Climate Change Reviews: 2022-2024

Climate Change Reviews: 2022-2024

Edited by

David Hemming

CABI

CABI is a trading name of CAB International

CABI
Nosworthy Way
Wallingford
Oxfordshire OX10 8DE
UK

Tel: +44 (0)1491 832111
E-mail: info@cabi.org
Website: www.cabi.org

CABI
200 Portland Street
Boston
MA 02114
USA

T: +1 (617)682-9015
E-mail: cabi-nao@cabi.org

A catalogue record for this book is available from the British Library, London, UK.

ISBN-13: 9781800629479 (hardback)
9781800629486 (ePDF)
9781800629493 (ePub)

DOI: 10.1079/9781800629493.0000

Commissioning Editor: David Hemming
Editorial Assistant: Helen Elliott
Production Editor: James Bishop

Typeset by Exeter Premedia Services Pvt Ltd, Chennai, India
Printed and bound in the UK by CPI Group (UK) Ltd, Croydon, CR0 4YY

Contents

Contributors

Richard O. Akinwale: Department of Crop Production and Protection, Obafemi Awolowo University, P.O. Box 2586, Ile-Ife, Nigeria

Marcelo de Carvalho Alves: Agricultural Engineering Department (DEA), Federal University of Lavras (UFLA), Caixa Postal 3037, CEP 37200-900, Lavras, Brazil

Franklin Ore Areche: National University of Huancavelica, Huancavelica, Peru

Deb Raj Aryal: Consejo Nacional de Ciencia y Tecnología (CONACyT), Av. Insurgentes Sur 1582, C.P. 03940, Mexico City, Mexico; Facultad de Ciencias Agronómicas, Universidad Autónoma de Chiapas, Villaflores, Carr. Ocozocoautla—Villaflores km 84.5, C.P. 30470, Chiapas, Mexico

Jhon Adolfo Quincho Astete: Altoandina National Autonomous University of Tarma, Tarma, Peru

E.P. Axelsson: Department of Wildlife, Fish and Environmental Studies, Swedish University of Agricultural Sciences, Umeå, Sweden

Baffour Badu-Apraku: International Institute of Tropical Agriculture (IITA), PMB 5320, Ibadan, Nigeria

Tégawendé O. Bonkoungou: International Institute of Tropical Agriculture (IITA), PMB 5320, Ibadan, Nigeria

Johanne Brunet: Brunet Research, Madison, Wisconsin, USA

Guillermo Gomer Cotrina Cabello: Daniel Alcides Carrión National University, Cerro de Pasco, Peru

Gary S. Caldwell: School of Natural and Environmental Sciences Reception, 4th Floor, Ridley Building 2, Newcastle University, Newcastle Upon Tyne, NE1 7RU, UK

Becquer Frauberth Camayo-Lapa: National University of Central Peru, Huancayo, Peru

L.M. Cardenas: Rothamsted Research, North Wyke, Okehampton, Devon, UK

Vittoria Catara: Department of Agriculture, Food and Environment, University of Catania, Catania, 95124, Italy

Mizeck G.G. Changunda: University of Hohenheim, Institute of Agricultural Science in the Tropics, Germany

Renae Charalambous: School of Agriculture and Food Sustainability, The University of Queensland, Gatton Campus, QLD 4343, Australia

Charles Bwalya Chisanga: Department of Plant and Environmental Sciences, Copperbelt University, P.O. Box 21692, Off Jambo Drive, Kitwe 10101, Zambia

Christopher P. Cockel: Royal Botanic Gardens, Kew, Millennium Seed Bank, Ardingly, RH17 6TN, UK

Suseelendra Desai: School of Agricultural Sciences and Technology, Narsee Monji Institute of Management Studies Deemed to be University, Shirpur 425405, Maharashtra, India

M.S. Dhanoa: Centre for Nutrition Modelling, Department of Animal Biosciences, University of Guelph, Guelph ON, N1G 2W1, Canada

Vijay Kumar Didal: School of Agricultural Sciences and Technology, Narsee Monji Institute of Management Studies Deemed to be University, Shirpur 425405, Maharashtra, India

Guillermo Donoso: Departamento de Economía Agraria, Facultad de Agronomía e Ingeniería Forestal, Pontificia Universidad Católica de Chile, Santiago, Chile

Mukesh Dubey: Department of Forest Mycology and Plant Pathology, Uppsala Biocenter, Swedish University of Agricultural Sciences, Uppsala, 75651, Sweden

Kate Dutton-Register: School of Agriculture and Food Sustainability, The University of Queensland, Gatton Campus, QLD 4343, Australia

J.L. Ellis: Centre for Nutrition Modelling, Department of Animal Biosciences, University of Guelph, Guelph ON, N1G 2W1, Canada

Morakinyo A.B. Fakorede: Department of Crop Production and Protection, Obafemi Awolowo University, P.O. Box 2586, Ile-Ife, Nigeria

Denis Dante Corilla Flores: National University of Huancavelica, Huancavelica, Peru

Dylan Fox: School of Agriculture and Food Sustainability, The University of Queensland, Gatton Campus, QLD 4343, Australia

Fabiana P. Fragoso: Unaffiliated, Barra Mansa, Rio de Janeiro, Brazil

J. France: Centre for Nutrition Modelling, Department of Animal Biosciences, University of Guelph, Guelph ON, N1G 2W1, Canada

Aqarab Husnain Gondal: Institute of Soil and Environmental Sciences, University of Agriculture Faisalabad, Punjab, Pakistan

Farideh Ghadamgahi: Department of Plant Breeding, Swedish University of Agricultural Sciences, Lomma, SE-234 22, Sweden

Maraeva Gianella: Royal Botanic Gardens, Kew, Millennium Seed Bank, Ardingly, RH17 6TN, UK

Malali Gowda: The University of Trans Disciplinary Health Sciences and Technology, University, Bangalore 560064, India

K.C. Grady: School of Forestry, Northern Arizona University, Flagstaff, AZ, USA

Nachiket Gudi: Public Health Evidence South Asia, Department of Health Information, Prasanna School of Public Health, Manipal Academy of Higher Education, Manipal, 576104, India

Filippo Guzzon: Centre for Pacific Crops and Trees (CePaCT), Land Resources Division (LRD), Pacific Community (SPC), Private Mail Bag, Suva, Fiji

Abdul Jabbar: Department of Citrus Research Institute Sargodha, Sargodha, Punjab, Pakistan

Grahame Jackson: PestNet, 24 Alt Street, Queens Park, NSW, Australia

Jiwoo Jeon: School of Agriculture and Food Sustainability, The University of Queensland, Gatton Campus, QLD 4343, Australia

Joe Joseph: School of Agriculture and Food Sustainability, The University of Queensland, Gatton Campus, QLD 4343, Australia

Juliane Kaoh: VARTC, Vanuatu Agricultural Research and Technical Centre, PO Box 231, Luganville, Santo, Vanuatu

Bekele Gelena Kelbessa: Department of Plant Breeding, Swedish University of Agricultural Sciences, Lomma, SE-234 22, Sweden

Ning-Yuan Ko: School of Agriculture and Food Sustainability, The University of Queensland, Gatton Campus, QLD 4343, Australia

Birte Komolong: NARI, National Agricultural Research Institute, Papua New Guinea

Donald R. Kugonza: Department of Agricultural Production, School of Agricultural Sciences, College of Agricultural & Environmental Sciences, Makerere University, Box 7062, Kampala, Uganda

Jean-Pierre Labouisse: CIRAD UMR AGAP Institut, Montpellier 34398, France

Floriane Lawac: VARTC, Vanuatu Agricultural Research and Technical Centre, PO Box 231, Luganville, Santo, Vanuatu

Vincent Lebot: CIRAD-AGAP-VARTC,PO Box 946, Port-Vila, Vanuatu

S. López: Instituto de Ganadería de Montaña (IGM), CSIC-Universidad de León, Departamento de Producción Animal, Universidad de León, E-24007, León, Spain

Luciano Carlos da Maia: Faculdade de Agronomia Eliseu Maciel, Departamento de Fitotecnia, Universidade Federal de Pelotas,Campus Capão do Leão, 96010-610, Pelotas, RS, Brazil

El H.M. Mashad: Department of Animal Science, University of California, Davis, One Shields Ave., Davis, CA 95616-8521, USA

Rohini Mattoo: Divecha Centre for Climate Change, Indian Institute of Science, Bangalore 560012, India

C.J. McCabe: Department of Animal Science, University of California, Davis, One Shields Ave., Davis, CA 95616-8521, USA

Nitin Misal: School of Agricultural Sciences and Technology, Narsee Monji Institute of Management Studies Deemed to be University, Shirpur 425405, Maharashtra, India

F.M. Mitloehner: Department of Animal Science, University of California, Davis, One Shields Ave., Davis, CA 95616-8521, USA

Miyanda Moombe: Department of Soil Science, University of Zambia, Lusaka, Zambia

Oyekunle Muhyideen: Department of Plant Science, Institute for Agricultural Research/Ahmadu Bello University, Zaria, Nigeria

Jonas V. Müller: Royal Botanic Gardens, Kew, Millennium Seed Bank, Ardingly, RH17 6TN, UK

Richard M. Mulwa: Department of Crop Horticulture and Soil, Egerton University, P.O. Box 536 – 20115, Egerton, Kenya

Okeyo Mwai: International Livestock Research Institute, PO BOX 30709-00100, Nairobi, Kenya

Edward Narayan: School of Agriculture and Food Sustainability, The University of Queensland, Gatton Campus, QLD 4343, Australia; Queensland Alliance for Agriculture and Food Innovation, The University of Queensland, St Lucia, QLD 4072, Australia

Charles Nelimor: Savanna Agricultural Research Institute (CSIR-SARI), P.O. Box TL 52, Tamale, Ghana

Sneh Kashinath Nerurkar: School of Agriculture and Food Sustainability, The University of Queensland, Gatton Campus, QLD 4343, Australia

Dante Daniel Cruz Nieto: José Faustino Sánchez Carrión National University, Huacho, Peru

K.N. Nithin: School of Agricultural Sciences and Technology, Narsee Monji Institute of Management Studies Deemed to be University, Shirpur 425405, Maharashtra, India

Valeria Oliveira Nizolli: Faculdade de Agronomia Eliseu Maciel, Departamento de Fitotecnia, Universidade Federal de Pelotas,Campus Capão do Leão, 96010-610, Pelotas, RS, Brazil

Samuel Nyalala: Department of Crop Horticulture and Soil, Egerton University, P.O. Box 536 – 20115, Egerton, Kenya

Jae-Won Oh: Department of Pediatrics, College of Medicine, Hanyang University, Hanyang University Guri Hospital, 153 Gyeongchun-Ro, Guri Gyeonggi-Do Korea, Seoul, Korea

Julie M.K. Ojango: International Livestock Research Institute, PO BOX 30709-00100, Nairobi, Kenya

Lilian A. Okiro: Department of Crop Horticulture and Soil, Egerton University, P.O. Box 536 – 20115, Egerton, Kenya; Department of Biochemistry and Molecular Biology, Egerton University, P.O. Box 536 – 20115, Egerton, Kenya

Moses Okpeku: Discipline of Genetics, School of Life Sciences, University of Kwazulu-Natal, Durban South Africa

Antonio Costa de Oliveira: Faculdade de Agronomia Eliseu Maciel, Departamento de Fitotecnia, Universidade Federal de Pelotas,Campus Capão do Leão, 96010-610, Pelotas, RS, Brazil

Victoria Freitas de Oliveira: Faculdade de Agronomia Eliseu Maciel, Departamento de Fitotecnia, Universidade Federal de Pelotas,Campus Capão do Leão, 96010-610, Pelotas, RS, Brazil

Victor E. Olori: Research & Development Department, Aviagen Ltd, 11 Lochend Rd, Newbridge EH28 8SZ, Scotland, UK

Rodomiro Ortiz: Department of Plant Breeding, Swedish University of Agricultural Sciences, Lomma, SE-234 22, Sweden

Richard Osei-Amponsah: Department of Animal Science, School of Agriculture, University of Ghana, Legon, Accra, Ghana

Alimatu S. Osuman: International Institute of Tropical Agriculture (IITA), PMB 5320, Ibadan, Nigeria; West Africa Center for Crop Improvement (WACCI), University of Ghana, PMB 30 Legon, Accra, Ghana; Crops Research Institute, Ghana. P.O. Box 3785, 039-5028 Kumasi, Ghana

Ellen Ould: School of Natural and Environmental Sciences Reception, 4th Floor, Ridley Building 2, Newcastle University, Newcastle Upon Tyne, NE1 7RU, UK

Maurice E. Oyoo: Department of Crop Horticulture and Soil, Egerton University, P.O. Box 536 – 20115, Egerton, Kenya

Harsh Pahuja: School of Agriculture and Food Sustainability, The University of Queensland, Gatton Campus, QLD 4343, Australia

L.R. Panchami: Department of Health Information, Prasanna School of Public Health, Manipal Academy of Higher Education, Manipal, 576104, India

Divya Sussana Patil: Public Health Evidence South Asia, Department of Health Information, Prasanna School of Public Health, Manipal Academy of Higher Education, Manipal, 576104, India

Camila Pegoraro: Faculdade de Agronomia Eliseu Maciel, Departamento de Fitotecnia, Universidade Federal de Pelotas,Campus Capão do Leão, 96010-610, Pelotas, RS, Brazil

Elijah Phiri: Department of Soil Science, University of Zambia, Lusaka, Zambia

C.D. Powell: Centre for Nutrition Modelling, Department of Animal Biosciences, University of Guelph, Guelph ON, N1G 2W1, Canada

J.V.N.S. Prasad: ICAR-Central Research Institute for Dryland Agriculture, Santoshnagar, Hyderabad 500059, Telangana, India

Nishit Rao: School of Agriculture and Food Sustainability, The University of Queensland, Gatton Campus, QLD 4343, Australia

K. Venkateswara Rao: ICAR-Central Research Institute for Dryland Agriculture, Santoshnagar, Hyderabad 500059, Telangana, India

José Luis Riveros F.: Departamento de Ciencias Animales, Facultad de Agronomía e Ingeniería Forestal, Pontificia Universidad Católica de Chile, Santiago, Chile

Alfonso Ruiz Rodriguez: National University of Huancavelica, Huancavelica, Peru

Pablo Ronda-Borzone: Facultad de Agronomía e Ingeniería Forestal, Pontificia Universidad Católica de Chile, Santiago, Chile

William Herminio Salas-Contreras: National University of Huancavelica, Huancavelica, Peru

José Yovera Saldarriaga: Santiago Antúnez of Mayolo National University, Peru

Luciana Sanches: Sanitary and Environmental Department, UFMT Campus, Cuiabá, MT, Zip Code 78060-900, Brazil

R. Sanderson: Institute of Biological, Environmental and Rural Sciences, Aberystwyth University, Gogerddan, Aberystwyth, Ceredigion, SY23 3EB, UK

A. Shepherd: Institute of Biological and Environmental Sciences, University of Aberdeen, Aberdeen, Scotland, UK

Sharvari Avinash Sherekar: School of Agriculture and Food Sustainability, The University of Queensland, Gatton Campus, QLD 4343, Australia

Kyle Stice: Pacific Farmer Organizations, PO Box 9048, Nadi, Fiji Islands

Amit Sukal: Centre for Pacific Crops and Trees (CePaCT), Pacific Community (SPC), Private Mail Bag, Suva, Fiji

Mary Taylor: University of the Sunshine Coast, QLD, Australia

Geberemariyam Terefe: Ethiopian Institute of Agricultural Research, Animal Feeds and Nutrition Research, Pox 2003, Addis Ababa, Ethiopia

Krishna Chaitanya Tirunagaru: School of Agricultural Sciences and Technology, Narsee Monji Institute of Management Studies Deemed to be University, Shirpur 425405, Maharashtra, India

Lídia Raiza Sousa Lima Chaves Trindade: Department of Engineering, Federal University of Lavras, Campus Universitário, PO Box 3037, ZIP Code 37200-000, Lavras, Brazil

Rathnakaran Upasana: School of Agricultural Sciences and Technology, Narsee Monji Institute of Management Studies Deemed to be University, Shirpur 425405, Maharashtra, India

Ramesh Raju Vetukuri: Department of Plant Breeding, Swedish University of Agricultural Sciences, Lomma, SE-234 22, Sweden

Mulugeta Walelegne: Ethiopian Institute of Agricultural Research, Animal Feeds and Nutrition Research, Pox 2003, Addis Ababa, Ethiopia

Zhiheng Wang: School of Agriculture and Food Sustainability, The University of Queensland, Gatton Campus, QLD 4343, Australia

Logotonu Meleisea Waqainabete: Centre for Pacific Crops and Trees (CePaCT), Pacific Community (SPC), Private Mail Bag, Suva, Fiji

Yifei Yang: School of Veterinary Sciences, The University of Queensland, Gatton Campus, QLD 4343, Australia

Rafael Julian Malpartida Yapias: Altoandina National Autonomous University of Tarma, Tarma, Peru

Rasoul Yousefpour: Forestry Economics and Forest Planning, University of Freiburg, Tennenbacherstr. 4, D-79117 Freiburg, Germany

Introduction

Climate change is a reality that affects all aspects of agriculture, and is also impacted by agriculture. The International Panel on Climate Change estimates that in the next decade alone, climate change will drive 32–132 million more people into extreme poverty. Climate change impacts are stressing agriculture, forestry, fisheries and aquaculture, increasingly hindering efforts to meet human needs. Crop yields are compromised by surface ozone, and global warming is negatively affecting crop and grassland quality and harvest stability, according to the IPCC. It is evident that warmer and drier conditions have increased tree mortality and forest disturbances in many areas. The IPCC's work suggests that vulnerable groups, such as women, children, low-income households, Indigenous or other minority groups are the most exposed to risk.

This collection of articles surveys a diverse range of impacts, approaches for limiting the threats of climate change, building resilience and ways to address the contribution of agricultural practices to global warming.

The book includes topics such as impact on pollinators, key crops, farming systems, animal welfare and on humans, role of farmer organisations in extending use of climate-smart practices, genetic resources able to cope with climate change, including wild relatives and breeding for improved tolerance, how rhizobacteria can improve resilience, mitigation of livestock greenhouse gas emissions and the push for climate neutrality in the dairy industry and carbon storage in grasslands and seaweed. These articles have been published in the international peer-reviewed journal CABI Reviews.

1 What are the main reasons for the worldwide decline in pollinator populations?

Johanne Brunet[1]* and Fabiana P. Fragoso[2]
[1]*Brunet Research, Madison, Wisconsin, USA;* [2]*Unaffiliated, Barra Mansa, Rio de Janeiro, Brazil*

Abstract

Pollinator populations are declining worldwide and over 87% of flowering plant species and 87 of the leading global food crops rely on pollinators for seed production. Pollinator decline seriously impacts biodiversity conservation, reduces crop yield, and threatens food security. Understanding the drivers of pollinator decline can guide the development of strategies and action plans to protect and conserve pollinators and the essential ecosystem services they provide. This review introduces the diversity of pollinators, addresses the main drivers of pollinator decline, and presents strategies to reduce their negative impacts. We discuss how managed bees negatively affect wild bee species, and examine the impacts of habitat loss, pesticide use, pests and pathogens, pollution, and climate change on pollinator decline. Connections between humans and pollinator decline are addressed. Climate change emerges as the most prominent threat to pollinators and humans and the most difficult threat to control. The changes in water and temperature associated with climate change can lower the quantity and quality of resources available to pollinators, decrease the survival of larvae or adults, and modify suitable habitats. Climate change is a global issue and a serious threat to food security, and to world stability. Efforts to control the various factors that negatively impact pollinators must continue given the dire consequences.

Keywords: bees, climate change, habitat loss, managed pollinators, pathogens, pesticides, pollution

Introduction

Pollinator populations around the world have been declining at an alarming rate in recent decades (Potts *et al.*, 2010; Goulson *et al.*, 2015; Nath *et al.*, 2022). Approximately 16% of vertebrate pollinators, such as birds and bats, and 40% of invertebrate pollinators, such as bees and butterflies, are at risk of extinction (IPBES, 2016). Species loss and steep population declines of insects, the most important group of pollinators in most biomes, have been recorded in all regions of the world (Wagner, 2020). In North America, for example, the eastern population of the monarch butterfly (*Danaus plexippus*) has declined by 80% over the past decades (Semmens *et al.*, 2016), while the relative abundances of several bumble bee species have declined by up to 96% (Cameron *et al.*, 2011).

The significant decline in pollinator populations has become a growing concern for ecologists, farmers, and policy makers, as pollinators are essential to the reproduction of

*Corresponding author: brunetresearch@gmail.com

many flowering plants and crops worldwide. As over 87% of flowering plant species are dependent upon animal pollination for reproduction (Ollerton *et al.*, 2011; Tong *et al.*, 2023), pollinator decline affects the stability of plant-pollinator networks and threatens the conservation of ecosystem functions (Ramos-Jiliberto *et al.*, 2020). In agriculture, 87 of the leading global food crops rely on pollinators for their production (Klein *et al.*, 2007), which corresponds to 35% of the global production volume of crops grown for human consumption. The reliance on animal pollination will only increase, as more acreages of pollinator-dependent crops are being planted, especially in developing countries (Aizen *et al.*, 2008, 2009). Therefore, any decline in pollinator populations will significantly impact biodiversity conservation, reduce crop yields, increase food prices (Reilly *et al.*, 2020), and thus threaten food security (Kremen *et al.*, 2002; Gallai *et al.*, 2009).

In order to create strategies and action plans to protect and conserve pollinator populations and the essential ecosystem services they provide we need to understand the main reasons for their worldwide decline. A range of factors, including habitat loss, pesticide use, pests and pathogens, pollution, and climate change can adversely affect pollinators (Fig. 1.1). Synergistic interactions between these stressors can intensify the negative impacts on pollinators and accelerate species loss (Goulson *et al.*, 2015; Janousek *et al.*, 2023; Williams and Hemberger, 2023). More recently, managed bees have been identified as an extra factor that negatively impacts the survival of wild bee species.

Here, we first describe key pollinators and provide general information on their food source, nesting habits, and general life history traits. We then discuss the effect of managed pollinators on wild bee decline, before examining the potential consequence of each of the aforementioned factors on pollinator behavior, health, and survival. We discuss strategies that have or could be implemented to limit the negative impacts of distinct factors on pollinator

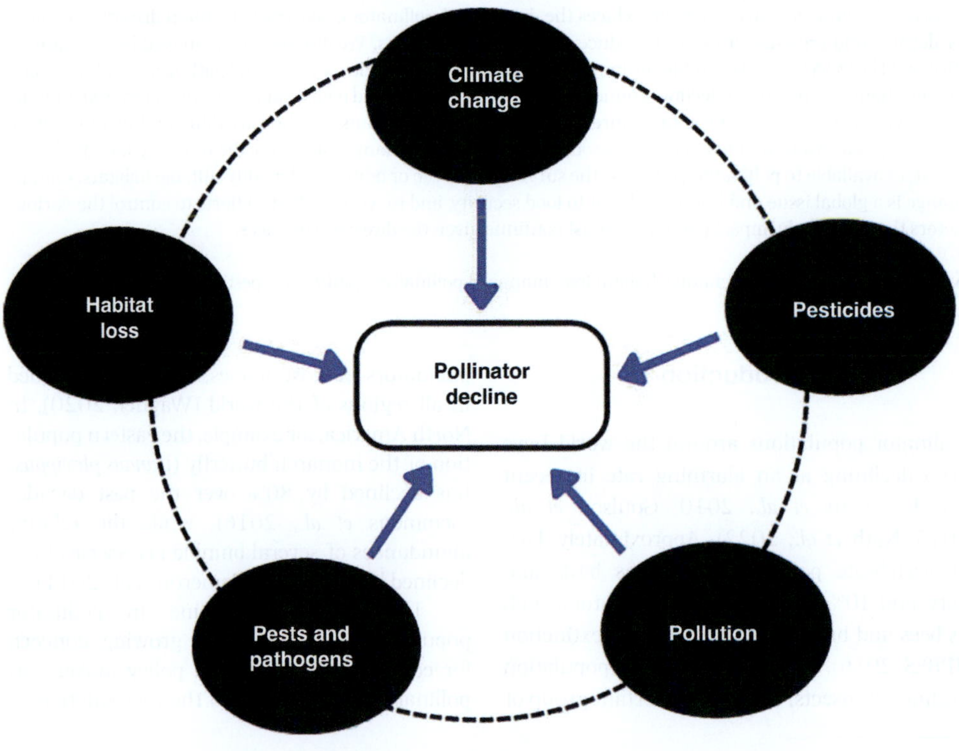

Fig. 1.1. Schematic representation of drivers of pollinator decline.

decline. Climate change emerged as the most prominent threat to pollinators, and the most difficult threat to control.

Review Methodology

We conducted systematic literature searches of peer-reviewed articles published in English using different databases such as Google Scholar (https://scholar.google.com/), Web of Science (https://www.webofscience.com/), PubMed (https://pubmed.ncbi. nlm.nih.gov/), and Scopus (https://www.scopus.com/). Initial search terms included "pollinator decline" and "bee decline"; and combinations of these two terms with "habitat loss"; "pesticides"; "climate change"; "pathogens"; "disease"; "invasive species"; and "genetic diversity." We reviewed relevant publications from these searches, and used the references from the articles obtained by this method, to check for additional relevant material.

Diversity and life history of key pollinators

Pollinators visit flowers to gather resources to feed themselves, their young, and in the case of eusocial bees, to maintain their colonies. Pollen grains adhere to their body while they forage and get transferred between flowers, which are incidentally pollinated. The great majority of pollinators are insects, which include bees, butterflies (Fig. 1.2A and 1.B), moths (i.e. hawkmoths Fig. 1.2C), flies (Fig. 1.2D), beetles, and wasps. Second to insects as pollinators come birds (Fig. 1.2E and 1.F), followed by bats. Some small mammals (Kress *et al.*, 1994) and lizards (Olesen and Valido, 2003) can also pollinate flowers. Bees are the most important pollinators and the only group that actively gathers pollen while foraging. Bees can collect pollen, nectar, or both during a foraging trip, while other pollinators forage exclusively for nectar. Bees have specialized structures for pollen collection: scopae or branched body hairs in most solitary bees (Fig. 1.3A), and corbiculae or pollen baskets, in eusocial and orchid bees (Euglossini) (Fig. 1.3B) (Thorp, 2000). Nectar is a source of carbohydrates and water, while pollen provides proteins, possibly microbes, and also contains lipids, vitamins, and minerals (Roulston and Cane, 2000; Nicolson, 2011; Dharampal *et al.*, 2019).

There exist approximately 20,000 bee species worldwide and native bees occur on every continent except Antarctica. While eusocial bees are best known, over 85% of bee species are solitary (Michener, 2007) (Fig. 1.3A, 1.C and 1.D). Solitary bees live for about a year, with most of that time spent between the larval and pupal stages, and their adult life lasts 3–8 weeks. Approximately 30% of solitary bees are cavity nesters and lay eggs in hollow stems or tree trunks. The remaining 70% nest in the ground where they create burrows and lay their eggs. All adult females are reproductive and provide each egg with pollen and nectar with no further parental care. Some species are aggregate nesters, as they nest in aggregation without cooperation.

Eusocial bees include honey bees, bumble bees and stingless bees. They form colonies with castes consisting of a queen, workers (females), and drones (males). The colony has a reproductive egg-laying female (the queen), division of labor, and cooperative brood care. The stingless bees (Fig. 1.3E) or Meliponini bees consist of over 500 bee species that inhabit tropical and subtropical regions, mostly in the Western Hemisphere (Michener, 2007). Stingless bee colonies can contain several thousand workers and are found in cavities above ground such as hollow tree trunks. Some stingless bee species make honey. The complexity of their communication system about resource location varies among species, with some species leaving scent trails, and others using sound and some form of zigzag dance to guide others to a food source (Nieh and Roubik, 1998; Jarau *et al.*, 2003).

Bumble bees have a cosmopolitan distribution and consist of approximately 250 species. They are absent in Australia but have been introduced to New Zealand and Tasmania. They are the only eusocial bees native to North America. Bumble bees build annual colonies in cavities, preferably underground (Fig. 1.3B). In late summer or early fall, males are produced to mate with the newly-emerged queens. The latter hibernate and the surviving mated queens start new colonies in spring or early summer. The queens gather nectar and pollen to nourish their initial brood, but after the first workers emerge,

Fig. 1.2. Pollinators (A) Monarch butterfly, (B) Skipper *Hesperiidae* sp., (C) White-lined sphinx hawkmoth, (D) Hoverfly (Syrphidae), and (E and F) Hummingbirds. Photos A, B, D, E, and F by Fabiana P. Fragoso and C by Brunet laboratory.

queens remain inside the colony to lay eggs. Depending on the bumble bee species, a colony may comprise 50 to 400–500 workers. Bumble bees store nectar in wax pots inside the hive; they do not make honey.

The best-known eusocial bees, the honey bees, include only eight currently living bee species worldwide (*Apis flora, A. andreniformis, A. dorsata, A. laboriosa, A. mellifera, A. cerana, A. koschevnikovi,* and *A. nigrocincta*). Besides *A. mellifera,* which is native to Europe, Africa, and the Middle East, the other *Apis* species occur exclusively in Asia. The western honey bee, *Apis mellifera,* has been widely introduced in South America, North America, and Australia, and now has populations on every continent

Fig. 1.3. Bee pollinators. (A) Solitary bee, (B) Bumble bee, *Bombus impatiens*, (C) Sweat bee *Augochlora* sp., (D) Sweat bee *Agapostemon* sp., (E) Stingless bee *Trigona spinipes*, and (F) Honey bee *Apis mellifera*. Photos by Fabiana P. Fragoso.

except Antarctica. It is the pollinator most widely managed for crop pollination and honey production (Fig. 1.3F). Honey bee colonies are perennial, that is, they persist for several years, like stingless bees. Established wild honey bee colonies generally last 5–6 years, but when newly established (first-year) colonies are considered, then average colony survival is 2 years (Seeley, 2017). Workers live for approximately 6 weeks. In the wild, honey bee colonies are built in cavities above ground, and can reach 60,000–80,000 workers during peak season. Honey bees use the waggle dance, a sophisticated and unique communication system, to share information about resource quality and location (Shackleton *et al.*, 2023).

Birds are the second most important group of pollinators worldwide after insects, and they pollinate around 500 plant genera (Sekercioglu, 2006). Bird pollination is very common in Australia, where over 100 bird species have been reported visiting flowers of 250 plant species (Ford *et al.*, 1979). Hummingbirds and honeycreepers are important pollinators in North America and Hawaii, respectively, while brush-tongued parrots (New Guinea) and sunbirds (Old World tropics) can serve as pollen vectors in the tropics. Bat pollination follows bird pollination in order of importance worldwide and is more prevalent in tropical and desert climates. Bats pollinate close to 528 flowering plant species from at least 67 plant families (Fleming *et al.*, 2009). Nectar-feeding bats belong to two bat families, the family Pteropodidae and the family Phyllostomidae. The former occurs in tropical and subtropical parts of Africa, Asia, Australia, Papua New Guinea, and the Pacific Islands, and the latter inhabits tropical and subtropical regions of the Americas.

Managed bees as a threat to wild bees

Honey bees can threaten wild bee survival. Honey bees are highly valued as pollinators, and Colony Collapse Disorder (CCD), an affliction of honey bees responsible for an annual loss of 20–30% of colonies in the United States, raised public attention toward pollinator decline. Colony losses can, however, be attributed to multiple factors, including *Varroa* mite infestations (Steinhauer *et al.*, 2018). To counteract honey bee colony decline, a growing number of companies and private citizens have begun maintaining honey bee hives in urban and residential areas. While this practice has increased the abundance of honey bees in these communities, it has also intensified competition for resources with wild bees, negatively impacting the survival and health of some wild bee species (Thomson, 2004; Cane and Tepedino, 2016; Angelella *et al.*, 2021; Weaver *et al.*, 2022).

Managed bumble bees can impact wild bumble bee health. The use of bumble bees as managed pollinators, mainly for greenhouse pollination and research purposes, is expanding. Managed bumble bee colonies not only compete for resources with wild bumble bee species (Goulson *et al.*, 2008), but these colonies can harbor high levels of pathogens (Murray *et al.*, 2013), which can be transmitted to wild bumble bee populations (Colla *et al.*, 2006; Otterstatter and Thomson, 2008). The frequency of pathogens shared by managed and wild bumble bee colonies is in fact higher in wild colonies located close to managed hives and decreases with increasing distance from the commercial hives (Murray *et al.*, 2013). These pathogens are suspected to be one factor contributing to the decline of some wild bumble bee species (Cameron *et al.*, 2011; Cordes *et al.*, 2012; Murray *et al.*, 2013).

Managed stingless bees can displace native bees. In the southern hemisphere, the use of stingless bees as managed pollinators for crop pollination is also increasing. Hives are often relocated beyond the natural range of these bee species, raising concern about the biological introduction of these bees in new areas, potentially leading to invasiveness and displacement of native bees (dos Santos *et al.*, 2022).

Bee decline and base-line data

Bees, the most important pollinators, are declining. A study using data available from 30% of the 4337 North American wild bee species (including Hawaii), concluded that over 50% of these species were declining while 24% were in peril (Kopec and Burd, 2017). For bumble bees, a group for which more data are available, diversity has declined in the UK (Williams, 1982), over much of Europe (Kosior *et al.*, 2007), in the United States (Cameron *et al.*, 2011; Bartomeus *et al.*, 2013), and in Asia (Williams and Osborne, 2009). Recent studies indicate that 28% of bumble bee species are threatened in Canada, the United States, and Mexico.

Data on bee decline vary between countries and for distinct bee genera. Efforts to determine the rates and drivers of pollinator loss internationally (Potts *et al.*, 2010) have highlighted strong differences among countries in the availability of pollinator studies (Archer *et al.*, 2014). Moreover, bumble bees are overrepresented in the available pollinator data, evidencing taxonomic bias (De Palma *et al.*, 2016). These geographic and taxonomic biases raise concern about generalizing and extrapolating study results to countries where pollinator data are sparse, or to more poorly studied bee taxa (Archer *et al.*, 2014; De Palma *et al.*, 2016).

The rising interest in pollinator decline has led to the realization that baseline data on bees and pollinators are sparse. This realization has generated increased research efforts to quantify bee diversity and richness in different communities and geographic regions and to examine the importance of bee diversity for crop pollination (Westphal *et al.*, 2008; Garibaldi *et al.*, 2013; Lewanski *et al.*, 2022). A recent study on bee richness, performed at a global scale, has identified peaks of bee richness in the northern and southern hemispheres, but not at the equator, a pattern that contrasts with many other pollinator groups whose richness peaks at the equator (Orr *et al.*, 2021). In the United States, the awareness of bee decline has led to increased research funding for bee health research, and to a USDA-funded effort to monitor native bees nationally (Woodard *et al.*, 2020).

Obtaining base-line data on pollinators generates more intense bee sampling. Given the increase in sampling intensity, a main emphasis should include the development and adoption of non-lethal methods of bee sampling. These collection methods should also aim at distinguishing between pollinators and mere floral visitors (Boyer *et al.*, 2020). Pan traps are commonly used and are often favored due to their ease of implementation on a large scale and the minimal personnel training required (Boyer *et al.*, 2020). Pan trap is also an efficient method to estimate bee species richness (Westphal *et al.*, 2008). Unfortunately, pan traps are inherently lethal, and over 10,000 bee specimens can be trapped and killed in a single study (Westphal *et al.*, 2008). While lethal bee sampling does not always have a negative impact on pollinator populations (Gezon *et al.*, 2015), few studies have examined this question. There is a pressing need to develop and adopt non-lethal methods of bee collection in order to minimize the potential negative impacts of intense sampling on bee populations.

Habitat loss and changes in land configuration

Primary drivers of change in pollinator populations are habitat loss and changes in land configuration, caused by human activities such as intensive agriculture, urbanization, and industrialization. Even though some bee guilds are positively affected by habitat conversion (Cane *et al.*, 2006; Winfree *et al.*, 2007), and others can tolerate moderate levels of disturbance (Carré *et al.*, 2009), most pollinator communities experience cascading adverse effects of habitat degradation and deforestation (Ricketts *et al.*, 2008; Winfree *et al.*, 2009).

Habitat loss and degradation decrease the availability of nesting sites and resources. It reduces the abundance and richness of flowering plants, which in turn decreases the availability of pollen and nectar resources necessary for pollinators' survival, growth, and reproduction (Carvell *et al.*, 2006; Cohen *et al.*, 2020). It can also increase isolation from native habitats which impacts pollinator species nesting above ground more negatively than species nesting below ground, likely due to loss of nesting sites (Williams *et al.*, 2010). Agricultural practices such as intensive grazing significantly reduce the abundance of food plants and bee diversity (Xie *et al.*, 2008), and the loss of a preferred host species can be a key factor in the decline of bee populations (Scheper *et al.*, 2014).

Habitat loss and changes in land configuration can lead to pollinator malnutrition. Bees with limited or low-quality resources have several disadvantages compared to those with access to abundant and nutritious pollen. High-quality pollen affects bee physiology and tolerance to parasites, increasing bee survival (Di Pasquale *et al.*, 2013; Glavinic *et al.*, 2017). Dietary phytochemicals enable bees to better fight infection, thus increasing their tolerance to pathogens and their longevity (Bernklau *et al.*, 2019). The amino acid composition of pollen significantly affects the growth rate of bumble bee larvae (Moerman *et al.*, 2016). Honey bee workers reared in pollen-stressed colonies become poor foragers and waggle dancers as adults (Scofield and Mattila, 2015). Thus, a reduced quality of available floral resources can weaken pollinators and make their populations more susceptible to other threats.

Habitat fragmentation is likely to affect pollinator movement (Hadley and Betts, 2011). Social and solitary bees consider both patch size and isolation distance when moving between patches, and typically prefer large, nearby patches (Fragoso *et al.*, 2021; Fragoso and Brunet, 2023). In fragmented landscapes, visiting small isolated patches is less profitable (Hadley and Betts, 2011). It increases energy expenditure, as greater search and travel times are required to travel between them, for reduced resource acquisition which can negatively impact bee health.

Habitat loss impacts both specialist and generalist bees. Although specialist bee species are generally considered to be more vulnerable to habitat loss, recent evidence shows that widespread generalist species are also at risk (Janousek *et al.*, 2023). In addition, social bee species are more strongly affected by isolation from native habitats than solitary bee species, which may be explained by differences in nest requirements (Williams *et al.*, 2010). Ensuring adequate and sufficient habitats for pollinators is paramount in their conservation strategies.

Potential solutions include the creation of urban gardens with native plants and fruit and vegetable crops attractive to bees to ensure that the land loss to urban development does not deprive bees of floral resources. Urban gardens should include some bare grounds for nesting sites and nesting cavities for different cavity nesters. Bees can survive in an urban environment as long as floral resources and nesting sites are available. Similarly, vacant lands could be transformed into gardens using plant species attractive to different bee species.

For land converted into agriculture, the maintenance of wild habitats around agricultural fields has proven profitable for crop pollination. Integrating agriculture with the surrounding landscape should become a priority and would benefit the farmers and the bees and other wild pollinators.

Pesticides and other agrochemicals

Pesticides are detrimental to pollinators and can reach bees via multiple routes. Pesticides, such as neonicotinoids and pyrethroids, present very high toxicity to a wide range of insects, particularly pollinators (Goulson, 2013; Pisa *et al.*, 2015; Cabezas and Farinós, 2022). Adverse effects of pesticides include behavioral changes, impaired foraging and learning abilities, altered physiological processes, suppressed immunity, reduced reproductive success, and overall population decline (Desneux *et al.*, 2007; Goulson, 2013; Gill and Raine, 2014; Pisa *et al.*, 2015; Czerwinski and Sadd, 2017). Pollinators can be exposed to pesticides through multiple routes, including seed coating and collected pollen (Girolami *et al.*, 2009; Krupke *et al.*, 2012; Rundlöf *et al.*, 2015). Chronic exposure to neonicotinoid pesticides can alter the interactions between bumble bees and wild plants (Stanley and Raine, 2016), and negatively impact the stability of plant-pollinator interactions and community persistence (Chattopadhyay *et al.*, 2023).

Pesticides can have lethal and sublethal effects on bees. Acute toxicity caused by direct contact or ingestion of pesticides can result in mortality, although sublethal effects are also detrimental. In honey bees, exposure to sublethal doses of pesticides can cause wing paralysis and impaired activity, locomotion, metabolism, and ontogenetic development (Girolami *et al.*, 2009; Yang *et al.*, 2012). Learning, memory, navigation, and homing ability are also negatively affected by exposure to sublethal doses of neonicotinoids (Henry *et al.*, 2012; Yang *et al.*, 2012). Impaired cognition and behavior will not only hinder pollinator foraging abilities (Gill and Raine, 2014), but it can reduce pollination effectiveness and efficiency, thus lowering plant reproductive success.

Other agrochemicals, herbicides, fungicides, and even some fertilizers, can indirectly affect pollinator populations by altering enzymes that affect pollinator behavior or impeding anther dehiscence of the host plant, thus reducing pollen availability (Londo *et al.*, 2014; Toledo-Hernández *et al.*, 2022).

Solutions include making sure to spray chemicals in the evening when bees are not active. This limits the lethal and sublethal impacts of agrochemicals on bees. Using chemicals with a shorter half-life would limit their buildup in the environment. Integrated pollinator management systems that minimize the impact on pollinators and the environment while optimizing crop yield should be implemented. These practices emphasize an integration of ecologically-based control strategies, supported by scientific data.

Pests and pathogens

Pests and pathogens negatively impact bee health. Insect pollinators can be infected by an array of viruses, bacteria, fungi, protozoans, nematodes, and external parasites (Yañez *et al.*, 2020; Toledo-Hernández *et al.*, 2022; Evans *et al.*, 2023; Figueroa *et al.*, 2023), which can cause disease and make pollinators more susceptible to other stressors. In the beekeeping industry, the spread of pests and pathogens have been considered the main driver of annual colony losses (Smith *et al.*, 2013; Freitas *et al.*, 2022; Peña-Chora *et al.*, 2023). *Varroa destructor* mites are particularly problematic because, in addition to feeding on honey bee larvae and pupae, they transmit numerous viruses, which further weaken their immune system (Wilfert *et al.*, 2016). Besides the widespread deformed wing virus (DWV) transmitted by *Varroa* (Wilfert *et al.*, 2016), at least 60 other viruses, such as the black queen cell virus (BQCV), acute bee paralysis virus (ABPV), and chronic bee paralysis virus (CBPV), have been identified in honey bees (Yañez *et al.*, 2020; Peña-Chora *et al.*, 2023). The detrimental effects of these viruses include compromised flight capabilities and reduced lifespan, ultimately leading to a decline in bee populations.

Horizontal transmission via shared floral resources plays a crucial role in virus transmission within and between bee species (Graystock *et al.*, 2015; Manley *et al.*, 2015; Budge *et al.*, 2020). Viruses that were first recorded in honey bees are now being detected in several bee species (Ravoet *et al.*, 2014; Yañez *et al.*, 2020). Viral infections by BQCV, ABPV, and IAPV (Israeli Acute Paralysis Virus), for example, have been registered in multiple species of stingless bees (reviewed by Toledo-Hernández *et al.*, 2022), and bumble bees (Figueroa *et al.*, 2023). Migratory beekeeping has exacerbated the risk of disease transmission among honey bee colonies (Martínez-López *et al.*, 2022), and the evidence of disease spillover from managed bees to wild bee species is accumulating (Fürst *et al.*, 2014; Tehel *et al.*, 2016), with a potential for spillback from the wild back to managed bees.

Several other invertebrate parasites are threatening bee populations. Phorid flies, wax moths (a common pest of honey bees), and several mites parasitize stingless bees (Toledo-Hernández *et al.*, 2022), and bumble bee hives (Evans *et al.*, 2023), and heavy infestations lead to colony decline. The invasive Argentine ant (*Linepithema humile*) increases DWV and BQCV loads in honey bees (Dobelmann *et al.*, 2023). The small hive beetle (SHB, *Aethina tumida*), a widespread parasite and scavenger of honeybee colonies native to sub-Saharan Africa, has been suggested as a potential biological vector of honey bee viruses (Eyer *et al.*, 2009). The SHB can also infest colonies of stingless bees (Nacko *et al.*, 2020; Pereira *et al.*, 2021), bumble bees (Hoffmann *et al.*, 2008), or complete their life cycle in nests of solitary bees (Gonthier *et al.*, 2019). Because of its opportunistic nature, the SHB presents a serious threat to bee communities.

Bacterial diseases such as American foulbrood (*Paenibacillus larvae*) and European foulbrood (*Melissococcus plutonius*) are highly destructive to honey bee colonies. European foulbrood has been recently recorded in stingless bees (Teixeira *et al.*, 2020). Little is known about bacterial infections in bumble bees (Figueroa *et al.*, 2023), but the wild bumble bee species decline in North America is closely associated with *Vairimorpha* (*Nosema*) infections (Cameron *et al.*, 2016; Orlova *et al.*, 2023). The combined impact of these pests and pathogens presents a critical challenge for bee populations.

Solutions include minimizing spill-ups of pests and pathogens between managed and wild bees by controlling pathogens in managed bees. Once the pest or pathogen has spread to wild bees it becomes extremely difficult to control.

Pollution

Metal pollutants are toxic to many organisms, but little is known about their effects on pollinators (Feldhaar and Otti, 2020). Besides air, soil and water pollution, bees can be exposed to heavy metals that accumulate in the nectar or pollen of plants (Meindl and Ashman, 2013; Sivakoff and Gardiner, 2017; Xun *et al.*, 2018). Plants grown in contaminated soils are shorter and have smaller flowers, leading to decreased visit duration by pollinators (Sivakoff and Gardiner, 2017). Under both laboratory and field conditions, the cognition and morphological development of bees chronically exposed to metals are impaired, and combined metals have additive negative effects on the cognitive processes (Monchanin, 2021). Exposure to metals can alter the composition of both the bumble bee and honey bee microbiomes, with negative consequences for bee health (Rothman, 2019). While bumble bees do not prefer flowers of plants grown in non-contaminated, as opposed to contaminated soils (Sivakoff and Gardiner, 2017), they spend less time on flowers containing contaminated nectar (Meindl and Ashman, 2013). The ability of honey bees to detect heavy metals in a sucrose solution varied depending on the type of metal and its concentration (Burden *et al.*, 2019). These studies suggest that bees respond to metal contaminants and these negatively affect their health.

Artificial light at night (ALAN) can negatively affect the behavior, foraging, and reproduction of nocturnal pollinators like moths and bats, although the impact varies across species (Knop *et al.*, 2017; MacGregor *et al.*, 2017; Straka *et al.*, 2021; Dzul-Cauich and Munguía-Rosas, 2022; Shivanna, 2022). ALAN may also impact diurnal plant-pollinator interactions (Giavi *et al.*, 2021). Tree cover can mitigate the effects of ALAN, as tree cover density positively affects species richness and abundance of nocturnal moths (Straka *et al.*, 2021).

Heavy metal pollution tends to be localized and can be remediated, often via the use of microorganisms (bioremediation) (Kapahi and Sachdeva, 2019). With respect to ALAN, careful consideration should be given when choosing artificial light, and preference given to movement detection technology to reduce its impact on pollinators. Solutions are available to lessen the impact of pollution on pollinators at a local scale.

Climate Change

Changes in the abiotic environment resulting from climate change will have direct and indirect effects on pollinators. Higher temperatures can alter the timing of pollinator emergence and the onset and duration of flowering in plants (their flowering phenology), leading to potential mismatches between plants and their pollinators (Memmott *et al.*, 2007; Hegland *et al.*, 2009). While plenty of evidence corroborates the impact of higher temperature on both the flowering phenology of plants and the emergence of pollinators (Inouye, 2022), the available data do not strongly support the predicted mismatch. Typically, plants flower earlier and pollinators emerge sooner at higher temperatures, and shifts in the plant flowering phenology are typically stronger than the changes in insect emergence (Forrest and Thomson, 2011; Kehrberger and Holzschuh, 2019). Moreover, pollinator groups can differ in their responses (Freimuth *et al.*, 2022). The empirical data does not currently support mismatches; higher temperature either improved synchrony between plants and their pollinators, or a complete mismatch was not observed (Forrest and Thomson, 2011; Ovaskainen *et al.*, 2013; Freimuth *et al.*, 2022).

Changes in water and temperature can affect both visual and olfactory plant traits involved in pollinator attraction. Reduced water availability and increased temperature often decrease flower size and the number of flowers on a plant thus reducing its floral display size (Gallagher and Campbell, 2017; Brunet and Van Etten, 2019; Rering *et al.*, 2020; Wilson Rankin *et al.*, 2020; Descamps *et al.*, 2021c;

Kuppler and Kotowska, 2021). Smaller floral displays provide fewer resources to pollinators (Makino and Sakai, 2007; Brunet *et al.*, 2015; Wilson Rankin *et al.*, 2020), with consequences for pollinator health and growth (Häussler *et al.*, 2017). Drought and higher temperature modify the emission of floral volatiles (Sagae *et al.*, 2008; Yuan *et al.*, 2009; Burkle and Runyon, 2016; Glenny *et al.*, 2018; Campbell *et al.*, 2019; Keefover-Ring *et al.*, 2022) but more studies are needed to determine the impact of such changes on pollinator foraging behavior and colony or individual fitness.

In addition, the quantity and quality of the rewards offered by plants, can both be affected by changes in water and temperature (Cnaani *et al.*, 2006; Wilson Rankin *et al.*, 2020). Limited water or higher temperature can lower the amount of pollen and nectar produced per flower (Carroll *et al.*, 2001; Takkis *et al.*, 2015; Waser and Price, 2016; Gallagher and Campbell, 2017; Brunet and Van Etten, 2019; Rering *et al.*, 2020). The amount of sucrose in nectar and sugar ratios are both affected by reduced water (Rering *et al.*, 2020; Wilson Rankin *et al.*, 2020; Descamps *et al.*, 2021a, b), while temperature can modify the relative proportion of the amino acids in pollen (Descamps *et al.*, 2021a). Moreover, increased temperature changes the abundance of some bacteria in nectar, leading to a decrease in the amount of fructose (Russell and McFrederick, 2021). Future work should examine the effect of reduced water on the microbiomes of nectar and pollen, and more studies should quantify the impacts of these changes on the foraging behavior and health of pollinators.

Changes in temperature and water related to climate change also have direct effects on pollinators. Higher temperature during development significantly increases larval or adult mortality (da Silva *et al.*, 2017; Ostap-Chec *et al.*, 2021), which can lead to reduced pollinator diversity (Ramos-Jiliberto *et al.*, 2020). Drier conditions can favor drought-tolerant plants (Feeley *et al.*, 2011) and generalist pollinators (Pardee *et al.*, 2022). Specialist pollinators have few host plants, and become at risk when these plants become rare or disappear (Ramos-Jiliberto *et al.*, 2020).

A warming climate can lead to a change in species distribution, as species move to higher elevations or higher latitudes (Pyke *et al.*, 2016; Marshall *et al.*, 2020).

Unlike measures to control land loss, use of agrochemicals, and spillover of pests and pathogens between managed and wild bees, controlling climate change requires efforts at a global scale. It also represents the factor with the most varied impacts. The changes in temperature and water availability associated with climate change are expected to decrease the quantity and quality of floral resources available to pollinators. Increased temperature can create asynchrony between plant flowering phenology and pollinator emergence, further limiting resource availability to pollinators. It can also directly affect the viability of many pollinators by increasing larval or adult mortality. These abiotic changes can affect the distribution of plant and pollinator species, including crops, leading to issues of food security. Given its multiple impacts, climate change is the most prominent threat to pollinators, and the most difficult to control.

Humans and pollinators decline

Pollinators are negatively impacted by human actions. Habitat loss and degradation, application of agrochemicals, climate change, and pollution, are all consequences of human activities. The negative impacts of managed bees on wild bees, and the more intense sampling of bees to obtain baseline data also result from human actions. While humans are responsible for many of the factors affecting pollinator decline, the loss of pollinators will affect several aspects of human lives.

In the absence of pollinators, the human diet will shift toward a preponderance of grains (e.g. wheat, rice, oat) and corn, as these are wind-pollinated crops. It will also include crops that were bred to reproduce mostly or entirely via self-fertilization and these include some varieties of beans, peas, lettuce, and tomatoes, chard and spinach, among others. Crops that reproduce vegetatively, such as bananas, will also be maintained. Self-fertilization does, however, reduce the heterozygosity of individual plants, rendering them more vulnerable to pests and diseases. Clonally reproduced crops exhibit lower levels of genetic variation, affecting their ability to adjust to environmental changes. Many fruits and vegetables will need to be hand-pollinated for their production and persistence. Hand-pollination is a very costly and labor-intensive process. In addition, the availability of self-pollinating crop varieties is uneven across the world. We thus expect sharp increases in food prices, and geographic variation in food availability, with dire consequences for food stability across the world.

A decline of pollinators will change plant communities and the many insects and animals they harbor. The plant species that rely on pollinators to reproduce will decline and wind-pollinated plants and obligate selfers, together with plants that mostly reproduce vegetatively, will start to dominate plant communities. This will change the community of insects and animals associated with the plant community. Such changes could impact not only recreational activities for humans but also future discoveries of new plant compounds for pharmaceuticals or other usage.

Steps must be taken to lessen the negative impacts of human activities on pollinators. Some initiatives are already underway. Strips of wild flowers are planted around crop fields, and there is a strong movement to develop integrated pollinator management practices to reduce the negative impacts of agrochemicals on pollinators. Studies demonstrating the importance of wild pollinators on crop pollination have increased the attention given to wild pollinators. Bees forage within larger landscapes containing agricultural and wild habitats, and resources available in such landscapes are shared by both managed and wild bees. Thus, management practices to help conserve pollinators must integrate agricultural land and wild habitats and consider both managed and wild bees.

Funding must remain available to encourage and sustain practices to help conserve wild pollinators. The setup of strips of wild flowers is expensive and such practices will only be sustained if funding remains available to farmers to pursue them. Funding must continue to encourage the further development of eco-friendly management practices for pollinators in agriculture. Education of the general public must persist and people should be encouraged to grow gardens with wild plants and crops on their properties to provide floral resources and nesting sites to pollinators in urban areas.

Climate change is the most critical threat to pollinators, and research must continue to study its impacts and develop solutions. Public education on the negative impacts of climate change on plants and their pollinators and the implications for human diet and lives must take place at a global scale. Climate change is a global problem and a political one with strong implications for food security and world stability.

Conclusion

Widespread use of sustainable practices in agriculture, and further development of integrated pollinator management strategies, eco-friendly strategies including reduction of pesticide use, will help preserve pollinators. Potential adverse effects of managed bees on the local wild bee populations must be mitigated. Non-lethal collection methods should be developed and adopted globally in response to the increasing need for baseline pollinator data collection. Adopting a more holistic approach to pollinator conservation, with management strategies that integrate natural habitats and agricultural systems, together with managed and wild bees, should become a high priority worldwide. Measures must keep being implemented to reduce climate change and prevent its serious negative impacts on pollinators. Climate change has the most diverse negative impacts on pollinators and is the threat most difficult to control. However, its consequences threaten food security and world stability thus efforts to control it must be prioritized on a global scale.

Author Contributions

The authors contributed equally to the development and writing of the manuscript.

References

Aizen, M., Garibaldi, L.A., Cunningham, S. and Klein, A.M. (2008) Long-term global trends in crop yield and production reveal no current pollination shortage but increasing pollinator dependency. *Current Biology* 18, 1572–1575. DOI: 10.1016/j.cub.2008.08.066.

Aizen, M., Garibaldi, L.A., Cunningham, S. and Klein, A.M. (2009) How much does agriculture depend on pollinators? Lessons from long-term trends in crop production. *Annals of Botany* 103, 1579–1588. DOI: 10.1093/aob/mcp076.

Angelella, G.M., McCullough, C.T. and O'Rourke, M.E. (2021) Honey bee hives decrease wild bee abundance, species richness, and fruit count on farms regardless of wildflower strips. *Scientific Reports* 11, 3202. DOI: 10.1038/s41598-021-81967-1.

Archer, C.R., Pirk, C.W.W., Carvalheiro, L.G. and Nicolson, S.W. (2014) Economic and ecological implications of geographic bias in pollinator ecology in the light of pollinator declines. *Oikos* 123, 401–407. DOI: 10.1111/j.1600-0706.2013.00949.x.

Bartomeus, I., Ascher, J.S., Gibbs, J. and Winfree, R. (2013) Historical changes in northeastern US bee pollinators related to shared ecological traits. *Proceedings of the National Academy of Science* 110, 4656–4660. DOI: 10.1073/pnas.1218503110.

Bernklau, E., Bjostad, L., Hogeboom, A., Carlisle, A. and Arathi, H.S. (2019) Dietary phytochemicals, honey bee longevity and pathogen tolerance. *Insects* 10, 14. DOI: 10.3390/insects10010014.

Boyer, K.J., Fragoso, F.P., Dieterich Mabin, M.E. and Brunet, J. (2020) Netting and pan traps fail to identify the pollinator guild of an agricultural crop. *Scientific Reports* 10, 13819. DOI: 10.1038/s41598-020-70518-9.

Brunet, J. and Van Etten, M.L. (2019) The response of floral traits associated with pollinator attraction to environmental changes expected under anthropogenic climate change in high-altitude habitats. *International Journal of Plant Sciences* 180, 954–964. DOI: 10.1086/705591.

Brunet, J., Thairu, M.W., Henss, J.M., Link, R.I. and Kluever, J.A. (2015) The effects of flower, floral display, and reward sizes on bumblebee foraging behavior when pollen is the reward and plants are dichogamous. *International Journal of Plant Sciences* 176, 811–819. DOI: 10.1086/683339.

Budge, G.E., Simcock, N.K., Holder, P.J., Shirley, M.D., Brown, M.A. *et al.* (2020) Chronic bee paralysis as a serious emerging threat to honey bees. *Nature Communications* 11, 1–9. DOI: 10.1038/s41467-020-15919-0.

Burden, C.M., Morgan, M.O., Hladun, K.R., Amdam, G.V., Trumble, J.J. *et al.* (2019) Acute sublethal exposure to toxic heavy metals alters honey bee (*Apis mellifera*) feeding behavior. *Scientific Reports* 9, 4253. DOI: 10.1038/s41598-019-40396-x.

Burkle, L.A. and Runyon, J.B. (2016) Drought and leaf herbivory influence floral volatiles and pollinator attraction. *Global Change Biology* 22, 1644–1654. DOI: 10.1111/gcb.13149.

Cabezas, G. and Farinós, G.P. (2022) Sensitivity of buff-tailed bumblebee (*Bombus terrestris* L.) to insecticides with different modes of action. *Insects* 13, 184. DOI: 10.3390/insects13020184.

Cameron, S.A., Lozier, J.D., Strange, J.P., Koch, J.B., Cordes, N. *et al.* (2011) Patterns of widespread decline in North American bumble bees. *Proceedings of the National Academy of Sciences* 108, 662–667. DOI: 10.1073/pnas.1014743108.

Cameron, S.A., Lim, H.C., Lozier, J.D. and Thorp, R. (2016) Test of the invasive pathogen hypothesis of bumble bee decline in North America. *Proceedings of the National Academy of Science* 113, 4386–4391. DOI: 10.1073/pnas.1525266113.

Campbell, D.R., Sosenski, P. and Raguso, R.A. (2019) Phenotypic plasticity of floral volatiles in response to increasing drought stress. *Annals of Botany* 123, 601–610. DOI: 10.1093/aob/mcy193.

Cane, J.H. and Tepedino, V.J. (2016) Gauging the effect of honey bee pollen collection on natural bee communities. *Conservation Letters* 10, 205–210. DOI: 10.1111/conl.12263.

Cane, J.H., Minckley, R.L., Kervin, L.J., Roulston, T.H. and Williams, N.M. (2006) Complex responses within a desert bee guild (*Hymenoptera:* Apiformes) to urban habitat fragmentation. *Ecological Applications* 16, 632–644. DOI: 10.1890/1051-0761(2006)016[0632:crwadb]2.0.co;2.

Carré, G., Roche, P., Chifflet, R., Morison, N., Bommarco, R, *et al.* (2009) Landscape context and habitat type as drivers of bee diversity in European annual crops. *Agriculture, Ecosystems & Environment* 133, 40–47. DOI: 10.1016/j.agee.2009.05.001.

Carroll, A.B., Pallardy, S.G. and Galen, C. (2001) Drought stress, plant water status, and floral trait expression in fireweed, *Epilobium angustifolium* (Onagraceae). *American Journal of Botany* 88, 438–446. DOI: 10.2307/2657108.

Carvell, C., Roy, D.B., Smart, S.M., Pywell, R.F., Preston, C.D. *et al.* (2006) Declines in forage availability for bumblebees at a national scale. *Biological Conservation* 132, 481–489. DOI: 10.1016/J.biocon.2006.05.008.

Chattopadhyay, A., Samadder, A. and Bhattacharya, S. (2023) Pesticide mediated critical transition in plant-pollinator networks, *arXiv*: 2303.08495. DOI: 10.48550/arXiv.2303.08495.

Cnaani, J., Thomson, J.D. and Papaj, D.R. (2006) Flower choice and learning in foraging bumblebees: Effects of variation in nectar volume and concentration. *Ethology* 112, 278–285. DOI: 10.1111/j.1439-0310.2006.01174.x.

Cohen, H., Philpott, S.M., Liere, H., Lin, B.B. and Jha, S. (2020) The relationship between pollinator community and pollination services is mediated by floral abundance in urban landscapes. *Urban Ecosystems* 24, 275–290. DOI: 10.1007/s11252-020-01024-z.

Colla, S.R., Otterstatter, M.C., Gegear, R.J. and Thomson, J.D. (2006) Plight of the bumble bee: pathogen spillover from commercial to wild populations. *Biological Conservation* 129, 416–467. DOI: 10.1016/j.%20biocon.2005.11.013.

Cordes, N., Huang, W.F., Strange, J.P., Cameron, S.A., Griswold, T.L. *et al.* (2012) Interspecific geographic distribution and variation of the pathogens *Nosema bombi* and *Crithidia* species in United States bumble bee populations. *Journal of Invertebrate Pathology* 109, 209–216. DOI: 10.1016/j.jip.2011.11.005.

Czerwinski, M.A. and Sadd, B.M. (2017) Detrimental interactions of neonicotinoid pesticide exposure and bumblebee immunity. *Journal of Experimental Zoology* 327, 273–283. DOI: 10.1002/jez.2087.

da Silva, M.A., Ferreira, N.S., Teixeira-Souza, V.H.S., Maia-Silva, C., de Assis de Oliveira, F. *et al.* (2017) On the thermal limits for the use of stingless bees as pollinators in commercial greenhouses. *Journal of Apicultural Research* 56, 8190. DOI: 10.1080/00218839.2016.1260380.

De Palma, A., Abrahamczyk, S., Aizen, M.A. and Albrecht, M. (2016) Predicting bee community responses to land-use changes: Effects of geographic and taxonomic biases. *Scientific Reports* 6, 31153. DOI: 10.1038/srep31153.

Descamps, C., Quinet, M. and Jacquemart, A.-L. (2021a) Climate change–induced stress reduce quantity and alter composition of nectar and pollen from a bee-pollinated species (*Borago officinalis*, Boraginaceae). *Frontiers in Plant Science* 12, 755843. DOI: 10.3389/fpls.2021.755843.

Descamps, C., Boubnan, N., Jacquemart, A.L. and Quinet, M. (2021c) Growing and flowering in a changing climate: Effects of higher temperatures and drought stress on the bee-pollinated species *Impatiens glandulifera* Royle. *Plants* 10, 988. DOI: 10.3390/plants10050988.

Descamps, C., Quinet, M. and Jacquemart, A.-L. (2021b) The effects of drought on plant–pollinator interactions: What to expect? *Environmental and Experimental Botany* 182, 104297. DOI: 10.1016/j.envexpbot.2020.104297.

Desneux, N., Decourtye, A. and Delpuech, J.M. (2007) The sublethal effects of pesticides on beneficial arthropods. *Annual Review of Entomology* 52, 81–106. DOI: 10.1146/annurev.ento.52.110405.091440.

Dharampal, P.S., Carlson, C., Currie, C.R. and Steffan, S.A. (2019) Pollen-borne microbes shape bee fitness. *Proceedings of the Royal Society B: Biological Sciences* 286(1904), 20182894. DOI: 10.1098/rspb.2018.2894.

Di Pasquale, G., Salignon, M., Le Conte, Y., Belzunces, L.P., Decourtye, A. *et al.* (2013) Influence of pollen nutrition on honey bee health: Do pollen quality and diversity matter? *PLoS One* 8, e72016. DOI: 10.1371/journal.pone.0072016.

Dobelmann, J., Felden, A. and Lester, P.J. (2023) An invasive ant increases deformed wing virus loads in honey bees. *Biology Letters* 19, 2022041. DOI: 10.1098/rsbl.2022.0416.

dos Santos, C.F., Acosta, A.L., Halinski, R., Souza-Santos, P.D., Borges, R.C. *et al.* (2022) The widespread trade in stingless beehive may introduce them into novel places and could threaten species. *Journal of Applied Ecology* 59, 965–981. DOI: 10.1111/1365-2664.14108.

Dzul-Cauich, H.F. and Munguía-Rosas, M.A. (2022) Negative effects of light pollution on pollinator visits are outweighed by positive effects on the reproductive success of a bat-pollinated tree. *Naturwissenschaften* 109, 12. DOI: 10.1007/s00114-021-01783-5.

Evans, E., Strange, J., Sadd, B., Tripodi, A., Figueroa, L, *et al.* (2023) Parasites, parasitoids, and hive products that are potentially deleterious to wild and commercially raised bumble bees (*Bombus* spp.) in North America. *Journal of Pollination Ecology* 32, 37–53. DOI: 10.26786/1920-7603(2023)710.

Eyer, M., Chen, Y.P., Schäfer, M.O., Pettis, J. and Neumann, P. (2009) Small hive beetle, *Aethina tumida*, as a potential biological vector of honeybee viruses. *Apidologie* 40, 419–428. DOI: 10.1051/apido:2008051.

Feeley, K.J., Davies, S.J., Perez, R., Hubbell, S.P. and Foster, R.B. (2011) Directional changes in the species composition of a tropical forest. *Ecology* 92, 871–882. DOI: 10.1890/10-0724.1.

Feldhaar, H. and Otti, O. (2020) Pollutants and their interaction with diseases of social Hymenoptera. *Insects* 11, 153. DOI: 10.3390/insects11030153.

Figueroa, L., Sadd, B., Tripodi, A., Strange, J., Colla, S., *et al.* (2023) Endosymbionts that threaten commercially raised and wild bumble bees (*Bombus* spp.). *Journal of Pollination Ecology* 32, 14–36. DOI: 10.26786/1920-7603(2023)713.

Fleming, T.H., Geiselman, C. and Kress, W.J. (2009) The evolution of bat pollination: A phylogenetic perspective. *Annals of Botany* 104, 1017–1043. DOI: 10.1093/aob/mcp197.

Ford, H.A., Paton, D.C. and Forde, N. (1979) Birds as pollinators of Australian plants. *New Zealand Journal of Botany* 17, 509–519. DOI: 10.1080/0028825X.1979.10432566.

Forrest, J.R.K. and Thomson, J.D. (2011) An examination of synchrony between insect emergence and flowering in Rocky Mountain meadows. *Ecological Monographs* 81, 469–491. DOI: 10.1890/10-1885.1.

Fragoso, F.P. and Brunet, J. (2023) The decision-making process of leafcutting bees when selecting patches. *Biology Letters* 9, 20220411. DOI: 10.1098/rsbl.2022.0411.

Freimuth, J., Bossdorf, O., Scheepens, J.F. and Willems, F.M. (2022) Climate warming changes synchrony of plants and pollinators. *Proceedings of the Royal Society B: Biological Sciences* 289(1971), 20212142. DOI: 10.1098/rspb.2021.2142.

Fragoso, F.P., Jiang, Q., Clayton, M.K. and Brunet, J. (2021) Patch selection by bumble bees navigating discontinuous landscapes. *Scientific Reports* 11, 8986. DOI: 10.1038/s41598-021-88394-2.

Freitas, C.D., Oki, Y., Resende, F.M., Zamudio, F., Freitas, G.S, *et al.* (2022) Impacts of pests and diseases on the decline of managed bees in Brazil: A beekeeper perspective. *Journal of Apicultural Research* 62, 969–982. DOI: 10.1080/00218839.2022.2099188.

Fürst, M.A., McMahon, D.P., Osborne, J.L., Paxton, R.J. and Brown, M.J.F. (2014) Disease associations between honeybees and bumblebees as a threat to wild pollinators. *Nature* 506, 364–366. DOI: 10.1038/nature12977.

Gallagher, M.K. and Campbell, D.R. (2017) Shifts in water availability mediate plant–pollinator interactions. *New Phytologist* 215, 792–802. DOI: 10.1111/nph.14602.

Gallai, N., Salles, J.-M., Settele, J. and Vaissière, B.E. (2009) Economic valuation of the vulnerability of world agriculture confronted with pollinator decline. *Ecological Economics* 68, 810–821. DOI: 10.1016/j.ecolecon.2008.06.014.

Garibaldi, L.A., Steffan-Dewenter, I., Winfree, R., Aizen, M.A., Bommarco, R. *et al.* (2013) Wild pollinators enhance fruit set of crops regardless of honey bee abundance. *Science* 340, 1608–1611. DOI: 10.1126/science.1230200.

Gezon, Z.J., Wyman, E.S., Ascher, J.S., Inouye, D.W. and Irwin, R.E. (2015) The effect of repeated, lethal sampling on wild bee abundance and diversity. *Methods in Ecology and Evolution* 6, 1044–1054. DOI: 10.1111/2041-210X.12375.

Giavi, S., Fontaine, C. and Knop, E. (2021) Impact of artificial light at night on diurnal plant-pollinator interactions. *Nature Communications* 12, 1690. DOI: 10.1038/s41467-021-22011-8.

Gill, R.J. and Raine, N.E. (2014) Chronic impairment of bumblebee natural foraging behaviour induced by sublethal pesticide exposure. *Functional Ecology* 28, 1459–1471. DOI: 10.1111/1365-2435.12292.

Girolami, V., Mazzon, L., Squartini, A., Mori, N., Marzaro, M. *et al.* (2009) Translocation of neonicotinoid insecticides from coated seeds to seedling guttation drops: A novel way of intoxication for bees. *Journal of Economic Entomology* 102, 1808–1815. DOI: 10.1603/029.102.0511.

Glavinic, U., Stankovic, B., Draskovic, V., Stevanovic, J., Petrovic, T. *et al.* (2017) Dietary amino acid and vitamin complex protects honey bee from immunosuppression caused by *Nosema ceranae*. *Plos One* 12(11), e0187726. DOI: 10.1371/journal.pone.0187726.

Glenny, W.R., Runyon, J.B. and Burkle, L.A. (2018) Drought and increased CO_2 alter floral visual and olfactory traits with context-dependent effects on pollinator visitation. *New Phytologist* 220, 785–798. DOI: 10.1111/nph.15081.

Gonthier, J., Papach, A., Straub, L., Campbell, J.W., Williams, G.R. *et al.* (2019) Bees and flowers: How to feed an invasive beetle species. *Ecology and Evolution* 9, 6422–6432. DOI: 10.1002/ece3.5217.

Goulson, D., Lye, G.C. and Darvill, B. (2008) Decline and conservation of bumble bees. *Annual Review of Entomology* 53, 191–208. DOI: 10.1146/annurev.ento.53.103106.093454.

Goulson, D. (2013) Review: An overview of the environmental risks posed by neonicotinoid insecticides. *Journal of Applied Ecology* 50, 977–987. DOI: 10.1111/1365-2664.12111.

Goulson, D., Nicholls, E., Botías, C. and Rotheray, E.L. (2015) Bee declines driven by combined stress from parasites, pesticides, and lack of flowers. *Science* 347, 1255957. DOI: 10.1126/science.1255957.

Graystock, P., Goulson, D. and Hughes, W.O.H. (2015) Parasites in bloom: Flowers aid dispersal and transmission of pollinator parasites within and between bee species. *Proceedings of the Royal Society B: Biological Sciences* 282, 20151371. DOI: 10.1098/rspb.2015.1371.

Hadley, A.S. and Betts, M.G. (2011) The effects of landscape fragmentation on pollination dynamics: Absence of evidence not evidence of absence. *Biological Reviews* 87, 526–544. DOI: 10.1111/j.1469-185X.2011.00205.x.

Häussler, J., Sahlin, U., Baey, C., Smith, H.G. and Clough, Y. (2017) Pollinator population size and pollination ecosystem service responses to enhancing floral and nesting resources. *Ecology and Evolution* 7, 1898–1908. DOI: 10.1002/ece3.2765.

Hegland, S.J., Nielsen, A., Lázaro, A., Bjerknes, A.-L. and Totland, Ø. (2009) How does climate warming affect plant-pollinator interactions? *Ecology Letters* 12, 184–195. DOI: 10.1111/j.1461-0248.2008.01269.x.

Henry, M., Beguin, M., Requier, F., Rollin, O., Odoux, J.-F. *et al.* (2012) A common pesticide decreases foraging success and survival in honeybees. *Science* 336, 350–351. DOI: 10.1126/science.1215039.

Hoffmann, D., Pettis, J.S. and Neumann, P. (2008) Potential host shift of the small hive beetle (*Aethina tumida*) to bumblebee colonies (*Bombus impatiens*). *Insect Sociaux* 55, 153–162. DOI: 10.1007/s00040-008-0982-9.

Inouye, D.W. (2022) Climate change and phenology. *WIREs Climate Change* 13, e764. DOI: 10.1002/wcc.764.

IPBES (2016) The assessment report of the intergovernmental science-policy platform on biodiversity and ecosystem services on pollinators, pollination and food production. In: Potts, S.G., Imperatriz-Fonseca, V.L. and Ngo, H.T. (eds) *Secretariat of the Intergovernmental Science-Policy Platform on Biodiversity and Ecosystem Services*. IPBES, Bonn, Germany, p. 552. DOI: 10.5281/zenodo.3402856.

Janousek, W.M., Douglas, M.R., Cannings, S., Clément, M.A., Delphia, C.M, *et al.* (2023) Recent and future declines of a historically widespread pollinator linked to climate, land cover, and pesticides. *Proceedings of the National Academy of Sciences* 120, e2211223120. DOI: 10.1073/pnas.2211223120.

Jarau, S., Hrncir, M., Schmidt, V.M., Zucchi, R. and Barth, F.G. (2003) Effectiveness of recruitment behavior in stingless bees (Apidae, Meliponini). *Insectes Sociaux* 50, 365–374. DOI: 10.1007/s00040-003-0684-2.

Kapahi, M. and Sachdeva, S. (2019) Bioremediation options for heavy metal pollution. *Journal of Health and Pollution* 9, 191203. DOI: 10.5696/2156-9614-9.24.191203.

Keefover-Ring, K., Hetherington, M.C. and Brunet, J. (2022) Population-specific responses of floral volatiles to abiotic factors in changing environments. *American Journal of Botany* 109, 676–688. DOI: 10.1002/ajb2.1846.

Kehrberger, S. and Holzschuh, A. (2019) Warmer temperatures advance flowering in a spring plant more strongly than emergence of two solitary spring bee species. *Plos One* 14, e0218824. DOI: 10.1371/journal.pone.0218824.

Klein, A.M., Vaissière, B.E., Cane, J.H., Steffan-Dewenter, I., Cunningham, S.A. *et al.* (2007) Importance of pollinators in changing landscapes for world crops. *Proceedings of the Royal Society B Biological Sciences* 274, 303–331. DOI: 10.1098/rspb.2006.3721.

Knop, E., Zoller, L., Ryser, R., Gerpe, C., Hörler, M. *et al.* (2017) Artificial light at night as a new threat to pollination. *Nature* 548, 206–209. DOI: 10.1038/nature23288.

Kopec, K. and Burd, L.A. (2017) Pollinators in peril. A systematic status review of North American and Hawaiian native bees. Center for Biodiversity. Available at: https://www.biologicaldiversity.org/campaigns/native_pollinators/pdfs/Pollinators_in_Peril.pdf (accessed 17 October 2023).

Kosior, A., Celary, W., Olejniczak, P., Fijal, J., Król, W, *et al.* (2007) The decline of the bumble bees and cuckoo bees (Hymenoptera: Apidae: Bombini) of western and central Europe. *Oryx* 41, 79–88. DOI: 10.1017/S0030605307001597.

Kremen, C., Williams, N.W. and Thorp, R.W. (2002) Crop pollination from native bees at risk from agricultural intensification. *Proceedings of the National Academy of Sciences* 99, 16812–16816. DOI: 10.1073/pnas.262413599.

Kress, W.J., Shatz, G.E., Andrianifahanana, M. and Morland, H.S. (1994) Pollination of *Ravenala madagascariensis* (Strelitziaceae) by lemurs in Madagascar: Evidence for an archaic coevolutionary system? *American Journal of Botany* 81, 542–551. DOI: 10.1002/j.1537-2197.1994.tb15483.x.

Krupke, C.H., Hunt, G.J., Eitzer, B.D., Andino, G. and Given, K. (2012) Multiple routes of pesticide exposure for honey bees living near agricultural fields. *Plos One* 7, e29268. DOI: 10.1371/journal.pone.0029268.

Kuppler, J. and Kotowska, M.M. (2021) A meta-analysis of responses in floral traits and flower–visitor interactions to water deficit. *Global Change Biology* 27, 30953108. DOI: 10.1111/gcb.15621.

Lewanski, N.J., Williams, N.W. and Winfree, R. (2022) Greater bee diversity is needed to maintain crop pollination over time. *Nature Ecology & Evolution* 6, 1516–1523. DOI: 10.1038/41559-022-01847-3.

Londo, J.P., McKinney, J., Schwartz, M., Bollman, M., Sagers, C. *et al.* (2014) Sub-lethal glyphosate exposure alters flowering phenology and causes transient male-sterility in *Brassica* spp. *BMC Plant Biology* 14, 70. DOI: 10.1186/1471-2229-14-70.

MacGregor, C.J., Evans, D.M., Fox, R. and Pocock, M.J.O. (2017) The dark side of street lighting: Impacts on moths and evidence for the disruption of nocturnal pollen transport. *Global Change Biology* 23, 697–707. DOI: 10.1111/gcb.13371.

Makino, T.T. and Sakai, S. (2007) Experience changes pollinator responses to floral display size: From size-based to reward-based foraging. *Functional Ecology* 21, 854–863. DOI: 10.1111/j.1365-2435.2007.01293.x.

Manley, R., Boots, M. and Wilfert, L. (2015) Emerging viral disease risk to pollinating insects: Ecological, evolutionary and anthropogenic factors. *Journal of Applied Ecology* 52, 331–340. DOI: 10.1111/1365-2664.12385.

Marshall, L., Perdijk, F., Dendoncker, N., Kunin, W., Roberts, S. *et al.* (2020) Bumblebees moving up: shifts in elevation ranges in the Pyrenees over 115 years. *Proceedings of the Royal Society B: Biological Sciences* 287, 20202201. DOI: 10.1098/rspb.2020.2201.

Martínez-López, V., Ruiz, C. and De la Rúa, P. (2022) Migratory beekeeping and its influence on the prevalence and dispersal of pathogens to managed and wild bees. *International Journal for Parasitology: Parasites and Wildlife* 18, 184–193. DOI: 10.1016/j.ijppaw.2022.05.004.

Meindl, G.A. and Ashman, T.L. (2013) The effects of aluminum and nickel in nectar on the foraging behavior of bumblebees. *Environmental Pollution* 177, 78–81. DOI: 10.1016/j.envpol.2013.02.017.

Memmott, J., Craze, P.G., Waser, N.M. and Price, M.V. (2007) Global warming and the disruption of plant-pollinator interactions. *Ecology Letters* 10, 710–717. DOI: 10.1111/j.1461-0248.2007.01061.x.

Michener, C.D. (2007) *The Bees of the World*. 2nd ed. The Johns Hopkins University Press, Baltimore, USA.

Moerman, R., Vanderplanck, M., Roger, N., Declèves, S., Wathlet, B. *et al.* (2016) Growth rate of bumblebee larvae is related to pollen amino acids. *Journal of Economic Entomology* 109, 25–30. DOI: 10.1093/jee/tov279.

Monchanin, C. (2021) Impacts of metallic pollutants on honey bees: From the colony to the brain. PhD Dissertation, Université de Toulouse 3 Paul Sabatier.

Murray, M.C., Coffey, M.F., Kehoe, E. and Horgan, F.G. (2013) Pathogen prevalence in commercially reared bumble bees and evidence of spillover in conspecific populations. *Biological Conservation* 159, 269–276. DOI: 10.1016/j.biocon.2012.10.021.

Nacko, S., Hall, M., Duncan, M., Cook, J., Riegler, M, *et al.* (2020) Scientific note on small hive beetle infestation of stingless bee (*Tetragonula carbonaria*) colony following a heat wave. *Apidologie* 51, 1199–1201. DOI: 10.1007/s13592-020-00799-6.

Nath, R., Singh, H. and Mukherjee, S. (2022) Insect pollinators decline: An emerging concern of Anthropocene epoch. *Journal of Apicultural Research* 62, 23–38. DOI: 10.1080/00218839.2022.2088931.

Nicolson, S.W. (2011) Bee food: The chemistry and nutritional value of nectar, pollen and mixtures of the two. *African Zoology* 46, 197–204. DOI: 10.3377/004.046.0201.

Nieh, J.C. and Roubik, D.W. (1998) Potential mechanisms for the communication of height and distance by a stingless bee, *Melipona panamica*. *Behavioral Ecology and Sociobiology* 43, 387–399. DOI: 10.1007/s002650050506.

Olesen, J.M. and Valido, A. (2003) Lizards as pollinators and seed dispersers: An island phenomenon. *Trends in Ecology and Evolution* 18, 177–181. DOI: 10.1016/S0169-5347(03)00004-1.

Ollerton, J., Winfree, R. and Tarrant, S. (2011) How many flowering plants are pollinated by animals? *Oikos* 120, 321–326. DOI: 10.1111/j.1600-0706.2010.18644.x.

Orlova, M., Porter, M., Hines, H.M. and Amsalem, E. (2023) Symptomatic infection with *Vairimorpha* spp. decreases diapause survival in a wild bumble bee species (*Bombus griseocollis*). *Animals* 13, 1656. DOI: 10.3390/ani13101656.

Orr, M.C., Hughes, A.C., Chesters, D., Pickering, J., Zhu, C.D. *et al.* (2021) Global patterns and drivers of bee distribution. *Current Biology* 31, 451–458. DOI: 10.1016/j.cub.2020.10.053.

Ostap-Chec, M., Kierat, J., Kuszewska, K. and Woyciechowski, M. (2021) Red mason bee (*Osmia bicornis*) thermal preferences for nest sites and their effects on offspring survival. *Apidologie* 52, 707–719. DOI: 10.1007/s13592-021-00858-6.

Otterstatter, M.C. and Thomson, J.D. (2008) Does pathogen spillover from commercially reared bumble bees threaten wild pollinators? *Plos One* 3, e2771. DOI: 10.1371/journal.pone.0002771.

Ovaskainen, O., Skorokhodova, S., Yakovleva, M., Sukhov, A., Kutenkov, A. *et al.* (2013) Community level phenological response to climate change. *Proceedings of the National Academy of Science* 110, 13434–13439. DOI: 10.1073/pnas.1305533110.

Pardee, G.L., Griffin, S.R., Stemkovski, M., Harrison, T., Portman, Z.M. *et al.* (2022) Life-history traits predict responses of wild bees to climate variation. *Proceedings of the Royal Society B: Biological Sciences* 289, 20212697. DOI: 10.1098/rspb.2021.2697.

Peña-Chora, G., Toledo-Hernández, E., Sotelo-Leyva, C., Damian-Blanco, P., Villanueva-Flores, A.G. *et al.* (2023) Presence and distribution of pests and diseases of *Apis mellifera* (Hymenoptera: Apidae) in Mexico: A review. *The European Zoological Journal* 90, 224–236. DOI: 10.1080/24750263.2023.2182920.

Pereira, S.N., Alves, L.H.S., da Costa, R.F.R., Prezoto, F. and Teixeira, E.W. (2021) Occurrence of the small hive beetle (*Aethina tumida*) in *Melipona rufiventris* colonies in Brazil. *Sociobiology* 68, e6021. DOI: 10.13102/sociobiology.v68i1.6021.

Pisa, L.W., Amaral-Rogers, V., Belzunces, L.P., Bonmatin, J.M., Downs, C.A. *et al.* (2015) Effects of neonicotinoids and fipronil on non-target invertebrates. *Environmental Science and Pollution Research* 22, 68–102. DOI: 10.1007/s11356-014-3471-x.

Potts, S.G., Biesmeijer, J.C., Kremen, C., Neumann, P., Schweiger, O. *et al.* (2010) Global pollinator declines: Trends, impacts and drivers. *Trends in Ecology and Evolution* 26, 345–353. DOI: 10.1016/j.tree.2010.01.007.

Pyke, G.H., Thomson, J.D., Inouye, D.W. and Miller, T.J. (2016) Effects of climate change on phenologies and distributions of bumble bees and the plants they visit. *Ecosphere* 7, e01267. DOI: 10.1002/ecs2.1267.

Ramos-Jiliberto, R., Moisset de Espanés, P. and Vázquez, D.P. (2020) Pollinator declines and the stability of plant–pollinator networks. *Ecosphere* 11, e03069. DOI: 10.1002/ecs2.3069.

Ravoet, J., De Smet, L., Meeus, I., Smagghe, G., Wenseleers, T. *et al.* (2014) Widespread occurrence of honey bee pathogens in solitary bees. *Journal of Invertebrate Pathology* 122, 55–58. DOI: 10.1016/j.jip.2014.08.007.

Reilly, J.R., Artz, D.R., Biddinger, D., Bobiwash, K., Boyle, N.K. *et al.* (2020) Crop production in the USA is frequently limited by a lack of pollinators. *Proceedings of the Royal Society B: Biological Sciences* 287, 20200922. DOI: 10.1098/rspb.2020.0922.

Rering, C.C., Franco, J.G., Yeater, K.M. and Mallinger, R.E. (2020) Drought stress alters floral volatiles and reduces floral rewards, pollinator activity, and seed set in a global plant. *Ecosphere* 11, e03254. DOI: 10.1002/ecs2.3254.

Ricketts, T.H., Regetz, J., Steffan-Dewenter, I., Cunningham, S.A., Kremen, C. *et al.* (2008) Landscape effects on crop pollination services: Are there general patterns? *Ecology Letters* 11, 499–515. DOI: 10.1111/j.1461-0248.2008.01157.x.

Rothman, J. (2019) Effects of toxicant exposure on honey bee and bumble bee microbiomes and impacts on host health. PhD Dissertation, University of California, Riverside.

Roulston, T.H. and Cane, J.H. (2000) Pollen nutritional content and digestibility for animals. *Plant Systematics and Evolution* 222, 187–209. DOI: 10.1007/BF00984102.

Rundlöf, M., Andersson, G., Bommarco, R., Fries, I., Hederström, V. *et al.* (2015) Seed coating with a neonicotinoid insecticide negatively affects wild bees. *Nature* 521, 77–80. DOI: 10.1038/nature14420.

Russell, K.A. and McFrederick, Q.S. (2021) Elevated temperature may affect nectar microbes, nectar sugars, and bumble bee foraging preference. *Microbial Ecology* 84, 473–482. DOI: 10.1007/s00248-021-01881-x.

Sagae, M., Oyama-Okubo, N., Ando, T., Marchesi, E. and Nakayama, M. (2008) Effect of temperature on the floral scent emission and endogenous volatile profile of *Petunia axillaris*. *Bioscience, Biotechnology, and Biochemistry* 72, 110–115. DOI: 10.1271/bbb.70490.

Scheper, J., Reemer, M., van Kats, R., Ozinga, W.A., van der, Linde, *et al.* (2014) Museum specimens reveal loss of pollen host plants as key factor driving wild bee decline in The Netherlands. *Proceedings of the National Academy of Sciences* 111, 17552–17557. DOI: 10.1073/pnas.1412973111.

Scofield, H.N. and Mattila, H.R. (2015) Honey bee workers that are pollen stressed as larvae become poor foragers and waggle dancers as adults. *Plos One* 10, e0121731. DOI: 10.1371/journal.pone.0121731.

Seeley, T.D. (2017) Life-history traits of wild honey bee colonies living in forests around Ithaca, NY, USA. *Apidologie* 48, 743–754. DOI: 10.1007/s13592-017-0519-1.

Sekercioglu, C.H. (2006) Increasing awareness of avian ecological function. *Trends in Ecology and Evolution* 21, 464–471. DOI: 10.1016/j.tree.2006.05.007.

Semmens, B.X., Semmens, D.J., Thogmartin, W.E., Wiederholt, R., López-Hoffman, L. *et al.* (2016) Quasi-extinction risk and population targets for the Eastern, migratory population of monarch butterflies (*Danaus plexippus*). *Scientific Reports* 6, 23265. DOI: 10.1038/srep23265.

Shackleton, K., Balfour, N.J., Al Toufailia, H., Emma James, E. and Ratnieks, F.L.W. (2023) Honey bee waggle dances facilitate shorter foraging distances and increased foraging aggregation. *Animal Behaviour* 198, 11e19. DOI: 10.1016/j.anbehav.2023.01.009.

Shivanna, K.R. (2022) Impact of light pollution on nocturnal pollinators and their pollination services. *Proceedings of the Indian National Science Academy* 88, 626–633. DOI: 10.1007/s43538-022-00134-w.

Sivakoff, F.S. and Gardiner, M.M. (2017) Soil lead contamination decreases bee visit duration at sunflowers. *Urban Ecosystems* 20, 1221–1228. DOI: 10.1007/s11252-017-0674-1.

Smith, K.M., Loh, E.H., Rostal, M.K., Zambrana-Torrelio, C.M., Mendiola, L. *et al.* (2013) Pathogens, pests, and economics: drivers of honey bee colony declines and losses. *EcoHealth* 10, 434–445. DOI: 10.1007/s10393-013-0870-2.

Stanley, D.A. and Raine, N.E. (2016) Chronic exposure to a neonicotinoid pesticide alters the interactions between bumblebees and wild plants. *Functional Ecology* 30, 1132–1139. DOI: 10.1111/1365-2435.12644.

Steinhauer, N., Kulhanek, K., Antúnez, K., Human, H., Chantawannakul, P. *et al.* (2018) Drivers of colony losses. *Current Opinion in Insect Science* 26, 42–148. DOI: 10.1016/j.cois.2018.02.004.

Straka, T.M., von der Lippe, M., Voigt, C.C., Gandy, M., Kowarik, I. *et al.* (2021) Light pollution impairs urban nocturnal pollinators but less so in areas with high tree cover. *Science of the Total Environment* 778, 146244. DOI: 10.1016/j.scitotenv.2021.146244.

Takkis, K., Tscheulin, T., Tsalkatis, P. and Petanidou, T. (2015) Climate change reduces nectar secretion in two common Mediterranean plants. *AoB Plants* 7, lv111. DOI: 10.1093/aobpla/plv111.

Tehel, A., Brown, M.J.F. and Paxton, R.J. (2016) Impact of managed honey bee viruses on wild bees. *Current Opinion in Virology* 19, 16–22. DOI: 10.1016/j.coviro.2016.06.006.

Teixeira, E.W., Ferreira, E.A., da Luz, C.F.P., Martins, M.F., Ramos, T.A. *et al.* (2020) European foulbrood in stingless bees (Apidae: Meliponini) in Brazil: Old disease, renewed threat. *Journal of Invertebrate Pathology* 172, 107357. DOI: 10.1016/j.jip.2020.107357.

Thomson, D. (2004) Competitive interactions between the invasive European honey bee and native bumble bees. *Ecology* 85, 458–470. DOI: 10.1890/02-0626.

Thorp, R.W. (2000) The collection of pollen by bees. *Plant Systematics and Evolution* 222, 211–223. DOI: 10.1007/BF00984103.

Toledo-Hernández, E., Peña-Chora, G., Hernández-Velázquez, V.M., Lormendez, C.C., Toribio-Jiménez, J. *et al.* (2022) The stingless bees (Hymenoptera: Apidae: Meliponini): A review of the current threats to their survival. *Apidologie* 53, 8. DOI: 10.1007/s13592-022-00913-w.

Tong, Z.-Y., Wu, L.-Y., Feng, H.-H., Zhang, M., Ambruster, W.S. *et al.* (2023) New calculations indicate that 90% of flowering plant species are animal-pollinated. *National Science Review* 10, wad219. DOI: 10.1093/nsr/nwad219.

Wagner, D.L. (2020) Insect declines in the Anthropocene. *Annual Review of Entomology* 65, 457–480. DOI: 10.1146/annurev-ento-011019-025151.

Waser, N.M. and Price, M.V. (2016) Drought, pollen and nectar availability, and pollination success. *Ecology* 97, 1400–1409. DOI: 10.1890/15-1423.1.

Weaver, J.R., Ascher, J.S. and Mallinger, R.E. (2022) Effects of short-term managed honey bee deployment in a native ecosystem on wild bee foraging and plant-pollinator networks. *Insect Conservation and Diversity* 15, 634–644. DOI: 10.1111/icad.12594.

Westphal, C., Bommarco, R., Carre, G., Lamborn, E., Morison, N. *et al.* (2008) Measuring bee diversity in different European habitats and biogeographical regions. *Ecological Monographs* 78, 653–671. DOI: 10.1890/07-1292.1.

Wilfert, L., Long, G., Leggett, H.C., Schmid-Hempel, P., Butlin, R. *et al.* (2016) Deformed wing virus is a recent global epidemic in honeybees driven by *Varroa* mites. *Science* 351, 594–597. DOI: 10.1126/science.aac9976.

Williams, N.M. and Hemberger, J. (2023) Climate, pesticides, and landcover drive declines of the western bumble bee. *Proceedings of the National Academy of Sciences* 120, e2221692120. DOI: 10.1073/pnas.2221692120.

Williams, N.M., Crone, E.E., Roulston, T.H., Minckley, R.L., Packer, L. *et al.* (2010) Ecological and life-history traits predict bee species responses to environmental disturbances. *Biological Conservation* 143, 2280–2291. DOI: 10.1016/j.biocon.2010.03.024.

Williams, P.H. (1982) The distribution and decline of British bumble bees (*Bombus* Latr.). *Journal of Apicultural Research* 21, 236–245. DOI: 10.1080/00218839.1982.11100550.

Williams, P.H. and Osborne, J.L. (2009) Bumblebee vulnerability and conservation world-wide. *Apidologie* 40, 367–387. DOI: 10.1051/apido/2009025.

Wilson Rankin, E.E., Barney, S.K. and Lozano, G.E. (2020) Reduced water negatively impacts social bee survival and productivity via shifts in floral nutrition. *Journal of Insect Science* 20, 15. DOI: 10.1093/jisesa/ieaa114.

Winfree, R., Griswold, T. and Kremen, C. (2007) Effect of human disturbance on bee communities in a forested ecosystem. *Conservation Biology* 21, 213–223. DOI: 10.1111/j.1523-1739.2006.00574.x.

Winfree, R., Aguilar, R., Vázquez, D.P., LeBuhn, G. and Aizen, M.A. (2009) A meta-analysis of bees' responses to anthropogenic disturbance. *Ecology* 90, 2068–2076. DOI: 10.1890/08-1245.1.

Woodard, S.H., Federman, S., James, R.R., Danforth, B.N., Griswold, T.L. *et al.* (2020) Towards a U.S. national program for monitoring native bees. *Biological Conservation* 252, 108821. DOI: 10.1890/07-1292.1.

Xie, Z., Williams, P.H. and Tang, Y. (2008) The effect of grazing on bumblebees in the high rangelands of the eastern Tibetan Plateau of Sichuan. *Journal of Insect Conservation* 12, 695–703. DOI: 10.1007/s10841-008-9180-3.

Xun, E., Zhang, Y., Zhao, J. and Guo, J. (2018) Heavy metals in nectar modify behaviors of pollinators and nectar robbers: Consequences for plant fitness. *Environmental Pollution* 242, 1166–1175. DOI: 10.1016/j.envpol.2018.07.128.

Yañez, O., Piot, N., Dalmon, A., de Miranda, J.R., Chantawannakul, P. *et al.* (2020) Bee viruses: Routes of infection in Hymenoptera. *Frontiers in Microbiology* 11, 943. DOI: 10.3389/fmicb.2020.00943.

Yang, E.C., Chang, H.C., Wu, W.Y. and Chen, Y.W. (2012) Impaired olfactory associative behavior of honeybee workers due to contamination of imidacloprid in the larval stage. *Plos One* 7, e49472. DOI: 10.1371/journal.pone.0049472.

Yuan, J.S., Himanen, S.J., Holopainen, J.K., Chen, F. and Stewart, C.N. (2009) Smelling global climate change: Mitigation of function for plant volatile organic compounds. *Trends in Ecology and Evolution* 24, 323–331. DOI: 10.1016/j.tree.2009.01.012.

2 How membership of a farmer organization can support farmers in adapting to climate change in the Pacific Island region

Mary Taylor[1] and Kyle Stice[2*]
[1]University of the Sunshine Coast, QLD, Australia; [2]Pacific Farmer Organizations, PO Box 9048, Nadi, Fiji Islands

Abstract
Climate projections for the Pacific Island region serve to emphasize the climate unpredictability, variability and intensity that farmers will have to adapt to in the future. The impact that climate change is having on agriculture in the Pacific is evident from the data around agriculture loss and damage from extreme weather events such as flooding and cyclones. It is clear that food systems will have to be resilient and farmers will have to adapt, if food security in the region is not to be seriously threatened. The factors considered as essential for building adaptive capacity are: (a) social capital; (b) ability of communities to engage effectively with external agents; (c) access to knowledge including how knowledge is generated, shared and exchanged; (d) merging of local and external knowledge; (e) space for farmers to interact, communicate, experiment and learn from each other; (f) trust in the adaptation measure(s)being promoted; (g) effective capacity building; (h) decentralized research; and (i) supportive policy. This article explores these factors and proposes that membership of a farmer organization (FO) provides an enabling platform for the effective and efficient delivery of these conditions, thereby providing farmers with the tools essential for adaptation to a changing climate. Examples illustrating how farmers around the Pacific Island region are working together to put in place measures that will strengthen the resilience of their farming systems are provided from national and regional farmer organizations in the region.

Keywords: farmer organizations, adaptation, Pacific Islands, climate change, food security, grassroots, family farmers

Review Methodology

Information was obtained from the following sources:

1. Pacific Islands Farmer Organization Network based in Fiji and Hawaii – an umbrella organization for national FOs. PIFON began informally operating in 2008 and registered as a not-for-profit company in 2013. Its reach covers 13 Pacific Island countries, 30 national FOs and 95,000 farmer livelihoods. I work with this organization.
2. CABI Direct using the search terms farmer organization, climate change, Pacific Islands

*Corresponding author: manager@pacificfarmers.com

3. Google Scholar using the search terms farmer organization, climate change, Pacific Islands.
4. PAIS – Pacific Agriculture Information System using the search terms farmer organization, climate change, Pacific Islands.
5. Personal contacts – I worked for 22 years in the Pacific Islands region, based for 8 years in Samoa and 14 years in Fiji. Towards the end of my time in Fiji I produced a book Taylor *et al.* (2016).

Pacific Farmers and Agriculture

Pacific Island farmers have a deep experience in managing the effects of climate variability on agriculture, largely due to the region's exposure to the El Niño Southern Oscillation (ENSO). Traditional farming systems, which centre on agroecological approaches, have been relatively resilient against external shocks and have helped to maintain food security. Local knowledge sustained over generations, through a range of traditional and cultural practices, has been the foundation of this resilience, and community cooperation and collaboration have provided the social safety net (Sisifa *et al.*, 2016; Mcleod *et al.*, 2018; Mcleod *et al.*, 2019).

Today, however, these resilient traditional farming systems are less common, and as a result, farming systems are now more vulnerable to climate change. Other pressures exist such as an expanding population, growing urbanization, labour migration, land degradation, such as soil nutrient depletion and soil loss, deforestation, loss of biodiversity, depletion of freshwater resources through saline incursions and contamination from urban, agricultural and industrial sources, and inadequate investment (Barnett, 2019). Further, changing aspirations and value systems have contributed to an under-valuing of traditional food systems resulting in a lack of interest from youth to engage in agriculture (Thaman, 1982; SPC, 2010; FAO, 2015; Sisifa *et al.*, 2016) (Fig. 2.1).

Climate change adds another dimension to these pressures, that of extreme variability and unpredictability. Climate change is affecting, and will continue to affect food systems in the Pacific Island region, including the supply of food from agriculture and fisheries, the ability of countries to import food (because of increasing costs, shortages in supply and export bans (Sharma, 2023)), distribution systems, and the ability of households to purchase and utilize food (Bell *et al.*, 2016; Taylor *et al.*, 2016).

Impact of Climate Change on Agriculture in the Pacific

Current and projected climate change

Climate projections are available for most of the Pacific Islands. The 'Next Generation Climate Projections for the Western Tropical Pacific', (RCCAP, n.d.) launched in October 2021, concluded that temperatures have increased, sea level has risen, and cyclones have become less frequent but more intense. Projections confirm continued warming, and increasing frequency of extreme weather, such as rainfall intensity. Cyclone impacts are likely to increase as a result of the projected increase in average cyclone intensity combined with sea level rise and increased heavy rainfall intensity. Sea level rise is occurring at a faster rate in the region (approximate increase of 4 mm annually in certain regions, which is higher than the worldwide average of 3.4 mm per year (WMO, 2022). A decrease in the return times for extreme weather events (the period of time that passes between extreme events) will affect the ability of food systems to recover, and changes in extremes will be compounded by changes in mean sea-level, temperature, and rainfall, posing significant challenges for Pacific farmers (Handmer *et al.*, 2012).

The ENSO has been influencing climate in the Pacific Islands for centuries, and remains a major source of climate variability, affecting rainfall, cyclone frequency and intensity and occurrence of extreme events (Murphy and Power, 2014; Lough *et al.*, 2016). How climate change will affect ENSO in the future is a common question in the region. Cai *et al.* (2023) concluded that after 1960, there is very strong variability in the ENSO and this strong variability has contributed to more extreme and frequent droughts, floods, heatwaves, bushfires and storms around the world. Projections for

Fig. 2.1. Indigenous root and nut crops proven to have more climate resilience than many introduced crops.

the future suggest more intense and frequent El Niño and La Niña events and also more frequent swings from a strong El Niño to a strong La Niña the following year (Cai, 2023). These projections serve to emphasize the climate unpredictability, variability and intensity that Pacific farmers will have to adapt to in the future.

Managing climate change in the Pacific Island region is made more complicated by the weather pattern differences that exist between the islands and within the islands. As noted by Lebot (2013), the temperature and rainfall can change significantly between different locations, for example, between the windward and leeward sides of the same island. This variation in weather patterns within and between the islands reinforces the importance of research that is relevant to the local context, that is, decentralized research.

Current impact of climate change on agriculture in pacific

Cyclones are a significant cause of lost agricultural production, for example, Tropical Cyclone (TC) Winston (2016) caused over US$200 million in crop damage and loss in Fiji (Mansur *et al.*, 2017). TC Pam devastated Vanuatu in 2015 and caused losses and damages to the agriculture sector valued at US$56.5 million

(Government of Vanuatu, 2015). The waves and strong winds from TC Pam destroyed about 30–90% of crops on many islands of Tuvalu with the economic impact estimated to be 25% of Tuvalu's projected GDP in 2015 (Katea, 2016). The drought which followed TC Pam destroyed vegetation across Vanuatu, and affected food security in PNG and the Republic of the Marshall Islands (Iese *et al.*, 2021). Drought presents problems to agriculture everywhere in the region, particularly given the lack of irrigation (Fig. 2.2).

Flooding in river catchments threatens food production. Flooding in Guadalcanal Island, Solomon Islands (2014), affected over 9000 households, destroying more than 75% of household food gardens (Government of the Solomon Islands, 2014). TC Cody (January 2022), brought significant rain affecting the whole of Fiji. An initial assessment estimated damages to the agricultural sector at over US$4 million, and more than US$1.3 million in relief was ultimately paid in assistance to the farmers (Duncan *et al.*, 2022).

Saltwater intrusion affects crop production on low-lying islands, either directly by impacting crop growth and/or indirectly through its impact on groundwater reserves (Tekinene,

Fig. 2.2. A papaya farm devastated by extreme rain events, waterlogging and disease pressure.

2014). In Tonga, wells located on the low-lying coastal areas showed increasing salinity because of saltwater intrusion (Government of Tonga, 2012). Several low-lying islands in the Solomon Islands and Micronesia have already been lost – including Kale and Rapita in the northern Solomon Islands – and more are experiencing severe erosion due to sea-level rise (Thomas *et al.*, 2020).

Urgency of the situation

'Adaptation opportunities will be reduced and the risks of unavoidable damages increased (medium confidence) in vulnerable regions, including small islands, that are projected to experience higher multiple inter-related climate risks at 1.5°C global warming compared to today, with risks increasing further with warming of 2°C (high confidence).' IPCC (2018)

There is no doubt that climate change is already affecting Pacific farmers. Current global projections suggest that there is 66% likelihood that the annual average near-surface global temperature between 2023 and 2027 will be more than 1.5°C above pre-industrial levels for at least 1 year. There is 98% likelihood that at least one of the next 5 years, and the 5-year period as a whole, will be the warmest on record (WMO, 2024). This surge in temperatures is fuelled by greenhouse gas emissions (GHG) and a naturally occurring El Niño event. IPCC's Sixth Assessment Report (AR6) states that unless there are immediate, rapid and large-scale reductions in GHG emissions, limiting warming to close to 1.5°C or even 2°C will be beyond reach (IPCC, 2021).

Strengthening the resilience of food production systems to these projected changes is vitally important – and not just resilience to short-term shocks but an enduring resilience that will be able to absorb the extremes of climate variability as well as the long-term changes, such as an increase in mean temperature. The challenge is to support and enable farmers to continuously adapt their farming practices to unpredictably changing environmental conditions. An important part of this challenge will be to identify and predict where adaptation limits are likely to occur and who is most likely to be affected, to enable better planning for climate impacts (Dow *et al.*, 2013; Mcleod *et al.*, 2019).

Factors that favour successful climate change adaptation by farmers

Family farms (FAO and IFAD, 2019) are increasingly being acknowledged as critical to food security and nutrition around the globe. The United Nations Decade of Family Farming 2019–2028 seeks to place family farming at the centre of national public policies and investments (FAO and IFAD, 2019). It is vital therefore that these family farms – which are crucial global assets – are supported in their efforts to improve the resilience of their farming systems to enable adaptation to a changing climate. Farmer organizations are considered to be an effective means of supporting family farms (Thierry, 2023).

Resilience and adaptive capacity are closely linked. In general, resilience refers to 'the ability of a system to absorb disturbance and reorganize while undergoing changes so as to still retain essentially the same function, structure, identity and feedbacks' (Walker *et al.*, 2004). Resilience is therefore considered as essential to strengthening the sustainability of food systems which is necessary if the increasing complexity and uncertainty associated with climate change is to be managed. At the farm level, resilience refers to the ability of farms to adapt to climatic, social and market shocks. Farmer adaptive capacity is therefore a prerequisite for building farm resilience to climate change and is linked to increasing the options for managing climate change and improving decision making under the uncertainty of climate change (Taylor *et al.*, 2016; Mcleod *et al.*, 2019; Kangogo *et al.*, 2020).

Various studies have considered what factors influence adaptive capacity. In the application of the Pacific Adaptive Capacity Analysis Framework (PACAF), social capital was identified as a critical adaptive capacity determinant, with leadership, collective action and engaging effectively with external agents as the most important (USP, 2011). The ability to engage effectively with external agents in sourcing and using adaptation resources (such as finance and technology) in a way that responds to their own immediate and future needs is an essential determinant of adaptive capacity, especially with the influx of adaptation investment in the Pacific (Taylor *et al.*, 2016).

Access to knowledge is considered vital in building adaptive capacity. How knowledge is generated, shared and exchanged is important, and learning platforms through participatory action research (PAR), farmer field schools and community-based initiatives have been found to be particularly effective. Merging of local and external knowledge is critical for widening and diversifying farmers' knowledge base (Silici *et al.*, 2021; Hainzer *et al.*, 2022). Pelling *et al.* (2008) also emphasized the important role of organizational structures for improving adaptive capacity in that they provide space for farmers to interact, communicate, experiment and learn from each other. Cinner *et al.* (2018) identified: (a) the flexibility to change strategies; (b) the ability to organize and act collectively; (c) learning to recognize and respond to change; and (d) the agency to determine when and how to change, as important assets in building adaptive capacity for resilience.

McNamara *et al.* (2022) discuss the challenges for adaptation in the Pacific Islands and put forward four mutually reinforcing adaptation pathways that could lead to more equitable, sustainable, and effective adaptation futures. They stress the importance of locally led adaptation, so that local knowledge, local resources and local realities are central to any adaptation measure or strategy that is implemented. Locally appropriate alternative entry points for adaptation are also considered important, including the use of traditional forms of governance.

Cvitanovic *et al.* (2016) in their study of climate adaptation in the Pacific Islands emphasized the importance of adaptation science, local social networks, key actors (influential and trusted individuals) and relevant forms of knowledge exchange in overcoming the barriers to climate adaptation. Their findings suggested that the trust is best developed through participatory research approaches, which allow for the inclusion of traditional knowledge into research.

Membership of a farmer organization (FO) provides a platform which supports many of the factors identified as critical for influencing adaptive capacity, such as, access to knowledge, technology etc. and the space for farmers to interact (Tompkins, 2005; Kangogo *et al.*, 2020; Barokatuminalloh and Setiarso, 2022; Ma *et al.*, 2023; White *et al.*, 2023). In its 2008 World Development Report, the World Bank identified producer organizations as a 'fundamental building block' of its agriculture-for-development agenda, and prioritized support to enhance producer organizations' performance as a key strategy to increase the productivity and sustainability of smallholder farming (Oxfam, 2008).

FO membership supports farmers to better adapt to climate change

Factors essential for enabling successful adaptation have been identified through numerous studies and have been discussed in Section 'Factors that favour successful climate change adaptation by farmers'. These include: (a) social capital – an ability to organize and act collectively; (b) ability of communities to engage effectively with external agents; (c) access to knowledge including how knowledge is generated, shared and exchanged; (d) merging of local and external knowledge; (e) space for farmers to interact, communicate, experiment and learn from each other; (f) trust in the adaptation measure(s)being promoted; (g) effective capacity building; (h) decentralized research; and (i) supportive policy (Fig. 2.3).

Membership of a FO provides an enabling platform and mechanism for all of these factors. FOs are good sources of information, and they provide a space in which farmers can interact, generating and sharing knowledge (Vermeulan and Wynte, 2014; Bingen and Simpson, 2015: PIFON, 2018; Bizikova *et al.*, 2020; Kangogo *et al.*, 2020). FOs can facilitate farmer exchange programmes across countries and regions so that less experienced farmers can learn from those with more experience. This farmer-to-farmer learning can be very beneficial if farmers have to diversify in order to adapt to climate change, that is, adopt crops and farming systems that are new to them. Learning from farmers with experience of these crops can be very effective in

Fig. 2.3. Bula Agro 'Tel-A-Woman' program training in backyard gardening.

the transfer of knowledge and skills (Robinson, 2020).

FOs can be the mechanism for the effective communication of external knowledge and the integration of local knowledge. Knowledge based on local practices may not be sufficient to elicit more transformative adaptation actions which are likely to be needed in the long-term management of climate change. Technical training and skills transfer are also critical enabling factors for improving the uptake of adaptation (Frank and Penrose Buckley, 2012; Taylor *et al.*, 2016; Mcleod *et al.*, 2019; Race *et al.*, 2023).

Innovation in agriculture is clearly an important response for effective and equitable climate change adaptation and mitigation (Borda-Rodriguez and Vicari, 2015). Innovation relies on access to finance, information and other resources, which can be acquired through participation in FOs, especially for resource-constrained farmers. Not all individuals will be innovators, hence the need for platforms and mechanisms that enable successful innovations to be shared more widely. FOs facilitate experimentation and in so doing support innovation and a more proactive approach to climate change challenges, essential for managing the unpredictability of climate change (Kangogo *et al.*, 2020). A diversity of knowledge available from different sources (farmers, scientists, advisory services, agricultural companies) is often a prerequisite for innovation. FOs can bring together these actors, acting as 'innovation intermediaries' (Iyabano *et al.*, 2022).

Decentralization allows for policies and practices specific to the local environmental needs, which are essential for effective climate change adaptation (Mcleod *et al.*, 2019; Fischer, 2021; Carter and Hollinsworth, 2022; McNamara *et al.*, 2022). It is supportive of innovation, and also of outcomes that are more useful at the local level (McGregor *et al.*, 2016; van Zonneveld *et al.*, 2020). It is an approach that promotes the use of local knowledge, leads to more relevant research questions and ensures that the results and skills are shared with those who will use them. Importantly with a decentralized approach to research, farmers are doing the research themselves – taking their ideas, combining and integrating them with good science, while at the same time, tailoring the technologies and methods to meet their needs (Stice and McGregor, 2016; Ensor and de Bruin, 2022).

FOs are best placed to implement decentralized research (PAPP, n.d.), especially in regions such as the Pacific Islands, where membership covers farmer associations from different islands within an archipelago, and from islands dispersed across the vast Pacific ocean (Chand and Kumar, 2019; Norton and Alwang, 2020). With the implementation of decentralized research, there is an opportunity to evaluate crops and systems across the diverse ecosystems found in the region (Palanivel and Shah, 2021). Decentralized research paves the way for participatory research and extension which are considered the major drivers behind agricultural innovation, mainly because of the multi-directional flow of knowledge and technology between farmers, extension providers and researchers (NRI, 2010), ensuring that appropriate and relevant research is conducted and then effectively 'translated' and disseminated among farmers (Beyer, 2015). The policy brief 'Agricultural Research and Farmer Organizations in the Pacific' (PIFON, n.d.-b) summarizes the advantages of the decentralized research model for Pacific Island farmers, emphasizing how it is an approach that allows for the efficient collection of diverse and widespread data, which in turn leads to higher farmer uptake (Fig. 2.4).

Additionally, FOs can facilitate the marketing of farm produce, which can give farmers the opportunity to access more lucrative markets, thereby generating improved revenues which can be reinvested in adaptation measures. Economies of scale may also support investment in communal resources, such as the commercial hot water dipping treatment available to Fiji papaya exporters through the Nature's Way Cooperative (NWC) (see Section 'Case studies illustrating how FOs can support farmers in adapting to climate change through improving the resilience of farming systems and value chains, and increasing market opportunities'). Investing in storage, transport and processing facilities can add value to their products and, with access to the right market information, can enable farmers to wait for better prices rather than selling to the first buyer (Frank and Penrose

Fig. 2.4. Farmers helping farmers in marcotting of breadfruit.

Buckley, 2012; Bizikova *et al.*, 2020; Kangogo *et al.*, 2020).

FOs can also assist farmers in accessing resources from governments, development agencies and private sector (Bizikova *et al.*, 2020; Othman *et al.*, 2020). Members of FOs may also enjoy significantly greater access to services owing to the cost savings that service providers enjoy by working with large groups including access to financial services such as credit and insurance institutions. Some FOs can support their members directly with access to financial capital (Frank and Penrose Buckley, 2012).

The ability to plan ahead is an essential characteristic of successful adaptation. Strong, forward-looking decision making from FOs will support their members to adapt successfully (Frank and Penrose Buckley, 2012). Further, participation in decision making within an FO can encourage local ownership and support community empowerment which all work together to strengthen trust in the decisions made and their consequences (Ma *et al.*, 2023).

A supportive and enabling policy is vital for effective climate change adaptation. FOs can support farmers to be advocates for policy change and through their greater collective voice, are better placed to bring about policy change. FOs can lobby for the needs and preferences of farmers with evidence based on local experiential knowledge. For example, they can survey members to assess how a specific policy will impact livelihoods and/or carry out financial analysis to show the impact in economic terms, and then provide this evidence to the policy makers (PIFON, 2018; Bizikova *et al.*, 2020; Iyabano *et al.*, 2022).

Case studies illustrating how FOS can support farmers in adapting to climate change through improving the resilience of farming systems and value chains, and increasing market opportunities

In the Pacific Island region, the Pacific Island Farmers Organization Network (PIFON, 2020a) serves as an umbrella organization for national FOs. PIFON began informally operating in 2008 and registered as a not-for-profit company in 2013. Its reach covers 13 Pacific Island countries, 30 national FOs and 95,000 farmer livelihoods. PIFON's mission is to make Pacific Islands' FOs more vibrant, viable and sustainable organizations. PIFON has been hugely successful in organizing farmer-to-farmer learning, for example: (a) the first Pacific Farmers Open Pollinated Seed Learning Exchange (2016), which involved 60 farmers from Samoa, Tonga, Vanuatu, Timor-Leste, Fiji, PNG and the Solomon Islands (PIFON, 2016); (b) farmer-to-farmer exchange programme in Vanuatu (2017) where farmers visited and inspected various agriculture and agro-tourism productions streams (PIANGO, 2017); and (c) Farmers Forum and Farmer to Farmer Learning Exchange (2017) (APFP, n.d.), – the latter, according to the feedback received, was the most favoured part of the Farmers Forum, where farmers learned from other farmers (Fig. 2.5).

Climate models suggest that breadfruit will grow well across the tropics for many decades to come. Planting more breadfruit trees could significantly contribute to food security (McGregor *et al.*, 2016; Yang *et al.*, 2022). PIFON has been supporting breadfruit farmers in the Pacific (PIFON, n.d.-a). Farmer-led trials evaluating the performance of trees derived from different propagation types have resulted in a package of best practices for mass propagation of breadfruit. In addition, data collection from farmer-led trials on breadfruit intercropping systems has identified crops that work well in providing an initial cash flow from a breadfruit orchard. Through this work farmer-owned demonstration orchards are now coming into production some 18 months ahead of expectations, greatly improving the expected viability of breadfruit as a sustainable, commercial crop (APFF, 2018).

The Nature's Way Cooperative (NWC), Fiji, a member of PIFON works directly with 140 members, made up of growers and exporters, for all applied research work. NWC have achieved a number of major successes using this model including: (a) the establishment of a certified producer's scheme for Fiji Red papaya; (b) investment in a commercial hot water dipping treatment available to Fiji papaya exporters through NWC (potential to save the industry FJD2mn); and (c) development of technologies supporting sea freight of papaya from Fiji to New Zealand.

Fig. 2.5. Participants at the Tutu Rural Training centre learn about propagation and breadfruit agroforestry.

Research findings indicate a 50% saving in freight costs with no reduction in fruit quality (NWC, n.d.). Much of the NWC applied research work on papaya was carried out under the Fiji Papaya Project (FPP) funded by the Australian Centre for International Agricultural Research (Clarke and Powell, 2022). NWC has recently begun processing fruits into value-added products as part of key plans to assist members, reduce post-harvest losses and promote consumption of local fruits (PIFON, 2020b). Best practice recommendations include selection of 'seed trees' based on performance under local conditions; bulking of seed stocks to quickly recover from natural disasters, and research into pre- and post-cyclone farm activities, such as defoliation, ratooning and sunburn protection (PIFON, n.d.-c).

NWC have also been the key in developing best practice for producing healthy planting material – an essential first step for sustainable, resilient farming – including developing disaster mitigation strategies for nurseries. Farmers are given training in establishing disaster-ready nurseries, including the use of containers for seedling storage as a response to cyclones (Tora *et al.*, 2022). Availability of and access to seed can be a challenge in the Pacific Island region, especially in remote areas, yet a secure seed supply is vitally important for farmers if they are to manage climate change and disaster impacts. In response to this need, PIFON has provided training for its members on open-pollinated and hybrid seed production (PIFON, n.d.-c).

In Samoa, the Samoa Farmers Association (SFA) has been involved in subsector consultations to discuss the challenges facing the agriculture sector from climate change. Stakeholders, experts, and SFA members identified strategies and actions to build resilience and enhance the productivity and profitability of the crop subsector (Ariu, 2023).

Diversification, such as the development of alternative markets, is an important adaptation strategy. In an effort to establish an off-season pineapple production SFA arranged technical exchanges with the Fiji farmer organization, Tei Tei Taveuni. Despite only a limited number of farmers engaging in off-season production practices, pineapple is now available nearly all year round in selected outlets (PIFON, 2020c).

The Tutu Rural Training Centre (TRTC) (Tutu, 2023) and Teitei Taveuni (2023) in Fiji are foundation members of PIFON and have been actively involved in farmer-to-farmer exchanges throughout the region. In an effort to address problems with decreasing fertility affecting taro and kava production in Fiji's Cakaudrove Province farmers cleared new forest areas for planting, clearly a move that was unsustainable. However, farmers, after being involved in trials conducted by the two FOs, are now exploring agroforestry initiatives, including the use of nitrogen-fixing *Mucuna pruriens*. The involvement of the two FOs was a significant factor in the uptake of the agroforestry initiatives. The positive results from the trials are now being disseminated to farmers across Fiji and the region through farmer-to-farmer exchanges organized through PIFON. Agroforestry increases adaptive capacity, reduces vulnerability, and thus helps farmers reduce climate risk (Quandt *et al.*, 2023)

Jiwaka Youth in Agriculture Association (Papua New Guinea) formed as a result of a training event in Fiji on off-season cropping of fruit and vegetables, such as capsicum, watermelon, papaya, and cabbage. The training highlighted the importance of diversification as an adaptation strategy and the potential of off-season production and marketing. The association has grown to a membership of 250 and have contracts for the supply of capsicum and watermelon. The success of the Association has helped the young farmers to build proper houses for their families and to raise the profile and status of youth farmers (PIFON, 2020e).

In Vanuatu, collaboration between the Vanuatu Farm Support Association (FSA) (PIFON, 2020d) and Vanuatu Agricultural Research Centre (VARTC) saw the distribution of different taro and yam varieties to 10 villages in different locations. Two years after distribution, monitoring of the diversity in the villages showed an 86% gain in yam diversity and a 61% gain in taro diversity. Importantly, this gain in 'new' varieties was not associated with the loss of traditional varieties (Camus and Lebot, 2010). Without the support of the FSA the improved diversity available from VARTC would not have made it into farmers' fields. Enriching farmers' varietal portfolios strengthens the resilience of their food production systems. Improving agricultural biodiversity is seen as underpinning resilient farm ecosystems (Frison *et al.*, 2011).

Conclusion

FOs can play an increasingly important role in supporting farmers to adapt to a changing climate through facilitating information sharing and learning, promoting sustainable agricultural practices and tools, and strengthening policy advocacy and development. Importantly, they facilitate locally led adaptation, so that local knowledge, local resources and local realities are central to any adaptation measure or strategy that is implemented. Effectively addressing climate change in a region as diverse as the Pacific Islands requires a focus on local conditions. Through a regional FO, such as PIFON, adaptation measures can be evaluated in other localities, and best practices scaled up, enabling a portfolio of adaptation measures to be established that work across a range of different environments.

Despite the successes of FOs in the Pacific and elsewhere, continued and strengthened support for FOs is necessary so as to enhance their effectiveness and sustainability, and to enable efficient and effective delivery of the key services necessary to build farmers' adaptive capacity. Governments and development partners must acknowledge the central role of farmers and FOs in addressing climate change and ensure that FOs are involved in the setting and delivery of agricultural research priorities so that farmers' needs are met. Financial support is vital, including proportional access to climate funds. Partnerships with FOs should be established to address the knowledge gaps around adaptation to a changing climate, including understanding of how farmers can continuously adapt their farming practices to unpredictably changing environmental conditions and the limits to adaptation. Governments, specifically, can provide support for FOs by providing enabling policy, regulatory and legal frameworks for FO operation which aim to enhance service delivery and strengthen the impact of FOs.

Author Contributions

Kyle Stice conceptualized the review, provided information and images, and reviewed the article. Mary Taylor carried out the research and the writing.

Acknowledgements

We extend our sincere gratitude to the following contributors who played pivotal roles in the development of this publication: Angela Birch, Ian Kennedy and Tony Jansen of Pacific Farmer Organizations. Their expertise, commitment, and valuable insights significantly enriched the content and quality of this work.

Furthermore, we would like to express our deep appreciation to the farmers of the Pacific Islands. Their unwavering dedication to sustainable agriculture, coupled with their willingness to share knowledge and experiences, has been instrumental in shaping the foundation of this publication. We also acknowledge the farmer organizations that represent and support these communities, as their collaboration has been indispensable to the success of this research.

Funding

With an implementation period of 54 months, the Farmers Organizations for Africa, Caribbean and the Pacific (FO4ACP) is expected to directly benefit 150,000 farmers in the (Pacific) region. The Program is a joint partnership between the European Union, the African, Caribbean and Pacific Group of States, the International Fund for Agricultural Development and the Pacific Island Farmers Organizations Network.

Conflicts of Interest

The authors have no conflicts of interest to declare.

References

APFF (Asia-Pacific Farmers Forum) (2018) Investing in the Future: The Breadfruit Story. Available at: http://www.asiapacificfarmersforum.net/pacific-pifon-investing-in-the-future-the-breadfruit-story/ (accessed 20 October 2023).

APFP (n.d.). Available at: https://asiapacificfarmersforum.net/fiji-farmers-forum/ (accessed 18 October 2023).

Ariu (2023). Available at: https://www.samoagovt.ws/2023/03/media-releaseenhancing-climate-resilience-of-the-agriculture-sector/ (accessed 17 October 2023).

Barnett, J. (2019) Climate change and food security in the Pacific Islands. In: Connell, J. and Lowitt, K. (eds) *Food Security in Small Island States*. Springer Nature, Singapore, pp. 25–38.

Barokatuminalloh, N.W. and Setiarso, O. (2022) Role of institutions to improve farmer's resilience in facing climate change. *IOP Conference Series: Earth and Environmental Science* 1114, 012087. DOI: 10.1088/1755-1315/1114/1/012087.

Bell, J., Taylor, M., Amos, M. and Andrew, N. (2016) Climate Change and Pacific Island Food Systems the Future of Food, Farming and Fishing in the Pacific Islands under a Changing Climate. Available at: https://cgspace. cgiar.org/rest/bitstreams/79049/retrieve (accessed 10 October 2023).

Beyer (2015). Available at: https://pacificfarmers.com/resource/background-paper-farmer-led-participatory-action-research/ (accessed 12 October 2023).

Bingen, R.J. and Simpson, B.M. (2015) *Farmer Organizations and Modernizing Extension and Advisory Services: A Framework and Reflection on Cases from Sub-Saharan Africa*. MEAS Discussion Paper.

Bizikova, L., Nkonya, E., Minah, M., Hanisch, M., Turaga, R.M.R. *et al.* (2020) A scoping review of the contributions of farmers' organizations to smallholder agriculture. *Nature Food* 1, 620–630. DOI: 10.1038/s43016-020-00164-x.

Borda-Rodriguez, A. and Vicari, S. (2015) Coffee cooperatives in Malawi: Building resilience through innovation. *Annals of Public and Cooperative Economics. Wiley Blackwell* 86(2), 317–338.

Cai, W. (2023). Available at: https://www.csiro.au/en/news/all/articles/2023/may/climate-change-affecting-el-nino (accessed 15 October 2023).

Cai, W., Ng, B., Geng, T., Jia, F., Wu, L. *et al.* (2023) Anthropogenic impacts on twentieth-century ENSO variability changes. *Nature Reviews Earth & Environment* 4, 407–418. DOI: 10.1038/s43017-023-00427-8.

Camus, P. and Lebot, V. (2010) On-farm assessment of clonal introduction of root crops diversity in Vanuatu. *Melanesia. Experimental Agriculture* 46(4), 541–559. DOI: 10.1017/S0014479710000591.

Carter, J. and Hollinsworth, D. (2022) Governing extension and extending governance for Pacific organic farming. *Geographical Research* 60(3), 498–510. DOI: 10.1111/1745-5871.12542.

Chand, A. and Kumar, S. (2019) Changing agriculture extension models in pacific island countries. *Journal of Agricultural Extension* 23(4), 30–38. DOI: 10.4314/jae.v23i4.4.

Cinner, J., Adger, W.N., Allison, E.H., Barnes, M.L., Brown, K. *et al.* (2018) Building adaptive capacity to climate change in tropical coastal communities. *Nature Climate Change* 8, 117–123.

Clarke, M. and Powell, L. (2022) Strengthening the Fiji papaya industry through applied research and information dissemination, ACIAR Impact Assessment no.104, Australian Centre for International Agricultural Research, Canberra. *ACIAR Impact Assessment Series No.104 (IAS104)*.

Cvitanovic, C., Crimp, S., Fleming, A., Bell, J., Howden, M. *et al.* (2016) Linking adaptation science to action to build food secure Pacific Island communities. *Climate Risk Management* 11, 53–62.

Dow, K., Berkhout, F., Preston, B.L., Klein, R.J., Midgley, G. *et al.* (2013) Limits to adaptation. *Nature Climate Change* 3, 305.

Duncan, J., Varea, R., Boruff, B. and Davies, K. (2022). Available at: https://www.climatechange.ai/blog/2022-09-06-grants-fiji-flood (accessed 12 October 2023).

Ensor, J. and de Bruin, A. (2022) The role of learning in farmer-led innovation. *Agricultural Systems* 197. DOI: 10.1016/j.agsy.2021.103356.

FAO (2015) *Developing the Knowledge, Skills and Talent of Youth to Further Food Security and Nutrition*. Food and Agriculture Organization of the United Nations/International Fund for Agricultural Development World Food Programme, Rome. Available at: http://www.fao.org/3/i5024e/i5024e. pdf (accessed 12 October 2023).

FAO and IFAD (2019) *United Nations Decade of Family Farming 2019–2028*. Global Action Plan, Rome. Available at: https://www.fao.org/3/ca4672en/ca4672en.pdf (accessed 10 October 2023).

Fischer, H. (2021) Decentralization and the governance of climate adaptation: Situating community-based planning within broader trajectories of political transformation. *World Development* 140(3).

Frank, J. and Penrose Buckley, C. (2012) Small-scale farmers and climate change. How can farmer organisations and Fairtrade build the adaptive capacity of smallholders? IIED, London. Available at: https://www.iied.org/sites/default/files/pdfs/migrate/16518IIED.pdf (accessed 10 October 2023).

Frison, E.A., Cherfas, J. and Hodgkin, T. (2011) Agricultural biodiversity is essential for a sustainable improvement in food and nutrition security. *Sustainability* 3(1), 238–253. DOI: 10.3390/su3010238.

Government of the Solomon Islands (2014) *Rapid Assessment of the Macro and Sectoral Impacts of Flash Floods in the Solomon Islands*. World Bank, Washington, DC. Available at: https://www.gfdrr.org/sites/default/files/publication/pda-2014-solomonislands.pdf (accessed 11 October 2023).

Government of Tonga (2012) Second National Communication. In response to its obligations under the United Nations Framework Convention on Climate Change. Second National Communication on Climate Change Project. Ministry of Environment and Climate Change, Nuku'alofa, Tonga. Available at: https://tonga-data.sprep.org/system/files/2nd%20Tonga%20National%20Communication%20Report%20-%20 2012.pdf (accessed 11 October 2023).

Government of Vanuatu (2015) Vanuatu Agriculture Sector Policy 2015–2030. Available at: https://www.nab.vu/sites/default/files/documents/Vanuatu%20Agriculture%20Sector%20Policy%202015-2030.pdf (accessed 11 October 2023).

Hainzer, K., O'Mullan, C. and Brown, P.H. (2022) The use of local knowledge in agricultural extension: A systematic review of the literature. *Journal of Agricultural Extension and Rural Development* 14(1), 1–12.

Handmer, J., Honda, Y., Kundzewicz, Z.W., Arnell, N., Benito, G. *et al*. (2012) Changes in impacts of climate extremes: Human systems and ecosystems. In: Field, C.B., Barros, V., Stocker, T.F., Qin, D. and Dokken, D.J. (eds) *Managing the Risks of Extreme Events and Disasters to Advance Climate Change Adaptation*. A Special Report of Working Groups I and II of the Intergovernmental Panel on Climate Change (IPCC). Cambridge University Press, Cambridge, UK, and New York, pp. 231–290.

Iese, V., Kiem, A.S., Mariner, A., Malsale, P., Tofaeono, T. *et al*. (2021) Historical and future drought impacts in the Pacific islands and atolls. *Climatic Change* 166, 19. DOI: 10.1007/s10584-021-03112-1.

IPCC (2018) Summary for Policymakers of the Special Report on Global Warming of 1.5°C (SR15). Available at: https://www.ipcc.ch/sr15/chapter/spm/ (accessed 11 October 2023).

IPCC (2021). Available at: https://www.ipcc.ch/2021/08/09/ar6-wg1-20210809-pr/ (accessed 10 October 2023).

Iyabano, A.I., Klerkx, L., Faure, G. and Toillier, A. (2022) Farmers' Organizations as innovation intermediaries for agroecological innovations in Burkina Faso. *International Journal of Agricultural Sustainability* 20(5), 857–873. DOI: 10.1080/14735903.2021.20020.

Kangogo, D., Dentoni, D. and Bijman, J. (2020) Determinants of farm resilience to climate change: The role of farmer entrepreneurship and value chain collaborations. *Sustainability* 12, 1–15. DOI: 10.3390/su12030868.

Katea, T. (2016) Tuvalu Progress Report. 16th RAV Tropical Cyclone Committee meeting. Honiara, Solomon Islands. Available at: https://www.wmo.int/pages/prog/www/tcp/documents /RAV_TCC-16_Tuvalu_ ProgressReport.pdf (accessed 13 October 2023).

Lebot, V. (2013) Coping with insularity: The need for crop genetic improvement to strengthen adaptation to climatic change. *Environment, Development and Sustainability* 15(6), 1405–1423. DOI: 10.1007/s10668-013-9445-1.

Lough, J., Sen Gupta, A., Power, S.B., Grose, M.R. and McGree, S. (2016) Observed and projected changes in surface climate of tropical Pacific Islands. In: Taylor, M., McGregor, A. and Dawson, B. (eds) *Vulnerability of Pacific Agriculture and Forestry to Climate Change*. Secretariat of the Pacific Community, Noumea.

Ma, W., Marini, M.A. and Rahut, D.B. (2023) Farmers' organizations and sustainable development: An introduction. *Annals of Public and Cooperative Economics* 94(3), 683–700. DOI: 10.1111/apce.12449.

Mansur, A., Doyle, J. and Ivaschenko, O. (2017) *Social Protection and Humanitarian Assistance Nexus for Disaster Response: Lessons Learnt from Fiji's Tropical Cyclone Winston*. World Bank, Washington, DC. Available at: https://elibrary.worldbank.org/doi/epdf/10.1596/26408 (accessed 14 October 2023).

McGregor, A.M., Tora, L.D. and Lebot, V. (2016) Planting breadfruit orchards as a climate change adaptation strategy for the Pacific islands. *Acta Horticulturae* 1128, 55–66. DOI: 10.17660/ActaHortic.2016.1128.8.

Mcleod, E., Arora-Jonsson, S., Masuda, Y.J., Bruton-Adams, M., Emaurois, C.O. *et al.* (2018) Raising the voices of Pacific Island women to inform climate adaptation policies. *Marine Policy* 93, 178–185.

Mcleod, E., Bruton-Adams, M., Förster, J., Franco, C., Gaines, G. *et al.* (2019) Lessons from the Pacific Islands – adapting to climate change by supporting social and ecological resilience. *Frontiers in Marine Science* 6, 289. DOI: 10.3389/fmars.2019.00289.

McNamara, K.E., Westoby, R. and Clissold, R. (2022) Lessons for adaptation pathways in the Pacific Islands. *PLOS Climate* 1(2), 1–4. DOI: 10.1371/journal.pclm.0000011.

Murphy, F.M. and Power, S.B. (2014) The varied impacts of el Ninô–southern oscillation on Pacific Island climates. *Journal of Climate* 27(11), 4015–4036. DOI: 10.1175/JCLI-D-13-00130.1.

Norton, G.W. and Alwang, J. (2020) Changes in agricultural extension and implications for farmer adoption of new practices. *Applied Economic Perspectives and Policy* 42(1), 8–20. DOI: 10.1002/aepp.13008.

NRI (2010) Agricultural Extension, Advisory Services and Innovation. Natural Resources Institute (NRI). University of Greenwich, London. Available at: https://www.nri.org/publications/thematic-papers/7-agricultural-extension-advisory-services-and-innovation/file (accessed 16 October 2023).

NWC (n.d.). Available at: https://nwcfiji.com/research-extension/ (accessed 18 October 2023).

Othman, M.S., Oughton, E. and Garrod, G. (2020) Significance of farming groups for resource access and livelihood improvement of rural smallholder women farmers. *Development in Practice* 30(5), 586–598. DOI: 10.1080/09614524.2020.1764502.

Oxfam (2008) Turning up the Heat: Climate Change and Poverty in Uganda. Oxfam GB, Kampala. Available at: https://reliefweb.int/report/uganda/turning-heat-climate-change-and-poverty-uganda (accessed 17 October 2023).

Palanivel, H. and Shah, S. (2021) Unlocking the inherent potential of plant genetic resources: Food security and climate change adaptation strategy in Fiji and the Pacific. *Environment, Development and Sustainability* 23, 14264–14323.

PAPP (n.d.). Available at: https://pafpnet.spc.int/resources/523-policy-brief-agricultural-research-and-farmer-organisations-in-the-pacific (accessed 11 October 2023).

Pelling, M., High, C., Dearing, J. and Smith, D. (2008) Shadow Spaces for social learning: A relational understanding of adaptive capacity to climate change within organisations. *Environmental Planning* A40, 867–884. DOI: 10.1068/a39148.

PIANGO (2017). Available at: https://pacific.scoop.co.nz/2017/10/farmer-to-farmer-exchange-to-strengthen-networks/ (accessed 15 October 2023).

PIFON (2016). Available at: https://pacificfarmers.com/wp-content/uploads/2016/11/Trip-Report-for-the-Open-Pollinated-Seed-Learning-Exchange.pdf (accessed 14 October 2023).

PIFON (2018) Farmers Having Their Say: A Guide to Help Pacific Farmers Organisations 'Build' and 'Promote' Their Case to Influence Policy. PIFON. (accessed 18 October 2023).

PIFON (2020a). Available at: https://pacificfarmers.com (accessed 13 October 2023).

PIFON (2020b). Available at: https://pacificfarmers.com/natures-way-cooperative-expands-market-base/ (accessed 18 October 2023).

PIFON (2020c). Available at: https://pacificfarmers.com/pineapples-in-paradise/ (accessed 12 October 2023).

PIFON (2020d). Available at: https://pacificfarmers.com/story/farm-supportassociation/ (accessed 14 October 2023).

PIFON (2020e). Available at: https://pacificfarmers.com/young-farmers-lead-the-way-off-season-vegetable-production/ (accessed 14 October 2023).

PIFON (n.d.-a). Available at: https://pacificfarmers.com/resource-category/crops/breadfruit/ (accessed 18 October 2023).

PIFON (n.d.-b). Available at: https://pacificfarmers.com/wp-content/uploads/2016/03 /Agricultural-Research-and-Farmer-Organisations-in-the-Pacific-1.pdf (accessed 13 October 2023).

PIFON (n.d.-c). Available at: https://www.unescap.org/sites/default/files/Role%20of%20Farmer%20 Organizations%20in%20CCA%20by%20the%20Pacific%20 Islands%20Farmer %20Organisations%20Network.pdf (accessed 19 October 2023).

Quandt, A., Neufeldt, H. and Gorman, K. (2023) Climate change adaptation through agroforestry: Opportunities and gaps. *Current Opinion in Environmental Sustainability* 60, 101244. DOI: 10.1016/j.

Race, D., Gentle, P. and Mathew, S. (2023) Living on the margins: Climate change impacts and adaptation by remote communities living in the Pacific Islands, the Himalaya and desert Australia. *Climate Risk Management* 40, 1–7. DOI: 10.1016/j.crm.2023.100503.

RCCAP (n.d.). Available at: https://www.rccap.org/climate-change-update-for-the-pacific/ (accessed 10 October 2023).

Robinson, V. (2020) *Farmer to Farmer Knowledge Exchange: Relevance and Challenge during Change*.

Sharma, S. (2023). Available at: https://www.aljazeera.com/economy/2023/8/16/how-indias-ban-on-some-rice-exports-is-ricocheting-around-the-world (accessed 10 October 2023).

Silici, L., Rowe, A., Suppiramaniam, N. and Knox, J.W. (2021) Building adaptive capacity of smallholder agriculture to climate change: Evidence synthesis on learning outcomes. *Environmental Research Communications* 3, 122001. DOI: 10.1088/2515-7620/ac44df.

Sisifa, A., Taylor, M. and Dawson, B. (2016) Pacific communities, agriculture and climate change. In: McGregor, A. and Dawson, B. (eds) *Vulnerability of Pacific Agriculture and Forestry to Climate Change*. Secretariat of the Pacific Community, Noumea. Available at: http://www.pacificfarmers.com/wp-content/uploads/2016/07/Vulnerability-of-Pacific-Island-agriculture-and-forestry-to-climate-change.pdf (accessed 9 October 2023).

SPC (2010) Pacific Youth in Agriculture Strategy 2011–2015 Echoing the Voices of Pacific Youth. SPC, Noumea, New Caledonia.

Stice, K. and McGregor, A. (2016) Vulnerability of high-value horticultural crops to climate change. In: Taylor, M., McGregor, A. and Dawson, B. (eds) *Vulnerability of Pacific Agriculture and Forestry to Climate Change*. Secretariat of the Pacific Community, Noumea.

Taylor, M., Crimp, S., Dawson, B., McGregor, A., Cvitanovic, C. *et al.* (2016) Adapting Pacific agriculture and forestry to climate change: Management measures and investments. In: Taylor, M., McGregor, A. and Dawson, B. (eds) *Vulnerability of Pacific Agriculture and Forestry to Climate Change*. Secretariat of the Pacific Community, Noumea. Available at: http://www.pacificfarmers.com/wp-content/uploads/2016/07/Vulnerability-of-Pacific-Island-agriculture-and-forestry-to-climate-change (accessed 9 October 2023).

Teitei Taveuni (2023). Available at: https://teiteitaveuni.com/ (accessed 13 October 2023).

Tekinene, M. (2014) An assessment of the impacts of climate change on cultivated pulaka (*Cyrtosperma chamissonis*) in Tuvalu, Master of Science Thesis, University of the South Pacific, Fiji.

Thaman, R.R. (1982) Deterioration of traditional food systems, increasing malnutrition and food dependency in the Pacific Islands. *Journal of Food and Nutrition* 39, 109–121.

Thierry (2023). Available at: https://moderndiplomacy.eu/2023/05/09/strengthening-professional-farmers-organizations-to-improve-family-farming-income/ (accessed 15 October 2023).

Thomas, A., Baptiste, A., Martyr-Koller, R., Pringle, P. and Rhiney, K. (2020) Climate change and small island developing states. *Annual Review of Environment and Resources* 45, 1–27. DOI: 10.1146/annurev-environ-012320-083355.

Tompkins, E. (2005) Planning for climate change in small islands: Insights from national hurricane preparedness in the Cayman Islands. *Global Environmental Change* 15, 139–149. DOI: 10.1016/j.gloenvcha.2004.11.002.

Tora, L., Kumar, S., Stice, K., Birch, A. and McGregor, A. (2022) Module 5: Seeds and Seedlings for Sustainable Fruit and Vegetable Production. Available at: https://pacificfarmers.com/wp-content/uploads/2022/09/Module-5_-Seeds-and-Seedlings_250822.pdf (accessed 14 October 2023).

Tutu (2023). Available at: https://tutufiji.com/ (accessed 13 October 2023).

USP (2011) Pacific Adaptive Capacity Analysis Framework (PACAF). Compiled by the Institute of Applied Science of the University of the Pacific, Suva Fiji in Partnership with the Secretariat of the Pacific Community and Australian Red Cross. Available at: https://www.dcceew.gov.au/sites/default/files/documents/usp-pacific-adaptive-capacity-analysis-framework-12.pdf (accessed 16 October 2023).

van Zonneveld, M., Turmel, M.-S. and Hellin, J. (2020) Decision-making to diversify farm systems for climate change adaptation. *Frontiers in Sustainable Food Systems* 4, 32. DOI: 10.3389/fsufs.2020.00032.

Vermeulan, S. and Wynte, A. (2014) Working with farmers for agricultural innovation and climate adaptation. CCAFS Working Paper no. 77. CGIAR Research Program on Climate Change, Agriculture and Food Security (CCAFS). Copenhagen, Denmark. Available at: www.ccafs.cgiar.org (accessed 17 October 2023).

Walker, B., Hollin, C.S., Carpenter, S.R. and Kinzig, A. (2004) Resilience, adaptability and transformability in social-ecological systems. *Ecology and Society* 9(5). Available at: http://www.ecologyandsociety.org/vol9/iss2/art5/ (accessed 16 October 2023).

White, A., Faulkner, J.W., Niles, M.T., Conner, D. and Mendez, V.E. (2023) The role of farmer networks in supporting adaptive capacity: Opening the door for innovation and transformation in the Northeastern United States. *Elementa:Science of the Anthropocene* 11, 1. DOI: 10.1525/elementa.2022.00039.

WMO (2022) State of the Climate in South-West Pacific 2021. World Meteorological Organization (WMO), Geneva. Available at: https://library.wmo.int/idurl/4/58225 (accessed 18 October 2023).

WMO (2024). Available at: https://wmo.int/news/media-centre/climate-change-indicators-reached-record-levels-2023-wmo (accessed 1 May 2024).

Yang, L., Zerega, N., Montgomery, A. and Horton, D.E. (2022) Potential of breadfruit cultivation to contribute to climate-resilient low latitude food systems. *PLOS Climate* 1(8), 1–19. DOI: 10.1371/journal.pclm.0000062.

3 The danger of the spread of *Ralstonia solanacearum* on potato crops worldwide and potential mitigation options

Lilian A. Okiro[1,2]*, Richard M. Mulwa[1], Maurice E. Oyoo[1] and Samuel Nyalala[1]
[1]Department of Crop Horticulture and Soil, Egerton University, P.O. Box 536 – 20115, Egerton, Kenya; [2]Department of Biochemistry and Molecular Biology, Egerton University, P.O. Box 536 – 20115, Egerton, Kenya

Abstract
Potato is a major crop that is consumed by over one billion people worldwide. However, production is hampered by diseases such as viral infections, late blight, and bacterial wilt. Bacterial wilt disease caused by *Ralstonia solanacearum* is the second most prevalent disease after blight, and its effects are exacerbated by both global trade and climate change. The extent of the disease devastation varies by country, with more serious consequences evident in major potato-producing countries. The *Ralstonia* species complex (RSSC) is a group of pathogenic strains that spread through the use of infected seed potato, soils, water runoffs, plant debris, or contaminated farm equipment and propagation material. The elements that contribute to the rapid spread of bacterial wilt disease in potato are discussed in this review article, and a disease-control strategy is proposed. Understanding how *R. solanacearum* spreads through international trade is critical for developing effective mitigation and management techniques to control and limit the spread of this damaging phytopathogen.

Keywords: bacterial wilt, mitigation, potato, *Ralstonia solanacearum*

Introduction

The potato crop (*Solanum tuberosum* L.), is cultivated and regularly consumed by more than one billion people globally (Mickiewicz *et al.*, 2022). It is the third most significant food crop after wheat and rice (Spooner *et al.*, 2005; Li *et al.*, 2018). Potato is presently cultivated on approximately 18 million hectares of agricultural land worldwide, with the overall potato production standing at 376 million tons (FAOSTAT, 2021). China, India, and the United States of America are the leading potato-producing nations, with most of the countries having averagely lower yields. Potato stands out as a remarkable source of energy, essential vitamins, crucial minerals, high-quality proteins, and beneficial lipids and are increasingly being harnessed as foundational raw materials for a diverse range of industrial products (Wang *et al.*, 2019; Khlestkin *et al.*, 2020; Devaux *et al.*, 2021).

Potato production has not achieved its full potential due to diseases that lead to a reduction in yield. The most frequently identified potato diseases comprise bacterial wilt, late blight, and viral infections. Bacterial wilt, caused by the

*Corresponding author: lilian.okiro@egerton.ac.ke

bacterium *R. solanacearum*, is the second most widespread disease of potato crops globally, after late blight (Kaguongo *et al.*, 2008). The disease is spreading at an unprecedented rate, primarily driven by escalating global trade and the consequences of climate change, which have facilitated the migration of the pathogen from its indigenous habitats to newly favored climatic zones (Caruso *et al.*, 2005). Strains of phylotype II sequevar 1(Race 3 Biovar 2), better suited to colder temperatures (<20°C) (Elphinstone, 2005), and still thrive in warmer temperatures ranging from 28 to 35°C (Huerta *et al.*, 2015). The magnitude of damage inflicted by this disease differs from one country to another, with more substantial repercussions evident in major potato-producing nations. This situation poses a significant obstacle to achieving household food self-sufficiency in different countries. Bacterial wilt disease inflicts devastating economic losses of approximately US$1 billion annually on potato cultivation, reverberating across 1.5 million hectares of farmland and impacting the lives of approximately 3 million farming households in over 80 nations (Elphinstone, 2005).

The RSSC is a collection of genetically and phenotypically diverse pathogenic strains (Boucher *et al.*, 2001; Fegan and Prior, 2005; Genin and Denny, 2012). Historically, RSSC was separated into distinct races depending on their host preferences (Hayward, 1991) and distinct biochemical characteristics (Ajanga, 1993). A system of classification that relies on the partial sequences of the *egl*, *hrpB*, 16S rRNA, and genes suggested by Fegan and Prior (2005). Based on the classification system employed, the *Ralstonia* spp. complex has been divided into four distinct phylotypes depending on their respective geographic origins: Phylotype I (Asia), Phylotype II (America), Phylotype III (Africa), and Phylotype IV (Indonesia). Phylotypes can be further divided into sequevars, determined by variations in a 750 bp section of the endoglucanase (egl) gene.

The classification of RSSC has recently been updated, taking into account the *egl* gene, the DNA-DNA hybridization analyses, 16S–23S rRNA ITS region, and genome sequence analysis (Safni *et al.*, 2014), namely *Ralstonia solanacearum*, encompassing all phylotype II strains; *Ralstonia pseudosolanacearum*, covering all phylotype I and phylotype III strains; and *Ralstonia syzygyi*, comprising all phylotype IV strains. Strains within phylotype II sequevar 1 and 2, also known as Race 3 Biovar 2 (R3bv2), are responsible for causing bacterial wilt disease in potato (Ailloud *et al.*, 2015; Yuan *et al.*, 2015; Muthoni *et al.*, 2020).

Ralstonia solanacearum (R3bv2) is believed to have evolved in South America and spread worldwide on latently infected propagation material and infested soil samples (Mark and Hayward, 2014). The disease is known as bacterial wilt as well as potato brown rot, and the most important host plants for R3bv2 are potato and tomato (Madden and Wheelis, 2003; Champoiseau *et al.*, 2009). The phytopathogen has also been reported in pelargonium, eggplant, pepper, and asymptomatic weeds (Champoiseau *et al.*, 2009; Gutarra *et al.*, 2017). The bacterium is subject to quarantine regulations in North America and Europe and is classified as a biological weapon by the United States of America (Madden and Wheelis, 2003).

Bacterial wilt disease is tuber-borne and spreads through the use of infected seed potato, infested soils, water run-offs, plant debris, or contaminated farm equipment and propagation material (van der Wolf and De Boer, 2007; Abdurahman *et al.*, 2017; Sharma *et al.*, 2022). The bacterium inhabits the vascular tissue, lenticels, and outer surface of seed potato tubers (van der Wolf and De Boer, 2007; Khokhani *et al.*, 2018). Its effectiveness as a crop pathogen can be attributed to its remarkable resilience to diverse habitats and capacity to endure unfavorable conditions over prolonged durations (Caruso *et al.*, 2005). The propagation and outbreak of bacterial wilt disease of potato is further exacerbated by a combination of factors, including appropriate agricultural practices, unrestricted access by animals and humans to agricultural fields, a dearth of knowledge regarding proper procedures for field sanitation, the repeated use of seed potato for planting, the high cost of certified seed potato, relaxed seed quarantine measures, soil acidity, and fluctuations in ecological conditions (Muthoni *et al.*, 2013; Tafesse *et al.*, 2021; Tessema and Seid, 2023). In addition, limited size of land parcels due to land fragmentation resulting from the traditional practice of dividing land among subsequent generations in many African countries has led

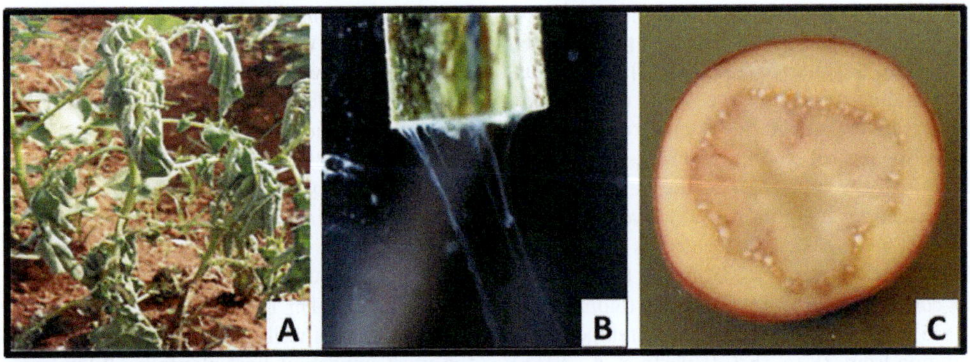

Fig. 3.1. Symptoms of bacterial wilt of potato caused by *Ralstonia solanacearum* in potato plant (A), bacterial ooze from infected stem (B), and potato tuber (C).

to the challenge of implementing effective crop rotation to manage the disease.

This review article examines the factors driving the rapid propagation of bacterial wilt disease in potato across continents and offers a strategic path for disease control. While the existence of the pathogen has been verified in numerous countries, comprehensive data detailing the extent and characteristics of bacterial wilt within the global potato value chain remains notably absent (EPPO, 2023). The regulations governing potato seed systems and the international movement of propagation materials remain disparate and unharmonized worldwide. Consequently, the transportation of propagation materials presents a substantial risk for the transmission of various strains within the *Ralstonia* species complex to regions unaffected by the disease. Thus, it is imperative to gain a profound understanding of how *R. solanacearum* spreads through the international trade of seed potato and propagation materials. Such insight is pivotal for implementing effective mitigation and management strategies aimed at controlling and curtailing the propagation of this destructive phytopathogen (Fig. 3.1).

Review Methodology

We searched databases including FAO, CAB Abstracts, Google Scholar, and ScienceDirect (Keyword search terms used: *Ralstonia solanacearum*, bacterial wilt disease, mitigation, potato). We used the references from the articles obtained by this method to check for additional relevant material.

Global distribution of *Ralstonia solanacearum*

Ralstonia solanacearum is a pathogen that inhabits soil and water. It is spread through infested soil, contaminated equipment or personnel, infected plant material, including latent propagation material and seed potato, irrigation, and surface water (Lebeau *et al.*, 2011; Mansfield *et al.*, 2012). There are numerous historical accounts of brown rot outbreaks in potato that have been linked to *R. solanacearum* (Phylotype IIB or R3bv2), particularly in the Americas (Norman *et al.*, 2020), Europe (Tjou-tam-sin *et al.*, 2017; Bergsma-Vlami *et al.*, 2018; EFSA, 2019), Asia (Jiang *et al.*, 2017; He *et al.*, 2020), the Mediterranean, and Africa. *Ralstonia solanacearum* (R3bv2), recognized for their capacity to flourish in cold conditions, has gained acknowledgement due to their substantial influence on the international potato sector. Their repercussions are felt across all continents, especially in the highland areas of temperate, tropical, and subtropical regions around the globe (Hayward, 1991). The impact of the bacterium on potato farming has been extensively documented, and it affects an estimated 3.76 million acres of land across 239 nations worldwide (FAOSTAT, 2021; EPPO, 2023). Despite the fact that the pathogen has been confirmed in a number of countries, there is no systematic data on the scope and severity of bacterial wilt disease in the potato value chain in the majority of countries (EPPO, 2023).

In Africa, RSSC has been reported to be widely spread in Cameroon, Côte d'Ivoire, Madagascar, Malawi, Mali, Mauritius, Nigeria, Uganda, Kenya, Ethiopia, Rwanda, and Kenya (Mahbou *et al.*, 2009; N'Guessan *et al.*, 2012; Abdurahman *et al.*, 2017, 2019; Sharma *et al.*, 2020; Okiro *et al.*, 2022; EPPO, 2023). The phylotype IIB-1 strain causing potato brown rot disease was found to be widely spread in most African potato-growing countries. However, knowledge of the impact on distribution and economic significance of the RSSC in Africa continues to be limited (Abdurahman *et al.*, 2017), as the diversity of the phytopathogen in the majority of the countries remains scanty. Diversity has been reported in areas where extensive sampling efforts have been made, resulting in the identification of three different strains of the RSSC found in multiple African nations. In particular, phylotype I strains are common in lowland areas, whereas strains of phylotypes II and III are more frequently found in the elevated regions of Cameroon, Rwanda, Kenya, Tanzania, Uganda, Ethiopia, and the Ivory Coast (N'Guessan *et al.*, 2012; Abdurahman *et al.*, 2017; Abdurahman *et al.*, 2019; Sharma *et al.*, 2022) (Table 3.1).

The main routes of transmission are the use of uncertified seed potato, imported plants, infected planting material, and infected irrigation water (Champoiseau *et al.*, 2009). Rwanda has an extensive history of *R. solanacraerum* phylotype IIB in potato, with yield losses estimated at 20–100% (Mutimawurugo *et al.*, 2019; Sharma *et al.*, 2020; Uwamahoro *et al.*, 2020). In Kenya, the phytopathogen causing potato brown rot has been identified in all the significant potato cultivation areas of the Rift Valley, Central, Western, and Coastal regions of the country, causing lower yields (Okiro *et al.*, 2022; Sharma *et al.*, 2022). The prevalence of bacterial wilt disease in potato in Ethiopia is 97%, with an overall disease incidence of 0.06–96.6%. Production of disease-free seed potato is difficult, especially in regions of Ethiopia where bacterial wilt of potato is widespread (Abdurahman *et al.*, 2017). In Cameroon, the phylotype IIB strains of RSSC were found to be most prevalent in the highlands of West Cameroon, known for potato farming (Mahbou *et al.*, 2009). In the southern African region, cases of bacterial wilt have been documented in Zambia, Mozambique, Angola, South Africa, Malawi, Swaziland, and Zimbabwe. However, there is currently lack of up-to-date records regarding the specific species responsible for causing the bacterial wilt of potato in southern Africa (Muthoni *et al.*, 2012). Nevertheless, it is presumed that *R. solanacearum* and *R. pseudosolanacearum* are both present (van der Waals and Krüger, 2020).

In the Americas, RSSC Phylotypes I and II have been reported in the USA and neighboring countries such as Canada, Peru, Brazil, and Colombia, among other countries in the region. Phylotype IIB, which causes potato brown rot disease, was introduced through imported infected geranium from Kenya, Guatemala, and Costa Rica and was later labeled as a bioterrorism agent in that country (Madden and Wheelis, 2003; Swanson *et al.*, 2005). Cuttings from propagative planting materials, including geranium, potato, tomato, heliconia, anthurium, and pothos, imported into the USA, were discovered to be infected with RSSC. The transportation of such propagated materials carries a significant risk, especially when dealing with asymptomatic hosts. The potential consequences of introducing these infections into a new environment could be profoundly destructive (Norman *et al.*, 2009). Strict measures have been put in place, especially in the USA and Canada, to monitor the introduction of new strains of the phytopathogen from imported propagative materials.

In Asian countries, brown rot, induced by R3bv2, has been documented in China, India, Iran, the Philippines, Indonesia, Nepal, and Thailand (Mark and Hayward, 2014). Bacterial wilt was detected in China in over 90 plant species representing at least 39 different families in 30 southern Chinese provinces, in many crops, including ornamental plants, medicinal plants, shrubs, and trees. Some of the economically important crops affected include peanut, tomato, mulberry, tobacco, potato, eggplant, pepper, ginger, and olive (Jiang *et al.*, 2017). In contrast to the situation in China, a comprehensive investigation of potato field infections in the warm and temperate areas of south-western Japan indicated that while there is a prevalence of RSSC from other variants, R3bv2 was not

Table 3.1. Global distribution of *Ralstonia* species complex in selected countries.

Country	RSSC Strains	Crops	Reference	Region
Cameroon	Phylotypes I, II, and III	Tomato, Potato, Huckleberry, Pepper, Amaranthus	Mahbou *et al.* (2009)	Africa
Côte d'Ivoire	Phylotypes I, II, and III	Tomato, Eggplant, Pepper	N'Guessan *et al.* (2012)	
Madagascar	Phylotypes I, II, and III	Potato, Tomato, Eggplant, Bean, weed, Black night shade, Pepper, Cape gooseberry	Ravelomanantsoa *et al.* (2018)	
Nigeria	Phylotype I and II	Tomato	Popoola *et al.* (2015)	
Kenya	Phylotype I and II	Potato, Tomato, Nightshade, Sodom apple, Pepper	Okiro *et al.* (2022), Sharma *et al.* (2022)	
Uganda	Phylotypes I, II, and III	Potato, Tomato, Eggplant, Chilli, Pepper, Garden egg, Night shade, Sodom apple, Nakati	Abdurahman *et al.* (2019)	
South Africa	Phylotype I and II	Tomato	Shutt *et al.* (2018)	
USA	Phylotype I and II	Potato, Tomato, Geranium, Eggplant, Anthurium, Pothos, Heliconia	Norman *et al.* (2009)	Americas
Peru	Phylotype I and II	Tomato, Potato, Pepper, plantain, Egg plant	Gutarra *et al.* (2017)	
Canada	Phylotype II	Geranium	EPPO (2020)	
Brazil		Tomato, Potato, Aubergine, Eggplant, Bell pepper, Long pepper, Geranium, Banana, and Eucalyptus	Santiago *et al.* (2017)	
Colombia	Phylotype II	Plantain, Taro, Pumpkin, Guava	Romo *et al.* (2012), Ramírez *et al.* (2020)	
United Kingdom	Phylotype II	Potato, Tomato	Parkinson *et al.* (2013)	Europe
Germany	Phylotype II	Pelargonium, Ginger, Potato	Janse *et al.* (2004); EPPO (2017)	
Netherlands	Phylotype I and II	Potato, Roses	Van Elsas *et al.* (2000), Bergsma-Vlami *et al.* (2018)	
China	Phylotype I and II	Potato, Tobacco, Peanut, Mulberry, Sesame, Chilli, Pepper Eggplant, Ginger, Tomato	Jiang *et al.* (2017), He *et al.* (2020)	Asia

Continued

Table 3.1. Continued

Country	RSSC Strains	Crops	Reference	Region
India	Phylotype I and II	Ginger, Mulberry, Tomato, Potato, Brinjal, and Pepper	Singh *et al.* (2021)	
Pakistan		Tomato, Brinjal, Pepper, Potato	Begum *et al.* (2012)	

found (Horita and Tsuchiya, 2001). This is because only locally produced seed potato have been allowed for cultivation due to strict regulations prohibiting the importation of seed potato from other countries (Mark and Hayward, 2014). Indonesia has been reported to contain Phylotype IV strains of RSSC that were initially identified in India and were probably introduced to Mauritius through contaminated potato tubers. Sequevar IV-10 isolates, which share a close relationship with strains from Indonesia and are distinct from sequevar IV-8 strains from Japan, exhibit significant virulence on potato, tobacco, and groundnut, and has spread to more than 10 provinces (Yahiaoui *et al.*, 2017).

In Oceania, phylotypes I and II occur in Papua New Guinea, Indonesia, Sri Lanka, and Australia, with the Philippines reporting all four phylotypes (Ajanga, 1993). The study by Dela Cueva *et al.* (2019) classified potato bacterial wilt pathogens as phylotypes I and II, with phylotype I inciting symptoms earlier in tomato and potato host plants. This suggests pathogenic variability, with non-certified planting materials likely to aggravate the spread of the pathogen. Previously, it was reported that phylotype IV was present in Japan, Australia, and Indonesia (Fegan and Prior, 2005). The disease was absent in Western Australia; however, several outbreaks were reported in the south-west of Perth, mainly attributed to the use of contaminated seed (Mark and Hayward, 2014).

The first report of bacterial wilt of potato in Sweden was in Olsson, 1976, and the disease then spread to other parts of Europe (Elphinstone *et al.*, 1998). Since the late 1980s, isolated instances of potato brown rot outbreaks have been documented in 11 European member states, including Germany, France, Greece, Belgium, Italy, Portugal, Spain, the Netherlands, the United Kingdom, Hungary, and Turkey. Elphinstone *et al.* (1998) have documented latent infections in *Solanum dulcamara* in Norway, the Netherlands, and England. Eradication of *R. solanacearum* causing potato brown rot disease in Sweden has been achieved through the systematic removal of the alternative host, *S. dulcamara* weed, from rivers (Elphinstone *et al.*, 1998) (Fig. 3.2).

Dangers of the spread of bacterial wilt disease of potato

According to Elphinstone (2005), the bacterial wilt disease has caused economic losses in potato production on a global scale that exceed $1 billion. Propagation materials harboring latent infections, such as seed potato and cuttings, represent a significant source of long-distance pathogen spread (Dickstein *et al.*, 2023). This poses a dual threat to crop production and contributes significantly to the dispersion of the pathogen internationally through trade (Norman *et al.*, 2020). In numerous countries where the organism is under quarantine status, significant losses ensue due to lack of enforcement of eradication protocols and the implementation of limitations on additional cultivation in affected areas. Every year, the United States imports more than 4 billion propagative cuttings, primarily contributing to the introduction of *R. solanacearum* (Drotleff, 2018; Norman *et al.*, 2020). Identifying the disease can prove especially challenging since plants imported often do not display visible symptoms (Weibel *et al.*, 2016; Álvarez *et al.*, 2019; Vargas *et al.*, 2023). The past few years have seen the emergence of new distinctive populations of *Ralstonia* on blueberry, mandevilla, and tea roses in the USA and Europe. These populations spread rapidly through the trading of latently infected, propagative stock (Weibel *et al.*, 2016;

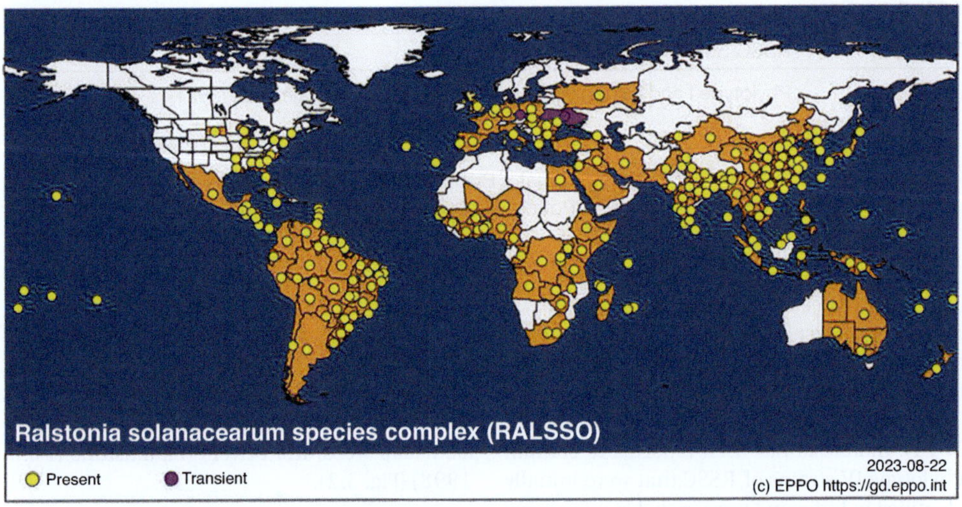

Fig. 3.2. Map displaying the worldwide distribution of the *Ralstonia solanacearum* species complex. Regions highlighted in yellow and orange represent presence, while purple indicates instances of transient presence as reported (Source: EPPO, 2023).

Tjou-tam-sin *et al.*, 2017; Bergsma-Vlami *et al.*, 2018; Conner *et al.*, 2022).

Ralstonia solanacearum stands as a notable environmental concern, instigating outbreaks within water reservoirs and the roots of semi-aquatic weeds (Caruso *et al.*, 2005, 2017; Elphinstone, 2005). This issue presents a formidable challenge for growers, particularly in regions with limited water resources and the European Union countries that have imposed irrigation restrictions (EFSA, 2019). Surface water used in irrigation is susceptible to contamination through runoff from agricultural fields infected with *R. solanacearum* or via exposure to wastewater from the potato processing industry (EFSA, 2019). Several weed species such as *Tropaeolum majus*, *Portulaca oleracea*, *Rumex* sp., *Tagetes* sp., and *Ipomoea* sp., *Solanum nigrum*, *Solanum dulcamara*, *Solanum cinereum*, *Physalis angulate*, and *B. pilosa* have been found to act as carriers of *R. solanacearum*, often without displaying symptoms, offering an additional means for the bacteria to persist (Ajanga, 1993; Pradhanang *et al.*, 2000; Janse *et al.*, 2004; EFSA, 2019; He *et al.*, 2020). *Ralstonia solanacearum* depends on infecting perennial host plants to persist and survive during the winter months in temperate climates. Bittersweet (*Solanum dulcamara*) is a persistent weed that has been observed to serve as a reservoir for its resilience and propagation. In the natural environment, *R. solanacearum* has the ability to inhabit the roots and vascular system of *S. dulcamara* plants, subsequently discharging the bacterium into surface water via their submerged roots (Elphinstone, 1996; Caruso *et al.*, 2005). Furthermore, it has been observed that the *R. solanacearum* phylotype IIB sequevar one strain can establish hidden infections in weeds, which then serve as a source of inoculation for infecting potato fields (Olsson, 1976; Lowe-Power *et al.*, 2018). In Uganda, many herbaceous weeds were latently infected with *R. solanacearum* (Tusiime *et al.*, 1998). Recently, bacterial wilt disease was detected in *Bidens pilosa*, a plant native to Central America that is now widely spread in China. This weed competes with crops, leading to agricultural and ecological harm. While crop rotation is a useful control method, latent infections in the weed hinder its effectiveness (He *et al.*, 2020).

The spread of plant diseases to new environments is attributed to global warming, which results from increases in atmospheric CO_2 and temperatures (Luck *et al.*, 2011). Temperature ranks as a critical factor affecting the spread of numerous plant-pathogenic bacteria. Climate change in Bolivia has led to an increase in the spread of brown rot due to rising temperatures and erratic weather events, potentially favoring *R. solanacearum* survival in colder climates and increasing its dispersion beyond border areas (José and Plata, 2016). *Ralstonia solanacearum* strains have been reported to adjust to temperature and oligotrophy, with low temperatures indicating water nutrient content protection. In temperate and warm conditions, adaptations encompass responses to survive in times of scarcity and alterations in structural characteristics influenced by temperature. *Ralstonia solanacearum* cells remain pathogenic in all conditions and can be resuscitated by reversing temperature (Belén *et al.*, 2022). Climate change-related weather patterns, such as erratic rainfall, higher temperatures, heavier rainfall, and flooding, are increasing the danger of bacterial infections in potato, including bacterial wilt disease (Luck *et al.*, 2011; José and Plata, 2016). Climate change is poised to trigger substantial shifts in the geographical distribution of both plant pathogens and their host plants. This transformative process holds the potential to profoundly impact ecosystems and agricultural systems (Kůdela, 2009). Nevertheless, factors beyond climate, such as the composition of microbial communities and variations in soil types, can also play a pivotal role in the restructuring of plant diseases (Huerta *et al.*, 2015).

Mitigation measures of bacterial wilt disease in potato

Managing *R. solanacearum* is challenging due to its soil-borne nature that enables it to endure in the soil for extended periods. The pathogen has a long host range and wide biological variability (Hayward, 1991). Different strategies are employed to control bacterial wilt, but none has proven to be 100% effective alone; hence, different control methods are applied together (Muthoni *et al.*, 2012). Current strategies for control include cultural practices, soil disinfection, chemical control, biological control, and host resistance. To date, breeding for resistance remains the most promising strategy (Muthoni *et al.*, 2013).

Cultural practices

Crop rotation offers several advantages, including the preservation of soil structure and organic content, as well as a notable reduction in soil erosion often linked to uninterrupted row planting (Yuliar *et al.*, 2015). In regions where RSSC has already established a widespread presence, implementing crop rotation emerges as the most effective control strategy (Dickstein *et al.*, 2023). Implementing crop rotation involving plants that are not host plants or less susceptible plants can result in a decrease in particular pathogenic populations (Muthoni *et al.*, 2012). This decline occurs as a result of the antagonistic actions and natural mortality of other microorganisms (Janvier *et al.*, 2007). Crop rotation inhibits the occurrence of specific plant pathogenic populations and often reduces soil-borne pathogen-induced plant diseases resulting from cropping the same susceptible host plant for successive harvests. In microbial diversity studies in relation to intercropping in Egypt, *R. solanacearum* pathogen was undetectable after onion planting but increased after cowpea planting. Onion soil and rhizosphere had the highest antagonist ratio, suggesting brown rot control. A higher ratio of endospores was observed, but no antagonists were detected. The pathogen was undetectable in wheat and second-year corn soil and rhizosphere due to long life span, fluorescent pseudomonad density, and antagonistic *Streptomyces* spp. ratio (Messiha *et al.*, 2019). Crop rotation of potato and maize was reported to increase potato yield more than planting only potato in a field infested by *R. solanacearum* (European and Mediterranean Plant Protection Organization, 2004; Yuliar *et al.*, 2015). For multi-cropping, reports indicate that growing Chinese chive and tomato has been shown to decrease the occurrence of bacterial wilt by 60%. This reduction is attributed to the fact that the root exudates of Chinese chive act as a preventative measure, inhibiting the infection of tomato plants by *R. solanacearum* (Yuliar *et al.*, 2015). However, this method may not be effective in some countries due to inadequate land for long rotational periods as a result of land fragmentation (Muthoni *et al.*, 2012).

Previous studies indicate that certain fertilizers have the potential to decrease both the frequency and intensity of bacterial wilt. Higher doses of calcium in plants reduce the severity of bacterial wilt and *R. solanacearum* populations in the stems of tomatoes (Yuliar *et al.*, 2015). Several elements in plant cell walls, such as silica, play a role in determining whether plants are vulnerable or resistant to pathogen infections. Mixing a specific combination of rock dust and commercially available organic fertilizer resulted in a reduction in the occurrence of bacterial wilt in tomato plants (Kiirika *et al.*, 2013). Field experiments carried out in regions of Ethiopia where bacterial wilt affects potato cultivation have shown that the application of lime has a significant impact on reducing bacterial wilt incidence. After 90 days, the incidence rate was only 10.8% in plots treated with 12 t/ha of lime, whereas it was as high as 40% in the control plots (Tafesse *et al.*, 2021).

Positive selection entails recognizing and collecting seed potato that exhibits a healthy phenotype around 6–7 weeks after planting, prior to flowering, and removing plants that manifest bacterial wilt symptoms later in the growing season. An enhancement of this technique involves combining positive selection and disease indexing for improved outcomes. This entails collecting seeds from visibly healthy plants and examining tubers for latent infections (Muthoni *et al.*, 2012). These methods are not popular with potato farmers in developing countries, especially in Africa, because they are expensive since they require specialized testing (Muthoni *et al.*, 2012). Negative selection involves pulling out visibly diseased plants from a relatively pristine potato field. This is achieved by completely digging out infected plants with all their root systems, stolons, tubers, and the soil around them. Ash or lime is mixed with the soil where rogue plants were removed to elevate the pH and eliminate bacteria (Gildemacher *et al.*, 2009; Muthoni *et al.*, 2012).

Solarization and hot water treatment have proven to be successful in combating *R. solanacearum* (Vinh *et al.*, 2005). The utilization of transparent plastic mulch for soil solarization for a period of 60 days before planting tomatoes resulted in a decreased occurrence of bacterial wilt (Yuliar *et al.*, 2015). Heat treatment of soil in incubators containing bacterial suspensions of *R. solanacearum* at 45 and 60°C for 2 days and 2 hour, respectively, prior to tomato planting inhibited infections by 100% and reduced bacterial wilt occurrence by 50–75% (Kongkiattikajorn and Thepa, 2007).

Biological soil disinfection (BSD), is a techniques that boosts soil microbial respiration by incorporating decomposable inexpensive organic matter while limiting oxygen availability through the use of sealed plastic sheets (Messiha *et al.*, 2007; Momma *et al.*, 2013). BSD resulted in a reduction of over 99.9% in *R. solanacearum* population in the soil. Polymerase chain reaction denaturing gradient gel electrophoresis (PCR-DGGE) investigations demonstrated that BSD had no effect on bacterial diversity but altered the composition of the bacterial community (Messiha *et al.*, 2007). A study assessed the efficacy of *C. recemoca* and *C. myrica* in dried algal biomass positively impacted soil microbial diversity, potentially serving as a preventative measure against bacterial wilt (Hamed *et al.*, 2023). However, the primary impediments to widespread adoption of *B. subtilis* among low-income farmers in tropical regions seem to be the high costs associated with plastic and labor.

Chemical control

Controlling brown rot and diseases caused by RSSC in the field poses a formidable challenge due to their broad host range, strain diversity, and the absence of effective chemical treatments. The chemicals that have been examined for their efficacy in managing bacterial wilt diseases encompass Actigard (acibenzolar-Smethyl, ASM), BABA (DL-3-aminobutyric acid), phosphorous acid SBP (calcium hypochlorite, also known as stable bleaching powder), and a mixture of urea and lime (Dickstein *et al.*, 2023). Acibenzolar-S-methyl (ASM) effectively decreases bacterial wilt in tomato cultivars with moderate resistance but does not have the same effect in susceptible ones when high concentrations of *R. solanacearum* are employed (Pradhanang *et al.*, 2005). Treatment of tomato plants using ASM reduces pathogen movement in stem tissues, thereby preventing *R. solanacearum* spread. A mixture involving methyl bromide, 1,3-dichloropropene, or sodium metal in combination with chloropicrin resulted in a significant decrease in bacterial wilt occurrence in the agricultural field, leading to yield enhancements of up to 100% for both tobacco and tomato crops (Yuliar *et al.*, 2015). In the group treated with pesticides, the tomato yield was 1.7–2.5 times higher than the yield of the untreated control group (Yuliar *et al.*, 2015). It is, however, important to note that the utilization of methyl bromide was discontinued in compliance with the 1997 Montreal Amendment to the 1987 Montreal Protocol, which addresses substances that contribute to ozone layer depletion.

Drenching of phosphorous acid salts into the soil was found to be an effective method for safeguarding geranium plants against *R. solanacearum* infections. Drenching proved to be a successful method for managing bacterial spots in tomatoes in both greenhouse and field trials (Wen *et al.*, 2009). Hayes *et al.* (2022) tested three antimicrobial substances on water-borne plant-pathogenic *Ralstonia* cells. Results showed Huwa-San, hydrogen peroxide, and bleach led to reduced populations of *Ralstonia* cells, while active chlorine was highly effective. A study by Hassan and Abo-Elyousr (2013) investigated the role of polyphenol oxidase and catalase in tomato protection against *R. solanacearum* bacterial wilt. It was found that BABA (DL-3-aminobutyric acid) enhanced polyphenol oxidase activity in both infected and control plants. Polyphenol oxidase is involved in resistance and defense through phenolic compound oxidation. The study suggests that increased phenolic acid may increase resistance. However, pesticides have been associated with health and environmental concerns, leading to a decline in their use in agricultural production. Exposure can cause negative effects, including dermatological, gastrointestinal, neurological, and reproductive issues (Nicolopoulou-Stamati *et al.*, 2016).

Plant health measures

Implementing phytosanitary measures is the primary and most effective approach for controlling bacterial wilt in agricultural fields. In areas where potato bacterial wilt is widespread or in regions where *R. solanacearum* is present but has not fully established, these techniques can be effective in specific circumstances (Muthoni *et al.*, 2012). In the absence of any available chemical control methods for remediation, the primary approach to preventing bacterial wilt hinges on two key factors: the accessibility of clean planting material and the implementation of robust surveillance and monitoring measures to safeguard regions devoid of the bacteria (Muthoni *et al.*, 2012; Gregory *et al.*, 2020).

Potato growers in Europe and North America primarily use certified seed, while farmers in the USA, particularly organic farmers, use informal sources, highlighting commodity-driven seed systems. Low-income countries often have informal seed systems with low-certified seed use, but efforts to establish certified seed programs and recent innovations aim to improve resource-poor farmers' seed systems. The choice between certified and informal seed sources in potato cultivation depends on various factors, including regional agricultural practices, the scale of farming, and economic considerations (Gregory *et al.*, 2020). While certified seed is widely adopted in developed countries, the low-income countries are still struggling to meet this requirement. Ultimately, the goal is to strike a balance between meeting the needs of commodity-driven production and supporting resource-poor farmers in their quest for improved seed quality and crop resilience.

Conducting routine post-harvest testing of seed potato tubers is generally imperative in order to mitigate the inadvertent dissemination of latent infections. Similarly, in the case of other crops that are propagated through vegetative means, it is crucial to ensure that the planting materials have undergone thorough testing to ascertain their freedom from any form of infection. Additionally, stringent controls must be implemented to restrict the transfer of planting materials from regions with infections to regions that are free from diseases.

The primary natural dissemination mechanism of brown rot is predominantly attributed to polluted irrigation water. Surface water employed for irrigation can easily become contaminated through contact with runoff from agricultural fields infected by *R. solanacearum* or through exposure to effluent water discharged from potato processing facilities (EFSA, 2019; Eisfeld *et al.*, 2022). *Ralstonia solanacearum* is a significant environmental concern, causing outbreaks in water reservoirs and semiaquatic weed roots. This presents a challenge for farmers, particularly in regions with limited water resources and within European Union countries that have imposed restrictions on water for irrigation (EFSA, 2019).

Biological control

The control of *R. solanacearum* is primarily achieved through the utilization of antagonistic interactions between different bacterial species, the utilization of genetically modified variants of *R. solanacearum*, and the implementation of bacteriophages as potential biological agents. (Yuliar *et al.*, 2015; Álvarez *et al.*, 2019; Cai *et al.*, 2021). Biological control mechanisms involve interactions among biological agents such as parasitism, nutrient competition, antibiosis, and induced systemic resistance. In India, studies on biological control methods to manage *R. solanacearum* revealed that local fields contain plant-growth-promoting rhizobacteria (PGPR) isolates that can be harnessed to counteract the harmful impact of *R. solanacearum* (Maji and Chakrabartty, 2014). A study revealed that rhizobacterial strains can inhibit both pathogen and helper strains, making indirect effects on helper strains more effective in reducing *R. solanacearum* (Li *et al.*, 2022). Other biocontrol studies indicate that various strains of rhizobacteria were tested against an Ethiopian *R. solanacearum* strain (RP87, B2G, APF1, APF2, APF3, and APF4) and exhibited effective inhibition (Maji and Chakrabartty, 2014) against bacterial wilt disease. Significantly, APF1 emerged as the most promising strain for both disease suppression and enhancing growth, leading to a remarkable 63% increase in dry weight when compared to untreated controls. Another study by Mohamed *et al.* (2020) reported that *E. cloacae* PS14 and *T. asperellum* T34 bioagents significantly reduced wilt disease severity in potato plants under greenhouse conditions, protecting them from pathogen damage and increasing crop productivity by 20.44–40.96% by the release of antibacterial compounds and the stimulation of innate resistance in potato plants. However, the success of disease control relies heavily on the local edaphic adaptation of the biocontrol agents. Limited success can be attributed to the inadequate competitive abilities and poor adaptation of introduced strains. It is imperative to screen antagonistic strains that can adapt effectively to the specific soil conditions in the local environment, as this is pivotal for successful biocontrol (Alamer *et al.*, 2020).

Lytic or lysogenic bacteriophages have been demonstrated to be effective against RSSC strains (Bhunchoth *et al.*, 2015; Wei *et al.*, 2017; Ahmad *et al.*, 2018; Alamer *et al.*, 2020). Three new lytic phages, vRsoP-WR2, vRsoP-WM2, and vRsoP-WF2, were shown to be effective and eco-friendly in the biocontrol of *R. solanacearum* populations in host plants and irrigation water. Their long-term persistence and lytic activity make them suitable for field or greenhouse use (Alamer *et al.*, 2020). A study involving 12 lytic bacteriophages sourced from Kenya and China was undertaken by Wei *et al.* (2017) to assess their potential in combating potato bacterial wilt. This research led to the development of a potent phage cocktail, P1, which displayed remarkable efficiency and rapidity in lysing *R. solanacearum* (Wei *et al.*, 2017). In another study by Umrao *et al.* (2021) in India it was found that phage PE204, a bacterium, can delay wilt development and completely inhibit disease in tomato plants. The lytic phage cocktail and a mix of podoviridae bacteriophages were found to have a variety of inhibition patterns of up to 98% against *R. solanacearum* that causes potato bacterial wilt disease. However, some unknown aspects of bacteriophages, like survival capacity and lytic activity maintenance, remain uncertain, especially after prolonged conservation using lyophilization. The viability and stability of these phages against different *R. solanacearum* strains should be thoroughly evaluated (Álvarez *et al.*, 2022). The use of bacteriophages for plant disease control faces challenges due to environmental factors like sunlight, high temperatures, and desiccation. However, rhizosphere application eliminates sunlight-related issues, according to studies by Iriarte *et al.* (2012). Another legitimate concern that revolves around bacteriophages is the potential for bacteria to undergo mutations, leading to resistance against individual phages (Jones *et al.*, 2012).

Host resistance

The available potato germplasm offers only a restricted selection of resistant options, and conventional breeding has achieved only a moderate level of resistance in most solanaceous crops (Lebeau *et al.*, 2013; Dickstein *et al.*, 2023). Resistance to bacterial wilt in solanaceous crops has been found to be dependent on regional climatic conditions (Hayward, 1991; Huet, 2014). Previous studies in Cameroon revealed that some strains of *R. solanacearum* preferentially establish themselves in different climatic conditions (N'Guessan *et al.*, 2012). For instance, potato resistant to bacterial wilt in one location was found to be susceptible in another geographical location due to local strain variation. Potato cultivars with resistance have been developed by utilizing wild potato species that are closely related to *Solanum tuberosum* (Carputo *et al.*, 2009; Narancio *et al.*, 2013). Previous studies have documented the presence of resistance to bacterial wilt in various *Solanum* species, including use of *Solanum stenotomum* through protoplast fusion (Fock *et al.*, 2001), somatic hybridization of *S. tuberosum* and S. *chacoense* (Chen *et al.*, 2013), *S. phureja* (Lopes *et al.*, 2021), and *S. commersonii* (Carputo *et al.*, 2009; Ferreira *et al.*, 2017; Otieno *et al.*, 2021), which provide new sources of resistance for developing commercially resistant potato cultivars.

A study by Boschi *et al.* (2017) discovered that the activation of the AtEFR gene led to enhanced resistance against wild *S. commersonii*, greater survival rates, and a decrease in disease symptoms. A study on 11 potato accessions from the Colombian Central Collection identified five highly resistant and tolerant for potato breeding (Vargas *et al.*, 2023). The potato variety Cruza 148, which is currently considered the most resistant, has demonstrated its ability to withstand wilt caused by race 3 strains of *R. solanacearum* in various regions, including the Philippines, Peru, Brazil, Uganda, Burundi, Rwanda, and the Democratic Republic of the Congo (Gutarra *et al.*, 2015; Abdurahman *et al.*, 2019; Uwamahoro *et al.*, 2020; Jose *et al.*, 2023). Therefore, Cruza and the other wild potato cultivars, having demonstrated resistance to bacterial wilt in potato, could be a valuable recommendation for fields with a history of bacterial wilt infestations.

Nanotechnology

The use of nanoparticles either alone or as a composite, have become increasingly favored across different sectors, particularly in agriculture (Chen *et al.*, 2019). Numerous research efforts have explored the application of nanotechnology in delivering various agrochemicals such as fertilizers, pesticides, herbicides, and the detection of pathogens and diseases (Ray *et al.*, 2023). A variety of metals, metal oxides, carbon based, and biocomposites have been extensively studied and proven to be effective fungicides and bactericides when developed into nanoparticles (Khot *et al.*, 2012; Shojaei *et al.*, 2016; Chen *et al.*, 2019; Khan *et al.*, 2021). According to earlier studies, nanoparticles have the potential to address the resistance of bacteria and fungi to conventional fungicides and bactericides (Ray *et al.*, 2023). Nanotechnology is currently being used to generate antimicrobial compounds to combat phytopathogens including bacteria and fungi (Shojaei *et al.*, 2016).

Previous research has shown that nanoparticles can kill bacteria by degrading the cell wall or producing large quantities of reactive oxygen species (Wang *et al.*, 2016). According to Khan *et al.* (2021), using zinc oxide nanoparticles to artificially infect tomato plants dramatically increased plant development by lowering bacterial cell densities and disease severity compared to control. Zinc oxide nanoparticles increase the permeability of the bacterial cell membrane triggering membrane disruption and potential changes at the cellular level such as interference with cell metabolism leading to bacterial cell death (Khan *et al.*, 2021). A study by Jiang *et al.* (2022) demonstrated that metallic nanoparticles (CuO, ZnO, and FeO) diminished the occurrence of bacterial wilt variably and influenced the composition and structure of the bacterial community in the rhizosphere. Fabricated nanocomposite composed of SiO_2/Ag has been used for antibacterial agent to combat *R. solanacearum* in agricultural settings (Zaki *et al.*, 2022). It was reported that SiO_2/Ag nanocomposites adhere to the surface of *R. solanacearum*, resulting in membrane impairment, formation of pits, and alteration of ion transportation. Reactive oxygen species (ROS) generation harms intracellular components such as DNA, plasmids, and organelles, ultimately inducing cellular toxicity attributed to oxidative stress. However, there are presently no set guidelines or regulatory measures for the application of nanotechnology, specifically in soil and aquatic settings. Furthermore, a significant portion of farmers and agricultural stakeholders still lack awareness regarding the potential impacts of nanoparticles (Ray *et al.*, 2023).

Conclusions

Ralstonia solanacearum (R3bv2) causes potato brown rot and can survive and spread in water, soil, and plant hosts. Its interactions with these environments are complex and not well understood. In developed countries, quarantine measures and disease-free planting material can prevent the disease. In developing countries, farmers face challenges in accessing and affording these resources. Replanting on infested land also increases the risk of infection. The disease can affect crop production and international trade by contaminating potato cuttings. It can also harm the environment by infecting water reservoirs and weeds. Global warming may favor the bacterium by changing the host and pathogen distribution. Enhancing disease management under these conditions presents a significant challenge for the future.

Conflict of Interest

The authors have no conflict of interests.

References

Abdurahman, A., Griffin, D., Elphinstone, J., Struik, P.C., Schulz, S. *et al.* (2017) Molecular characterization of *Ralstonia solanacearum* strains from Ethiopia and tracing potential source of bacterial wilt disease outbreak in seed potatoes. *Plant Pathology* 66(5), 826–834. DOI: 10.1111/ppa.12661.

Abdurahman, A., Parker, M.L., Kreuze, J., Elphinstone, J.G., Struik, P.C. *et al.* (2019) Molecular epidemiology of *Ralstonia solanacearum* species complex strains causing bacterial wilt of potato in Uganda. *Phytopathology* 109(11), 1922–1931. DOI: 10.1094/PHYTO-12-18-0476-R.

Ahmad, A.A., Elhalag, K.M., Addy, H.S., Nasr-Eldin, M.A., Hussien, A.S. *et al.* (2018) Sequencing, genome analysis and host range of a novel Ralstonia phage, RsoP1EGY, isolated in Egypt. *Archives of Virology* 163(8), 2271–2274. DOI: 10.1007/s00705-018-3844-4.

Ailloud, F., Lowe, T., Cellier, G., Roche, D., Allen, C. *et al.* (2015) Comparative genomic analysis of *Ralstonia solanacearum* reveals candidate genes for host specificity. *BMC Genomics* 16(1), 270. DOI: 10.1186/s12864-015-1474-8.

Ajanga, S. (1993) Status of bacterial wilt of potato in Kenya. In: Hartman, G.L. and Hayward, A.C. (eds) *Proceedings of International Conference on Bacterial Wilt*, Australian Centre for International Agricultural Research, pp. 338–340.

Alamer, I.S.A., Tomah, A.A., Li, B. and Zhang, J.Z. (2020) Isolation, identification and characterization of rhizobacteria strains for biological control of bacterial wilt (*Ralstonia solanacearum*) of eggplant in China Iman. *Agriculture (Switzerland)* 10(2), 1–10. DOI: 10.3390/agriculture10020037.

Álvarez, B., López, M.M. and Biosca, E.G. (2019) Biocontrol of the major plant pathogen *Ralstonia solanacearum* in irrigation water and host plants by novel waterborne lytic bacteriophages. *Frontiers in Microbiology* 10(Dec), 1–17. DOI: 10.3389/fmicb.2019.02813.

Álvarez, B., Gadea-Pallás, L., Rodríguez, A., Vicedo, B., Figàs-Segura, À. *et al.* (2022) Viability, stability and biocontrol activity in planta of specific *Ralstonia solanacearum* bacteriophages after their conservation prior to commercialization and use. *Viruses* 14(2), 183. DOI: 10.3390/v14020183.

Begum, N., Haque, M.I., Mukhtar, T., Naqvi, S.M. and Wang, J.F. (2012) Status of bacterial wilt caused by *Ralstonia solanacearum* in Pakistan. *Pakistan Journal of Phytopathology* 24(1), 37–39.

Belén, Á., María, M.L. and Elena, G.B. (2022) *Ralstonia solanacearum* facing spread-determining climatic temperatures, sustained starvation, and naturally induced resuscitation of viable but non-culturable cells in environmental water. *Microorganisms* 10(12), 2503. DOI: 10.3390/microorganisms10122503.

Bergsma-Vlami, M., van de Bilt, J.L.J., Tjou-Tam-Sin, N.N.A., Westenberg, M., Meekes, E.T.M. *et al.* (2018) Phylogenetic assignment of *Ralstonia pseudosolanacearum* (*Ralstonia solanacearum* Phylotype I) isolated from *Rosa* spp. *Plant Disease* 102(11), 2258–2267. DOI: 10.1094/pdis-09-17-1345-re.

Bhunchoth, A., Phironrit, N., Leksomboon, C., Chatchawankanphanich, O., Kotera, S. *et al.* (2015) Isolation of *Ralstonia solanacearum*-infecting bacteriophages from tomato fields in Chiang Mai, Thailand, and their experimental use as biocontrol agents. *Journal of Applied Microbiology* 118(4), 1023–1033. DOI: 10.1111/jam.12763.

Boschi, F., Schvartzman, C., Murchio, S., Ferreira, V., Siri, M.I. *et al.* (2017) Enhanced bacterial wilt resistance in potato through expression of Arabidopsis EFR and Introgression of quantitative resistance from Solanum commersonii. *Frontiers in Plant Science* 8(Sep), 1–11. DOI: 10.3389/fpls.2017.01642.

Boucher, C., Genin, S. and Arlat, M. (2001) Current concepts of pathogenicity in plant pathogenic bacteria. Concepts actuels sur la pathogénie chez les bactéries phytopathogènes. *Comptes Rendus de l'Académie Des Sciences-Series III-Sciences de La Vie* 324(10), 915–922. DOI: 10.1016/S0764-4469(01)01375-0.

Cai, Q., Zhou, G., Ahmed, W., Cao, Y., Zhao, M. *et al.* (2021) Study on the relationship between bacterial wilt and rhizospheric microbial diversity of flue-cured tobacco cultivars. *European Journal of Plant Pathology* 160(2), 265–276. DOI: 10.1007/s10658-021-02237-4.

Carputo, D., Aversano, R., Barone, A., Di Matteo, A., Iorizzo, M. *et al.* (2009) Resistance to *Ralstonia solanacearum* of sexual hybrids between *Solanum commersonii* and *S. tuberosum*. *American Journal of Potato Research* 86, 196–202. DOI: 10.1007/s12230-009-9072-4.

Caruso, P., Palomo, J.L., Bertolini, E., Álvarez, B., López, M.M. *et al.* (2005) Seasonal variation of *Ralstonia solanacearum* biovar 2 populations in a Spanish river: Recovery of stressed cells at low temperatures. *Applied and Environmental Microbiology* 71(1), 140–148. DOI: 10.1128/AEM.71.1.140-148.2005.

Caruso, P., Biosca, E.G., Bertolini, E., Marco-Noales, E., Gorris, M.T. *et al.* (2017) Genetic diversity reflects geographical origin of *Ralstonia solanacearum* strains isolated from plant and water sources in Spain. *International Microbiology* 20(4), 155–164. DOI: 10.2436/20.1501.01.298.

Champoiseau, P.G., Jones, J.B. and Allen, C. (2009) *Ralstonia solanacearum* race 3 biovar 2 causes tropical losses and temperate anxieties. *Plant Health Progress* 10, 35. DOI: 10.1094/php-2009-0313-01-rv.

Chen, L., Guo, X., Xie, C., He, L., Cai, X. *et al.* (2013) Nuclear and cytoplasmic genome components of *Solanum tuberosum* + *S. chacoense* somatic hybrids and three SSR alleles related to bacterial wilt resistance. *Theoretical and Applied Genetics* 126(7), 1861–1872. DOI: 10.1007/s00122-013-2098-5.

Chen, J., Mao, S., Xu, Z. and Ding, W. (2019) Various antibacterial mechanisms of biosynthesized copper oxide nanoparticles against soilborne *Ralstonia solanacearum*. *RSC Advances* 9(7), 3788–3799. DOI: 10.1039/c8ra09186b.

Conner, C.J., Bocsanczy, A.M., Spakes-Richter, B. and Norman, D.J. (2022) Bacterial wilt resistance in blueberry species. *International Journal of Fruit Science* 22(1), 852–859. DOI: 10.1080/15538362.2022.2137275.

Dela Cueva, F.M., Balendres, M.A.O., Justo, V.P. and Pathania, N. (2019) Phylotypes of the potato bacterial wilt pathogen in the Philippines and their relationship to pathogen aggressiveness. *Journal of Plant Pathology* 101(1), 23–29. DOI: 10.1007/s42161-018-0133-8.

Devaux, A., Goffart, J.P., Kromann, P., Andrade-Piedra, J., Polar, V. *et al.* (2021) The potato of the future: Opportunities and challenges in sustainable agri-food systems. *Potato Research* 64(4), 681–720. DOI: 10.1007/s11540-021-09501-4.

Dickstein, E.R., Bocsanczy, A.M., Champoiseau, P., Jones, J., Norman, D. *et al.* (2023) Recovery plan for *Ralstonia solanacearum* race 3 biovar 2 (Phylotype IIB, sequevars 1 and 2) causing brown rot of potato, bacterial wilt of tomato, and southern wilt of geranium. *Plant Health Progress* 25(1), 98–139. DOI: 10.1094/php-03-23-0027-rp.

Drotleff, L. (2018) *Cuttings Producers Improve Quality and Focus on Streamlining the Supply Chain*. Greenhouse Grower. Available at: https://www.greenhousegrower.com/production /cuttings-producers-improve-quality-and-focus-on-streamlining-the-supply-chain/

EFSA (2019) Pest categorisation of the *Ralstonia solanacearum* species complex. *EFSA Journal* 17(2), e05618. DOI: 10.2903/j.efsa.2019.5618.

Eisfeld, C., Schijven, J.F., Kastelein, P., van Breukelen, B.M., Medema, G. *et al.* (2022) Dose-response relationship of *Ralstonia solanacearum* and potato in greenhouse and in vitro experiments. *Frontiers in Plant Science* 13(Dec), 1–17. DOI: 10.3389/fpls.2022.1074192.

Elphinstone, J.G. (1996) Survival and possibilities for extinction of *Pseudomonas solanacearum* (Smith) Smith in cool climates. *American Journal of Potato Research* 39(1996), 403–410.

Elphinstone, J.G., Stanford, H. and Stead, D.E. (1998) Detection of *Ralstonia solanacearum* in potato tubers, *Solanum dulcamara* and associated irrigation water. In: Prior, P., Allen, C. and Elphinstone, J.G. (eds) *Bacterial Wilt Disease: Molecular and Ecological Aspects*. Springer, Berlin, Heidelberg, pp. 133–139.

Elphinstone, J.G. (2005) The current bacterial wilt situation: A global overview. In: Allen, C., Prior, P. and Hayward, A.C. (eds) *Bacterial Wilt Disease and the Ralstonia solanacearum Species Complex*. American Phytopathological Society, St. Paul, MN, pp. 9–28.

EPPO (2017) Current Pest Situation Evaluated by EPPO. Available at: https://gd.eppo.int/taxon /RALSSL /distribution/DE (accessed 23 November 2023).

EPPO (2020) *Ralstonia solanacearum* Detected in Pelargonium in Canada. Available at: https://gd.eppo.i nt/reporting/article-6776 (accessed 23 November 2023).

EPPO (2023) EPPO Datasheets on Pests Recommended for Regulation. Available at: https://gd.eppo.int (accessed 23 November 2023).

European and Mediterranean Plant Protection Organization (2004) Standards diagnostic protocols for regulated pests. *Bulletin, OEPP/EPPO* 34.

FAOSTAT (2021) Statistical Databases and Data-Sets of the Food and Agriculture Organization of the United Nations. Available at: https://www.fao.org/faostat/en/#data (accessed 27 November 2023).

Fegan, M. and Prior, P. (2005) How complex is the "*Ralstonia solanacearum* species complex"? In: Allen, C., Prior, P. and Hayward, A.C. (eds) *Bacterial Wilt Disease and the Ralstonia solanacearum Species Complex*. American Phytopathological Society, Saint Paul, MN, pp. 449–461.

Ferreira, V., Pianzzola, M.J., Vilaró, F.L., Galván, G.A., Tondo, M.L. *et al.* (2017) Interspecific potato breeding lines display differential colonization patterns and induced defense responses after *Ralstonia solanacearum* infection. *Frontiers in Plant Science* 8(Aug), 1–14. DOI: 10.3389/fpls.2017.01424.

Fock, I., Collonnier, C., Luisetti, J., Purwito, A., Souvannavong, V. *et al.* (2001) Use of *Solanum stenotomum* for introduction of resistance to bacterial wilt in somatic hybrids of potato. *Plant Physiology and Biochemistry* 39(10), 899–908. DOI: 10.1016/S0981-9428(01)01307-9.

Genin, S. and Denny, T.P. (2012) Pathogenomics of the *Ralstonia solanacearum* species complex. *Annual Review of Phytopathology* 50(Apr), 67–89. DOI: 10.1146/annurev-phyto-081211-173000.

Gildemacher, P.R., Kaguongo, W., Ortiz, O., Tesfaye, A., Woldegiorgis, G. *et al.* (2009) Improving potato production in Kenya, Uganda and Ethiopia: A system diagnosis. *Potato Research* 52, 173–205. DOI: 10.1007/s11540-009-9127-4.

Gregory, A.F., Amy, C., Jorge, A.-P., Monica, L.P. and Elmar, S.-G. (2020) The potato crop. In: *Potato Seed Systems*. Springer Nature, Switzerland, pp. 413–518. DOI: 10.1007/978-94-011-2340-2.

Gutarra, L., Kreuze, J., Lindqvist-Kreuze, H. and De Mendiburu, F. (2015) Variation of resistance to different strains of *Ralstonia solanacearum* in highland tropics adapted potato genotypes. *American Journal of Potato Research* 92(2), 258–265. DOI: 10.1007/s12230-014-9426-4.

Gutarra, L., Herrera, J., Fernandez, E., Kreuze, J. and Lindqvist-Kreuze, H. (2017) Diversity, pathogenicity, and current occurrence of bacterial wilt bacterium *Ralstonia solanacearum* in Peru. *Frontiers in Plant Science* 8(1221), 1–12. DOI: 10.3389/fpls.2017.01221.

Hamed, S.M., Kamal, M. and Messiha, N.A.S. (2023) Potential of algal-based products for the management of potato brown rot disease. *Botanical Studies* 64(29), 1–17. DOI: 10.1186/s40529-023-00402-y.

Hassan, M.A.E. and Abo-Elyousr, K.A.M. (2013) Activation of tomato plant defence responses against bacterial wilt caused by *Ralstonia solanacearum* using DL-3-aminobutyric acid (BABA). *European Journal of Plant Pathology* 136(1), 145–157. DOI: 10.1007/s10658-012-0149-4.

Hayes, M.M., Dewberry, R.J., Babujee, L., Moritz, R. and Allen, C. (2022) Validating methods to eradicate plant-pathogenic *Ralstonia* strains reveals that growth in planta increases bacterial stress tolerance. *Microbiology Spectrum* 10(6). DOI: 10.1128/spectrum.02270-22.

Hayward, A.C. (1991) Biology and epidemiology of bacterial wilt caused by *Pseudomonas solanacearum*. *Annual Review of Phytopathology Review of Phytopathology* 29, 65–87.

He, Y., Mo, Y., Zheng, D., Li, Q., Lin, W. *et al.* (2020) Different sequevars of *Ralstonia pseudosolanacearum* causing bacterial wilt of *Bidens pilosa* in China. *Plant Disease* 104(11), 2768–2773. DOI: 10.1094/PDIS-12-19-2738-SC.

Horita, M. and Tsuchiya, K. (2001) Genetic diversity of Japanese strains of *Ralstonia solanacearum*. *Phytopathology* 91(4), 399–407. DOI: 10.1094/PHYTO.2001.91.4.399.

Huerta, A.I., Milling, A. and Allen, C. (2015) Tropical strains of *Ralstonia solanacearum* outcompete race 3 biovar 2 strains at lowland tropical temperatures. *Applied and Environmental Microbiology* 81(10), 3542–3551. DOI: 10.1128/AEM.04123-14.

Huet, G. (2014) Breeding for resistances to *Ralstonia solanacearum*. *Frontiers in Plant Science* 5(715), 1–5. DOI: 10.3389/fpls.2014.00715.

Iriarte, F.B., Obradović, A., Wernsing, M.H., Jackson, L.E., Balogh, B. *et al.* (2012) Soil-based systemic delivery and phyllosphere in vivo propagation of bacteriophages. *Bacteriophage* 2(4), e23530. DOI: 10.4161/bact.23530.

Janse, J.D., Beld, H.E., Elphinstone, J.G., Simpkins, S. and Vaerenbergh, J.V. (2004) Introduction to Europe of *Ralstonia solanacearum* biovar 2, race 3 in *Pelargonium zonale* cuttings. *Journal of Plant Pathology* 86(2), 147–155.

Janvier, C., Villeneuve, F., Alabouvette, C., Edel-Hermann, V., Mateille, T. *et al.* (2007) Soil health through soil disease suppression: which strategy from descriptors to indicators? *Soil Biology and Biochemistry* 39(1), 1–23. DOI: 10.1016/j.soilbio.2006.07.001.

Jiang, G., Wei, Z., Xu, J., Chen, H., Zhang, Y. *et al.* (2017) Bacterial wilt in China: History, current status, and future perspectives. *Frontiers in Plant Science* 8(1549), 1–10. DOI: 10.3389/fpls.2017.01549.

Jiang, H., Lv, L., Ahmed, T., Jin, S., Shahid, M. *et al.* (2022) Effect of the nanoparticle exposures on the tomato bacterial wilt disease control by modulating the rhizosphere bacterial community. *International Journal of Molecular Sciences* 23(1), 414. DOI: 10.3390/ijms23010414.

Jones, J.B., Vallad, G.E., Iriarte, F.B., Obradović, A., Wernsing, M.H. *et al.* (2012) Considerations for using bacteriophages for plant disease control. *Bacteriophage* 2(4), e23857. DOI: 10.4161/bact.23857.

José, A.C. and Plata, G. (2016) The expansion of Brown Rot disease throughout Bolivia: Possible role of climate change José. *Canadian Journal of Microbiology* 256(0), 1–14.

Jose, J., Éva, C., Bozsó, Z., Hamow, K.Á., Fekete, Z. *et al.* (2023) Global transcriptome and targeted metabolite analyses of roots reveal different defence mechanisms against *Ralstonia solanacearum* infection in two resistant potato cultivars. *Frontiers in Plant Science* 13(Jan), 1–19. DOI: 10.3389/fpls.2022.1065419.

Kaguongo, W., Gildemacher, P., Demo, P., Wagoire, W., Kinyae, P. *et al.* (2008) Farmer practices and adoption of improved potato varieties in Kenya and Uganda. In: *International Potato Center (CIP), Lima, Peru. Social Sciences Working Paper* 2008-5. p. 85.

Khan, R.A.A., Tang, Y., Naz, I., Alam, S.S., Wang, W. *et al.* (2021) Management of *Ralstonia solanacearum* in tomato using ZnO nanoparticles synthesized through *Matricaria chamomilla*. *Plant Disease* 105(10), 3224–3230. DOI: 10.1094/PDIS-08-20-1763-RE.

Khlestkin, V.K., Erst, T.V., Rozanova, I.V., Efimov, V.M. and Khlestkina, E.K. (2020) Genetic loci determining potato starch yield and granule morphology revealed by genome-wide association study (GWAS). *PeerJ* 8, e10286. DOI: 10.7717/peerj.10286.

Khokhani, D., Tuan, T., Lowe-Power, T. and Allen, C. (2018) Plant assays for quantifying *Ralstonia solanacearum* virulence. *Bio-Protocol* 8(18), 1–19. DOI: 10.21769/bioprotoc.3028.

Khot, L.R., Sankaran, S., Maja, J.M., Ehsani, R. and Schuster, E.W. (2012) Applications of nanomaterials in agricultural production and crop protection: A review. *Crop Protection* 35, 64–70. DOI: 10.1016/j.cropro.2012.01.007.

Kiirika, L.M., Stahl, F. and Wydra, K. (2013) Phenotypic and molecular characterization of resistance induction by single and combined application of chitosan and silicon in tomato against *Ralstonia solanacearum*. *Physiological and Molecular Plant Pathology* 81, 1–12. DOI: 10.1016/j.pmpp.2012.11.002.

Kongkiattikajorn, J. and Thepa, S. (2007) Increased tomato yields by heat treatment for controlling *Ralstonia solanacearum*, in soil. *Proceedings of the 45th Kasetsart University Annual Conference, Kasetsart* 450–457.

Kůdela, V. (2009) Potential impact of climate change on geographic distribution of plant pathogenic bacteria in Central Europe. *Plant Protection Science* 45(Special Issue), S27–S32. DOI: 10.17221/2832-pps.

Lebeau, A., Daunay, M.-C., Frary, A., Palloix, A., Wang, J.-F. *et al.* (2011) Bacterial wilt resistance in tomato, pepper, and eggplant: Genetic resources respond to diverse strains in the *Ralstonia solanacearum* species complex. *Phytopathology* 101(1), 154–165. DOI: 10.1094/PHYTO-02-10-0048.

Lebeau, A., Gouy, M., Daunay, M.C., Wicker, E., Chiroleu, F. *et al.* (2013) Genetic mapping of a major dominant gene for resistance to *Ralstonia solanacearum* in eggplant. *Theoretical and Applied Genetics* 126(1), 143–158. DOI: 10.1007/s00122-012-1969-5.

Li, M., Pommier, T., Yin, Y., Wang, J., Gu, S. *et al*. (2022) Indirect reduction of *Ralstonia solanacearum* via pathogen helper inhibition. *ISME Journal* 16(3), 868–875. DOI: 10.1038/s41396-021-01126-2.

Li, Y., Colleoni, C., Zhang, J., Liang, Q., Hu, Y. *et al*. (2018) Genomic analyses yield markers for identifying agronomically important genes in potato. *Molecular Plant* 11, 473–484. DOI: 10.1016/j.molp.2018.01.009.

Lopes, C.A., Carvalho, A.D., Pereira, A.S., Azevedo, F.Q., Castro, C.M. *et al*. (2021) Performance of *Solanum phureja*-derived bacterial-wilt resistant potato clones in a field naturally infested with *Ralstonia solanacearum* in Central Brazil. *Horticultura Brasileira* 39(4), 411–416. DOI: 10.1590/s0102-0536-20210410.

Lowe-Power, T.M., Khokhani, D. and Allen, C. (2018) How *Ralstonia solanacearum* exploits and thrives in the flowing plant xylem environment. *Trends in Microbiology* 26(11), 929–942. DOI: 10.1016/j.tim.2018.06.002.

Luck, J., Spackman, M., Freeman, A., TreBicki, P., Griffiths, W. *et al*. (2011) Climate change and diseases of food crops. *Plant Pathology* 60(1), 113–121. DOI: 10.1111/j.1365-3059.2010.02414.x.

Madden, L.V. and Wheelis, M. (2003) The threat of plant pathogens as weapons againsst U.S. crops. *Annual Review of Phytopathology* 41, 155–176. DOI: 10.1146/annurev.phyto.41.121902.102839.

Mahbou, G., Toukam, S., Recherche, I.D., Maraîchères, C., Yaoundé, B.P. *et al*. (2009) Broad diversity of *Ralstonia solanacearum* strains in Cameroon. *Plant Disease* 93(11), 1123–1130. DOI: 10.1094/PDIS-93-11-1123.

Maji, S. and Chakrabartty, P.K. (2014) Biocontrol of bacterial wilt of tomato caused by *Ralstonia solanacearum* by isolates of plant growth promoting rhizobacteria. *Australian Journal of Crop Science* 8(2), 208–214.

Mansfield, J., Genin, S., Magori, S., Citovsky, V., Sriariyanum, M. *et al*. (2012) Top 10 plant pathogenic bacteria in molecular plant pathology. *Molecular Plant Pathology* 13(6), 614–629. DOI: 10.1111/j.1364-3703.2012.00804.x.

Mark, F. and Hayward, A.C. (2014) *Ralstonia solanacearum* Race 3, Biovar 2. In: *Manual of Security Sensitive Microbes and Toxins*. CRC Press, Boca Raton, FL, pp. 811–817.

Messiha, N.A.S., Van Diepeningen, A.D., Wenneker, M., Van Beuningen, A.R., Janse, J.D. *et al*. (2007) Biological soil disinfestation (BSD), a new control method for potato brown rot, caused by *Ralstonia solanacearum* race 3 biovar 2. *European Journal of Plant Pathology* 117(4), 403–415. DOI: 10.1007/s10658-007-9109-9.

Messiha, N.A.S., Elhalag, K.M.A., Balabel, N.M., Farag, S.M.A., Matar, H.A. *et al*. (2019) Microbial biodiversity as related to crop succession and potato intercropping for management of brown rot disease. *Egyptian Journal of Biological Pest Control* 29(1), 1–16. DOI: 10.1186/s41938-019-0185-x.

Mickiewicz, B., Volkova, E. and Jurczak, R. (2022) The global market for potato and potato products in the current and forecast period. *European Research Studies Journal* XXV(3), 740–751.

Mohamed, B.F.F., Sallam, N.M.A., Alamri, S.A.M., Abo-Elyousr, K.A.M., Mostafa, Y.S. *et al*. (2020) Approving the biocontrol method of potato wilt caused by *Ralstonia solanacearum* (Smith) using *Enterobacter cloacae* PS14 and *Trichoderma asperellum* T34. *Egyptian Journal of Biological Pest Control* 30(1). DOI: 10.1186/s41938-020-00262-9.

Momma, N., Kobara, Y., Uematsu, S., Kita, N. and Shinmura, A. (2013) Development of biological soil disinfestations in Japan. *Applied Microbiology and Biotechnology* 97(9), 3801–3809. DOI: 10.1007/s00253-013-4826-9.

Muthoni, J., Kabira, J., Shimelis, H. and Melis, R. (2013) Spread of bacterial wilt disease of potatoes in Kenya: Who is to blame? *African Crop Science Conference Proceedings* 11, 235–239.

Muthoni, J., Shimelis, H. and Melis, R. (2012) Management of bacterial wilt [*Rhalstonia solanacearum* Yabuuchi *et al*., 1995] of potatoes: Opportunity for host resistance in Kenya. *Journal of Agricultural Science* 4(9), 64–78. DOI: 10.5539/jas.v4n9p64.

Muthoni, J., Shimelis, H. and Melis, R. (2020) Conventional breeding of potatoes for resistance to bacterial wilt (*Ralstonia solanacearum*): Any light in the horizon? *Australian Journal of Crop Science* 14(3), 485–494. DOI: 10.21475/ajcs.20.14.03.p2144.

Mutimawurugo, M.C., Wagara, I.N., Muhinyuza, J.B. and Ogweno, J.O. (2019) Virulence and characterization of isolates of potato bacterial wilt caused by *Ralstonia solanacearum* (Smith) in Rwanda. *African Journal of Agricultural Research* 14(6), 311–320. DOI: 10.5897/AJAR2018.13686.

Narancio, R., Zorrilla, P., Robello, C., Gonzalez, M., Vilaró, F. *et al*. (2013) Insights on gene expression response of a characterized resistant genotype of *Solanum commersonii* Dun. against *Ralstonia solanacearum*. *European Journal of Plant Pathology* 136(4), 823–835. DOI: 10.1007/s10658-013-0210-y.

N'Guessan, C.A., Abo, K., Fondio, L., Chiroleu, F., Lebeau, A. *et al.* (2012) So near and yet so far: The specific case of *Ralstonia solanacearum* populations from Côte d'Ivoire in Africa. *Phytopathology* 102(8), 733–740. DOI: 10.1094/PHYTO-11-11-0300.

Nicolopoulou-Stamati, P., Maipas, S., Kotampasi, C., Stamatis, P. and Hens, L. (2016) Chemical pesticides and human health: The urgent need for a new concept in agriculture. *Frontiers in Public Health* 4(Jul), 1–8. DOI: 10.3389/fpubh.2016.00148.

Norman, D.J., Zapata, M., Gabriel, D.W., Duan, Y.P., Yuen, J.M.F. *et al.* (2009) Genetic diversity and host range variation of *Ralstonia solanacearum* strains entering North America. *Phytopathology* 99(9), 1070–1077. DOI: 10.1094/PHYTO-99-9-1070.

Norman, D.J., Yuen, J.M.F. and Bocsanczy, A.M. (2020) Threat of brown rot of potato and existing resistance. *American Journal of Potato Research* 97(3), 272–277. DOI: 10.1007/s12230-020-09775-3.

Okiro, L.A., Mulwa, R.M., Oyoo, M.E., Machuka, E.M., Parker, M.L. *et al.* (2022) Phylogenetic distribution of *Ralstonia solanacearum* species complex populations in potato in Kenya. *Plant Disease* 106(6), 1736–1742. DOI: 10.1094/PDIS-11-21-2556-RE.

Olsson, K. (1976) Experience of brown rot caused by *Pseudomonas solanacearum* (Smith) Smith in Sweden. *EPPO Bulletin* 6(4), 199–207. DOI: 10.1111/j.1365-2338.1976.tb01546.x.

Otieno, S.A., Collins, P., Coombs, J., Allen, C. and Douches, D.S. (2021) Screening for *Ralstonia solanacearum* resistance in *Solanum commersonii*. *American Journal of Potato Research* 98(1), 72–77. DOI: 10.1007/s12230-020-09819-8.

Parkinson, N., Bryant, R., Bew, J., Conyers, C., Stones, R. *et al.* (2013) Application of variable-number tandem-repeat typing to discriminate *Ralstonia solanacearum* strains associated with English watercourses and disease outbreaks. *Applied and Environmental Microbiology* 79(19), 6016–6022. DOI: 10.1128/AEM.01219-13.

Popoola, A., Ganiyu, S., Enikuomehin, O., Bodunde, J., Adedibu, O. *et al.* (2015) Isolation and characterization of *Ralstonia solanacearum* causing bacterial wilt of tomato in Nigeria. *Nigerian Journal of Biotechnology* 29(1), 1. DOI: 10.4314/njb.v29i1.1.

Pradhanang, P.M., Elphinstone, J.G. and Fox, R.T.V. (2000) Identification of crop and weed hosts of *Ralstonia solanacearum* biovar 2 in the hills of Nepal. *Plant Pathology* 49(4), 403–413. DOI: 10.1046/j.1365-3059.2000.00480.x.

Pradhanang, P.M., Ji, P., Momol, M.T., Olson, S.M., Mayfield, J.L. *et al.* (2005) Application of acibenzolar-S-methyl enhances host resistance in tomato against *Ralstonia solanacearum*. *Plant Disease* 89(9), 989–993. DOI: 10.1094/PD-89-0989.

Ramírez, M., Moncada, R.N., Villegas-Escobar, V., Jackson, R.W. and Ramírez, C.A. (2020) Phylogenetic and pathogenic variability of strains of *Ralstonia solanacearum* causing moko disease in Colombia. *Plant Pathology* 69(2), 360–369. DOI: 10.1111/ppa.13121.

Ravelomanantsoa, S., Vernière, C., Rieux, A., Costet, L., Chiroleu, F. *et al.* (2018) Molecular epidemiology of bacterial wilt in the Madagascar highlands caused by Andean (phylotype IIB-1) and African (Phylotype III) brown rot strains of the *Ralstonia solanacearum* species complex. *Frontiers in Plant Science* 8(2258), 1–17. DOI: 10.3389/fpls.2017.02258.

Ray, M.K., Mishra, A.K., Mohanta, Y.K., Mahanta, S., Chakrabartty, I. *et al.* (2023) Nanotechnology as a promising tool against phytopathogens: A futuristic approach to agriculture. *Agriculture (Switzerland)* 13(9), 1–41. DOI: 10.3390/agriculture13091856.

Romo, J.P., Morales, J.G., Ii, O., Salazar, M., Iii, Y. and Professor, Y. (2012) Identification of new hosts for *Ralstonia solanacearum* (Smith) race 2 from Colombia. *Revista de Protección Vegetal* 27(3), 151–161.

Safni, I., Cleenwerck, I., De Vos, P., Fegan, M., Sly, L. *et al.* (2014) Polyphasic taxonomic revision of the *Ralstonia solanacearum* species complex: Proposal to emend the descriptions of *Ralstonia solanacearum* and *Ralstonia syzygii* and reclassify current *R. syzygii* strains as *Ralstonia syzygii* subsp. *syzgii* subsp. nov., R. s. *International Journal of Systematic and Evolutionary Microbiology* 64, 3087–3103. DOI: 10.1099/ijs.0.066712-0.

Santiago, T.R., Lopes, C.A., Caetano-Anollés, G. and Mizubuti, E.S.G. (2017) Phylotype and sequevar variability of *Ralstonia solanacearum* in Brazil, an ancient centre of diversity of the pathogen. *Plant Pathology* 66(3), 383–392. DOI: 10.1111/ppa.12586.

Sharma, K., Kreuze, J., Abdurahman, A., Parker, M., Nduwayezu, A. *et al.* (2020) Molecular diversity and pathogenicity of *Ralstonia solanacearum* species complex associated with bacterial wilt of potato in Rwanda. *Plant Disease* 105(4), 770–779.

Sharma, K., Iruegas-Bocardo, F., Abdurahman, A., Alcalá-Briseño, R.I., Garrett, K.A. *et al.* (2022) *Ralstonia* strains from potato-growing regions of Kenya reveal two phylotypes and epidemic clonality of Phylotype II sequevar 1 strains. *Phytopathology* 112(10), 2072–2083. DOI: 10.1094/PHYTO-11-21-0455-R.

Shojaei, T.R., Salleh, M.A.M., Sijam, K., Rahim, R.A., Mohsenifar, A. *et al.* (2016) Fluorometric immunoassay for detecting the plant virus Citrus tristeza using carbon nanoparticles acting as quenchers and antibodies labeled with CdTe quantum dots. *Microchimica Acta* 183(7), 2277–2287. DOI: 10.1007/s00604-016-1867-7.

Shutt, V.M., Shin, G., van der Waals, J.E., Goszczynska, T. and Coutinho, T.A. (2018) Characterization of ralstonia strains infecting tomato plants in South Africa. *Crop Protection* 112(Sep 2017), 56–62. DOI: 10.1016/j.cropro.2018.05.013.

Singh, D., Chaudhary, G. and Yadav, D.K. (2021) Characterization and diversity of Indian isolates of *Ralstonia solanacearum* inciting bacterial wilt of tomato. *Indian Phytopathology* 74(2), 425–429. DOI: 10.1007/s42360-021-00365-9.

Spooner, D.M., McLean, K., Ramsay, G., Waugh, R. and Bryan, G.J. (2005) A single domestication for potato based on multilocus amplified fragment length polymorphism genotyping. *Proceedings of the National Academy of Sciences of the United States of America* 102(41), 14694–14699. DOI: 10.1073/pnas.0507400102.

Swanson, J.K., Yao, J., Tans-Kersten, J. and Allen, C. (2005) Behavior of *Ralstonia solanacearum* race 3 biovar 2 during latent and active infection of geranium. *Phytopathology* 95(2), 136–143. DOI: 10.1094/PHYTO-95-0136.

Tafesse, S., Braam, C., van Mierlo, B., Lemaga, B. and Struik, P.C. (2021) Association between soil acidity and bacterial wilt occurrence in potato production in Ethiopia. *Agronomy* 11(8), 1541. DOI: 10.3390/agronomy11081541.

Tessema, G.L. and Seid, H.E. (2023) Potato bacterial wilt in Ethiopia: history, current status, and future perspectives. *PeerJ* 11, e14661. DOI: 10.7717/peerj.14661.

Tjou-tam-sin, N.N.A., Bilt, J.L.J.V.D. and Westenberg, M. (2017) Assessing the pathogenic ability of *Ralstonia pseudosolanacearum* (*Ralstonia solanacearum* Phylotype I) from ornamental *Rosa* spp. *Plants* 8(Nov), 1–9. DOI: 10.3389/fpls.2017.01895.

Tusiime, G., Adipala, E., Opio, F. and Bhagsari, A.S. (1998) Weeds as latent hosts of *Ralstonia solanacearum* in highland Uganda: Implications to development of an integrated control package for bacterial wilt. In: Prior, P. (ed.) *Bacterial Wilt Disease*. Springer-Verlag, Berlin, Heidelberg, pp. 413–419. DOI: 10.1007/978-3-662-03592-4_63.

Umrao, P.D., Kumar, V. and Kaistha, S.D. (2021) Biocontrol potential of bacteriophage ϕsp1 against bacterial wilt-causing *Ralstonia solanacearum* in Solanaceae crops. *Egyptian Journal of Biological. Pest Control* 31(1), 1–12. DOI: 10.1186/s41938-021-00408-3.

Uwamahoro, F., Berlin, A., Bucagu, C., Bylund, H. and Yuen, J. (2020) *Ralstonia solanacearum* causing potato bacterial wilt: host range and cultivars' susceptibility in Rwanda. *Plant Pathology* 69(3), 559–568. DOI: 10.1111/ppa.13140.

van der Waals, J.E. and Krüger, K. (2020) Emerging potato pathogens affecting food security in southern Africa: Recent research. *South African Journal of Science* 116(11), 1–7. DOI: 10.17159/sajs.2020/8055.

van der Wolf, J.M. and De Boer, S.H. (2007) Bacterial pathogens of potato. *Potato Biology and Biotechnology: Advances and Perspectives* 595–617. DOI: 10.1016/B978-044451018-1/50069-5.

Van Elsas, J.D., Kastelein, P., Van Bekkum, P., van der Wolf, J.M., De Vries, P.M. *et al.* (2000) Survival of *Ralstonia solanacearum* biovar 2, the causative agent of potato brown rot, in field and microcosm soils in temperate climates. *Phytopathology* 90(12), 1358–1366. DOI: 10.1094/PHYTO.2000.90.12.1358.

Vargas, C.D., Soto-Suárez, M. and Zuluaga, P. (2023) Screening for resistance against *Ralstonia solanacearum* in commercially available Colombian potato varieties. *Ciencia y Tecnología Agropecuaria* 24(2). DOI: 10.21930/rcta.vol24_num2_art:2976.

Vinh, M., Tung, T. and Quang, H. (2005) Primary bacterial wilt study on tomato in vegetable areas of Ho Chi Minch city, Vietnam. In: Allen, C., Prior, P. and Hayward, A. (eds.) *Bacterial Wilt Disease and the Ralstonia solanacearum Species Complex*. American Phytopathological Society Press, St Paul, MN, pp. 177–184.

Wang, P., Lombi, E., Zhao, F.J. and Kopittke, P.M. (2016) Nanotechnology: A new opportunity in plant sciences. *Trends in Plant Science* 21(8), 699–712. DOI: 10.1016/j.tplants.2016.04.005.

Wang, Y., Rashid, M.A.R., Li, X., Yao, C., Lu, L. *et al.* (2019) Collection and evaluation of genetic diversity and population structure of potato landraces and varieties in China. *Frontiers in Plant Science* 10(139), 1–10. DOI: 10.3389/fpls.2019.00139.

Wei, C., Liu, J., Maina, A.N., Mwaura, F.B., Yu, J. *et al.* (2017) Developing a bacteriophage cocktail for biocontrol of potato bacterial wilt. *Virologica Sinica* 32(6), 476–484. DOI: 10.1007/s12250-017-3987-6.

Weibel, J., Tran, T.M. and Norman, D.J. (2016) A *Ralstonia solanacearum* strain from Guatemala infects diverse flower crops, including new asymptomatic hosts Vinca and Sutera, and causes symptoms in Geranium, mandevilla vine, and new host African daisy. *Plant Health Progress* 17(2), 114–121. DOI: 10.1094/PHP-RS-16-0001.

Wen, A., Balogh, B., Momol, M.T., Olson, S.M. and Jones, J.B. (2009) Management of bacterial spot of tomato with phosphorous acid salts. *Crop Protection* 28(10), 859–863. DOI: 10.1016/j.soilbio.2017.11.012.

Yahiaoui, N., Chéron, J.J., Ravelomanantsoa, S., Hamza, A.A., Petrousse, B. *et al.* (2017) Genetic diversity of the *Ralstonia solanacearum* species complex in the southwest Indian ocean islands. *Frontiers in Plant Science* 8(Dec), 2139. DOI: 10.3389/fpls.2017.02139.

Yuan, K. (Xiaoli), Cullis, J., Lévesque, C.A., Tambong, J., Chen, W. *et al.* (2015) Draft genome sequences of *Ralstonia solanacearum* race 3 biovar 2. *Genome Announcements* 3(4), 4–5. DOI: 10.1128/genomeA.00815-15.

Yuliar, N., Asi, Y. and Toyota, K. (2015) Recent trends in control methods for bacterial wilt diseases caused by *Ralstonia solanacearum*. *Microbes and Environments* 30(1), 1–11. DOI: 10.1264/jsme2.ME14144.

Zaki, A.G., Hasanien, Y.A. and El-Sayyad, G.S. (2022) Novel fabrication of SiO2/Ag nanocomposite by gamma irradiated *Fusarium oxysporum* to combat *Ralstonia solanacearum*. *AMB Express* 12(1). DOI: 10.1186/s13568-022-01372-3.

4 Assessing the impact of climate action mitigation strategies from livestock production: A review – Solution for greenhouse gases mitigation

Geberemariyam Terefe* and Mulugeta Walelegne
Ethiopian Institute of Agricultural Research, Animal Feeds and Nutrition Research, Pox 2003, Addis Ababa, Ethiopia

Abstract
The main sources of greenhouse gas (GHG) emissions are enteric fermentation from ruminants and manure management. An integrated strategy-combination approach incorporating nutrition and manure management is necessary to dramatically reduce GHGs and the carbon footprint of milk production. The aim of this review was to assess the effect of climate action mitigation strategies on the GHG production of ruminant animals. This review was based on the different references, which are textbooks, journals, proceedings, conferences, and abstracts accessed through Agricola, Agris, CAB Abstracts, Embase, MEDLINE, and the Web of Science. The research findings were done by using dairy cattle, sheep, and goat; the numbers of animals used in each experiment were given the majority of attention, and decisions were also made based on the method used to quantify methane (CH_4) and nitrous oxide. Enteric CH_4 per unit of milk was reduced by 6–8% when higher-starch corn silage was used in place of barley or alfalfa silage. Cows fed seaweed (1.77–2%, on dry matter [DM] basis) experienced a linear drop in enteric CH_4 emission from 383 to 400 g/d. Supplementation of vegetable fats (6%) and coconut oil (12 g/day) to ruminant diets can decrease CH_4 emissions by 21–40% and 50–61.2%, respectively. Similarly, the addition of 21.5 g nitrate/kg DM reduces CH_4 yield (17%) in sheep. Bovaer can reduce methane emissions (30%) for dairy cows. Feed processing and additives are alternative strategies to minimize GHG production in ruminants, but further studies on additives and their residual effect on livestock output is needed.

Keywords: enteric methane, feed additives, gas mitigation

Introduction

Enteric fermentation from the feed digestion in ruminants and manure management are the two main sources of greenhouse gas (GHG) emissions (Steinfeld *et al.*, 2006). Enteric methane emissions are the single largest source of direct GHG in beef and dairy value chains and a substantial contributor to anthropogenic methane emissions globally (Vijn *et al.*, 2020). An integrated strategy-combination approach incorporating nutrition and manure management is necessary to dramatically reduce GHGs and the carbon footprint of milk production (Jayasundara *et al.*, 2016). Methane (CH_4) and nitrous oxide are the most significant GHGs

*Corresponding author: geberemom@gmail.com

emitted by animal agriculture. The global warming potential of methane produced by enteric fermentation and manure storage, and nitrous oxide produced through manure storage and the application of organic/inorganic fertilizers were 28 and 265 times more than carbon dioxide, respectively (IPCC, 2013). As a means of reducing GHG emissions and enhancing feed conversion efficiency, it is preferable to minimize enteric CH_4 emissions from ruminants without changing animal production (Honan *et al.*, 2021). Dietary manipulation influences CH_4 production by directly influencing the rumen microbiome. There is the potential to affect the rumen fermentation profiles and microbiota community structure positively and meet sustainability goals by reducing CH_4 emissions from cattle production systems (Min *et al.*, 2022). Enteric CH_4 production is affected by forage quality and digestibility. Plant growth results in an increase in lignin content, which lowers plant digestibility. As a result, fodder (particularly grass) that is harvested for ensiling early in its maturation process has a higher amount of soluble carbohydrates and less lignification. Practices of improving pasture quality have the ability to reduce enteric CH_4 by roughly 5% every unit of milk (Knapp *et al.*, 2014). Physical feed processing, such as chopping, grinding, and steaming, also increases forage digestibility and reduces ruminant intestinal methane output (Gerber *et al.*, 2013). The application of anti-methanogenic chemical compounds is an effective strategy in reducing CH_4 emissions; however, their effects can be transitory and such compounds may have adverse impacts on both animal performance and the environment (Patra *et al.*, 2017). Marine macroalgae is an alternative feed resource that can largely decrease enteric CH_4 production from ruminants (Maia *et al.*, 2016). Saponin is a natural product extracted from plants which is important as a natural rumen manipulator and in altering rumen fermentation. Saponins may decrease CH_4 synthesis indirectly through defaunating actions and/or directly by lowering methanogen activities such as methanogenesis rate or expression of methane-producing genes (Patra and Saxena, 2009). The addition of saponin-rich extracts of *Yucca schidigera* did not reduce CH_4 production (Pen *et al.*, 2007). Tannin and flavonoids are secondary plant metabolite which can block methanogens directly and hydrogen-producing bacteria indirectly, and they are useful to reduce the production of CH_4 gas in the digestive tracts of ruminants (Kumar *et al.*, 2014; Formato *et al.*, 2022). Adding *Asparagopsis* and biochar to Rhodes grass hay at a rate of 5% on a DM basis can reduce cumulative CH_4 by 23.8% (Parra *et al.*, 2023). The red and brown algae stand out as having potential for GHG mitigation with the brown alga *Cystoseira* having the most prominent effect (Dubois *et al.*, 2013). This review provides comprehensive data that are currently available in high impact and indexed journals which provide information on the mitigation strategy of GHGs emission from livestock production. Even though a vast amount of research across the world has been done in GHGs mitigation strategies from livestock the information is not well organized and written separately, therefore, this review was intended to summarize and document the effect of feed physical processing and feed additives as GHGs mitigation strategies from the ruminant animals.

Review Methodology

To survey the current GHG mitigation strategies of the ruminants, this review was conducted by accessing different research papers (53 research articles and five review papers), which have sufficient information and were undertaken to document the effect of different feed processing and additives as mitigation strategies on the methane production of the ruminants. The detailed methodology of this review article is discussed in the following subheadings.

Inclusion criteria and aims

The primary aim of this review was to collect and present the available research evidence on existing methods of GHG mitigation strategies in livestock animals related to the title and abstract screening stage in a review, and this will reduce the waste of time by searching different literature on the topics. The different records were assessed against the following criteria based on their titles and abstracts: which must be relevant, full publication information, and a brief procedure of the work from the previous publication in the topic. The other search criteria adopted for this selection were "methane production from ruminants," "feed additives and management for greenhouse gas mitigation," and "feed processing for cattle." In brief, attention has been given primarily to studies that reported research findings from beef cattle, dairy cows, and sheep. After the number of animals per experiment was prioritized, the duration of the study was also taken into account, and finally, selection was made based on the technique adopted for quantification of methane and nitrous oxide. Additionally, we have also used references cited in the articles.

Search methods and criteria

The database and website searches were used as the two main methods of accessing the present mitigation strategies for GHGs from the ruminants. The details of the search method and the number of journal articles and case report publications included in this review are presented in the PRISMA flow diagram (Fig. 4.1), which was designed by Page *et al.* (2021). This article also included publications known to the team and those recommended by colleagues. This review was done based on a literature study that aimed to finalize the effect of different feed processing and additives as mitigation strategies on the methane production of ruminants. In the textbooks, journals, proceedings, conferences, and abstracts, a number of search terms and reference sources were used through Agricola, Agris, CAB Abstracts, Embase, Web of Science, Research Gate, and Google Scholar.

Methane Production in the Rumen

In the methanogenesis process, methanogenic archaea produce methane as a by-product of their metabolism (Króliczewska *et al.*, 2023). Methane is a colourless and odourless gas, and it is produced predominantly in the rumen (87%) and to a small extent (13%) in the large

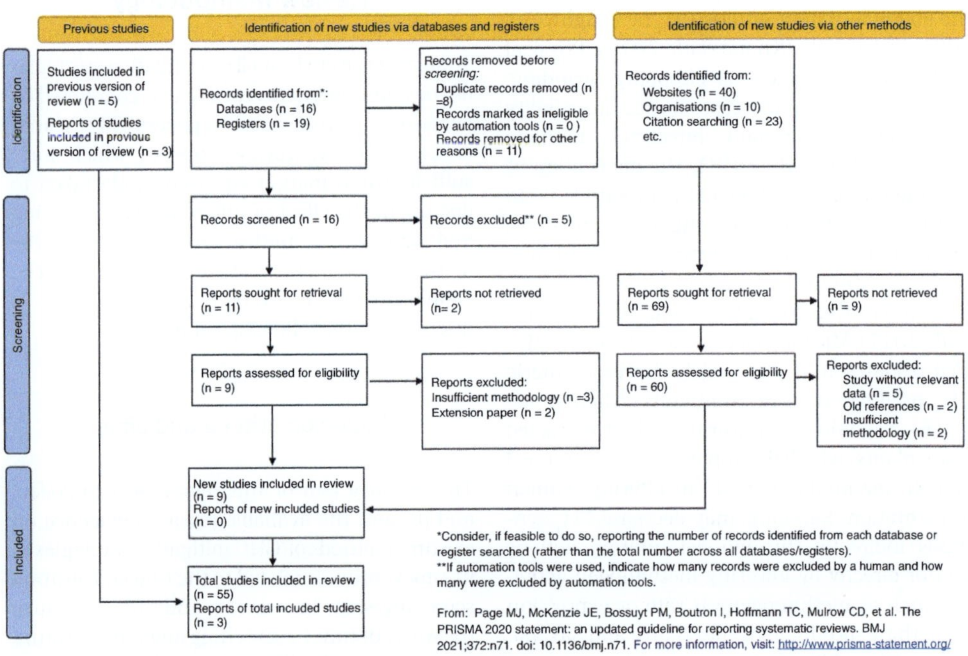

Fig. 4.1. PRISMA 2020 flow diagram for updated systematic reviews which included searches of databases, registers and other sources. From: Page *et al.* (2021). For more information, visit: http://www.prisma-statement.org/.

intestines through microbial fermentation of feed components (Murray *et al.*, 1976). The conversion of feed material to CH_4 in the rumen involves the integrated activities of different microbial species, with the final step carried out by methanogenic bacteria (Moss *et al.*, 2000). Proteins, starches, and plant cell wall polymers are hydrolysed by primary digestive microorganisms (bacteria, protozoa, and fungi) and changed into amino acids and sugars. After that, the primary and secondary digestive bacteria convert these by-products into volatile fatty acids (VFAs), hydrogen, and carbon dioxide. The VFAs, acetate, propionate, and butyrate are then absorbed and used by the host animal (Van Soest, 1988).

Greenhouse Gases Mitigation Strategies

According to the review paper published by Króliczewska *et al.* (2023), genetic animal selection and biological feed additives (secondary plant metabolites and probiotics) are viable methods for reducing enteric methane emissions. When developing strategies to reduce CH_4 emissions from the ruminants, it is important to consider how the rumen manages the fermentation and reducing H_2 liberating activities, promoting alternative H_2-using reactions, or promoting H_2 disposal pathways should be able to reduce CH_4 production (Joblin, 1999). The various management strategies, including nutrition, reproduction, or genetics, are essential for increasing animal productivity since they help to reduce the quantity of CH_4 produced per unit of meat or milk. The amount of feed energy associated with animal maintenance is about 70–75% in beef cattle and 50% in dairy cattle. Thus, as productivity increases CH_4 emissions go up slightly, but CH_4 emissions per unit of product decrease (Johnson *et al.*, 1996).

Feed processing

The feed efficiency and GHG emission levels have a substantial impact on animal productivity (Herrero *et al.*, 2013). Increasing forage digestibility and lowering intestinal CH_4 emission from the ruminant are the two benefits of physical feed processing, such as chopping, grinding, and steaming (Gerber *et al.*, 2013). The amount of milk produced significantly increases or decreases as fodder amounts change from smallest to largest and enteric CH_4 emissions per farm move in the same way. An extreme increase in the amount of grass fed on Normandy farm decreased GHG (7117 vs 5587 kg CO_2 eq./farm/year) and enteric CH_4 emissions (3870 vs 3296 kg CO_2 eq./farm/year, respectively) (Senga Kiessé *et al.*, 2022). Grinding or pellet forages for a ruminant diet are essential to reduce CH_4 losses per unit of feed intake by 20–40% (Johnson *et al.*, 1996). Cows grazing the pasture emitted significantly more CH_4 gas than cows consuming a total mixed ration (TMR) (Robertson and Waghorn, 2002). Reduced TMR consumption and the addition of grasses to dairy cow diets can maintain or slightly lower the C footprint in subtropical conditions (Ribeiro-Filho *et al.*, 2020).

Silage feeding

The silage-making process requires considerable fermentation, which reduces the rumen's capacity to absorb ensiled forages. The rumen fermentation of grass-silage-based feed was differentiated from hay-based diet by a higher molar proportion of butyrate and a smaller fraction of acetate (Shingfield *et al.*, 2002). The enhancement of fodder digestibility and feed intake is important to reduce GHGs emissions (Gerber *et al.*, 2013). Ensiled forage is a welcomed strategy for the mitigation of enteric CH_4 in the livestock industry, not only for reducing its impact as a GHG but also for increasing productivity because the energy contained in enteric CH_4 is not lost by the animal, which is accompanied by an increase in intake of ensiled forage. Another method that should be highlighted as a mitigation strategy is harvesting early cut swards for ensiling. Best practices for ensiling and feeding out must be followed for it to be effective. The ensiling fermentation process can be aided by *Lactobacillus* inoculants, which can improve the rumen's ecology and environment and help to mitigate CH_4 (Evans, 2018). Increasing the amount of maize silage in dairy cow diets increased intake and performance and decreased CH_4 generation, while the starch-to-fibre ratio in the concentrate had little effect on performance or CH_4 production (Hart *et al.*, 2015). It has been discovered that grains and whole plant silages made from maize improve the digestibility of cow diets and are effective approaches to lowering GHG emissions (Vellinga and Hoving, 2011). Enteric CH_4 per unit of milk was reduced by 6% when higher starch corn silage was used in place of barley or alfalfa silage (Jayasundara *et al.*, 2016). The emissions intensities were lowest for the silage (50% timothy, 35% meadow fescue, and 15% red clover), which was harvested early, and highest for the silage prepared at late maturity, independent of concentrate levels for both milk and beef (Åby *et al.*, 2019). Similarly, feeding corn silage to dairy cows generated 8% fewer emissions per kg of protein in food products as compared to alfalfa silage (Little *et al.*, 2017). The aerobic degradation of grass silage had a negative impact on net return to labour and management, reduced fermentation quality, and resulted in a modest rise in GHG emissions (Van Schooten and Philipsen, 2012).

Supplementation of microalgae

Several feed additives provide a promising option to increase the sustainability of animal-sourced foods and reduce enteric CH_4 emissions. Milk supplemented with dried seaweeds (*Ulva lactuca*, *Ascophyllum nodosum*, or *Saccharina latissima*, nearly 47 g/d) did not affect daily body weight gain of pre-weaning dairy calves (Samarasinghe and Sehested, 2021). With the exception of a slight decrease in milk protein and casein content, seaweed supplementation (9% *Laminaria digitata* and 91% *Ascophyllum nodosum*) has no effect on cow production or milk basic composition (Newton *et al.*, 2021). Feeding small amounts of seaweed to cattle had an impact on reducing the production of enteric methane (Vijn *et al.*, 2020). The alga (*A. taxiformis*) is an exotic species and from an environmental perspective, it is less favourable as feed supplement for the purpose of methane reduction in Swedish livestock (Silwer, 2018). The red and brown algae are particularly notable for their potential to reduce GHG emissions, with the brown alga (*Cystoseira*) having the most prominent effect (Dubois *et al.*, 2013). *A. taxiformis* is a good option for a biotic methane mitigation technique for dairy calves due to its ability to reduce methane during rumen fermentation (Roque *et al.*, 2019). The use of *A. taxiformis* supplemented to beef cattle diets reduced enteric CH_4 emissions for duration of 21 weeks without any loss in efficacy (Roque *et al.*, 2021). Cows fed huge levels of seaweed or oregano leaves (1.77%, on DM basis) experienced a linear drop in enteric methane emission from 383 to 351 g/d. Due to a significant link between feed intake and methane emissions, the observed decrease in DM intake was most likely the reason for this 8.4% decline. However, alternative seaweed diets had no impact on methane yield or intensity (Stefenoni *et al.*, 2021). When *B. hamifera* was added to the basal feed substrate (ryegrass), enteric CH_4 generation was completely eliminated at 6% inclusion per daily OM (organic matter) intake. Both *E. formissima* and *P. cirrhosum* have strong anti-methanogenic activity at 6% OM inclusion. Hence, these species may offer different anti-methanogenic seaweed targets that are less likely to lose volatile bioactives during processing (Mihaila *et al.*, 2022). In contrast, the supplementation of dried oregano plant (18–53 g/kg DM intake) did not lower the methane production of dairy cows over four consecutive days (Olijhoek *et al.*, 2019). The brief summary and the effect of feed processing and seaweed on GHGs mitigation from livestock is presented in Table 4.1.

Supplementation of oils and nitrate

Rumen inhibitors have the potential to reduce enteric CH_4 generation (up to 98%), despite their accessibility and risk to animal welfare are varied. Rumen modifiers, such as tannins, saponins, charcoal, and lipids, can be found all over the world, but their compositions and efficacy varies as well. Probiotics or direct-fed microorganisms have not provided enough evidence to be recognized as rumen modifiers to lower CH_4 production (Honan *et al.*, 2021). When the dietary lipids were increased from 3 to 6% of the DM, enteric CH_4 production was decreased by

Table 4.1. The effect of feed processing and seaweed on greenhouse gases mitigation.

Experimental animals	Animal number in the study	Duration of the study (day)	Diet (forage-concentrate ratio, % CP)	Average dry matter intake (kg/day)	Experimental feeds or Inclusion type	Level of inclusion (%)	Methane mitigation level (%)	References
Dairy cows	–	–	–	–	Improve pasture quality	–	5	Knapp *et al*. (2014)
Dairy cows	–	–	–	–	Physical processing	–	18	Senga Kiessé *et al*. (2022)
Dairy cows	–	–	–	–	Grinding & pelleting	–	20–40	Johnson *et al*. (1996)
Dairy cows	–	–	–	–	Maize silage	50	6	Jayasundara *et al*. (2016)
Dairy cows	160	–	56.4:43.6	–	Maize silage	56.4	8	Little *et al*. (2017)
Cattle	–	–	–	–	Seaweed	1	84.7	Silwer (2018)
Cattle	–	–	–	–	Green algae	8	80	Dubois *et al*. (2013)
Dairy cows	2		75:25	–	Red seaweed	5	67.2	Roque *et al*. (2019)
Beef steers	6	148	60:40,17.2	8.4	Red seaweed	0.5	68	Roque *et al*. (2021)
Cattle	–	–	–	–	Red seaweed	25	55	Maia *et al*. (2016)
Cattle	–	–	–	–	Red seaweed	2	55	Brooke *et al*. (2018)
Dairy cows	20	28	60:40,16	23.5	Red seaweed	0.25	35	Stefenoni *et al*. (2021)
Dairy cows	–	–	–	–	Seaweeds	6	95	Mihaila *et al*. (2022)

9% (Jayasundara *et al.*, 2016). Adding lipids or fatty acids to ruminant diets can reduce enteric CH_4 emissions by reducing the amount of energy provided from fermentable carbohydrates and altering the microbial community of the rumen (Llonch *et al.*, 2017). Nitrate and lipids are useful dietary supplements for lowering enteric CH_4 generation (Duthie *et al.*, 2017; Villar *et al.*, 2020). The combination dietary nitrate (20 g/kg DM) and canola oil (50 g/kg DM) additives allow high levels of CH_4 mitigation to be sustained with a lower dose of NO_3^- and therefore a reduced likelihood of NO_2^- toxicity, which is a practical constraint to inclusion of NO_3^- in diets as a no protein N source in ruminants (Villar *et al.*, 2020). While, feeding nitrate (18g nitrate/kg diet DM) and increasing the level of dietary lipid from 25 to 48 g/kg DM were not effective to reduce CH_4 yield and cannot be recommended for use within high concentrate diets (Duthie *et al.*, 2016). Similarly, an *in vitro*-based study shows that the addition of nitrate and biochar (8%, on a DM basis) to Rhodes grass hay has an impact on the reduction of methane emissions by 25.9% (Parra *et al.*, 2023).

Addition of supplementary fat (250–375 g/day) to the diet of ruminants effectively reduces methane production (Rasmussen and Harrison, 2011). Adding vegetable fats (6% on DM basis) to ruminant diets is a suitable approach to decrease methane emissions (21–28%) (Vargas *et al.*, 2020). Animals' diets were supplemented with fats (53g kg–1 DM of palm kernel oil, coconut oil, or canola oil) to reduce CH_4 generation by 34, 21, or 20%, respectively (Dohme *et al.*, 2000). Sheep diets containing 3.5 and 7% coconut oil had a 28 and 73% of daily CH_4 reduction, respectively (Machmüller and Kreuzer, 1999). Similarly, adding coconut oil (12 g/day) to the diet of Tibetan sheep can reduce methane emissions (by 61.2%) and this persists for at least the 4 week feeding period (Ding *et al.*, 2012). Feeding canola oil (4.6% of DMI) to beef cattle decreased daily methane emissions (32%) and decreased methane emissions as a percentage of gross energy intakes by 21% but, adding fumaric acid (75 g/day) to the diet of beef cattle had no measurable effect on CH_4 emissions (Beauchemin and McGinn, 2006). Similarly, the inclusion of coconut oil (350 g/day) in forage and concentrate ratio (60:40%) mixed feed to finishing heifers showed a reduction of CH_4 emissions without affecting the production (Lovett *et al.*, 2003). Pasture based diet for dairy cows supplemented with concentrates containing linseed oil emitted 18% less CH_4 per kg of milk (Boland *et al.*, 2020). An additive effect of nitrate (3%) and linseed oil (4%) in concentrate feed (50%, DM basis) reduced methanogenesis in the dairy cows without altering diet digestibility (Guyader *et al.*, 2015). A supply of linseed (35%) and nitrate (6%) in bull feed failed to decrease enteric methane yield and impaired bull live weight gain but without adverse effects for animal health or food safety (Doreau *et al.*, 2018). Linseed (3.5%) plus nitrate (1.8%) may have a long-term CH_4-mitigating effect in dairy cows and consuming milk products from cows fed nitrate may be safe in terms of nitrate and nitrite residuals (Guyader *et al.*, 2016). The addition of 21.5g nitrate/kg DM reduced the CH_4 yield of finishing steers by 17%, while increasing the oil content (rapeseed cake) of the same diet from 27 to 53g oil/kg DM reduced the CH_4 yield by 7.5% (Troy *et al.*, 2015). Although NO_2 toxicity is a practical barrier to using NO_3 in diets as a nonprotein N source in ruminants, the combination of NO_3 and canola oil additives enables high levels of CH_4 mitigation to be sustained with a lower dose of NO_3 and a reduced likelihood of NO_2 toxicity (Villar *et al.*, 2020). 3-Nitrooxypropanol (3-NOP) is involved in inhibiting the enzyme (Methyl-coenzyme M reductase), which is responsible for CH_4 production by methanogenic archaea in the rumen. The addition of 3-NOP (60 mg/kg of DM) to dairy cow diets can reduce enteric CH_4 emissions (20–35%) (60 mg/kg DM) (Duin *et al.*, 2016; Bampidis *et al.*, 2021). The brief summary of dietary lipids and nitrate supplementation for GHGs mitigation from ruminant is presented in Table 4.2.

Feeding brewery by-products

Limiting the risk of environmental damage from brewery by-product disposal is important to make it valuable for the food sector and it can be assured with effective management (Rachwał *et al.*, 2020). The spent craft yeast is currently becoming a value-added product that increases livestock efficiency. Supplemental craft yeast decreased the amount of CH_4

Table 4.2. The effect of dietary lipids and nitrate for greenhouse gases mitigation.

Experimental animals	Animal number in the study	Duration of the study (day)	Diet (forage-concentrate ratio, % CP)	Average dry matter intake (kg/day)	Methodology used to measure CH4 production	Experimental feeds or Inclusion type	Level of inclusion (%)	Methane mitigation level (%)	References
Sheep	6	42	71.2:28.8,15.2	1.08	Chamber	Coconut oil	3.5	28	Machmüller and Kreuzer (1999)
Sheep	6	42	55.4:54.6,18.4	0.89	Chamber	Coconut oil	7	73	Machmüller and Kreuzer (1999)
Dairy cows	–	–	–	–	*In vitro*	Palm kernel oil	5.3	34	Dohme *et al.* (2000).
Cattle/steers	4	82	40:60, 13.8	6.86	Chamber	Canola oil	5.0	6	Villar *et al.* (2020)
Dairy heifer	36	77	65:35,15.8	–	–	Coconut oil	35	34	Lovett *et al.* (2003)
Beef cattle	16	84	75:25, 14.6	5.64	Chamber	Canola oil	4.6	32	Beauchemin and McGinn (2006)
Tibetan sheep	12	28	60:40, –	0.79	Tunnel	Coconut oil	1.2	61.2	Ding *et al.* (2012)
Finishing steers	76	91	50:50, 14	10	Chamber	Rapeseed cake	2.7–5.3	7.5	Troy *et al.* (2015)
Dairy cows	45	49	79:21, –	15.2	Calibrated tracer	Linseed	16	18	Boland *et al.* (2020)
Sheep/ewes	32	21	70:30, 14.5	–	*In vitro*	Olive oil	6	18	Vargas *et al.* (2020)
Sheep/ewes	32	21	70:30, 14.5	–	*In vitro*	Sunflower oil	6	20	Vargas *et al.* (2020)
Sheep/ewes	32	21	70:30, 14.7	–	*In vitro*	Linseed	6	21	Vargas *et al.* (2020)
Finishing steers	76	91	50:50, 14	10	Chamber	Nitrate	2.15	17	Troy *et al.* (2015)
Finishing steers	4	82	40:60, 16.7	6.58	–	Nitrate	2	8.6	Villar *et al.* (2020)

Continued

Table 4.2. Continued

Experimental animals	Animal number in the study	Duration of the study (day)	Diet (forage-concentrate ratio, % CP)	Average dry matter intake (kg/day)	Methodology used to measure CH4 production	Experimental feeds or Inclusion type	Level of inclusion (%)	Methane mitigation level (%)	References
Finishing steers	32	84	52:48,14.8	12	—	Nitrate	1.8	0	Duthie *et al.* (2016)
Beef cattle	80	91	55:45,14.1	11.43	Chamber	Nitrate	2.5	11	Duthie *et al.* (2017)
Dairy cows	4	35	50:50,11.7	12.2	Chamber	Lipid & nitrate	Nitrate (3) & LSO (4)	32.85	Guyader *et al.* (2015)
Dairy cows	16	120	60:40,16	17.3	Chmber	Lipid & nitrate	LSO(3.5) & nitrate (1.8)	48.41	Guyader *et al.* (2016)
Bulls	16	60	70:30, 16.7	10.82	GreenFeed	Lipid & nitrate	LSO (35) & nitrate (6)	10.78	Doreau *et al.* (2018)
Cattle/steers	4	82	40:60, 16.6	6.36	Chamber	Lipid & nitrate	Nitrate (2) & lipid (5)	25	Villar *et al.* (2020)

LSO = Linseed oil.

produced by bovine and caprine rumen fermentation (Pszczolkowski *et al.*, 2016). Yeast (*Saccharomyces cerevisiae*) products are a class of feed additives known as direct-fed microbials that are advised for use in cattle to reduce rumen gas production and enhance feed efficiency, cow health, and production performance (McAllister *et al.*, 2011). Feeding yeast for dairy cows is important to reduce CH_4 emissions (Chung *et al.*, 2011; Bayat *et al.*, 2015). The inclusion of liquid brewery yeast (>30%) into cassava by-product and sundry brewery spent yeast (30–50% on DM basis) into wheat straw improves its chemical composition (increasing CP and digestibility), while decreasing gas production, without decreasing fermentation quality (Ding *et al.*, 2014; Kamphayae *et al.*, 2017; Terefe *et al.*, 2023). Some agro-industrial by-products (cottonseed cake, brewer's grains, brewers spent yeast and cold-pressed canola meal) are useful to reduce enteric fermentation and improve dairy cow milk yield (Moate *et al.*, 2011; Geberemariyam *et al.*, 2024). An *in vitro* based result show that the addition of yeast (1067 mg/kg DM) in ruminant diets has an impact on the reduction of methane emission (Cagliari *et al.*, 2023).

Conclusion

A substantial source of feed energy loss during production and GHG emissions is enteric CH_4 emissions. In the last 10 years, significant amounts of research on enteric CH_4 emissions from ruminants have been conducted. Therefore, the different mitigation strategies for CH_4 emissions for the ruminants were organized in this review. The previous findings indicated that increasing dietary silage, feeding physically processed forages (grinding and pelleting), microalgae, dietary lipid, and nitrate supplementation to the ruminant's diet can reduce enteric CH_4 emissions from ruminants by 6%, 20–40%, 35–95%, 6–61.2%, and 12%, respectively.

Conflict of Interest

The authors declare that there is no conflict of interest.

References

Åby, B.A., Randby, Å.T., Bonesmo, H. and Aass, L. (2019) Impact of grass silage quality on greenhouse gas emissions from dairy and beef production. *Grass and Forage Science* 74(3), 525–534.

Bampidis, V., Azimonti, G., Bastos, M.D.L., Christensen, H., Dusemund, B. *et al.* (2021) Safety and efficacy of a feed additive consisting of 3-nitrooxypropanol (Bovaer® 10) for ruminants for milk production and reproduction (DSM Nutritional Products Ltd.). *EFSA Journal* 19, e06905.

Bayat, A.R., Kairenius, P., Stefański, T., Leskinen, H., ComtetMarre, S. *et al.* (2015) Effect of Camelina oil or live yeasts (*Saccharomyces cerevisiae*) on ruminal methane production, rumen fermentation, and milk fatty acid composition in lactating cows fed grass silage diets. *Journal of Dairy Science* 98, 3166–3181. DOI: 10.3168/jds.2014-7976.

Beauchemin, K.A. and McGinn, S.M. (2006) Methane emissions from beef cattle: Effects of fumaric acid, essential oil, and canola oil. *Journal of Animal Science* 84(6), 1489–1496.

Boland, T.M., Pierce, K.M., Kelly, A.K., Kenny, D.A., Lynch, M.B. *et al.* (2020) Feed Intake, methane emissions, milk production and rumen methanogen populations of grazing dairy cows supplemented with various c 18 fatty acid sources. *Animals* 10(12), 2380.

Brooke, C.G., Roque, B.M., Najafi, N., Gonzalez, M., Pfefferlen, A. *et al.* (2018) Evaluation of the potential of two common pacific coast macroalgae for mitigating methane emissions from ruminants. *bioRxiv* 10, 434480. DOI: 10.1101/434480.

Cagliari, A.R., Magnani, E., Rigon, F., Loregian, K.E., Casagrande, A.C, *et al.* (2023) Evaluation of yeast-based additives, as an alternative to ionophores, on rumen fermentation of ruminant diets using an in vitro gas production system. *Frontiers in Animal Science* 4. DOI: 10.3389/fanim.2023.1233273.

Ding, G., Chang, Y., Zhou, Z., Ren, L. and Meng, Q. (2014) Effect of *Saccharomyces cerevisiae* on rumen fermentation characteristics, nutrient degradation and cellulase activity of steers fed diets with different concentrate to forage ratios. *World Journal of Agricultural Research* 2(6), 303–308.

Ding, X., Long, R., Zhang, Q., Huang, X., Guo, X. and Mi, J. (2012) Reducing methane emissions and the methanogen population in the rumen of Tibetan sheep by dietary supplementation with coconut oil. *Tropical Animal Health and Production* 44(7), 1541–1545.

Chung, Y.H., Walker, N.D., McGinn, S.M. and Beauchemin, K.A. (2011) Differing effects of 2 active dried yeast (*Saccharomyces cerevisiae*) strains on ruminal acidosis and methane production in nonlactating dairy cows. *Journal of Dairy Science* 94, 2431–2439. DOI: 10.3168/jds.2010-3277.

Dohme, F., Machmüller, A., Wasserfallen, A. and Kreuzer, M. (2000) Comparative efficiency of various fats rich in medium-chain fatty acids to suppress ruminal methanogenesis as measured with RUSITEC. *Canadian Journal of Animal Science* 80(3), 473–482. DOI: 10.4141/a99-113.

Doreau, M., Arbre, M., Popova, M., Rochette, Y. and Martin, C. (2018) Linseed plus nitrate in the diet for fattening bulls: Effects on methane emission, animal health and residues in offal. *Animal* 12(3), 501–507.

Dubois, B., Tomkins, N.W., Kinley, R.D., Bai, M., Seymour, S. *et al.* (2013) Effect of tropical algae as additives on rumen in vitro gas production and fermentation characteristics. *American Journal of Plant Sciences* 4, 34–43.

Duin, E.C., Wagner, T., Shima, S., Prakash, D., Cronin, B. *et al.* (2016) Mode of action uncovered for the specific reduction of methane emissions from ruminants by the small molecule 3-nitrooxypropanol. *Proceedings of the National Academy of Sciences of the USA* 113, 6172–6177.

Duthie, C.A., Rooke, J., Troy, S., Hyslop, J., Ross, D., Waterhouse, A. and Roehe, R. (2016) Impact of adding nitrate or increasing the lipid content of two contrasting diets on blood methaemoglobin and performance of two breeds of finishing beef steers. *Animal* 10, 786–795.

Duthie, C.-A., Troy, S.M., Hyslop, J.J., Ross, D.W., Roehe, R. and Rooke, J.A. (2017) The effect of dietary addition of nitrate or increase in lipid concentrations, alone or in combination, on performance and methane emissions of beef cattle. *Animal* 12(2), 280–287.

Evans, B. (2018) The role ensiled forage has on methane production in the rumen. *Animal Husbandry, Dairy and Veterinary Science* 2(4), 1–4.

Formato, M., Cimmino, G., Brahmi-Chendouh, N., Piccolella, S. and Pacifico, S. (2022) Polyphenols for livestock feed: Sustainable perspectives for animal husbandry? *Molecules* 27, 7752.

Geberemariyam, T., Mulugeta, W., Getu, K., Mesfin, D., Aemiro, K. *et al.* (2024) Supplementary value of sun dry brewer spent yeast as a replacement of cotton seed cake in the diet of lactating crossbred cows: Intake, digestibility, milk yield and quality. *Livestock Research for Rural Development* 36, Article #23. Available at: http://www.lrrd.org/lrrd36/03/3623gebe .html (accessed 17 February 2024).

Gerber, P.J., Henderson, B., Makkar, H.P.S. and et al, S. (2013) *Mitigation of Greenhouse Gas Emissions in Livestock Production: A Review of Technical Options for Non-CO2 Emissions*. (Issue 177) Food and Agriculture Organization of the United Nations (FAO).

Guyader, J., Eugène, M., Meunier, B., Doreau, M., Morgavi, D.P. *et al.* (2015) Additive methane-mitigating effect between linseed oil and nitrate fed to cattle. *Journal of Animal Science* 93(7), 3564–3577.

Guyader, J., Doreau, M., Morgavi, D.P., Gérard, C., Loncke, C. and Martin, C. (2016) Long-term effect of linseed plus nitrate fed to dairy cows on enteric methane emission and nitrate and nitrite residuals in milk. *Animal* 10(7), 1173–1181.

Hart, K.J., Huntington, J.A., Wilkinson, R.G., Bartram, C.G. and Sinclair, L.A. (2015) The influence of grass silage-to-maize silage ratio and concentrate composition on methane emissions, performance and milk composition of dairy cows. *Animal* 9(6), 983–991.

Herrero, M., Havlík, P., Valin, H., Notenbaert, A., Rufino, M.C. *et al.* (2013) Biomass use, production, feed efficiencies, and greenhouse gas emissions from global livestock systems. *Proceedings of the National Academy of Sciences of the United States of America* 110(52), 20888–20893. DOI: 10.1073/pnas.1308149110.

Honan, M., Feng, X., Tricarico, J.M. and Kebreab, E. (2021) Feed additives as a strategic approach to reduce enteric methane production in cattle: Modes of action, effectiveness and safety. *Animal Production Science* 62, 1303–1317.

IPCC (2013) Climate Change 2013: The Physical Science Basis. Contribution of Working Group I to the Fifth Assessment Report of the Intergovernmental Panel on Climate Change. In: Stocker, T.F., Qin, D., Plattner, G.-K., Tignor, M. and Allen, S.K. (eds). Cambridge University Press, Cambridge, UK and New York, NY, p. 1535.

Jayasundara, S., Ranga Niroshan Appuhamy, J.A.D., Kebreab, E. and Wagner-Riddle, C. (2016) Methane and nitrous oxide emissions from Canadian dairy farms and mitigation options: An updated review. *Canadian Journal of Animal Science* 96(3), 306–331.

Joblin, K.N (1999) Ruminal acetogens and their potential to lower ruminant methane emissions. *Australian Journal of Agricultural Research* 50(8), 1307–1314.

Johnson, D.E., Ward, G.W. and Ramsey, J.J. (1996) Livestock methane: Current emissions and mitigation potential. In: *Nutrient Management of Food Animals to Enhance and Protect the Environment*. CRC Press, Boca Raton, pp. 219–234.

Kamphayae, S., Kumagai, H., Angthong, W., Narmseelee, R. and Bureenok, S. (2017) Effects of different ratios and storage periods of liquid brewers yeast mixed with cassava pulp on chemical composition, fermentation quality and in vitro ruminal fermentation. *Asian-Australasian Journal of Animal Sciences* 30(4), 470.

Knapp, J.R., Laur, G.L., Vadas, P.A., Weiss, W.P. and Tricarico, J.M. (2014) Invited review: Enteric methane in dairy cattle production: Quantifying the opportunities and impact of reducing emissions. *Journal of Dairy Science* 97(6), 3231–3261.

Króliczewska, B., Pecka-Kiełb, E. and Bujok, J. (2023) Strategies used to reduce methane emissions from ruminants: Controversies and issues. *Agriculture* 13(3), 602. DOI: 10.3390/agriculture13030602.

Kumar, S., Choudhury, P.K., Carro, M.D., Griffith, G.W., Dagar, S.S. *et al*. (2014) New aspects and strategies for methane mitigation from ruminants. *Applied Microbiology and Biotechnology* 98, 31–44.

Little, S.M., Benchaar, C., Janzen, H.H., Kröbel, R., McGeough, E.J. and Beauchemin, K.A. (2017) Demonstrating the effect of forage source on the carbon footprint of a Canadian dairy farm using whole-systems analysis and the Holos model: Alfalfa silage vs. corn silage. *Climate* 5(4), 87.

Llonch, P., Haskell, M.J., Dewhurst, R.J. and Turner, S.P. (2017) Current available strategies to mitigate greenhouse gas emissions in livestock systems: An animal welfare perspective. *Animal* 11(2), 274–284.

Lovett, D., Lovell, S., Stack, L., Callan, J., Finlay, M., Conolly, J. and O'Mara, F.P. (2003) Effect of forage/concentrate ratio and dietary coconut oil level on methane output and performance of finishing beef heifers. *Livestock Production Science* 84(2), 135–146.

Machmüller, A. and Kreuzer, M. (1999) Methane suppression by coconut oil and associated effects on nutrient and energy balance in sheep. *Canadian Journal of Animal Science* 79(1), 65–72. DOI: 10.4141/A98-079.

Maia, M.R.G., Fonseca, A.J.M., Oliveira, H.M., Mendonça, C. and Cabrita, A.R.J. (2016) The potential role of seaweeds in the natural manipulation of rumen fermentation and methane production. *Scientific Reports* 6(1), 32321.

McAllister, T.A., Beauchemin, K.A., Alazzeh, A.Y., Baah, J., Teather, R.M. and Stanford, K. (2011) The use of direct fed microbials to mitigate pathogens and enhance production in cattle. *Canadian Journal of Animal Science* 91(2), 193–211.

Mihaila, A.A., Glasson, C.R.K., Lawton, R., Muetzel, S., Molano, G. and Magnusson, M. (2022) New temperate seaweed targets for mitigation of ruminant methane emissions: An in vitro assessment. *Applied Phycology* 3(1), 1–11.

Min, B.-R., Lee, S., Jung, H., Miller, D.N. and Chen, R. (2022) Enteric methane emissions and animal performance in dairy and beef cattle production: Strategies, opportunities, and impact of reducing emissions. *Animals* 12(8), 948.

Moate, P.J., Williams, S.R.O., Grainger, C., Hannah, M.C., Ponnampalam, E.N. and Eckard, R.J. (2011) Influence of cold-pressed canola, brewers grains and hominy meal as dietary supplements suitable for reducing enteric methane emissions from lactating dairy cows. *Animal Feed Science and Technology* 166, 254–264.

Moss, A.R., Jouany, J.-P. and Newbold, J. (2000). Methane production by ruminants: Its contribution to global warming. *Annales de Zootechnie* 49(3), 231–253.

Murray, R.M., Bryant, A.M. and Leng, R.A. (1976) Rates of production of methane in the rumen and large intestine of sheep. *British Journal of Nutrition* 36(1), 1–14.

Newton, E.E., Pétursdóttir, Á.H., Ríkharðsson, G., Beaumal, C., Desnica, N. *et al*. (2021) Effect of dietary seaweed supplementation in cows on milk macrominerals, trace elements and heavy metal concentrations. *Foods* 10(7), 1526.

Olijhoek, D.W., Hellwing, A.L.F., Grevsen, K., Haveman, L.S., Chowdhury, M.R. *et al*. (2019) Effect of dried oregano (*Origanum vulgare* L.) plant material in feed on methane production, rumen fermentation, nutrient digestibility, and milk fatty acid composition in dairy cows. *Journal of Dairy Science* 102(11), 9902–9918.

Page, M.J., McKenzie, J.E., Bossuyt, P.M., Boutron, I., Hoffmann, T.C. *et al.* (2021) The PRISMA 2020 statement: An updated guideline for reporting systematic reviews. *BMJ* 372(1). DOI: 10.1136/bmj.n71.

Parra, M.C., Forwood, D.L., Chaves, A.V. and Meale, S.J. (2023) In vitro screening of anti-methanogenic additives for use in Australian grazing systems. *Frontiers in Animal Science* 4, 1123532. DOI: 10.3389/fanim.2023.1123532.

Patra, A.K. and Saxena, J. (2009) The effect and mode of action of saponins on the microbial populations and fermentation in the rumen and ruminant production. *Nutrition Research Reviews* 22(2), 204–219.

Pen, B., Takaura, K., Yamaguchi, S., Asa, R. and Takahashi, J. (2007) Effects of *Yucca schidigera* and *Quillaja saponaria* with or without β 1–4 galacto-oligosaccharides on ruminal fermentation, methane production and nitrogen utilization in sheep. *Animal Feed Science and Technology* 138(1), 75–88.

Patra, A., Park, T., Kim, M. and Yu, Z. (2017) Rumen methanogens and mitigation of methane emission by anti-methanogenic compounds and substances. *Journal of Animal Science and Biotechnology* 8(1), 13.

Pszczolkowski, V.L., Bryant, R.W., Harlow, B.E., Aiken, G.E., Martin, L.J. *et al.* (2016) Effects of spent craft brewers' yeast on fermentation and methane production by rumen microorganisms. *Advances in Microbiology* 6(9), 716.

Rachwał, K., Waśko, A., Gustaw, K. and Polak-Berecka, M. (2020) Utilization of brewery wastes in food industry. *PeerJ* 8, 1–28. DOI: 10.7717/peerj.9427.

Rasmussen, J. and Harrison, A. (2011) The benefits of supplementary fat in feed rations for ruminants with particular focus on reducing levels of methane production. *International Scholarly Research Notices* 2011(613172). DOI: 10.5402/2011/613172.

Ribeiro-Filho, H.M.N., Civiero, M. and Kebreab, E. (2020) Potential to reduce greenhouse gas emissions through different dairy cattle systems in subtropical regions. *PLoS ONE* 15(6), 1–18. DOI: 10.1371/journal.pone.0234687.

Robertson, L.J. and Waghorn, G.C. (2002) New Zealand society of animal production online archive dairy industry perspectives on methane emissions and production from cattle fed pasture Greenhouse Gas (GHG) issues and the New Zealand dairy. *Proceedings of the New Zealand Society of Animal Production* 62, 213–218.

Roque, B.M., Brooke, C.G., Ladau, J., Polley, T., Marsh, L.J. *et al.* (2019) Effect of the macroalgae *Asparagopsis taxiformis* on methane production and rumen microbiome assemblage. *Animal Microbiome* 1(1), 1–14.

Roque, B.M., Venegas, M., Kinley, R.D., de Nys, R., Duarte, T.L. *et al.* (2021) Red seaweed (*Asparagopsis taxiformis*) supplementation reduces enteric methane by over 80 percent in beef steers. *PLoS One* 16(3), e0247820.

Samarasinghe, M.B. and Sehested, J. (2021) Milk supplemented with dried seaweed affects the systemic innate immune response in preweaning dairy calves. *Journal of Dairy Science* 104, 3575–3584. DOI: 10.3168/jds.2020-19528.

Senga Kiessé, T., Corson, M.S. and Wilfart, A. (2022) Analysis of milk production and greenhouse gas emissions as a function of extreme variations in forage production among French dairy farms. *Journal of Environmental Management* 307, 114537.

Shingfield, K.J., Jaakkola, S. and Huhtanen, P. (2002) Effect of forage conservation method, concentrate level and propylene glycol on intake, feeding behaviour and milk production of dairy cows. *Animal Science* 74(2), 383–397.

Silwer, H. (2018) Macroalgae as feed supplement for reduction of methane emission in livestock.

Stefenoni, H.A., Räisänen, S.E., Cueva, S.F., Wasson, D.E., Lage, C.F.A. *et al.* (2021) Effects of the macroalga *Asparagopsis taxiformis* and oregano leaves on methane emission, rumen fermentation, and lactational performance of dairy cows. *Journal of Dairy Science* 104(4), 4157–4173.

Steinfeld, H., Gerber, P., Wassenaar, T.D., Castel, V., Rosales, M. *et al.* (2006) *Livestock's Long Shadow: Environmental Issues and Options*. FAO.

Terefe, G., Walelgne, M., Fekadu, D., Kitaw, G., Dejene, M., Kehaliu, A. and Habteyesus, Y. (2023) Effect of sun dry brewer spent yeast on chemical composition, in vitro digestibility, and ruminal degradation kinetics of wheat straw. *CABI Agriculture and Bioscience* 4(1), 1–8.

Troy, S.M., Duthie, C.A., Hyslop, J.J., Roehe, R., Ross, D.W. *et al.* (2015) Effectiveness of nitrate addition and increased oil content as methane mitigation strategies for beef cattle fed two contrasting basal diets. *Journal of Animal Science* 93(4), 1815–1823.

Van Schooten, H. and Philipsen, B. (2012) Proceedings XVI International Silage Conference, Helsinki, Finland, 2-4 July 2012, pp 126-127. Available at: https://edepot.wur.nl/265367

Van Soest, P.J. (1988) Effect of environment and quality of fibre on the nutritive value of crop residues. In: *Plant Breeding and the Nutritive Value of Crop Residues. International Livestock Centre for Africa (ILCA), Addis Ababa, Ethiopia*. pp. 71–94.

Vargas, J.E., Andrés, S., López-Ferreras, L., Snelling, T.J., Yáñez-Ruíz, D.R., García-Estrada, C. and López, S. (2020) Dietary supplemental plant oils reduce methanogenesis from anaerobic microbial fermentation in the rumen. *Scientific Reports* 10(1), 1–9.

Vellinga, T.V. and Hoving, I.E. (2011) Maize silage for dairy cows: Mitigation of methane emissions can be offset by land use change. *Nutrient Cycling in Agroecosystems* 89(3), 413–426.

Vijn, S., Compart, D.P., Dutta, N., Foukis, A., Hess, M. *et al.* (2020) Key considerations for the use of seaweed to reduce enteric methane emissions from cattle. *Frontiers in Veterinary Science* 7, 1135.

Villar, M.L., Hegarty, R.S., Nolan, J.V., Godwin, I.R. and McPhee, M. (2020) The effect of dietary nitrate and canola oil alone or in combination on fermentation, digesta kinetics and methane emissions from cattle. *Animal Feed Science and Technology* 259, 114294. DOI: 10.1016/j.anifeedsci.

5 Dorper sheep in Africa: A review of their use and performance in different environments

Julie M.K. Ojango*[1], Moses Okpeku[2], Richard Osei-Amponsah[3], Donald R. Kugonza[4], Okeyo Mwai[1], Mizeck G.G. Chagunda[5] and Victor E. Olori[6]
[1]International Livestock Research Institute, PO BOX 30709-00100, Nairobi, Kenya; [2]Discipline of Genetics, School of Life Sciences, University of Kwazulu-Natal, Durban South Africa; [3]Department of Animal Science, School of Agriculture, University of Ghana, Legon, Accra, Ghana; [4]Department of Agricultural Production, School of Agricultural Sciences, College of Agricultural & Environmental Sciences, Makerere University, Box 7062, Kampala, Uganda; [5]University of Hohenheim, Institute of Agricultural Science in the Tropics, Germany; [6]Research & Development Department, Aviagen Ltd, 11 Lochend Rd, Newbridge EH28 8SZ, Scotland, UK

Abstract

The Dorper breed developed in South Africa is used either as a pure breed or crossbred with existing indigenous breeds by many countries across the African continent to improve sheep production. This article presents documented information on the adoption, use, and performance of Dorper sheep across the continent of Africa and opportunities for their more sustainable production under the changing climatic conditions in Africa. Apart from the well-documented information on the Dorper sheep in South Africa, published information on the performance of the sheep is mainly from Eastern Africa. Most countries initially retained purebred Dorpers in nationally owned institutions for multiplication and crossbreeding trials with different indigenous breeds prior to distributing the crossbreds to diverse livestock keepers. The offspring produced through crossbreeding programs with the Dorper have better growth rates than indigenous breeds in the different countries; however, the performance of Dorper sheep in South Africa has not been achieved in any of the other countries. Genomic studies including Dorper sheep have identified regions of interest for resistance to brucellosis and Mycoplasma ovipneumonia that imply adaptability to challenging environments within Dorper sheep. Unfortunately, limitations in systems for guided breeding and monitoring of sheep productivity in Africa have resulted in haphazard crossbreeding of the Dorper. Targeted efforts are required across the different countries to develop breeding programs for improving locally adapted Dorper sheep populations and their crosses with indigenous breeds. New science and technologies need to be innovatively packaged and used to identify and propagate more productive and resilient Dorper and Dorper-based breed-types for the increasingly challenging tropical African range environments.

Keywords: Dorper sheep, productivity, genetic parameters, crossbreeding

*Corresponding author: j.ojango@cgiar.org

Introduction

The Dorper sheep is a hardy, early maturing composite breed developed in the semi-arid Karoo region of South Africa from the 1930s (Cloete *et al.*, 2000; Milne, 2000; AGTR, 2011; Zishiri *et al.*, 2013). The breed was developed by the South African Department of Agriculture and cooperating farmers in the region in response to the need for a hardy productive sheep breed able to survive the arid, harsh environmental elements of the Karoo and produce good-quality meat (Ramsay *et al.*, 2000). Through selective crossbreeding of the Dorset Horn sheep (known for quality meat production), with the Persian sheep (well adapted to arid environments), two variants of the breed were produced, one with a characteristic black head and white body and the other that is entirely white in color (Cloete *et al.*, 2000; Milne, 2000; Schoeman, 2000; Wanjala *et al.*, 2023a). The Dorper is an embodiment of adaptation to various climatic conditions, has high fertility and a good growth rate under extensive management in the semi-arid region of South Africa, and produces a high-quality carcass with well-marbled meat, making it a choice breed for meat production (Schoeman, 2000). Although developed for the arid regions, the attributes of Dorper sheep in South Africa enabled the breed to be exported to many countries on the African continent, North America (United States and Canada), Australia, and in more temperate environments of Europe, New Zealand, and Tasmania (Cloete *et al.*, 2000; Milne, 2000; Schoeman, 2000; Cosgrove, 2023).

In Africa, Dorper sheep are used for crossbreeding with various indigenous breeds raised in semi-arid environments (Getachew *et al.*, 2016; Zonabend König *et al.*, 2017a; Oyieng *et al.*, 2022; Besufkad *et al.*, 2023).

Characteristics and performance of Dorper sheep in South Africa are well documented (Ramsay *et al.*, 2000; Cloete *et al.*, 2012; Zishiri *et al.*, 2013; Hoffman *et al.*, 2020; Kao *et al.*, 2022); however, recent scientific reports on Dorper sheep in other countries of Africa are few and scattered. This review presents documented information on the adoption, use, and performance of Dorper sheep and their crosses with indigenous breeds in different countries of Africa and opportunities for their more sustainable production under changing climatic conditions.

Review Methodology

When scoping information to get a general overview of what has been documented on Dorper sheep in Africa, we used "Chat GPT" to generate an initial list of topics to prioritize in the review. The general framework used in implementing the review is presented in Table 5.1.

A series of search strings were developed that combined components representing the scope of the proposed review including: (i) Background and distribution of Dorper sheep in Africa: Origin/Genesis of the Dorper; (ii) Breeding programs involving Dorper sheep in different countries of Africa; (iii) Genetic diversity

Table 5.1. The framework adopted for this review.

Steps	Outcome
Protocol	Define scope for the review: Determine key topics to be covered in the review
Search	Define the search strategy to be adopted Prioritize digital libraries to source content
Appraisal and quality assessment	Define criteria for inclusion and exclusion of publications/documents
Synthesis and review	Synthesize and iteratively categorize information collated in line with topics defined
Writing of publication	Develop narrative of organized information, iteratively review drafts to identify gaps, compare results, and derive conclusions

Adapted from Mengist *et al.* (2020).

in Dorper sheep populations across Africa: Phenotypic and Genetic parameter estimates for Dorper sheep and their crosses in different environments of Africa; (iv) Opportunities and strategies for future breeding programs involving Dorper sheep.

Search strings comprising a combination of key terms in the prioritized topics using the Boolean operators "OR" and "AND" were applied to CABI, Scopus, PubMed, ScienceDirect, Google Scholar, and a web search of Google to target grey literature, conference proceedings, and Graduate thesis that may not be in the academic databases. Only publications in English were included in the search. Considering the reviews on Dorper sheep in Africa that were published in 2000 (Cloete *et al.*, 2000; Milne, 2000; Schoeman, 2000), the publication date for our search was set to publications after 1999. The search was implemented between May and August 2023.

Utilization of the Dorper in Breeding Programs in Africa

The exact number of Dorper sheep within the different countries of Africa is not known as such data are limited and generally not captured at country levels. The information available through the Food and Agricultural Organization (FAO) statistical database (FAOSTAT, 2023) on sheep populations is based on all sheep breeds that are found within each country and therefore reported at the country level. The Southern Africa region has the largest population of Dorper sheep, with the population in South Africa alone estimated to be more than 5 million by the Dorper Sheep Breeders' Society (Available at: https://dorpersa.co.za/breed-history/), accessed 7 October 2023. Dorper sheep were introduced in Eastern Africa around 1952 (Kariuki *et al.*, 2010) and in Ethiopia in the 1980s (Getachew *et al.*, 2016). The breed was subsequently established through additional importations to flocks owned by national organizations and used both as purebreds and for crossbreeding with existing indigenous breeds (Kosgey and Okeyo, 2007; Kariuki *et al.*, 2010; Getachew *et al.*, 2016; Ayichew, 2019).

Multiplication, breeding, and sheep improvement programs using Dorper sheep that are documented from different countries of Africa (Table 5.2) are managed either by government-owned or privately owned ranches. The Ethiopian Sheep and Goat Productivity Improvement Program (ESGPIP) re-introduced Dorper sheep in 2007 to be multiplied through government stations and subsequently used for crossbreeding with local indigenous sheep breeds such as the Tumele, Hararghe Highland, Afar, and Menz (Lakew *et al.*, 2014; Ayichew, 2019; Tesema *et al.*, 2022). In Kenya, a Dorper sheep breeder's society was formed in 1970 and formerly registered in 2012 to promote the genetic improvement of a national Dorper flock (DSBSK, 2023). Though the society in Kenya has set breed standards, documentation of the performance of the registered animals is scarce. The Dorper is however widely used for crossbreeding with the indigenous Red Maasai sheep breed in Kenya (Muigai *et al.*, 2009; Kariuki *et al.*, 2010; Zonabend König *et al.*, 2017a).

Crossbred animals resulting from mating Dorper sheep with indigenous breeds are popular because of the fast growth rate and good market demand as meat animals (Zonabend König *et al.*, 2016; Lakew *et al.*, 2021; Tesema *et al.*, 2022). Reports from crossbreeding Dorper sheep with the Red Maasai sheep breed in Kenya (Zonabend König *et al.*, 2017a; Oyieng *et al.*, 2022; Wanjala *et al.*, 2023b) and with Tumele sheep in Ethiopia (Tesema *et al.*, 2020) indicate that the use of the Dorper improved the reproductive performance and growth rates of sheep under semi-arid production environments. The indigenous breeds in the different countries however display better resistance to internal parasites compared with the Dorper (Tembely *et al.*, 1998; Baker *et al.*, 1999; Matika *et al.*, 2003; Baker *et al.*, 2004). The offspring from crossbreeding programs that use Dorper sheep have been multiplied by livestock keepers rearing sheep under extensive management in semi-arid to arid environments in Ethiopia and Kenya (Kosgey *et al.*, 2008; Ayele Abebe *et al.*, 2015; Ayichew, 2019). Published reports on the commercialization of Dorper crossbred populations and their performance across the different farming communities are mainly from animals reared in Ethiopia and Kenya. Results from the crossbreeding of Dorper sheep with indigenous Ethiopian sheep breeds

Table 5.2. Documented sheep improvement programs using the Dorper breed in different countries of Africa.

Region country	Organization/institution affiliated	Indigenous breed used in crossbreeding	Reference
Southern Africa			
Botswana	Farmers in different districts	Tswana	Bolowe *et al.* (2022)
Namibia	Extensive farming systems	N/A	Kandiwa *et al.* (2020)
South Africa[a]	Research institutions such as Nortier experimental farm and University of Limpopo	South African Merino, N	Cloete *et al.* (2000); Milne (2000); Ramsay *et al.* (2000); Schoeman (2000); Zishiri *et al.* (2013)
Swaziland	Research institutes such as University of Free State and Malotwana Silvopastoral farm	N/A	Patricia Walker (2013); Nxumalo *et al.* (2022)
Zambia	Mazabuka Regional Research Station; Mansa Research station	Indigenous breeds	Mwenya (1994); Goma *et al.* (2007)
Zimbabwe	Research Institutes, e.g. Grassland Research Station, Matopos Research Station	Sabi	Matika *et al.* (2003); Assan and Makuza (2005)
Eastern Africa			
Ethiopia	Ethiopian Sheep and Goat Productivity Improvement Program (ESGPIP)	Hararghe Highland, Blackhead Ogaden, Menz	Ayele Abebe *et al.* (2015); Getachew *et al.* (2016); Weldeyesus (2017); Ayichew (2019); Tesema *et al.* (2020); Lakew *et al.* (2021); Tesema *et al.* (2022); Besufkad *et al.* (2023)
Kenya	Ol'Magogo Research station; International Livestock Research Institute, Kapiti; pastoral production systems and private ranches	Red Maasai Sheep	Baker *et al.* (2004); Mugambi *et al.* (2005); Kariuki *et al.* (2010); Liljestrand (2012); Zonabend König *et al.* (2017b); Oyieng *et al.* (2022); Wanjala *et al.* (2023c)

[a]Original developers of breed.

(Getachew *et al.*, 2016; Ayichew, 2019) indicate that the introduction of the new breeds has had a positive effect on sheep production. However, the collection of data on the performance of the introduced breeds and their crosses in the local farming communities remains limited. The practice of ubiquitous dissemination and selling of crossbred rams for breeding to individual farmers may however dilute efforts of developing stabilized crossbred animals with improved performance (Getachew *et al.*, 2016).

Performance of Dorper Sheep and Their Crosses in Africa

The economic returns from rearing non-wool sheep including the Dorper depend on the offspring produced and marketed, which in turn is influenced by the reproduction and growth performance of the animals (Cloete *et al.*, 2000; Cloete *et al.*, 2012; Zishiri *et al.*, 2013; Deribe *et al.*, 2021). A review of the performance of Dorper sheep and its crosses with other breeds under different production systems in Africa (Schoeman, 2000) noted that the Dorper was superior to the other breeds in reproductive and growth traits; however, it was noted that mortality rates of up to 49% were reported in Kenya under semi-arid conditions (Inyangala *et al.*, 1992). Though the high mortality rate could make the breed unsuitable for commercial use in the more humid African Tropics, observation of flocks owned by pastoralist livestock keepers in the southern rangelands of Kenya and anecdotal information points to the contrary as Dorper sheep and their crosses are herded in large numbers and their population is increasing in Kenya (Muigai *et al.*, 2009; Zonabend König *et al.*, 2017a). Corroborating this observation, subsequent publications from different countries in Africa illustrate the widespread adoption of Dorper sheep (Table 5.2) accompanied by changes in productivity within the specified environments as outlined below.

Reproduction

Cloete *et al.* (2000) reviewed published literature on the reproduction of Dorper sheep which indicated a broad range of performance across environments. The review reported age at first oestrus to range from 213 to 328 days, with reproductive rates of 0.99–1.40 lambs per ewe mated. Subsequent studies on the fertility of Dorper sheep within Africa present a broader range in the age at first lambing (AFL) (Table 5.3). Zishiri *et al.* (2013) evaluated the overall reproduction, longevity, and survival of Dorper sheep in South Africa by deriving several fertility parameters. These included the number of lambing chances afforded per ewe lifetime (LC/EL = 2.45), the average litter size (1.23), the

Table 5.3. Average reproductive performance (mean ± standard deviation) of pure and crossbred Dorper sheep in different countries of Africa.

Age at first lambing (days)	Lambing interval (days)	Lambs weaned per ewe mated (LW/EM)	Number of ewes lambing per ewe mated (EL/EM)	Breed	Country	References
555 ± 6.25	287 ± 2.38			50% Dorper ×indigenous	Ethiopia	Lakew *et al.* (2014)
786 ± 57.8	380 ± 19		0.85 ± 0.01	Dorper ×Menz 50%	Ethiopia	Goshme *et al.* (2021)
570 ± 54	398 ± 14		0.68 ± 0.03	Dorper	Ethiopia	
346–588	198–231	0.99–1.40		Dorper	South Africa	Cloete *et al.* (2000)
	298 ± 44.4	0.80 ± 0.26		Dorper	South Africa	Zishiri *et al.* (2013)
756 ± 162	418 ± 92			Dorper	Kenya	Ojango *et al.* (2023)
720 ± 183	411 ± 88			75% Dorper ×Red Maasai	Kenya	
690 ± 170	422 ± 91			50% Dorper ×50% Red Maasai	Kenya	
		0.882	1.008	Dorper	Kenya	Wanjala *et al.* (2023b)

number of lambs born per ewe per lifetime (NLB/EL = 2.14), and the number of lambs weaned per ewe per lifetime (NLE/EL = 1.83). From this study, the reproductive performance of ewes defined as the total weight of lamb weaned per ewe lambing (TWW/L) was proposed as a good measure for both reproduction and fitness of ewes that can be improved through either direct or indirect selection.

Studies on Dorper sheep raised in national stations of Kenya and Ethiopia have reported the number of ewes lambing per ewe mated (EL/EM) as presented in Table 5.3, and the number of lambs weaned per ewe lambing. The number of ewes lambing per ewe mated reported for Dorper sheep within a mixed flock in Kenya (Wanjala *et al.*, 2023c) was higher than that reported for purebred Dorper sheep at the Debre Berhan station in Ethiopia (Goshme *et al.*, 2021), reflecting differences in the production environment on the fertility.

Individual fertility traits including the AFL and lambing intervals (LI) reported for pure and crossbred Dorper sheep raised in the arid rangelands of Kenya have been shown to fluctuate with variations in precipitation over the seasons and years (Ojango *et al.*, 2023). From previously reviewed studies on fertility (Cloete *et al.*, 2000; Deribe *et al.*, 2021), it is evident that the inter-lambing period is greatly influenced by the management practices adopted by each breeder rather than the genetic potential of the animals, while post-partum anoestrus of ewes strongly depends on the lambing season and feeding level of the animals.

Survival

Lamb survival to weaning for Dorper sheep in South Africa derived as the proportion of the total number of lambs weaned over the total number of lambs born is reported to be 78% by Kao *et al.* (2022) and 88% by Zishiri *et al.* (2013) with a direct heritability estimate of 0.07 ± 0.01. In Ethiopia, Tesema *et al.* (2020, 2021) reported mortality rates for crossbred Dorper lambs of 14% from birth to weaning, and 32.1% from birth to yearling. The main causes of mortality in the flocks were noted to be gastrointestinal parasites and pneumonia (Tesema *et al.*, 2020). The results on mortality indicate the importance of managing the environment for better survival of young animals. Evaluations of the length of productive life of Dorper ewes are limited, yet the ability of ewes to remain productive over several years enhances the efficiency of production and reduces replacement costs. Cloete *et al.* (2000) reported that Dorpers remained in breeding flocks for an average of 4.7 seasons in Zimbabwe, while Zishiri *et al.* (2013) reported the heritability for stayability to 5 years of age in Dorper sheep to be less than 0.05.

Growth

Reports on the growth performance of Dorper sheep and their crosses with indigenous breeds in Africa include information on lamb performance from birth to weaning, and from weaning to maturity which is either reported at 9 months of age or 1 year of age. Results from published documents on the growth performance of Dorper's and their crosses from various studies on populations in Africa are presented in Table 5.4.

The growth of Dorper sheep and their crosses varies greatly across countries and production environments (Table 5.4). In South Africa, Dorper sheep are crossed with the South African Mutton Merino for meat production (Kao *et al.*, 2022), and the Namaqua Africana an adapted indigenous breed for their adaptability (Cloete *et al.*, 2021). Several studies present results on the growth performance of Dorper sheep and their crosses in Ethiopia (Table 5.4). Among these results, the study by Besufkad *et al.* (2023) presented a general decline in growth performance of the pure-bred Dorper (2012–2021) reared at a higher altitude in Debre Birhan Agricultural Research Center of Ethiopia, mainly attributed to management and nutrition of the animals post-weaning. In Kenya, Oyieng *et al.* (2022) reported improved growth rates in Dorper's and their crosses with Red Maasai sheep over 17 years under a semi-arid environment (Table 5.4). The crosses comprising 75% Dorper and 25% Red Maasai breed exhibited the best productivity in the

Table 5.4. Growth performance of Dorper and their crosses (50 and 70% Dorper) with other breeds reported for different countries of Africa.

Breed	Birth	Weaning (3 months)	9 months	12 months	Country	Source
Pure-bred Dorper	4.19 ± 0.09	32.7 ± 0.6			South Africa	Cloete *et al*. (2021)
		30.9 ± 0.013	45.1 ± 0.034		South Africa	Zishiri *et al*. (2013)
	4.16 ± 0.10	31.6 ± 0.04			South Africa	Kao *et al*. (2022)
	3.33 ± 0.10				Ethiopia	Goshme *et al*. (2021)
	3.39 ± 0.08	16.18 ± 0.35		34.43 ± 0.79	Ethiopia	Ayele Abebe *et al*. (2015)
	3.8 ± 0.02	19.38 ± 0.10	29.86 ± 0.16	36.6 ± 0.19	Kenya	Kariuki *et al*. (2010)
	4.11 ± 0.14			38.16 ± 1.13	Kenya	Zonabend König *et al*. (2017a)
	3.8 ± 0.02	17.5 ± 0.14	27.3 ± 0.21		Kenya	Oyieng *et al*. (2022)
50% Dorper crossed to breed listed						
Namaqua Afrikaner	4.40 ± 0.11	33.1 ± 0.7			South Africa	Cloete *et al*. (2021)
Menz	2.51 ± 0.11				Ethiopia	Goshme *et al*. (2021)
	2.77 ± 0.04	12.3 ± 0.25		31.33 ± 0.56	Ethiopia	Ayele Abebe *et al*. (2015)
Blackhead Ogaden	3.0 ± 0.1	15.1 ± 0.3			Ethiopia	Teklebrhan *et al*. (2014)
Afar	2.57 ± 0.06	9.45 ± 0.87		24.96 ± 3.77	Ethiopia	Ayele Abebe *et al*. (2015)
Hararghe Highland	2.9 ± 0.1	14.9 ± 0.1			Ethiopia	Teklebrhan *et al*. (2014)
Indigenous	3.24 ± 0.04	14.59 ± 0.21		31.37 ± 0.38	Ethiopia	Lakew *et al*. (2014)
	3.03 ± 0.03	14.4 ± 0.18	20.3 ± 0.28	24.6 ± 0.32	Ethiopia	Tesema *et al*. (2022)
Red Maasai	3.79 ± 0.12			39.37 ± 0.97	Kenya	Zonabend König *et al*. (2017a)
	3.4 ± 0.02	16.6 ± 0.11	26.6 ± 0.15		Kenya	Oyieng *et al*. (2022)
75% Dorper crossed to breed listed						
Blackhead Ogaden	3.1 ± 0.1	20.7 ± 1.1			Ethiopia	Teklebrhan *et al*. (2014)
Hararghe Highland	3.1 ± 0.2	19.2 ± 1.1			Ethiopia	Teklebrhan *et al*. (2014)
Dorper ×indigenous	3.00 ± 0.07	15.2 ± 0.59	21.4 ± 0.68	26.5 ± 0.53	Ethiopia	Tesema *et al*. (2022)
Red Maasai	4.05 ± 0.09			38.59 ± 0.77	Kenya	Zonabend König *et al*. (2017a)
Red Maasai	3.72 ± 0.02	17.64 ± 0.11	28.82 ± 0.16		Kenya	Oyieng *et al*. (2022)

semi-arid environment (Oyieng *et al.*, 2022). It is however notable that Dorper sheep in South Africa exhibit the highest weights at different ages (Table 5.4), reflecting the potential of the breed under optimal conditions, and the gaps in performance levels attained in the different countries.

In addition to documenting weights at different ages of the Dorper sheep and their crosses in Ethiopia, the authors have calculated Kleiber ratios (KR: weight gain per unit metabolic weight) as an indicator of feed efficiency by the animals in different phases of growth (Deribe *et al.*, 2021; Tesema *et al.*, 2022; Besufkad *et al.*, 2023). The KR for sheep in Ethiopia differ significantly depending on the proportion of Dorper genes in the animals, and depending on the phase of growth. Sheep comprising 50% Dorper and 50% indigenous breeds had a KR from birth to weaning of 6.49 relative to 5.79 for those with 75% Dorper and 25% indigenous breeds. Additionally, a higher KR is reported between weaning and 6 months of age (5.01–5.76) relative to that between 6 and 9 months of age (3.04–3.6) (Deribe *et al.*, 2021; Tesema *et al.*, 2022; Besufkad *et al.*, 2023). Though the management and feeding of the animals are noted to influence their growth, the impacts of the different temperatures and humidity on the performance of the different breeds are not yet documented.

Phenotypic and Genetic Parameters of Dorper Sheep and Their Crosses in Africa

The genetic relationships across Dorper sheep populations in Africa are poorly documented. Genetic parameter estimates for some traits are published for populations of Dorper sheep in South Africa, Ethiopia, and Kenya. Those published prior to the year 2000 are well documented in Cloete *et al.* (2000).

Heritability estimates reported for survival to weaning for Dorper sheep and their crosses with indigenous breeds in Africa are presented in Table 5.5. Parameter estimates for the composite fertility traits defined by Zishiri *et al.* (2013) ranged from 0.09 ± 0.01 for NLW/EL to 0.23 ± 0.01 for LC/EL, indicating that genetic gains in the defined fertility traits are realizable. Their study also reviewed previously reported heritability estimates for the number of lambs born per ewe joined across different parities in various sheep breeds, noting that the reported estimates tended to be low (below 0.10).

Heritability estimates for weaning weight and post-weaning weight for Dorper's across different countries of Africa are moderate (Table 5.5), illustrating the propensity of the sheep to transmit reasonable genetic variation that will result in genetic gains in growth traits. From the estimates published, except in the study by Kariuki *et al.* (2010), the maternal effects that are important to weaning diminish with lamb age post-weaning (Table 5.5). Maternal effects and influence reflect the milk-producing ability and mothering behavior of ewes which support early lamb growth and survival. The total weight of lambs weaned per year as a function of the number of lambs born, their survival, and individual lamb weights at weaning as proposed in Zishiri *et al.* (2013) provides a good measure of flock productivity. The differences in genetic parameter estimates within the sub-populations of Dorper sheep in the different countries reflect significant levels of genetic diversity of the sub-populations which is yet to be exploited through structured national and regional breeding programs.

Positive and moderate genetic correlations are reported between growth traits and composite fertility traits for Dorper sheep in South Africa (Zishiri *et al.*, 2013). The same study also reported positive genetic correlations between growth traits and lamb survival to weaning (0.36 ± 0.05) and post-weaning (0.24 ± 0.03), implying that selection for liveweight will not compromise the fitness of sheep in the South African production environments. The genetic correlation between lamb survival to weaning and litter size was however unfavorable (-0.41 ± 0.15). Additional studies in the different production environments across the continent are however required to enhance the understanding of the relationship between survival and production traits.

Table 5.5. Genetic parameter estimates for growth and fertility traits in Dorper sheep and their crosses with indigenous breeds from different countries of Africa.

	Growth traits (kg)				Functional traits				
	Birth weight	Weaning weight	9-month weight	Yearling weight	Age at first lambing	Survival to weaning	Breed	Country	Source
Heritability estimates		0.21 ± 0.01	0.27 ± 0.02			0.07 ± 0.01	Dorper	South Africa	Zishiri *et al*. (2013)
	0.24 ± 0.09	0.10 ± 0.07	0.08 ± 0.06				Dorper	Kenya	Oyieng *et al*. (2022)
	0.18 ± 0.01	0.25 ± 0.05	0.14 ± 0.05	0.29 ± 0.09			Dorper	Kenya	Kariuki *et al*. (2010)
	0.33 ± 0.07	0.38 ± 0.07	0.40 ± 0.07				50% Dorper ×Red Maasai	Kenya	Oyieng *et al*. (2022)
	0.37 ± 0.12	0.21 ± 0.12					50% Dorper ×indigenous	Ethiopia	Tesema *et al*. (2022)
Direct maternal heritability		0.05 ± 0.01	0.04 ± 0.02				Dorper	South Africa	Zishiri *et al*. (2013)
	0.16 ± 0.01	0.19 ± 0.04		0.18 ± 0.06			Dorper	Kenya	Kariuki *et al*. (2010)
	0.08 ± 0.06	0.05 ± 0.03	0.00				Dorper	Kenya	Oyieng *et al*. (2022)
	0.12 ± 0.05	0.00	0.00				50% Dorper ×Red Maasai	Kenya	Oyieng *et al*. (2022)
	0.20 ± 0.06	0.13 ± 0.07					50% Dorper ×indigenous	Ethiopia	Tesema *et al*. (2022)
Dam permanent environment		0.08 ± 0.01	0.04 ± 0.02				Dorper	South Africa	Zishiri *et al*. (2013)

Genomic Studies with Dorper Sheep

Genomic studies on sheep across the world have included Dorper's in a variety of ways. Analyses of the population structure of African sheep breeds using principal component analysis (PCA) and ADMIXTURE analyses show Dorper sheep to be clustered between Dorset and Blackhead Persian (Molotsi *et al.*, 2017; Dzomba *et al.*, 2021). The black head coloration of the Dorper and similarly colored sheep breeds have been associated with a haplotype that covers the melanocortin receptor 1 (MC1R) gene (Wang *et al.*, 2021).

A study on the genomic structure of the Dorper sheep variants in South Africa and Hungary using the ovine 50K single nucleotide polymorphism (SNP) chip reported a strong genetic relationship between the white Dorper in Hungary and South Africa (Wanjala *et al.*, 2023a). The Dorper variants from South Africa had a higher level of genetic diversity within the population. The genetic differentiation between the Dorper populations was noted to be related to environmental adaptations. The first report and analysis of the full sequence of the Dorper sheep reference genome by Qiao *et al.* (2022) revealed allele-specific expression (ASE) genes related to the immune system as well as lipid metabolism and adaptation of the breed to the environment. Runs of homozygosity (RoH) of South African breeds of sheep including the White Dorper sheep using the Ovine SNP50 Illumina® bead chip array showed a majority of short, i.e., 1–6 Mb RoH in the Dorper and other tested breeds (Dzomba *et al.*, 2021). They found that chromosome 10, which harbours a genomic region associated with horns in sheep (Kijas *et al.*, 2012), had the highest incidence of common RoH per SNP, indicating a significant change in genetic architecture due to selection for pollness in sheep. The White Dorper sheep had the smallest mean number of RoH compared to the other breeds in the study.

In a study using Oxford Nanopore Technology (ONT) sequencing and Hi-C (chromatin conformation capture) approaches, ASE genes involved in the immune system, lipid metabolism, and environment adaptation were identified in Dorper sheep (Qiao *et al.*, 2022). A study by Lv *et al.* (2023) identified candidate genes associated with immunity, reproductive efficiency, and meat quality in Dorper sheep. Zhao *et al.* (2022) using a label-free proteomics approach investigated the genetic basis of body weight in the Dorper and Hu breeds of sheep. They reported several differentially abundant proteins (DAPs) associated with muscle development and fat synthesis that were significantly enriched in kinase activity, metabolism of arachidonic acid, and biosynthesis of steroid hormones. They concluded that body weight is regulated via a variety of pathways in the Dorper and Hu sheep breeds.

Dorper sheep and their crosses have also been studied for their resistance to different diseases. Marshall *et al.* (2013) undertook a genome-wide scan to detect quantitative trait loci (QTL) for resistance to gastrointestinal nematodes in the Dorper sheep. Interesting regions were found in chromosome 26 and chromosome two with putative QTLs for traits of interest. They concluded that favourable QTLs for resistance to disease originated in the Dorper and the Red Maasai sheep, but these were not fixed. A study of genetic mechanisms of Brucellosis disease by Li *et al.* (2021) using whole genome sequences of samples from a population of Dorper × Hu sheep revealed suitable molecular markers for resistance to brucellosis in sheep that can be used in sheep breeding. Another study of polymorphisms associated with Mycoplasma ovipneumonia (MO) by Wang *et al.* (2020) identified that MvaI *bb* and HaeIII *ee* genotypes were predominant in breeds that were susceptive to MO, while resistant breeds like the Dorper sheep carried the MvaI *cc* and HaeIII *dd* genotypes. This indicates that MvaI *cc* and HaeIII *dd* genotypes are associated with MO resistance in sheep.

Results from the diverse studies imply that the Dorper is a resilient sheep breed with the potential to adapt to a variety of environments. In reference to the use of the Dorper in Kenya as an example, most of the local Dorper populations have been established through the initial crossing of imported Dorper rams with local fat-tailed ewes, notably the Red Maasai sheep, and subsequently back-crossing to pure Dorper rams. The Kenya Dorper sheep populations thus carry a degree of DNA inheritance from the local but more resilient Red Maasai sheep (Baker *et al.*, 2003; Muigai *et al.*, 2009). The diversity in the breed reflected in the relatively high degree of resilience in Dorper sheep raised in the harsh Northern and Southern rangelands of Kenya needs to be characterized and exploited through appropriately designed breeding programs.

Challenges and Opportunities for Breeding Programs Involving Dorper Sheep in Africa

Results from studies on use of Dorper sheep across Africa reflect sub-optimal exploitation of the attributes of the Dorper mainly resulting from: (i) *Lack of structured breeding programs*: The absence of organized breeding programs that facilitate sustainable utilization of indigenous breeds in combination with Dorper sheep is a compelling limitation (Zonabend König *et al.*, 2016). Without a structured breeding program, the establishment of clear breeding objectives and selection becomes very challenging. (ii) *The absence of well-defined breeding objectives and selection indices* prevents breeders from making informed decisions regarding crossbreeding with the Dorper breed and hinders progress in archiving desired traits and genetic improvement (Kosgey and Okeyo, 2007; Getachew *et al.*, 2016; Abebe and Alemayehu, 2019). (iii) *Variation in environmental factors:* The impact of diverse environmental conditions, notably effects of high altitudes and large variations in diurnal temperatures on Dorper sheep and their crosses, has not been rigorously evaluated in Africa. Apart from studies in South Africa on the performance of the Dorper under variable environments and feeding systems, in-depth comparative studies on the adaptability and productive potential of Dorper crosses under changing climatic conditions are rarely adopted to drive strategic sheep breeding programs for improved productivity. (iv) *Genetic erosion:* There are concerns about the possible genetic erosion of indigenous breeds due to crossbreeding with the Dorper breed and other introduced breeds (Molotsi *et al.*, 2020). Genetic profiles of sheep populations in pastoral communities of Kenya show massive introgression of Dorper genes in the indigenous populations resulting from haphazard crossbreeding (Muigai *et al.*, 2009), which may also have diluted the adaptive capacity of the indigenous breeds in different communities.

The role of the emerging sciences (Rege *et al.*, 2011) notably, innovative application of genomic technologies, emerging phenotyping methods, and reproductive technologies such as the use of surrogate sires (Gottardo *et al.*, 2019) provide new opportunities for attaining faster genetic improvements in Dorper crosses in pastoral systems. Smart application of the new science would help overcome the technical, organizational, and institutional bottlenecks to genetic improvement of sheep and goat productivity in low input systems.

Establishing structured Dorper breeding programs in different countries of Africa with guidelines on options for exploiting cross-country genetic resources can help facilitate effective and sustainable crossbreeding and improvement of the animals. Through engaging the relevant livestock producers in determining breeding objectives, the risk of undervaluing and potentially deteriorating traits that may contribute in the long term to the overall efficiency and resilience of future production systems is reduced. Community-based approaches to developing and implementing sheep improvement programs (Kosgey and Okeyo, 2007; Mueller *et al.*, 2015; Haile *et al.*, 2019; Mueller *et al.*, 2019; Haile *et al.*, 2020) combined with collaborative use of reference sires could be a more viable means of introducing and propagating more productive and resilient sheep. However, this would require targeted efforts to develop capacity within communities to adopt guided animal monitoring and selection practices for flock productivity gains.

Conclusions

The performance of Dorper sheep and their crosses with indigenous breeds in Eastern Africa and South Africa is extensively documented. Outside of South Africa however, well-defined breeding objectives for using Dorper sheep guided by structured selection indices are limited. High genetic correlations reported between growth traits and net reproductive rates indicate that environments that promote improvements in live weight for young animals are beneficial for their survival. Composite traits such as the total weight of lambs weaned per year as a function of the number of lambs born, their survival, and individual lamb weights at weaning provide a good measure of flock productivity and should be considered when selecting to improve reproduction and fitness in sheep.

Results from genomic studies including Dorper sheep present evidence of markers for resistance to diseases that could be further exploited in developing breeding programs for low-input extensive production environments. Community-based breeding programs supported by smart application of emerging genomic, reproductive, and phenotyping technologies offer new opportunities for realizing faster genetic improvements. There is opportunity for new science and technologies to be innovatively packaged and used to continuously identify and propagate more productive and resilient Dorper and Dorper-based breed-types for the increasingly challenging tropical African range environments.

Author Contributions

The general structure and outline for this review was developed by J.M.K. Ojango who also undertook the lead role in presenting the content for all the sections. M. Okpeku and O. Mwai contributed to the background and the challenges and opportunities for utilizing Dorper Sheep in Africa. R. Osei Amponsah and D.R. Kugonza provided inputs on phenotypic and genetic parameter estimates, while V.E. Olori, D.R Kugonza, and M.G.G. Chagunda provided the content on genomic studies on Dorper sheep. All Authors reviewed the paper for clarity of content.

Acknowledgments

This review was written with support through the CGIAR Research Initiative on Livestock, Climate and Systems Resilience (LCSR) and is supported by contributors to the CGIAR Trust Fund. CGIAR is a global research partnership for a food-secure future dedicated to transforming food, land, and water systems in a climate crisis.

Conflict of Interest

The authors have no conflicts of interest to declare.

References

Abebe, S.A. and Alemayehu, K. (2019) Insights into sheep genetic improvement practices through crossbreeding in Ethiopia: Achievements, challenges and future prospects. *Advances in Life Science and Technology* 75, 1–8. DOI: 10.7176/alst/75-01.

AGTR (2011) Breed information: Dorper sheep. In: Ojango, J.M., Malmfors, B. and Okeyo, A.M. (eds) *Animal Genetics Training Resource, Version 3, 2011*. International Livestock Research Institute, Nairobi, Kenya, and Swedish University of Agricultural Sciences, Uppsala, Sweden. Available at: http://ago.dagris.info/dorper

Assan, N. and Makuza, S.M. (2005) The effect of non-genetic factors on birth weight and weaning weight in three sheep breeds of Zimbabwe. *Asian-Australasian Journal of Animal Sciences* 18(2), 151–157. DOI: 10.5713/ajas.2005.151.

Ayele Abebe, A.A., Solomon Gizaw, S.G., Asfaw Bisrat, A.B., Shenkute Goshme, S.G., Shambel Besufekad, S.B. *et al.* (2015) Growth performance of Dorper and its F 1 Crossbreds at Debre-Birhan Agricultural Research Center. *International Institute for Science, Technology and Education (IISTE): E-Journals* 5(13), 90–98. Available at: https://core.ac.uk/works/70345678 (accessed 7 October 2023).

Ayichew, D. (2019) Dorper sheep cross breeding with Indigenous sheep breed in Ethiopia. *Journal of Applied and Advanced Research* 4, 36–41. DOI: 10.21839/jaar.2019.v4i1.250.

Baker, R.L., Mwamachi, D.M., Audho, J.O., Aduda, E.O. and Thorpe, W. (1999) Genetic resistance to gastro-intestinal nematode parasites in Red Maasai, Dorper and Red Maasai X Dorper ewes in the sub-humid tropics. *Animal Science* 69(2), 335–344. Available at: https://www.cambridge.org/core/product/identifier/S1357729800050906/type/journal_article.

Baker, R.L., Nagda, S., Rodriguez-Zas, S.L., Southey, B.R., Audho, J.O. *et al.* (2003) Resistance and resilience to gastro-intestinal nematode parasites and relationships with productivity of Red Maasai, Dorper and Red Maasai X Dorper crossbred lambs in the sub-humid tropics. *Animal Science* 76(1), 119–136. DOI: 10.1017/s1357729800053388.

Baker, R.L., Mugambi, J.M., Audho, J.O., Carles, A.B. and Thorpe, W. (2004) Genotype by environment interactions for productivity and resistance to gastro-intestinal nematode parasites in Red Maasai and Dorper sheep. *Animal Science* 79(3), 343–353. DOI: 10.1017/S1357729800090214.

Besufkad, S., Goshme, S., Abebe, A., Bisrat, A., Abebe, A. *et al.* (2023) Genetic evaluation of growth rate and efficiency-related traits in Dorper sheep. *Ecological Genetics and Genomics* 28(April), 100186. DOI: 10.1016/j.egg.2023.100186.

Bolowe, M.A., Thutwa, K., Kgwatalala, P.M., Monau, P.I. and Malejane, C. (2022) Production characteristics and management practices of indigenous Tswana sheep in southern districts of Botswana. *Animals* 12(7), 830. DOI: 10.3390/ani12070830.

Cloete, J.J.E., Hoffman, L.C., Claasen, B. and Cloete, S.W.P. (2012) Effect of production system on the growth rate, carcass characteristics and carcass composition of Dorper lambs. *Livestock Research for Rural Development* 24(6), 8.

Cloete, S.W.P., Snyman, M.A. and Herselman, M.J. (2000) Productive performance of Dorper sheep. *Small Ruminant Research* 36(2), 119–135. DOI: 10.1016/S0921-4488(99)00156-X.

Cloete, S.W.P., Thutwa, K., Scholtz, A.J., Cloete, J.J.E., Dzama, K. *et al.* (2021) Breed effects and heterosis for weight traits and tick count in a cross between an indigenous fat-tailed and a commercial sheep breed. *Tropical Animal Health and Production* 53(1), 165. DOI: 10.1007/s11250-021-02612-7.

Cosgrove, N. (2023) Dorper Sheep: Care Guide, Varieties, Pictures and More. Available at: https://petkeen.com/dorper-sheep/ (accessed 25 July 2023).

Deribe, B., Tesema, Z., Lakew, M., Zegeye, A., Alemu, K. *et al.* (2021) Reproductive Performance of Indigenous Sheep and Survival of Their Dorper Crossbred Lambs under Extensive Management in Raya Kobo Area, Ethiopia. Available at: https://www.researchgate.net/publication/352222752

DSBSK (2023) Brief Profile of Dorper Sheep Breeders Society of Kenya. Available at: https://dorperkenya.org/dbsk-profile/ (accessed 7 October 2023).

Dzomba, E.F., Chimonyo, M., Pierneef, R. and Muchadeyi, F.C. (2021) Runs of homozygosity analysis of South African sheep breeds from various production systems investigated using OvineSNP50k data. *BMC Genomics* 22(2021), 7. DOI: 10.1186/s12864-020-07314-2.

FAOSTAT (2023) Population of Sheep and Goats on Sub-Saharan Africa. FAO Statistical Database. Available at: https://www.fao.org/faostat/en/#data/QCL (accessed 21 September 2023).

Getachew, T., Haile, A., Wurzinger, M., Rischkowsky, B., Gizaw, S. *et al.* (2016) Review of sheep crossbreeding based on exotic sires and among indigenous breeds in the tropics: An Ethiopian perspective. *African Journal of Agricultural Research* 11(11), 901–911. DOI: 10.5897/AJAR2013.10626.

Goma, F.Y., Geurden, T., Siwila, J., Phiri, I.G.K., Gabriel, S. *et al.* (2007) The prevalence and molecular characterisation of *Cryptosporidium* spp. in small ruminants in Zambia. *Small Ruminant Research* 72(1), 77–80. DOI: 10.1016/j.smallrumres.2006.08.010.

Goshme, S., Besufekad, S., Abebe, A., Bisrat, A. and Abebe, A. (2021) Reproductive and productive performance of Dorper sheep and their crossbreds with local highland sheep at Debre Birhan agricultural research center, Ethiopia. *Livestock Research for Rural Development* 33, Article #112. Available at: http://www.lrrd.org/lrrd33/9/33112senk.html

Gottardo, P., Gorjanc, G., Battagin, M., Gaynor, R.C., Jenko, J. *et al.* (2019) A strategy to exploit surrogate sire technology in livestock breeding programs. *G3: Genes, Genomes, Genetics* 9(1), 203–215. DOI: 10.1534/g3.118.200890.

Haile, A., Gizaw, S., Getachew, T., Mueller, J.P., Amer, P. *et al.* (2019) Community-based breeding programmes are a viable solution for Ethiopian small ruminant genetic improvement but require public and private investments. *Animal Breeding and Genetics* 136, 319–328. DOI: 10.1111/jbg.12401.

Haile, A., Getachew, T., Mirkena, T., Duguma, G., Gizaw, S. *et al.* (2020) Community-based sheep breeding programs generated substantial genetic gains and socioeconomic benefits. *Animal* 14(7), 1362–1370. DOI: 10.1017/S1751731120000269.

Hoffman, L.C., Claasen, B., van der Merwe, D.A., Petrus Cloete, S.W. and Erasmus Cloete, J.J. (2020) The effects of production system and sex on the sensory quality characteristics of Dorper lamb. *Foods* 9(6), 725. DOI: 10.3390/FOODS9060725.

Inyangala, B.A.O., Rege, J.E.O. and Itulya, S. (1992) Growth traits of the Dorper sheep.II. Genetic and phenotypic parameters-Paramètres de croissance du mouton Dorper. II. Paramètres génétiques et phénotypiques. Available at: https://cgspace.cgiar.org/handle/10568/70832 (accessed 7 October 2023).

Kandiwa, E., Nguarambuka, U., Chitate, F., Samkange, A., Madzingira, O. *et al.* (2020) Production performance of sheep and goat breeds at a farm in a semi-arid region of Namibia. *Tropical Animal Health and Production* 52(5), 2621–2629. DOI: 10.1007/s11250-020-02283-w.

Kao, M.A., Van Wyk, J.B., Scholtz, A.J., Cloete, J.J.E., Matebesi, P.A. *et al.* (2022) Breed and crossbreeding effects on growth, fitness and reproduction of commercial sheep in South Africa. *Small Ruminant Research* 212(1–6), 106705. DOI: 10.1016/j.smallrumres.2022.106705.

Kariuki, C.M., Ilatsia, E.D., Kosgey, I.S. and Kahi, A.K. (2010) Direct and maternal (co)variance components, genetic parameters and annual trends for growth traits of Dorper sheep in semi-arid Kenya. *Tropical Animal Health and Production* 42(3), 473–481.

Kijas, J.W., Lenstra, J.A., Hayes, B., Boitard, S., Porto Neto, L.R. *et al.* (2012) Genome-wide analysis of the world's sheep breeds reveals high levels of historic mixture and strong recent selection. *PLoS Biology* 10(2), e1001258. DOI: 10.1371/journal.pbio.1001258.

Kosgey, I.S. and Okeyo, A.M. (2007) Genetic improvement of small ruminants in low-input, smallholder production systems: Technical and infrastructural issues. *Small Ruminant Research* 70(1), 76–88. DOI: 10.1016/j.smallrumres.2007.01.007.

Kosgey, I.S., Rowlands, G.J., van Arendonk, J.A.M. and Baker, R.L. (2008) Small ruminant production in smallholder and pastoral/extensive farming systems in Kenya. *Small Ruminant Research* 77(1), 11–24. DOI: 10.1016/j.smallrumres.2008.02.005.

Lakew, M., Haile-Melekot, M. and Mekuriaw, G. (2014) Evaluation of Growth Performance of Local and Dorper × Local Crossbred Sheep in Eastern Amhara Region, Ethiopia. *Iranian Journal of Applied Animal Science* 4, 787–794.

Lakew, M., Tessema, Z. and Deribe, B. (2021) Sheep production system and farmers' perception towards Dorper crossbred lambs in Eastern Amhara, Ethiopia. Scientific Papers: Animal Science and Biotechnologies 54(2), 26–37. Available at: https://www.researchgate.net/publication/ 355807599.

Liljestrand, J. (2012) Breeding Practices of Red Maasai sheep in Maasai Pastoralist Communities. SLU, Department of Animal Breeding and Genetics, Uppsala. Available at: https://cgspace.cgiar.org/handle/10568/32708.

Li, X., Wu, Q., Zhang, X., Li,, C., Zhang, D. *et al.* (2021) Whole-genome resequencing to study brucellosis susceptibility in sheep. *Frontiers in Genetics* 12. DOI: 10.3389/fgene.2021.653927.

Lv, X., Chen, W., Wang, S., Cao, X., Yuan, Z. *et al.* (2023) Whole-genome resequencing of Dorper and Hu sheep to reveal selection signatures associated with important traits. *Animal Biotechnology* 34(7), 3016–3026. DOI: 10.1080/10495398.2022.2127409.

Marshall, K., Mugambi, J.M., Nagda, S., Sonstegard, T.S., Van Tassell, C.P. *et al.* (2013) Quantitative trait loci for resistance to *Haemonchus contortus* artificial challenge in Red Maasai and Dorper sheep of East Africa. *Animal Genetics* 44(3), 285–295. DOI: 10.1111/j.1365-2052.2012.02401.x.

Matika, O., Nyoni, S., Van Wyk, J.B., Erasmus, G.J. and Baker, R.L. (2003) Resistance of Sabi and Dorper ewes to gastro-intestinal nematode infections in an African semi-arid environment. *Small Ruminant Research* 47(2), 95–102. DOI: 10.1016/S0921-4488(02)00251-1.

Mengist, W., Soromessa, T. and Legese, G. (2020) Method for conducting systematic literature review and meta-analysis for environmental science research. *MethodsX* 7, 100777. DOI: 10.1016/j.mex.2019.100777.

Milne, C. (2000) The history of the Dorper sheep. *Small Ruminant Research* 36(2), 99–102. DOI: 10.1016/S0921-4488(99)00154-6.

Molotsi, A.H., Dube, B. and Cloete, S.W.P. (2020) The current status of indigenous ovine genetic resources in Southern Africa and future sustainable utilisation to improve livelihoods. *Diversity* 12(1), 14. DOI: 10.3390/d12010014.

Molotsi, A.H., Taylor, J.F., Cloete, S.W., Muchadeyi, F., Decker, J.E. *et al.* (2017) Genetic diversity and population structure of South African smallholder farmer sheep breeds determined using the OvineSNP50 beadchip. *Tropical Animal Health and Production* 49(8), 1771–1777. DOI: 10.1007/s11250-017-1392-7.

Mueller, J.P., Rischkowsky, B., Haile, A., Philipsson, J., Mwai, O. *et al.* (2015) Community-based livestock breeding programmes: Essentials and examples. *Journal of Animal Breeding and Genetics* 132, 155–168.

Mueller, J.P., Haile, A., Getachew, T., Rekik, M. and Rischkowsky, B. (2019) Genetic progress and economic benefit of community-based breeding programs for sheep out- and upscaling options in Ethiopia. *Small Ruminant Research* 177, 124–132. DOI: 10.1016/j.smallrumres.2019.06.025.

Mugambi, J.M., Audho, J.O. and Baker, R.L. (2005) Evaluation of the phenotypic performance of a Red Maasai and Dorper double backcross resource population: Natural pasture challenge with gastrointestinal nematode parasites. *Small Ruminant Research* 56(1–3), 239–251.

Muigai, A.W.T., Okeyo, A.M., Kwallah, A.K., Mburu, D. and Hanotte, O. (2009) Characterization of sheep populations of Kenya using microsatellite markers: Implications for conservation and management of indigenous sheep populations. *South African Journal of Animal Science* 39, 93–96.

Mwenya, W.N.M. (1994) Sheep production in Zambia: A review of sheep breeding. In: *In Small Ruminant Research and Development in Africa: Proceedings of the Second Biennial Conference of the African Small Ruminant Research Network: AICC*. Arusha, Tanzania, pp. 7–11.

Nxumalo, K.S., Grobler, P., Ehlers, K., Nesengani, L.T. and Mapholi, N.O. (2022) The genetic assessment of South African Nguni sheep breeds using the ovine 50K chip. *Agriculture (Switzerland)* 12(5), 663. DOI: 10.3390/agriculture12050663.

Ojango, J.M.K., Oyieng, E., Oloo, R.D., Sila, W., Audho, J. *et al.* (2023) Evaluating fertility in a breeding program for improved growth of indigenous sheep in the arid lands of Kenya. In: *8th All Africa Conference on Animal Agriculture*. Gabarone, Botswana. Available at: https://cgspace.cgiar.org/handle/10568/132001 (accessed 31 October 2023).

Oyieng, E., Mrode, R., Ojango, J.M.K., Ekine-Dzivenu, C.C., Audho, J. *et al.* (2022) Genetic parameters and genetic trends for growth traits of the Red Maasai sheep and its crosses to Dorper sheep under extensive production system in Kenya. *Small Ruminant Research* 206, 106588. DOI: 10.1016/J.SMALLRUMRES.2021.106588.

Patricia Walker, K. (2013) Performance of Dorper lambs fed buffel grass and browse as basal diet and sorghum bran as supplement. *International Journal of Plant and Animal Sciences* 1(2), 37–41.

Qiao, G., Tingting, G., Pan, X., Xu, P., Wu, Y. *et al.* (2022) Genetic Basis of Dorper Sheep (*Ovis aries*) Revealed by Long-Read De Novo Genome Assembly. *Frontiers in Genetics* 13, 2022. DOI: 10.3389/fgene.2022.846449.

Ramsay, K., Swart, D., Olivier, B. and Hallowell, G. (2000) An evaluation of the breeding strategies used in the development of the Dorper sheep and the improved Boer goat of South Africa. *ICAR Technical Series* 3(3), 339–346.

Rege, J.E.O., Marshall, K., Notenbaert, A., Ojango, J.M.K. and Okeyo, A.M. (2011) Pro-poor animal improvement and breeding – What can science do? *Livestock Science* 136(1), 15–28. DOI: 10.1016/j.livsci.2010.09.003.

Schoeman, S.J. (2000) A comparative assessment of Dorper sheep in different production environments and systems. *Small Ruminant Research* 36(2), 137–146. DOI: 10.1016/S0921-4488(99)00157-1.

Teklebrhan, T., Urge, M., Mekasha, Y. and Baissa, M. (2014) Pre-weaning growth performance of crossbred lambs (Dorper × indigenous sheep breeds) under semi-intensive management in eastern Ethiopia. *Tropical Animal Health and Production* 46(2), 455–460. DOI: 10.1007/s11250-013-0513-1.

Tembely, S., Lahlou-Kassi, A., Rege, J.E., Mukasa-Mugerwa, E., Anindo, D. *et al.* (1998) Breed and season effects on the peri-parturient rise in nematode egg output in indigenous ewes in a cool tropical environment. *Veterinary Parasitology* 77(2–3), 123–132.

Tesema, Z., Deribe, B., Kefale, A., Lakew, M., Tilahun, M. *et al.* (2020) Survival analysis and reproductive performance of Dorper × Tumele sheep. *Heliyon* 6(4). DOI: 10.1016/j.heliyon.2020.e03840.

Tesema, Z., Deribe, B., Kefale, A., Lakew, M., Tilahun, M. *et al.* (2021) In: *13th Annual Regional Conference on Completed Livestock Research activities. Ethiopia*, pp. 138–152.

Tesema, Z., Deribe, B., Lakew, M., Getachew, T., Tilahun, M. *et al.* (2022) Genetic and non-genetic parameter estimates for growth traits and Kleiber ratios in Dorper × indigenous sheep. *Animal* 16(6), 100533. DOI: 10.1016/j.animal.2022.100533.

Wang, K., Liu, X., Li, Q., Wan, K., Gao, R. *et al.* (2020) MHC-DRB1 exon 2 polymorphism and its association with mycoplasma ovipneumonia resistance or susceptibility genotypes in sheep. *Journal of Genetics* 99, 22. PMID 32366733.

Wang, S., Zhang, D., Zhou, H., Pan, J., Zhang, Y. *et al.* (2021) Screening of genes coupled to heat response in Mongolian and Dorper sheep breeds. *Biologia* 76(3), 949–958. DOI: 10.2478/s11756-020-00616-6.

Wanjala, G., Astuti, P.K., Bagi, Z., Kichamu, N., Strausz, P. *et al.* (2023a) Assessing the genomics structure of Dorper and white Dorper variants, and Dorper populations in South Africa and Hungary. *Biology* 12(3), 386. DOI: 10.3390/biology12030386.

Wanjala, G., Kichamu, N., Cziszter, L.T., Astuti, P.K. and Kusza, S. (2023b) An on-station analysis of factors affecting growth traits of pure red Maasai and Dorper sheep breeds under an extensive production system. *Animals* 13(2), 300. DOI: 10.3390/ani13020300.

Wanjala, G., Kichamu, N., Strausz, P., Astuti, P.K. and Kusza, S. (2023c) On-station comparative analysis of reproductive and survival performance between Red Maasai, Dorper, and Merino sheep breeds. *Animal* 17(3), 100715. DOI: 10.1016/j.animal.2023.100715.

Weldeyesus, G. (2017) Improving live body weight gain of local sheep through crossbreeding with high yielding exotic Dorper sheep under smallholder farmers. *International Journal of Livestock Production* 8(5), 67–71. DOI: 10.5897/ijlp2016.0316.

Zhao, L., Zhang, D., Li, X., Zhang, Y., Zhao, Y. *et al.* (2022) Comparative proteomics reveals genetic mechanisms of body weight in Hu sheep and Dorper sheep. *Journal of Proteomics* 267, 104699. DOI: 10.1016/j.jprot.2022.104699.

Zishiri, O.T., Cloete, S.W.P., Olivier, J.J. and Dzama, K. (2013) Genetic parameters for growth, reproduction and fitness traits in the South African Dorper sheep breed. *Small Ruminant Research* 112(1), 39–48.

Zonabend König, E., Strandberg, E., Ojango, J.M.K., Mirkena, T., Okeyo, A.M. *et al.* (2017) Purebreeding of Red Maasai and crossbreeding with Dorper sheep in different environments in Kenya. *Journal of Animal Breeding and Genetics* 134(6), 531–544. DOI: 10.1111/jbg.12260.

Zonabend König, E., Mirkena, T., Strandberg, E., Audho, J., Ojango, J. *et al.* (2016) Participatory definition of breeding objectives for sheep breeds under pastoral systems—The case of Red Maasai and Dorper sheep in Kenya. *Tropical Animal Health and Production* 48(1), 9–20. DOI: 10.1007/s11250-015-0911-7.

Zonabend König, E., Ojango, J.M.K., Audho, J., Mirkena, T., Strandberg, E. *et al.* (2017) Live weight, conformation, carcass traits and economic values of ram lambs of Red Maasai and Dorper sheep and their crosses. *Tropical Animal Health and Production* 49, 121–129. DOI: 10.1007/s11250-016-1168-5.

6 Cultivars of Arabica coffee: Characterization and interception of solar radiation

Lídia Raiza Sousa Lima Chaves Trindade[1]*, Marcelo de Carvalho Alves[1] and Luciana Sanches[2]

[1]Department of Engineering, Federal University of Lavras, Campus Universitário, PO Box 3037, 37200-000, Lavras, Brazil; [2]Department of Environmental and Sanitary Engineering, Federal University of Mato Grosso, UFMT Campus, Cuiabá, Mato Grosso, 78060-900, Brazil

Abstract

The importance of coffee crops in the global agricultural scenario goes beyond economic aspects, also encompassing political and sociocultural factors. Brazil stands out in this scenario as the largest producer and exporter of coffee in the world, with a particular emphasis on the state of Minas Gerais. Coffee production is mainly focused on the *Coffea arabica* L. species. However, several factors directly impact coffee crop productivity, resulting in significant losses. Among the most relevant factors are adverse climatic conditions, nutritional deficiencies, and the presence of pests and diseases. Therefore, it is essential to seek an efficient metabolic response from coffee crops in the face of biotic and abiotic stresses, developing plants that are more resistant to climate variations. Based on the literature, this review is structured as follows: first, the characteristics and growth habits of coffee crops are presented; second, 18 cultivars of Arabica coffee analyzed in the study are discussed, providing specific agronomic information for each of them; subsequently, the coffee's phenological cycle and productivity are addressed; finally, radiometric variables such as photosynthetically active radiation (PAR), fraction of absorbed photosynthetically active radiation (f_{APAR}), transmittance, reflectance, and leaf area index (LAI) are treated. The evidence gathered in this review unanimously points to the need for intensified studies on solar radiation interception and LAI in different cultivars of Arabica coffee. The literature highlights the importance of using field-obtained datasets to complement information on the characteristics and growth patterns of varieties, as well as their phenological cycle, in order to understand the plant's behavior before and after harvesting and its relation to productivity.

Keywords: coffee crop, phenological cycle, radiometric variables

Introduction

The coffee crop (*Coffea arabica* L.) originates from the mountainous forests of Ethiopia, south-eastern Sudan, and northern Kenya (Aerts *et al.*, 2017). Its domestication took place in Yemen, and its spread to Asia and Latin America occurred during the Renaissance

*Corresponding author: lidiaraiza@hotmail.com

period (1300–1700), leading to the expansion of coffee cultivation throughout Europe and the Americas (Aerts *et al.*, 2017). Coffee was introduced to the country in 1727, starting in the state of Para, from where it spread across the entire national territory (Alves *et al.*, 2017).

The habit of drinking coffee was transmitted by the Arabs, and its importance in the global agricultural scenario goes beyond just economic aspects, encompassing political and sociocultural factors as well (Aerts *et al.*, 2017). Coffee is one of the most consumed and appreciated products worldwide, making a significant contribution to the economies of coffee-producing and exporting countries (Toledo *et al.*, 2017). Brazil stands out in this scenario, being considered the largest producer and exporter of coffee in the world. In 2023, coffee cultivation spans a total area of 2.24 million hectares in Brazil, with 1.88 million hectares dedicated to active plantations, marking a 1.9% increase from the previous year, while 362.5 thousand hectares are under development. In the current harvest, the production of Arabica coffee in the country is projected to reach 38.16 million bags, representing 70.2% of the total, which signifies a 16.6% increase compared to the 2022 harvest. This growth is the result of a 2.4% expansion in the cultivation area, along with an anticipated 13.9% increase in productivity, thanks to more favorable weather conditions compared to the last two harvests (Companhia Nacional de Abastecimento (CONAB), 2023).

For 2023, there was an increase of 28.8% in comparison to the total quantity harvested in the previous harvest, which can be explained by a 6.5% expansion in the cultivation area, a substantial 21% increase in production efficiency, and, above all, due to the improvement in the conditions of the plantations after the last harvests, which were affected by unfavorable weather conditions (Companhia Nacional de Abastecimento (CONAB), 2023).

Coffee cultivation in Minas Gerais is responsible for over 50% of the national production and is present in four main regions: Cerrado (Triângulo Mineiro/Alto Paranaíba), Sul de Minas (Southern/Southwestern), Chapadas de Minas (Vale do Jequitinhonha/Mucuri), and Matas de Minas (Zona da Mata/Rio Doce) (Companhia Nacional de Abastecimento (CONAB), 2023).

The coffees produced in Minas Gerais have distinct flavor and aroma characteristics, influenced by the peculiarities of climate, altitude, and production systems adopted in each region (Barbosa *et al.*, 2010). Coffee belongs to the *Coffea* genus, and two species stand out in the global market: *Coffea arabica* and *Coffea canephora*, which have distinct characteristics.

In Minas Gerais, coffee production is primarily focused on cultivating Arabica coffee, which represents approximately 69% of the national production of this type of coffee. The state has a coffee-growing area of 1210 thousand hectares (Companhia Nacional de Abastecimento (CONAB), 2023). Arabica coffee beans are highly valued in the market due to their balance of desirable chemical compounds and their ability to offer a beverage of superior quality, making them more appreciated by consumers (Ribeiro *et al.*, 2014).

Several causes directly impact coffee crop productivity, leading to significant losses. Among the most relevant factors are adverse climatic conditions, nutritional deficiencies, and the presence of pests and diseases (Oestreich-Janzen, 2010). Some of the significant coffee diseases include orange rust caused by *Hemileia vastatrix* Berk. & Br., cercospora leaf spot caused by *Cercospora coffeicola* Berk & Cooke, coffee berry anthracnose caused by *Colletotrichum kahawae* Waller & Bridge (considered a quarantine pest A1 in Brazil), *Meloidogyne* spp., Phoma spot caused by *Phoma tarda* (Stewart) Boerema & Bollen, and aureolate spot caused by *Pseudomonas syringae* pv. *garcae* Young, Dye & Wilkie, among others (Zambolim *et al.*, 2005; Patrício *et al.*, 2008).

The climatic conditions of the environment where coffee crops are cultivated can affect productivity due to variations in photosynthetic activity (Souza *et al.*, 2017). There are numerous coffee genotypes cultivated in different countries worldwide. The selection of plants adapted to different conditions has been crucial for choosing the varieties used (Gomes *et al.*, 2008). The *Coffea arabica* L. genus exhibits high phenotypic plasticity, enabling its adaptation to diverse climatic conditions, including temperature and radiation.

According to the fifth report of the IPCC (Intergovernmental Panel on Climate Change), the global average temperature has increased by 0.61°C since pre-industrial time, and this increase is expected to continue in the coming years (Intergovernmental Panel on Climate Change (IPCC), 2014). The projected climate changes would have severe impacts on coffee bean production and quality, especially for the species *Coffea arabica* L. (Botelho *et al.*, 2015; Läderach *et al.*, 2017), which is adapted to average annual temperatures of 18–22°C (Gesisky *et al.*, 2008; Camargo, 2010), and experiences water deficit of less than 100 mm annually (Zullo *et al.*, 2011).

In Minas Gerais, the climatological average of monthly rainfall totals varies from 70 to 310 mm, with average minimum and maximum temperatures ranging from 16 to 18°C and 32 to 34°C, respectively. In the mountainous regions of the state, the average minimum and maximum temperatures range from 16 to 18°C and 26 to 29°C. The occurrence of dry spells in the months of January, February, and March has been common, especially in the Central North and East regions of the state (Companhia Nacional de Abastecimento (CONAB), 2023). Climate zoning studies have shown that the increase in global temperature may render many regions, especially in south-eastern Brazil, unsuitable for the cultivation of *Coffea arabica* L., or require the adoption of irrigation practices (Assad *et al.*, 2004; Bragança *et al.*, 2016; Tavares *et al.*, 2018).

A 38% reduction in bioclimatic areas favorable for Arabica coffee production is predicted by 2080 in Ethiopia due to the climate changes that have occurred in recent decades. The rise in temperature, changes in the distribution of rainfall throughout the year, and increased carbon dioxide (CO_2) emissions have impacted commercial coffee farming (Miniussi *et al.*, 2015). In recent years, global minimum temperatures have increased twice as much as maximum temperatures, posing a threat to tropical crops like Arabica coffee (Craparo *et al.*, 2015). It is necessary to seek an excellent metabolic response from coffee crops to biotic and abiotic stresses, developing more resilient plants to climate variations (Aerts *et al.*, 2017).

Based on this perception, there is a clear research need that the current study aims to address: to analyze existing research that explores the characteristics and behavior of different Arabica coffee cultivars regarding solar radiation interception. This knowledge gap is crucial, both from a productive and ecological perspective, as understanding the impact of solar radiation throughout the coffee phenological cycle is essential to maintain and optimize its productivity.

This study, structured based on the literature, specifically addresses this research need. Initially, the characteristics and growth habits of the coffee crop are presented. Next, different cultivars widely used commercially are analyzed in the study, detailing the agronomic information and characteristics of each. Subsequently, the focus shifts to the coffee's phenological cycle and its productivity, with the aim of thoroughly understanding the various stages of plant development and growth over time. Finally, radiometric variables, including photosynthetically active radiation (PAR), fraction of absorbed photosynthetically active radiation (f_{APAR}), transmittance, reflectance, and leaf area index (LAI), are meticulously explored to unravel how solar radiation interception influences plant growth and development, as well as the relationship of these variables with climatic factors and productivity.

Review Methodology

The objective of this study is to present a literature review on the agronomic characteristics and growth habits of different cultivars of Arabica coffee. Additionally, the biophysical parameters that can help understand the temporal behavior of the plant throughout its phenological cycle were discussed. The research was conducted using various databases, including Scopus (Elsevier), Web of Science (WoS) from Clarivate Analytics, Google Scholar, ScienceDirect (Elsevier), DOAJ (Directory of Open Access Journals), Springer Science and Scientific Electronic Library Online (SciELO). The article was organized into topics that address the characteristics and growth habits of the plant, different cultivars of Arabica coffee, the coffee's phenological cycle and productivity, as well as issues related to radiation, radiometric variables such as transmittance, reflectance, photosynthetically active radiation (PAR), and fraction of absorbed photosynthetically active radiation (f_{APAR}), in addition to the leaf area index (LAI).

Additionally, the references from the articles were used to identify additional relevant material. This strategy allowed for the aggregation of complementary and relevant information related to the subject at hand.

Characteristics of the plant and growth habits

The coffee crop is a perennial tropical dicotyledonous shrub with a more upright stem structure. Its height can vary from 4 to 6m without pruning or from two to three meters with pruning (Carvalho *et al.*, 2010).

Regarding carbon metabolism, coffee is considered a C3 plant, and most of these plants show photosynthetic saturation between 600 and 800 μmol m^{-2} s^{-1}. According to Charbonnier *et al.* (2017), when atmospheric CO_2 concentrations increase above the compensation point without saturation, photosynthesis is stimulated, and the photon flux occurs linearly in relation to the photosynthetic rate and biomass, resulting in an increase in plant productivity. Coffee reaches its maximum yield about 4 years after planting. It is worth noting that crops are typically renewed every 16 years, although some plantations remain productive for decades (Frederico, 2014).

There are differences among coffee genotypes, which can manifest as small bushes or large trees with hard and thick stems. All species in the *Coffea* genus are diploids, except for *C. arabica*, which is tetraploid and the only species with four sets of the genus's basic chromosomal number (n=11), totaling 44 chromosomes. The stem is woody, rigid, and erect, with *C. arabica* having a single stem (monocaul) and *C. canephora* having multiple stems (multicaul), with diameters ranging from 8 to 10 cm (Rodrigues *et al.*, 2012).

There is dimorphism in the growth pattern of the branches, known as orthotropic branches when they grow vertically and plagiotropic branches when they grow horizontally. The leaves have short petioles, are elliptical, and typically range from dark green to greenish-brown, with 9–12 veins and a leaf surface area of 12–20 cm^2 in mature plants (Damatta, 2004).

The coffee plant's root system consists of a taproot that can reach up to 1m deep. Additionally, there are auxiliary roots that can extend up to 3m deep. The coffee crop has several lateral roots parallel to the ground, located about 2m from the trunk, as well as root hairs. Approximately 75% of the secondary and adventitious roots are less than 30 cm deep and are concentrated in the protruding regions of the canopy, performing functions of plant support and water and nutrient absorption (Covre *et al.*, 2015).

The coffee fruit has an oval shape and usually contains two seeds. The maturation of the fruit occurs over a period of 7–9 months, and the seeds have a dark red or yellow color when ripe. The fruit is composed of the peduncle, crown, exocarp, mesocarp, endocarp, and seed. The seed is surrounded by an inner shell called parchment, which is coated with the endosperm (silver skin), mesocarp (pulp), and exocarp (skin). The seed consists of the embryo, endosperm, and tegument (Damatta, 2004).

Years of low productivity are associated with the allocation of photosynthetic resources to branching (Mendonça *et al.*, 2011). Genetic variation is one of the determining factors for differences among coffee varieties (Hein and Gatzweiler, 2006; Kitzberger *et al.*, 2014). However, there is a growing demand for genotypes that can adapt to specific conditions and demonstrate satisfactory productive results.

It is known that the coffee crop has a significant adaptive capacity, leading to physiological, biochemical, morphological, and anatomical changes (Silveira, 2014). Therefore, there is an increasing demand for cultivars adapted to different environmental stress conditions, climatic factors, and resistance to pests and diseases, as global climate change and increased atmospheric carbon dioxide may pose risks to coffee cultivation (Chemura *et al.*, 2017).

Arabica Coffee Cultivars

Brazil is a country known for its diversity of Arabica coffee cultivars available for commercial use. Currently, there are 142 cultivars registered in the National Registry of Varieties (RNC) of the Ministry of Agriculture, Livestock, and Supply

(MAPA), which is responsible for the registration and protection of these cultivars. However, only about 40 cultivars are widely used commercially. This is partly due to a lack of understanding of the benefits, agronomic performance, and characteristics of the new cultivars available for cultivation (Carvalho *et al.*, 2022).

It is essential to know the characteristics and agronomic information of Arabica coffee cultivars to make an appropriate decision in choosing the most suitable variety for planting. Table 6.1 presents some characteristics, such as productivity, plant size, canopy diameter, vegetative vigor, maturation period, seed size, and tolerance to water deficit, of the cultivars used in this study. This information is valuable in assisting coffee growers in selecting the most suitable cultivar for their needs and environmental cultivation conditions.

Phenological Cycle of Coffee and Productivity

Most plants bloom in the spring and bear fruit in the same phenological year. Arabica coffee is a special plant that takes two years to complete its phenological cycle. First, during the months of long days, vegetative branches with axillary buds form at the nodes. In January, when the days start to shorten, the axillary vegetative buds are stimulated by photoperiods in the reproductive buds (Gouveia, 1984).

The second phenological year begins with flowering, and the small fruits precede the inflation of the beans until they reach their normal size. Then comes the grain filling and ripening phase. And finally, aging, death of terminal plagiotropic branches, and familiar self-pruning. In the spring of the following civil year, new vegetative buds germinate and begin to spread, allowing for new late growth production in the following year (Camargo and Camargo, 2001).

Fig. 6.1 presents a detailed diagram of the phenological stages of Arabica coffee (*Coffea arabica* L.) according to Camargo and Camargo (2001). For the tropical conditions of Brazil, the phenological cycle is divided into six distinct stages, two vegetative and four reproductive: (1) plant formation and leaf budding; (2) induction and maturation of floral buds; (3) inflorescence (4) grain filling; (5) fruit ripening; and (6) dormancy and aging of the third and fourth branches.

Studies conducted in Campinas (Sondahl and Sharp, 1979) and Chinchiná, Colombia (Gomez-Gomez, 1977) have shown that Arabica coffee fruits mature approximately in the 32nd week after anthesis. However, the time required to reach full maturity is influenced by climatic conditions and the genetic composition of the coffee crop (Sondahl and Sharp, 1979).

Arabica coffee is native to shaded environments, where it exhibits lower photosynthetic rates, even under optimal cultivation conditions. However, the spatial and temporal fluctuations in photosynthesis and the reasons behind the low photosynthetic rates are not fully understood (Damatta, 2018). Therefore, it is of utmost importance to study the biophysical parameters and their impact on Arabica coffee.

Radiation

The accumulation of dry matter in plants of different crops is directly related to solar interception, as pointed out by Monteith (1972) and Monteith (1977). The efficiency of light interception can be a crucial factor influencing crop performance under different climatic conditions, including within forests. The amount of solar energy captured by plants depends on the characteristics of the incident solar radiation and canopy properties. To understand and model crop growth and production, it is essential to use light interception or radiative transfer models, which combine vegetation characteristics with how radiation is attenuated by the canopy (Sinclair and Muchow, 1999).

In general terms, interception can be understood as the process by which solar radiation strikes a certain surface. For the leaf surface to be intercepted at any point in the canopy, the solar radiation incident on the top of the canopy must be transmitted. Radiative transfer along the canopy depends on factors such as the type of radiation (direct or diffuse), optical properties of the plants, and canopy structure. According to Ross (1975), the optical properties of plants are characterized by the coefficients of reflection

Table 6.1. Characteristics and agronomic information about Arabica coffee cultivars.

Characteristics	**Acaua**	**Acaua Novo**	**Aranas RV**	**Araponga MG1**
Plant size	Low	Low	Low	Low
Canopy diameter	Medium	Medium	Medium	Medium
Color of young leaves	Bronze	Green	Green/Bronze	Green/Bronze
Fruit color at maturity	Red	Red	Red	Red
Seed size	Medium	Medium	Large	Medium
Maturation period	Late	Med./Late	Med./Late	Medium
Rust resistance	Resistant	Resistant	Part. resist	Resistant
Nematode resistance	Resistant to *M. exigua*	Resistant to *M. exigua*	Resistant to *M. exigua*	Susceptible
Other diseases	None	None	None	None
Vegetative vigor	High	High	High	High
Cup quality	Regular	Regular	Regular	Regular
Productivity	High	High	High	Medium
Characteristics	**Catigua MG1**	**Catigua MG2**	**Catigua MG3**	**Catucai Amarelo 2SL**
Plant size	Low	Low	Low	Low
Canopy diameter	Medium	Medium	Medium	Medium
Color of young leaves	Bronze	Green/Bronze	Light Bronze	Green/Bronze
Fruit color at maturity	Red	Red	Red	Yellow
Seed size	Medium	Small	Medium	Medium
Maturation period	Medium	Med./Late	Medium	Medium
Rust resistance	Resistant	Resistant	Resistant	Part. resistant
Nematode resistance	Susceptible	Susceptible	Resistant to *M. exigua*	Susceptible
Other diseases	None	Resistant to *Pseudomonas*	None	Partially to *Phoma* spot
Vegetative vigor	High	High	High	High
Cup quality	Regular	Distinguished	Regular	Regular
Productivity	Medium	Medium	High	High
Characteristics	**Iapar 59**	**IPR 100**	**IPR 102**	**IPR 103**
Plant size	Low	Low	Low	Low
Canopy diameter	Small	Medium	Medium	Medium
Color of young leaves	Bronze	Bronze	Green/Bronze	Bronze
Fruit color at maturity	Red	Red	Red	Red
Seed size	Medium	Medium	Large	Medium

Continued

Table 6.1. Continued

Maturation period	Early/Medium	Late	Late	Late
Rust resistance	Resistant	Susceptible	Resistant	Part. resistant
Nematode resistance	Resistant to *M. exigua*	Resistant to *M. paranaensis* and *M. exigua*	Susceptible to *M. paranaensis* and *M. incognita*. Not tested for *M. exigua*	Susceptible to *M. paranaensis*, *M. incognita*. Not tested for *M. exigua*
Other diseases	Partially to bacterial halo blight	None	Resistant to aureolate spot and *Phoma* spot	None
Vegetative vigor	Low	High	Medium	High
Cup quality	Distinguished	Regular	Regular	Regular
Productivity	High	Very high	High	Very high
Characteristics	**Rubi – MG 1192**	**Siriema**	**Saira**	**Arara**
Plant size	Low	Low	Low	Low
Canopy diameter	Medium	Small	Medium	Large
Color of young leaves	Bronze (predominant)	Green	Green/Bronze	Green
Fruit color at maturity	Red	Red/Yellow	Red	Yellow
Seed size	Medium	Small	Medium	Large
Maturation period	Medium	Early	Med./Late	Late
Rust resistance	Susceptible	Part. resistant	Resistant	Resistant
Nematode resistance	Susceptible	Susceptible	Susceptible	Susceptible
Other diseases	None	Resistant to leaf miner in about 35% of plants	None	Resistant to *Pseudomonas*
Vegetative vigor	High	High	High	High
Cup quality	Regular	Regular	Regular	Distinguished
Productivity	Medium	Medium	High	Very high
Characteristics	**Guara**	**Pau Brasil MG1**		
Plant size	Low	Low		
Canopy diameter	Large	Small		
Color of young leaves	Green/Bronze	Green		
Fruit color at maturity	Red	Red		
Seed size	Large	Medium		
Maturation period	Medium	Medium		
Rust resistance	Part. resistant	Resistant		

Continued

Table 6.1. Continued

Nematode resistance	Moderately resistant to *M. exigua*	Susceptible
Other diseases	None	None
Vegetative vigor	High	Medium
Cup quality	Regular	Regular
Productivity	High	Very high

Adapted from Carvalho *et al.* (2022).

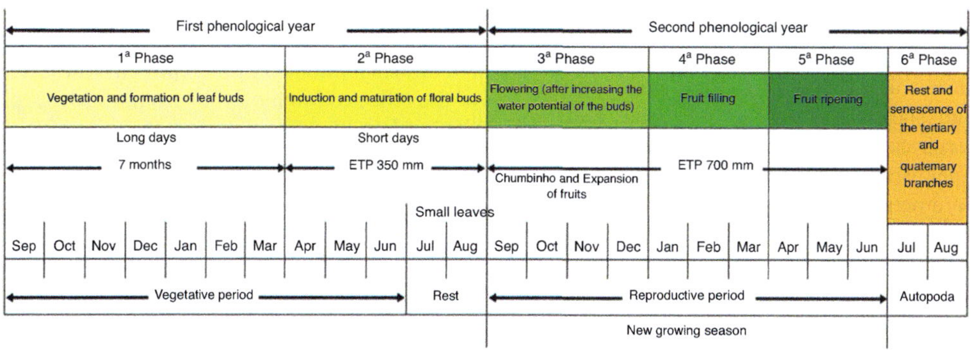

Fig. 6.1. The six phenological stages of *Coffea arabica* L., over 24 months, in the tropical climatic conditions of Brazil. Source: Adapted from Camargo and Camargo (2001).

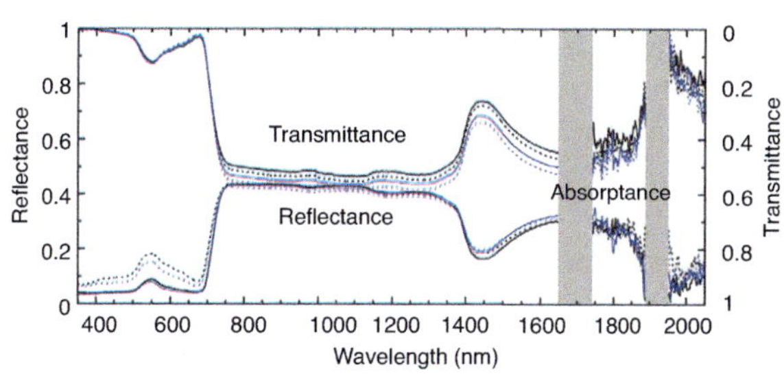

Fig. 6.2. Relationship between reflection, transmission, and absorption of a green leaf systematized. Source: Adapted from Monteith (2013).

(ρ), transmission (τ), and absorption (α), as shown in the following Eqn 1:

$$\alpha = 1 - \rho - \tau \quad (1)$$

Furthermore, when it comes to radiation transfer, it is also common to use the term scattering coefficient (ω), which can be calculated using Eqn 2:

$$\omega = \rho + \tau \quad (2)$$

Although the visual characteristics of leaves may vary with age, structure, and the presence of diseases, it is possible to create a generic idealization of the optical behavior of healthy green leaves (Fig. 6.2).

Fig. 6.2 shows a significant absorption of radiation in the visible range (0.4–0.7 μm), reaching around 90% absorption. From 0.7 μm, the absorption decreases drastically, reaching a minimum of 10% between 0.7 and 1.2 μm. From 1.2 μm, the absorption increases again at 1.5, 2.1, and 2.5 μm (Monteith, 2013). In addition to optical properties, the transmission of radiation also depends on the canopy structure. Ross and Ross (1981) defined 'architecture' as the set of characteristics that define the shape, size, geometry, and external structure of a plant.

The calculation of solar radiation can be performed at any point in the Earth's outer atmosphere for each day of the year, based on the latitude and solar declination. However, gases and clouds cause changes in the amplitude and spectral composition of solar radiation. Empirical models are used to estimate global solar irradiance, which is a useful tool when parameters can be calibrated at different locations. These models have the advantage of using widely available meteorological data (Meza and Varas, 2000).

According to Gröll (2008), crop reflectance has been studied in the visible region (400–700 nm) and near-infrared (760–1300 nm) for plant health detection. In healthy plants, in the visible region, chlorophyll absorbs higher amounts of blue and red radiation in the spectrum, while green is more reflected, giving healthy plants their natural green appearance. In the near-infrared, radiation is also strongly reflected due to the internal leaf structure.

Radiometric Quantities

Radiation measurements are essential when analyzing the interaction of electromagnetic radiation with the Earth's surface. The quantities that describe the radiation field resulting from the interaction with various surfaces are reflectance, transmittance, and absorption.

Transmittance

Understood as the fraction of radiant energy that penetrates to reach parallel levels within the medium. It can also be defined as the ratio of the incoming solar radiation flux (Rad_0) within the canopy to the incoming radiation flux above the canopy (Rad_1), according to Eqn 3.

$$t = \frac{Rad_0}{Rad_1} \tag{3}$$

Transmittance is related to the canopy openness fraction and is directly affected by the zenith angle and the amount of diffuse radiation (Gower *et al.*, 1999).

Reflectance

Derived from the Latin word "reflectu + ância," it is a term in photometry that represents the ratio of the reflected luminous flux from a surface to the incident luminous flux, i.e., the ratio of the radiant flux reflected from the surface (Rad_r) to the incident radiant flux (Rad_i), according to Eqn 4.

$$r = \frac{Rad_r}{Rad_i} \tag{4}$$

Reflectance represents the fraction of the total incident solar radiation that is reflected by a surface. It is a variable used to study the interaction of solar radiation with the Earth. Changes caused by human activities in land cover and albedo have an impact on the proportion of solar radiation reflected back into space, making it an important radiative climate variable (Kaufman *et al.*, 1998).

The albedo of vegetated surfaces varies according to solar elevation angle, vegetation type, atmospheric conditions, surface moisture, soil type, humidity, and cloud amount and type (Azevedo *et al.*, 1990). In the case of forest vegetation, when the foliage is clustered in the canopy, the canopy surface presents crevices and depressions, allowing the penetration of a significant amount of incident solar radiation before being reflected (Shuttleworth *et al.*, 1984).

Photosynthetically Active Radiation (PAR)

It should be considered that a canopy consists of various elements of the plant itself, such as leaves, branches, fruits, flowers, among others. The radiation flux that strikes these elements will undergo processes of transmission and absorption. The spectral characteristics of leaves and multiple reflections between them result in a high degree of absorption of PAR in the visible spectrum (0.4–0.72 µm) and moderate reflectance in the near-infrared region (0.72–4.0 µm). When the ground is not covered by vegetation, it generally shows a gradual increase in radiation reflectance between 0.4 and 4.0 µm (Sellers *et al.*, 1997).

The amount of PAR depends primarily on two factors at any given moment: the position of the sun and atmospheric transmissivity. The position of the sun follows a regular pattern of altitude, which can be used to estimate the maximum potential PAR for a given time and location under clear sky conditions (Nöjd and Hari, 2001).

Fraction of Absorbed Photosynthetically Active Radiation (f_{APAR})

The evolution of terrestrial ecosystems is characterized by physical, biochemical, and physiological variables, some of which can be extracted from remotely sensed data collected from space, with a high level of acceptance. This makes the f_{APAR} an excellent indicator of canopy status (Gobron *et al.*, 2000). Thus, f_{APAR} is a reliable variable for determining the presence of vegetation on a global scale.

The amount of PAR absorbed by green leaves can be used to estimate carbon sequestration by the canopy (Gallo *et al.*, 1993). Fortunately, vegetation remote sensing indices show a stronger correlation with f_{APAR} (Hall *et al.*, 1990).

The determination of f_{APAR} depends on canopy structure, LAI, zenith angle, diffuse radiation rate, and soil albedo (Mariscal *et al.*, 2000; Nouvellon *et al.*, 2014). Leaf area index (LAI) is often estimated, and f_{APAR} is calculated assuming that the standard canopy model is suitable (Gower *et al.*, 1999).

According to Sellers *et al.* (1997), f_{APAR} is a key parameter in several ecosystem productivity models, as well as in global climate, hydrology, biochemical, and ecological models.

The study conducted by Costa *et al.* (2019) emphasizes the importance of analyzing the absorption of PAR in the context of parameterizing physiological models. The researchers examined the behavior of the LAI and the radiation extinction coefficient in coffee crops subjected to different irrigation levels, using *in situ* sensors. They observed significant variations related to the irrigation treatments and the plant canopy structure.

In the case of plants subjected to the L130% irrigation level, it was found that approximately 90% of PAR was absorbed in the upper canopy quadrants. In the middle and lower quadrants, these values were approximately 35% and 4%, respectively. However, plants treated with L100%, i.e., with the ideal amount of water to fully meet the crop's needs, exhibited a similar pattern, with approximately 90% PARi absorption, comparable to the L130% treatment (Costa *et al.*, 2019).

Another study, like that of Thao *et al.* (2022), in the context of climate change, developed a method for predicting coffee production on a regional scale. In this study, f_{APAR} was used as one of the vegetation variables obtained through satellite remote sensing. The authors employed the Crop Growth Monitoring System Statistical Tool (CGMSstatTool – CST) to analyze these vegetation variables derived from satellite remote sensing (SPOT-VEGETATION and PROBA-V). They emphasized the significance of these variables in the accurate prediction of coffee production, which is crucial for all farmers in the coffee industry.

Leaf Area Index (LAI)

It is known that the leaf area of crops is an indicative parameter of yield, as photosynthesis depends on the capture of light energy and its conversion into chemical energy. According to Sellers *et al.* (1997), photosynthetic efficiency depends on the rate of photosynthesis per unit leaf area and the absorption of solar radiation, which is affected, among other things, by canopy structural characteristics and system size.

Thus, the leaf surface area of a plant is the basis for the potential yield of the crop. In addition, knowledge of the leaf area of the plant allows for the estimation of water loss, as leaves are the main organs involved in the transpiration process, and responsible for gas exchange with the environment (Leong, 1980). Therefore, understanding the temporal variation of LAI in perennial crops can be useful in evaluating various cultural practices such as pruning, fertilization, irrigation, spacing, and pesticide application, among others.

According to Pereira *et al.* (1997), the use of LAI in coffee cultivation was initially suggested by Castillo in 1961, in Colombia. However, Valencia (1973) was the first researcher to conduct studies on this topic. Pereira *et al.* (1997) conducted a study correlating LAI with coffee productivity based on planting density. The results indicate that the highest crop productivity was achieved when LAI reached a value of eight, which occurred 3 years after planting for a population of 10,000 plants ha^{-1} and only after the 4th year following planting when the population was 5000 plants ha^{-1}.

LAI is a variable of great importance used to estimate water, carbon, and energy fluxes. This index plays a relevant role in studies related to the understanding of phenomena at different scales, from the leaf to the canopy, as well as in the calculation of the photosynthetically active radiation extinction coefficient (kPAR), providing crucial information for the parameterization of physiologically based models (Huerta and Alvim, 1962).

It is expected that the agricultural sector will be substantially affected by climate change, due to the sensitivity of crops to increasing temperature and water scarcity (Mendelsohn, 2008; Taugourdeau *et al.*, 2014). The negative effects are evident and include declines in crop production and quality, as well as increased pest and disease infestation, resulting in a reduction in global agricultural production (Intergovernmental Panel on Climate Change (IPCC), 2014).

Ramirez-Villegas and Challinor (2012) reported that, in 2006, about 57% of coffee producers suffered production losses, with 26% due to water stress and 27% due to excessive rainfall. These challenges represent significant difficulties for small-scale farmers, many of whom depend on rainfed cultivation and have limited access to financial and technical support (Van Westen *et al.*, 2008; Cohn *et al.*, 2017), which could help them to cope with constantly changing weather conditions.

In the study conducted by Costa *et al.* (2019), which aimed to assess the variation in LAI in coffee crops subjected to different irrigation levels using a linear radiation bar sensor (SunScan-Delta-T Devices Ltd, London), it was observed that the average LAI values (for the entire plant) were 4.7, 4.2, 3.7, and 3.5 for the L130, L100, L70, and L40 treatments, respectively. The average LAI values were higher in plants subjected to the L130 and L100 treatments compared to the values observed in coffee plantations receiving deficit irrigation levels (70 and 40%).

The study conducted by Dos Santos *et al.* (2020) monitored the evolution of the LAI in a coffee plantation. This monitoring was based on plant measurements derived from 3D point clouds generated in conjunction with the application of the SfM algorithm to digital images captured by a camera attached to an unmanned aerial vehicle (UAV). The results revealed LAI values exceeding 3.0 in various blocks during the months of January through May 2018.

The LAI is a crucial indicator of ecosystem services and management practices, as evidenced in a study conducted by Taugourdeau *et al.* (2014) on a coffee agroforestry farm in Costa Rica. In this study, LAI results derived from MODIS showed seasonal variations ranging from 2.4 to 4.4 $m^2\ m^{-2}$. The authors emphasized the significant role of LAI as an important co-predictor of production and discussed how it is influenced by natural factors such as phenology and climate, as well as management.

Research like the one conducted by Thao *et al.* (2022) emphasized the importance of the LAI, revealing that the development of coffee production prediction models based on multiple linear regression of satellite-derived vegetation biophysical variables, such as LAI, resulted in coffee production forecast models with satisfactory accuracy.

Conclusion

Overall, the synthesis of evidence presented in this review unanimously points to the conclusion that there is an urgent need to deepen research on solar radiation interception and LAI in different Arabica coffee cultivars. This becomes particularly relevant considering that many coffee cultivation regions are already subject to climate variability, which may worsen in the future. Despite the abundance of studies addressing solar radiation and LAI, these analyses often focus on plantations as a whole, lacking, however, investigations that delve into the biophysical behavior of different cultivars.

A detailed understanding of solar radiation interception is essential not only to identify cultivars more resilient to climatic adversities but also to select the most suitable varieties for different regions. Although the literature includes various studies using *in situ* sensors and satellite data, there is a clear need for research dedicated to evaluating the individual behavior of each cultivar. This effort aims to gain a comprehensive understanding of plant behavior both before and after harvesting, as well as its relationship with productivity.

By gaining knowledge of the temporal behavior of these varieties and their interactions with solar radiation, it is possible to potentially develop cultivars more resilient to periods of drought, better equipped to face adversities, and with higher productivity, with the goal of providing support not only to coffee producers in their practices but also playing a fundamental role in promoting food security.

Conflicts of Interest

The authors have no conflicts of interest to declare.

References

Aerts, R., Geeraert, L., Berecha, G., Hundera, K., Muys, B. *et al.* (2017) Conserving wild Arabica coffee: Emerging threats and opportunities. *Agriculture, Ecosystems and Environment* 237, 75–79. DOI: 10.1016/j.agee.2016.12.023.

Alves, E.L., Pereira, F.A.C. and Dalchiavon, F.C. (2017) Potencial econômico da utilização de microterraceamento em lavouras de café: um estudo de caso. *Revista iPecege* 3, 24. DOI: 10.22167/r.ipecege.2017.1.24.

Assad, E.D., Pinto, H.S., Zulla Junior, J. and Ávila, A.M.H. (2004) Impacto das mudanças climáticas no zoneamento agroclimático do café no Brasil. *Pesquisa Agropecuária Brasileira* 39, 1057–1064.

Azevedo, P.V., Leitão, M.M.V.B.R., Souza, I.F. and Maciel, G.F. (1990) Balanço de radiação sobre culturas irrigadas no semi-árido do Nordeste do Brasil. *Revista de Meteorologia* 5(1), 403–410.

Barbosa, J.N., Borém, F.M., Alves, H.M.R., Volpato, M.M.L., Vieira, T.G.C. *et al.* (2010) Spatial distribution of coffees from Minas Gerais state and their relation with quality. *Coffee Science* 5(3), 237–250.

Botelho, C.E., Carvalho, J.P.F., Venturin, R.P., Silva, V.A., da Cunha, R.L. *et al.* (2015) Cafeicultura: convivência do arábica com a seca e introdução do café Conilon em regiões alternativas de Minas Gerais. *Informe Agropecuário* 36(285), 58–66.

Bragança, R., dos Santos, A.R., de Souza, E.F., de Carvalho, A.J.C., Luppi, A.S.L. *et al.* (2016) Impactos das mudanças climáticas no zoneamento agroclimatológico do café Arábica no Espírito Santo. *Revista Agro@mbiente On-Line* 10(1), 77–82.

Camargo, Â.P. de and Camargo, M.B.P. de. (2001) Definição e esquematização das fases fenológicas do cafeeiro arábica nas condições tropicais do Brasil. *Bragantia* 60, 65–68.

Camargo, M.B.P. (2010) The impact of climatc variability and climate change on Arabic coffee crop in Brazil. *Bragantia* 69, 239–247.

Carvalho, A.M.D., Mendes, A.N.G., Carvalho, G.R., Botelho, C.E., Gonçalves, F.M.A. *et al.* (2010) Correlação entre crescimento e produtividade de cultivares de café em diferentes regiões de Minas Gerais, Brasil. *Pesquisa Agropecuária Brasileira* 45, 269–275.

Carvalho, C.H.S.D., Bartelega, L., Sera, G.H., Matiello, J.B., Almeida, S.R.D, *et al.* (2022) Catálogo de cultivares de café arábica. Available at: https://www.infoteca.cnptia.embrapa.br/infoteca/handle/doc/1145722

Charbonnier, F., Roupsard, O., Maire, G.L., Guillemot, J., Casanoves, F. *et al.* (2017) Increased light-use efficiency sustains net primary productivity of shaded coffee plants in an agroforestry system. *Plant, Cell and Environment* 40(9), 1592–1608.

Chemura, A., Mutanga, O. and Dube, T. (2017) Remote sensing leaf water stress in coffee (*Coffea arabica*) using secondary effects of water absorption and random forests. *Physics and Chemistry of the Earth, Parts A/B/C* 100, 317–324.

Cohn, A.S., Newton, P., Gil, J.D., Kuhl, L., Samberg, L. *et al.* (2017) Smallholder agriculture and climate change. *Annual Review of Environment and Resources* 42(1), 347–375.

Companhia Nacional de Abastecimento (CONAB) (2023) Acompanhamento da safra brasileira – segundo levantamento 10(2), 1–44.

Costa, J.D.O., Coelho, R.D., Barros, T.H.D.S., Fraga, E.F. and Fernandes, A.L.T. (2019) Leaf area index and radiation extinction coefficient of a coffee canopy under variable drip irrigation levels. *Acta Scientiarum Agronomy* 41.

Covre, A.M., Partelli, F.L., Gontijo, I. and Zucoloto, M. (2015) Distribuição do sistema radicular de cafeeiro conilon irrigado e não irrigado. *Pesquisa Agropecuária Brasileira* 50, 1006–1016.

Craparo, A.C.W., Van Asten, P.J.A., Läderach, P., Jassogne, L.T.P. and Grab, S.W. (2015) Coffea arabica yields decline in Tanzania due to climate change: Global implications. *Agricultural and Forest Meteorology* 207, 1–10.

Damatta, F.M. (2004) Ecophysiological constraints on the production of shaded and unshaded coffee: A review. *Field Crops Research* 86, 99–114.

Damatta, F.M. (2018) Coffee tree growth and environmental acclimation. In: *Achieving Sustainable Cultivation of Coffee*. Burleigh Dodds Science Publishing, London, pp. 39–66.

Dos Santos, L.M., De Souza Barbosa, B.D., Diotto, A.V., Andrade, M.T., Conti, L, *et al.* (2020) Determining the leaf area index and percentage of area covered by coffee crops using UAV RGB images. *IEEE Journal of Selected Topics in Applied Earth Observations and Remote Sensing* 13, 6401–6409.

Frederico, S. (2014) Circuito espacial produtivo do café e o jogo de escalas. *Mercator* 13(1), 37–48.

Gallo, K.P., Daughtry, C.S.T. and Wiegand, C.L. (1993) Errors in measuring absorbed radiation and computing Canopy crop radiation use efficiency. *Agronomy Journal* 85(6), 1222–1228.

Gesisky, J., Girardi, G., Pinto, H.S., Zullo Júnior, J., Macedo Júnior, C. *et al.* (2008) Aquecimento global e a nova geografia da produção agrícola no Brasil. Embrapa/Unicamp, Campinas.

Gobron, N., Pinty, B., Verstraete, M.M. and Widlowski, J.L. (2000) Advanced vegetation indices optimized for up-coming sensors: Design, performance, and applications. *IEEE Transactions on Geoscience and Remote Sensing* 38(6), 1350–1360.

Gomes, I.A.C., de Castro, E.M., Soares, A.M., Alves, J.D., Alvarenga, M.I.N. *et al.* (2008) Alterações morfofisiológicas em folhas de *Coffea arabica* L. cv. "Oeiras" sob influência do sombreamento por Acacia mangium willd. *Ciência Rural* 38(1), 109–115. DOI: 10.1590/S0103-84782008000100018.

Gomez-Gomez, L. (1977) Influência de los factores climáticos sobre la periodicidad de crecimiento del cafeto. *Cenicafé* 28(1), 3–17.

Gouveia, N.M. (1984) Estudo da diferenciação e crescimento das gemas florais de *Coffea arabica* L.: observações sobre antese e maturação dos frutos. Campinas, 237p, Dissertação (Mestrado em Biologia) – Instituto de Biologia, UNICAMP.

Gower, S.T., Kucharik, C.J. and Norman, J.M. (1999) Direct and indirect stimation of leaf área index, fapar, and net primary production of terrestrial ecosystems. *Remote Sensing of Environment* 70, 29–51.

Gröll, K. (2008) Use of sensor technologies to estimate and assess the effects of various plant diseases on crop growth and development, Dissertation (Dr. sc. agr./Ph.D. in Agricultural Sciences) – Faculty of Agricultural Science of the University of Hohenheim, Bietigheim-Bissingen.

Hall, F.G., Huemmrich, F. and Goward, S.N. (1990) Use of narrow-band spectra to estimate the fraction of absorbed photosynthetic active radiation. *Remote Sensing of Environment* 32(1), 47–54.

Hein, L. and Gatzweiler, F. (2006) The economic value of coffee (*Coffea arabica*) genetic resources. *Ecological Economics* 60(1), 176–185.

Huerta, S.A. and Alvim, P. de T. (1962) Índice de área foliar y su influencia en la capacidad fotosintética del cafeto. *Cenicafe, Caldas* 13(2), 75–84.

Intergovernmental Panel on Climate Change (IPCC) (2014) Summary for policymakers. In: Field, C.B., Barros, V.R., Dokken, K.J., Mach, M.D., Mastrandrea, M.D. *et al.* (eds) *Climate Change 2014: Impacts,*

Adaptation, and Vulnerability. Part A: Global and Sectoral Aspects. Contribution of Working Group II to the Fifth Assessment Report of the Intergovernmental Panel on Climate Change. Cambridge University Press, Cambridge, UK and NY, pp. 1–32.
Kaufman, Y.J., Hobbs, P.V., Kirchhoff, V.W.J.H., Artaxo, P., Remer, L.A. *et al*. (1998) Smoke, clouds, and radiation-brazil (SCAR-B) experiment. *Journal of Geophysical Research* 103(D24), 31783–31808.
Kitzberger, C.S.G., Scholz, M.B., Dos, S. and Benassi, M.D.T. (2014) Bioactive compounds content in roasted coffee from traditional and modern Coffea Arabica cultivars grown under the same edapho-climatic conditions. *Food Research International* 61, 61–66.
Läderach, P., Ramirez-Villegas, J., Navarro-Racines, C., Zelaya, C., Martinez-Valle, A. *et al*. (2017) Climate change adaptation of coffee production in space and time. *Climatic Change* 141(1), 47–62.
Leong, W. (1980) Canopy modification and its effects on the growth and yield of *Hevea brasiliensis* Muell, Arg. Thesis (Ph.D.), Faculty of Agriculture Sciences of Ghent, Ghent.
Mariscal, M.J., Orgaza, F. and Villalobos, F.J. (2000) Modelling and measurement of radiation interception by olive canopies. *Agricultural and Forest Meteorology* 100, 183–197.
Mendelsohn, R. (2008) The impact of climate change on agriculture in developing countries. *Journal of Natural Resources Policy Research* 1, 5–19.
Mendonça, R.F.D., Rodrigues, W.N., Martins, L.D. and Tomaz, M.A. (2011) Abordagem Sobre a Bienalidade de Produção em Plantas de Café. *Enciclopédia Biosfera* 7, 1–9.
Meza, F. and Varas, E. (2000) Estimation of mean monthly solar global radiation as a function of temperature. *Agricultural Forest Meteorology* 100, 231–241.
Miniussi, M., Terra, L.D., Savi, T., Pallavicini, A. and Nardini, A. (2015) Aquaporins in *Coffea arabica* L.: Identification, expression, and impacts on plant water relations and hydraulics. *Plant Physiology and Biochemistry* 95, 92–102.
Monteith, J.L. (1972) Solar radiation and productivity in tropical ecosystems. *Journal of Applied Ecology* 9, 747–766.
Monteith, J.L. (1977) Climate and efficiency of crop production in Britain. *Philosophical Transactions of the Royal Society: B* 281, 277–294.
Monteith, J.L. (2013) *Principles of Environmental Physics*, 4th edn. Edward Arnold, London, p. 401.
Nöjd, P. and Hari, P. (2001) Instantaneous PAR estimated using long records of daily temperature and rainfall. *Agricultural and Forest Meteorology* 109, 47–59.
Nouvellon, Y., Bégué, A., Moran, M.S. and Seen, D.L. (2014) Leaf area index as an indicator of ecosystem services and management practices: An application for coffee agroforestry. *Agriculture, Ecosystems & Environment* 192, 11–37.
Oestreich-Janzen (2010) Chemistry of coffee in comprehensive Natural Products II: Chemistry and biology. The Science of Coffee, Illy. Available at: http://www2.illy.com/wps/wcm/connect/us/illy (accessed 22 July 2022).
Patrício, F.R.A., Almeida, I.M.G., Barros, B.C., Santos, A.S. and Frare, P.M. (2008) Effectiveness of acibenzolar-S-methyl, fungicides and antibiotics for the control of brown eye spot, bacterial blight, brown leaf spot and coffee rust in coffee. *Annals of Applied Biology* 152(1), 29–39.
Pereira, A.R., Villa Nova, N.A. and Sediyama, G.C. (1997) *Evapotranspiração*. Fealq, Piracicaba.
Ramirez-Villegas, J. and Challinor, A. (2012) Assessing relevant climate data for agricultural applications. *Agricultural and Forest Meteorology* 161, 26–45.
Ribeiro, B.B., Mendonça, L.M.V.L., Assis, G.A., Mendonça, J.M.A.D., Malta, M.R. *et al*. (2014) Avaliação química e sensorial de blends de Coffea canephora e Coffea arabica L. *Coffee Science* 9(2), 178–186.
Rodrigues, W.P., Vieira, H.D., Barbosa, D.H.S.G. and Vittorazzi, C. (2012) Growth and yield of *Coffea arabica* L. in Northwest Fluminense: 2nd harvest. *Revista Ceres* 59(6), 809–815.
Ross, I. and Ross, J. (1981) *The Radiation Regime and Architecture of Plant Stands*, Vol. 3. Springer Science, Business Media, Boston.
Ross, J. (1975) Radiative transfer in plant communities. In: Monteith, J.L. (ed.) *Vegetation and the Atmosphere, I*. Academic Press, California, pp. 13–55.
Sellers, P.J., Dickinson, R.E., Randall, D.A., Betts, A.K., Hall, F.G. *et al*. (1997) Modeling the exchanges of energy, water, and carbon between continents and the atmosphere. *Science* 275, 502–509.
Shuttleworth, W.J., Gash, J.H., Lloyd, C.R., Moore, C.J., Roberts, J. *et al*. (1984) Observations of radiation exchange above and below Amazonian forest. *Quarterly Journal of the Royal Meteorological Society* 97, 541–564.
Silveira, H.R. de. O. (2014) Excesso de água em mudas de café (Coffea arabica L.): Efeitos fisiológicos e anatômicos. Dissertação Mestrado, UFLA, 78 p, Lavras – MG.

Sinclair, T.R. and Muchow, R.C. (1999) Radiation use efficiency. *Advances in Agronomy* 65, 215–265.
Sondahl, M.R. and Sharp, W.R. (1979) Research in Coffea spp. and applications of tissue culture methods. In: Paddock, E.F. and Raghavan, V. (eds) *Plant Cell and Tissue Culture: Principles and Applications*. Ohio State University Press, Columbus, pp. 527–584.
Souza, J.D.C., Resende, M.L.V., Botelho, D.M.S., Monteiro, A.C.A., Silva, J.A.G. *et al.* (2017) Tolerância à seca em cultivares de *Coffea arabica* resistentes à ferrugem.
Taugourdeau, S., Le Maire, G., Avelino, J., Jones, J.R., Ramirez, L.G. *et al.* (2014) Leaf area index as an indicator of ecosystem services and management practices: An application for coffee agroforestry. *Agriculture, Ecosystems & Environment* 192, 19–37. DOI: 10.1016/j.agee.2014.03.%20042.
Tavares, P.D.S., Giarolla, A., Chou, S.C., Silva, A.J.D.P. and Lyra, A.D.A. (2018) Climate change impact on the potential yield of Arabica coffee in South-east Brazil. *Regional Environmental Change* 18, 873–883.
Thao, N.T.T., Khoi, D.N., Denis, A., Viet, L.V., Wellens, J. *et al.* (2022) Early prediction of coffee yield in the central highlands of Vietnam using a statistical approach and satellite remote sensing vegetation biophysical variables. *Remote Sensing* 14(13), 2975.
Toledo, P.R., de Melo, M.M., Pezza, H.R., Toci, A.T., Pezza, L. *et al.* (2017) Discriminant analysis for unveiling the origin of roasted coffee samples: A tool for quality control of coffee related products. *Food Control* 73, 164–174. DOI: 10.1016/j.foodcont.2016.08.001.
Valencia, G.A. (1973) Relación entre el índice de area foliar y la productividad del cafeto. *Cenicafe, Caldas* 24(4), 79–89.
Van Westen, C.J., Castellanos, E. and Kuriakose, S.L. (2008) Spatial data for landslide susceptibility, hazard, and vulnerability assessment: An overview. *Engineering Geology* 102(3–4), 112–131.
Zambolim, L., Vale, F.X.R. and Zambolim, E.M. (2005) Doenças do cafeeiro. In: Kimati, A. (ed.) *Manual de Fitopatologia*, 4th edn, Vol. 2. Agronômica Ceres, São Paulo, pp. 165–180.
Zullo, J., Pinto, H.S., Assad, E.D. and de Ávila, A.M.H. (2011) Potential for growing Arabica coffee in the extreme south of Brazil in a warmer world. *Climatic Change* 109(3–4), 535–548.

7 Using intraspecific molecular and phenotypic variation to promote multi-functionality of reforestation during climate change – A review of tropical forest case studies in South-east Asia

K.C. Grady[1] and E.P. Axelsson[2*]

[1]*School of Forestry, Northern Arizona University, Flagstaff, AZ, USA;* [2]*Department of Wildlife, Fish and Environmental Studies, Swedish University of Agricultural Sciences, Umeå, Sweden*

Abstract

The study of intraspecific genetic variation in plant traits for use in tropical forest restoration has broad potential for increasing our ability to achieve multi-functional objectives during this era of climate change. Developing seed-sourcing guidelines that optimize phenotypic characteristics best suited to a particular planting site as well as to future conditions imposed by environmental change could be useful for effective reforestation. Because evolution operates differently across tree species, this is an especially cumbersome task in tropical forests that contain thousands of species. Partially due to this high plant diversity, research and application of intraspecific variation in genetics, plant traits, and plant function in tropical forests wane far behind less diverse forest biomes. To examine the potential for improving reforestation efforts in tropical forests by considering intraspecific variation in plant traits and functions, we review the state of knowledge on intraspecific variation in South-east Asia as a case study. We focus on the dipterocarp family (*Dipterocarpaceae*), a highly diverse family of 16 genera with approximately 695 known species that often dominate lowland tropical rainforests of South-east Asia with many of these forests in a degraded state and in need of restoration. We found that there is research accumulating to understand genetic variation in approximately 10% of these 695 species. Intraspecific molecular variation exists at different spatial scales among species with 74% of species having moderate to high population differentiation (Fst > 0.10) and 92% of species with evidence of fine-scale genetic structure. Although this suggests a high potential for trait variation, few studies associated molecular with phenotypic variation. Seventeen tree species across 11 studies revealed intraspecific variation in traits or functions. Research indicates that intraspecific variation in growth may vary two-fold and drought tolerance four-fold among genotypes highlighting the possibility to pre-adapt trees to climate change during reforestation and to use intraspecific variation to promote the use of native species in commercial forestry. Our review presents opportunities and ideas for developing seed-sourcing guidelines to take advantage of intraspecific variation in traits and function by identifying how to locate this variation, which species would benefit, and how to test for trait variation. We also highlight an emerging area of research on local adaptation, common garden studies, and adaptive drought conditioning to improve reforestation during climate change.

Keywords: tropical forest restoration, climate pre-adaptation, assisted migration, local adaptation, fine-scale genetic structure, seed source guidelines, "climate-adjusted" provenancing

*Corresponding author: petter.axelsson@slu.se

Introduction

The application of evolutionary-based research to tropical reforestation efforts has broad potential to increase our ability to achieve multi-functional objectives during this era of climate change. Specifically, intraspecific variation in plant traits such as wood density, specific leaf area (SLA), foliar nitrogen (N) content, leaf lifespan, photosynthetic and transpiration rates, leaf-level water-use efficiency, and carbon isotope discrimination are often related to key ecosystem functions such as growth rate, promotion of canopy arthropod diversity, carbon sequestration potential, drought tolerance, nutrient-use, and economic values such as timber and fruit production (Grady *et al.*, 2013; Gregorio *et al.*, 2017; Axelsson *et al.*, 2022). Such trait-function relationships can be identified and incorporated into seed collection protocols for use in reforestation to match appropriate phenotypes to specific environmental conditions of the planting site to maximize growth, survivorship, and desired functionality. However, the functionality of different genotypes/phenotypes may be impacted by environmental changes such as global warming, drought, or deforestation. As such, site-matching should include both appropriate phenotypes for current and future environmental conditions. While much tropical research has focused on intraspecific variation in traits of commercially valuable species that are largely non-native to areas where they are grown, there is a growing need to assess how genetic variation in a broad range of native tree species can be utilized during reforestation for both commercial and conservation values (Thomas *et al.*, 2014; Prober *et al.*, 2015; Gregorio *et al.*, 2017; Jalonen *et al.*, 2018; Axelsson *et al.*, 2020). Such research has a high potential to diversify the portfolio of tree species that can be used in both commercial forestry and ecosystem restoration during climate change. It is widely recognized that restoration of native ecosystems is fundamentally necessary for achieving multiple benefits such as rebuilding native flora and fauna while at the same time promoting the livelihoods of local people. Achieving multi-functional benefits as a goal of restoration is highlighted by The Intergovernmental Science-Policy Platform on Biodiversity and Ecosystem Services (IPBES, 2019) and the United Nations Declaration of the Decade of Restoration (FAO, 2019).

Improved seed-sourcing guidelines of native tropical tree species during reforestation is a crucial step toward meeting reforestation objectives and contributing to solving the global biodiversity and climate crises (Gregorio *et al.*, 2017; Bosshard *et al.*, 2021). Using inappropriate seed sources that do not achieve desired plant functions can increase the costs (Gregorio *et al.*, 2017; Jalonen *et al.*, 2018) and reduce the long-term benefits of reforestation (Gardiner *et al.*, 2019; Andivia *et al.*, 2021). Furthermore, as climate change is expected to make locally adapted populations maladapted to future climates (Alberto *et al.*, 2013; Isaac-Renton *et al.*, 2018), reforestation efforts may benefit by selecting genetic material adapted to future climatic conditions (Prober *et al.*, 2015; Axelsson *et al.*, 2020). In non-tropical ecosystems, much research has illustrated high-intraspecific variation in functional traits that corresponds to environmental variation (Grady *et al.*, 2015; Zucchi *et al.*, 2017; Pastorino and Marchelli, 2020); much less research on intraspecific trait variation has been conducted in the tropics (though see Cruz *et al.*, 2019; Barton *et al.*, 2020). As such, in contrast to the wide application of evolutionary-based themes in non-tropical systems such as assisted migration (Aitken and Whitlock, 2013; Grady *et al.*, 2015), pre-adapting forests to climate change during reforestation (Axelsson *et al.*, 2020), or using a "climate-adjusted" provenancing strategy when conducting restoration (Prober *et al.*, 2015), few such efforts are being tested or used in tropical ecosystems. Unless the forests that are established through restoration today can maintain functionality in the climate of tomorrow, restoration efforts might be ineffective. Understanding the magnitude of intraspecific trait variation at multiple scales, and examining if that variation is related to climatic and/or biotic factors and/or is locally adapted, is an important first step toward identifying and applying the correct trait-function genotypes that promote functional resilience in trees undergoing climate change. A review of the research literature related to intraspecific trait variation in tropical forests of Southeast Asia research will promote the development of germplasm/seed collection strategies for reforestation.

Review Methodology

We conducted a rapid review on the available literature pertaining to spatial scales of intraspecific molecular and phenotypic variation, intraspecific variation in trait-to-function relationships, and local adaptation of the dipterocarps. Rapid reviews are the recommended review type in cases such as this when there are time constraints that need consideration (Grant and Booth, 2009). To find literature relevant to our objective, we built a search string consisting of two components (Appendix S1). The first component contained nine terms in relation to intraspecific variation, phenotypic variation, genetics, population differentiation, and local adaptation plus relevant synonyms. To delimit our search to dipterocarps we added a second component with synonyms to *Dipterocarpaceae* as well as each of the genera within the family. Using this search string, we compiled a database of articles extracted from Web-of-Science using the "all databases" and "Topic" features on May 25, 2023. We delimited our search to empirical articles published in peer-reviewed journals (but kept review articles for reference).

Using this search, we found 275 articles. We then reviewed titles and abstracts to determine if studies were pertinent to our main research objectives. Inclusion at this stage was based on if there was a genetic focus of relevant taxa (i.e. dipterocarps) and if studies were written in English. Studies with a purely phylogenetic focus or technical papers with a focus on developing molecular- or forensic tools (i.e. verify the origin of timber products) were not considered. Review papers addressing dipterocarps with a genetic focus were kept as reference but not included in the final list of papers. This generated 97 studies that were accessible online (Appendix S2).

Using molecular data to locate intraspecific trait variation

Understanding the spatial and ecological drivers underlying intraspecific variation in tropical trees is daunted by the high species diversity in tropical forests, making the development of seed-sourcing guidelines challenging. One of the most commonly used measures of genetic structure is the fixation index Fst, which measures genetic differentiation among populations across large spatial scales. Fst essentially measures the level of genetic differentiation between two or more populations and ranges from 0 (i.e. no genetic differentiation) to 1 (complete genetic differentiation). For many tree species, high Fst (e.g. high population differentiation) across broad spatial scales is related to high environmental variation that reflects natural selection to different environmental factors (Carvalho *et al.*, 2019). For species with high levels of Fst, we would expect high intraspecific variation in plant traits/functions as genetic variation generally correlates with phenotypic variation. For such species, intraspecific variation in traits could be examined among populations using greenhouse and common garden experiments to identify desirable plant characteristics for use in reforestation. In addition, tree species with high Fst would be potential candidates to screen for evidence of local adaptation. Where local adaptation exists, strategies such as assisted migration (Aitken and Whitlock, 2013) may be useful to apply in a reforestation context. Seed-sourcing recommendations could be developed accounting for which populations are best suited to current and future environmental conditions.

In tropical forests, we expect many species to also exhibit low within population genetic differentiation where neighboring individuals are highly genetically related. Such related groups of individuals can be genetically distinct from nearby groups of other individuals (e.g. at distances less than 1 km; Tito de Morais *et al.*, 2015) where there are limits to gene flow among them. This genetic aggregation and isolation is referred to as fine-scale genetic structure (FSGS). High FSGS is thought to be found where there are strong barriers to gene flow such as where trees are pollinated by wingless insects that cannot travel long distances or from limited seed or pollen dispersal (Harata *et al.*, 2012; Tito de Morais *et al.*, 2015; Lee *et al.*, 2016). Limited gene flow can lead to closely genetic-related neighboring individuals with a high occurrence of inbreeding and subsequent high potential for reduced fitness and low ability to adapt to changing environments. FSGS may be common in highly diverse forests where a high diversity of tree species inherently creates obstacles to gene flow. For example, in a forest with 100 or more species per hectare, the probability is low of a tree being in close proximity to a conspecific tree neighbor that would promote pollination. For species with high levels of FSGS, we would expect intraspecific variation in traits at relatively short distances (e.g. trees separated by 50 m–1200 m; Harata *et al.*, 2012; Tito de Morais *et al.*, 2015). Greenhouse and common garden experiments could be used to screen species with FSGS for desirable phenotypes for use in reforestation.

Anthropogenic factors such as deforestation that cause population fragmentation are also known to influence the genetic structure of species (Pandey and Geburek, 2010; Ismail *et al.*, 2014; Dai *et al.*, 2018). Both Fst and FSGS and disturbance-related genetic impacts are likely to vary among species such that seed-sourcing guidelines would vary widely depending on the species being used in reforestation. The interaction of deforestation with naturally occurring intraspecific trait variation could be incorporated into the development of a strategic plan for seed sourcing during reforestation (Thomas *et al.*, 2014).

Local adaptation and pre-adapting forests to future climates

In tropical forests, species selection to climatic conditions may be weak where homogenous atmospheric temperatures and consistently high precipitation do not result in natural selection pressures great enough to result in adaptation (Pimm and Sugden, 1994). In such forests, water may not be a limiting resource, and selection to it may be absent considering that soil saturation is often present at both high (i.e. 300 cm per year) or very high (600 cm per year) levels of precipitation. While intraspecific variation that is related to variation in climatic conditions has been rarely studied in wet tropical forests (though see Barton *et al.*, 2020), there is evidence emerging highlighting the evolutionary consequences of drought in seasonal tropical rainforests (Ng *et al.*, 2021). Furthermore, the amount and seasonality of rainfall varies across altitudinal gradients, and plant species distribution patterns are correlated with rainfall even in wet tropical forests (Engelbrecht *et al.*, 2005). Species turnover across environmental gradients and differences among species in distribution ranges suggest that evolution influences the interspecific variation of plant species in different ways. There is evidence of variation among species in response to drought (Bartlett *et al.*, 2016; Axelsson *et al.*, 2021) suggesting that drought tolerance is an important factor that could be under selection in tropical forests (Axelsson *et al.*, 2020; Ng *et al.*, 2021). Knowledge regarding functional genetic variation across geographical gradients is commonly used in non-tropical systems to enhance restoration success, e.g., via assisted migration (Grady *et al.*, 2015; Zucchi *et al.*, 2017) or low-intensity breeding (Pastorino and Marchelli, 2020). We acknowledge that other factors such as nutrient limitation, inter- and intraspecific competition among trees (Grady *et al.*, 2017), insect and pathogen impacts on trees (Grady *et al.*, 2015), among others, could also have evolutionary impacts such as have been found in non-tropical systems and that these factors may interact in complex ways with adaptation to climate. Improved understanding of the extent and evolutionary causes of functional and adaptive intraspecific variation of native tree species in tropical forests could potentially be used to direct seed collection campaigns to support reforestation during global warming and other environmental changes.

A review of South-east Asian dipterocarps

Here we review the current knowledge about intraspecific molecular and phenotypic variation in tree species in the dipterocarp family (*Dipterocarpaceae*), a highly diverse family of 16 genera and about 695 known species typically dominating lowland tropical rainforests of South-east Asia. We were explicitly interested in assessing the extent of, and knowledge about, intraspecific variation in traits and function across different spatial scales in order to examine the potential to incorporate such variation into commercial forestry using native species and conservation-based reforestation efforts where adapting to future environmental conditions is important. We specifically reviewed the literature where Fst and FSGS values were reported as these measures can be used to distinguish among species variation in the scale of population differentiation. That is, as high Fst is often correlated to high trait variation among populations and high FSGS values would suggest low intraspecific variation at very small scales (e.g. trees less than 1 km apart), and high intraspecific variation at intermediate scales (e.g. trees greater than 1 km apart; Tito de Morais *et al.*, 2015). We also reviewed research that evaluated intraspecific variation in plant traits, and research that correlated such variation to environmental variables such as drought, nutrient limitation, and atmospheric temperature. Finally, we reviewed the literature for studies on local adaptation within dipterocarp tree species. Compiling this knowledge is the first step to examine the application of intraspecific variation in plant traits and function to native species commercial forestry and conservation-based reforestation efforts during climate change in dipterocarp-dominated forests of South-east Asia. Through this review, we specifically aimed to address the following questions of relevance for seed-sourcing guidelines and reforestation success in the face of climate change.

a. How does intraspecific molecular variation of dipterocarp species vary among different spatial scales and what are the potential drivers of this variation? Can we use such molecular data to catalyze efforts to understand phenotypic variation and apply it to reforestation?
b. What amount of phenotypic variation in plant characteristics and function have been found that could be incorporated into reforestation to achieve multi-function objectives?
c. What is the evidence that such genetic/phenotypic variation is locally adapted or associated with variation in response to a changing climate?

Based on the state of knowledge of intraspecific variation in plant traits, we then highlight opportunities for developing seed-sourcing guidelines to (1) enhance forest functionality and achievement of multi-functional objectives during reforestation, and (2) enhance plant fitness under changing climatic conditions.

Material and Methods

Dipterocarp forests

Mixed-dipterocarp forests are a dominant forest ecosystem throughout South-east Asia. These forests are subjected to considerable degradation, and restoration is urgently required (Kettle, 2009). In addition to being a dominant component of overall plant community composition in South-east Asia, dipterocarps are emergent species, occupying the highest places in the forest canopy with some species reaching over 80m in height. As such, they have a key role in carbon sinks and are thus very important from a climate change perspective. Dipterocarps also promote a wide array of other diverse taxa such as insects (Sakai *et al.*, 1999; Axelsson *et al.*, 2022), birds (Engstrom *et al.*, 2020), and mammals (Charles, 1996; Chapman *et al.*, 2018). Furthermore, tropical forests of South-east Asia have been managed by local communities for thousands of years (Barker *et al.*, 2017) and many tree species are culturally significant to local communities (Barker *et al.*, 2017; Axelsson *et al.*, 2021). Due to their foundational role in these dipterocarp-dominated forests (Axelsson *et al.*, 2022), a range of species within this family are commonly included in reforestation efforts in the region (Kettle, 2009; Gustafsson *et al.*, 2016; Axelsson *et al.*, 2020). Many dipterocarp species are categorized as near threatened or endangered as a result of exploitation and massive population reduction (Guan and Yen, 2000; BGCI, 2021) highlighting an evident need to strengthen research related to genetic conservation of dipterocarps. Being a highly diverse family of 16 genera with approximately 695 known species, the dipterocarp family has species with varying mating systems that, in turn, generate large differences among species in genetic structure (Tito de Morais *et al.*, 2015). This in combination with reports of both variations in Fst and FSGS suggests that dipterocarps are a likely candidate taxonomic group for evaluating the potential

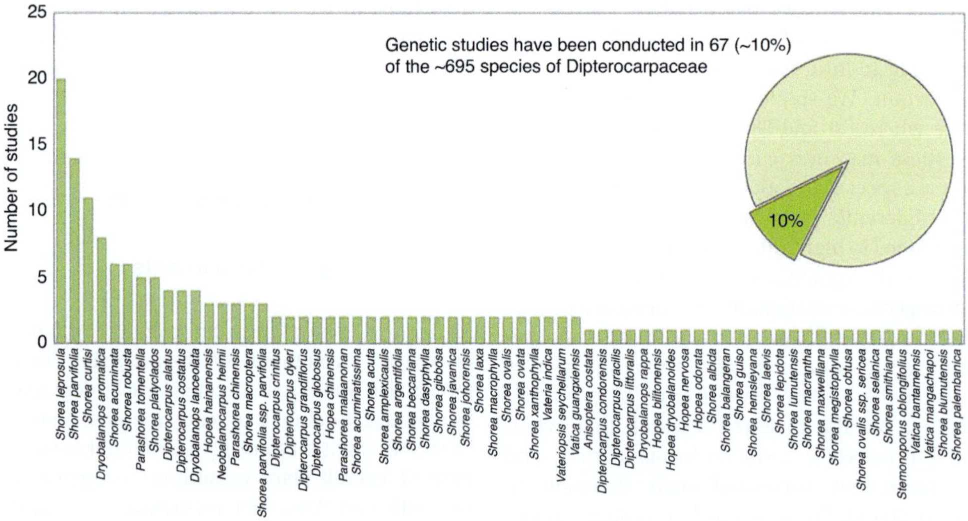

Fig. 7.1. Genetic aspects have been studied in 67 species of dipterocarps which represents roughly 10% of the ~695 known species in the family. A few species are well studied as represented in the literature whereas a majority of the species have only been included in one or two studies.

of intraspecific trait-matching to current and future conditions during reforestation.

Results

In total, we found 97 studies that covered 67 species and 2 subspecies (Fig. 7.1). These 67 species represent roughly 10% of the known species in the dipterocarp family (i.e. 67/695). The most commonly studied species were *Shorea leprosula*, *Shorea parvifolia*, and *Shorea curtisii* that occurred in 22, 15, and 11 studies, respectively. Most species were, however, represented in very few studies, i.e., 75% of the species or subspecies were only represented in one or two studies (Fig. 7.1). A large majority of the studies were molecular studies focusing on the genetic variation among populations (i.e. population differentiation, Fst) or genetic variation within populations (i.e. fine-scale genetic structure, FSGS) and spatial genetic structure in natural- or semi-natural populations (i.e. effects of forest management, degradation, etc.). In a subset of such studies, molecular data were used to link species variation in spatial genetic structure and gene flow with life history traits, breeding system, and pollination syndromes (Harata *et al.*, 2012; Tani *et al.*, 2012; Tito de Morais *et al.*, 2015). Studies were also found that linked spatial genetic structure to landscape features such as elevation and topography (Lee *et al.*, 2013; Lee *et al.*, 2016; Smith *et al.*, 2018). Molecular genetic information has also been used to enhance *ex situ* (Lee *et al.*, 2000; Li *et al.*, 2002; Kusuma *et al.*, 2019) and *in situ* conservation (Ng *et al.*, 2019). We found a small number of nine studies addressing phenotypic variation or genetic variation in plant functioning but few that linked molecular to phenotypic data. The phenotypic traits and functions addressed included variation in seed size and germination (Naito *et al.*, 2005; Lee *et al.*, 2006; Lee *et al.*, 2013; Takeuchi and Diway, 2021), seedling performance and growth rate (Lee *et al.*, 2006; Ismail *et al.*, 2014; Nutt *et al.*, 2016; Axelsson *et al.*, 2020; Ng *et al.*, 2021; O'Brien and Escudero, 2022). Only two studies included research on intraspecific variation in response to climatic factors (Axelsson *et al.*, 2020; Ng *et al.*, 2021), three studies examined intraspecific variation in response to soil nutrients, forest type, or insect herbivory (Indriani *et al.*, 2019; Li *et al.*, 2021; Ng *et al.*, 2022), and one common garden study examined local adaptation to abiotic and biotic conditions (O'Brien and Escudero, 2022).

Population differentiation (FST) and Fine-Scale Genetic Structure (FSGS)

A large majority of the studies included in this review used molecular tools to assess intraspecific genetic variation in species of dipterocarps. These studies reveal that intraspecific genetic variation ranges from small to large and that the extent of population differentiation (Fst) varies widely among species. Mean population differentiation estimates (Fst) range from 0 (no significant differentiation) to 0.519 and differ across species and vary across different spatial scales (Table 7.1). Although these molecular-based studies never included a phenotypic component, the majority of the 34 species that presented population structure data generally showed moderate to high population structure (Table 7.1). For example, of the 34 species included, 25 species (74%) had at least one study where moderate to high population differentiation was reported (Fst, Gst, or Rst above 0.10) while for 6 species (18%), at least one study confirmed a low level of genetic differentiation (Fst, Gst, or Rst between 0.05 and 0.099). In sum, 74% of species had at least one study showing moderate to high levels of population differentiation. Moderate to high genetic differentiation among populations suggests the potential to identify populations with desirable traits/functions as well as to explore the potential for local adaptation.

Our case studies with two species (*Shorea macrophylla* and *Dipterocarpus alatus*) in which pairwise Fst values have been reported across a multitude of populations show that species vary in how spatial scale influences genetic differentiation. In *S. macrophylla* genetic differentiation follows a linear fit and Fst increases mainly across large spatial scales. In *D. alatus*, genetic differentiation follows a logarithmic fit and Fst increases quickly across relatively small spatial scales (Fig. 7.2). The Fst to spatial distance relationship also has a better fit in *S. macrophylla* ($R^2 = 0.66$) than in *D. alatus* ($R^2 = 0.52$) suggesting that *D. alatus* may have a larger proportion of variation explained by environmental factors other than distance (Fig. 7.2). Such information can be used to guide the scope of seed sourcing to increase the likelihood of finding phenotypic variation that can be tested in common gardens to infer local adaptation to environmental conditions.

In addition to moderate to high overall population structure among dipterocarp species (74% of species with a moderate to high level of population differentiation), we also found that 92% of species studied had at least one case where fine-scale genetic (FSGS) structure was present (Table 7.2). Only 2 out of 26 species investigated across a total of 13 studies did not have FSGS. The distance at which genetically related individuals were aggregated varied among species (Table 7.2) and multiple studies on the same species generally followed the same trends for evidence, strength, and distance of FSGS (Table 7.2). Intraspecific variation in phenotypic traits for any of these species was not reported. Table 7.2 illustrates that for 92% of dipterocarp species, collecting seeds from different mother trees spaced at 1200m would ensure genetic differentiation for 92% of dipterocarps. Using this seed-sourcing guideline could be useful for developing studies to examine phenotypic variation and future applications of using this phenotypic variation in reforestation efforts. In addition, such a protocol could be established for screening for phenotypic variation where seeds from different mother trees at least 1200m apart could be grown in greenhouse or common garden environments (Axelsson *et al.*, 2020).

Genetic isolation appears to be driven by low gene flow in trees pollinated by less mobile insects such as wingless thrips or small beetles compared to larger pollinators such as giant honeybees (Apis dorsata; Harata *et al.*, 2012; Tito de Morais *et al.*, 2015). Tree species with smaller flowers tend to be pollinated by smaller insects that only disperse short distances and both flower size and pollinator size have been correlated to FSGS. In addition, seed dispersal by wind is common for dipterocarps but dipterocarps have heavy seeds that do not disperse far, are actively fed on by a variety of herbivores, have long periods between flowering events (some species up to 20 years), and are quickly decomposed and unviable if germination is not immediate; taken together, seed dispersal is limited and limits gene flow (Harata *et al.*, 2012; Tito de Morais *et al.*, 2015).

A range of studies also highlight how landscape features such as landscape connectivity

Table 7.1. Population differentiation values (Fst, Rst, and Gst) reported for dipterocarp tree species in studies conducted across different spatial scales.

Species	Populations (n)	Life stage	Fst min	Fst	Fst max	Gst	Spatial scale (km)	Authors and year
Dipterocarpus alatus	7	Adults	–	–	0.135	–	30	Chokthaweepanicha *et al.* (2022)
Dipterocarpus alatus	5	Unspecified	0.317	0.274	0.492	–	150	Vu *et al.* (2019)
Dipterocarpus alatus	10	Adults	–	0.266	0.492	–	350	Tam *et al.* (2014)
Dipterocarpus alatus	5	Adults	–	0.274	0.492	–	150	Vu *et al.* (2019)
Dipterocarpus condorensis	8	Adults	0.024	0.122	0.138	–	200	Nguyen *et al.* (2022)
Dipterocarpus costatus	3	Unspecified	0.053	0.347	0.403	–	150	Vu *et al.* (2019)
Dipterocarpus costatus	4	Mature trees	0.315	0.405	0.481	–	209	Duc *et al.* (2016)
Dipterocarpus costatus	3	Adults	–	0.347	0.403	–	150	Vu *et al.* (2019)
Dipterocarpus dyeri	6	Adults	–	0.132	0.192	0.135	300	Nguyen *et al.* (2020)
Dipterocarpus dyeri	2	Adults	–	0.123	0.123	–	50	Tam *et al.* (2019)
Dipterocarpus dyeri	2	Juveniles	–	0.188	0.188	–	50	Tam *et al.* (2019)
Dipterocarpus gracilis	4	Seedlings	–	0.251	0.251	–	2	Romadini *et al.* (2021)
Dipterocarpus littoralis	2	Adults	0.05	0.05	0.05	–	1	Dwiyanti *et al.* (2014)
Dipterocarpus littoralis	2	Saplings	0.036	0.036	0.036	–	1	Dwiyanti *et al.* (2014)
Dryobalanops aromatica	10	Unspecified	–	–	–	0.036	400	Lee *et al.* (2000)
Dryobalanops aromatica	5	Seedlings	–	–	–	0.067	450	Lim *et al.* (2002)
Dryobalanops aromatica	3	Adults	–	–	–	0.126	150	Ritonga *et al.* (2018)

Continued

Table 7.1. Continued

Species	Populations (n)	Life stage	Fst min	Fst	Fst max	Gst	Spatial scale (km)	Authors and year
Hopea bilitonensis	2	Unspecified	–	–	–	0.116	3	Lee *et al*. (2013)
Hopea chinensis	4	Adults	–	–	–	0.0735	400	Tang *et al*. (2015)
Hopea hainanensis	7	Unspecified	–0.046	–	0.295	–	100	Chen *et al*. (2022)
Hopea hainanensis	10	Adults	0.001	0.23	0.271	0.229	250	Wang *et al*. (2020)
Hopea odorata	3	Adults	0.17	0.251	0.31	0.193	1000	Nguyen *et al*. (2014)
Neobalanocarpus heimii	32	Adults	–	0.127	0.127	0.39	700	Tnah *et al*. (2013)
Parashorea chinensis	15	Unspecified	–	0.32	0.3202	–	1.2	Li *et al*. (2022)
Parashorea tomentella	3	Adults	–	0.078	0.078	–	6	Kettle *et al*. (2011)
Shorea acuminata	3	Unspecified	–0.003	–	0.093	–	200	Ishiyama *et al*. (2008)
Shorea albida	11	Adults	0.0003	0.04	0.1374	–	50	Ogasahara *et al*. (2023)
Shorea balangeran	2	Unspecified	–	–	–	0.03	30	Indriani *et al*. (2019)
Shorea curtisii	3	Unspecified	–0.01	–	0.083	–	200	Ishiyama *et al*. (2008)
Shorea curtisii	2	Adults and saplings	–	0.519	0.519	–	1400	Kamiya *et al*. (2012)
Shorea curtisii	6	Adults and saplings	–	0.033	0.033	–	500	Kamiya *et al*. (2012)
Shorea curtisii	2	Adults and saplings	–	0.035	0.035	–	125	Kamiya *et al*. (2012)
Shorea guiso	4	Adults	–	0.05	0.107	–	700	Tinio *et al*. (2014)

Continued

Table 7.1. Continued

Species	Populations (n)	Life stage	Fst min	Fst	Fst max	Gst	Spatial scale (km)	Authors and year
Shorea javanica	6	Adults	–	0	0	–	600	Rachmat *et al.* (2012a)
Shorea javanica	6	Adults	–0.064	0.076	0.158	–	500	Rachmat *et al.* (2012b)
Shorea laevis	7	Adults	–	–	–	0.736	1000	Zulfahmi *et al.* (2015)
Shorea leprosula	7	Adults and saplings	–	0.25	0.471	–	1900	Cao *et al.* (2006)
Shorea leprosula	2	Adults and saplings	–	0.404	0.404	–	1500	Cao *et al.* (2009)
Shorea leprosula	8	Adults and seedlings	–	–	–	0.117	900	Lee *et al.* (2000)
Shorea leprosula	7	Adults and seedlings	–	–	–	0.085	600	Lee *et al.* (2000)
Shorea leprosula	10	Seeds	0.00012	–	0.023	–	1	Lee *et al.* (2000)
Shorea lumutensis	10	>1 cm DBH	–	–	–	0.058[a]	300	Lee *et al.* (2006)
Shorea macrophylla	13	Adults	–	0.093	0.166	–	1000	Utomo *et al.* (2018)
Shorea obtusa	5	Adults	–	0.03	0.056	–	500	Senakun *et al.* (2011)
Shorea parvifolia	6	Adults and saplings	–	0.31	0.717	–	1900	Cao *et al.* (2006)
Shorea parvifolia	2	Adults and saplings	–	0.336	0.336	–	1500	Cao *et al.* (2009)
Shorea parvifolia	18	Adults	–	0.15	0.1555	–	2000	Ohtani *et al.* (2021)

Continued

Table 7.1. Continued

Species	Populations (n)	Life stage	Fst min	Fst	Fst max	Gst	Spatial scale (km)	Authors and year
Shorea parvifolia	7	Adults	–	–	–	0.582	2000	Zulfahmi *et al.* (2015)
Shorea platyclados	3	Unspecified	0.01	–	0.11	–	250	Muhammad *et al.* (2016)
Shorea platyclados	27	Unspecified	–	0.060	0.060	–	900	Ng *et al.* (2019)
Shorea robusta	15	Adults	–	0.043	0.047	–	500	Pandey and Geburek (2010)
Shorea selanica	3	Adults	0.394	0.702	0.907	–	300	Rachmat *et al.*, 2012aRachmat *et al.*, 2012b
Stemonoporus oblongifolius	4	Adult and subadult	–	–	–	0.163	10	Murawski and Bawa (1994)
Vateria indica	2	Adults	–	0.071	0.071	–	2.3	Ismail *et al.* (2014)
Vateria indica	2	Wildings	–	0.098	0.0979	–	2.3	Ismail *et al.* (2014)
Vateriopsis seychellarum	7	Adults	–	0.2	0.27	–	20	Finger *et al.* (2012)
Vateriopsis seychellarum	7	Juveniles	–	0.38	0.66	–	20	Finger *et al.* (2012)
Vatica guangxiensis	3	Unspecified	–	–	–	0.3764	900	Li *et al.* (2002)
Vatica mangachapoi	11	DBH ≥1.5 cm	0.156	–	0.257	–	200	Dai *et al.* (2018)

[a]Rst.

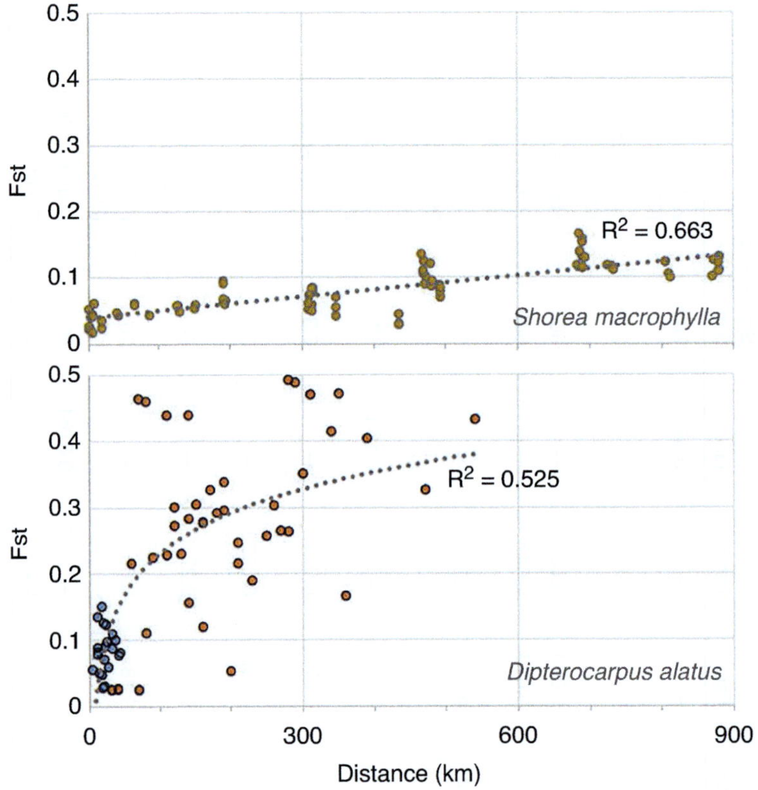

Fig. 7.2. Illustration of how genetic differentiation (Fst) is influenced by spatial scale differently for two dipterocarp tree species. The blue data points are from Chokthaweepanicha *et al.* (2022), red from Tam *et al.* (2014) and yellow from Utomo *et al.* (2018). In *S. macrophylla*, genetic differentiation follows a linear fit and Fst increases mainly across large spatial scales. In *D. alatus,* genetic differentiation follows a logarithmic fit and Fst can increase quickly across relatively small spatial scales.

(Pandey and Geburek, 2010; Ismail *et al.*, 2014), and logging and forest degradation may impact intraspecific genetic variation, inbreeding and gene flow within and among populations (Obayashi *et al.*, 2002; Wickneswari *et al.*, 2004; Ng *et al.*, 2009; Ang *et al.*, 2016; Widiyatno *et al.*, 2016; Widiyatno *et al.*, 2017) and that such effects can vary in extent depending on mating system (Ng *et al.*, 2009). Widiyatno *et al.* (2016) found that the genetic diversity of logged tropical forests gradually decreases depending on logging rotation times, especially with respect to sensitive genetic parameters such as the numbers of rare alleles, and that enrichment planting can be used to increase the genetic variation of logged forests. Molecular genetic information has also been used to identify genetically unique populations that may be particularly interesting for conservation (Lee *et al.*, 2000; Li *et al.*, 2002; Kusuma *et al.*, 2019; Ng *et al.*, 2019). Incorporating genetically unique populations into reforestation as well as

Table 7.2. Dipterocarp tree species variation in patterns of fine-scale genetic structure (FSGS) as indicated by F1 (mean pairwise kinship coefficient among geographically close individuals); Sp (index of the strength of fine-scale genetic structure); and Spatial Scale in meters (distance beyond which kinship relatedness is non-significant). Multiple values correspond to statistics reported for different studies. Where individual studies included more than two values, we report the highest value. We report studies that found non-significant F1 (*ns*). Only a Spatial Scale is reported where F1 or another indicator of FSGS was statistically significant. In several cases, neither F1 nor Sp were reported but other statistical procedures were used to evaluate FSGS – in such cases, only a Spatial Scale is reported. The region where the study was conducted is indicated by Bo = Borneo; PM = Peninsular Malaysia; and SI = South India.

Species	F1 (FSGS)	Sp (FSGS)	Spatial scale (m)	Region	References
Dipterocarpus crinitus	ns; ns	–	–	Bo	Harata *et al*. (2012); Tito de Morais *et al*. (2015)
Dipterocarpus globosus	0.069	0.007; 0.044	75	Bo	Harata *et al*. (2012); Tito de Morais *et al*. (2015)
Dipterocarpus grandiflorus	0.020; 0.029	0.002; 0.002	20, 100	Bo	Kettle *et al*. (2011); Tito de Morais *et al*. (2015)
Dryobalanops aromatica	0.067	0.010; 0.032	200	Bo	Tito de Morais *et al*. (2015); Harata *et al*. (2012)
Dryobalanops lanceolata	ns	0.035	–	Bo	Tito de Morais *et al*. (2015); Harata *et al*. (2012)
Hopea bilitonensis	–	–	30	PM	Lee *et al*. (2013)
Hopea dryobalanoides	–	–	300	PM	Takeuchi and Diway (2021)
Parashorea tomentella	0.051; 0.083; 0.094	0.013; 0.013; 0.025	25, 140, 300	Bo	Kettle *et al*. (2011); Tito de Morais *et al*. (2015); Smith *et al*. (2018)
Shorea acuminatissima	0.001	0.015	200	Bo	Tito de Morais *et al*. (2015)
Shorea acuta	0.094	0.015; 0.051	200	Bo	Harata *et al*. (2012); Tito de Morais *et al*. (2015)
Shorea amplexicaulis	ns; ns	–	–	Bo	Tito de Morais *et al*. (2015)
Shorea argentifolia	0.142	0.03	–	Bo	Tito de Morais *et al*. (2015)
Shorea beccariana	0.083	0.019; 0.041	300	Bo	Harata *et al*. (2012); Tito de Morais *et al*. (2015)

Continued

Table 7.2. Continued

Species	F1 (FSGS)	Sp (FSGS)	Spatial scale (m)	Region	References
Shorea curtisii	0.075	0.026; 0.033	250, 250	Bo	Ng *et al*. (2006); Harata *et al*. (2012); Tito de Morais *et al*. (2015)
Shorea gibbosa	0.053	0.007	50	Bo	Tito de Morais *et al*. (2015)
Shorea laxa	–	–	1250	Bo	Takeuchi and Diway (2021)
Shorea leprosula	0.058	0.015	25, 35, 250, 250	Bo, PM	Lee *et al*. (2001); Ng *et al*. (2004, 2006); Smith *et al*. (2018)
Shorea lumutensis	–	–	100	PM	Lee *et al*. (2006)
Shorea macroptera	–	–	250	PM	Ng *et al*. (2006)
Shorea ovalis	–	–	35	PM	Ng *et al*. (2004)
Shorea ovata	0.179	0.052; 0.102	200	Bo	Harata *et al*. (2012); Tito de Morais *et al*. (2015)
Shorea parvifolia	0.072; ns	0.009; 0.010	25, 240, 300	Bo, PM	Takeuchi *et al*. (2004); Harata *et al*. (2012); Tito de Morais *et al*. (2015); Lee *et al*. (2016); Smith *et al*. (2018)
Shorea smithiana	0.067	0.012	300	Bo	Tito de Morais *et al*. (2015)
Shorea xanthophylla	0.046; 0.052	0.007; 0.007	100, 120	Bo	Kettle *et al*. (2011); Tito de Morais *et al*. (2015)
Vateria indica	0.122	0.028; 0.031	640, 1200	SI	Ismail *et al*. (2014); Tito de Morais *et al*. (2015)
Vateriopsis seychellarum	0.187	0.041	1200	Bo	Tito de Morais *et al*. (2015)

avoiding populations with high inbreeding such as those found in highly degraded forests will promote genetic diversity in reforestation efforts.

Phenotypic variation/Genetic variation in plant functioning

A limited number of 11 studies addressed intraspecific variation in traits and functions within a total of 17 species of dipterocarps (Table 7.3). These studies show that seeds from different mother trees can vary in size, germination rate, and seedling survival (Ismail *et al.*, 2014; Nutt *et al.*, 2016; Tito de Morais *et al.*, 2020; Takeuchi and Diway, 2021). For example, Tito de Morais *et al.* (2020) found that low genetic diversity increased the probability of seedling mortality over 30 months in three out of four species. Nutt *et al.* (2016) found that seedling survival, but not growth, of seedlings from different mother trees of *Parashorea tomentella* over 18 months increased with greater heterozygosity. Furthermore, Ismail *et al.* (2014) found that the growth of selfed progeny was approximately 75% of the growth of outbred progeny in *Vateria indica*. Axelsson *et al.* (2020) studied seedling performance in a nursery experiment and found that basic characteristics such as growth and susceptibility to herbivory commonly varied among half-sib families originating from different mother trees. Variation was shown to be quite substantial, i.e., in some cases growth varied two-fold among families (Fig. 7.3). Also, in two species, seedling performance was influenced by water availability which indicated intraspecific variation in drought tolerance (Axelsson *et al.*, 2020). Ng *et al.* (2021) exposed seedlings to drought and were able to detect the upregulation of genes associated with drought tolerance. The same "drought tolerance" genes were also shown to be associated with landscape-level variation in precipitation (Ng *et al.*, 2021) providing molecular evidence for the evolutionary relevance of drought in some dipterocarp species despite a general lack of an annual dry season in these systems. Identifying and deploying drought-tolerant genotypes could be useful during reforestation as tropical forests in South-east Asia are confronting longer and more severe droughts as well as record-high temperatures.

Although local adaptation to biotic conditions has seldom been reported, we did find that common garden studies are becoming increasingly used to evaluate local adaptation and population variation in important plant traits. In Sabah, Malaysia, half-sib families from six mother trees of six dipterocarp species from either riparian or upland ecotones were grown in replicated common gardens and compared (O'Brien and Escudero, 2022). While genotypes from riparian and upland ecotones did not appear to be locally adapted to abiotic conditions, when they were exposed to varying levels of tree competition there was variation in growth rates among genotypes from riparian compared to upland ecotones. Such genetic by environmental (G × E) interaction was found for all six species (*Shorea leprosula, Shorea johorensis, Shoea parvifolia, Parashorea malaanonan, Hopea nervosa, Dryobalanops lanceolata*). This suggests that competition has differently impacted the riparian and upland genotypes and may indicate natural selection to the biotic environment; however, given the data provided, it was not possible to tease apart the genetics by multiple interacting environmental factors.

In another common garden experiment, nine populations of *Shorea leprosula* were grown in a common garden in Peninsular Malaysia (Ng *et al.*, 2022). Intraspecific variation in herbivory by a scale insect and growth rate were detected with the fastest growing populations being those that were most resistant to the scale insect. Local adaptation to environmental factors was not evaluated and molecular data was not included. However, a few studies did link molecular data with landscape features in attempts to address how selection to different environments may affect genetic variation (Indriani *et al.*, 2019; Li *et al.*, 2021). For example, Indriani *et al.* (2019) explored to what extent populations of *Shorea balangeran* occurring on wet peat forest and dry heath forest differed genetically but found no evidence of genetic clustering of populations according to forest type. However, the study by Li *et al.* (2021) provides some evidence of the existence of adaptive genetic variation in *Parashorea chinensis* related to soil nutrient availability.

Table 7.3. Evidence of functional variation in plant performance and traits related to genetic variation in the dipterocarps.

Type of study	Species	Level of genetic variation	Life stage	Response	Main findings	References
Functional	*Shorea leprosula, Shorea parvifolia, Parashorea malaanonan, Shorea johorensis, Dryobalanops lanceolata, Hopea nervosa*	Topography of seed-origin	Seedlings	Growth and survival	Within species, topographic seed-origin determined the response of seedling growth and survival.	O'Brien and Escudero (2022)
Molecular +Function	*Vateria indica*	Selfed vs Outbred	Seedlings	Growth	Growth of selfed projeny ~75% of outbred	Ismail *et al*. (2014)
Molecular +Function	*Hopea bilitonensis*	"Seed batches" from two different populations	Seeds	Germination	Germination after 15 days varied ~62–88% among "seed batches."	Lee *et al*. (2013)
Molecular +Function	*Shorea lumutensis*	Half-sib families	Seeds to seedlings	Germination and growth	Variation among mother trees in seed size and germination. Not necessarily translating to performance in the field.	Lee *et al*. (2006)
Molecular +Function	*Neobalanocarpus heimii*	Selfed vs Outbred half-sib families	Seeds	Seed mass and germination	Significant effect of mother tree and selfing on seed mass +germination was influenced by seed mass and mother tree.	Naito *et al*. (2005)
Molecular +Function	*Shorea leprosula*		Seedlings	Upregulation	Drought causes upregulation of "drought tolerance" genes.	Ng *et al*. (2021)
Molecular +Function	*Parashorea tomentella*	Heterozygosity	Seed to seedlings	Gemination, growth rate, and survival	Seedling survival over 18 months increased with heterozygosity.	Nutt *et al*. (2016)
Molecular +Function	*Shoreo laxa*	Parental relatadness	Seed	Seed size	Seed size was decreased with increasing pairwise relatedness of seed parents.	Takeuchi and Diway (2021)
Molecular +Function	*Shorea acuminatissima, Shorea argentifolia, Shorea gibbosa, Parashorea tomentella*		Seedlings	Seedling performance	In three of four species, lower genetic diversity increased the probability of mortality over 30 months.	Tito de Morais *et al*. (2020)
Functional	*Dryobalanops keithii, Dryobalanops lanceolata, Shorea leprosula, Shorea parvifolia, Parashorea smythiesii*	Half-sib families	Seedlings	Growth and drought tolerance	Genetic variation in basic traits such as growth and susceptibility to herbivory. Two out of five species show genetic variation in drought tolerance.	Axelsson *et al*. (2020)
Molecular +Function	*Parashorea chinensis*	Genetic diversity	Overstory trees	Plant community composition	Genetic diversity of *P. chinensis* promotes the genetic diversity of one understory-abundant species.	Li *et al*. (2022)

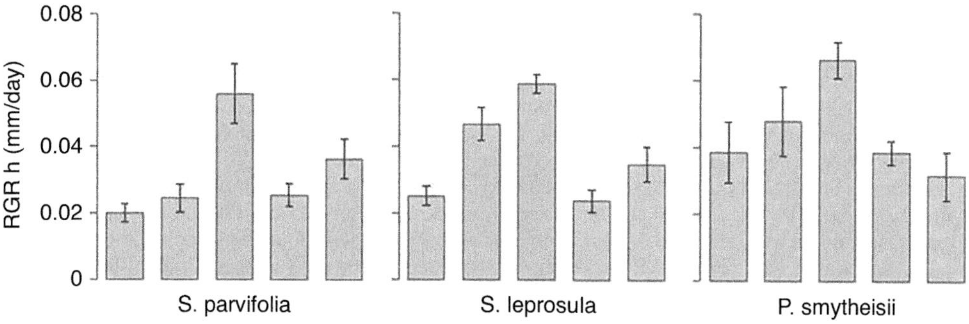

Fig. 7.3. Variation in height growth among five half-sib families of seedlings of different dipterocarp tree species growing in controlled settings in a nursery in the town of Luasong in Sabah, Malaysia on the island of Borneo. The graphs are reproduced from data in Axelsson *et al.* (2020).

Discussion

The aim of this article was to review what we currently know about intraspecific variation in traits and functioning of dipterocarp species, a family of trees dominating mixed dipterocarp forests of South-east Asia and in high need for conservation and restoration (Kettle, 2009), and its potential for application to reforestation. We found a wide range of studies presenting values of both low and high population variation (Fst, Gst, Rst) among tree species (Table 7.1) with a majority of species with moderate to high levels of Fst (74%). A high degree of population differentiation makes such species obvious candidates for screening traits and functions appropriate for site-matching to current and future environmental conditions. Although only a few studies estimated FSGS, there was broad congruence across these studies in the occurrence of FSGS (92% of study species revealed FSGS). A few studies explained high levels of FSGS by differences in gene flow as influenced by species variation in mating systems and pollination syndromes (Takeuchi *et al.*, 2004; Harata *et al.*, 2012; Tani *et al.*, 2012; Tito de Morais *et al.*, 2015). The spatial scale of aggregation of genetically related individuals for species with FSGS differed among species between 50 and 1200m (Tito de Morais *et al.*, 2015). This means that intraspecific variation in traits would be expected for some species at very close distances between trees (i.e. 50 m), and for other species at slightly larger scales (i.e. 1200m between trees). Using seed-sourcing strategies for enhanced functional diversity could potentially benefit from considering the genetic structure of the species. This is so because spatially determined population differentiation is likely to predict how seed sourcing across different spatial scales may impact the level of genetic and phenotypic variation covered (Box 7.1).

The frequent evidence of variation in genetic structure (Fst or FSGS) found in our review was rarely linked to differing phenotypes or plant function. Significant research gaps in research on intraspecific variation in traits and function limit our ability to effectively incorporate genetic variation into reforestation strategies despite the great opportunities that such research has elucidated in non-tropical systems. However, these research gaps highlight the need for future research toward developing better intraspecific site-matching guidelines. Below, we discuss how a better understanding of intraspecific variation in plant traits and functions can improve achieving multi-functional restoration objectives, and we discuss experimental design strategies to screen likely candidate species (i.e. species with high Fst and high FSGS) to identify intraspecific variation in traits and functions. Second, there is an incomplete understanding of local adaptation in dipterocarps and only few studies considering how to conduct reforestation to match appropriate phenotypes to specific environmental conditions of either current or future conditions imposed by environmental change. Following, we discuss the potential for incorporating molecular and phenotypic data to pre-adapt forests to future environmental conditions.

Box 7.1.

A framework for identifying how phenotypic variation in seed sourcing can be maximized based on knowledge about the underlying genetic structure of tree species assuming that genetic variation is linked to phenotypic variation. In species with fine-scale genetic structure (FSGS), where patches of trees can be highly inbreed because of limited pollen transfer and seed dispersal, phenotypic variation is predicted to increase quickly from seed sourcing across even quite small spatial scales. In species with high population structure (Fst), where most genetic variation is occurring across populations, maximising phenotypic variation is most effectively achieved by seed sourcing across populations.

For species A, with high FSGS and high Fst, genetic variation will increase rapidly from seed sourcing across quite small spatial scales (e.g. within a population) and continues to increase with increasing spatial scale (across populations). Focusing seed sourcing from multiple trees within as well as across populations will obtain maximal phenotypic and functional diversity.

For species B, with high FSGS and low Fst, genetic diversity will increase rapidly from seed sourcing across small spatial scales (within population) but does not increase much from seed sourcing on large spatial scales (across populations). For such a species, phenotypic and functional diversity are most effectively enhanced by collecting from as many trees as possible within a single population.

For species C, with low FSGS and High Fst, genetic diversity increase slowly from seed sourcing across small spatial scales (within populations) but will increase from seed sourcing across larger scales (across populations). For such a species, maximal phenotypic and functional diversity will be captured by incorporating as many populations as possible (eg. collecting seeds from across a broad geographic area).

For species D, with low FSGS and low Fst, there is very little increase of genetic variation from seed sourcing across either small scales (within populations) or large scales (across populations). In such species most phenotypic and functional diversity is captured even within a very local population.

Genetic variation in phenotypes and plant function to improve multi-functionality

Only a few studies using a limited number of 17 species confirm intraspecific variation in plant traits and functions such as growth (Ismail *et al.*, 2014; Axelsson *et al.*, 2020) and drought tolerance (Axelsson *et al.*, 2020) in these dipterocarp forests. While these traits/functions are important, there are a wide array of other plant traits and functions that could be selected to improve reforestation effectiveness and to achieve multiple objectives or forest multi-functionality. Understanding intraspecific variation in growth rates and other traits is extremely important for achieving timber and carbon sequestration objectives (Jayawickrama, 2001) and as a starting point for developing native tree breeding programs. The majority of tropical forest restoration efforts continue to rely on exotic non-native tree species due to their rapid growth rates under a variety of environmental conditions. For example, forest landscape restoration (FLR) programs, large-scale projects across thousands of hectares such as those implemented in China (Xu, 2011) and South Korea (Temperton *et al.*, 2014), focus almost exclusively on the use of exotic tree species to increase tree cover in deforested landscapes. However, the selection of these non-native timber species has benefited tremendously from genetic testing for genetic improvement; i.e. the performance of commonly used species such as eucalypts, acacias, and teak, has been greatly improved by sourcing seeds from genotypes and populations with fast-growth characteristics followed by breeding programs using these sources. The high genetic/molecular variation found in many of the species reviewed here suggests that there is great potential for native species to compete with exotic timber using genetic and phenotypic testing. Incorporating a wide array of genetic variation into tree planting programs could promote the use of more native tree species in these programs to thereby enhance multi-functionality. Native species could compete not only on the timber market, but they are also more likely than exotic species to support desired functions such as supporting native food webs and the immense diversity found in the tree canopy and rhizosphere of tropical forests (Charles, 1996; Sakai *et al.*, 1999; Chapman *et al.*, 2018; Engstrom *et al.*, 2020; Axelsson *et al.*, 2022). With the new era defined as the Insect Apocalypse and the 6th Mass Extinction, we suggest that a reforestation framework that supports a large amount of diversity should be prioritized and that developing knowledge of genetic variation in native tree species could help to achieve this objective.

Using a genetic-based trait-to-function framework is an effective way to increase forest functionality during reforestation to obtain multiple evolutionary, ecological, economic, and cultural objectives. In addition to growth rate as a desirable trait that can be selected to achieve multi-functional goals, genetically based trait correlations are commonly used to predict a wide array of other functions. Examples of such functions include; soil microbial diversity (Cowan *et al.*, 2022; Senior *et al.*, 2022), canopy arthropod diversity (Gosney *et al.*, 2017), natural regeneration potential (Mamo *et al.*, 2006), carbon sequestration potential (Jayawickrama, 2001), tolerance to drought and limited soil nutrients (Depardieu *et al.*, 2020; Ismael *et al.*, 2022), bird and mammal abundance and richness (Mann *et al.*, 2012), duration of flowering to support pollinators, duration of fruiting to support animals and human communities (Nyoka *et al.*, 2015), resistance to fires (Hernández *et al.*, 2022), and production of fibers and craft-wood for artisans (Axelsson and Grady, 2022), among others. What our review clearly illustrates is that while such genetic-based trait-to-function frameworks are well established in temperate ecosystems, such a framework has not been well developed in ecosystems dominated by dipterocarps despite evidence of high genetic variation that could be employed to achieve forest restoration objectives.

Climate-based reforestation and seed source selection strategies

While multi-functionality is often a goal of reforestation, there is a need to consider how to maintain functionality during climate change (Prober *et al.*, 2015). Even in tropical forests, drought events, intense rain events, and seasonality is changing as a consequence of global warming and these changes are having large impacts on forest ecosystems (Axelsson *et al.*, 2021; Browne *et al.*, 2021; Miyamoto *et al.*, 2021). Maintaining the fitness of tree populations during global warming may require interventions. How do we identify genetic resources to stabilize functionality during climate change?

One approach is to examine within population variation in functions such as drought tolerance as increasing drought events from climate change are impacting South-east Asian forests (Axelsson *et al.*, 2020). Species with moderate to high FSGS are likely to have low genetic variation at small spatial scales but with increasing scale, genetic differentiation increases. Thus, if seeds are collected from mother trees separated by 1200 m, distinct phenotypes are likely to be encountered (Tito de Morais *et al.*, 2015; Table 7.2). Such phenotypic variation may include different responses to drought as has been found in prior research (Axelsson *et al.*, 2020). While FSGS may result in inbreeding at the local population level, it also implies that there is considerable genetic variation across different patches of trees and that the variation across the landscape may be high. Axelsson *et al.* (2020) found large phenotypic variation in response to water limitation among half-sib-families of *Dryobalanops keitthii* collected from mother trees within a 45 km seed collection diameter. According to this review, a large fraction of dipterocarp tree species contains moderate to high FSGS. One potential method to screen species for the presence of FSGS is to start by looking for species that are likely to have gene flow limitations, i.e., species with small flower size, small pollinators, or large seed mass (*sensu* Harata *et al.* (2012) and Tito de Morais *et al.* (2015)). For species suspected to have high FSGS, which our review indicates is a large fraction of them (92%; Table 7.2) simple nursery progeny experiments can identify drought tolerant genotypes and such genotypes can be increased in frequency in reforestation efforts (Axelsson *et al.*, 2020). In addition to finding within population variation in drought tolerance, there is also potential for populations to vary in response to increasing aridity. Differentiation across larger geographical scales and across populations are more likely to be determined by environmental conditions than mating systems (Carvalho *et al.*, 2019). We found one such study by Ng *et al.* (2021) that characterized drought-tolerant genes in *Shorea leprosula* and found that the same "drought-tolerant" genes were associated with landscape-level variation in precipitation. By incorporating drought tolerant phenotypes at high frequencies relative to the overall phenotypic composition of a seed source used in reforestation (e.g. pre-adaptation *sensu* Axelsson *et al.*, 2020) and/or enhancing overall population genetic diversity as a bet-hedging strategy to allow populations to adapt to a variety of changing environmental conditions (Axelsson *et al.*, 2020), we may promote restoration goals during climate change.

Another method for maintaining plant fitness during climate change is assisted migration - to move populations from drier or warmer areas to wetter or cooler areas in anticipation of climate change (e.g. if climate warming and drought are the anticipated eventuality). Similarly, Prober *et al.* (2015) propose a "climate-adjusted" provenancing strategy combining genetic diversity and adaptability, targeting projected climate change directions while allowing for uncertainty in such projections as well as unforeseen selective agents. However, for this method to be effective, we need to understand if populations are locally adapted to environmental conditions and/or if some populations are simply inherently more tolerant to resource stress. Our review reveals a paucity of studies on local adaptation to abiotic and biotic conditions in the dipterocarps (though see O'Brien and Escudero, 2022). Species with moderate to high Fst may be more likely to show patterns of local adaptation, i.e., populations that do not differ molecularly are less likely to have functional trait variation. Although common garden studies where multiple populations from differing climates are planted in home and away environments (i.e. where away environments may include drier areas that represent future conditions) have been very valuable for investigating local adaptation and the potential for assisted migration in other systems (Grady *et al.*, 2015); such experiments are lacking not only in South-east Asia but are also rare around the global tropics (Chen and Schemske, 2015; Cruz *et al.*, 2019; Barton *et al.*, 2020; Muehleisen *et al.*, 2020). These studies have provided evidence of local adaptation of tropical tree species to climatic factors (Chen and Schemske, 2015; Cruz *et al.*, 2019; Barton *et al.*, 2020; Muehleisen *et al.*, 2020) while a lack of local adaptation has also been found (Fetched *et al.*, 2000). Considering the great wealth of biodiversity in the tropics, it is surprising that more attention has not been paid to natural selection to the environment in the tropics. Summing up the few studies that do exist provides evidence for or against natural selection of tropical trees to climate for less than 20 species.

Developing an understanding of the extent and causes of intraspecific variation in traits and function within the dipterocarps would help in designing seed-sourcing campaigns in a strategic way. For example, including a range of genetic material in reforestation could hedge bets that genetic material suited for future conditions is included in restoration and hence "pre-adapt" developing forests for future conditions (Axelsson *et al.*, 2020). Furthermore, in cases where genetic variation occurs as a consequence of evolutionary adaptation to certain environmental conditions (Li *et al.*, 2021; Ng *et al.*, 2021) such knowledge could be used to direct seed-sourcing campaigns (Prober *et al.*, 2015).

This is currently being done in non-tropical forests; e.g., via assisted migration (Grady *et al.*, 2015; Zucchi *et al.*, 2017) and low-intensity breeding (Pastorino and Marchelli, 2020). Despite good evidence that evolutionary knowledge can increase the effectiveness of reforestation (Thomas *et al.*, 2014), a strategic use of genetic resources is not currently being widely applied to the restoration of native tropical forests. This review suggests that developing such strategies would require a better knowledge of what type of selection pressures are influencing intraspecific variation in these forests and how they are distributed across abiotic and biotic environmental gradients.

Seed sourcing for improving genetic diversity and genetic connectivity

Given that many species of dipterocarps are shown to have FSGS, there are likely special considerations needed for effective seed sourcing for restoration in the face of climate change. Forest degradation commonly results in a loss of genetic diversity in tree populations that can cause inbreeding as has been demonstrated in several dipterocarp tree species (Obayashi *et al.*, 2002; Wickneswari *et al.*, 2004; Ng *et al.*, 2009; Ang *et al.*, 2016; Widiyatno *et al.*, 2016, 2017). Populations of species with FSGS are more likely to experience inbreeding as neighboring individuals tend to be genetically related. As a consequence, we often find decreased performance and fitness of subsequent generations of trees (Ismail *et al.*, 2014; Nutt *et al.*, 2016; Tito de Morais *et al.*, 2020; Takeuchi and Diway, 2021). For example, the study by Widiyatno *et al.* (2016) demonstrates that genetic diversity decreases with degradation from logging, especially with respect to sensitive genetic parameters such as the numbers of rare alleles. Furthermore, species with known limitation in dispersal or that are pollinated by immobile insects (e.g. species with high FSGS) may be extra sensitive to isolation and forest disturbance and hence need extra care in restoration. Designing restoration projects that explicitly consider genetic connectivity may help minimize inbreeding and also promote adaptation to environmental change by maintaining gene flow (Ismail *et al.*, 2014; Nutt *et al.*, 2016; Tito de Morais *et al.*, 2020; Takeuchi and Diway, 2021). This may be especially important in cases when competing land use objectives are causing fragmentation that limits gene flow.

Conclusions

We conclude that there are broad differences among tree species in their levels of intraspecific molecular and phenotypic variation with a high proportion of dipterocarp species having fine-scale genetic structure (92%). The occurrence of FSGS suggests that trees aggregated in a forest patch are highly genetically related to each other with high inbreeding potential. Studies of FSGS also suggest that intraspecific variation occurs across relatively short distances (e.g. from 50 to 1200m between patches). This would also mean that species with FSGS might need special consideration in seed sourcing to prevent inbreeding but also that such species may contain large phenotypic and functional variation on quite small spatial scales. If mother trees for species with FSGS are separated by at least 1200 m, then for 92% of dipterocarps, we would expect phenotypic variation. We suggest that species with high genetic differentiation (high Fst or occurrence of FSGS) are those that are most likely to be usefully employed for increasing multi-functionality and being used in an adaptive capacity either through within population identification of desirable traits for species with high FSGS or through identification of populations that may be pre-adapted to future conditions and used in assisted migration for species with high Fst. By selecting traits such as high tolerance to drought and/or using assisted migration, we can help maintain fitness during climate change. Maintaining genetic diversity and high genetic connectivity also has likely fitness benefits during climate change. However, our major finding that we emphasize here is that there is a major research gap in the understanding of functional variation and the presence of local adaptation within the dipterocarps – this lack of research is consistent globally across tropical forests where very little evolutionary research is applied to ecological restoration despite the clear benefits that have been demonstrated in non-tropical forests. A strategic use of evolutionary science applied to tropical reforestation can improve forest multi-functionality, adaptation to climate change, and the prioritization of native species over exotic species for wood production. While there is still a long way to go to understand how natural selection and other evolutionary factors drive intraspecific genetic variation in the tropics (with research on 20 out of 40,000 known tropical tree species), we suggest that the overwhelming majority of dipterocarp species have both moderate to high Fst and FSGS and that finding species that have useful intraspecific traits to improve reforestation should be relatively easy and immediately useful.

Acknowledgements

This research was funded by FORMAS (Grant Number: 2016–20005), the Swedish research council (Grant Number: 2022–04565) and the United States, Department of State, Fulbright award. We are grateful to two anonymous reviewers for their inputs on earlier versions of this manuscript.

Conflict of interest

The authors have no conflicts of interest to declare.

References

Aitken, S.N. and Whitlock, M.C. (2013) Assisted gene flow to facilitate local adaptation to climate change. *Annual Review of Ecology, Evolution, and Systematics* 44, 367–388.

Alberto, F.J., Aitken, S.N., Alía, R., González-Martínez, S.C., Hänninen, H. *et al.* (2013) Potential for evolutionary responses to climate change – evidence from tree populations. *Global Change Biology* 19, 1645–1661.

Andivia, E., Villar-Salvador, P., Oliet, J.A., Puértolas, J., Dumroese, R.K. *et al.* (2021) Climate and species stress resistance modulate the higher survival of large seedlings in forest restorations worldwide. *Ecological Applications* 31, e02394.

Ang, C.C., O'Brien, M.J., Ng, K.K.S., Lee, P.C., Hector, A. *et al.* (2016) Genetic diversity of two tropical tree species of the *Dipterocarpaceae* following logging and restoration in Borneo: High genetic diversity in plots with high species diversity. *Plant Ecology & Diversity* 9, 459–469.

Axelsson, E.P. and Grady, K.C. (2022) Symphony for the native wood(s): Global reforestation as an opportunity to develop a culture of conservation. *People and Nature* 4, 576–587.

Axelsson, E.P., Abin, J.V., Lardizabal, M.L.T., Ilstedt, U. and Grady, K.C. (2022) A trait-based plant economic framework can help increase the value of reforestation for conservation. *Ecology and Evolution* 12, e8855.

Axelsson, E.P., Franco, F.M., Lussetti, D., Grady, K.C. and Ilstedt, U. (2021) Mega El Niño's change the playing field for culturally important tree species and hence the foundation for human-nature interactions in tropical forests. *Trees, Forests and People* 5, 100109.

Axelsson, E.P., Grady, K.C., Lardizabal, M.L.T., Nair, I.B.S., Rinus, D. *et al.* (2020) A pre-adaptive approach for tropical forest restoration during climate change using naturally occurring genetic variation in response to water limitation. *Restoration Ecology* 28, 156–165.

Barker, G., Hunt, C., Barton, H., Gosden, C., Jones, S. *et al.* (2017) The 'cultured rainforests' of Borneo. *Quaternary International* 448, 44–61.

Bartlett, M.K., Zhang, Y., Yang, J., Kreidler, N., Sun, S. *et al.* (2016) Drought tolerance as a driver of tropical forest assembly: Resolving spatial signatures for multiple processes. *Ecology* 97, 503–514.

Barton, K.E., Jones, C., Edwards, K.F., Shiels, A.B. and Knight, T. (2020) Local adaptation constrains drought tolerance in a tropical foundation tree. *Journal of Ecology* 108, 1540–1552.

BGCI (2021) *State of the World's Trees*. BGCI, Richmond, UK.

Bosshard, E., Jalonen, R., Kanchanarak, T., Yuskianti, V., Tolentino, E. *et al.* (2021) Are tree seed systems for forest landscape restoration fit for purpose? An analysis of four Asian countries. *Diversity* 13, 575.

Browne, L., Markesteijn, L., Engelbrecht, B.M.J., Jones, F.A., Lewis, O.T. *et al.* (2021) Increased mortality of tropical tree seedlings during the extreme 2015–16 El Niño. *Global Change Biology* 27, 5043–5053.

Cao, C.-P., Finkeldey, R., Siregar, I.Z., Siregar, U.J. and Gailing, O. (2006) Genetic diversity within and among populations of *Shorea leprosula* Miq. and *Shorea parvifolia* Dyer (Dipterocarpaceae) in Indonesia detected by AFLPs. *Tree Genetics & Genomes* 2, 225–239.

Cao, C.-P., Gailing, O., Siregar, I.Z., Siregar, U.J. and Finkeldey, R. (2009) Genetic variation in nine *Shorea* species (Dipterocarpaceae) in Indonesia revealed by AFLPs. *Tree Genetics & Genomes* 5, 407–420.

Carvalho, Y.G.S., Vitorino, L.C., de Souza, U.J.B. and Bessa, L.A. (2019) Recent trends in research on the genetic diversity of plants: Implications for conservation. *Diversity* 11, 62.
Chapman, P.M., Wearn, O.R., Riutta, T., Carbone, C., Rowcliffe, J.M. *et al.* (2018) Inter-annual dynamics and persistence of small mammal communities in a selectively logged tropical forest in Borneo. *Biodiversity and Conservation* 27, 3155–3169.
Charles, J.K. (1996) Small mammal diversity in riparian and dipterocarp habitats in Belalong forest, Brunei Darussalam. In: Edwards, D.S., Booth, W.E. and Choy, S.C. (eds) *Tropical Rainforest Research – Current Issues: Proceedings of the Conference held in Bandar Seri Begawan, April 1993*, Springer Netherlands, Dordrecht, pp. 175–182.
Chen, G.F. and Schemske, D.W. (2015) Ecological differentiation and local adaptation in two sister species of *Neotropical costus* (Costaceae). *Ecology* 96(2), 440–449.
Chen, Y., Zhang, H.L., Zhang, L., Nizamani, M.M., Zhou, T. *et al.* (2022) Genetic diversity assessment of *Hopea hainanensis* in Hainan Island. *Frontiers in Plant Science* 13, 1075102.
Chokthaweepanicha, H., Butdavapeeb, S. and Phongkaewb, P. (2022) Genetic diversity and relationships of *Dipterocarpus alatus* Roxb. (Dipterocarpaceae) on island in freshwater lake and nearby mainland. *Agriculture and Natural Resources* 056, 351–362.
Cowan, J.A., Gehring, C.A., Axelsson, E.P., Ilstedt, U., Subbiah, V.K. *et al.* (2022) Ectomycorrhizal (dipterocarp) and arbuscular mycorrhizal (non-dipterocarp) tree hosts and their relative distribution in a tropical forest predict soil bacterial communities. *Journal of Ecology n/a*.
Cruz, M.V., Mori, G.M., Signori-Müller, C., da Silva, C.C., Oh, D.-H. *et al.* (2019) Local adaptation of a dominant coastal tree to freshwater availability and solar radiation suggested by genomic and ecophysiological approaches. *Scientific Reports* 9, 19936.
Dai, Z., Si, C., Zhai, D., Huang, P., Qi, S. *et al.* (2018) Genetic effects of historical anthropogenic disturbance on a long-lived endangered tropical tree *Vatica mangachapoi*. *Journal of Forestry Research* 29, 291–299.
Depardieu, C., Girardin, M.P., Nadeau, S., Lenz, P., Bousquet, J. *et al.* (2020) Adaptive genetic variation to drought in a widely distributed conifer suggests a potential for increasing forest resilience in a drying climate. *New Phytologist* 227, 427–439.
Duc, N.M., Duy, V.D., Xuan, B.T., Thang, B.V., Ha, N.T. *et al.* (2016) Genetic structure of the threatened *Dipterocarpus costatus* populations in lowland tropical rainforests of Southern Vietnam. *Genetics and Molecular Research* 15.
Dwiyanti, F.G., Harada, K., Siregar, I.Z. and Kamiya, K. (2014) Population genetics of the critically endangered species *Dipterocarpus littoralis* blume (Dipterocarpaceae) endemic on Nusakambangan Island, Indonesia. *Biotropia: The Southeast Asian Journal of Tropical Biology* 21, 1–12.
Engelbrecht, B.M.J., Kursar, T.A. and Tyree, M.T. (2005) Drought effects on seedling survival in a tropical moist forest. *Trees* 19, 312–321.
Engstrom, R.T., Edenius, L., Thapa, T.B., Bidari, B., Gurung, A. *et al.* (2020) Bird communities of two forest types in Chitwan Valley, Nepal. *Ornithological Science* 19, 29–40.
FAO (2019) The UN decade on ecosystem restoration 2021–2030 "Prevent, halt and reverse the degradation of ecosystems worldwide. Available at: https://documents.un.org/doc/undoc/gen/n19/060/16/pdf/n1906016.pdf
Fetched, N., Cordero, R.A. and Voltzow, J. (2000) Lack of ecotypic differentiation: Plant response to elevation, population origin, and wind in the Luquillo Mountains, Puerto Nico1. *Biotropica* 32, 225–234.
Finger, A., Kettle, C.J., Kaiser-Bunbury, C.N., Valentin, T., Mougal, J. *et al.* (2012) Forest fragmentation genetics in a formerly widespread island endemic tree: *Vateriopsis seychellarum* (Dipterocarpaceae). *Molecular Ecology* 21, 2369–2382.
Gardiner, R., Shoo, L.P. and Dwyer, J.M. (2019) Look to seedling heights, rather than functional traits, to explain survival during extreme heat stress in the early stages of subtropical rainforest restoration. *Journal of Applied Ecology* 56, 2687–2697.
Gosney, B., O'Reilly-Wapstra, J., Forster, L., Whiteley, C. and Potts, B. (2017) The extended community-level effects of genetic variation in foliar wax chemistry in the forest tree eucalyptus globulus. *Journal of Chemical Ecology* 43, 532–542.
Grady, K.C., Kolb, T.E., Ikeda, D.H. and Whitham, T.G. (2015) A bridge too far: Cold and pathogen constraints to assisted migration of riparian forests. *Restoration Ecology* 23, 811–820.
Grady, K.C., Wood, T.E., Kolb, T.E., Hersch-Green, E.I., Shuster, S.M. *et al.* (2017) Local biotic adaptation of trees and shrubs to plant neighbors. *Oikos* 126, 583–593.

Grady, K.C., Laughlin, D.C., Ferrier, S.M., Kolb, T.E., Hart, S.C., Allan, G.J. and Whitham, T.G. (2013) Conservative leaf economic traits correlate with fast growth of genotypes of a foundation riparian species near the thermal maximum extent of its geographic range. *Functional Ecology* 27, 428–438.

Grant, M.J. and Booth, A. (2009) A typology of reviews: An analysis of 14 review types and associated methodologies. *Health Information & Libraries Journal* 26, 91–108.

Gregorio, N., Herbohn, J., Harrison, S., Pasa, A. and Ferraren, A. (2017) Regulating the quality of seedlings for forest restoration: Lessons from the national greening program in the Philippines. *Small-Scale Forestry* 16, 83–102.

Guan, S.L. and Yen, S.Y. (2000) Conservation of dipterocarpaceae in Peninsular Malaysia. *Journal of Tropical Forest Science* 12(3), 593–615.

Gustafsson, M., Gustafsson, L., Alloysius, D., Falck, J., Yap, S. *et al.* (2016) Life history traits predict the response to increased light among 33 tropical rainforest tree species. *Forest Ecology and Management* 362, 20–28.

Harata, T., Nanami, S., Yamakura, T., Matsuyama, S., Chong, L. *et al.* (2012) Fine-scale spatial genetic structure of ten dipterocarp tree species in a Bornean rain forest. *Biotropica* 44, 586–594.

Hernández, M.A., Butler, J.B., Ammitzboll, H., Freeman, J.S., O'Reilly-Wapstra, J. *et al.* (2022) Genetic variation in fire recovery and other fire-related traits in a global eucalypt species. *Tree Genetics & Genomes* 18, 42.

Indriani, F., Siregar, U., Matra, D. and Siregar, I. (2019) Ecological aspects and genetic diversity of *Shorea balangeran* in two forest types of Muara Kendawangan Nature Reserve, West Kalimantan, Indonesia. *Biodiversitas* 20, 482–488.

IPBES (2019) Global assessment report on biodiversity and ecosystem services of the intergovernmental science-policy platform on biodiversity and ecosystem services Bonn, Germany.

Isaac-Renton, M., Montwé, D., Hamann, A., Spiecker, H., Cherubini, P. *et al.* (2018) Northern forest tree populations are physiologically maladapted to drought. *Nature Communications* 9, 5254.

Ishiyama, H., Inomata, N., Yamazaki, T., Shukor, N.A.A. and Szmidt, A.E. (2008) Demographic history and interspecific hybridization of four *Shorea* species (Dipterocarpaceae) from peninsular Malaysia inferred from nucleotide polymorphism in nuclear gene regions. *Canadian Journal of Forest Research* 38, 996–1007.

Ismael, A., Xue, J., Meason, D., Klápště, J., Gallart, M. *et al.* (2022) Genetic variation in drought-tolerance traits and their relationships to growth in *Pinus radiata* D. Don under water stress. *Frontiers Plant Science* 12.

Ismail, S., Ghazoul, J., Ravikanth, G., Kushalappa, C., Shaanker, R. *et al.* (2014) Fragmentation genetics of *Vateria indica*: Implications for management of forest genetic resources of an endemic dipterocarp. *Conservation Genetics* 15, 533–545.

Jalonen, R., Valette, M., Boshier, D., Duminil, J. and Thomas, E. (2018) Forest and landscape restoration severely constrained by a lack of attention to the quantity and quality of tree seed: Insights from a global survey. *Conservation Letters* 11, e12424.

Jayawickrama, K.J.S. (2001) Potential genetic gains for carbon sequestration: A preliminary study on radiata pine plantations in New Zealand. *Forest Ecology and Management* 152, 313–322.

Kamiya, K., Nanami, S., Kenzo, T., Yoneda, R., Diway, B. *et al.* (2012) Demographic history of *Shorea curtisii* (Dipterocarpaceae) inferred from chloroplast DNA sequence variations. *Biotropica* 44, 577–585.

Kettle, C. (2009) Ecological considerations for using dipterocarps for restoration of lowland rainforest in Southeast Asia. *Biodiversity and Conservation* 19, 1137–1151.

Kettle, C.J., Hollingsworth, P.M., Burslem, D.F., Maycock, C.R., Khoo, E. *et al.* (2011) Determinants of fine-scale spatial genetic structure in three co-occurring rainforest canopy trees in Borneo. *Perspectives in Plant Ecology, Evolution and Systematics* 13, 47–56.

Kusuma, Y.W.C., Ariati, S.R., Risna, R.A., Mitsuyuki, C., Suyama, Y. *et al.* (2019) Seedling selection using molecular approach for ex situ conservation of critically endangered tree species (*Vatica bantamensis* (Hassk.) Benth. & Hook. ex Miq.) in Java, Indonesia. *Tropical Conservation Science* 12, 1940082919849506.

Lee, S., Ang, K.C. and Muhammad, N. (2000) Genetic diversity of *Dryobalanops aromatica* Gaertn. F. (Dipterocarpaceae) in Peninsular Malaysia and its pertinence to genetic conservation and tree improvement. *Forest Genetics* 7, 211–219.

Lee, S., Wickneswari, R., Mahani, M. and Zakri, A.H. (2001) Comparative genetic diversity studies of *Shorea leprosula* (Dipterocarpaceae) using rapd and allozyme markers. *Journal of Tropical Forest Science* 13, 202–215.

Lee, S.L., Chua, L.S.L., Ng, K.K.S., Hamidah, M., Lee, C.T. *et al.* (2013) Conservation management of rare and predominantly selfing tropical trees: An example using *Hopea bilitonensis* (Dipterocarpaceae). *Biodiversity and Conservation* 22, 2989–3006.

Lee, S.L., Ng, K.K.S., Ng, C.H., Tnah, L.H., Lee, C.T. *et al.* (2016) Spatial studies of *Shorea pavifolia* spp. *parvifolia* (Dipterocarpaceae) in lowland and hill Dipterocarp forests. *Journal of Tropical Forest Science* 28, 309–317.

Lee, S.L., Ng, K.K.S., Saw, L.G., Lee, C.T., Muhammad, N. *et al.* (2006) Linking the gaps between conservation research and conservation management of rare dipterocarps: A case study of *Shorea lumutensis*. *Biological Conservation* 131, 72–92.

Li, Q., Xu, Z. and He, T. (2002) Ex situ genetic conservation of endangered *Vatica guangxiensis* (Dipterocarpaceae) in China. *Biological Conservation* 106, 151–156.

Li, Q.-M., Cai, C.-N., Xu, W.-M., Cao, M., Sha, L.-Q. *et al.* (2021) Adaptive genetic diversity of dominant species contributes to species coexistence and community assembly. *Plant Diversity* 44(3), 271–278.

Li, Q.-M., Cai, C.-N., Xu, W.-M., Cao, M., Sha, L.-Q. *et al.* (2022) Adaptive genetic diversity of dominant species contributes to species coexistence and community assembly. *Plant Diversity* 44, 271–278.

Lim, L.-S., Wickneswari, R. and Lee, A.J. (2002) Genetic variation of *Dryobalanops aromatica* Gaertn F. (Dipterocarpaceae) in Peninsular Malaysia using microsatellite DNA markers. *Forest Genetics* 9(2), 125–136.

Mamo, N., Mihretu, M., Fekadu, M., Tigabu, M. and Teketay, D. (2006) Variation in seed and germination characteristics among *Juniperus procera* populations in Ethiopia. *Forest Ecology and Management* 225, 320–327.

Mann, A.N., O'Reilly-Wapstra, J.M., Iason, G.R., Sanson, G., Davies, N.W. *et al.* (2012) Mammalian herbivores reveal marked genetic divergence among populations of an endangered plant species. *Oikos* 121, 268–276.

Miyamoto, K., Aiba, S.-I., Aoyagi, R. and Nilus, R. (2021) Effects of El Niño drought on tree mortality and growth across forest types at different elevations in Borneo. *Forest Ecology and Management* 490, 119096.

Muehleisen, A., Engelbrecht, B., Jones, F., Manzane, E. and Comita, L. (2020) Local adaptation to herbivory within tropical tree species along a rainfall gradient. *Ecology* 101(11), e03151. DOI: 10.1002/ecy.3151.

Muhammad, A.J., Cannon, C.H. and Ratnam, W. (2016) Cross-specific amplification of microsatellite DNA markers in *Shorea platyclados*. *Journal of Forestry Research* 27, 27–32.

Murawski, D.A. and Bawa, K.S. (1994) Genetic structure and mating system of *Stemonoporus oblongifolius* (Dipterocarpaceae) in Sri Lanka. *American Journal of Botany* 81, 155–160.

Naito, Y., Konuma, A., Iwata, H., Suyama, Y., Seiwa, K. *et al.* (2005) Selfing and inbreeding depression in seeds and seedlings of *Neobalanocarpus heimii* (Dipterocarpaceae). *Journal of Plant Research* 118, 423–430.

Ng, C.-H., Lee, S.-L., Tnah, L.-H., Ng, K.K.S., Lee, C.-T. *et al.* (2019) Genetic diversity and demographic history of an upper hill Dipterocarp (Shorea platyclados): Implications for conservation. *Journal of Heredity* 110, 844–856.

Ng, C.H., Ng, K.K.S., Lee, S.L., Suwa, R., Lee, C.T. *et al.* (2022) Growth performance and scale insect infestation of *Shorea leprosula* in a common garden experimental plot. *Journal of Forestry Research* 34, 781–792.

Ng, K.K., Lee, S.L. and Koh, C.L. (2004) Spatial structure and genetic diversity of two tropical tree species with contrasting breeding systems and different ploidy levels. *Molecular Ecology* 13, 657–669.

Ng, K.K.S., Kobayashi, M.J., Fawcett, J.A., Hatakeyama, M., Paape, T. *et al.* (2021) The genome of *Shorea leprosula* (Dipterocarpaceae) highlights the ecological relevance of drought in aseasonal tropical rainforests. *Communications Biology* 4, 1166.

Ng, K.K.S., Lee, S.L. and Ueno, S. (2009) Impact of selective logging on genetic diversity of two tropical tree species with contrasting breeding systems using direct comparison and simulation methods. *Forest Ecology and Management* 257, 107–116.

Ng, K.K.S., Lee, S.L., Saw, L.G., Plotkin, J.B. and Koh, C.L. (2006) Spatial structure and genetic diversity of three tropical tree species with different habitat preferences within a natural forest. *Tree Genetics and Genomes* 2, 121–131.

Nguyen, D.M., Nguyen, H.L.P. and Nguyen, T.M. (2022) Genetic structure of the endemic *Dipterocarpus condorensis* revealed by microsatellite markers. *AoB PLANTS* 14, plac007.

Nguyen, T.M., Vu, D.D., Nguyen, D.M., Dang, H.P., Phan, L.K. *et al.* (2020) Microsatellite analysis reveals genetic diversity of the endangered species *Dipterocarpus dyeri*. *Journal of Forest Research* 25, 198–201.

Nguyen, T.M.P., Tran, T.H., Nguyen, M.D., Sierens, T. and Triest, L. (2014) Genetic population of threatened *Hopea odorata* Roxb. in the protected areas of Vietnam. *Journal of Vietnamese Environment* 6, 69–76.

Nutt, K.S., Burslem, D.F.R.P., Maycock, C.R., Ghazoul, J., Khoo, E. *et al.* (2016) Genetic diversity affects seedling survival but not growth or seed germination in the Bornean endemic dipterocarp *Parashorea tomentella*. *Plant Ecology & Diversity* 9, 471–481.

Nyoka, B.I., Chanyenga, T., Mng'omba, S.A., Akinnifesi, F.K. and Sagona, W. (2015) Variation in growth and fruit yield of populations of *Sclerocarya birrea* (A. rich.) Hochst. *Agroforestry Systems* 89, 397–407.

Obayashi, K., Tsumura, Y., Ihara-Ujino, T., Niiyama, K., Tanouchi, H. *et al.* (2002) Genetic diversity and outcrossing rate between undisturbed and selectively logged forests of *Shorea curtisii* (Dipterocarpaceae) using microsatellite DNA analysis. *International Journal of Plant Sciences* 163, 151–158.

O'Brien, M.J. and Escudero, A. (2022) Topography in tropical forests enhances growth and survival differences within and among species via water availability and biotic interactions. *Functional Ecology* 36, 686–698.

Ogasahara, M., Cobb, A.R., Sukri, R.S., Metali, F. and Kamiya, K. (2023) Genetic structure and population history of a peat swamp forest tree species, *Shorea albida* (Dipterocarpaceae), in Brunei Darussalam. *Genes and Genetic Systems* 98, 35–44.

Ohtani, M., Tani, N., Ueno, S., Uchiyama, K., Kondo, T. *et al.* (2021) Genetic structure of an important widely distributed tropical forest tree, *Shorea parvifolia*, in Southeast Asia. *Tree Genetics and Genomes* 17, 44.

Pandey, M. and Geburek, T. (2010) Genetic differences between continuous and disjunct populations: Some insights from sal (*Shorea robusta* Roxb) in Nepal. *Conservation Genetics* 11, 977–984.

Pastorino, M.J. and Marchelli, P. (2020) *Low Intensity Breeding of Native Forest Trees in Argentina: Genetic Basis for their Domestication and Conservation*. Springer International Publishing.

Pimm, S.L. and Sugden, A.M. (1994) Tropical diversity and global change. *Science* 263, 933–934.

Prober, S., Byrne, M., McLean, E., Steane, D., Potts, B. *et al.* (2015) Climate-adjusted provenancing: A strategy for climate-resilient ecological restoration. *Frontiers in Ecology and Evolution* 3. DOI: 10.3389/fevo.2015.00065.

Rachmat, H.H., Kamiya, K. and Harada, K. (2012a) Contrasting cpDNA variation in two Indonesian endemic lowland dipterocarp species and implications for their conservation. *Pakistan Journal of Biological Sciences: PJBS* 15(16), 783–788.

Rachmat, H.H., Kamiya, K. and Harada, K. (2012b) Genetic diversity, population structure and conservation implication of the endemic Sumatran lowland dipterocarp tree species (*Shorea javanica*). *International Journal of Biodiversity and Conservation* 4, 573–583.

Ritonga, F.N., Dwiyanti, F.G., Kusmana, C., Siregar, U.J. and Siregar, I.Z. (2018) Population genetics and ecology of Sumatran camphor (*Dryobalanops aromatica*) in natural and community-owned forests in Indonesia. *Biodiversitas Journal of Biological Diversity* 19(6), 2175–2182.

Romadini, N.P., Indrioko, S., Widiyatno, W., Faridah, E. and Ratnaningrum, Y.W.N. (2021) Genetic diversity in seedling populations of *Dipterocarpus gracilis* in Kecubung Ulolanang nature conservation reserve,Indonesia. *Biodiversitas* 22.

Sakai, S., Momose, K., Yumoto, T., Kato, M. and Inoue, T. (1999) Beetle pollination of *Shorea parvifolia* (Section Mutica, Dipterocarpaceae) in a general flowering period in Sarawak, Malaysia. *American Journal of Botany* 86, 62–69.

Senakun, C., Changtragoon, S., Pramual, P. and Prathepha, P. (2011) Genetic structure and diversity of *Shorea obtusa* (Dipterocarpaceae) in Thailand. *Journal of Systematics and Evolution* 49, 120–125.

Senior, J.K., Gundale, M.J., Iason, G.R., Whitham, T.G. and Axelsson, E.P. (2022) Progeny selection for enhanced forest growth alters soil communities and processes. *Ecosphere* 13, e3943.

Smith, J.R., Ghazoul, J., Burslem, D.F.R.P., Itoh, A., Khoo, E. *et al.* (2018) Are patterns of fine-scale spatial genetic structure consistent between sites within tropical tree species? *Plos One* 13, e0193501.

Takeuchi, Y. and Diway, B. (2021) Long pollen dispersal prevents biparental inbreeding depression in seeds in a natural population of the tropical tree *Shorea laxa*. *Forest Ecology and Management* 489, 119063.

Takeuchi, Y., Ichikawa, S., Konuma, A., Tomaru, N., Niiyama, K. *et al.* (2004) Comparison of the fine-scale genetic structure of three dipterocarp species. *Heredity* 92, 323–328.

Tam, N.M., Duy, V.D., Duc, N.M., Giap, V.D. and Xuan, B.T. (2014) Genetic variation in and spatial structure of natural populations of *Dipterocarpus alatus* (Dipterocarpaceae) determined using single sequence repeat markers. *Genetics and Molecular Research* 13, 5378–5386.

Tam, N.M., Duy, V.D., Duc, N.M., Thanh, T.T.V., Hien, D.P. *et al.* (2019) Genetic variation and outcrossing rates of the endangered tropical species *Dipterocarpus dyeri*. *Journal of Tropical Forest Science* 31, 259–268.

Tang, W.-X., Dai, W.-J., Hu, X.-H. and Huang, S.-X. (2015) ISSR-based genetic diversity of wild and ex situ conserved population of *Hopea chinensis* (Merr.) Hand.-Mazz.) *Bangladesh Journal of Botany* 44, 407–413.

Tani, N., Tsumura, Y., Fukasawa, K., Kado, T., Taguchi, Y. *et al.* (2012) Male fecundity and pollen dispersal in hill dipterocarps: Significance of mass synchronized flowering and implications for conservation. *Journal of Ecology* 100, 405–415.

Temperton, V.M., Higgs, E., Choi, Y.D., Allen, E., Lamb, D. *et al.* (2014) Flexible and adaptable restoration: An example from South Korea. *Restoration Ecology* 22, 271–278.

Thomas, E., Jalonen, R., Loo, J., Boshier, D., Gallo, L. *et al.* (2014) Genetic considerations in ecosystem restoration using native tree species. *Forest Ecology and Management* 333, 66–75.

Tinio, C., Finkeldey, R., Prinz, K. and Fernando, E. (2014) Genetic variation in natural and planted populations of *Shorea guiso* (Dipterocarpaceae) in the Philippines revealed by microsatellite DNA markers. *Asia Life Sciences* 23, 75–91.

Tito de Morais, C., Ghazoul, J., Maycock, C.R., Bagchi, R., Burslem, D. *et al.* (2015) Understanding local patterns of genetic diversity in dipterocarps using a multi-site, multi-species approach: Implications for forest management and restoration. *Forest Ecology and Management* 356, 153–165.

Tito de Morais, C., Kettle, C.J., Philipson, C.D., Maycock, C.R., Burslem, D.F.R.P. *et al.* (2020) Exploring the role of genetic diversity and relatedness in tree seedling growth and mortality: A multispecies study in a Bornean rainforest. *Journal of Ecology* 108, 1174–1185.

Tnah, L.H., Lee, S.L., Ng, K.K., Lee, C.T., Bhassu, S. *et al.* (2013) Phylogeographical pattern and evolutionary history of an important Peninsular Malaysian timber species, *Neobalanocarpus heimii* (Dipterocarpaceae). *Journal of Heredity* 104, 115–126.

Utomo, S., Uchiyama, K., Ueno, S., Matsumoto, A., Widiyatno, S.I. *et al.* (2018) Effects of pleistocene climate change on genetic structure and diversity of *Shorea macrophylla* in Kalimantan rainforest. *Tree Genetics and Genomes* 14, 44.

Vu, D.-D., Bui, T.T.-X., Nguyen, M.-D., Shah, S.N.M. and Vu, D.-G. (2019) Genetic diversity and conservation of two threatened dipterocarps (Dipterocarpaceae) in Southeast Vietnam. *Journal of Forestry Research* 30, 1823–1831.

Wang, C., Ma, X., Ren, M. and Tang, L. (2020) Genetic diversity and population structure in the endangered tree *Hopea hainanensis* (Dipterocarpaceae) on Hainan Island. *Plos One* 15, e0241452.

Wickneswari, R., Ho, W.-S., Lee, K. and Lee, C.T. (2004) Impact of disturbance on population and genetic structure of tropical forest trees. *Forest Genetics* 11(3), 193.

Widiyatno, S.I., Na'iem, M., Uchiyama, K., Numata, S., Ohtani, M. *et al.* (2016) Effects of different silvicultural systems on the genetic diversity of *Shorea parvifolia* populations in the tropical rainforest of Southeast Asia. *Tree Genetics and Genomes* 12, 73.

Widiyatno, W., Indrioko, S., Na'iem, M., Purnomo, S., Hosaka, T. *et al.* (2017) Effects of logging rotation in a lowland dipterocarp forest on mating system and gene flow in *Shorea parvifolia*. *Tree Genetics and Genomes* 13.

Xu, J. (2011) China's new forests aren't as green as they seem. *Nature* 477, 371.

Zucchi, M.I., Sujii, P.S., Mori, G.M., Viana, J.P.G., Grando, C. *et al.* (2017) Genetic diversity of reintroduced tree populations in restoration plantations of the Brazilian Atlantic forest. *Restoration Ecology* 26, 694–701. DOI: 10.1111/rec.12620.

Zulfahmi, Z., Mahfira, U., Siregar, U., Siregar, I. and Yunanto, T. (2015) Comparison of levels of chloroplast DNA diversity of two *Shorea* species with contrasting geographical distribution. *Asia-Pacific Journal of Molecular Biology and Biotechnology* 23, 291–302.

8 Sustainable agricultural practices in South Asia: A comprehensive review

L.R. Panchami[1], Nachiket Gudi[2] and Divya Sussana Patil[2]*

[1]Department of Health Information, Prasanna School of Public Health, Manipal Academy of Higher Education, Manipal, 576104, India; [2]Public Health Evidence South Asia, Department of Health Information, Prasanna School of Public Health, Manipal Academy of Higher Education, Manipal, 576104, India

Abstract

Overpopulation is a root cause that impedes food production and supply to the continuously growing population. Since agriculture is the primary source of food production, producers are inclined towards various unsustainable agriculture practices contributing to climate change. Therefore, sustainable agricultural practices have gained the spotlight to reduce the risk of climate change and increase crop yields. Specifically, South Asian countries take a significant hit with respect to agriculture and climate change because much of the population is earning their livelihood through agriculture. We conducted a comprehensive review to map different sustainable agriculture practices and understand how these practices impact climate change, the livelihood of farmers, and food security in South Asia. A systematic search carried out on PubMed (NCBI) and Scopus (Elsevier), resulted in the inclusion of 42 articles for analysis. The review identified agroforestry and organic farming as the most commonly used practices in the South Asian region, followed by zero tillage practices, use of hybrid seeds, residue and integrated pest management, and cropping systems. These sustainable agriculture practices ensure that sufficient food can be produced to feed the rising population with efficient use of natural resources. In addition, sustainable agriculture gains economic profitability for farmers and improved employment opportunities for women. Since sustainable agricultural practices are the best way forward, we recommend encouraging it by incentivizing the farmers. Due to contextual variations, future research can focus on identifying best practices for specific contexts.

Keywords: sustainable agriculture, climate change, agricultural practices, climate-smart agriculture, South Asia

Introduction

Globally, food security is a concern due to the increasing population. To balance the food supply along with the rising population, unsustainable methods of food production are adopted leading to environmental damage (Rasul, 2021; Stray Dog Institute, 2021). However, to achieve food security and reduce environmental threats, the concept of sustainable food systems can be employed for environmentally-friendly techniques (Boix-Fayos and de Vente, 2023). Sustainable food system is 'a food system that delivers food security and nutrition for all in such a way that the economic, social, and environmental bases to generate food security and nutrition for future generations

*Corresponding author: divya.patil@manipal.edu

are not compromised' (Dupouy and Gurinovic, 2020). Food systems refer to 'the entire range of actors and their interrelated value-adding activities involved in the production, aggregation, processing, distribution, consumption, and disposal of food products that originate from agriculture, forestry, or fisheries, as well as the wider economic, social, and natural environments within which they are embedded' (Braun *et al.*, 2023). However, this article focuses only on food production originating from agriculture and forestry (agroforestry).

Review Methodology

Information sources: A systematic search was carried out on PubMed (NCBI) and Scopus (Elsevier) from the inception till April 2023. Keywords were identified by referring to previous literature (Knox *et al.*, 2012; Smith and Gregory, 2013; Aryal *et al.*, 2020a), and a search strategy was developed. An example of a search strategy on PubMed is given in Appendix 1 (supplementary file). The search query was used for PubMed (NCBI) and later adapted for the Scopus (Elsevier) database accordingly. The terms were combined using Boolean operators such as 'AND' or 'OR.' The search was limited to English language literature only.

Study selection: All identified reports were exported to Rayyan software (Ouzzani *et al.*, 2016), and duplicates were removed. The articles were screened for titles and abstracts by the first author and cross-checked by another co-author based on the inclusion and exclusion criteria given in Table 8.1. Further, full-text articles were retrieved and screened for inclusion in the analysis by two reviewers.

Data extraction: Data was extracted using a predetermined data extraction sheet by the first author and cross-checked by another co-author. We extracted data from the following fields: author, title, place, objective, practices mentioned, findings from the study, author conclusion, and outcome.

Transforming Agricultural Practices

As stated by the Food and Agriculture Organization (FAO), sustainable food and agriculture have the potential to produce nutritious foods without compromising on future needs and natural resources (Food and Agriculture Organization of the United Nations, 2023). Food security is also one of the aims of the sustainable development goals (SDGs) adopted in 2015 to end hunger, improve nutrition, and achieve food security by 2030 (United Nations, 2023). Effective utilization of natural resources such as sustainable land and water use management, cutting down on synthetic fertilizers to enhance the soil properties, and mitigating global warming was ensured in the European farming system. Furthermore, modifying agricultural practices that support biodiversity, purchasing organic products, using eco-friendly food packaging, and food waste minimization were followed in Europe for efficient food production (Zaharia *et al.*, 2021). Additionally, farmers in the European Union used precision farming to ensure that plant health is maintained at its best using fewer resources (Barnes *et al.*, 2019). Such techniques can give maximum yield and have proven to be environmentally safe (Ashley *et al.*, 2020). For example, tomatoes were grown in a greenhouse using the hydroponics system, where the greenhouse roof is made of glass as it captures heat during the day and keeps the greenhouse warm when it gets cold at night. LED lights allow the plants to grow all night, and pesticides and herbicides are not used (Os *et al.*, 2016). In Sweden, food waste was converted into biogas, which was used as fuel for public transport (Liu *et al.*, 2022). In 2016, to minimize overall greenhouse gas (GHGs) emissions, a total of 195 countries signed the Paris Agreement on climate change, where a pledge was made to minimize GHG emissions (Seo, 2017). Likewise, in India, phosphorus from industries is being leached off heavily into local water bodies causing eutrophication which is a major environmental issue. Practices such as precipitation can reduce phosphorous runoff and sustainably maintain its levels, as phosphorous is an essential nutrient for crop development. Struvite precipitation can recover around 80–90% of phosphorous from wastewater and is a sustainable practice as it slowly provides phosphorous to the soil compared to phosphorous fertilizer (Vasa and Pothanamkandathil Chacko, 2021). In Nepal, adopting good agricultural practices (GAP)

Table 8.1. Inclusion and exclusion criteria.

	Inclusion	Exclusion
Context	South Asian countries (World Bank classification): India, Sri Lanka, Bangladesh, Pakistan, Nepal, Afghanistan, Bhutan, Maldives (World Bank Group, 2023).	Countries other than those under the South Asian region
Intervention	Sustainable agricultural and food system practices: Sustainable food production, global food systems, sustainable food preparation, food security, food insecurity, food supply chain, sustainable agriculture practices, agro-ecology, regenerative agriculture, organic farming, natural farming, less resource-intensive farming, sustainable agriculture system, agroforestry, precision farming, integrated pest management, biodynamic agriculture, partial conservation agriculture, vermicompost, permaculture, farm productivity, nutrition security, food security policy, agricultural sustainability, sustainable agricultural development, sustainable farming, environmentally sustainable, feeding system, food production system, food processing, agricultural production, food supply, food availability, food accessibility.	Aquatic ecosystem, marine fisheries, livestock farming, animal husbandry, food delivery system, food web, fish supply, animals, agricultural biotechnology, household cooking.
Outcome	Primary outcomes: Climate change, global warming, climate emergency, greenhouse gas emissions, food related GHG emissions, climate crisis, global heating, temperature change, environmental impacts, global emissions, environmental pollution, green revolution, biodiversity loss, resource depletion, water quality, ecosystem, atmospheric conditions, climate variability, climate uncertainty, meteorological conditions, extreme weather events, air pollution, deforestation, irregular rainfall. Secondary outcomes: malnutrition, hunger, rural employment, livelihood, women's employment, women's empowerment, women's income.	Water pollution, obesity, energy use, monsoon variability, groundwater depletion, green revolution, blue economy, indoor air pollution, coastal hazards.
Study design	Peer reviewed primary studies such as surveys, impact evaluations, and qualitative research.	Systematic reviews, rapid reviews, umbrella reviews, scoping reviews, narrative reviews, literature reviews, meta-analysis, meta-synthesis, book chapters, and studies where full text was not available.

helped improve crop yield by 36%, and further enhanced farmers' income by more than 100% since they could use the natural resources more effectively, enhancing soil richness and ensuring food security (Kharel *et al.*, 2022).

Climate Change and Agriculture

Agriculture is affected by climate change, particularly in South Asian countries, where employment share in the agricultural sector is

41% of total employment in 2021 (Elder and Viegelahn, 2022). Approximately 57% of the land in South Asia is deemed appropriate for cultivation, offering employment to around 40% of the people in South Asia in the agriculture sector (Bibi *et al.*, 2021). With South Asia having the second-highest population in the world, farming plays a crucial role in maintaining the economy and providing nutritious food for a large population (Asian Development Bank, 2021). The agriculture sector is both a contributor and a victim of climate change. Improper agriculture techniques contribute to the emission of GHG, leading to global warming, demonstrating the interlinkage between agriculture and climate change (Bhatt *et al.*, 2019). Such threats to the environment can be tackled by using climate-smart agricultural practices (Das *et al.*, 2022). Climate-smart agriculture aims to help farmers adopt sustainable techniques that have a favourable impact on soil and its properties. It assists in effective land and water management, balanced nutrient intake, and safe handling practices during pest control. Besides, it helps lower GHG emissions and develops climate-tolerant crop species (Bhatt *et al.*, 2019). Traditional form of rice cultivation is a significant source of methane emissions, due to churning of the soil with water. Therefore, employing direct seeded rice cultivation can reduce GHG emissions (Aryal *et al.*, 2020a). A study reported in Punjab state of India successfully showed that the direct seeded rice technique reduced the GHG emissions compared to puddling of the rice field (Pathak *et al.*, 2013). Additionally, using organic fertilizers enhances soil properties by increasing the soil's organic carbon by 2.25% and helping mitigate GHG emissions (Haque *et al.*, 2021).

Sustainable agriculture can improve nutritional intake by improving productivity and combining nutrient-rich foods with efficient resource use (Pradhan *et al.*, 2021). Agroforestry, a sustainable practice, merges crops with trees that provide benefits such as fuelwood, fruits, and medicinal plants, ensuring the livelihood security of the farmers (Islam *et al.*, 2022). However, in addition to the lack of knowledge about the practices among farmers, factors such as age, education, gender, and income influenced the adoption of sustainable agriculture practices in South Asia (Begho *et al.*, 2022).

Previous research has focused on various sustainable agriculture practices that are practiced across South Asian countries and their impact on climate change (Aryal *et al.*, 2020b), climate resilient agriculture practices (Aryal *et al.*, 2021), the impact of climate disasters on agriculture (Chhogyel and Kumar, 2018; Habib-ur-Rahman *et al.*, 2022; Nor Diana *et al.*, 2022), the influence of agricultural practices on livelihoods of farmers (Ali and Erenstein, 2017), and traditional farming practices in India (Aich *et al.*, 2022). However, there is limited knowledge of how various sustainable agriculture practices influence climate change, livelihood, and food security of the farmers in South Asia. Therefore, in this comprehensive review, we aim to map available evidence on different types of sustainable agriculture practices in South Asia and understand how these practices influence climate change, the livelihood of farmers, and food security.

Results

A total of 4482 results were exported to Rayyan. On de-duplication (n=74), 4408 articles were screened for title and abstract based on the eligibility criteria, of which (n=111) were included for full-text screening. Finally, 42 articles were selected for data extraction. The results of the review are presented narratively under six sustainable agriculture practices namely agroforestry, tillage practices, organic farming, hybrid seeds, cropping systems, and residue management.

There were 25 studies conducted in India (Yadvinder-Singh *et al.*, 2004; Agoramoorthy *et al.*, 2012; Ramesh *et al.*, 2015; Sharma *et al.*, 2016; Sapkota *et al.*, 2017; Das *et al.*, 2018; Eyhorn *et al.*, 2018; Parihar *et al.*, 2018; Arora *et al.*, 2019; Jat *et al.*, 2019; Kumar *et al.*, 2019; Rizvi *et al.*, 2019; Adhikari *et al.*, 2020; Meetei *et al.*, 2020; Chaudhary *et al.*, 2021; Chavan *et al.*, 2021; Kumar *et al.*, 2021; Pradhan *et al.*, 2021; Ahmad *et al.*, 2022; Das *et al.*, 2022; Dawar *et al.*, 2022; Dhanya *et al.*, 2022; Jinger *et al.*, 2022; Dubey *et al.*, 2023; Fagodiya *et al.*, 2023), eleven studies were conducted in Bangladesh (Rasul and Thapa, 2004; Rahman *et al.*, 2008; Nath *et al.*, 2011; Bala and Hossain,

2013; Rahman, 2013; Acharjee *et al.*, 2019; Emran *et al.*, 2019; Shew *et al.*, 2019; Haque *et al.*, 2021; Aravindakshan *et al.*, 2022; Islam *et al.*, 2022), three studies in Pakistan (Imran *et al.*, 2022; Raza *et al.*, 2022; Becker *et al.*, 2023), two studies were conducted in Nepal (Atreya *et al.*, 2008; Bajracharya *et al.*, 2021), and one study was conducted in Sri Lanka (Williams and Carrico, 2017).

Agroforestry practices were reported in ten studies (Rahman *et al.*, 2008; Nath *et al.*, 2011; Ramesh *et al.*, 2015; Rizvi *et al.*, 2019; Adhikari *et al.*, 2020; Chavan *et al.*, 2021; Kumar *et al.*, 2021; Ahmad *et al.*, 2022; Islam *et al.*, 2022; Jinger *et al.*, 2022), seven studies reported findings on zero tillage practices (Atreya *et al.*, 2008; Sapkota *et al.*, 2017; Parihar *et al.*, 2018; Kumar *et al.*, 2019; Chaudhary *et al.*, 2021; Aravindakshan *et al.*, 2022; Fagodiya *et al.*, 2023), eleven studies on organic farming (Rasul and Thapa, 2004; Yadvinder-Singh *et al.*, 2004; Agoramoorthy *et al.*, 2012; Bala and Hossain, 2013; Eyhorn *et al.*, 2018; Emran *et al.*, 2019; Haque *et al.*, 2021; Pradhan *et al.*, 2021; Das *et al.*, 2022; Dawar *et al.*, 2022; Dubey *et al.*, 2023), three studies on practices using hybrid seeds (Sharma *et al.*, 2016; Williams and Carrico, 2017; Shew *et al.*, 2019), four studies on residue management and integrated pest management (Rahman, 2013; Arora *et al.*, 2019; Bajracharya *et al.*, 2021; Raza *et al.*, 2022) and cropping system practices were reported by six studies (Das *et al.*, 2018; Acharjee *et al.*, 2019; Jat *et al.*, 2019; Meetei *et al.*, 2020; Imran *et al.*, 2022; Becker *et al.*, 2023). A detailed study characteristic table is given in Appendix 2 (supplementary file).

Agroforestry

Agroforestry is defined as 'the incorporation of trees with crops on the same piece of land'. The review identified that this practice helps in carbon sequestration by storing atmospheric carbon in the soil by 35–70% (Rizvi *et al.*, 2019) and prevents the emission of GHG and global warming (Ramesh *et al.*, 2015). In India (Uttar Pradesh, Madhya Pradesh, and Rajasthan), agroforestry decreased emissions by about 5.24% from the agricultural sector (Rizvi *et al.*, 2019). In hilly regions of north-east India, it was found that Alder trees had a huge potential to improve the sustainability of the agroforestry system by increasing total organic carbon by 52% in the surface soil (Ramesh *et al.*, 2015). In Uttarakhand, it was found that Q. *leucotrichophora* (bang oak tree) based agroforestry systems can isolate huge amounts of carbon and guide in mitigating climate change because it helped store 30–45% of carbon in their biomass compared to a forest (Kumar *et al.*, 2021). Agroforestry opts for zero tillage, which is essential to recover the ravine lands and improve productivity (Jinger *et al.*, 2022). Evidence indicated that landscapes in Jharkhand were best suited for agroforestry practices, suggesting a need for policies to expand agroforestry, particularly in Jharkhand, to secure the livelihoods of farmers and eliminate poverty (Ahmad *et al.*, 2022). In the central Himalayas, both Agri-silviculture and Agri-horticulture systems increased crop yield, crop productivity, and carbon sequestration in India (Adhikari *et al.*, 2020). Policy interventions such as 'Green India Mission 2003', 'National Agroforestry Policy 2014', and 'Green Highway Policy 2015' were undertaken to expand agroforestry in India (Chavan *et al.*, 2021).

In Bangladesh, two common agroforestry practices were practiced i.e., homestead agroforestry, which is mainly practiced by people for everyday consumption, and cropland agroforestry, which includes intercropping of crops and trees on the same land. Cropland agroforestry is further divided into two types, i.e., boundary plantation (hedgerows) and alley cropping, of which alley cropping was practiced extensively in Bangladesh because it helped improve farmers' income (Rahman *et al.*, 2008). The practice of agroforestry was highlighted in Bangladesh since it helped increase soil organic matter, maintain soil quality, balance nutrient content, and reduce air pollution and groundwater depletion (Islam *et al.*, 2022). In addition, areca and betel-based agroforestry in Bangladesh was identified as a common practice under homestead agroforestry. One of the significant advantages of this practice is that it can grow on land unsuitable for cultivation. In the face of the climate crisis, poor farmers in Bangladesh carry out this practice to sustain their livelihoods (Nath *et al.*, 2011).

Tillage practices

Traditional practices involve intensive tillage that degrades the soil quality and produces high nitrous oxide emissions compared to conservation agriculture, which has a higher greenhouse effect in comparison to carbon dioxide (Parihar *et al.*, 2018). However, zero tillage practice involves keeping residues rather than burning them, positively impacting the environment, and enhancing soil quality. The review identified that in India, conservation agriculture was practiced using zero tillage, which was soil-friendly and helped mitigate global warming (Sapkota *et al.*, 2017; Parihar *et al.*, 2018; Chaudhary *et al.*, 2021). Relay cropping that does not follow tillage practice was beneficial in cultivating winter crops such as Lathyrus (wild pea) and lentils in Patna, India. This practice released lower GHG, increased production, and improved soil quality (Kumar *et al.*, 2019). The drip irrigation and sprinklers in conservational tillage and residue management practices in the western Indo-Gangetic plain of India consumed less water and energy, positively impacting groundwater depletion and global warming (Fagodiya *et al.*, 2023). In Nepal and Bangladesh, mulching and reduced tillage had a positive impact on the soil properties in comparison with conventional tillage (Atreya *et al.*, 2008; Aravindakshan *et al.*, 2022).

Organic farming

Our review identified that organic amendments such as farmyard manure (FYM), green manure (GM), biochar, and vermicompost play a vital role in building sustainable agriculture by enhancing soil properties and crop development (Yadvinder-Singh *et al.*, 2004; Dawar *et al.*, 2022). In India, organic farming was identified as a successful option for small-scale farmers because more prominent companies tend to buy food products grown organically directly from farmers and provide decent pay for them (Eyhorn *et al.*, 2018). Another practice called double transplantation used farmyard manure to help increase crop yields. In addition, it provided more job opportunities for women because in double transplantation, primarily seeds are sown in the primary nursery, followed by implantation in a secondary nursery, and finally transplanted to the main rice field. Double transplantation requires more labour since it involves transplanting the rice seedling onto two lands (Vijayakumar *et al.*, 2019; Dubey *et al.*, 2023). By using organic manure for plants, women were able to cultivate vegetables, which made them financially independent (Agoramoorthy *et al.*, 2012). This improved their affordability for healthcare and other services. The farming system for nutrition (FSN) was practiced in India, which aims to create awareness about nutrition through group discussions on growing nutritionally sound crops, livestock farming, and encouraging homestead farming. This helped to meet people's nutritional requirements (Pradhan *et al.*, 2021).

In Bangladesh, ecological agriculture was employed as it consumed fewer pesticides and chemicals and encouraged the use of organic manure to deliver good quality food for the consumers, improved water retention capacity, and reduced soil erosion (Rasul and Thapa, 2004). Another study in Bangladesh reported that using recommended fertilizers and organic material improved overall crop development and was more efficient in reducing methane gas emissions (Haque *et al.*, 2021). With increasing risks to food security and climate change, cultivating horticultural crops is the best method to meet the food demand with a favourable impact on nature (Bala and Hossain, 2013). Climate-smart agricultural practice involving natural resource management using organic manure was practiced in India and Bangladesh, which increased agricultural production with minimal impact on the environment (Emran *et al.*, 2019; Das *et al.*, 2022).

Hybrid seeds

The review highlights that aerobic rice cultivation requires less water than flood-transplanted (FT) rice producing high yields (Sharma *et al.*, 2016). Adding organic amendments to it further helped reduce the emission of methane and nitrous oxide, minimizing the risk of global warming. Farmers in Sri Lanka employed kakulama and parachute methods to overcome water shortages. In the kakulama method, dry seeding consumes less irrigated water. In the parachute method, seedlings are planted inside flexible trays and transplanted into muddy fields, which can help to produce high yields (Williams and Carrico, 2017). In Bangladesh, high-yielding rice varieties were more convenient than traditional ones, creating more yields in all three seasons and positively impacting the environment by reducing global warming and water consumption (Shew *et al.*, 2019).

Residue management and integrated pest management

Burning of crop residues releases huge amounts of GHGs that harm nature and human health. In the interest of nature and human health, sustainable residue management practices (SRMPs) such as residue decomposition were practiced in Pakistan that improved soil properties, reduced soil erosion, and balanced plant nutrient availability (Raza *et al.*, 2022). The study further reported that crop residues were used to feed livestock and as beds in animal husbandry during winter, increasing the milk produced by the animals (Raza *et al.*, 2022). Integrated pest management (IPM) was practiced in Bangladesh to reduce the harmful effects of pesticides on human health and the environment (Rahman, 2013). It was identified that IPM could protect the environment and biodiversity and can control weeds. In contrast, non-integrated pest management (NIPM) seriously threatens the environment and biodiversity and hampers crop development (Arora *et al.*, 2019).

Cropping systems

In India, crop rotation was identified to impact soil organic carbon (SOC) significantly. Leguminous plants such as peas and French beans have a beneficial impact on soil properties because of their ability to store nutrients in root nodules (Meetei *et al.*, 2020). Another study identified a modified system of rice intensification (MSRI) that proved effective in crop development by using less water, reducing the use of chemical fertilizers, and increasing the SOC, which increased carbon sequestration (Das *et al.*, 2018). In Pakistan, to battle against heat stress, employing heat-resistant crops and alteration in planting crops was proven more effective than water-associated tactics because cultivation was mainly affected by changing temperatures (Becker *et al.*, 2023). However, another study in Pakistan reported that adopting water-innovative practices in climate-smart agriculture resulted in better profits (Imran *et al.*, 2022).

Discussion

This review provides an overview of sustainable food production practices in South Asia to the existing literature. When the population and demand for food are on the rise, keeping the land healthy and replenished can be of great significance. Sustainable agricultural practices can reduce environmental pollution by limiting the use of chemicals, therefore conserving the environment. The findings from this review will help researchers in the field to explore further the suitable practices for sustainable food production in different contexts within the South Asian region. Our findings showed that agroforestry can reduce GHG emissions by storing carbon in the soil and increasing SOC. Similar results were seen in Belgium and other parts of Europe, where SOC was increased due to agroforestry (Aertsens *et al.*, 2013; Pardon *et al.*, 2017). Alley cropping, an agroforestry practice, was widespread in Bangladesh that helped to improve farmers' income. A report from the project 'AGFORWARD' in Europe also highlighted similar results, where alley cropping was practiced experimentally in Germany (Mosquera-Losada *et al.*, 2016). Though agroforestry has been practiced in India for a very long time, its adoption by farmers has been low in India (Sharma *et al.*, 2017). Since agroforestry has the potential to mitigate global warming and help farmers to improve their income levels, it is important to identify the challenges faced by farmers at the ground level and find solutions to overcome them and adopt agroforestry practices. In 2022, 198 countries have pledged to reduce carbon emissions and become carbon-negative countries. Of the 198 countries, 4.5% (Liberia, Benin, Guyana,

Bhutan, Suriname, Guinea-Bissau, Cambodia, Madagascar, and Gabon) have succeeded in achieving carbon neutrality, 10.6% (Nigeria, Israel, Australia, Eswatini, Bahrain, Congo, Saudi Arabia, Vietnam, United Arab Emirates, India, Andorra, Russian Federation, South Africa, Ghana, Malaysia, Thailand, Zimbabwe, Côte d'Ivoire, Cameroon, Kazakhstan, and Estonia) have said to achieve carbon neutrality between 2050 and 2070, carbon neutrality was legislated by 8.6%, policies have been regulated by 29.3%, and the remaining 47% are still in the operational level (Chen *et al.*, 2022).

The review highlighted that zero tillage did not disturb soil properties and had a beneficial impact on the environment. Similar findings on soil conservation through zero tillage and conservation agriculture techniques were reported in Brazil (De Freitas and Landers, 2014). Another review conducted in Asia (China and Mongolia) reported that mulching combined with zero tillage can minimize residue burning and air pollution and improve crop yield (Ramírez *et al.*, 2022), which aligns with our results. Among South Asian countries, India has the highest residue production, resulting in the burning of residue, causing air pollution (Lin and Begho, 2022). Therefore, using zero tillage practices may help countries similar to India in reducing air pollution due to residue burning. Sustainable residue management by the government of India involves practices favourable to both nature and human health (Havrysh *et al.*, 2021). Educating the farmers about using residue for biogas production through local stakeholders and providing incentives for not burning the residue may benefit both the environment and farmers (Zhang *et al.*, 2008).

Organic amendments such as farmyard manure, green manure, crop residues, biochar, and vermicompost retained water content in the soil and reduced soil destruction. Our findings were congruent with a review where biochar and vermicompost ameliorate the soil-building properties in Australia (Quilty and Cattle, 2011). Evidence from China observed reduced GHG emissions by 53.9–62.8% after using biochar soil amendments (Zhang *et al.*, 2015). The use of organic amendments in north-east Japan was found to balance the nutrient concentrations in soil (Hasegawa *et al.*, 2005). However, in Sri Lanka, there was a rapid decline in food production, leading to food insecurity and hindering the livelihood of the farmers. Despite the ecological benefits of organic farming, it took much work to produce enough animal manure to provide the required nitrogen to the plants compared to synthetic fertilizers (Nordhaus and Shah, 2022). Though chemical fertilizers were banned, lack of incentives to buy organic fertilizers, inadequate supply of organic manure, and lack of knowledge regarding the usage of organic amendments resulted in reduced use of organic manure. In addition, reduced yields compared to conventional systems, soil nutrient management challenges, certification, and market barriers, added to challenges in using organic farming in Sri Lanka (Jouzi *et al.*, 2017). To make organic farming work, allowing the soil to adapt to organic manure is essential. Organic farming is not just the replacement of chemical fertilizers with organic manure; it is about slowly changing the crop management method. In addition to using organic manure, one must inhibit pest infestation, restrict the use of non-renewable resources, and promote crop rotation (Pagliarino *et al.*, 2020). In India, multiple schemes (e.g. National Project on Organic Farming, Rashtriya Krishi Vikas Yojana, National Programme for Organic Production, etc.) to boost organic farming were made available to farmers that increased the area under cultivation used for organic farming by 17-fold between 2004 and 2014 (Mukherjee *et al.*, 2022). Despite the efforts, the adoption of organic farming has been low in India due to high costs, low yields, and lack of knowledge and training among the farmers (Azam and Shaheen, 2019). This demands substantial support from the government with respect to implementation of policies and training for farmers to promote organic farming.

Climate-smart agriculture was beneficial in sub-Saharan African regions as well, where the use of organic fertilizers resulted in better yield and protected natural soil resources (Hammed *et al.*, 2019). On the contrary, Hellin *et al.* (2023) argue that a failure to contextualize climate-smart agriculture with respect to socio-economic factors can increase the vulnerability to climate change for some farmers. This defeats the purpose of climate-smart practices, which focus on reducing rural poverty and increasing equity. Therefore, the authors argue that climate-resilient agricultural practices may be more beneficial due to their adaptive nature and focus on social equity (Hellin *et al.*, 2023). Agricultural resilience and sustainability require effective management of natural resources and harnessing of renewable energy to address climate-related disasters. Further, findings from our review revealed that in India, organic vegetable farming was practiced by women, which helped them acquire financial freedom, which complies with a study conducted in Thailand (Thapa and Rattanasuteerakul, 2011). Therefore, by promoting the use of organic amendments, we will be able to achieve the targets of SDG 5, which focuses on gender equality and empowerment of women and giving them an opportunity for employment. In addition, organic manure was used in the cultivation of aerobic rice to reduce GHG emissions. This practice was followed in regions with water scarcity in China (International Rice Research Institute and Plant Research International, 2002). It also increased crop productivity in Japan (Kato *et al.*, 2009) and benefited the farmers economically in Brazil (Pinheiro *et al.*, 2006).

The review brought to light the importance of IPM to control pests in an environmentally safe approach. It is beneficial for both the society as well as the environment. Studies conducted in Sweden, Poland, and Africa reported the key elements of IPM, which align with our review findings (Banwo and Adamu, 2010; Sundgren, 2014; Havrysh *et al.*, 2021). Farmers in India benefitted through training on IPM with the help of collaborating partners such as CropLife Asia, CropLife International, CropLife India, and local Indian organizations (CropLife International, 2014). Emphasis on such trainings is the need of the hour to adapt practices for safe and sustainable food production. Additionally, the review highlights that MSRI reduced water usage and was employed to maintain soil fertility, increasing rice productivity and rising incomes. Similar findings were seen in Tanzania, where MSRI positively impacted crop growth and increased farmers' income (Nakano *et al.*, 2018). The cultivation of rice and wheat crops in South Asia has proven unsustainable due to the traditional method of cultivation, i.e., flood-transplanted rice, as it releases massive amounts of methane gas into the environment, causing global warming (Ullah *et al.*, 2021). In addition, this method requires more labour, water, and the application of nitrogen fertilizer, causing damage to the soil and biodiversity. To overcome these problems, farmers in South Asia should be encouraged to adopt practices such as direct-seeded rice plantation, which uses water efficiently and reduces methane emissions. Since rice is more commonly grown in the South Asian region, adapting this system can offer opportunities to reduce water demand in water-scarce areas.

Strengths and Limitations

This comprehensive review mapped the evidence for sustainable agricultural practices (food production) in South Asia. Owing to access and time limitations, we searched only two databases and did not explore grey literature, which may have resulted in missing a few reports/ articles. Future research may focus on climate, soil, and economic variations within individual countries and identify the best practices suitable for specific regions in South Asia. This can help relevant stakeholders educate farmers to adopt sustainable practices.

Conclusion

To feed a growing population, various unsustainable practices are put into action resulting in ecological consequences. Agriculture in South Asian countries is the hardest hit by climate change. Due to deforestation, extensive tillage, residue burning, and the use of synthetic fertilizers, climate change eventually leads to the deterioration of soil properties, land use patterns, and natural resources. The agricultural practices highlighted in this review positively impact the environment, farmers' livelihood, and food security. It is time we make use of local solutions available for food production. Policies to support climate-smart agricultural practices can benefit the environment and help economic progress. Nature-based solutions can be promoted through innovative financing mechanisms. Sustainable agriculture is the best alternative to achieve food security without hampering the needs of the next generation. Therefore, it is important for farmers to understand the advantages of sustainable agriculture. It is recommended that the government may repurpose existing incentives for farmers to boost sustainable agriculture and restore every farmland, thus helping sustain humankind and the natural environment right now and in the future.

Author Contributions

LRP and DSP were involved in formulating the research question, designing, and carrying out the study, analysing and interpreting data, and writing the article. NG reviewed and edited the manuscript.

Acknowledgements

The authors would like to thank Mr Ravitejas Patil, who generously shared his time in providing valuable insights to this project. We would also like to thank the Department of Health Information, Prasanna School of Public Health for the institutional support for this review.

Conflict of Interest

The authors have no conflicts of interest to declare.

References

Acharjee, T.K., van Halsema, G., Ludwig, F., Hellegers, P. and Supit, I. (2019) Shifting planting date of Boro rice as a climate change adaptation strategy to reduce water use. *Agricultural Systems* 168, 131–143. DOI: 10.1016/j.agsy.2018.11.006.

Adhikari, B., Lodhiyal, N. and Lodhiyal, L.S. (2020) Assessment of crop yield, productivity and carbon sequestration in agroforestry systems in Central Himalaya, India. *Agroforestry Systems* 94, 281–296. DOI: 10.1007/s10457-019-00388-2.

Aertsens, J., De Nocker, L. and Gobin, A. (2013) Valuing the carbon sequestration potential for European agriculture. *Land Use Policy* 31, 584–594. DOI: 10.1016/j.landusepol.2012.09.003.

Agoramoorthy, G., Hsu, M.J. and Shieh, P. (2012) India's women-led vegetable cultivation improves economic and environmental sustainability. *Scottish Geographical Journal* 128(2), 87–99. DOI: 10.1080/14702541.

Ahmad, F., Talukdar, N.R., Goparaju, L., Biradar, C., Dhyani, S.K. *et al.* (2022) GIS-based assessment of land-agroforestry potentiality of Jharkhand state, India. *Regional Sustainability* 3(3), 254–268. DOI: 10.1016/j.regsus.2022.10.004.

Aich, A., Dey, D. and Id, A.R. (2022) Climate change resilient agricultural practices: A learning experience from indigenous communities over India. *PLOS Sustainability and Transformation* 1(7), –e0000022. DOI: 10.1371/JOURNAL.PSTR.0000022.

Ali, A. and Erenstein, O. (2017) Assessing farmer use of climate change adaptation practices and impacts on food security and poverty in Pakistan. *Climate Risk Management* 16, 183–194. DOI: 10.1016/J.CRM.2016.12.001.

Aravindakshan, S., Al Qahtany, A., Arshad, M., Manjunatha, A.V. and Krupnik, T.J. (2022) A metafrontier approach and fractional regression model to analyze the environmental efficiency of alternative tillage practices for wheat in Bangladesh. *Environmental Science and Pollution Research International* 29(27), 41231–41246. DOI: 10.1007/s11356-021-18296-3.

Arora, S., Sehgal, M., Srivastava, D.S., Arora, S. and Sarkar, S.K. (2019) Rice pest management with reduced risk pesticides in India. *Environmental Monitoring and Assessment* 191(4), 241. DOI: 10.1007/s10661-019-7384-5.

Aryal, J.P., Rahut, D.B., Sapkota, T.B., Khurana, R. and Khatri-Chhetri, A. (2020a) Climate change mitigation options among farmers in South Asia. *Environment, Development and Sustainability* 22(4), 3267–3289. DOI: 10.1007/s10668-019-00345-0.

Aryal, J.P., Sapkota, T.B., Khurana, R., Khatri-Chhetri, A., Rahut, D.B. *et al.* (2020b) Climate change and agriculture in South Asia: Adaptation options in smallholder production systems. *Environment, Development and Sustainability* 22(6), 5045–5075. DOI: 10.1007/s10668-019-00414-4.

Aryal, J.P., Sapkota, T.B., Rahut, D.B., Marenya, P. and Stirling, C.M. (2021) Climate risks and adaptation strategies of farmers in East Africa and South Asia. *Scientific Reports* 11(1), 1–14. DOI: 10.1038/s41598-021-89391-1.

Ashley, L., Carter, R., Ferdinand, T., Choularton, R., Appadurai, N. *et al.* (2020) Applying Climate Services to Transformative Adaptation in Agriculture. Available at: www.wri.org/publication/applyingclimate-services (accessed 11 October 2023).

Asian Development Bank (2021) Transforming agriculture in Asia. In: *Asian Development Outlook 2021 Update*. Available at: https://www.adb. org/sites/default/files/publication/726556/ ado2021-update-theme-chapter.pdf (accessed 10 October 2023).

Atreya, K., Sharma, S., Bajracharya, R.M. and Rajbhandari, N.P. (2008) Developing a sustainable agro-system for central Nepal using reduced tillage and straw mulching. *Journal of Environmental Management* 88(3), 547–555. DOI: 10.1016/j.jenvman.2007.03.017.

Azam, M.S. and Shaheen, M. (2019) Decisional factors driving farmers to adopt organic farming in india: A cross-sectional study. *International Journal of Social Economics* 46(4), 562–580. DOI: 10.1108/IJSE-05-2018-0282.

Bajracharya, S.B., Mishra, A. and Maharjan, A. (2021) Determinants of crop residue burning practice in the Terai region of Nepal. *PloS One* 16(7), e0253939. DOI: 10.1371/journal.pone.0253939.

Bala, B.K. and Hossain, M.A. (2013) Modeling of ecological footprint and climate change impacts on food security of the hill tracts of Chittagong in Bangladesh. *Environmental Modeling and Assessment* 18, 39–55. DOI: 10.1007/s10666-012-9326-2.

Banwo, O.O. and Adamu, R.S. (2010) Insect pest management in African agriculture: Challenges in the current millenium. *Archives of Phytopathology and Plant Protection* 36(1), 59–68. DOI: 10.1080/0323540031000106723.

Barnes, A.P., Soto, I., Eory, V., Beck, B., Balafoutis, A. *et al.* (2019) Exploring the adoption of precision agricultural technologies: A cross regional study of EU farmers. *Land Use Policy* 80, 163–174. DOI: 10.1016/j.landusepol.2018.10.004.

Becker, R., Schüth, C., Merz, R., Khaliq, T., Usman, M. *et al.* (2023) Increased heat stress reduces future yields of three major crops in Pakistan's Punjab region despite intensification of irrigation. *Agricultural Water Management* 281, 108243. DOI: 10.1016/j.agwat.2023.108243.

Begho, T., Glenk, K., Anik, A.R. and Eory, V. (2022) A systematic review of factors that influence farmers' adoption of sustainable crop farming practices: Lessons for sustainable nitrogen management in South Asia. *Journal of Sustainable Agriculture and Environment* 1(2), 149–160. DOI: 10.1002/sae2.12016.

Bhatt, R., Kaur, R. and Ghosh, A. (2019) Strategies to practice climate-smart agriculture to improve the livelihoods under the rice-wheat cropping system in South Asia. *Sustainable Management of Soil and Environment* 29–71. DOI: 10.1007/978-981-13-8832-3_2.

Bibi, Z., Khan, D. and Haq, I. ul. (2021) Technical and environmental efficiency of agriculture sector in South Asia: A stochastic frontier analysis approach. *Environment, Development and Sustainability* 23(6), 9260–9279. DOI: 10.1007/s10668-020-01023-2.

Boix-Fayos, C. and de Vente, J. (2023) Challenges and potential pathways towards sustainable agriculture within the European green deal. *Agricultural Systems* 207, 103634. DOI: 10.1016/j.agsy.2023.103634.

Braun, J. von, Afsana, K., Fresco, L.O., Hassan, M.H.A. and Torero, M. (2023) Food system concepts and definitions for science and political action. In: *Science and Innovations for Food Systems Transformation*. Springer International Publishing, Switzerland, AG, pp. 11–17.

Chaudhary, V.P., Chandra, R., Chaudhary, R. and Bhattacharyya, R. (2021) Global warming potential and energy dynamics of conservation tillage practices for different rabi crops in the Indo-Gangetic Plains. *Journal of Environmental Management* 296, 113182. DOI: 10.1016/j.%20 jenvman.2021.113182.

Chavan, S.B., Newaj, R., Rizvi, R.H., Ajit, P.R., Alam, B. *et al.* (2021) Reduction of global warming potential vis-à-vis greenhouse gases through traditional agroforestry systems in Rajasthan, India. *Environment, Development and Sustainability* 23, 4573–4593. DOI: 10.1007/s10668-020-00788-w.

Chen, L., Msigwa, G., Yang, M., Osman, A.I., Fawzy, S. *et al.* (2022) Strategies to achieve a carbon neutral society: A review. *Environmental Chemistry Letters* 20(4), 2277–2310. DOI: 10.1007/ s10311-022-01435-8.

Chhogyel, N. and Kumar, L. (2018) Climate change and potential impacts on agriculture in Bhutan: A discussion of pertinent issues. In: *Agriculture and Food Security*, Vol. 7. Springer Nature, Switzerland, AG. DOI: 10.1186/s40066-018-0229-6.

CropLife International (2014) *Integrated Pest Management*. CropLife International Aisbl. Available at: https://croplife.org/wp-content/uploads/pdf_files/Integrated-pest-management.pdf (accessed 15 June 2023).

Das, A., Layek, J., Ramkrushna, G.I., Patel, D.P., Choudhury, B.U. *et al.* (2018) Modified system of rice intensification for higher crop and water productivity in Meghalaya, India: Opportunities for improving livelihoods for resource-poor farmers. *Paddy and Water Environment* 16, 23–34. DOI: 10.1007/ s10333-017-0609-8.

Das, U., Ansari, M.A. and Ghosh, S. (2022) Effectiveness and upscaling potential of climate smart agriculture interventions: Farmers' participatory prioritization and livelihood indicators as its determinants. *Agricultural Systems* 203, 103515. DOI: 10.1016/J.AGSY.2022.103515.

Dawar, K., Khan, A., Mian, I.A., Khan, B., Ali, S. *et al.* (2022) Maize productivity and soil nutrients variations by the application of vermicompost and biochar. *PloS One* 17(5), e0267483. DOI: 10.1371/journal. pone.0267483.

De Freitas, P.L. and Landers, J.N. (2014) The transformation of agriculture in Brazil through development and adoption of Zero Tillage conservation agriculture. *International Soil and Water Conservation Research* 2(1), 35–46.

Dhanya, P., Ramachandran, A. and Palanivelu, K. (2022) Understanding the local perception, adaptation to climate change and resilience planning among the farmers of semi-arid tracks of South India. *Agricultural Research* 1–18. DOI: 10.1007/s40003-021-00560-0.

Dubey, P.K., Chaurasia, R., Pandey, K.K., Bundela, A.K., Singh, A. *et al.* (2023) Double transplantation as a climate resilient and sustainable resource management strategy for rice production in eastern Uttar Pradesh, North India. *Journal of Environmental Management* 329, 117082. DOI: 10.1016/j. jenvman.2022.117082.

Dupouy, E. and Gurinovic, M. (2020) Sustainable food systems for healthy diets in Europe and Central Asia: Introduction to the special issue. *Food Policy* 96, 101952. DOI: 10.1016/j.foodpol.2020.101952.

Elder, S. and Viegelahn, C. (2022) *Asia–Pacific Employment and Social Outlook 2022: Rethinking Sectoral Strategies for a Human-Centred Future of Work*. International Labour Office, Geneva, Switzerland. DOI: 10.54394/EQNI6264.

Emran, S.A., Krupnik, T.J., Kumar, V., Ali, M.Y. and Pittelkow, C.M. (2019) Agronomic, economic, and environmental performance of nitrogen rates and source in Bangladesh's coastal rice agroecosystems. *Field Crops Research* 241, 107567. DOI: 10.1016/j.fcr.2019.107567.

Eyhorn, F., van den Berg, M., Decock, C., Maat, H. and Srivastava, A. (2018) Does organic farming provide a viable alternative for smallholder rice farmers in India? *Sustainability (Switzerland)* 10(12), 4424. DOI: 10.3390/su10124424.

Fagodiya, R.K., Singh, A., Singh, R., Rani, S., Kumar, S, *et al.* (2023) The food-energy-water-carbon nexus of the rice-wheat production system in the western indo-gangetic plain of India: An impact of irrigation system, conservational tillage and residue management. *The Science of the Total Environment* 860, 160428. DOI: 10.1016/j.scitotenv.2022.160428.

Food and Agriculture Organization of the United Nations (2023) Sustainable Food and Agriculture. Available at: https://www.fao.org/sustainability/background/en/ (accessed 10 October 2023).

Habib-ur-Rahman, M., Ahmad, A., Raza, A., Hasnain, M.U., Alharby, H.F. *et al.* (2022) Impact of climate change on agricultural production; issues, challenges, and opportunities in Asia. *Frontiers in Plant Science* 13, 925548. DOI: 10.3389/FPLS.2022.925548.

Hammed, T.B., Oloruntoba, E.O. and Ana, G.R.E.E. (2019) Enhancing growth and yield of crops with nutrient-enriched organic fertilizer at wet and dry seasons in ensuring climate-smart agriculture. *International Journal of Recycling of Organic Waste in Agriculture* 8, 81–92. DOI: 10.1007/S40093-019-0274-6.

Haque, M.M., Datta, J., Ahmed, T., Ehsanullah, M., Karim, M.N. *et al.* (2021) Organic amendments boost soil fertility and rice productivity and reduce methane emissions from paddy fields under sub-tropical conditions. *Sustainability* 13(6), 3103. DOI: 10.3390/SU13063103.

Hasegawa, H., Furukawa, Y. and Kimura, S.D. (2005) On-farm assessment of organic amendments effects on nutrient status and nutrient use efficiency of organic rice fields in Northeastern Japan. *Agriculture, Ecosystems and Environment* 108(4), 350–362. DOI: 10.1016/J.AGEE.2004.12.015.

Havrysh, V., Kalinichenko, A., Brzozowska, A. and Stebila, J. (2021) Agricultural residue management for sustainable power generation: The Poland case study. *Applied Sciences* 11(13), 5907. DOI: 10.3390/APP11135907.

Hellin, J., Fisher, E., Taylor, M., Bhasme, S. and Loboguerrero, A.M. (2023) Transformative adaptation: From climate-smart to climate-resilient agriculture. *CABI Agriculture and Bioscience* 4(1), 30. DOI: 10.1186/s43170-023-00172-4.

Imran, M.A., Ali, A., Culas, R.J., Ashfaq, M., Baig, I.A. *et al.* (2022) Sustainability and efficiency analysis w.r.t adoption of climate-smart agriculture (CSA) in Pakistan: A group-wise comparison of adopters and conventional farmers. *Environmental Science and Pollution Research International* 29(13), 19337–19351. DOI: 10.1007/s11356-021-17181-3.

International Rice Research Institute and Plant Research International (2002) *Water_Wise_Rice_Production*, Vol. 1.

Islam, M.A., Biswas, R., Sharmin, A., Dey, T., Ashaduzzaman, M. *et al.* (2022) Sustainable livelihoods and household participation in agroforestry: A case study adjacent to the Sundarban reserve forest in Bangladesh. *GeoJournal* 88(3), 3059–3078. DOI: 10.1007/S10708-022-10796-0.

Jat, S.L., Parihar, C.M., Dey, A., Nayak, H.S., Ghosh, A, *et al.* (2019) Dynamics and temperature sensitivity of soil organic carbon mineralization under medium-term conservation agriculture as affected by residue and nitrogen management options. *Soil and Tillage Research* 190, 175–185. DOI: 10.1016/j.still.2019.02.005.

Jinger, D., Kumar, R., Kakade, V., Dinesh, D., Singh, G. *et al.* (2022) Agroforestry for controlling soil erosion and enhancing system productivity in ravine lands of Western India under climate change scenario. *Environmental Monitoring and Assessment* 194(4), 267. DOI: 10.1007/s10661-022-09910-z.

Jouzi, Z., Azadi, H., Taheri, F., Zarafshani, K., Gebrehiwot, K. *et al.* (2017) Organic farming and small-scale farmers: Main opportunities and challenges. *Ecological Economics* 132, 144–154. DOI: 10.1016/j.ecolecon.2016.10.016.

Kato, Y., Okami, M. and Katsura, K. (2009) Yield potential and water use efficiency of aerobic rice (*Oryza sativa* L.) in Japan. *Field Crops Research* 113(3), 328–334. DOI: 10.1016/j.fcr.2009.06.010.

Kharel, M., Dahal, B.M. and Raut, N. (2022) Good agriculture practices for safe food and sustainable agriculture in Nepal: A review. *Journal of Agriculture and Food Research* 10, 100447. DOI: 10.1016/j.jafr.2022.100447.

Knox, J., Hess, T., Daccache, A. and Wheeler, T. (2012) Climate change impacts on crop productivity in Africa and South Asia. *Environmental Research Letters* 7(3), 034032. DOI: 10.1088/1748-9326/7/3/034032.

Kumar, R., Mishra, J.S., Rao, K.K., Bhatt, B.P., Hazra, K.K. *et al.* (2019) Sustainable intensification of rice fallows of eastern India with suitable winter crop and appropriate crop establishment technique. *Environmental Science and Pollution Research International* 26(28), 29409–29423. DOI: 10.1007/s11356-019-06063-4.

Kumar, S., Bijalwan, A., Singh, B., Rawat, D., Yewale, A.G. *et al.* (2021) Comparison of carbon sequestration potential of *Quercus leucotrichophora*–based agroforestry systems and natural forest in Central Himalaya, India. *Water, Air, and Soil Pollution* 232(9), 350. DOI: 10.1007/s11270-021-05294-x.

Lin, M. and Begho, T. (2022) Crop residue burning in South Asia: A review of the scale, effect, and solutions with a focus on reducing reactive nitrogen losses. *Journal of Environmental Management* 314, 115104. DOI: 10.1016/j.jenvman.2022.115104.

Liu, T., Zhang, Q., Kang, X., Hou, J., Luo, T. *et al*. (2022) Household food waste to biogas in Västerås, Sweden: A comprehensive case study of waste valorization. *Sustainability* 14(19), 11925. DOI: 10.3390/SU141911925.

Meetei, T.T., Kundu, M.C. and Devi, Y.B. (2020) Long-term effect of rice-based cropping systems on pools of soil organic carbon in farmer's field in hilly agroecosystem of Manipur, India. *Environmental Monitoring and Assessment* 192(4), 209. DOI: 10.1007/s10661-020-8165-x.

Mosquera-Losada, R., Javier, J., Freijanes, S., Pisanelli, A., Rois, M. *et al*. (2016) Extent and success of current policy measures to promote agroforestry across Europe. Available at: www.agforward.eu (accessed 15 June 2023).

Mukherjee, K., Konar, A. and Ghosh, P. (2022) Organic farming in India: A brief review. *International Journal of Research in Agronomy* 5(2), 113–118. DOI: 10.33545/2618060X.2022.v5.i2b.120.

Nakano, Y., Tanaka, Y. and Otsuka, K. (2018) Impact of training on the intensification of rice farming: Evidence from rainfed areas in tanzania. *Agricultural Economics (United Kingdom)* 49(2), 193–202. DOI: 10.1111/AGEC.12408.

Nath, T.K., Inoue, M., Pradhan, F.E. and Kabir, M.A. (2011) Indigenous practices and socio-economics of *Areca catechu* L. and *Piper betel* L. based innovative agroforestry in northern rural Bangladesh. *Forests Trees and Livelihoods* 20(2–3), 175–190. DOI: 10.1080/14728028.2011.9756705.

Nor Diana, M.I., Zulkepli, N.A., Siwar, C. and Zainol, M.R. (2022) Farmers' adaptation strategies to climate change in Southeast Asia: A systematic literature review. *Sustainability* 14(6), 3639. DOI: 10.3390/su14063639.

Nordhaus, T. and Shah, S. (2022) In Sri Lanka, organic farming went catastrophically wrong. Foreign Policy. Available at: https://foreignpolicy.com/2022/03/05/sri-lanka-organic-farming-crisis/

Os, E. van, Waked, L. and Blok, C. (2016) *Hydroponic Training Visit*. Wageningen UR Greenhouse Horticulture. Available at: https://research. wur.nl/en/publications/hydroponic-training-visit (accessed 15 June 2023).

Ouzzani, M., Hammady, H., Fedorowicz, Z. and Elmagarmid, A. (2016) Rayyan—A web and mobile app for systematic reviews. *Systematic Reviews* 5(1), 210. DOI: 10.1186/s13643-016-0384-4.

Pagliarino, E., Orlando, F., Vaglia, V., Rolfo, S. and Bocchi, S. (2020) Participatory research for sustainable agriculture: The case of the Italian agroecological rice network. *European Journal of Futures Research* 8(1), 7. DOI: 10.1186/s40309-020-00166-9.

Pardon, P., Reubens, B., Reheul, D., Mertens, J., De Frenne, P. *et al*. (2017) Trees increase soil organic carbon and nutrient availability in temperate agroforestry systems. *Agriculture, Ecosystems and Environment* 247, 98–111. DOI: 10.1016/j.agee.2017.06.018.

Parihar, C.M., Parihar, M.D., Sapkota, T.B., Nanwal, R.K., Singh, A.K. *et al*. (2018) Long-term impact of conservation agriculture and diversified maize rotations on carbon pools and stocks, mineral nitrogen fractions and nitrous oxide fluxes in inceptisol of India. *Science of the Total Environment* 640, 1382–1392. DOI: 10.1016/j.scitotenv.2018.05.405.

Pathak, H., Sankhyan, S., Dubey, D.S., Bhatia, A. and Jain, N. (2013) Dry direct-seeding of rice for mitigating greenhouse gas emission: Field experimentation and simulation. *Paddy and Water Environment* 11(1–4), 593–601. DOI: 10.1007/s10333-012-0352-0.

Pinheiro, B.D.S., de Castro, E.D.M. and Guimarães, C.M. (2006) Sustainability and profitability of aerobic rice production in Brazil. *Field Crops Research* 97(1), 34–42. DOI: 10.1016/j.fcr.2005.08.013.

Pradhan, A., Raju, S., Nithya, D.J., Panda, A.K., Wagh, R.D. *et al*. (2021) Farming system for nutrition-a pathway to dietary diversity: Evidence from India. *PLoS ONE* 16(3), e0248698. DOI: 10.1371/journal.pone.0248698.

Quilty, J.R. and Cattle, S.R. (2011) Use and understanding of organic amendments in Australian agriculture: A review. *Soil Research* 49(1), 1. DOI: 10.1071/SR10059.

Rahman, S. (2013) Pesticide consumption and productivity and the potential of IPM in Bangladesh. *Science of the Total Environment* 445, 48–56. DOI: 10.1016/j.scitotenv.2012.12.032.

Rahman, S.A., Imam, M.H., Wachira, S.W., Farhana, K.M., Torres, B. *et al*. (2008) Land use patterns and the scale of adoption of agroforestry in the rural landscapes of padma floodplain in Bangladesh. *Forests Trees and Livelihoods* 18(2), 193–207. DOI: 10.1080/14728028.

Ramesh, T., Manjaiah, K.M., Mohopatra, K.P., Rajasekar, K. and Ngachan, S.V. (2015) Assessment of soil organic carbon stocks and fractions under different agroforestry systems in subtropical hill agroecosystems of North-east India. *Agroforestry Systems* 89, 677–690. DOI: 10.1007/s10457-015-9804-z.

Ramírez, D.A., Silva-Diaz, C., Ninanya, J., Carbajal, M., Rinza, J. *et al.* (2022) Potato zero-tillage and mulching is promising in achieving agronomic gain in Asia. *Agronomy* 12(7), 1494.

Rasul, G. (2021) A framework for addressing the twin challenges of COVID-19 and climate change for sustainable agriculture and food security in South Asia. *Frontiers in Sustainable Food Systems* 5, 679037. DOI: 10.3389/fsufs.2021.679037.

Rasul, G. and Thapa, G.B. (2004) Sustainability of ecological and conventional agricultural systems in Bangladesh: An assessment based on environmental, economic and social perspectives. *Agricultural Systems* 79(3), 327–351. DOI: 10.1016/S0308-521X(03)00090-8.

Raza, M.H., Abid, M., Faisal, M., Yan, T., Akhtar, S. *et al.* (2022) Environmental and health impacts of crop residue burning: Scope of sustainable crop residue management practices. *International Journal of Environmental Research and Public Health* 19(8), 4753. DOI: 10.3390/ijerph19084753.

Rizvi, R.H., Newaj, R., Chaturvedi, O.P., Prasad, R., Handa, A.K. *et al.* (2019) Carbon sequestration and CO_2 absorption by agroforestry systems: An assessment for Central Plateau and Hill region of India. *Journal of Earth System Science* 128(3), 56. DOI: 10.1007/s12040-019-1071-3.

Sapkota, T.B., Jat, R.K., Singh, R.G., Jat, M.L., Stirling, C.M. *et al.* (2017) Soil organic carbon changes after seven years of conservation agriculture in a rice–wheat system of the eastern Indo-Gangetic plains. *Soil Use and Management* 33(1), 81–89. DOI: 10.1111/sum.12331.

Seo, S.N. (2017) Beyond the paris agreement: Climate change policy negotiations and future directions. *Regional Science Policy and Practice* 9(2), 121–140. DOI: 10.1111/RSP3.12090.

Sharma, P., Tiwari, P., Verma, K. and Singh, M.K. (2017) Agroforestry systems: Opportunities and challenges in India. *Journal of Pharmacognosy and Phytochemistry* 6(6S), 953–957. Available at: https://www.phytojournal.com/special-issue/2017.v6.i6S.3005/agroforestry-systems-opportunities-and-challenges-in-india

Sharma, S.K., Singh, Y.V., Tyagi, S. and Bhatia, A. (2016) Influence of rice varieties, nitrogen management and planting methods on methane emission and water productivity. *Paddy and Water Environment* 14, 325–333. DOI: 10.1007/s10333-015-0502-2.

Shew, A.M., Durand-Morat, A., Putman, B., Nalley, L.L. and Ghosh, A. (2019) Rice intensification in Bangladesh improves economic and environmental welfare. *Environmental Science and Policy* 95, 46–57. DOI: 10.1016/j.envsci.2019.02.004.

Smith, P. and Gregory, P.J. (2013) Climate change and sustainable food production. *Proceedings of the Nutrition Society* 72(1), 21–28. DOI: 10.1017/S0029665112002832.

Stray Dog Institute (2021) *Sustainable Food Production Offers Multiple Global Benefits*. Environment. Stray Dog Institute, Kansas.

Sundgren, A. (2014) Experiences with implementation and adoption of integrated pest management strategies in Sweden. *Integrated Pest Management* 4, 555–570. DOI: 10.1007/978-94-007-7802-3_22.

Thapa, G.B. and Rattanasuteerakul, K. (2011) Adoption and extent of organic vegetable farming in Mahasarakham province, Thailand. *Applied Geography* 31(1). DOI: 10.1016/j.apgeog.2010.04.004.

Ullah, A., Nawaz, A., Farooq, M. and Siddique, K.H.M. (2021) Agricultural innovation and sustainable development: A case study of rice–wheat cropping systems in South Asia. *Sustainability* 13(4), 1965. DOI: 10.3390/su13041965.

United Nations (2023) *Take Action for the Sustainable Development Goals*. United Nations.

Vasa, T.N. and Pothanamkandathil Chacko, S. (2021) Recovery of struvite from wastewaters as an eco-friendly fertilizer: Review of the art and perspective for a sustainable agriculture practice in India. *Sustainable Energy Technologies and Assessments* 48, 101573. DOI: 10.1016/j.seta.2021.101573.

Vijayakumar, S., Nayak, A.K., Satapathy, B.S., Munda, S., Tripathi, R. *et al.* (2019) Double transplanting—A contingent strategy for rice cultivation in low-lying areas. *Indian Farming* 69(5), 36–38.

Williams, N.E. and Carrico, A. (2017) Examining adaptations to water stress among farming households in Sri Lanka's dry zone. *Ambio* 46(5), 532–542. DOI: 10.1007/s13280-017-0904-z.

World Bank Group (2023) *World Bank Open Data*. The World Bank. Available at: https://data.worldbank.org/ (accessed 24 June 2023).

Yadvinder-Singh, Bijay-Singh, Ladha, J.K., Khind, C.S., Gupta, R.K. *et al.* (2004) Long-term effects of organic inputs on yield and soil fertility in the rice-wheat rotation. *Soil Science Society of America*

Journal 68(3), 845–853. Available at: https://www.scopus.com/inward/record.uri?eid=2-s2.0-2442646838&partnerID=40&md5=74f94058a7f0320a523cb20ff0865a44

Zaharia, A., Diaconeasa, M.C., Maehle, N., Szolnoki, G. and Capitello, R. (2021) Developing sustainable food systems in Europe: National policies and stakeholder perspectives in a four-country analysis. *International Journal of Environmental Research and Public Health* 18(14), 7701. DOI: 10.3390/IJERPH18147701.

Zhang, A., Bian, R., Li, L., Wang, X., Zhao, Y. *et al*. (2015) Enhanced rice production but greatly reduced carbon emission following biochar amendment in a metal-polluted rice paddy. *Environmental Science and Pollution Research International* 22(23), 18977–18986. DOI: 10.1007/s11356-015-4967-8.

Zhang, Q., Yang, Z. and Wu, W. (2008) Role of crop residue management in sustainable agricultural development in the North China plain. *Journal of Sustainable Agriculture* 32(1), 137–148. DOI: 10.1080/10440040802121502.

9 Agricultural impacts of climate change in India and potential adaptations

Suseelendra Desai[1]*, Krishna Chaitanya Tirunagaru[1], K. Venkateswara Rao[2], Rathnakaran Upasana[1], J.V.N.S. Prasad[2], K.N. Nithin[1], Vijay Kumar Didal[1] and Nitin Misal[1]

[1]School of Agricultural Sciences and Technology, Narsee Monji Institute of Management Studies Deemed to be University, Shirpur 425405, Maharashtra, India; [2]ICAR-Central Research Institute for Dryland Agriculture, Santoshnagar, Hyderabad 500059, Telangana, India

Abstract

The research and development efforts coupled with suitable policies over decades have resulted in India becoming self-reliant in food grain production. However, the rising incidence of extreme weather events due to climate change and climatic variability over the last few decades could impact the food security of the country. The impact of these events could result in increased frequency of dry spells and floods; multiple stresses in a given season; emergence of new pests/biovars; soil health deterioration; reduced animal and poultry productivity; changes in fish species diversity; and disrupted market services. Efforts at the regional/agro-ecology level to minimize the adverse impacts through adaptation strategies have shown positive results. These efforts included a demonstration of selected technologies, enhancing awareness about impacts to the stakeholders, and operationalization of agricultural contingency plans. New investments and research-extension-policy networks over the last decade by the government of India have shown impressive results with a possibility of developing 'Climate smart village clusters'. To ensure resilience coupled with sustainable profitability, thrust should be laid to investing on infrastructure for advanced research, packaging of farming system/agro-ecology-specific adaptation strategies, state-of-the-art knowledge management systems, skill development programs and enabling policy framework.

Keywords: climate change impacts, crops, soil health, pests, livestock, poultry, adaptation, contingency plans

Introduction

The challenge of food and nutritional security for the growing populations is a daunting task considering the limited arable land for horizontal spread. The outcome of the meta-analysis of anticipated world food demand vis-à-vis population at hunger risk shows that while the total food demand could increase by 35 to 56% between 2010 and 2050, the population at risk of hunger is expected to touch +8% (van Dijk *et al.*, 2021). If animal populations are also considered, then the overall demand could be far higher. In recent decades, the farmers are bearing the brunt of expensive input costs denting farm profitability (Srivastava *et al.*, 2017), and severe crop losses due to recurring extreme weather events (Swaminathan and Rengalakshmi,

*Corresponding author: desai1959@yahoo.com

2021). Though the initial outcomes of coping strategies are encouraging, packaging of these adaptation strategies is often missing to derive the anticipated benefits. Since, the research on climate change is at its infancy in India, understanding the issues from a systems perspective could help to reap the benefits of interventions and their synergies in a holistic way. In this backdrop, the review focuses on (i) the impacts of climate change on various agricultural sectors, (ii) available adaptation strategies, (iii) current gaps, and (iv) steps required to bridge these gaps to make agriculture climate resilient.

Indian agricultural scenario

Agriculture is the primary occupation covering more than 50% population of India. As per the Ministry of Statistics and Program Implementation, Government of India, the Gross Added Value (GAV) from agriculture and allied sectors was estimated at about USD 1.28 trillion during 2020–2021 accounting for about 20.19% at current rates (statisticstimes.com). More than 50% of the households are dependent on agriculture and live in rural India. Over the years, the research and development coupled with timely policy interventions have resulted in making India self-reliant in food grain production. It is on an upward trend during the last decade with a phenomenal food grain production of 327 million tonnes in 2020–2021 (Fig. 9.1). In 2020–2021, the Indian livestock population accounted for about 31% of the global livestock population with milk production of about 198 million tonnes.

Review Methodology

About 85% of Indian farmers are resource-poor small and marginal farm holders who are highly vulnerable to climate-driven risks in agriculture. The frequent occurrences of extreme weather events in recent decades leading to severe crop losses across agriculture and allied sectors prompted the authors to look into the research status on (i) impacts of climate change and climatic variability on various sectors of Indian agriculture; (ii) current status of adaptation strategies; and (iii) future strategies. Initially, sources such as google scholar helped to understand the trending research papers published using broad search terms. Subsequently, specific data was collected from databases of FAOSTAT, INDIASTAT, and various ministry websites. An all-inclusive comprehensive literature including

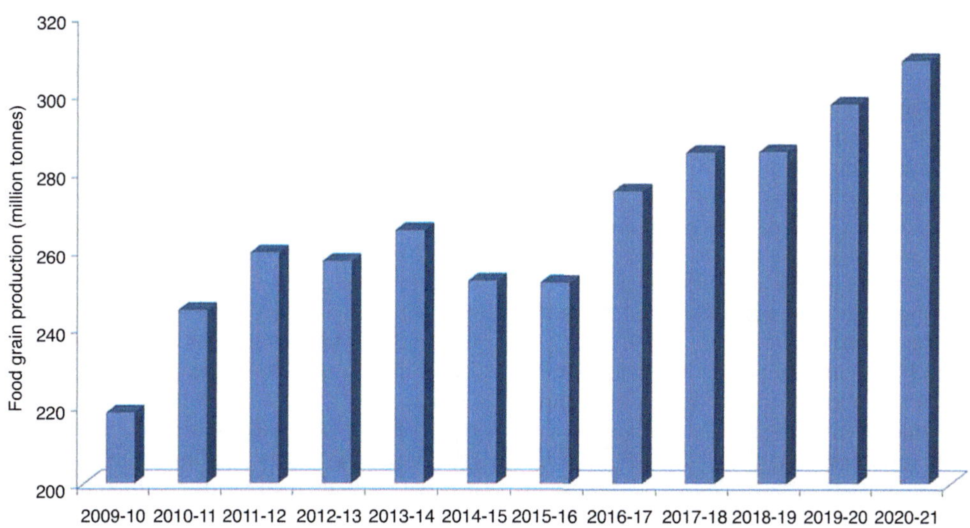

Fig. 9.1. Total food grain production trend in India during 2009–2010 and 2020–2021. Data source: Ministry of Agriculture & Farmers' Welfare, Government of India.

published and unpublished information were collected and systematically screened. The short-listed relevant potential studies provided a coherent lens to make sense of extant knowledge and avoid biases or mistakes. The scientific quality of the primary studies was assessed based on the rigour of their research designs and methods. Sometimes, contrasting literature was considered to arrive at unbiased inferences. Wherever necessary, the data were transformed, synthesized, and analysed for meaningful interpretations.

Climate Change Scenarios and Vulnerability of Indian Agriculture

Climate change projections

The Greenhouse Gas (GHG) emissions include carbon dioxide (CO_2), methane, and nitrous oxide which are considered as major drivers of climate change. A sharp rise in extreme weather events in the last two decades worldwide and their indomitable impacts on humans, animals, and agriculture has prompted the global scientific communities to look into the role of GHGs on climate change. The Inter-governmental Panel on Climate Change (IPCC) has recently published a summary report for policy makers on climate change in 2018 (IPCC, 2021). The highlights of the same are as follows:

- Climate-related risks to health, livelihoods, food security, water supply, human security, and economic growth are projected to increase with global warming of 1.5°C. However, the risks depend on the rate and magnitude of warming, geographic location, levels of development and vulnerability, and adaptation and mitigation interventions.
- Climate models project that global mean sea level rise would be around 0.1m lower than current levels and global warming will be 1.5°C by 2100.
- To impacts on biodiversity and ecosystems, including species loss and extinction, ocean acidity and ocean oxygen levels could be minimized if the temperature rises are contained at 1.5°C.
- To reduce the impacts due to future climate-related risks, upscaling and acceleration of far-reaching, multilevel and cross-sectoral climate mitigation, and stepped-up transformational adaptation strategies will be required.

Climate change projections for India

The projected climate change impacts on India are (i) temperature increase by about 1.1°C under the RCP2.6 and by about 4°C by 2080–2099 under the RCP8.5 emissions pathways, respectively; (ii) warming with 18–21% higher rise in annual minimum temperatures than the mean temperatures; (iii) higher emission scenarios are likely to influence the winter and pre-monsoon months; (iv) local long-term precipitation trends exhibit considerable uncertainty which is further compounded because of little understanding of the relationship between El Niño Southern Oscillation (ENSO) and the monsoon; and (v) the projected country-level changes in the seasonality of monthly precipitation are also unclear though the intensity of extreme precipitation events in South India and in Himalayas region could be more (Climate Knowledge Portal, 2021).

Agricultural and socio-economic vulnerability

Indian farming community by and large comprises resource-poor small and marginal farm holders. Agriculture in about 35% of districts was highly vulnerable to climate change whereas nearly 20% of districts fell into the very high-risk category (Rama Rao *et al.*, 2019). Late and very late sown wheat may suffer about 23% yield losses by 2050 due to climate change (Kumar *et al.*, 2012). Radhika *et al.* (2017) reported that Karnataka state could be one of the vulnerable states to climate change. On average, 30% of crop yields could reduce due to climate change by the mid-21st century in South Asian countries (Rao *et al.*, 2019). The projections of climate change impacts for 2080 on different crops varied across 15 agro-climatic zones of India with more reductions in yields in crops like sugarcane, cotton, and groundnut (Singh *et al.*, 2019).

Climate change impacts on agriculture could in turn affect the food-, nutritional-, and livelihood securities. The rate of climate refugees is 1.5 million people per year in India and 19.3 million people per year globally and the poor accommodation of millions of climate refugees in Bangladesh forced their migration to India (Wilson, 2021). Migrants suffered from very high rates of acute illness (31%), chronic illness (9%), and vector-borne diseases (44%) due to vulnerabilities.

Among the migrants also, children were more vulnerable (Santha, 2015). Disaster-hit countries experienced direct economic losses to the tune of USD 2908 billion during 1998–2017 out of which about 77% losses were due to climate-related disasters (Pascaline and Rowena, 2018). Climate change could reduce farm incomes by 15–18% in irrigated areas and by 20–25% in unirrigated areas (Hari *et al.*, 2018). Elevated temperatures could elevate the heat load of the livestock and thus impact the productivity of meat, wool, milk, and draught power (Upadhyay *et al.*, 2008). Economic and environmental factors impacted the fisheries sector and impacted the livelihoods of fishermen in Tamil Nadu (Johnson *et al.*, 2016). Kantamaneni *et al.* (2020) proposed a novel index for the evaluation of agricultural vulnerability to climate change in coastal areas of Andhra Pradesh and observed that climate change affects farmers' incomes and socio-cultural activities such as marriages. Letha Devi *et al.* (2021) developed the Livelihood Vulnerability Index (LVI) to estimate the climate vulnerability of livestock farming in Karnataka and reported that dry districts were more vulnerable whereas those who adopted an integrated farming system were less vulnerable. Mahapatra *et al.* (2021) reported that district-level degree of vulnerability to climate change was positively associated with malnutrition among children.

Climate Change Impacts

The impacts of climate change are multifarious and affect various components of agricultural systems. The impacts on each of the systems depend on how intricately they are linked to other systems. The climate change impacts could be either positive or negative. The key climate change impacts on Indian agriculture include frequent recurrence of mid- and late-season droughts in parts of Rajasthan, Maharashtra, Uttar Pradesh, Haryana, Gujarat, Andhra Pradesh, Karnataka, and Bihar; floods due to very high intense rainfall events; heat waves in parts of Rajasthan; damage to field and horticultural crops especially at reproductive phases; altered pest scenarios; changes in milk and meat production in livestock; and altered growth and reproduction of fishes (Rao *et al.*, 2019).

Soils and nutrient dynamics

Climate change has direct and indirect impact on nutrient dynamics, land, and soil systems. Soil is the second largest carbon sink after oceans. The soil processes such as chemical weathering, organic matter decomposition, microbial activity, and nutrient releases could be accelerated by warm temperatures and also affect the type of vegetation (Hamidev *et al.*, 2018). Climate change drivers affect organic matter status, carbon and nutrient cycling in the soils, though soil pH is not drastically altered (Reth *et al.*, 2005). Due to faster reduction in organic matter content and C:N ratio under warm soil temperature regimes, slow organic matter decomposition, reduced plant root growth, and increased vulnerability to soil erosion have been reported (Chatterjee, 2021). Rengasamy and Churchman (1999) reported that high soil temperatures decreased the size of clay fractions and organic matter thereby leading to a reduction in the cation exchange capacity of the soil impacting soil structure, stability, nutrient availability, and water-holding capacity. While the regions with high-intensity rainfall could contribute the increased methane release, the deficient rainfall regions could experience CO_2 loss, deficit water supply, changes in foraging patterns, and altered soil invertebrate populations (Chander, 2012). Increased soil temperature models predicted 2.0–11.5% soil organic carbon reduction in the north-central regions of the United States and the addition of organic residues facilitated CO_2 liberation (Navarro-Pedreno *et al.*, 2021).

Interestingly, an increase in CO_2 levels and temperature resulted in an enhanced soil C:N ratio resulting in better availability of nitrogen and enhanced carbon uptake which in turn reduced the net mineralization process (Hungate *et al.*, 2003). However, An *et al.* (2005) reported that nitrogen mineralization effects were realized only for 1 year. The elevated CO_2 levels increased the symbiotic nitrogen fixations whereas the free-living and associate N_2 fixers were non-responsive (Reich *et al.*, 2006). However, the most challenging issue could be to establish the relationship between climate change and soil nitrogen vis-à-vis carbon sequestration. The low temperature hindered the release of phosphorus from organic material thus impacting phosphorus availability (Gahoonia and Nielsen, 2003). Hou *et al.* (2018) confirmed that the availability of phosphorus was inversely proportional to increasing annual precipitation and temperature.

An increase in soil temperature could result in increased microbial populations resulting in enhanced enzymatic activity (Kaur *et al.*, 2014). However, its impact on populations of plant growth-promoting bacteria and endophytic fungi was either positive or negative or neutral, depending on the temperature range (Compant *et al.*, 2010). DeAngelis *et al.* (2015) reported that relative abundance tilted towards soil bacteria from fungal communities when soil temperatures increased by 5°C. In a long-term experiment, Gao *et al.* (2020) found that the effect of eCO_2 on soil respiration at low N supply during latter years resulted in changed microbial functional traits associated with C and N cycling which in turn could affect temporal shifts in soil and plant N availability. Guo *et al.* (2018) showed that warming affected community structure and divergent succession of bacteria and fungi in grasslands. Stochastic processes played a larger role in explaining taxonomic and phylogenetic compositions under normal conditions while warming elicited a potential deterministic environmental filtering. Climatic warning resulted in robust molecular ecological networking coupled with enhanced stability of the microbial communities and further strengthened relationships between network structure to community functional potentials and key ecosystem functioning Yuan *et al.* (2021). They further suggested preserving microbial interactions for ecosystem management. In an detailed review on climate change and microbes, Tiedje *et al.* (2022) summarized that climate change impacts stimulated soil respiration and SOM decomposition; accelerates microbial temporal scaling, reduces microbial diversity; lowered respiratory temperature sensitivity but does not affect soil carbon storage. They emphasized the need for innovative tools and integrated climate models to understand these implications. Extreme heatwaves and droughts during 2018–2019 in central Europe impacted the structure, assembly, and function of soil microbiomes in cropland and grassland resulting in the relative abundance of actinobacteria, fungi, and viruses (Bei *et al.*, 2023). They further reported homogeneous selection to bacterial community assembly increased from 40.0% in normal summers to 51.9% in extreme summers.

Agricultural water systems

The climate change on agricultural water systems could impact surface and groundwater availability; water quality; seasonal streamflow patterns; crop water demand; and groundwater recharge. The changes in precipitation patterns in terms of a number of rainy days, intensity of rain events, and dry spells could impact availability of surface and sub-surface water for drinking purposes as well as for agriculture. The effects could aggravate the runoff potential; availability of water and irrigation requirement of crops especially when coupled with modified land use patterns, *in situ* and *ex situ* conservation measures. The effects of climate change on water resources differ substantially among different regions and river basins of India (Goyal and Surampalli, 2018). In several river basins across India, the precipitation is projected to increase up to 30% by 2040–2069 and 50% by 2070–2099 (Mishra and Lilhare, 2016). Kumar *et al.* (2015) studied the use of an ensemble of hydrological models compared eight popular hydrological models and concluded that the best-performing ensemble outperformed all eight individual models in simulating the observed discharge and flow volume. Dubey *et al.* (2020) used a bias-corrected ensemble Soil

Water Assessment Tool (SWAT) to predict water yield and evapotranspiration for 2011–2040, 2041–2070, and 2071–2099 under RCP4.5 and RCP8.5 scenarios.

Insect pests and diseases

The incidence and intensity of insect pests and diseases are highly correlated with the weather conditions. The climate change impacts on insect pests and pathogens could be (i) a shift in major/minor status of the pest/pathogen; (ii) a new host range within a geographical area (iii) an emergence of new biovars/pathovars, and (iv) a spread to new areas leading to biosecurity related issues.

Shift in major and minor pest status

A rise of 1–5°C could increase insect survival due to low-winter mortality, increased population build-up, early infestations, and higher crop damages (Sharma *et al.*, 2010). Thus, a global increase in temperature within the favourable range may pose an enhanced threat to tropical and subtropical insect pests due to their accelerated rate of development, reproduction, and survival. The green-veined butterfly, *Pieris napi* reacted to warming-mediated poor-quality foliage in *Brassicaceae* and consumed significantly higher amounts of plant tissue (Bauerfeind and Fischer, 2013). On the contrary, *Myzus persicae* and *Brevicoryne brassicae* were not affected despite being fed with oilseed rape plants grown under varied temperatures and thus different in nutritional parameters (Himanen *et al.*, 2008). High temperatures reduced Bt toxin production in cotton plants resulting in enhanced susceptibility to *Heliothis virescens*, *Helicoverpa armigera*, and *H. punctigera* (Kaiser, 1996).

Climate change may lead to more incidence of plant diseases transmitted by insects due to the rapid multiplication of vectors (Sharma, 2014). Research has shown that host plants such as wheat and oats become more susceptible to rust diseases with increased temperature; but some forage species become more resistant to fungi with increased temperature (Coakley *et al.*, 1999). The changes in vector overwintering and oversummering are certain when a new vector species enters the crop fields and may influence pathogen survival, movement, and reproduction (Newman, 2005). A change in temperature could directly affect the spread of infectious diseases and their survival between seasons. Evidence suggests that sunlight affects plant pathogens due to the accumulation of phytoalexins or protective pigments in host tissue. Milus *et al.* (2006) reported that at higher temperatures, isolates of *Puccinia striiformis* showed increased aggressiveness. Huot *et al.* (2017) showed that elevated temperature significantly increased the susceptibility of Arabidopsis to *Pseudomonas syringae* pv. Tomato (Pst) DC3000, independent of the phyB/PIF thermo-sensing pathway. Debela and Tola (2018) noted that elevated temperature and CO_2 significantly influenced disease development due to a clear increment of growth of the pathogen and fecundity. Soil temperature variability could impact pathogens such as *Plasmodiophora brassicae* and *Verticillium longisporum* (Priyanka *et al.*, 2020) and prolonged dry spells could increase the frequency of tree pathogens.

Expanded new host range/geographical area

With rise in temperature, the insect-pests are expected to extend their geographic range from the tropics and subtropics to temperate regions along with shifts in cultivation areas of their host plants (Sharma, 2014). The general prediction is that if global temperatures increase, the species will shift their geographical ranges closer to the poles or to higher elevations and increase their population size (Sutherst, 2000). Outbreak of sugarcane woolly aphid *Ceratovacuna lanigera* in the sugarcane belt of Karnataka and Maharashtra states during 2002–2003 resulted in 30% yield losses (Rahul *et al.*, 2015). Most warm-loving plant pathogenic bacteria like *Acidovorax avenae* subsp. *avenae*, *Ralstonia solanacearum*, and *Burkholderia glumae* have emerged as serious problem worldwide (Schaad, 2008). Weather and dry root rot disease data collected in India for one decade showed a higher incidence of dry root rot in chickpea varieties that resist Fusarium wilt in years when soil temperatures exceeded 33°C. On the contrary, cooler temperatures coupled with wet conditions were associated with increased incidence of *Sclerotinia* stem rot on soybean, *Ascochyta* blight in chickpea, lentil, pea; and anthracnose in chickpea and lentil (Pande *et al.*, 2010). Interestingly, a rise in global temperature of 2°C will reduce the risk of late blight in warmer areas while it will increase in cooler areas and early onset of the epidemics (Singh *et al.*, 2013). Srinivasarao *et al.* (2015) predicted the high incidence of *Spodoptera litura* on groundnuts due to increased generations under eCO_2 conditions.

Outbreak of new pests/diseases

The South American tomato pinworm was recorded for the first time in India in 2014 (Sankarganesh *et al.*, 2017). The fall armyworm was first noticed in the Indian subcontinent in Bengaluru Rural and Chikkaballapur districts from May to June 2018 (Ganiger *et al.*, 2018) and South Karnataka during the first fortnight of July 2018 (ICAR-NBAIR Pest Alert, 2018). Later the pest was recorded in Chikkamagaluru, Chitradurga, and Davangere with an infestation of 40–70% (ICAR-NBAIR Pest Alert, 2018). Changes in temperature regimes may lead to outbreaks of epidemics due to new races of pathogens hither to unknown (Reddy, 2013). Temperature and duration of exposure are important in determining the effect of climate change on disease intensity (Chakraborty, 2013; Ferrocino *et al.*, 2013).

Livestock systems

The livestock sector brings resilience to Indian agriculture and is an integral part of the socio-economic development of rural households. As per the 20th livestock census, the total livestock population was 535.78 million which contributed nearly 4.11% to India's gross domestic product. Though livestock have in-built resilience to climatic stresses offered by the body's anatomy and physiology, very high temperatures could affect growth and productivity, leading to a rise in respiratory rates, body temperature, and pulse rate (Pereira *et al.*, 2008). The climate change impacts on the livestock sector could be discussed under the broad heads viz. (i) milk production, (ii) meat production, (iii) egg production, (iv) growth and reproduction of the animals, (v) fodder quality and yields, and (vi) animal health based on published literature and are summarized in Fig. 9.2.

Breeds of milch animals with high productivity produce high-metabolic heat under

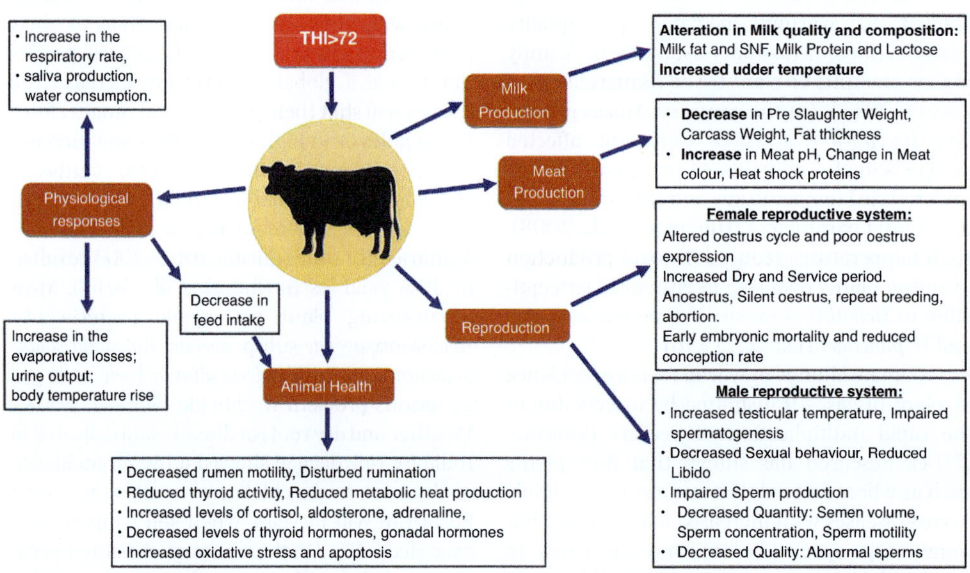

Fig. 9.2. Summary impacts of climate change on livestock sector. Data source: Ministry of Agriculture & Farmers' Welfare, Government of India.

adverse climatic conditions and their feed intake gets reduced which ultimately reflects in reduced milk yields (Marchesini *et al.*, 2018). In dairy animals, Temperature Humidity Index (THI) is widely used to measure heat tolerance, and values above 72 caused a sharp decline in milk yield by 0.2 kg per unit increase of THI (Ravagnolo and Misztal, 2000). The negative impact of temperature rise on total milk production has been estimated to be more than 15 million tonnes by 2050 in India (Das *et al.*, 2016). The proportion of milk fat, solid not fat, and milk protein significantly decreased by 39.7, 18.9, and 16.9%, respectively (Kadzere *et al.*, 2002). The incidence of clinical mastitis in dairy animals is usually high during hot climatic conditions because of increased udder temperature and high vector fly populations (Prathap *et al.*, 2017). Apart from the length and severity of heat stress, the genetic potential, age, nutrition and geographical habitat of animals are also important traits that impact milk production.

India stands first in buffalo meat production in the world. Climate change directly impacts meat production by affecting meat quality and organoleptic parameters (Spehar *et al.*, 2009). Mitloehner *et al.* (2001) observed that heat-stressed feedlot cattle showed physiological and behavioural responses that negatively affected productivity. Climatic stresses during pre-slaughter resulted in dark, firm and dry beef (Miller, 2007) whereas climatic stress just prior to slaughter produced high pH that caused changes in the meat's colour and consistency. (Warner *et al.*, 2010).

Layers are extremely sensitive to temperature-associated issues, especially heat stress. Environmental temperatures above 30°C adversely affected the performance of birds by causing a reduction in feed consumption and hence, disturbed enzyme activity required for nutrient absorption (Mashaly *et al.*, 2004). Insufficient nutrient intake also affected egg production and egg weight in layers (Tankson *et al.*, 2001).

Thermal stress decreased the dry matter intake and growth rate and a rise in temperature (above the thermo-comfort zone of 15–29°C) negatively affected the growth rate of young animals resulting in their daily weight gain and body condition score leading to delayed puberty and sexual maturity (Padodara and Jacob, 2013). Due to prolonged climatic stress, alterations at the hypothalamo-hypophyseal gonadal axis affected the endocrine function, leading to a disturbance in the secretory impulses of follicle-stimulating hormone (Daley *et al.*, 1999). The secretory impulses were responsible for the production of hormones related to oestrus expression and pregnancy (Naqvi *et al.*, 2004). In the case of males, the heat stress significantly affected the quantity and quality of sperm production and thereby decreased the fertility rate (Balic *et al.*, 2012).

The crop residues and their by-products are majorly used as feed supplements for the livestock and the quality and quantity of feed are directly proportional to the productivity of the animals. Kumar *et al.* (2011) predicted a shortage of dry fodder availability under future climate change scenarios with a huge impact on livestock production systems. The increased atmospheric CO_2 concentration significantly affected crop growth and its quality, altered the biomass of pasture land to change over to legumes, and increased lignification of plant tissues as well as other cell wall components (Minson, 1990). These changes in the forage components could lead to a reduction in the digestibility and nutrient availability. The research in the animal sciences sector is emerging in India and the methodology could be used for generating similar datasets for Indian breeds.

Animal health

The extreme weather events could disturb the natural balance of co-evolution of animals and their pathogens resulting in the emergence and re-emergence of infectious zoonotic and vector-borne diseases (Nardone *et al.*, 2010), especially when the animals were relocated. The diseases such as haemorrhagic septicaemia, anthrax, leptospirosis, cryptosporidiosis, fascioliasis, haemonchosis, etc. will also succeed as the temperature, relative humidity, soil moisture, and rainfall favour the growth of the organism, thereby causing significant economic loss due to decreased livestock production and survivability. The increased climate stress may also cause alteration in the transmission cycle of the vectors of diseases like lumpy skin, foot and mouth disease, malaria, blue tongue, and babesiosis (Basu and Bandhyopadhyay, 2004;

Thornton *et al.*, 2009) and aggravate their infestation in livestock.

Market-related aspects

Climate changes result in a higher frequency of extreme variations and lengthier growing seasons, which ultimately affect negatively the income of farmers (Chakrabarty, 2016). Declining agricultural productivity, rising sea levels, and negative health were estimated to impact 3% of the GDP at 1°C of global warming pushing the national poverty rate to go up by 3.5% by the year 2040 resulting in around 50 million more poor people (Picciariello *et al.*, 2021). The Economic Survey of India for 2017–2018 predicted that the income of farmers may decrease by 15–18% in general and up to 25% for unirrigated areas due to climate change (Sarkar, 2018). Food supply in India is sensitive to the effect of unevenness in monsoon (distribution and intensity) and could result in a 121–194% price rise by 2050 (Davis *et al.*, 2019). As a consequence of climate change and price rise, cereal consumption is projected to decline in Asia by 24% by 2050; ~15% decline in the average calorie availability; ~4.6% rise in food consumer prices; and ~6.2% fall in food purchasing power, especially in the poor rural households (FAO, 2018). The projected yield fluctuation of −2.3 to 43.2% in groundnuts by 2050 due to climate alone was reduced to −0.9 to 16.2% with the addition of market (elasticities, trade etc.) and economic (population and income) variables under an optimistic scenario (Kadiyala *et al.*, 2021). Out of 34 districts of Maharashtra in India, 11 were categorized as highly vulnerable, 14 as moderately vulnerable, and remaining as less vulnerable to climate change (Adhav *et al.*, 2021). Further, based on socioeconomic vulnerability (consisting of crop and dairy indicators) the highest and lowest exposure were recorded for Nandurbar (0.91) and Pune (0.41) districts, respectively.

Adaptation Strategies

To cope with the climate change impacts on Indian agriculture, basic and applied research is required to understand the dynamics of interactions of various sectors of agriculture. Such studies should further be corroborated by developments in other related areas to get a holistic understanding. Some of the present adaptation strategies are discussed here.

Soil dynamics

Modified soil and land management practices could minimize the adverse effects of temperature and precipitation or extreme weather conditions including crop rotation, minimum tillage, mulching, crop residues, and agronomic management strategies such as cropping pattern, planting density, planting time, and row spacing. Such practices could help in carbon sequestration and reduced GHG emissions. Zero tillage system and other practices like crop rotation, mulching, and cover crops improved carbon sequestration (Hobbs and Govaerts, 2010; Alvaro-Fuentes and Paustian, 2011). Conservation agriculture practices such as reduced tillage and maintaining soil temperature and moisture through mulching have enhanced carbon sequestration (Pratibha *et al.*, 2016). Biochar application has improved soil organic carbon levels and reduced N_2O and CH_4 emissions from soil (Xiao *et al.*, 2018). Adoption of agroforestry systems have been found to provide land cover and thereby helped in increased carbon sequestration, water holding capacity, and soil microbial biomass. While the land cover could minimize soil erosion and eutrification of rivers, enhanced soil microbial activity has improved the available phosphorus in the soil (Garske and Ekardt, 2021). Site-specific nutrient management could decrease total emissions over conventional fertilization application methods (Zhang *et al.*, 2017). Neem-coated urea has helped in the slow release of nitrogen and thus contributed to reduce N_2O emissions (Hala *et al.*, 2014). The use of value-added potassium fertilizers have enhanced the resilience in terms of enhanced crop productivity and helped in the uptake of Zn resulting in up to 75% increment in Zn content in grains (Hussain *et al.*, 2020).

Water systems

It is established that the total amount of rainfall may not decrease but its distribution and intensities may vary across agro-ecologies. Hence, there is a need to revisit the cropping seasons and patterns including crop diversifications. The short-term adaptation strategies could include rainwater harvesting systems coupled with water-saving cropping systems (drought tolerant varieties/crops), prevention of soil erosion, proper drainage, water-saving irrigation techniques, and improved weather forecasts and insurance (FAO, 2021). The medium-term strategies are drainage and flood management, adaptation of SOPs of dam operation rules, integrated water management, efficient irrigation schemes, re-prioritization of water uses (FAO, 2021). The long-term strategies emphasize developing national adaptation plans, improved monitoring systems, strengthening rights of access to water, new irrigation areas, desalination and wastewater recycling, and developing climate-resilient crops (FAO, 2021). Machine transplantation, direct seeding, system of rice intensification, and alternate wetting and drying methods in rice could reduce water requirement by 20–25% and also reduce CH_4 emissions. Deficit irrigation has been suggested as a way to increase system benefits, at the cost of individual benefits, by decreasing the crop water allocation and increasing the total irrigated land (Garg and Dadhich, 2014). Small, modular, and decentralized check dams can improve rainwater use and climate resilience while minimizing the environmental impacts of large irrigation infrastructure (Rockstrom *et al.*, 2017).

Pests and diseases

The threat due to climate change impacts on pests and plant pathogens could seriously affect the profitability and livelihood security of farmers in the absence of suitable adaptation strategies. Further, rainfed production systems in India are fragile, and hence affordable adaptation strategies such as change of planting time, use of climate-resilient varieties, intercropping systems, use of organic manures, and crop diversification should be developed, with due care to trade-offs (Desai *et al.*, 2021). At present, during the rainy season, sesamum suffers crop losses due to end-of-season rains whereas groundnut is contaminated by mycotoxins due to end-of-season drought. If these crops could be accommodated in a post-rainy cropping system, such losses could be avoided. Similarly, yield penalty due to changing planting dates should be compensated by short-duration high-yielding varieties with market preference (Wolfe *et al.*, 2008). As the tritrophic interactions are impacted due to climate change, selecting an appropriate biocontrol agent could be a valuable option. Melloy *et al.* (2010) highlighted the role of crop residue management in pest management. A cohesive farming systems approach would reduce crop losses due to pests (Reidsma *et al.*, 2015). Speed breeding techniques could help to develop pest-resistant/tolerant varieties at a faster pace (Watson *et al.*, 2018). Strong survey and surveillance systems coupled with early detection and warning systems have been suggested to meet the biosecurity threats from alien invasive species (Desai *et al.*, 2021). Carvajal-Yepes *et al.* (2019) proposed a global surveillance system to strengthen and interconnect crop biosecurity systems. Similarly, GIS-based pest risk mapping will help to emphasize management strategies (Vennila *et al.*, 2016). Maintaining the special and temporal agro-diversity could help in balancing the pests and their natural enemies in a given agro-ecology (Altieri, 1999). Development of pesticides with novel mode of action and using it in a mixture to counter multiple insect pest attacks could reduce the risk of development of resistance (Ma *et al.*, 2021).

Livestock systems

Modification of the environment, nutritional management, and genetic improvements were suggested as key areas in developing the adaptation strategies in livestock systems. The modification of the micro-environment could dissipate the heat and thus alleviate the heat stress. The common methods adopted to combat the negative effects of heat stress were (i) providing shade, (ii) provision of evaporative cooling systems (foggers/sprinklers), (iii) use of fans, and (iv) proper selection of building site of the animal shelters with proper ventilation (Sejian *et al.*, 2012a, b). Choosing the right location and direction coupled with the provision of shade is the simplest and most economical adaptation strategy against heat stress in livestock (Singh and Upadhyay, 2009). Roofing modification with thermal insulators, mud plasters, roof paints or solar panels, etc.; provision of fans, mist, foggers, sprinklers; or wallowing tanks in sheds reduced the thermal heat stress and enhanced animal productivity (Chauhan, 2004; Kumar *et al.*, 2007; Narwaria *et al.*, 2017). Rearing of animals with high production traits in lower numbers would provide more open space for animals and also reduce GHG emissions (Batima, 2007).

Modification in the feed composition and improved feeding regimen could help to increase the dry matter intake in heat-stressed animals. A high-quality feed and a balanced ration should be designed for quicker digestion with less heat production (Tahmasbi *et al.*, 2012). Supplementation of niacin and antioxidants (Vitamin C and E) in the diet during heat stress helped in improving the fertility rate and milk production in livestock (El-Tarabany and Nasr, 2015) and egg production in poultry (Lin *et al.*, 2006). As lactating animals consume 120–200% more water during heat stress (Sirohi *et al.*, 2010), provision should be made for a continuous supply of clean and fresh water abundantly. Organizing animal camps during summer months has been found to be effective in reducing mortality due to heat stress.

On the basis of phenotypic characteristics, the genetic selection of livestock would help to increase the heat tolerance ability in animals. Advanced breeding strategies to select progeny with high production traits even under harsh environmental conditions such as heat stress helped to improve the production and growth rate of livestock (Kimothi and Ghosh, 2005; Henry *et al.*, 2012). Kadzere *et al.* (2002) reported that the seaweed supplemented at 0.25% of the diet prevented a decline in milk production during heat stress, especially for larger cows, which are more susceptible to heat stress because of the higher metabolic rate.

Market-related aspects

Agricultural insurance helps farmers to protect themselves from the financial losses suffered due to extreme climate, stabilizes the farm income, and enhances farmers' willingness to adapt and make use of innovative technologies. Accessible and affordable credit and crop insurance help the more vulnerable small and marginal farmers to cope with the ill effects of climate change and the schemes could be improved through tools such as GIS and remote sensing for real-time loss assessment and quick settlements (Kean, 2021). However, the instruments should be designed with climatically sensitive triggers to ensure that the benefit always marginally tilts towards the farmers. Knowledge and technology management plays a crucial role in the adaptation of the farmers for sustainable farm incomes, food security of households, and dietary diversity (Ghosh *et al.*, 2021). The mobile technology coupled with artificial intelligence and machine learning tools could help to spread tailor-made adaptation and mitigation strategies (Talaviya *et al.*, 2020) to stakeholders and thus reach a new economic equilibrium. Selection of agro-climatically suitable alternative crops such as finger millet, maize, pearl millet, and sorghum reduced vulnerability coupled with potential co-benefits and tradeoffs (Davis *et al.*, 2019). Agro-forestry systems act as vulnerability neutralizers as they ensure food and livelihood security, enhance farm income, reduce migrations, and sequester carbon (Batini, 2019). Managing post-harvest losses could be an adaptation strategy to ensure food and nutritional security and indirectly reduce emissions (Chegere, 2018). The exploitation of GPS and satellite imaging and sensors and the use of nano-packaging, eco-friendly storage, and supply chain management practices can make the agri-supply chain robust and resilient (Devi *et al.*, 2020). Blockchain technology applications can improve transparency and accountability in the supply chain and support farmers' adaptation to climate change, create climate finance opportunities and track investments as well as results of enriched management practices (UNDP, 2021). A global carbon database of impacts and adaptation/mitigation strategies will help for effective monitoring and evaluation of climate change adaptation/

mitigation activities and support carbon market development (FAO, 2021). Adaptive social protection practices such as mobility, diversification of livelihoods, risk-pooling, and informal/social entrepreneurship activities help in building resilience among the poor and vulnerable populations (Santha *et al.*, 2015).

Mitigation Strategies

The agriculture sector in India contributed 14% of total emissions in 2016, of which, 54% was from enteric fermentation, 19% from agricultural soils, 17% from rice cultivation, 6% from manure management, and 2% due to the burning of agricultural residues (MoEFCC, 2021). Adoption of cheaper mitigation options can contribute up to 80% of the total technical mitigation potential in Indian agriculture (Sapkota *et al.*, 2019). For instance, judicious water management in rice and mid-season drainage (Pathak *et al.*, 2012); adoption of a System of Rice Intensification (SRI), modified SRI, aerobic rice, and direct seeding of rice (Pathak, 2015). Similarly, leaf colour chart-based urea application in rice-wheat cropping system (Bhatia *et al.*, 2012); slow release of nitrogen fertilizers like neem-coated urea and polymer-coated urea and the use of sulphate forms of nitrogen and the right timing of fertilizer application could reduce emissions. Micro-irrigation systems can significantly increase the water use efficiency and sub-surface drip, surface drip, and sprinkler methods could help in conserving the water and energy resources (Gao *et al.*, 2019; Mehmood *et al.*, 2021). Globally, renewable energy technologies are being promoted as potential mitigation strategies and solar pumps can minimize emissions to the extent of 0.891t CO_2 e/ha/year (MoEFCC, 2021) for providing irrigation water. Conservation Agriculture (CA) with reduced tillage or zero-tillage contributed to resource conservation, improving soil properties, water and nutrient use efficiencies, and ecosystem services leading to more stable yields and a reduction in GHG emissions (Sapkota *et al.*, 2015). The adoption of zero-tillage for land preparation and crop establishment in rice–wheat systems could minimize 93kg CO_2 emission/ha/year in Indo-Gangetic regions (Gupta *et al.*, 2016). Laser land leveling reduced the water pumping time in rice-wheat areas resulting in mitigation of 0.30t CO_2 eq/ha/year (Jat *et al.*, 2015). Agro-forestry systems acted as above and below ground carbon sinks and sequestered up to 14.42 tonnes of C/ha/year depending on the tree species, location, tree density and crown growth (Prasad *et al.*, 2012). Such systems also benefit to conserve biodiversity and maintain soil health. Arresting land degradation through afforestation through the National Afforestation Programme could result in a potential sink of emission 2.14 CO_2eq/year (MoEFCC, 2021). Manipulating the diet of livestock to manipulate enteric fermentation has reduced the CH_4 emissions by 19% (Hossain *et al.*, 2017). Optimizing the use of protein supplements within the ruminant system by feeding bypass proteins has led to an emission reduction of 0.078 tCO_2eq/animal during 2017–2018 (MoEFCC, 2021). Minimizing the inorganic fertilizers by promoting organic manure will also reduce emissions (Pathak, 2015).

Government of India Initiatives

Climate change research financing

In India, several research projects/programmes were initiated by government research organizations, NGOs, and private enterprises to assess the impacts of climate change and develop suitable adaptation and mitigation strategies. Under the National Action Plan for Climate Change (NAPCC), the Government of India launched eight national missions to address the climate change-related issues in a holistic way. While the National Mission on Sustainable Agriculture (NMSA) directly addresses the issues related to agriculture, other missions indirectly contribute to addressing the impacts of climate change. Since 2011–2012, many projects were funded by DST. Projects are also being funded by the Department of Biotechnology, Ministry of Environment, Forestry & Climate Change and NABARD. Similarly, international agencies such as DFID, CIDA, CDKN, GEF trust, World Bank, U.K., Canada, Norway, Aga Khan Foundation, GIZ, USAID, IDRC, Asian Development Bank, and other agencies are also implementing climate change research programmes.

However, as these programs were scattered, the outcomes were negligible. To formalize the research at national level for dedicated outcomes, ICAR had initiated an umbrella project viz. Network Project on Climate Change (NPCC). The initial research outcomes of this project demonstrated the need for further investments and thus the Government of India, through ICAR has launched "National Initiative on Climate Resilient Agriculture (NICRA)" in 2011 with ICAR-Central Research Institute for Dryland Agriculture as lead institute and an outlay of about 40 million USD for 2 years. It has continued with enhanced outlays and participating institutions and has transformed into National Innovations in Climate Resilient Agriculture subsequently. The project has four components viz. (i) Strategic research, (ii) Technology demonstration, (iii) Capacity building, and (iv) Sponsored/competitive grants. Under strategic research component, state-of-the art research facilities have been established to work on developing stress tolerant genotypes, natural resource conservation, and technologies for management of farming systems.

The Technology Demonstration Component focused on the participatory demonstration of adaptation and mitigation technologies in 121 village clusters to make them climate resilient. The four modules of the TDC include (i) Natural resources, (ii) Crop production, (iii) Livestock and fisheries, and (iv) Institutional interventions. Some success stories of this component include (a) rainwater harvesting coupled with efficient irrigation systems resulting in 11–63% yield enhancement; (b) *in-situ* moisture conservation measures for minimizing the impact of dry spells and for safe discharge of excess water by promoting conservation furrow, trench-cum-bunding, compartmental bunding, mulching, ridge and furrow, broad bed furrow, contour bunding, etc. resulting in 17–64% yield increase; (c) promoting stress escaping/tolerant cultivars in rice, pigeonpea, castor beans, pearl millet, wheat, chickpea, finger millet, black gram, groundnut, sorghum, foxtail millet and mustard; (d) resilient intercropping systems for sustained income such as maize +soybean (1:2) with 121% maize equivalent yield and pigeonpea +setaria (5:1) in Rayalaseema region even under 33–75% deficit rainfall; (e) Livestock technologies such as cultivating improved fodder cultivars with drought tolerance, year-round fodder production of cultivars with the available moisture during winter, and promotion of perennial fodders could serve as fodder banks; and (f) improved feed management through silaging, urea-molasses mineral blocks, dry fodder enrichment, concentrate feeds, and mineral mixture supplementation, etc. Since inception in 2011, 467 awareness programmes for 60,704 farmers on climate change impacts and 18,183 training programmes on climate resilient technologies for 443,656 farmers were conducted. Village Climate Risk Management Committees (VCRMCs) have been established and nurtured to enable communities to respond to climate stresses in a continuous manner beyond the project period.

The sponsored and competitive grants component focuses on specific areas of research not covered under the strategic research programme to bridge the gap between both strategic and TD components. The Capacity Building program was intended to hone the skills of the National Agricultural Research and Education Systems (NARES) in emerging areas of research such as systems simulation modelling, phenotyping, and other areas.

After successful implementation of the programme for two plan periods, it was renamed as National Innovations in Climate Resilient Agriculture and taken into fold many more villages and research organizations to systematically characterize the impacts and develop adaptation and mitigation strategies.

District Agricultural Contingency Plans (DACPS)

Another adaptation approach is Agricultural Contingency planning for enhanced preparedness of the stakeholders to cope with aberrant weather-related risks *viz.*, droughts, floods, heat waves, cold waves, untimely and high-intensity rainfall, frost, hailstorms, pest, and disease outbreaks. DACPs have been prepared with information on (i) district agricultural scenario, (ii) stage-wise vulnerability of various agricultural systems, and (iii) recommended adaptation

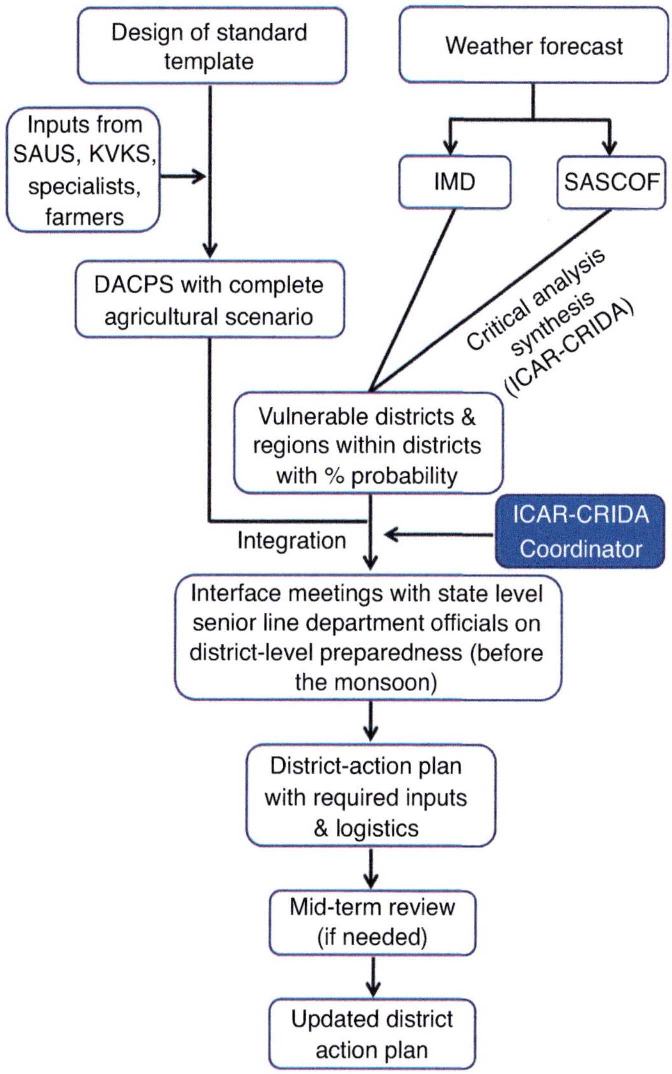

Fig. 9.3. A flow diagram of the preparation and operationalization of district agricultural contingency plans. Data source: Ministry of Agriculture & Farmers' Welfare, Government of India.

measures. These plans are being updated with funding from NMSA, the Government of India to incorporate recent scenarios and technologies.

The DACPs have been operationalized since 2013 by integrating the long-range weather forecast with the agricultural calendar of the district with an emphasis on the agro-ecologies supposed to face weather aberrations. The step-wise operationalization is shown in Fig. 9.3.

Impacts of contingency plan implementation

The impacts of contingency plans were summarized by Srinivasarao *et al.* (2020) and other publications. Some of the highlights are:

- During 2014–2015, despite −20 to −59% deficit rainfall in 30 out of 51 districts in Madhya Pradesh, the total sown area and food grain production increased by 15 and 13%, respectively.

- Though most of Gujarat state received 18–23% deficit rainfall in 2014–2015, the impact was negligible in many crops except for groundnut, sorghum, and pearl millet.
- In Karnataka, the deficit rainfall situation was managed by (a) encouraging the sowing of maize, cotton, pearl millet in unsown areas of sorghum, pigeon pea, groundnut, and sunflower; (b) encouraging the farming community to take up land configuration, soil management practices; (c) utilizing the common property resources for fodder production; and (d) promoting drought safety net programmes i.e. crop insurance.
- Promotion of community nurseries for rice seedlings to overcome the delayed sowing of paddy on a village basis to meet the demand of the entire village.
- Lining of ponds and strengthening of bunds were promoted for effective water storage in Jharkhand.
- Additional power supply to irrigate standing maize crop at critical growth stage; adoption of *in situ* moisture conservation measures, short duration rice varieties like MTU 1011, direct seeded rice, crop diversification, and micro irrigation in cotton as water saving mechanism were taken up in Telangana.

Current Gaps

Current gaps in combating the impacts of climate change on agriculture could be classified under research, development, and policy framework components. On the research front, despite huge investments, considering the wide agro-ecological variability, the outcomes of the research front do not match with the quantum of the impacts in this important sector. Efforts are underway to bring all researchers onto one platform to harmonize research priorities and programmes. However, there is a huge gap in formal functional networking of the research teams leading to duplication of the efforts. There is an urgent need for huge investments in establishing infrastructure, skilling of the human resources and establishment of database networks. While the information sector is growing at an enormous pace, the creation of databases and their exchange is still at its nascent stage, primarily due to non-collation of cured datasets. Interactions among the biophysical resource diversity of a given agro-ecology vis-à-vis agro-enterprise need to be characterized using systems biology approach. Standardized research methodologies for climate change research are another major limiting factor.

On the development front, there is an acute shortage of the trained human resource to scale up and scale out best bet climate resilient technologies. Agricultural enterprises spill across multiple development departments and the benefits of the adaptation and mitigation strategies have not been harnessed by all stakeholders due to lack of coherence among them (Byravan and Rajan, 2012). Investments are another hurdle to take up large-scale demonstrations of climate-resilient technologies under farmers' conditions and their refinement, if needed.

On the policy front, 'one-size fits all' strategy has obstructed the development of tailor-made agro-ecology policies that can effectively address regional dimensions of climate change impacts. The inter-state water disputes pose both water-surplus and water-deficit conflicts denting the ultimate interest of the farmers. In India, agriculture is the State subject thus limiting the role of the central government to make major policy changes. It has sometimes led to conflicts in implementation of schemes of the central government in some states.

Way Forward

The agriculture sector in India is expected to generate better momentum in the coming decades due to increased investment in agricultural infrastructure such as irrigation facilities, warehousing, and cold storage. The all-embracing National Actin Plan for Climate Change of 2008 was prepared under the guidance of the Prime Minister's Council on Climate Change to reap the 'co-benefits' by implementing effective and appropriate climate change adaptation and mitigation strategies (MOEF, 2008). The States in turn have prepared State Actin Plan for Climate Change (SAPCC) for effective implementation of climate change adaptation and mitigation plans to climate-proof through development investments and infrastructure planning. The suggested way forward thus should focus on:

- developing short-term adaptation strategies to address immediate needs of the farmers while medium and long-term adaptation strategies to fit to the future climate scenarios;
- preparing a roadmap for research and development activities for sustainable agricultural systems with adequate coping mechanisms to climate change;
- making ecosystem services as an integral part of the research programmes and assess their role in agriculture and allied sectors in relation to climate change;
- establishing databases and data sharing mechanisms for seamless exchange of information and research methodologies among research groups;
- inter-disciplinary and inter-institutional networks at regional/national/international level for methodologies, databases, and knowledge sharing mechanisms for exchange of ideas, datasets, and methodologies;
- measurable impact assessment strategies and timelines for monitoring the programmes on climate change and their benefits and suggest mid-course corrections, where needed;
- human resource development and skill enhancement to conduct climate change research and development programmes;
- organizing awareness and training programmes for all stakeholders for effective implementation of the adaptation strategies;
- address the disparity in government investments between irrigated and rainfed agriculture, especially in climate change research programmes to benefit the most vulnerable rainfed small and marginal farming communities;
- agro-ecology specific differential functional policy framework with suitable enabling mechanisms to ensure climate-proofing of the vulnerable agricultural systems;
- large-scale climate finance to significantly reduce emissions and to adapt to the adverse effects of changing climate;
- amicable policy to resolve the inter-state water disputes to develop long-term strategies for scientific management of water;
- incentivizing the farmers to adopt climate-resilient strategies such as crop diversification, modified crop calendars, etc. to motivate non-adopters
- mainstreaming district agricultural contingency plans by all states

Conclusion

In the Climate Change Performance Index rankings, India stands at 10th position and with the timely research and policy interventions being taken by the government, the ranking could go higher. The climate change impacts are a reality and hence, need to be addressed on priority to ensure food-, nutritional-, and livelihood securities. A multi-pronged approach addressing research, value addition, development, marketing, knowledge management, and enabling policy framework will help to address the regional and national concerns in a holistic manner. The adaptation and mitigation strategies need to be customized and mainstreamed to reap maximum benefits. For instance, rainwater harvesting must be revisited based on soil types, rainfall projections, and cropping patterns to ensure optimized harvesting and storage of rainwater from highly intense rainfall events to protect crops at later critical phenophases. While updation of the DACPs will help to chalk out strategies at the administration unit level, downscaling of the district agricultural contingency plans to the sub-district level will help to implement customized adaptation strategies for the given agro-ecology. The capital and knowledge investments will help to reduce the disasters and also prepare the nation to meet SDGs 2, 3, 12, and 13 directly, whereas 1, 14, 15, and 17, indirectly.

Conflicts of interest

The authors have no conflicts of interest to declare.

References

Adhav, C.A., Sendhil, R., Chandel, B.S., Gunjan, B., Ponnusamy, K. *et al.* (2021) Socio-economic vulnerability to climate change – index development and mapping for districts in, Maharashtra, India 25.

Altieri, M.A. (1999) The ecological role of biodiversity in agroecosystems. *Agriculture Ecosystems and Environment* 74, 19–31.

Alvaro-Fuentes, J. and Paustian, K. (2011) Potential soil carbon sequestration in a semiarid mediterranean agroecosystem under climate change: Quantifying management and climate effects. *Plant and Soil* 338, 261–272.

An, Y., Wan, S., Zhou, X., Subedar, A.A., Wallace, L.A. *et al.* (2005) Plant nitrogen concentration, use efficiency, and contents in a tallgrass prairie eco-system under experimental warming. *Global Change Biology* 11, 1733–1744.

Balic, I.M., Milinkovic-Tur, S., Samardzija, M. and Vince, S. (2012) Effect of age and environmental factors on semen quality, glutathione peroxidase activity and oxidative parameters in Simmental bulls. *Theriogenology* 78, 423–431.

Basu, A.K. and Bandhyopadhyay, P.K. (2004) The effect of season on the incidence of ticks. *Animal Health and Production in Africa* 52, 39–42.

Batima, P. (2007) *Climate Change Vulnerability and Adaptation in the Livestock Sector of Mongolia. Assessments of Impacts and Adaptations to Climate Change*. International START Secretariat, Washington DC, p. 105.

Batini, N. (2019) Reaping what we sow- smart changes to how we farm and eat can have a huge impact on our planet. *Finance and Development* 4(56), 30–33.

Bauerfeind, S.S. and Fischer, K. (2013) Increased temperature reduces herbivore hostplant quality. *Global Change Biology* 19, 3272–3282.

Bei, Q., Reitz, T., Schnabel, B., Eisenhauer, N., Schädler, M. *et al.* (2023) Extreme summers impact cropland and grassland soil microbiomes. *The ISME Journal* 17, 1589–1600.

Bhatia, A., Aggarwal, P.K., Jain, N. and Pathak, H. (2012) Greenhouse gas emission from rice-and wheat-growing areas in India: Spatial analysis and upscaling. *Greenhouse Gases: Science and Technology* 2, 115–125.

Byravan, S. and Rajan, S.C. (2012) *An Evaluation of India's National Action Plan on Climate Change*. IFMR – Centre for Development Finance and IIT Chennai, India.

Carvajal-Yepes, M., Cardwell, K., Nelson, A., Garrett, K.A., Giovani, B.D. *et al.* (2019) A global surveillance system for crop diseases. *Science* 364, 1237–1239.

Chakrabarty, M. (2016) Climate Change and Food Security in India. Issue Briefs and Special Reports. Available at: https://www.orfonline. org/research/climate-change-and-food-security-in-india/ (accessed 1 November 2023).

Chakraborty, S. (2013) Migrate or evolve: Options for plant pathogens under climate change. *Global Change Biology* 19, 1985–2000.

Chander, S. (2012) Impact of climate change on insects. In: Pathak, H. and Aggarwal, P.K. (eds) *Climate Change Impact, Adaptation and Mitigation in Agriculture: Methodology for Assessment and Application*. Indian Agricultural Research Institute, New Delhi, India, p. 131.

Chatterjee, A. (2021) Climate Change and Indian Agriculture. Available at: https://www.devalt.org/newsletter/nov03/of_5.htm (accessed 1 November 2023).

Chauhan, T.R. (2004) Feeding strategies for sustainable buffalo production. In: *Proceedings XI Animal Nutrition Conference*, pp. 74–83.

Chegere, M.J. (2018) Post-harvest losses reduction by small-scale maize farmers: The role of handling practices. *Food Policy* 77, 103–115.

Climate Knowledge Portal (2021). Available at: https://climateknowledgeportal. worldbank.org/country/india (accessed 1 November 2021).

Coakley, S.M., Scherm, H. and Chakraborty, S. (1999) Climate change and disease management. *Annual Review of Phytopathology* 37, 399–426.

Compant, S., Heijden, M.G.A.V.D. and Sessitsch, A. (2010) Climate change effects on beneficial plant-microorganism interactions. *FEMS Microbiology Ecology* 73, 197–214.

Daley, C.D., Sakurai, H., Adams, B.M. and Adams, T.E. (1999) Effect of stress like concentrations of cortisol of gonadotroph function in orchidectomized sheep. *Biology of Reproduction* 60, 158–163.

Das, R., Sailo, L., Verma, N., Bharti, P., Saikia, J. *et al.* (2016) Impact of heat stress on health and performance of dairy animals: A review. *Veterinary World* 9, 260–268.

Davis, K.F., Chhatre, A., Rao, N.D., Singh, D. and Defries, R. (2019) Sensitivity of grain yields to historical climate variability in India. *Environmental Research Letters* 14, 064013.

DeAngelis, K.M., Pold, G., Topcuoglu, B.D., Van Diepen, L.T.A., Varney, R.M. *et al.* (2015) Long-term forest soil warming alters microbial communities in temperate forest soils. *Frontiers in Microbiology* 6, 104. DOI: 10.3389/fmicb.2015.00104.

Debela, C. and Tola, M. (2018) Effect of elevated CO_2 and temperature on crop-disease interactions under rapid climate change. *International Journal of Environmental Sciences & Natural Resources* 13, 555–851.

Desai, S., Dubey, S.C., Meenakshi, T., Uzma, S. and Sravani, P. (2021) Crop disease management strategies for rainfed cropping systems under changing climate scenarios. *Indian Phytopathology* 74, 485–494.

Devi, T.S., Biswas, S., Gaibimei, P., Arjun, A.D., Tarafdar, A. *et al.* (2020) Climate smart post-harvest agriculture and food systems. In: Srinivasarao, C., Srinivas, T., Rao, R.V.S., Srinivasa Rao, N., Senthil Vinayagam, S. *et al.* (eds) *Climate Change and Indian Agriculture: Challenges and Adaptation Strategies*. ICAR- National Academy of Agricultural Research Management, Hyderabad, India, pp. 359–384.

Dubey, S.K., Sharma, D., Babel, M.S. and Mundetia, N. (2020) Application of hydrological model for assessment of water security using multi-model ensemble of CORDEX-South Asia experiments in a semi-arid river basin of India. *Ecological Engineering* 143, 105641.

El-Tarabany, M.S. and Nasr, M.A.F. (2015) Reproductive performance of brown Swiss, holstein and their crosses under subtropical environmental conditions. *Theriogenology* 84, 559–565.

FAO (2018) The State of Agricultural Commodity Markets—Climate Change and Food Security Agricultural Commodity Markets. Available at: https://www.fao.org/3/I9542EN/i9542en.pdf (accessed 1 November 2023).

FAO (2021) Water Management Options for Climate Change Adaptation. Available at: https://www.fao.org/climate-smart-agriculture-sourcebook/production-resources/module-b6-water/chapter-b6-4/en/ (accessed 1 November 2023).

Ferrocino, I., Chitarra, W., Pugliese, M., Gilardi, G., Gullino, M. *et al.* (2013) Effect of elevated atmospheric CO_2 and temperature on disease severity of *Fusarium oxysporum* f.sp. *lactucae* on lettuce plants. *Applied Soil Ecology* 72, 1–6.

Gahoonia, T.S. and Nielsen, N.E. (2003) Phosphorus uptake and growth of root hairless barley mutant (bald root barley) and wild type in low and high–p soils. *Plant, Cell and Environment* 26, 1759–1766.

Ganiger, P.C., Yeshwanth, H.M., Muralimohan, K., Vinay, N., Kumar, A.R.V. *et al.* (2018) The occurrence of the new invasive pest, fall armyworm, *Spodoptera frugiperda* (J.E. smith) (Lepidoptera: Noctuidae), in the maize fields of Karnataka, India. *Current Science* 115, 25.

Gao, S., Hendratna, A., Cai, Z., Duan, Y., Qin, R. *et al.* (2019) Subsurface drip irrigation reduced nitrous oxide emissions in a pomegranate orchard. *International Journal of Environmental Science and Development* 10, 79–85.

Gao, Q., Wang, J., Xue, K., Yang, Y., Xie, J. *et al.* (2020) Stimulation of soil respiration by elevated CO_2 is enhanced under nitrogen limitation in a decade-long grassland study. *Proceedings of National Academy of Sciences USA* 117, 33317–33324.

Garg, N.K. and Dadhich, S.M. (2014) Integrated non-linear model for optimal cropping pattern and irrigation scheduling under deficit irrigation. *Agricultural Water Management* 140, 1–13.

Garske, B. and Ekardt, F. (2021) Economic policy instruments for sustainable phosphorus management: Taking into account climate and biodiversity targets. *Environmental Sciences Europe* 33, 56.

Ghosh, J.S., Kapoor, R., Ghosh, U., Singh, A., Downs, S. *et al.* (2021) Pathways of climate change impact on agroforestry, food consumption pattern, and dietary diversity among indigenous subsistence farmers of Sauria Paharia tribal community of India: A mixed methods study. *Frontiers in Sustainable Food Systems* 5, 174.

Goyal, M.K. and Surampalli, R.Y. (2018) Impact of climate change on water resources in India. *Journal of Environmental Engineering* 144, 04018054.

Guo, X., Feng, J., Shi, Z., Zhou, X., Yuan, M. *et al.* (2018) Climate warming leads to divergent succession of grassland microbial communities. *Nature Climate Change* 8, 813–818.

Gupta, D.K., Bhatia, A., Kumar, A., Das, T.K., Jain, N. *et al.* (2016) Mitigation of greenhouse gas emission from rice–wheat system of the Indo-Gangetic plains through tillage, irrigation and fertilizer management. *Agriculture, Ecosystems & Environment* 230, 1–9.

Hala, Y., Jumadi, O., Muis, A. and Hartati, H. (2014) Development of urea coated with neem (*Azadirachta indica*) to increase fertilizer efficiency and reduce greenhouse gases emission. *Jurnal Teknologi* 69, 11–15.

Hamidev, A., Helming, K., Bellochi, G., Bojar, W., Dalgaard, T. *et al.* (2018) Impacts of climate change adaptation options on soil functions: A review of European case-studies. *Land Degradation and Development* 29, 2378–2389.

Hari, S., Khare, P. and Subramanian, A. (2018) Climate change and Indian agriculture. Available at: https://www.ideasforindia.in/topics/agriculture/climate-change-and-indian-agriculture.html (accessed 1 November 2023).

Henry, B., Charmley, E., Eckard, R., Gaughan, J.B. and Hegarty, R. (2012) Livestock production in a changing climate: Adaptation and mitigation research in Australia. *Crop and Pasture Science* 63, 191–202.

Himanen, S.J., Nissinen, A., Dong, W.X., Nerg, A.M., Stewart, C.N. *et al.* (2008) Interactions of elevated carbon dioxide and temperature with aphid feeding on transgenic oilseed rape. *Global Change Biology* 14, 1437–14354.

Hobbs, P.R. and Govaerts, B. (2010) How conservation agriculture can contribute to buffering climate change. In: Reynolds, M.P. (ed.) *Climate Change and Crop Production*. CPI Antony Rowe, Chippenham, UK, pp. 177–199.

Hossain, S.A., Sherasia, P.L., Phondba, B.T., Pathan, F.K. and Garg, M.R. (2017) Effect of feeding green fodder based diet in lactating buffaloes: Milk production, economics and methane emission. *Indian Journal of Dairy Science* 70, 767–773.

Hou, E.Q., Chen, C.R., Luo, Y.Q., Zhou, G.Y., Kuang, Y.W. *et al.* (2018) Effects of climate on soil phosphorus cycle and availability in natural terrestrial ecosystems. *Global Change Biology* 24, 3344–3356.

Hungate, B.A., Dukes, J.S., Shaw, M.R., Luo, Y. and Field, C.B. (2003) Nitrogen and climate change. *Science* 302, 1512–1513.

Huot, B., Castroverde, M.D.C., Velásquez, C.A., Hubbard, E., Jane, A. *et al.* (2017) Dual impact of elevated temperature on plant defence and bacterial virulence in Arabidopsis. *Nature Communications* 8, 1808.

Hussain, S., Shah, M.A.A., Khan, A.M., Ahmad, F. and Hussain, M. (2020) Potassium enhanced grain zinc accumulation in wheat grown on a calcareous saline-sodic soil. *Pakistan Journal of Botany* 52, 69–74.

ICAR (Indian Council for Agricultural Research)-NBAIR Pest Alert (2018) *Spodotera frugiperda (J.E. Smith)*.

IPCC (2021) Summary for policymakers. In: Masson-Delmotte, V., Zhai, P., Pirani, A., Connors, S.L., Péan, C. *et al.* (eds) *Climate Change 2021: The Physical Science Basis. Contribution of Working Group I to the Sixth Assessment Report of the Intergovernmental Panel on Climate Change*. Cambridge University Press, Cambridge, UK and New York, pp. 3–32. DOI: 10.1017/9781009157896.001.

Jat, M.L., Singh, Y., Gill, G., Sidhu, H., Aryal, J.P. *et al.* (2015) Laser assisted precision land leveling: Impacts in irrigated intensive production systems of South Asia. *Advances in Soil Science* 22, 323–352.

Johnson, B., Shyam, S.S., Kripa, V., Zachriah, P.U., Swathilekshmi, P.S. *et al.* (2016) Vulnerability assessment of coastal fishermen households in Tamil nadu, India: A climate change perspective. *Indian Journal of Extension Education* 52(1 & 2), 20–26.

Kadiyala, M.D.M., Nedumaran, S., Padmanabhan, J., Gumma, S.K., Gummadi, S. *et al.* (2021) Modeling the potential impacts of climate change and adaptation strategies on groundnut production in India. *Science of the Total Environment* 776, 145996.

Kadzere, C.T., Murphy, M.R., Silanikove, N. and Maltz, E. (2002) Heat stress in lactating dairy cows: A review. *Livestock Production Science* 77, 59–91.

Kaiser, J. (1996) Pests overwhelm Bt cotton crop. *Nature* 273, 423.

Kantamaneni, K., Rice, L., Yenneti, K. and Campos, L.C. (2020) Assessing the vulnerability of agriculture systems to climate change in coastal areas: A novel index. *Sustainability* 12, 1–24.

Kaur, J., Gosal, S.K. and Kaur, P. (2014) Effects of climate change on plant associated microbial communities and enzyme activities. *African Journal of Microbiology Research* 8, 3087–3093.

Kean, N.G. (2021) Paving the way for digital green economy in Indian agriculture. India Matters. Available at: https://www.orfonline.org/expert-speak/paving-way-digital-green-economy-indian-agriculture/ (accessed 1 November 2023).

Kimothi, S.P. and Ghosh, C.P. (2005) Strategies for ameliorating heat stress in dairy animals. *Dairy Year Book* 371–377.

Kumar, A., Singh, R., Jena, P.P., Chatterjee, C. and Mishra, A. (2015) Identification of the best multi-model combination for simulating river discharge. *Journal of Hydrology* 525, 313–325.

Kumar, R., Kamra, D.N., Agarwal, N. and Chaudhary, L.C. (2007) *In vitro* methanogenesis and fermentation of feeds containing oilseed cakes with rumen liquor of buffalo. *Asian-Australasian Journal of Animal Science* 20, 1196–2000.

Kumar, S.B.V., Kumar, A. and Meena, K. (2011) Effect of heat stress in tropical livestock and different strategies for its amelioration. *A Review Journal of Stress Physiology and Biochemistry* 7, 45–54.

Kumar, S.N., Singh, A.K., Aggarwal, P.K., Rao, V.U.M. and Venkateswarlu, B. (2012) Climate change and Indian agriculture: Impact, adaptation and vulnerability – salient achievements from ICAR network project 1–33.

Letha Devi, G., Adhiguru, G., Mech, P., Kataktalware, M.A., Chaithra, G. *et al.* (2021) Livelihood vulnerability analysis to climate variability and change risks of livestock farming in Karnataka. *Indian Journal of Extension Education* 57, 1–5.

Lin, H., Jiao, H.C., Buyse, J. and Decuypere, E. (2006) Strategies for preventing heat stress in poultry. *World's Poultry Science Journal* 62, 71–86.

Ma, C., Zhang, W., Peng, Y., Zhao, F., Chang, X. *et al.* (2021) Climate warming promotes pesticide resistance through expanding overwintering range of a global pest. *Nature Communications* 12, 5351.

Mahapatra, B., Walia, M., Rao, C.A.R., Raju, B.M.K. and Saggurti, N. (2021) Vulnerability of agriculture to climate change increases the risk of child malnutrition: Evidence from a large-scale observational study in India. *PLoS ONE* 16(6), e0253637.

Marchesini, G., Cortese, M., Mottaran, D., Ricci, R., Serva, L. *et al.* (2018) Effects of axial and ceiling fans on environmental conditions, performance and rumination in beef cattle during the early fattening period. *Livestock Science* 214, 225–230.

Mashaly, M.M., Hendricks, G.L., Kamala, M.A., Gehad, A.E., Abbas, A.O. *et al.* (2004) Effect of heat stress on production parameters and immune responses of commercial laying hens. *Poultry Science* 83, 889–894.

Mehmood, F., Wang, G., Gao, Y., Liang, Y., Zain, M. *et al.* (2021) Impacts of irrigation managements on soil CO_2 emission and soil CH_4 uptake of winter wheat field in the North China plain. *Water* 13, 2052.

Melloy, P., Hollaway, G., Luck, J., Norton, R., Aitken, E. *et al.* (2010) Production and fitness of *Fusarium pseudograminearum* inoculum at elevated carbon dioxide in FACE. *Global Change Biology* 16, 3363–3373.

Miller, M. (2007) *Dark, Firm and Dry Beef*. Centennial, Co. Beef Facts-Product Enhancement, pp. 1–4.

Milus, E.A., Seyran, E. and McNew, R. (2006) Aggressiveness of *Puccinia striiformis* f. sp. *tritici* isolates in the south-central United States. *Plant Disease* 90, 847–852.

Minson, D.J. (1990) *Forage in Ruminant Nutrition*. Academic Press, San Diego, p. 483.

Mishra, V. and Lilhare, R. (2016) Hydrologic sensitivity of Indian sub-continental river basins to climate change. *Global and Planetary Change* 139, 78–96.

Mitloehner, F.M., Morrow, J.L., Dailey, J.W., Wilson, S.C., Galyean, M.L. *et al.* (2001) Shade and water misting effects on behaviour, physiology, performance and carcass traits of heat-stressed feedlot cattle. *Journal of Animal Science* 79, 2327–2335.

MOEF (2008) National Action Plan on Climate Change, 2008. Available at: https://moef.gov.in/wp-content/uploads/2018/04/Pg01-52_2.pdf (accessed 1 November 2023).

MoEFCC (2021) *Annual Report 2020–2021*. Government of India Ministry of Environment, Forest and Climate Change, New Delhi.

Naqvi, S.K., Maurya, V.P., Gulyani, R., Joshi, A. and Mittal, J.P. (2004) Effect of thermal stress on superovulatory response and embryo production in Bharat merino ewes. *Small Ruminant Research* 55, 57–63.

Nardone, A., Ronchi, B., Lacetera, N., Ranieri, M.S. and Bernabucci, U. (2010) Effects of climate changes on animal production and sustainability of livestock systems. *Livestock Science* 130, 57–69.

Narwaria, U.S., Singh, M., Verma, K.K. and Bharti, P.K. (2017) Amelioration of thermal stress using modified roof in dairy animals under tropics: A review. *Journal of Animal Research* 7, 801–812.

Navarro-Pedreno, J., Almendro-Candel, M.B. and Zorpas, A.A. (2021) The increase of soil organic matter reduces global warming, myth or reality. *Science* 3, 18.

Newman, J.A. (2005) Climate change and the fate of cereal aphids in Southern Britain. *Global Change Biology* 11, 940–944.

Padodara, R.J. and Jacob, N. (2013) Climate change: Effect on growth of animals. *Basic Research Journal of Agricultural Science and Review* 2, 85–95.

Pande, S., Sharma, M., Mangala, U.N., Ghosh, R. and Sundaresan, G. (2010) Phytophthora blight of Pigeonpea [*Cajanus cajan* (L.) Millsp.]: An updating review of biology, pathogenicity and disease management. *Crop Protection* 30, 951–956.

Pascaline, W. and Rowena, H. (2018) Economic losses, poverty and disasters: 1998 – 2017. Centre for Research on the Epidemiology of Disasters, United Nations Office for Disaster Risk Reduction.

Pathak, H., Chakrabarti, B., Bhatia, A., Jain, N. and Aggarwal, P.K. (2012) Potential and cost of low carbon technologies in rice and wheat systems: A case study for the Indo-Gangetic Plains.

Pathak, H. (2015) Greenhouse gas emission from Indian agriculture: Trends, drivers and mitigation strategies. *Proceedings of the Indian National Science Academy* 81, 1133–1149.

Pereira, A.M.F., Baccari, J.R., Titto, F. and Almeida, J.A.A. (2008) Effect of thermal stress on physiological parameters, feed intake and plasma thyroid hormones concentration in Alentejana, Mertolenga, Frisian and Limousine cattle breeds. *International Journal of Biochemistry* 52, 199–208.

Picciariello, A., Colenbrander, S., Bazaz, A. and Roy, R. (2021) The costs of climate change in India: A review of the climate-related risks facing India, and their economic and social costs. *ODI Literature Review* 31.

Prasad, J.V.N.S., Srinivas, K., Rao, C.S., Ramesh, C., Venkatravamma, K. *et al.* (2012) Biomass productivity and carbon stocks of farm forestry and agroforestry systems of *Leucaena* and *Eucalyptus* in Andhra Pradesh, India. *Current Science* 103, 536–540.

Prathap, P., Archana, P.R., Joy, A., Veerasamy, S., Krishnan, G. *et al.* (2017) Heat stress and dairy cow: Impact on both milk yield and composition. *International Journal of Dairy Science* 12, 1–11.

Pratibha, G., Srinivas, I., Rao, K.V., Shanker, A.K., Raju, B.M.K. *et al.* (2016) Net global warming potential and greenhouse gas intensity of conventional and conservation agriculture system in rainfed semi-arid tropics of India. *Atmospheric Environment* 145, 239–250.

Priyanka, M.A.K., Varma, S., Kumar, V. and Sharma, R.S. (2020) Impact of climate change on plant diseases and management strategies: A review. *International Journal of Chemical Studies* 8, 2968–2973.

Radhika, V.S., Priyadarshini, C.G. and Patil, B.L. (2017) Socioeconomic vulnerability of climate change in Karnataka. *International Journal of Agriculture, Environment and Bioresearch* 2, 196–198.

Rahul, K., Keval, R. and Yadav, A. (2015) Effect of climate change on insect pests and management adaptations: Review. *Indian Journal of Agriculture and Allied Sciences* 1, 45–46.

Rama Rao, C.A., Raju, B.M.K., Islam, A., Subba Rao, A.V.M., Rao, K.V. *et al.* (2019) *Risk and Vulnerability Assessment of Indian Agriculture to Climate Change*. ICAR-Central Research Institute for Dryland Agriculture, Hyderabad.

Rao, S., Prasad, R.S. and Mohapatra, T. (2019) *Climate Change and Indian Agriculture: Impacts, Coping Strategies, Programmes and Policy. Technical Bulletin/Policy Document*. Indian Council of Agricultural Research, Ministry of Agriculture and Farmers' Welfare and Ministry of Environment, Forestry and Climate Change, Government of India, New Delhi.

Ravagnolo, O. and Misztal, I. (2000) Genetic component of heat stress in dairy cattle, parameter estimation. *Journal of Dairy Science* 83, 2126–2130.

Reddy, P. (2013) Impact of climate change on insect pests, pathogens and nematodes. *Pest Management in Horticultural Ecosystems* 19, 225–228.

Reich, P.B., Hungate, B.A. and Luo, Y. (2006) Carbon-nitrogen interactions in terrestrial ecosystems in response to rising atmospheric carbon dioxide. *Annual Review of Ecology Evolution and Systematics* 37, 611–636.

Reidsma, P., Wolf, J., Kanellopoulos, A., Schaap, B.F., Mandryk, M. *et al.* (2015) Climate change impact and adaptation research requires integrated assessment and farming systems analysis: A case study in the Netherlands. *Environmental Research Letters* 10, 045004.

Rengasamy, P. and Churchman, G.J. (1999) Cation exchange capacity, exchangeable cations and solidity. In: Peverrilli, K.I., Sparrow, L.A. and Reuter, D.J. (eds) *Soil Analysis. An Interpretation Manual Collingwood*. CSIRO Publishing, Australia, pp. 147–157.

Reth, S., Reichstein, M. and Falge, E. (2005) The effect of soil water content, soil temperature, soil pH-value and root mass on soil CO_2 efflux – A modified model. *Plant and Soil* 268, 21–33.

Rockstrom, J., Williams, J., Daily, G., Noble, A. and Matthews, N. (2017) Sustainable intensification of agriculture for human prosperity and global sustainability. *Ambio* 46, 4–17.

Sankarganesh, E., Firake, D.M., Sharma, B., Verma, V.K. and Behere, G.T. (2017) Invasion of the South American Tomato Pinworm, *Tuta absoluta*, in North-eastern India: A new challenge and biosecurity concerns. *Entomologia Generalis* 36, 335–345.

Santha, S., Jaswal, S., Sasidevan, D., Datta, K., Kuruvilla, A. *et al.* (2015) Adaptation strategies in the city: Recognizing livelihood struggles of migrant workers in India. In: *Asian Cities Climate Resilience Working Paper 19, IIED, London*. Available at: http://pubs.iied.org/10736IIED.html (accessed 1 November 2023).

Santha, S.D. (2015) Climate change and migrant workers in India: From vulnerability to adaptation. Asian Cities Climate Resilience Policy Brief. Available at: http://pubs.iied.org/10747IIED.html (accessed 1 November 2023).

Sapkota, T.B., Jat, M.L., Aryal, J.P., Jat, R.K. and Khatri-Chhetri, A. (2015) Climate change adaptation, greenhouse gas mitigation and economic profitability of conservation agriculture: Some examples from cereal systems of Indo-Gangetic Plains. *Journal of Integrative Agriculture* 14, 1524–1533.

Sapkota, T.B., Vetter, S.H., Jat, M.L., Sirohi, S., Shirsath, P.B. *et al.* (2019) Cost-effective opportunities for climate change mitigation in Indian agriculture. *Science of the Total Environment* 655, 1342–1354.

Sarkar, S. (2018) *Climate Change May Shrink India's Farm Incomes by 25%*. India Climate Dialogue. Available at: https://indiaclimatedialogue. net/2018/01/30/climate-change-mayshrink-indias-farm-incomes-25/ (accessed 1 November 2023).

Schaad, N.W. (2008) Emerging plant pathogenic bacteria and global warming. In: Fatmi, M., Collmer, A., Iacobellis, N.S., Mansfield, J.W. and Murilli, J. (eds) *Pseudomonas Syringae Pathovars and Related Pathogens-Identification, Epidemiology and Genomics*. Springer, Dordrecht, pp. 369–379. DOI: 10.1007/978-1-4020-6901-7_38.

Sejian, V., Maurya, V.P. and Naqvi, S.M.K. (2012b) Effect of walking stress on growth, physiological adaptability and endocrine responses in Malpura ewes under semi-arid tropical environment. *International Journal of Biometeorology* 56, 243–252.

Sejian, V., Maurya, V.P., Kumar, K. and Naqvi, S.M.K. (2012a) Effect of multiple stresses (thermal, nutritional and walking stress) on the reproductive performance of Malpura ewes. *Veterinary Medicine International* 2012, 471760. DOI: 10.1155/2012/471760.

Sharma, H.C. (2014) Climate change effects on insects: Implications for crop protection and food security. *Journal of Crop Improvement* 28, 229–259.

Sharma, M., Mangala, U.N., Krishnamurthy, M., Vadez, V. and Pande, S. (2010) Drought and dry root of chickpea. In: *5th International Food Legumes Research Conference (IFLRC V) & European Conference on Grain Legumes (AEP VII)*, Antalya, Turkey.

Singh, B.P., Dua, V.K., Govindakrishnan, P.M. and Sharma, S. (2013) Impact of climate change on potato. In: Singh, H.C.P., Rao, N.K.S. and Shivashankar, K.S. (eds) *Climate-Resilient Horticulture: Adaptation and Mitigation Strategies*. Springer, India, pp. 125–135.

Singh, N.P., Singh, S. and Anand, B. (2019) *Impact of Climate Change on Indian Agriculture: An Agro-Climatic Zone Level Estimation. Policy Brief 44*. ICAR-National Institute of Agricultural Economics and Policy Research, p. 8. Available at: http://www.nicra-icar.in/nicrarevised/images/publications/TDC/Impact%20of%20Climate%20Change%20on%20Indian%20 Agriculture%20An%20Agro-Climatic%20Zone%20Level%20Estimation.pdf (accessed 1 November 2023).

Singh, S.V. and Upadhyay, R.C. (2009) Impact of temperature rise on physiological function, thermal balance and milk production of lactating Karan fries and Sahiwal cows. *Indian Veterinary Journal* 86, 141–144.

Sirohi, S., Sirohi, S.K. and Pandey, P. (2010) Adapting small holder dairy production system to climate change natural and anthropogenic disasters. In: Jha, M.K. (ed.) *Natural and Anthropogenic Disasters*. Springer, Dordrecht, pp. 432–448.

Spehar, M., Vincek, D. and Zgur, S. (2009) Beef quality: Factors affecting tenderness and marbling. *Stocarstvo* 62, 463–478.

Srinivasarao, C., Rao, K.V., Gopinath, K.A., Prasad, Y.G., Arunachalam, A. *et al.* (2020) Agriculture contingency plans for managing weather aberrations and extreme climatic events: Development, implementation and impacts in India. *Advances in Agronomy* 159, 35–91.

Srinivasarao, M., Swathi, P., Rama Rao, C.A., Rao, K.V., Raju, B.M.K. *et al.* (2015) Model and scenario variations in predicted number of generations of *Spodoptera litura* Fab. on peanut during future climate change scenario. *PLoS ONE* 10(2), e0116762.

Srivastava, S.K., Chand, R. and Singh, J. (2017) Changing crop production cost in India: Input prices, substitution and technological effects. *Agricultural Economics Research Review* 30, 171–182.

Sutherst, R. (2000) Climate change and invasive species: A conceptual framework. In: Mooney, H.A. and Hobbs, R.J. (eds) *Invasive Species in a Changing World*. Island Press, Washington, DC, pp. 211–240.

Swaminathan, M.S. and Rengalakshmi, R. (2021) Impact of extreme weather events in Indian agriculture: Enhancing the coping capacity of farm families. *Mausam* 67, 1–4.

Tahmasbi, A.M., Kazemi, M., Moheghi, M.M., Bayat, J. and Shahri, A.M. (2012) Effects of selenium and vitamin E and night or day feeding on performance of holstein dairy cows during hot weather. *Journal of Cell and Animal Biology* 6, 33–40.

Talaviya, T., Shah, D., Patel, N., Yagnik, H. and Shah, M. (2020) Implementation of artificial intelligence in agriculture for optimisation of irrigation and application of pesticides and herbicides. *Artificial Intelligence in Agriculture* 4, 58–73.

Tankson, J.D., Vizzier-Thaxton, Y., Thaxton, J.P., May, J.D. and Cameron, J.A. (2001) Stress and nutritional quality of broilers. *Poultry Science* 80, 1384–1389.

Thornton, P.K., Van de, S.J., Notenbaert, A. and Herrero, M. (2009) The impacts of climate change onlive-stock and livestock systems in developing countries: A review of what we know and what we need to know. *Agricultural Systems* 101, 113–127.

Tiedje, J.M., Bruns, M.A., Casadevall, A., Eloe-Fadrosh, E., Karl, D.M. *et al.* (2022) Microbes and climate change: A research prospectus for the future. *Mbio* 13, e00800 – 22.

UNDP (2021) More Than Just Cryptocurrencies – Using Blockchain for Climate Action in Agriculture. Available at: https://www.adaptation-undp.org/using-blockchain-for-climate-action-in-agriculture (accessed 1 November 2023).

Upadhyay, R.C., Singh, S.V., Hooda, O.K. and Vaidya, M.M. (2008) Impact of climate change on livestock. In: *Proceedings of 36th Dairy Industry Conference*, Varanasi, India.

van Dijk, M., Morley, T., Rau, M. and Saghai, Y. (2021) A meta-analysis of projected global food demand and population at risk of hunger for the period 2010–2050. *Nature Food* 2, 494–501.

Vennila, S., Lokare, R., Singh, N., Ghadge, S. and Chattopadhyay, C. (2016) Crop pest surveillance and advisory project of Maharashtra. In: *A Role Model for an E-Pest Surveillance and Area Wide Implementation of Integrated Pest Management in India*. ICAR National Research Centre for Integrated Pest Management, New Delhi, India, p. 56.

Warner, R.D., Greenwood, P.L., Pethick, D.W. and Ferguson, D.M. (2010) Genetic and environmental effects on meat quality. *Meat Science* 86, 171–183.

Watson, A., Ghosh, S., Williams, M., Cuddy, W.S., Simmonds, J. *et al.* (2018) Speed breeding is a powerful tool to accelerate crop research and breeding. *Nature Plants* 4, 23–29.

Wilson, S. (2021) Recognition of Climate Refugees: What should be India's Stand? Available at: https://rli.blogs.sas.ac.uk/2021/03/18/recognition-of-climate-refugees-what-should-be-indias-stand/ (accessed 1 November 2023).

Wolfe, D.W., Ziska, L., Petzoldt, C., Seaman, A., Chase, L. and Hayhoe, K. (2008) Projected change in climate thresholds in the North-eastern U.S.: Implications for crops, pests, livestock, and farmers.*Mitigation and Adaptation Strategies for Global Change* 13, 555–575.

Xiao, Y., Yang, S., Xu, J., Ding, J., Sun, X. *et al.* (2018) Effect of biochar amendment on methane emissions from paddy field under water saving irrigation. *Sustainability* 10, 1371.

Yuan, M.M., Guo, X., Wu, L., Zhang, Y., Xiao, N. *et al.* (2021) Climate warming enhances microbial network complexity and stability. *Nature Climate Change* 11, 343–348.

Zhang, J.J., He, P., Xu, X.P., Ding, W.C., Ullah, S. *et al.* (2017) Nutrient expert improves nitrogen efficiency and environmental benefits for winter wheat in China. *Agronomy Journal* 110, 696–706.

10 Impacts of climate change on animal welfare

Joe Joseph[1], Renae Charalambous[1], Harsh Pahuja[1], Dylan Fox[1], Jiwoo Jeon[1], Ning-Yuan Ko[1], Nishit Rao[1], Zhiheng Wang[1], Sneh Kashinath Nerurkar[1], Sharvari Avinash Sherekar[1], Yifei Yang[2], Kate Dutton-Regester[1] and Edward Narayan[1,3*]

[1]School of Agriculture and Food Sustainability, The University of Queensland, Gatton Campus, QLD 4343, Australia; [2]School of Veterinary Sciences, The University of Queensland, Gatton Campus, QLD 4343, Australia; [3]Queensland Alliance for Agriculture and Food Innovation, The University of Queensland, St Lucia, QLD 4072, Australia

Abstract

Climate change has a significant impact on domesticated and wildlife animals globally. Projections for a 2°C rise in atmospheric temperature will result in catastrophic impacts globally. Animal welfare is an integral component of human-animal interaction and it is of paramount importance for wildlife rehabilitation and domestic animals. Animal welfare can be assessed using the five domains model, which includes nutrition, environment, physical health, behaviour, and mental attributes of an animal's response to environmental change. Researchers use some of these domain attributes to evaluate the effects of stress on their animals and to improve the management and welfare of animals. Although there are variations in how animals respond biologically to stress, in general, the five domains model provides a robust tool for research use, and to evaluate the proximate effects of climatic variability on animals. In this review, our research group (The Stress Lab; www.edwardstresslab.org) presents a series of wildlife and domesticated animal examples to showcase how climate change impacts animal welfare. We provide examples of animals from various countries, across both aquatic and terrestrial systems, and provide an overview of the impacts of climate change on each of the five domains of animal welfare. We hope that future researchers will apply the animal welfare domains to evaluate how climate change impacts animals, and further research will pave the way to the protection of animals from the catastrophic impacts of climate change.

Keywords: extinction crisis, global warming, human-animal-interaction, stress, welfare domains

Introduction

Climate change poses an unprecedented global crisis for humans and biodiversity. In recent years, extreme climatic events such as droughts and floods have profoundly impacted human populations. To address this, scientists have been enhancing weather prediction technology and preparing communities for the impending climate crisis (Hennessy *et al.*, 2005). While humans have support mechanisms to cope with climate change, animals require human

*Corresponding author: e.narayan@uq.edu.au

intervention to mitigate climate risks. Many wildlife species exhibit remarkable abilities to naturally forecast and adapt to the imminent dangers of climate change through behavioural and physiological changes. For instance, migrating sea turtles demonstrate these adaptive strategies (Pike *et al.*, 2006). According to the World Wildlife Fund (WWF), climate change drives nature loss through elevated temperatures, rising sea levels, and the exacerbation of extreme events (WWF, 2022). Global warming, primarily caused by human-induced factors such as increased levels of greenhouse gases (e.g. carbon dioxide, methane, and nitrous oxide) in the atmosphere, results from activities like agriculture, forestry, and other land use practices (Yoro and Daramola, 2020). The challenges posed by climate change are not limited to environmental concerns alone. Animals play a crucial role in meeting human needs, such as food security (e.g. animal-based protein, manufacturing, clothing) and fulfilling aesthetic desires like zoos and nature recreational parks (Argent, 2022). Consequently, climate change has detrimental effects on animal welfare, impacting their biology, ecology, and survival directly and indirectly, which will be discussed in this review.

In layman's terms, animal welfare can be defined as the genuine state of well-being and condition of the animal, independent of human expectations or desires for their prosperity. If climate stress is prolonged, it can have severe consequences on the health, well-being, and/or productivity of animals. There are various reviews on the topic of animal welfare, which can be referred here (Hamann *et al.*, 2007; Marsh *et al.*, 2007; Adams and Hayes, 2008; Adams, 2010; Narayan *et al.*, 2012; Huey and Kingsolver, 2019). While researchers have extensively examined the effects of climate change on animals, the direct correlation between climate change and animal welfare, particularly in the context of wild animals, remains relatively scarce in existing studies.

Therefore, the aim of this review is to utilize the five welfare domains (see Fig. 10.1) (Broom, 2014; Hemsworth *et al.*, 2015; Mellor *et al.*, 2020) to discuss the potential impacts of climate change on animal welfare. Our review provides a broad overview of the impacts of climate change on animal welfare, drawing on examples from various animal groups across terrestrial and aquatic habitats, encompassing both wildlife and domesticated species.

Methods

The data presented in this work were obtained through comprehensive searches conducted on various web and scientific search engines including Google Scholar, PubMed, and Science Direct. Primary search terms comprised the animal welfare domains (nutrition, environment, physical health, behaviour, and mental), environmental temperature as the surrogate for climate change, and the model animal species, which included invertebrates, fishes, amphibians, reptiles, birds, and small and large mammals. It is a non-systematic review. Table 10.1 provides an overview of the animal welfare domains and dimensions covered by the research.

Invertebrates

Diversity and importance

'Invertebrate' is a term of convenience and refers to a group of animals that do not develop a spine (vertebral column) and are not members of the Vertebrata (Agassiz, 2013). They are a diverse group, having a greater number and variety of species than all known vertebrates (Richards and Davies, 2013); it is estimated that they compose 90% of the existing animal species. They have a wide range of uses, such as research, aquaculture, farming, and displays in aquariums for conservation and educational purposes (Carere *et al.*, 2011). Many familiar invertebrates serve as major food sources and are economically viable, such as shrimps, clams, squids, locusts, termites, grasshoppers, and beetle grubs, as well as honey from bees. Other invertebrates are valued for their ability to make luxuries, such as silk, pearls, and shells (Kellert, 1993). Due to their diversity, abundance, accessibility, quick generation times, and large number of offspring compared to vertebrates, they are also of scientific significance. Despite this, invertebrates are frequently overlooked by

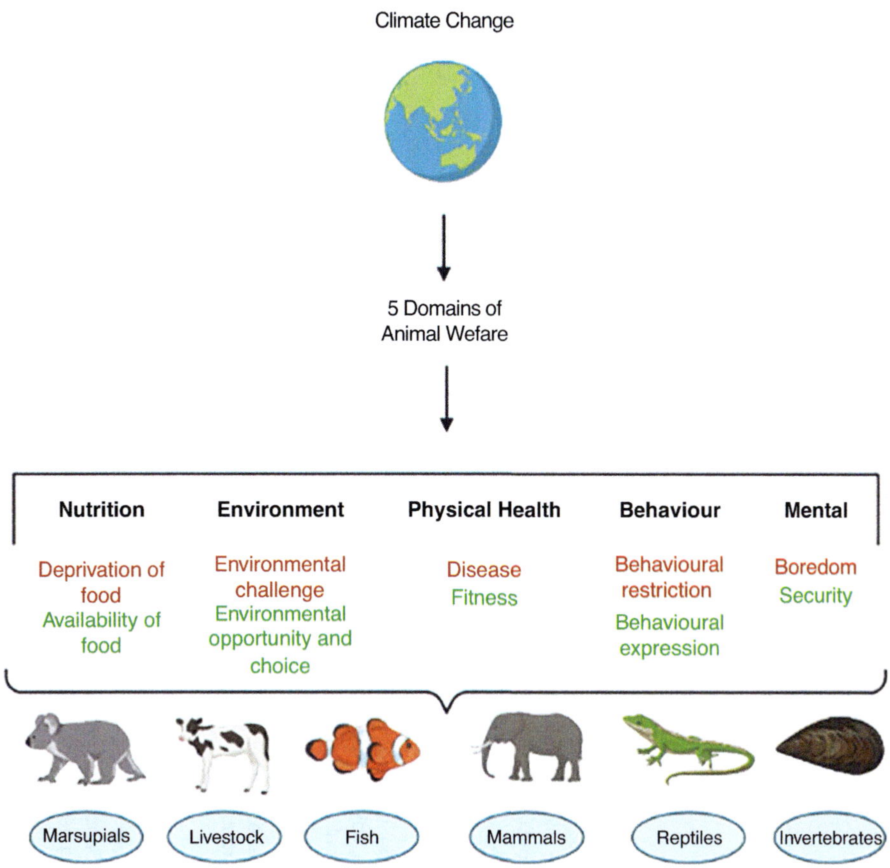

Fig. 10.1. The framework of the study which uses the five domains model (Mellor *et al.*, 2020) to explore the impacts of climate change on various animal groups.

the public and scientific community and often kept with limited animal care and supervision. Unfortunately, the majority of invertebrates are disliked or feared due to concerns of disease (Kellert, 1993). Consequently, the focus tends to be on charismatic species that capture the public attention, while the welfare of vital invertebrate species often goes unnoticed.

Invertebrates are ectotherms, and their growth rates are strongly affected by environmental factors such as temperature, humidity (Forrest, 2017), and water properties (Culler-Juarez and Onthank, 2021). Hence, there are substantial concerns surrounding the effects of global warming and the acidification of oceans on invertebrates (Raven *et al.*, 2005). Changes in temperature and rainfall patterns brought on by climate change are predicted to have a significant direct impact on the distribution of invertebrates and indirect effects on food plant ranges and phenologies. For instance, compared to data from a few decades ago, the first flight of butterflies has advanced by as much as a month (Roy and Sparks, 2000).

Many invertebrate species, particularly insects, are adapted to specific ecosystems all over the world. Many of these species are restricted to relatively small climatic ranges, and their response to climate change is dependent on whether individuals are able to adapt to the

Table 10.1. An overview of the animal welfare domains and dimensions covered by the research.

Animal Group and management system	Welfare dimension and dimension attribute	Methods used	Key results	Conclusion	References
Invertebrate; unspecified	Nutrition; food quantity	Review paper	Elevated temperatures weaken the symbiotic relationship between cnidarians and algae, resulting in the loss of their primary food source.	The loss of symbiotic algae decreases energy available to cnidarians resulting in death and the phenomenon known as coral bleaching.	Cziesielski *et al.* (2019)
Fish; intensive	Environment; temperature	Measured steroidogenesis markers, immunity status, and ovarian tissue of female Nile tilapia (*Oreochromis niloticus*) experiencing heat stress	Reproductive hormones and cortisol significantly declined due to heat stress; further, gene expression was down-regulated, and ovary tissues were injured.	Heat stress has harmful effects on female tilapia physiology through modulation of steroidogenesis-related gene expression and immunity	Mahmoud *et al.* (2020)
Fish; intensive	Nutrition; food quantity	Measured the effects of dietary nano-selenium on organizational structure, lipid changes, and biochemical parameters as well as heat-shock protein-and selenoprotein-related gene expression in rainbow trout (*Oncorhynchus mykiss*).	Stress began to increase significantly when trout experienced heat stress.	Supplementation with 5 mg/kg nano-Se effectively alleviates stress damage in rainbow trout.	Mahmoud *et al.* (2020)
Fish; unspecified	Environment; temperature	Measured transcriptomic profile, histopathology, and oxidative stress in the liver samples of pikeperch (*Sander lucioperca*) experiencing heat stress.	The most significant heat-stress-related pathway was the protein processing in the endoplasmic reticulum; liver histology was changed due to heat stress.	Heat stress at 29°C caused oxidative damage and dysfunction in the pikeperch liver between 3 hour and 48 hour.	Wang *et al.* (2019)

Continued

Table 10.1. Continued

Animal Group and management system	Welfare dimension and dimension attribute	Methods used	Key results	Conclusion	References
Fish; unspecified	Behaviour; temperature	Measured the performance of zebrafish (*Danio rerio*) exposed to different environmental temperatures and subject to four behavioural tests.	Heat treatment reduced anxiety and increases the boldness of zebrafish.	The increase in ambient temperature may compromise zebrafish survival rate in the natural environment.	Angiulli *et al*. (2020)
Amphibian; intensive	Health; disease	Measured urinary corticosterone levels of Chytridiomycosis positive and negative Stony Creek Frogs (*Litoria wilcoxii*) via enzyme-immunoassay.	Elevated corticosterone concentrations result in the reduction of eosinophils and lymphocytes.	Corticosterone elevation induced by heat stress increases susceptibility to diseases such as chytridiomycosis.	Kindermann *et al*. (2012)
Amphibian; intensive	Environment; temperature	Measured urinary corticosterone levels of Cane Toads (*Rhinella marina*) via enzyme-immunoassay.	Elevated environmental temperatures induce excessive corticosterone production in amphibians.	Excess production of corticosterone results in alterations to metabolic rate and other physiological traits in amphibians.	Narayan *et al*. (2012)
Reptile; unspecified	Nutrition; food quality	Review chapter	Variability in environmental temperature could influence the distribution and quality of seagrasses, which green sea turtles are reliant on as a food source.	Variability in the quality and distribution of seagrasses because of climate change can influence the ability of green sea turtles to satisfy their nutritional requirements.	Hamann et al. (2007)
Wildlife mammal; captive	Nutrition; food intake	Investigated whether koalas adjust the number and/or the size of their meals when offered eucalypt foliage with varying concentrations of formylated phloroglucinol compounds (FPC).	High FPC reduced the total intake of foliage.	Climate change results in increased toxic content and reduced nitrogen levels in eucalypts.	Marsh *et al*. (2007)

Continued

Table 10.1. Continued

Animal Group and management system	Welfare dimension and dimension attribute	Methods used	Key results	Conclusion	References
Wildlife mammal; free-ranging	Health; limitations	Trapping of wild bats and monitoring against climatic data, analysis utilized principal component analysis.	Elevated temperature and dry weather led to decreased reproductive performance in female bats.	Environmental conditions have a significant effect on bat reproduction.	Adams (2010)
Wildlife mammal; free-ranging	Nutrition; water intake	Capturing and tagging female bats to monitor activity and behaviour around an artificial water source.	Lactating females have a higher water requirement as compared to their non-lactating counterparts.	Increased temperatures will reduce water availability, increasing the chance of bat population reduction	Adams and Hayes (2008)
Wildlife mammal; free-ranging	Environment; choices	Measuring bat activity against environmental variables. Data analysed using generalized linear mixed model (GLMM).	Moonlight and temperature have the highest effect on insectivorous bat activity.	The ways in which bat species respond to environmental variables are predictable.	Appel *et al*. (2019)
Wildlife mammal; free-ranging	Environment; limitations	Trapping of bats, and comparing data with similar studies done on bats.	The number of reproducing female bats was lower in years with higher precipitation.	Higher precipitation delays or prevents reproduction in female bats.	Grindal *et al*. (1992)
Wildlife mammal; free-ranging	Behaviour; thermal comfort	Conducted behavioural observations of female bats in elevated temperatures.	Nursing female bats exposed to extreme temperatures demonstrated an increase in fanning and grooming behaviour concurrent with reduced resting behaviour. Further, nursing females were at increased risk of death.	Nursing female bats are more susceptible to heat-stress compared to their non-lactating counterparts.	Snoyman *et al*. (2013)

Continued

Table 10.1. Continued

Animal Group and management system	Welfare dimension and dimension attribute	Methods used	Key results	Conclusion	References
Wildlife mammal; free-ranging	Nutrition: water and food intake	Analysed elephant carcasses in Kenya to determine the cause of death.	Most elephant deaths could be classified under natural causes (i.e. nutritional deficiency), poaching or human conflict.	Drought and resource scarcity, as a result of climate change, are key contributors to elephant mortality rates.	Mukeka *et al.* (2022)
Wildlife mammal; unspecified	Mental; thermal comfort	Observation and analysis of the impacts of climate change on elephants.	Elephants perform evaporative water loss through their skin for thermoregulation, even at low temperatures. Consequently, they require large amounts of drinking water to support thermoregulatory processes within the body.	Reduced water availability and intake result in reduced effectiveness of elephants to thermoregulate, thereby increasing heat stress.	Fuller *et al.* (2016)
Domesticated species, captive	Health; disease	Autopsies were conducted on broiler chickens kept under hot conditions for 4 days.	Elevated temperatures result in greater concentrations of reactive oxygen, which can react with the cells of the animals and cause necrosis.	Higher cases of necrosis among broiler chickens will reduce the animals' quality of life and reduce the quality of their meat.	Santos *et al.* (2015)
Domesticated species; intensive	Environment; temperature	Measured feed intake and milk production of dairy cows.	Heat stress resulted in a 35% reduction in milk production.	The extent of milk yield decline observed in heat-stressed cows depends on various factors that interact with high air temperature.	Bernabucci *et al.* (2010)

new conditions and the genetic diversity among populations (Butterfield and Coulson, 1997). Additionally, the atmospheric carbon dioxide levels in the ocean, which have been increasing at a rate of around 0.5% per year primarily due to human activity, have recently experienced the most rapid change in CO_2 concentrations (Raven *et al.*, 2005; Fabry *et al.*, 2008). The ocean's acidity has increased by 26% after absorbing nearly one-third of the human CO_2 produced (Sabine *et al.*, 2004); a further 250% increase in acidity is expected by the end of this century (Pachauri and Meyer, 2014). The ocean absorbs surplus heat from a warming world and serves as a 'sink' for atmospheric carbon dioxide, with land and ocean surface temperatures already increasing by 0.85°C (Pachauri and Meyer, 2014). The coastal sea surface temperature is likely to rise 1–3°Cmore than what has already been recorded if more CO_2 emission mitigation measures are not taken (Parry *et al.*, 2007; Pachauri and Meyer, 2014).

Stress response in invertebrates

As invertebrates are important to many ecosystems and play a vital role in maintaining the balance of nature, understanding their stress physiology can help us to better protect these animals and their habitats. Invertebrates have a variety of mechanisms for responding to stress; the mechanism utilized depends on their specific physiology and the type of stress they are experiencing (Butterfield and Coulson, 1997). For example, some invertebrates may respond to physical stress, such as extreme temperatures or changes in salinity, by regulating their body fluids or by moving to a more favourable environment (Jin *et al.*, 2022). Others may respond to biological stress, such as the presence of predators or changes in their food supply, by changing their behaviour or by using chemical defences (Horwitz *et al.*, 2020). Overall, the stress response in invertebrates is like that of vertebrates, involving the activation of hormones and neurotransmitters that help the animal to cope with the stressor. However, the specific mechanisms involved may differ among different types of invertebrates. For example, insects and other arthropods may use hormones called ecdysteroids to help them cope with stress (Moen *et al.*, 2022), while molluscs may use molecules called opisthobranch peptides (Roggatz *et al.*, 2016).

Biogenic amines and heat stress in insects

The biosynthesis of biogenic amines and other compounds is a crucial process in insects. Biogenic amines, such as dopamine and serotonin, are neurotransmitters that play important roles in the nervous system (Neven, 2000). They are released in response to various stimuli, including heat stress. Other neuroendocrine factors, such as adipokinetic hormone, are also released in response to heat and serve to mobilize the insect's energetic reserves (González-Tokman *et al.*, 2020). However, the exact role of these hormones in the response to heat stress is not yet well understood. Heat stress can have synergistic effects when combined with other stressors, such as insecticides. For example, in the moth *Plodia interpunctella*, heat and insecticide exposure have a synergistic effect that greatly increases the stress response and mortality (González-Tokman *et al.*, 2020). These synergistic effects are common in nature, but the mechanisms underlying them are not yet fully understood.

Heat can have significant effects on insect physiology, including changes in metabolism, respiration, nervous and endocrine systems, and the elicitation of heat shock proteins (HSPs). In order to be effective, heat treatments used to disinfect fresh and stored products must take into account the sensitivity of insects to high temperatures (Neven, 2000). Insects can be transferred from one temperature to another in a step-wise or ramp-wise manner. The former demonstrates how quickly an insect can respond to a thermal challenge while the latter reveals mechanisms involved in thermal tolerance and indicates whether tolerance limits are wider in response to a ramp-wise or step-wise function (Neven, 2000). Heat treatments can also cause damage to insect cells and tissues, leading to death.

HSPs are a highly conserved class of proteins that display elevated transcription during periods of stress, playing important roles in protecting the cell and its components from damage. Different insects have different sets of HSPs, and the specific HSPs that are upregulated in response to heat vary among species (González-Tokman *et al.*, 2020). Heat stress can have a range of negative effects on insects, including reducing their ability to fly, reproduce, and feed (González-Tokman *et al.*, 2020). It can also affect their immune function, making them more susceptible to diseases. However, some insects have evolved mechanisms to tolerate or even benefit from heat stress. For example, desert locusts can increase their body temperature by sunbathing, which enhances their metabolism and immune function.

Conversely, heat shock has negative effects on male reproductive success in the springtail *Orchesella cincta* (Zizzari and Ellers, 2011). Researchers exposed male springtails to four temperature treatments: heat hardening (35.2°C for 1 hour), heat shock (37.2°C for 1 hour), heat hardening plus heat shock (35.2°C for 1 hour followed by 37.2°C for 1 hour), and control (20°C). Significant mortality was only found in the treatment with heat shock, with and without heat hardening. A direct effect of heat treatment was found for time to first reproduction, which was significantly longer after heat shock (with or without heat hardening) than in the control treatment. There was no difference among treatments in the number of spermatophores produced in the first reproductive instar. The mechanism of stress physiology in this study is the induction of HSPs in response to heat stress. Exposure to heat shock causes the expression of the heat shock gene Hsp70 to increase, indicating that the treatments induced thermal stress. This increase in HSPs may help protect the organism against the detrimental effects of heat stress, but it does not protect against all negative effects (Zizzari and Ellers, 2011).

Stress in crustaceans

Chang (2005) has developed methods for the quantification of stress in crustaceans. They have observed increases in crustacean hyperglycemic hormone (CHH) under conditions of acute hypoxia, elevated temperature, and altered salinity. In addition, elevated CHH concentrations were observed in Norway lobsters that were parasitized with the dinoflagellate *Hematodinium sp.* Further, they examined the influence of heat-shock, osmotic stress, and the moult cycle on HSP expression at the protein and mRNA levels. When comparing claw and abdominal muscles during moulting, a tissue-specific HSP response was observed (Chang, 2005). Quantification of these different stress responses may serve as early indicators of the degradation of environmental health. The CHH ELISA may also be useful for the quantification of circulating CHH levels following exposure to endocrine-disrupting compounds (Chang, 2005).

Chronic heat stress can negatively impact the growth, survival, and feeding of the sea urchin *Strongylocentrotus intermedius* (Zhan *et al.*, 2019). Researchers cultured the sea urchins in the northern Yellow Sea in China under two conditions: 20°C and 25°C. After 28 days, they found that the specific growth and ingestion rates of the sea urchins reared under high-temperature conditions decreased compared to those reared under optimal temperature conditions. In addition, the study identified 2125 differentially expressed genes in the sea urchins reared under high-temperature conditions compared to those subjected to optimal temperature conditions, which included 1015 upregulated and 1100 downregulated genes (Zhan *et al.*, 2019). The findings provide new insight into the molecular mechanism of echinoid species in response to heat stress (Zhan *et al.*, 2019).

Coral bleaching

Coral bleaching is a process that occurs when corals experience extreme and prolonged heat stress. This can cause the symbiotic relationship between corals and the algae of the family *Symbiodiniaceae* to break down (Cziesielski *et al.*, 2019). The algae provide most of the energetic requirements of the coral host, enabling them to effectively calcify and become the foundation of modern reefs. When the symbiotic relationship is disrupted, the coral experiences an energetic deficit that can lead to its death. The loss and degradation of coral reefs because of bleaching events is a major concern. Initial research into bleaching began in the early 1900s, but it was not until the 1990s that mass bleaching events were recorded (Cziesielski *et al.*, 2019). Since then, research into the mechanisms underlying coral bleaching has advanced rapidly, with the use of omics tools providing new insights. Despite these advances, the underlying molecular mechanisms of host and symbiont heat stress response are still not fully understood.

In conclusion, heat stress is a significant environmental factor that can have negative effects on insects. It can affect their behaviour, physiology, and immune function. However, the specific mechanisms underlying these effects are not yet fully understood, and further research is needed to better understand how insects respond to heat stress and other climatic stressors more broadly.

Fish

The welfare of fish is relatively novel in the field of animals. Indeed, the term 'fish welfare' was rarely used by fish researchers until 2000 (Kristiansen and Bracke, 2020), but the Medway report is believed to be the first report that specifically related to fish welfare (Medway, 1980). Today, fish have become an important food source for humans (Martos-Sitcha *et al.*, 2020). According to the Food and Agriculture Organization of the United Nations (FAO), finfish (including freshwater, diadromous, and marine fish) had the greatest share of the total fisheries and aquaculture production, reaching 135 million tonnes in 2019 (FAO, 2021). The main change in aquatic source production was that aquaculture has gradually replaced wild catch with 48% of contribution in global fisheries and aquaculture production in 2019 (FAO, 2021). Therefore, this section will focus on the impact of climate change on farmed fish because of the substantial number of them kept for food purposes; both marine and freshwater species are discussed.

In recent years, climate change has become one of the major threats to fish production. Climatic variables include cyclones, drought, flood, global warming, ocean acidification, salinity, rainfall variation, and rising sea level (Ahmed *et al.*, 2019). Temperature rise due to global warming can have a profound impact on fish farming as water temperature is a primary factor in fish welfare (Cochrane *et al.*, 2009). Increased water temperature (refer to heat stress in land animals) is a core stressor to ectothermic animals. Researchers have found that elevated water temperatures (> 30°C) have harmful effects on female Nile tilapia (*Oreochromis niloticus*) physiology through modulation of steroidogenesis markers, immunity function and ovarian tissue (Mahmoud *et al.*, 2020). More specifically, when the temperature exceeds 32°C, significant increases were found in serum stress indicators and enzyme activities, in addition to mRNA expression of HSP (Mahmoud *et al.*, 2020). Further, many cold-water fish species such as rainbow trout (*Oncorhynchus mykiss*) are even more vulnerable to heat stress (Wang *et al.*, 2019). The addition of appropriate dietary Selenium (Se) provides one solution to enhance heat resistance in rainbow trout (Li *et al.*, 2022). In addition to physiological effects, temperature can either directly or indirectly alter the swimming performance of fish, both directly or indirectly (Domenici *et al.*, 2019). For example, zebrafish (*Danio rerio*) were more mobile in warm water (34°C) compared to cold water (26°C), showing much higher swimming speed and total distance travelled (Angiulli *et al.*, 2020). Temperature also affects the selection of depths and habitats of marine fish. For example, Atlantic cod (*Gadus morhua*) swim to colder and deeper water areas during the summer when sea surface temperatures are high (Freitas *et al.*, 2021).

Cyclones, floods, ocean acidification, salinity changes, and sea level rise pose additional risks to coastal and marine aquaculture (Ahmed *et al.*, 2019). Climatic events affect not only infrastructure but also fish productivity, particularly outdoor operations. Recently, a land-based, indoor, and controlled farming system known as the 'Recirculating Aquaculture System' (RAS), has been developed to adapt to environmental challenges (Bueno and Soto, 2017; Ahmed *et al.*, 2019). The RAS, initially conceived in Japan in the 1950s, was first commercially used in Denmark, and then further developed in North America, Asia, Europe, and other countries

(Ying *et al.*, 2015). This fish-rearing facility is purpose-built for eco-friendly and intensive farming, which eliminates the effects of climate change on fish welfare (Ying *et al.*, 2015; Ahmed and Turchini, 2021). However, costly operations, greenhouse gas emissions, energy consumption, and engineering approaches are still the main challenges of the RAS (Ying *et al.*, 2015).

Global warming, water temperature rise, and any other alteration of water parameters can pose many risks to fish. As an ectotherm, fish are more vulnerable to climate change. Increased water temperatures not only have an impact on fish physiology but also their behaviour and habitat selection.

Amphibians

Amphibians are an ancient class of animals, vestigial of the first terrestrial vertebrates, and can be identified by their use of their skin as their primary organ for respiration (Wake and Koo, 2018). Currently, there are an estimated 8000 extant species, separated into three orders: (i) the Anurans, which includes frogs and toads. This is the most successful order, found on all continents excluding Antarctica and comprising approximately 7000 of the surviving amphibian species (Wake and Koo, 2018); (ii) Caudates, including salamanders and newts which are found throughout the northern hemisphere and South America; finally, (iii) Gymniophones which include caecilians. This is the least studied order which is constrained to the tropics (Wake and Koo, 2018; Stebbins and Cohen, 2021). This section will specifically focus on the impacts of climate change on the physical health, nutritional, and environmental domain of amphibian welfare.

Amphibians are far more sensitive to the quality of their environment than other terrestrial vertebrates due to their reliance on water for maintaining their skin moisture for respiration, reproduction, ion exchange, and a variety of other physiological processes. This makes them susceptible to rapid changes in the environment (Wake and Koo, 2018; Stebbins and Cohen, 2021). Environmental temperature has been shown to be integral in modulating metabolism, and performance and fitness traits such as locomotion and reproduction (Moore and Zoeller, 1985; Seebacher *et al.*, 2005; Narayan *et al.*, 2012; Narayan and Hero, 2014a). As such, increasing mean temperatures approaching and exceeding the thermal stress tolerances of many amphibian taxa, and erratic weather patterns resulting in extended periods of drought, flood, and fire instigated by climate change have resulted in the exposure of amphibians to the greatest risk of extinction of any vertebrate class (Bernardo and Spotila, 2006; Wake and Koo, 2018; CSIRO, 2020).

Amphibians are ectotherms; animals that are reliant on external sources of heat for the regulation of their body temperature which in turn controls their metabolic rate, behaviour, immunity, and other physiological qualities (Narayan *et al.*, 2012; Beltrán *et al.*, 2021). One pathway through which environmental temperature can influence the physiology of amphibians is through the synthesis of the stress hormone corticosterone. Corticosterone typically induces a short-term stress response; for example, corticosterone can increase lipid, amino acid, and glucose mobilization and availability, to provide additional energy and increase the organism's metabolic activity (Narayan *et al.*, 2012). In general, stress hormones are known to influence immune cell distribution and corticosterone is no exception as it decreases concentrations of lymphocytes and eosinophils (Kindermann *et al.*, 2012). Elevated environmental temperatures have been shown to elevate corticosterone concentrations in amphibians and these concentrations do not decrease unless the temperature declines (Narayan *et al.*, 2012). The positive relationship between mean temperature and corticosterone concentrations and the ongoing impacts of climate change on temperature may result in permanent impairment of amphibian health, nutrition requirements, and behaviour (Narayan *et al.*, 2012).

The resultant increase in metabolic activity from elevated corticosterone is detrimental to the nutritional and behavioural welfare of amphibians, as the mobilization of fats, glucose, and amino acids results in the reduction of the animals' energy reserves while simultaneously increasing the nutritional requirement of the animal, thereby increasing their susceptibility to periods of food scarcity (Narayan *et al.*, 2012;

Huey and Kingsolver, 2019). Furthermore, the impacts of mean temperature elevation are likely to increase amphibian susceptibility to disease (Narayan *et al.*, 2012). Absence of lymphocytes and eosinophils will weaken the adaptive immune system and increase the predisposition to infection from fungi and parasites (Narayan *et al.*, 2012; Figueiredo and Neves, 2018; Flajnik, 2018). Chytridiomycosis is an amphibian targeting, fungal disease caused by *Batrachochytrium dendrobatidis* (Bd) and has been responsible for declines in amphibian populations throughout the world (Kindermann *et al.*, 2012). As a disease of fungal origin, the reduction in eosinophil concentration further increases the vulnerability of amphibians to chytridiomycosis (Kindermann *et al.*, 2012; Figueiredo and Neves, 2018). Furthermore, extended exposure to heat stress can decrease the sensitivity of the endocrine system to acute stressors such as pathogens, resulting in greater loss of amphibians to chytridiomycosis and other diseases (Narayan and Hero, 2014b).

The prevalence of heat stress among amphibians due to climate change and growing mean temperatures has resulted in a decrease in amphibian welfare, with the health of amphibians being heavily impacted due to the suppression of their immune systems (Narayan *et al.*, 2012).

Reptiles

There are over 9000 species that form the class Reptilia. This includes more than 8000 species of snakes and lizards (order Squamata). Threats to these species include invasive species, habitat destruction, disease, pollution, and overexploitation, all of which have resulted in around 22% of reptiles' becoming threatened or extinct (Cabrelli, 2013). One of the most important challenges for reptiles, however, is climate change. Cabrelli (2013) reported that, over the past five decades there has been a global rise in temperature of around 0.46°C, along with variations in water temperature, sea level, rainfall patterns, and frequency and intensity of extreme climatic conditions. This section will specifically focus on the broad-headed snake (*Hoplocephalus bungaroides*) and the impact climate change has on its welfare regarding the environmental and behavioural domains.

Global temperature rise leads to an environment, which is extremely stressful for reptiles (Hall, 2020). Reptiles are commonly found in dry environments and can do so as their integuments are thick and scaly which reduces water loss. However, excessive heating of habitats could restrict lizards to cooler microhabitats and alter behavioural patterns such as periods of activity (Gunn, 1942; Cooper, 2000). Reptiles are ectotherms and are the most vulnerable species towards climate warming. The fluctuation in temperature can generate 'realized' thermal reactions i.e. fitness norms that are different from the fundamental norms in standard temperature conditions (Paaijmans *et al.*, 2013).

Reptiles, like all animals, have a thermal threshold within which they are comfortable. Physiological markers such as HSPs can determine the physiological responses of reptiles to thermal stress. In some reptiles, such as sea turtles, which rely on temperature for sex determination, the thermal environment becomes more important for gender-balanced population dynamics (Hamann *et al.*, 2007). By the year 2050, the global temperature range will increase by 1.1–1.2°C. The changes that take place in ultraviolet light and radiation will also have drawbacks on food availability and quality for species such as the green sea turtle, which is an herbivore. This consequence is a direct example of the impact of climate change on the nutritional domain of reptiles (see Fig. 10.1) (Hamann *et al.*, 2007).

The Broad-headed snake is the most endangered snake species in Australia (Penman *et al.*, 2010; Cogger, 2014). This species has specific habitat requirements for thermoregulation. For example, when temperatures are decreasing they are observed to reside in crevices and under rocks in locations which are exposed; conversely, in summer they move into hollow trees (Pringle *et al.*, 2003). The Broad-headed snake is distributed throughout climatically suitable areas, otherwise known as the bioclimatic envelope. These locations are restricted for species that are threatened by habitat degradation and collection from the wild (Penman *et al.*, 2010; Schlesinger and Shine, 1994; Shine *et al.*, 1998; Webb *et al.*, 2002a, b). The distribution of the Broad-headed snake has resulted in a loss of genetic variation (Cabrelli, 2013). These changes may alter various biological traits of the reptiles such as their reproductive cycle, sex ratio, breeding pattern, site selection, habitat use, and rates of survival (Forchhammer *et al.*, 1998; Crick and Sparks, 1999; Winkler *et al.*, 2002; Visser and Both, 2005; Pike *et al.*, 2006; Telemeco *et al.*, 2009; Penman *et al.*, 2010). The effects of these changes might benefit some animals by extending their ranges to unexplored areas, while others may suffer due to lack of connectivity to the area to provide suitable conditions for survival (Penman *et al.*, 2010). However, it was reported that climate change, wildfire, extinction of fauna, and modifications in anthropogenic landscapes drive changes in vegetation.

It can be concluded that rising temperatures adversely impact the Broad-headed snake. It alters biological traits such as site selection, habitat use, survival rates, reproductive cyclicity, and genetic variations. Climate change has impacted the ecology of reptiles in the wild and researchers have assessed the eco-physiological responses of species *in situ* using physiology, behavioural, and other methods. However, there has been limited work done to link the impacts directly to welfare domains. As climate change drives more wild populations to their ecological limits, there will be potential welfare consequences and considerations that need to be accounted for.

Birds

The impact of climate change on birds has been well documented. It is expected that one-eighth of species of birds will become extinct in the coming decades due to climate change. It is projected that increases in temperature by 3.5°C will lead to the extinction of approximately 900 bird species, with over 80% in the tropic region (Şekercioğlu *et al.*, 2012). Climate change impacts wild bird species in many ways such as habitat distribution, breeding season and migration, and the gross morphology of some bird species. Tropical birds such as northern snowbirds will be highly vulnerable to climate change due to limited access to high-elevation habitats. Birds are homeotherms; however, heat stress due to climate change can impact birds both physiologically and behaviourally. Excess heat dissipation due to exposure to acute heat stress can be costly for birds and trade-off with crucial life-history stages such as growth, development, and reproduction. Effects of heat stress can have significant consequences on breeding success and parental behaviour. Recent research has shown that seabirds are impacted by significant heat stress during breeding. For example, seabird breeding habitats and nests can be impacted by high coastal temperatures resulting in nest abandonment (Olin *et al.*, 2023). Researchers have highlighted the direct and indirect effects of climate change on food resources and disease transmission (e.g. avian influenza and malaria) will increase extinction risks for wild birds (Crick, 2004; Auer and Martin, 2013).

Marsupials

Marsupials are a subclass of mammals that are characterized by premature birth and continued development of the newborn while attached to the nipples on the mother's lower belly within a pouch (Kennedy, 1992; Encyclopaedia Britannica, 1999). Of the over 250 species of marsupials, most can be found in Australia and New Guinea, with the rest being found in America (Kennedy, 1992; Tyndale-Biscoe, 2005). Tooth dentition provides information relating to the type of food an animal eats and, therefore, can indicate their role within the ecosystem (Tyndale-Biscoe, 2005). Marsupial tooth dentition varies, indicating that some marsupials are herbivorous (i.e. the koala *Phascolarctos cinereus*), some are omnivorous (i.e. the common brushtail possum *Trichosurus vulpecula*), and others are carnivorous (i.e. the Tasmanian devil *Sarcophilus harrisii*) (Tyndale-Biscoe, 2005). Here, we will explore how climate change affects the welfare of marsupials, with a focus on the koala and their environmental and nutritional domain.

Climate change has been recognized as a major factor driving population decline across many species of marsupial, including the koala (Adams-Hosking *et al.*, 2011; Lunney *et al.*, 2014). Although Australian marsupials have been traditionally well adapted to the country's erratic climatic patterns, in modern times their survival is threatened by extreme weather events (Tyndale-Biscoe, 2005; CSIRO, 2020), including increasing mean temperatures, rainfall extremes such as prolonged rain and drought, and more frequent and longer fire seasons (Woinarski and Burbidge, 2016; CSIRO, 2020).

Koalas maintain their body temperature at a fairly consistent level (35.5°C) (Tyndale-Biscoe, 2005). They do this by expending the least amount of energy when the surrounding temperature is nearly the same as their body temperature (Tyndale-Biscoe, 2005). Essentially, when their surrounding temperature is lower, more energy is required to generate heat, and when their surrounding temperature is higher, more energy is expended to cool down (Tyndale-Biscoe, 2005). Koalas are a highly specialized species with a specific diet of around 40 eucalyptus species out of the almost 600 species that grow in Australia (Phillips and Callaghan, 2000; Phillips *et al.*, 2000). Eucalyptus is a highly tolerant tree species (Booth, 2013), and contains low levels of nitrogen and high toxic metabolites such as formylated phloroglucinol compounds (Moore *et al.*, 2005; Hovenden and Williams, 2010). However, current climate change projections indicate poor outcomes for eucalyptus leaf quality (Hovenden and Williams, 2010). Increasing mean temperatures influenced by climate change mean marsupial species such as the koala will be required to expend more energy to maintain their body temperature, using a food source that is reduced in quality.

Koalas use dietary water derived from foliage to meet their water requirements (Ellis *et al.*, 1995). Leaf water is normally sufficient to meet the moisture requirements of koalas; however, minimum thresholds are much higher when their surrounding temperature is higher (Hume and Esson, 1993; Clifton *et al.*, 2007). This means that as temperatures become more extreme, the leaf water content may not be sufficient to meet the needs of koalas (Mella *et al.*, 2019), and large-scale mortality events may occur due to dehydration (Gordon *et al.*, 1988; Seabrook *et al.*, 2011; Lunney *et al.*, 2012). Furthermore, eucalyptus is considered the most proficient and successful epicormic resprouter in the incidence of bushfires (Burrows, 2013), which potentially means only short-term loss of food for koalas (Matthews *et al.*, 2007). However, the recovery of eucalyptus post-fire is dependent on a number of factors such as forest composition, seasonal conditions before and after the fire, and the patchiness and intensity of the fire (Heath *et al.*, 2016). This means that the incidence of more frequent and longer fire seasons influenced by climate change could result in long-term loss of food for koalas (Beale *et al.*, 2022).

With the loss of quality food, and potentially food altogether, climate change has already begun to, and will continue to affect the welfare of marsupials, including the iconic koala.

Bats

Bats (Chiropterans) are considered to be the second most diverse mammalian order, with more than 1400 documented species, having a one-fifth contribution to the worldwide mammalian diversity (Simmons and Cirranello, 2020). Bats have a major role to play in the ecosystem, primarily as pollen transporters and seed disposers across many plant species in different parts of the world (Fujita and Tuttle, 1991); for example, bats act as a major pollinator of ferns in New Zealand (Brock and Collier, 2020).

Climate change is having a significant effect on the survival, population density, and reproductive performance of many different animal species (Mawdsley *et al.*, 2009). A study conducted by Appel *et al.* (2019) investigated the major environmental factors affecting insectivorous bat activity. They reported that moonlight and temperature had the highest impact on bat activity followed by rainfall, which affected the activity patterns of a very limited proportion of species. This section will explore the impacts of climate change on the environmental, nutritional, and behavioural domains of bat welfare.

Due to the absence of sweat glands and a high surface-volume ratio, bats are unable to dissipate heat efficiently making them extremely vulnerable to the ill effects of high temperatures (Downs *et al.*, 2012; Korine *et al.*, 2016). One such instance was observed in New South Wales when temperatures reached up to 42°C (Welbergen *et al.*, 2008). This led to the death of around 3500 bats across nine different species from separate colonies (Welbergen *et al.*, 2008).

The temperature increase was linked to a set of behaviours observed in bats, including fanning, increased saliva-spreading, and panting, which were presumably employed as mechanisms to dissipate heat (Welbergen *et al.*, 2008). However, these behaviours were followed by individuals falling from trees and subsequently dying (Welbergen *et al.*, 2008).

Insectivorous bats can be severely affected by climate change as the insects that they consume require very specific climatic conditions for their growth and development (Burles *et al.*, 2009). Temperatures below 10°C can reduce insect activity, which eventually leads to delayed development (Cryan *et al.*, 2000). Consequently, this disruption in the insect life cycle leaves insectivorous bats with less food, which can ultimately compromise their reproductive performance (Cryan *et al.*, 2000). Frick *et al.* (2012) observed that the time of emergence of bats from their caves for foraging was partially dependent on climatic conditions. Bats emerged earlier in years with extremely dry conditions as compared to the years with high moisture, which may increase their chances of predation. Conversely, when the surface temperature was higher the bats emerged later in search of food.

Highly variable habitat temperature because of climate change has an adverse effect on many different bat behavioural patterns. Downs *et al.* (2012) studied the effects of different seasons on the sleep patterns of the bats, observing that bats slept less in summer than in winter. They reported that high temperatures reduce sleep duration considerably as the bats spent most of their time trying to cool themselves. An increase in fanning behaviour was observed by Snoyman *et al.* (2013) in response to high temperatures. Additionally, females with young pups showed an increase in grooming behaviour compared to other individuals in the same group, indicating their vulnerability to high temperatures. Moreover, female bats consumed more water with increased ambient temperature (Adams and Hayes, 2008).

Finally, the effects of climate change are also seen in the reproductive success of female bats. Elevated temperatures and dry weather have an adverse effect on the reproductive patterns of bats (Adams, 2010). Increased rainfall has also been shown to reduce reproductive success in female bats due to reduced foraging activities and prey detection capacity (Grindal *et al.*, 1992; Lučan *et al.*, 2013). In summary, climate change proposes to be a significant threat to the survival of bats and will continue to have adverse effects if mitigation strategies are not established.

Production animal examples

Broiler chickens

Chicken meat has become a dietary staple for Australians due to its affordability and versatility (ACMF, 2022). The price per kilogram has remained stable since the mid-1990s, around AUS $5.50, making it the cheapest meat available for consumers (Phibbs *et al.*, 2021). While many Australians are reflecting on their meat consumption behaviours, ABARES have forecast a per capita consumption of 48.8 kg for FY 2021/22, with consumption expected to continue to increase year after year (ACMF, 2022). Decades of genetic selection of broilers has created highly efficient feed converters, which have a rapid growth rate (Phibbs *et al.*, 2021). Selective breeding has led to many morphological changes, especially in highly marketable carcase traits, namely more breast muscle. However, these breeding efforts have negatively impacted leg health, resulting in altered gait and an increase in leg-related issues (Danbury *et al.*, 2000). This raises significant animal welfare concerns for broiler chickens.

A survey by Taylor and Signal (2009) revealed that Australian consumers are aware of broiler welfare concerns; however, the public's overall knowledge of production methods is poorly understood. Additionally, they also show that while consumers advocate for more ethical production methods, consumers are typically unwilling to pay more. This situation gives rise to a complex dilemma, wherein producers lack sufficient incentives to embrace more ethical production methods. To address this issue, a deeper understanding of consumer motivations is needed to encourage the adoption of ethical production methods.

Commercial broiler systems typically opt for caged or floor-based methods (Shields and Greger, 2013). While uncommon in Australia, caged broilers are known to have higher instances of flesh wounds due to abrasion against the cage floor, which reduces meat quality (Slepukhin *et al.*, 2000; Shane, 2001). Caged systems often incur greater labour costs compared to floor systems (Reece *et al.*, 1971). Such systems significantly limit free space per bird and thus reduce the range of natural behaviours, such as dustbathing, that animals can partake in (Olsson and Keeling, 2005). As such, cage broilers are known to have weaker bones as a consequence of their inability to exercise (Wabeck and Littlefield, 1972; Travis *et al.*, 1983). Floor systems enable broilers to access deep litter, where they are free to express natural behaviours, which are otherwise limited in caged production systems (Shields and Greger, 2013). In floor systems, the birds are provided with ample space and therefore sufficient opportunity to exercise as they grow, and thus strengthen their bones, ultimately reducing bone health concerns. Free-range systems, where birds are raised in houses with voluntary access to the outdoors (Martínez-Pérez *et al.*, 2017), are becoming popular in Australia, and are seen as a more proactive approach to animal welfare and meat chicken production. Similar to floor systems, free-range enterprises enable birds to explore natural behaviours, which reduces stress, and improves health and animal performance (Abouelezz *et al.*, 2012).

In Australia, state and territory governments have implemented legislation that provides a broad scope of the minimum animal welfare standards required in broiler production systems (Phibbs *et al.*, 2021), while additional, best practice protocols exist at the federal level (Pisc, 2002). Animal welfare organizations such as the RSPCA offer accreditation programmes that producers can opt into, including their Approved Farming Scheme, and the Free Range Egg and Poultry Australia (FREPA) scheme (Phibbs *et al.*, 2021). Both accreditors provide elevated animal welfare standards to ensure producers outperform current minimum requirements, with the RSPCA encouraging the use of perches for enrichment and bone health (RSPCA, 2016). In return, successful producers are able to advertise their meat as 'RSPCA Approved' (RSPCA, 2020). About 70% of Australian chicken meat is produced according to the RSPCA Scheme, including chicken products sold by two major Australian supermarket chains (RSPCA, 2018). The FREPA standards offer fewer specific requirements for producers and account for ~20% of Australian meat production. The adoption of improved animal welfare practices into commercial farms has not significantly affected the producer's bottom line nor the cost to the consumer, with chicken meat prices remaining stable during the transition (Phibbs *et al.*, 2021).

Strong domestic and international demand for Australian chicken meat will inevitably result in a significant increase in resource requirements throughout the supply chain, including additional water and land for feed production, increased energy requirements for climate regulation within growing sheds, and labour costs (Izar-Tenorio *et al.*, 2020). Climatic changes, which are expected to result in elevated temperatures, translate to greater energy expenditure per unit of production. The welfare of birds, especially mature broilers, under warmer conditions, is of great concern. Birds have a limited capacity for heat regulation as they lack sweat glands, and regulate temperature by panting, limiting their activity, and drinking more (Fairchild, 2009; Izar-Tenorio *et al.*, 2020). Depending on the season, young chicks are provided with additional heating between 32 and 35°C for the first ~20 days, until they are of sufficient size to regulate their own body temperature (Xin *et al.*, 2001). Energy-expensive heating systems may be required during winter and cooler nights to ensure growing sheds are kept at ~21°C (Fairchild, 2009). During summer, mechanical ventilation and fans are commonly used to encourage air circulation and decrease environmental temperature (Izar-Tenorio *et al.*, 2020). As stated by Fairchild (2009), in warm climates, air conditioning units may be required to maintain an optimal production temperature.

In summary, chicken meat is a popular and affordable choice in Australia, but broiler welfare concerns persist. Consumers desire ethical production methods but are unwilling to pay higher prices, creating a dilemma for producers. Different production systems have varying impacts on animal welfare, with cage-free systems offering better conditions. Accreditation programmes and legislation aim to improve standards without significantly affecting costs. However, the industry faces resource challenges and climate change risks. Balancing ethics, affordability, and sustainability is crucial for the future of Australian chicken meat.

Ruminants

Climate change has a negative impact on livestock, affecting their productivity and physical health. Ruminants suffer from heat stress, which can alter their dry matter intake and impact forage crops and grazing practices. Climate change leads to variations in greenhouse gas concentrations and temperature, causing changes in forage growth, shifts in forage composition, and fluctuations in nitrogen content (Slepukhin *et al.*, 2000).

In Australia, the world's largest apparel wool growing country, climate change poses risks to livestock welfare, including reproductive potential and hormonal function. Rising temperatures and thermal stress can result in embryo loss and chronic stress in Merino sheep. Heat stress reduces feed intake and reproductive output, posing animal welfare concerns and affecting nutrition and health (Olsson and Keeling, 2005). Previous research from our lab revealed that heat stress (THI >75) and elevated glucocorticoid levels (indexed using faecal glucocorticoid metabolites) are linked to embryo loss in Merino ewes (Narayan *et al.*, 2018). Chronic stress can also be assessed in livestock such as Merino sheep using wool profiles of glucocorticoids (Wabeck and Littlefield, 1972). Heat stress has been shown to reduce feed intake and reproductive output in livestock, raising animal welfare concerns and impacting nutrition and health domains (Martínez-Pérez *et al.*, 2017).

Livestock's thermal comfort zone is challenged when temperatures surpass their optimal range, leading to changes in endocrine status, ruminant activity frequency, nutrient uptake, and feed intake behaviour. Energy use efficiency declines in animals experiencing heat stress, necessitating dietary modifications to minimize heat production. Maintaining consistent acid detergent fibre (ADF) and total plant fibre levels is crucial for optimal rumen function (Abouelezz *et al.*, 2012). According to RSPCA (2018), the efficiency of conversion to milk in diets containing 100%, 75%, or 50% hay is 54%, 61%, and 65%, respectively. Researchers have also found that the addition of alfalfa and dietary fibre both increase the heat load and exacerbate heat stress (Izar-Tenorio *et al.*, 2020). To ensure optimal rumen function, it is advisable to maintain ADF and total plant fibre at a consistent level of 18% and 28%, respectively (RSPCA, 2020).

Ruminants respond to climate change through behaviour adjustments such as seeking shade and increasing standing time for cooling. Heat stress significantly impacts lactation performance, immune function, and calf health. Monitoring cows' activity and rumination time helps detect heat stress, and appropriate heat dissipation strategies such as sunshades and sprinklers can mitigate its effects. Techniques like embryo transfer can increase conception rates during the hot season (Fairchild, 2009). In animal husbandry, the monitoring of cows' activity time and rumination time serves as an early warning system for detecting heat stress (3BL, 2022). During lactation, there is evidence of a correlation between the temperature-humidity index (THI) and conception rate (CR) in dairy cows (WHO, 2020). According to the relationship between mean THI and CR, the threshold at which heat stress can have the most significant effect on CR is when the THI value is 73 (WHO, 2020). Studies have also shown that cows living in milder climates are more susceptible to heat stress than those in tropical and subtropical climates (Castrica *et al.*, 2018). Further, cows are most susceptible to heat stress during the first 1–21 days before breeding. It is, therefore, important to develop appropriate heat dissipation strategies to avoid the effects of heat stress (German *et al.*, 2018). Heat stress can affect production levels and physical fitness of livestock, but breeders can provide equipment and techniques to help them recover (Miller *et al.*, 2018). It has also been shown that embryo transfer can be used during the hot season to increase the chances of conception and counteract the negative effects of heat stress (Packer *et al.*, 2019).

In conclusion, heat stress can affect production levels and physical fitness of livestock. But breeders can provide equipment and techniques to help them recover.

Companion animals

Companion animals, including cats and dogs, play a significant and influential role in societies across the world, with an estimated population of hundreds of millions serving as companions to people worldwide (AVMA, 2022).

In contemporary society, the realm of pet food has grown increasingly sophisticated, with a significant rise in expenditure, underscoring the paramount significance and elevated status bestowed upon pets in today's world. Several studies have highlighted the connection between companion animals and their environmental impact, revealing that the allocation of additional resources to these animals can potentially contribute to the escalation of climate change (Protopopova *et al.*, 2021). This connection stems from the often overlooked factor of food production for companion animals, which contributes to environmental damage (Su *et al.*, 2018). In the USA, the environmental impact of the pet dog and cat diets can reach 30% (Okin, 2017), and while these impacts may be lower on a global scale, the trend is continuing for pet food to be of increasingly high standards, almost comparable to that of human food (3BL, 2022). These high standards of pet food could also further exacerbate the global food crisis (WHO, 2020). Overfeeding companion animals can also have detrimental environmental consequences and jeopardize their own health (Su *et al.*, 2018). To alleviate environmental stress, companion animal food can be sourced from human food remnants or waste (Castrica *et al.*, 2018). By embracing these measures, companion animal owners can play an active role in reducing the ecological impact of their pets while simultaneously improving their welfare.

For dogs, exercise forms an important part of their health and welfare. About 50% of dogs in the UK are overweight. There are dietary reasons for this, but insufficient exercise is also an important cause of obesity and overweight status (German *et al.*, 2018). Factors such as weather conditions can be a potential hindrance to dog walking. For example, about 87% of owners report that they exercise their dogs less during hot weather (Hall *et al.*, 2021); therefore, extreme heat may contribute to increased rates of canine obesity. Further, heat stress can affect the performance and physical health of working dogs. For example, studies have shown that heat-related diseases are a leading cause of death in military working dogs (Miller *et al.*, 2018). Physical exercise is a major contributor to heat-related illness in the UK and Israel (Packer *et al.*, 2019). Therefore, the gradual increase in global temperature has the potential to decline the level of welfare for dogs, which may also lead to changes in the role of dogs in human society. In the case of cats and other animals in shelters, environmental changes can induce stress (Stella and Croney, 2016).

To safeguard the well-being of companion animals amidst the challenges posed by global warming, it is crucial to consider the behaviour and breeding practices of pet owners. For instance, one important measure is to refrain from leaving pets unattended in vehicles without proper air conditioning, as this can pose a significant risk. Additionally, certain breeds, particularly those with short-headed characteristics, are more susceptible to heat stroke despite their popularity (Hall *et al.*, 2020). Hence, responsible pet ownership entails being mindful of these factors to protect the health of companion animals in a warming world (Hall *et al.*, 2020).

The stark reality that the climate is moving in a direction that is detrimental to both humans and companion animals is already in front of human eyes, and it is necessary for humans to see companion animals as an important link in influencing climate change. This is because companion animals can, to some extent, contribute to climate warming, while climate warming can also affect the survival of companion animals, especially canines.

Large wildlife mammals

While most of this review has focused on endemic Australian species, it is important to acknowledge that climate change has a variety of differing influences in different regions, and ecosystems, and on organisms that cannot be represented with a restricted focus on Australian animals. An example of this is the African bush elephant (*Loxodonta africana*), a large herbivore endemic to many dryland environments in the African continent (Fuller *et al.*, 2021). This section will focus on the impacts of climate change on African elephant welfare, and compare these impacts to the red kangaroo (*Macropus rufus*), another herbivore endemic to dryland habitats (Fuller *et al.*, 2021).

As the largest, extant terrestrial land mammal, the African elephant has significant daily food and water requirements (Fuller *et al.*, 2021). However, with the effects of climate change compounding with every passing day, the ability of elephants to meet their nutritional needs decreases (Mukeka *et al.*, 2022). As droughts become more frequent and unpredictable, the availability of water and vegetation cover declines, increasing elephant heat and nutrition stress, contributing to the elevated elephant mortality we are currently witnessing (Mukeka *et al.*, 2022).

A mature African elephant requires up to 200l of water and 90 kg of dry matter per day to reach their nutritional and metabolic requirements (Ullrey *et al.*, 1997; Greene *et al.*, 2019). Elephants require large quantities of daily drinking water as a physiological adaptation to endure the heat of their environment, while maintaining their large body size. Due to their size, elephants are far more inefficient at maintaining homeostatic thermoregulation when compared to smaller animals. As a result, elephants have evolved adaptations such as highly vascularized ears and a high surface-to-volume ratio, to expedite their ability to cool themselves (Tattersall *et al.*, 2012). One of these adaptations utilizes their excess skin to increase evaporative water loss, a process similar to sweating (Fuller *et al.*, 2016). Even under climatic temperatures of 10°C, elephants have been reported to have a greater rate of cutaneous evaporative cooling than other animals (Mitchell *et al.*, 2018). With the current patterns of increased drought and elevated temperatures, elephants will have to find other methods to regulate their body temperatures, while also meeting their water and nutritional needs with reduced access to water and foliage.

The implications of increased drought frequency directly reduce the nutritional welfare of elephants through decreased water availability and thereby decreased foliage density (Mukeka *et al.*, 2022). The severity of nutritional and water stress is further exacerbated by the tendency of mammals to catabolize endogenous proteins to extract water, during periods of low water availability, resulting in greater degradation of body condition (McCue *et al.*, 2017). These nutritional issues develop into mental and environmental welfare issues as a result of reduced vegetation for utilization as shelter from the sun, as well as an increase in heat stress from the animals' inability to achieve their water requirements and thereby being unable to effectively thermoregulate (Dunkin *et al.*, 2013; Fuller *et al.*, 2016).

In comparison with elephants, kangaroos are far more adapted to erratic weather patterns and periodic drought (Tyndale-Biscoe, 2005; Mukeka *et al.*, 2022). Kangaroos undergo periods of population growth and decline, dictated by environmental factors, particularly the absence or abundance of food and water (Caughley *et al.*, 1985; McLeod *et al.*, 2021). However, despite this, ultimately both species are prone to elevated mortality rates as a result of drought (Caughley *et al.*, 1985; Douglas-Hamilton, 1987). From an ecological perspective, this process of drought-induced die-offs and rain-induced population booms is natural. However, complexity arises when the issue is examined through a welfare-centric lens (Wilson and Edwards, 2019; Zanker, 2021). The process of die-offs is induced by the starvation of animals as a result of insufficient vegetation to support the population, greatly impacting the nutritional welfare of the animals (Wilson and Edwards, 2019).

As a result of climatic variability, elephants are no longer able to acquire their nutritional requirements, and as a result their environmental and mental welfare suffers (Mukeka *et al.*, 2022). In comparison, the kangaroo while also suffering from nutritional disparity is involved in a complex balancing act between ecological balance and animal welfare maintenance, highlighting the complexities that are associated with the effects of climate change on animal welfare (Caughley *et al.*, 1985; Wilson and Edwards, 2019; Zanker, 2021).

Conclusion

As demonstrated in our review, climate change has impacted the welfare of animals in both the wild and domestic settings.

Researchers have assessed the eco-physiological responses of species using physiology, behavioural, and other methods. Although common themes regarding the methods and measurements are emerging, we believe that thus far, there has been limited work done to link the impacts of climate change directly to welfare domains, especially in wildlife. As climate change drives more wild populations to ecological limitations, there will be potential welfare consequences and considerations to explore; for example, when vulnerable species would need to be transferred to new environments (e.g. captive breeding), should food and habitat become limiting resources. Likewise, production animals and other domesticated species will be impacted by the extreme environmental changes with consequences on each of the dimensions within the five domains of animal welfare (Fig. 10.1). We believe that the impacts of climate change will be multi-faceted, whereby there could be cascading effects across all domains. Researchers need to carefully consider which domains are immediate and future priorities to safeguard animal longevity and welfare.

Author Contributions

All authors contributed to the conceptualization of the presented article, curation of data, analysis of data, conduction of the research process, formation of the methodology, presentation of the final work, construction of the final draft and review, and editing of the final work. EN was responsible for project administration and supervision.

Acknowledgements

We would like to thank all the team members of The Stress Lab who contributed to the initial discussion regarding the formation of the article.

Funding

This review article did not receive any form of funding from funding agencies in the public, commercial, or not-for-profit sectors.

Conflict of Interest

There are no conflicts of interest to disclose.

References

3BL (2022) Humanization of pet food' driving new capacity for manufacturers, CRB report finds. In: *3BL Blogs - 3BL Media*. Proquest.

Abouelezz, F.M.K., Sarmiento-Franco, L., Santos-Ricalde, R. and Solorio-Sanchez, F. (2012) Outdoor egg production using local forages in the tropics. *World's Poultry Science Journal* 68(4), 679–692.

ACMF (2022) Facts and Figures. Available at: https://www.chicken.org.au/facts-and-figures/ (accessed 8 December 2022).

Adams, R.A. (2010) Bat reproduction declines when conditions mimic climate change projections for western North America. *Ecology* 91(8), 2437–2445.

Adams, R.A. and Hayes, M.A. (2008) Water availability and successful lactation by bats as related to climate change in arid regions of western North America. *Journal of Animal Ecology* 77(6), 1115–1121.

Adams-Hosking, C., Grantham, H.S., Rhodes, J.S., McAlpine, C. and Moss, P.T. (2011) Modelling climate-change-induced shifts in the distribution of the koala. *Wildlife Research* 38(2), 122–130.

Agassiz, L. (2013) *Essay on Classification*. Dover Publications, Mineola, NY.

Ahmed, N. and Turchini, G.M. (2021) Recirculating aquaculture systems (RAS): Environmental solution and climate change adaptation. *Journal of Cleaner Production* 297, 126604.

Ahmed, N., Thompson, S. and Glaser, M. (2019) Global aquaculture productivity, environmental sustainability, and climate change adaptability. *Environmental Management* 63(2), 159–172.

Angiulli, E., Pagliara, V., Cioni, C., Frabetti, F., Pizzetti, F. *et al.* (2020) Increase in environmental temperature affects exploratory behaviour, anxiety and social preference in Danio rerio. *Scientific Reports* 10(1), 1–12.

Appel, G., López-Baucells, A., Magnusson, W.E. and Bobrowiec, P.E.D. (2019) Temperature, rainfall, and moonlight intensity effects on activity of tropical insectivorous bats. *Journal of Mammalogy* 100(6), 1889–1900.

Argent, G. (2022) Human-animal relationships and welfare in the Anthropocene: Pandemics, climate change, and other disasters. *Journal of Applied Animal Welfare Science* 25(2), 115–118.

Auer, S.K. and Martin, T.E. (2013) Climate change has indirect effects on resource use and overlap among coexisting bird species with negative consequences for their reproductive success. *Global Change Biology* 19(2), 411–419.

AVMA (2022) U.S. Pet Ownership Statistics. Available at: https://www. avma.org/resources-tools/reports-statistics/us-pet-ownership-statistics (accessed 14 December 2022).

Beale, P., Youngentob, K. and Marsh, K. (2022) *Effects of Fire on Koalas and Their Habitat*. The Australian National University, Canberra.

Beltrán, I., Herculano-Houzel, S., Sinervo, B. and Whiting, M.J. (2021) Are ectotherm brains vulnerable to global warming? *Trends in Ecology & Evolution* 36(8), 691–699.

Bernabucci, U., Lacetera, N., Baumgard, L.H., Rhoads, R.P., Ronchi, B. *et al.* (2010) Metabolic and hormonal acclimation to heat stress in domesticated ruminants. *Animal* 4(7), 1167–1183.

Bernardo, J. and Spotila, J.R. (2006) Physiological constraints on organismal response to global warming: Mechanistic insights from clinally varying populations and implications for assessing endangerment. *Biology Letters* 2(1), 135–139.

Booth, T.H. (2013) Eucalypt plantations and climate change. *Forest Ecology and Management* 301, 28–34.

Brock, J.M.R. and Collier, K. (2020) Bat dispersal of fern spores in New Zealand. *New Zealand Journal of Ecology* 44(2), 1–3.

Broom, D.M. (2014) *Sentience and Animal Welfare*. CAB International, Wallingford, UK.

Bueno, P.B. and Soto, D. (2017) Adaptation strategies of the aquaculture sector to the impacts of climate change12. *Fisheries & Aquaculture Circular* 99, 1–28.

Burles, D.W., Brigham, R.M., Ring, R.A. and Reimchen, T.E. (2009) Influence of weather on two insectivorous bats in a temperate pacific North-west rainforest. *Revue Canadienne De Zoologie* 87(2), 132–138.

Burrows, G. (2013) Buds, bushfires and resprouting in the eucalypts. *Australian Journal of Botany* 61(5), 331–349.

Butterfield, J.E. and Coulson, J.C. (1997) Terrestrial invertebrates and climate change: Physiological and life-cycle adaptations. In: *Past and Future Rapid Environmental Changes*. Springer, UK, pp. 401–412.

Cabrelli, A. (2013) *The Impacts of Climate Change on Australian Reptiles*. Faculty of Science, Department of Biological Sciences, Macquarie University.

Carere, C., Wood, J.B. and Mather, J. (2011) Species differences in captivity: Where are the invertebrates? *Trends in Ecology & Evolution* 26(5), 211.

Castrica, M., Tedesco, D.E., Panseri, S., Ferrazzi, G., Ventura, V. *et al.* (2018) Pet food as the most concrete strategy for using food waste as feedstuff within the European context: A feasibility study. *Sustainability* 10(6), 2035.

Caughley, G., Grigg, G.C. and Smith, L. (1985) The effect of drought on kangaroo populations. *The Journal of Wildlife Management* 679–685.

Chang, E.S. (2005) Stressed-out lobsters: Crustacean hyperglycemic hormone and stress proteins. *Integrative and Comparative Biology* 45(1), 43–50.

Clifton, I.D., Ellis, W.A., Melzer, A. and Tucker, G. (2007) Water turnover and the northern range of the koala (*Phascolarctos cinereus*). *Australian Mammalogy* 29(1), 85–88.

Cochrane, K., De Young, C., Soto, D. and Bahri, T. (2009) Climate change implications for fisheries and aquaculture. *FAO Fisheries and Aquaculture Technical Paper* 530, 212.

Cogger, H. (2014) *Reptiles and Amphibians of Australia*. CSIRO Publishing, Clayton, Victoria.

Cooper, W. (2000) Effect of temperature on escape behaviour by an ectothermic vertebrate, the keeled earless lizard (*Holbrookia propinqua*). *Behaviour* 137(10), 1299–1315.

Crick, H.Q.P. (2004) The impact of climate change on birds. *Ibis* 146(s1), 48–56.

Crick, H.Q. and Sparks, T.H. (1999) Climate change related to egg-laying trends. *Nature* 399(6735), 423–423.

Cryan, P.M., Bogan, M.A. and Altenbach, J.S. (2000) Effect of elevation on distribution of female bats in the black hills, South Dakota. *Journal of Mammalogy* 81(3), 719–725.

CSIRO (2020) State of the Climate Report 2020. Available at: https://www.csiro.au/en/research/environmental-impacts/climate-change/State-of-the-Climate/Previous/State-of-the-Climate-2020

Culler-Juarez, M.E. and Onthank, K.L. (2021) Elevated immune response in *Octopus rubescens* under ocean acidification and warming conditions. *Marine Biology* 168(9), 1–10.

Cziesielski, M.J., Schmidt-Roach, S. and Aranda, M. (2019) The past, present, and future of coral heat stress studies. *Ecology and Evolution* 9(17), 10055–10066.

Danbury, T.C., Weeks, C.A., Waterman-Pearson, A.E., Kestin, S.C. and Chambers, J.P. (2000) Self-selection of the analgesic drug carprofen by lame broiler chickens. *Veterinary Record* 146(11), 307–311.

Domenici, P., Allan, B.J., Lefrançois, C. and McCormick, M.I. (2019) The effect of climate change on the escape kinematics and performance of fishes: Implications for future predator–prey interactions. *Conservation Physiology* 7(1), coz078.

Douglas-Hamilton, I. (1987) African elephants: Population trends and their causes. *Oryx* 21(1), 11–24.

Downs, C.T., Zungu, M.M. and Brown, M. (2012) Seasonal effects on thermoregulatory abilities of the Wahlberg's epauletted fruit bat (*Epomophorus wahlbergi*) in Kwazulu-Natal, South Africa. *Journal of Thermal Biology* 37(2), 144–150.

Dunkin, R.C., Wilson, D., Way, N., Johnson, K. and Williams, T. (2013) Climate influences thermal balance and water use in African and Asian elephants: Physiology can predict drivers of elephant distribution. *Journal of Experimental Biology* 216(15), 2939–2952.

Ellis, W.A.H., Melzer, A., Green, B., Newgrain, K., Hindell, M.A. *et al.* (1995) Seasonal-variation in water flux, field metabolic-rate and food-consumption of free-ranging koalas (*Phascolarctos-cinereus*). *Australian Journal of Zoology* 43(1), 59–68.

Encyclopaedia Britannica (1999) *Marsupial*. Encyclopaedia Britannica, Chicago, IL.

Fabry, V.J., Seibel, B.A., Feely, R.A. and Orr, J.C. (2008) Impacts of ocean acidification on marine fauna and ecosystem processes. *ICES Journal of Marine Science* 65(3), 414–432.

Fairchild, B.D. (2009) *Environmental Factors to Control When Brooding Chicks*.

FAO (2021) *World Food and Agriculture – Statistical Yearbook 2021*. FAO, Rome.

Figueiredo, R.T. and Neves, J.S. (2018) Eosinophils in fungal diseases: An overview. *Journal of Leukocyte Biology* 104(1), 49–60.

Flajnik, M.F. (2018) A cold-blooded view of adaptive immunity. *Nature Reviews Immunology* 18(7), 438–453.

Forchhammer, M.C., Post, E. and Stenseth, N.C. (1998) Breeding phenology and climate. *Nature* 391(6662), 29–30.

Forrest, J.R. (2017) Insect pollinators and climate change. In: *Global Climate Change and Terrestrial Invertebrates*. John Wiley, Hoboken, NJ, pp. 69–91.

Freitas, C., Villegas-Ríos, D., Moland, E. and Olsen, E.M. (2021) Sea temperature effects on depth use and habitat selection in a marine fish community. *Journal of Animal Ecology* 90(7), 1787–1800.

Frick, W.F., Stepanian, P.M., Kelly, J.F., Howard, K.W., Kuster, C.M. *et al.* (2012) Climate and weather impact timing of emergence of bats. *PLoS One* 7(8), e42737.

Fujita, M.S. and Tuttle, M.D. (1991) Flying foxes (*Chiroptera: Pteropodidae*): Threatened animals of key ecological and economic importance. *Conservation Biology* 5(4), 455–463.

Fuller, A., Mitchell, D., Maloney, S.K. and Hetem, R.S. (2016) Towards a mechanistic understanding of the responses of large terrestrial mammals to heat and aridity associated with climate change. *Climate Change Responses* 3(1), 1–19.

Fuller, A., Mitchell, D., Maloney, S.K., Hetem, R.S., Fonsêca, V.F. *et al.* (2021) How dryland mammals will respond to climate change: The effects of body size, heat load and a lack of food and water. *Journal of Experimental Biology* 224(Suppl_1), jeb238113.

German, A.J., Woods, G.R., Holden, S.L., Brennan, L. and Burke, C. (2018) Dangerous trends in pet obesity. *The Veterinary Record* 182(1), 25.

González-Tokman, D., Córdoba-Aguilar, A., Dáttilo, W., Lira-Noriega, A., Sánchez-Guillén, R.A. *et al.* (2020) Insect responses to heat: Physiological mechanisms, evolution and ecological implications in a warming world. *Biological Reviews* 95(3), 802–821.

Gordon, G., Brown, A. and Pulsford, T. (1988) A koala (*Phascolarctos cinereus Goldfuss*) population crash during drought and heatwave conditions in south-western Queensland. *Australian Journal of Ecology* 13(4), 451–461.

Greene, W., Dierenfeld, E.S. and Mikota, S. (2019) A review of Asian and African elephant gastrointestinal anatomy, physiology, and pharmacology: Elephant gastrointestinal anatomy, physiology, and pharmacology. *Journal of Zoo and Aquarium Research* 7(1), 1–14.

Grindal, S.D., Collard, T.S., Brigham, R.M. and Barclay, R.M. (1992) The influence of precipitation on reproduction by *m*yotis bats in British Columbia. *The American Midland Naturalist* 128(2), 339–344.

Gunn, D. (1942) Body temperature in poikilothermal animals. *Biological Reviews* 17(4), 293–314.

Hall, J.M. (2020) *Heat Tolerance of Reptile Embryos in A Changing World: Physiological Mechanisms and Ecological Effects*. Nature Portfolio, UK.

Hall, E.J., Carter, A.J. and O'Neill, D.G. (2020) Incidence and risk factors for heat-related illness (heatstroke) in UK dogs under primary veterinary care in 2016. *Scientific Reports* 10(1), 1–12.

Hall, E.J., Carter, A.J. and Farnworth, M.J. (2021) Exploring owner perceptions of the impacts of seasonal weather variations on canine activity and potential consequences for human–canine relationships. *Animals* 11(11), 3302.

Hamann, M., Limpus, C.J. and Read, M.A. (2007) *Vulnerability of Marine Reptiles in the Great Barrier Reef to Climate Change*. The Great Barrier Reef Marine Park Authority, Townsville, Australia.

Heath, J.T., Chafer, C.J., Bishop, T.F. and Van Ogtrop, F.F. (2016) Post-fire recovery of eucalypt-dominated vegetation communities in the Sydney Basin, Australia. *Fire Ecology* 12(3), 53–79.

Hemsworth, P.H., Mellor, D.J., Cronin, G.M. and Tilbrook, A.J. (2015) Scientific assessment of animal welfare. *New Zealand Veterinary Journal* 63(1), 24–30.

Hennessy, K., Lucas, C., Nicholls, N., Bathols, J., Suppiah, R. *et al.* (2005) *Climate Change Impacts on Fire-Weather in South-East Australia*. Climate Impacts Group, CSIRO Atmospheric Research and the Australian Government Bureau of Meteorology, Aspendale.

Horwitz, R., Norin, T., Watson, S.A., Pistevos, J.C., Beldade, R. *et al.* (2020) Near-future ocean warming and acidification alter foraging behaviour, locomotion, and metabolic rate in a keystone marine mollusc. *Scientific Reports* 10(1), 5461.

Hovenden, M.J. and Williams, A.L. (2010) The impacts of rising CO_2 concentrations on Australian terrestrial species and ecosystems. *Austral Ecology* 35(6), 665–684.

Huey, R.B. and Kingsolver, J.G. (2019) Climate warming, resource availability, and the metabolic meltdown of ectotherms. *The American Naturalist* 194(6), E140–E150.

Hume, I. and Esson, C. (1993) Nutrients, antinutrients and leaf selection by captive koalas (*Phascolarctos-cinereus*). *Australian Journal of Zoology* 41(4), 379–392.

Izar-Tenorio, J., Jaramillo, P., Griffin, W.M. and Small, M. (2020) Impacts of projected climate change scenarios on heating and cooling demand for industrial broiler chicken farming in the Eastern US. *Journal of Cleaner Production* 255, 120306.

Jin, X., Wang, X., Tse, W.K.F. and Shi, Y. (2022) Editorial: Homeostasis and physiological regulation in the aquatic animal during osmotic stress. *Frontiers in Physiology* 13, 977185.

Kellert, S.R. (1993) Values and perceptions of invertebrates. *Conservation Biology* 7(4), 845–855.

Kennedy, M. (1992) *Australasian Marsupials and Monotremes: An Action Plan for Their Conservation*. IUCN, Switzerland.

Kindermann, C., Narayan, E.J. and Hero, J.-M. (2012) Urinary corticosterone metabolites and chytridiomycosis disease prevalence in a free-living population of male stony creek frogs (*Litoria wilcoxii*). *Comparative Biochemistry and Physiology Part A: Molecular & Integrative Physiology* 162(3), 171–176.

Korine, C., Adams, R., Russo, D., Fisher-Phelps, M. and Jacobs, D. (2016) Bats and water: anthropogenic alterations threaten global bat populations. In: *Bats in the Anthropocene: Conservation of Bats in a Changing World*. Springer, Cham, pp. 215–241.

Kristiansen, T.S. and Bracke, M. (2020) A brief look into the origins of fish welfare science. In: *The Welfare of Fish*. Springer, pp. 1–17.

Li, L., Liu, Z., Quan, J., Lu, J., Zhao, G. *et al.* (2022) Dietary nanoselenium supplementation for heat-stressed rainbow trout: Effects on organizational structure, lipid changes, and biochemical parameters as well as heat-shock-protein-and selenoprotein-related gene expression. *Fish Physiology and Biochemistry* 48(3), 707–722.

Lučan, R.K., Weiser, M. and Hanák, V. (2013) Contrasting effects of climate change on the timing of reproduction and reproductive success of a temperate insectivorous bat. *Journal of Zoology* 290(2), 151–159.

Lunney, D., Crowther, M.S., Wallis, I., Foley, W.J., Lemon, J. *et al.* (2012) Koalas and climate change: A case study on the Liverpool Plains, North-west new South Wales. In: Lunney, D. and Hutchings, P. (eds) *Wildlife and Climate Change: Towards Robust Conservation Strategies for Australian Fauna*. Royal Zoological Society of New South Wales, Mosman, Australia, pp. 150–168.

Lunney, D., Stalenberg, E., Santika, T. and Rhodes, J.R. (2014) Extinction in Eden: Identifying the role of climate change in the decline of the koala in south-eastern NSW. *Wildlife Research* 41(1), 22–34.

Mahmoud, S., Sabry, A., Abdelaziz, A. and Shukry, M. (2020) Deleterious impacts of heat stress on steroidogenesis markers, immunity status and ovarian tissue of nile tilapia (*Oreochromis niloticus*). *Journal of Thermal Biology* 91, 102578.

Marsh, K.J., Wallis, I.R. and Foley, W.J. (2007) Behavioural contributions to the regulated intake of plant secondary metabolites in koalas. *Oecologia* 154(2), 283–290.

Martínez-Pérez, M., Sarmiento-Franco, L., Santos-Ricalde, R.H. and Sandoval-Castro, C.A. (2017) Poultry meat production in free-range systems: Perspectives for tropical areas. *World's Poultry Science Journal* 73(2), 309–320.

Martos-Sitcha, J.A., Mancera, J.M., Prunet, P. and Magnoni, L.J. (2020) Welfare and stressors in fish: Challenges facing aquaculture. *Frontiers Media SA* 11, 162.

Matthews, A., Lunney, D., Gresser, S. and Maitz, W. (2007) Tree use by koalas (*Phascolarctos cinereus*) after fire in remnant coastal forest. *Wildlife Research* 34(2), 84–93.

Mawdsley, J.R., O'Malley, R. and Ojima, D.S. (2009) Review of climate-change adaptation strategies for wildlife management and biodiversity conservation. *Conservation Biology* 23(5), 1080–1089.

McCue, M.D., Sandoval, J., Beltran, J. and Gerson, A.R. (2017) Dehydration causes increased reliance on protein oxidation in mice: A test of the protein-for-water hypothesis in a mammal. *Physiological and Biochemical Zoology* 90(3), 359–369.

McLeod, S.R., Finch, N., Wallace, G. and Pople, A.R. (2021) Assessing the spatial and temporal organization of Red Kangaroo, Western Grey Kangaroo and Eastern Grey Kangaroo populations in eastern Australia using multivariate autoregressive state-space models. *Ecological Management and Restoration* 22, 106–123.

Medway, L. (1980) *Report of the Panel of Enquiry into Shooting and Angling (1976–79)*. Horsham, UK.

Mella, V.S., McArthur, C., Krockenberger, M.B., Frend, R. and Crowther, M.S. (2019) Needing a drink: Rainfall and temperature drive the use of free water by a threatened arboreal folivore. *PLoS One* 14(5), e0216964.

Mellor, D.J., Beausoleil, N.J., Littlewood, K.E. and McLean, A.N. (2020) The 2020 five domains model: Including human–animal interactions in assessments of animal welfare. *Animals* 10(10), 1870.

Miller, L., Pacheco, G.J., Janak, J.C., Grimm, R.C., Dierschke, N.A. *et al.* (2018) Causes of death in military working dogs during operation Iraqi freedom and operation enduring freedom, 2001–2013. *Military Medicine* 183(9–10), e467–e474.

Mitchell, D., Snelling, E.P., Hatem, R.S., Maloney, S.K., Strauss, W.M. *et al.* (2018) Revisiting concepts of thermal physiology: Predicting responses of mammals to climate change. *Journal of Animal Ecology* 87(4), 956–973.

Moen, C., Johnson, J.C. and Hackney Price, J. (2022) Ecdysteroid responses to urban heat island conditions during development of the western black widow spider (*Latrodectus hesperus*). *PLoS One* 17(4), e0267398.

Moore, F.L. and Zoeller, R.T. (1985) Stress-induced inhibition of reproduction: Evidence of suppressed secretion of LH-RH in an amphibian. *General and Comparative Endocrinology* 60(2), 252–258.

Moore, B.D., Foley, W.J., Wallis, I.R., Cowling, A. and Handasyde, K.A. (2005) Eucalyptus foliar chemistry explains selective feeding by koalas. *Biology Letters* 1(1), 64–67.

Mukeka, J.M., Ogutu, J.O., Kanga, E., Piepho, H.P. and Røskaft, E. (2022) Long-term trends in elephant mortality and their causes in Kenya. *Frontiers in Conservation Science* 3, 82.

Narayan, E.J. and Hero, J.-M. (2014a) Acute thermal stressor increases glucocorticoid response but minimizes testosterone and locomotor performance in the cane toad (*Rhinella marina*). *PLoS One* 9(3), e92090.

Narayan, E.J. and Hero, J.-M. (2014b) Repeated thermal stressor causes chronic elevation of baseline corticosterone and suppresses the physiological endocrine sensitivity to acute stressor in the cane toad (*Rhinella marina*). *Journal of Thermal Biology* 41, 72–76.

Narayan, E.J., Cockrem, J.F. and Hero, J.-M. (2012) Effects of temperature on urinary corticosterone metabolite responses to short-term capture and handling stress in the cane toad (*Rhinella marina*). *General and Comparative Endocrinology* 178(2), 301–305.

Narayan, E., Sawyer, G. and Parisella, S. (2018) Faecal glucocorticoid metabolites and body temperature in Australian merino ewes (*Ovis aries*) during summer artificial insemination (AI) program. *PloS One* 13(1), e0191961.

Neven, L.G. (2000) Physiological responses of insects to heat. *Postharvest Biology and Technology* 21(1), 103–111.

Okin, G.S. (2017) Environmental impacts of food consumption by dogs and cats. *PLoS One* 12(8), e0181301.

Olin, A.B., Dück, L., Berglund, P.A., Karlsson, E., Bohm, M. *et al.* (2023) Breeding failures and reduced nest attendance in response to heat stress in a high-latitude seabird. *Marine Ecology Progress Series heat* HEATav3.

Olsson, I.A.S. and Keeling, L.J. (2005) Why in earth? Dustbathing behaviour in jungle and domestic fowl reviewed from a Tinbergian and animal welfare perspective. *Applied Animal Behaviour Science* 93(3–4), 259–282.

Paaijmans, K.P., Heinig, R.L., Seliga, R.A., Blanford, J.I., Blanford, S. *et al.* (2013) Temperature variation makes ectotherms more sensitive to climate change. *Global Change Biology* 19(8), 2373–2380.

Pachauri, R. and Meyer, L. (2014) *Climate Change 2014: Synthesis Report. Contribution of Working Groups I, II and III to the Fifth Assessment Report of the Intergovernmental Panel on Climate Change*. Intergovernmental Panel on Climate Change (IPCC), Geneva.

Packer, R.M., O'Neill, D.G., Fletcher, F. and Farnworth, M.J. (2019) Great expectations, inconvenient truths, and the paradoxes of the dog-owner relationship for owners of brachycephalic dogs. *PLoS One* 14(7), e0219918.

Parry, M.L., Canziani, O.F., Palutikof, J.P., van der Linden, P.J. and Hanson, C.E. (eds) (2007) *Climate Change 2007-Impacts, Adaptation and Vulnerability: Working Group II Contribution to the Fourth Assessment Report of the IPCC*, Vol. 4. Cambridge University Press, Cambridge, UK.

Penman, T.D., Pike, D.A., Webb, J.K. and Shine, R. (2010) Predicting the impact of climate change on Australia's most endangered snake, *Hoplocephalus bungaroides*. *Diversity and Distributions* 16(1), 109–118.

Phibbs, D., Groves, P. and Muir, W. (2021) Leg health of meat chickens: Impact on welfare, consumer behaviour, and the role of environmental enrichment. *Animal Production Science* 61(12), 1203–1212.

Phillips, S. and Callaghan, J. (2000) Tree species preferences of koalas (*Phascolarctos cinereus*)in the Campbelltown area south-west of Sydney, New South wales. *Wildlife Research* 27(5), 509–516.

Phillips, S., Callaghan, J. and Thompson, V. (2000) The tree species preferences of koalas (*Phascolarctos cinereus*) inhabiting forest and woodland communities on quaternary deposits in the Port Stephens area, New South Wales. *Wildlife Research* 27(1), 1–10.

Pike, D.A., Antworth, R.L. and Stiner, J.C. (2006) Earlier nesting contributes to shorter nesting seasons for the loggerhead seaturtle, *Caretta caretta*. *Journal of Herpetology* 40(1), 91–94.

Pisc (2002) *Model Code of Practice for the Welfare of Animals: Domestic Poultry*, Vol. 83. CSIRO Publishing, Clayton, Victoria.

Pringle, R.M., Webb, J.K. and Shine, R. (2003) Canopy structure, microclimate, and habitat selection by a nocturnal snake, *Hoplocephalus bungaroides*. *Ecology* 84(10), 2668–2679.

Protopopova, A., Ly, L.H., Eagan, B.H. and Brown, K.M. (2021) Climate change and companion animals: Identifying links and opportunities for mitigation and adaptation strategies. *Integrative and Comparative Biology* 61(1), 166–181.

Raven, J., Caldeira, K., Elderfield, H., Hoegh-Guldberg, O., Liss, P. *et al.* (2005) *Ocean Acidification Due to Increasing Atmospheric Carbon Dioxide*. The Royal Society, London.

Reece, F.N., Deaton, J.W., May, J.D. and May, K.N. (1971) Cage versus floor rearing of broiler chickens. *Poultry Science* 50(6), 1786–1790.

Richards, O.W. and Davies, R.G. (2013) *Imms' General Textbook of Entomology: Volume 2: Classification and Biology*. Springer Science & Business Media, Zurich, Switzerland.

Roggatz, C.C., Lorch, M., Hardege, J.D. and Benoit, D.M. (2016) Ocean acidification affects marine chemical communication by changing structure and function of peptide signalling molecules. *Global Change Biology* 22(12), 3914–3926.

Roy, D. and Sparks, T. (2000) Phenology of British butterflies and climate change. *Global Change Biology* 6(4), 407–416.

RSPCA (2016) *RSPCA Approved Farming Scheme Impact Report 2016*. RSPCA, Canberra.

RSPCA (2018) *RSPCA Approved Farming Scheme*. RSPCA, Canberra.

RSPCA (2020) *RSPCA Impact Report 2020*. RSPCA, Canberra.

Sabine, C.L., Feely, R.A., Gruber, N., Key, R.M., Lee, K. *et al.* (2004) The oceanic sink for anthropogenic CO2. *Science* 305(5682), 367–371.

Santos, R.R., Awati, A., Roubos-van den Hil, P.J., Tersteeg-Zijderveld, M.H., Koolmees, P.A. *et al.* (2015) Quantitative histo-morphometric analysis of heat-stress-related damage in the small intestines of broiler chickens. *Avian Pathology* 44(1), 19–22.

Schlesinger, C.A. and Shine, R. (1994) Choosing a rock: Perspectives of a bush-rock collector and a saxicolous lizard. *Biological Conservation* 67(1), 49–56.

Seabrook, L., McAlpine, C., Baxter, G., Rhodes, J., Bradley, A. *et al.* (2011) Drought-driven change in wildlife distribution and numbers: A case study of koalas in south west Queensland. *Wildlife Research* 38(6), 509–524.

Seebacher, F., Davison, W., Lowe, C.J. and Franklin, C.E. (2005) A falsification of the thermal specialization paradigm: Compensation for elevated temperatures in antarctic fishes. *Biology Letters* 1(2), 151–154.

Şekercioğlu, Ç.H., Primack, R.B. and Wormworth, J. (2012) The effects of climate change on tropical birds. *Biological Conservation* 148(1), 1–18.

Shane, S. (2001) Versatility of broiler growing cages. *Zootecnica International* 8, 18–19.

Shields, S. and Greger, M. (2013) Animal welfare and food safety aspects of confining broiler chickens to cages. *Animals* 3(2), 386–400.

Shine, R., Webb, J.K., Fitzgerald, M. and Sumner, J. (1998) The impact of bush-rock removal on an endangered snake species, *Hoplocephalus bungaroides* (*Serpentes: Elapidae*). *Wildlife Research* 25(3), 285–295.

Simmons, N. and Cirranello, A. (2020) Bat species of the world: A taxonomic and geographic database. *Accessed On* 7(10), 2020.

Slepukhin, V., Galpern, I. and Cherepanov, S. (2000) Breeding Russian broilers to adapt them to the cage environment. *World Poultry* 16(10), 25–27.

Snoyman, S., Muhic, J. and Brown, C. (2013) Corrigendum to "nursing females are more prone to heat stress: Demography matters when managing flying-foxes for climate change" [Appl. Anim. Behav. Sci. 142 (December (1–2)) (2012)90–97]. *Applied Animal Behaviour Science* 145(3–4), 138–138.

Stebbins, R.C. and Cohen, N.W. (2021) *A Natural History of Amphibians*. Princeton University Press, Princeton, NJ.

Stella, J.L. and Croney, C.C. (2016) Environmental aspects of domestic cat care and management: Implications for cat welfare. *Scientific World Journal* 2016, 6296315.

Su, B., Martens, P. and Enders-Slegers, M.-J. (2018) A neglected predictor of environmental damage: The ecological paw print and carbon emissions of food consumption by companion dogs and cats in China. *Journal of Cleaner Production* 194, 1–11.

Tattersall, G.J., Sinclair, B.J., Withers, P.C., Fields, P.A., Seebacher, F. *et al.* (2012) Coping with thermal challenges: Physiological adaptations to environmental temperatures. *Comprehensive Physiology* 2(3), 2151–2202.

Taylor, N. and Signal, T.D. (2009) Willingness to pay: Australian consumers and 'on the farm' welfare. *Journal of Applied Animal Welfare Science* 12(4), 345–359.
Telemeco, R.S., Elphick, M.J. and Shine, R. (2009) Nesting lizards (*Bassiana duperreyi*) compensate partly, but not completely, for climate change. *Ecology* 90(1), 17–22.
Travis, D., Sloan, D. and Hughes, B. (1983) Bone fragility in broilers as affected by pen height, sex, and a comparison of left and right humeri. *Poultry Science* 62(10), 2117–2119.
Tyndale-Biscoe, C.H. (2005) *Life of Marsupials*. CSIRO Publishing, Clayton, Victoria.
Ullrey, D.E., Crissey, S.D. and Hintz, H.F. (1997) *Elephants: Nutrition and Dietary Husbandry*. Nutrition Advisory Group East Lansing, Brookfield, IL.
Visser, M.E. and Both, C. (2005) Shifts in phenology due to global climate change: The need for a yard-stick. *Proceedings of the Royal Society B: Biological Sciences* 272(1581), 2561–2569.
Wabeck, C. and Littlefield, L. (1972) Bone strength of broilers reared in floor pens and in cages having different bottoms. *Poultry Science* 51(3), 897–899.
Wake, D.B. and Koo, M.S. (2018) Amphibians. *Current Biology* 28(21), R1237–R1241.
Wang, Y., Li, C., Pan, C., Liu, E., Zhao, X. *et al.* (2019) Alterations to transcriptomic profile, histopathology, and oxidative stress in liver of pikeperch (*Sander lucioperca*) under heat stress. *Fish & Shellfish Immunology* 95, 659–669.
Webb, J.K., Brook, B.W. and Shine, R. (2002a) Collectors endanger Australia's most threatened snake, the broad-headed snake *Hoplocephalus bungaroides*. *Oryx* 36(2), 170–181.
Webb, J.K., Brook, B.W. and Shine, R. (2002b) What makes a species vulnerable to extinction? Comparative life-history traits of two sympatric snakes. *Ecological Research* 17(1), 59–67.
Welbergen, J.A., Klose, S.M., Markus, N. and Eby, P. (2008) Climate change and the effects of temperature extremes on Australian flying-foxes. *Proceedings of the Royal Society B* 275(1633), 419–425.
WHO (2020) *The State of Food Security and Nutrition in the World 2020: Transforming Food Systems for Affordable Healthy Diets*, Vol. 2020. FAO.
Wilson, G.R. and Edwards, M. (2019) Professional kangaroo population control leads to better animal welfare, conservation outcomes and avoids waste. *Australian Zoologist* 40(1), 181–202.
Winkler, D.W., Dunn, P.O. and McCulloch, C.E. (2002) Predicting the effects of climate change on avian life-history traits. *Proceedings of the National Academy of Sciences* 99(21), 13595–13599.
Woinarski, J. and Burbidge, A. (2016) *Phascolarctos cinereus*. *The IUCN Red List of Threatened Species 2016* e. T16892A21960344.
WWF (2022) *Effects of Climate Change*. WWF, Gland, Switzerland.
Xin, H., Berry, I.L., Tabler, G.T. and Costello, T.A. (2001) Heat and moisture production of poultry and their housing systems: Broilers. *Transactions of the ASAE* 44(6), 1851.
Ying, L., Baoliang, L., Ce, S. and Guoxiang, S. (2015) Recirculating aquaculture systems in China-current application and prospects. *Fisheries and Aquaculture Journal* 6(3), 1.
Yoro, K.O. and Daramola, M.O. (2020) CO_2 emission sources, greenhouse gases, and the global warming effect. In: *Advances in Carbon Capture*. Elsevier, pp. 3–28.
Zanker, L. (2021) Kangaroos, drought and a frustrated landowner. *Ecological Management & Restoration* 22, 66.
Zhan, Y., Li, J., Sun, J., Zhang, W., Li, Y. *et al.* (2019) The impact of chronic heat stress on the growth, survival, feeding, and differential gene expression in the sea urchin *Strongylocentrotus intermedius*. *Frontiers in Genetics* 10, 301.
Zizzari, Z.V. and Ellers, J. (2011) Effects of exposure to short-term heat stress on male reproductive fitness in a soil arthropod. *Journal of Insect Physiology* 57(3), 421–426.

11 Conventional breeding of Pacific Island staple crops: A paradox

Vincent Lebot[1]*, Birte Komolong[2], Jean-Pierre Labouisse[3], Floriane Lawac[4], Juliane Kaoh[4], Logotonu Meleisea Waqainabete[5], Amit Sukal[5] and Grahame Jackson[6]

[1]CIRAD-AGAP-VARTC, PO Box 946, Port-Vila, Vanuatu; [2]NARI, National Agricultural Research Institute, Papua New Guinea; [3]CIRAD UMR AGAP Institut, Montpellier 34398, France; [4]VARTC, Vanuatu Agricultural Research and Technical Centre, PO Box 231, Luganville, Santo, Vanuatu; [5]Centre for Pacific Crops and Trees (CePaCT), Pacific Community (SPC), Private Mail Bag, Suva, Fiji; [6]PestNet, 24 Alt Street, Queens Park, NSW, Australia

Abstract

The Pacific Islands (Melanesia, Micronesia, Polynesia) were the last region in the world to be colonized by modern humans. In their migrations, they dispersed asexually propagated food plants that were first domesticated in Melanesia and exposed them to further selection in diverse insular environments. Later, staple food crops were introduced from South America. In our search for research on genetic resources characterization and breeding conducted between 1950 and 2022, we retrieved 338 references on 13 crop species. All studies were examined, and a review is presented here. Relatively few were on genetic improvement (92 references) compared to those on morphological diversity (246 references). The paradox is that the Pacific is considered a region exposed to the climatic crisis, perhaps more vulnerable than any other region in the world. This paradox is surprising when we consider that most people in the Pacific are growing staple food crops, for household use and/or for local markets, and occasionally for export. In some countries, it involves more than 75% of the population with a considerable contribution to gross domestic product (GDP). Plant breeding has not been considered a priority to mitigate against forthcoming changes in climate, which is likely to be detrimental to crops with a narrow genetic base. We suggest several possible reasons and that conventional plant breeding should be given high priority based on new approaches that include geographical distribution of allelic diversity.

Keywords: asexually propagated crops, breeding, genetic improvement, recurrent selection, Oceania

Introduction

The countries in this review are the 22 Pacific island states and territories (PICTs) of the Pacific Community (Pacific Community, 2020). They form three ethnogeographic groupings within Oceania: Melanesia, Polynesia and Micronesia (Fig. 11.1). The population is approximately 12.8 million on a total land mass of 541,000 km^2. The countries differ greatly in size and populations, but have several common features, including decreasing amounts of arable land

*Corresponding author: lebot@vanuatu.com.vu

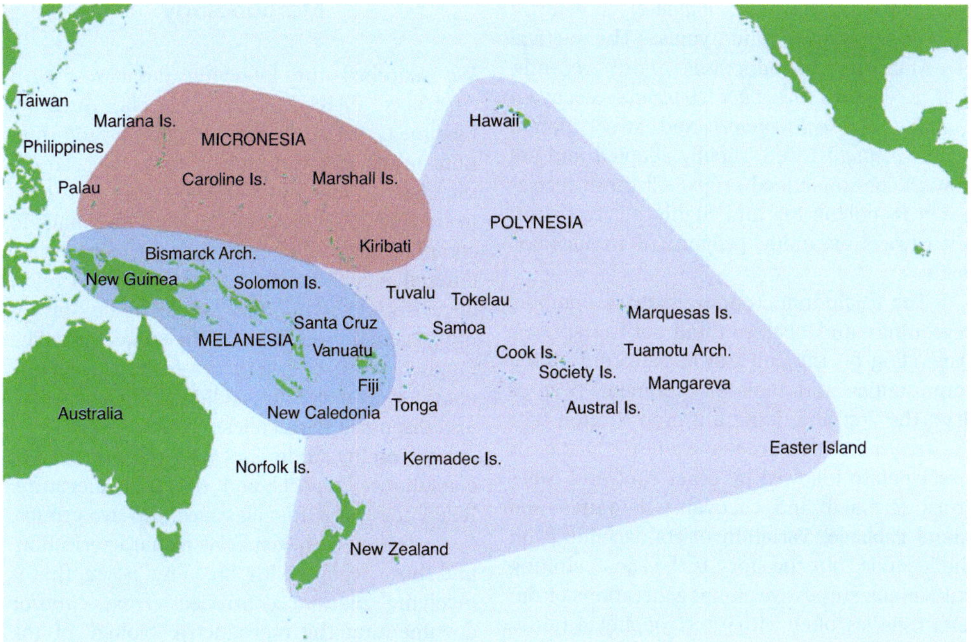

Fig. 11.1. Map of Oceania (downloaded from the internet, public domain).

because of rapid population growth, increasing urbanization and vulnerability to changing climates (Georgeou *et al.*, 2022).

Oceania was the last region in the world to be inhabited by modern humans. The people came by way of the *Sahul* plate connecting Australia, New Guinea and Tasmania at least 45,000 BP (Choin *et al.*, 2021). These hunters-gatherers from South-east Asia probably found the tropical rainforests of New Guinea familiar to them. However, many of the plants were new. On the northern part of the plate, wild bananas (*Musa* spp.) and roots and tubers were among the first plant species domesticated. They could be roasted without the need for cooking containers, stored *in situ* by delaying harvest, and provided a degree of tolerance to extreme weather events, not being destroyed completely by cyclones. Direct selection from wild forms occurred and, by vegetative propagation, domestication was rapid.

Stone tools recovered in New Guinea, and dated to about 46,000 BP, have been shown to have abundant starch grains of yams (*Dioscorea alata*, *D. bulbifera*, *D. esculenta*, *D. nummularia* and *D. pentaphylla*) (Summerhayes *et al.*, 2010) indicating that humans were already exploiting their tubers at that time. Further east, in the northern Solomon Islands, starch granules of aroids (*Colocasia esculenta* and *Alocasia macrorrhizos*) have been found on stone tools dated 28,000 BP (Loy *et al.*, 1992). Much later, about 4000 BP, Lapita People from Asia (Taiwan and northern Philippines) colonized the Bismarck Archipelago, a group of islands off the north-east coast of New Guinea, and from there penetrated Remote Oceania, sailing eastwards across Polynesia to reach Tahiti, the Marquesas and Easter Island, north to Hawaii and south to New Zealand (Kirch, 2017). In their migrations, they dispersed food plants that were first domesticated in Melanesia and in these new locations exposed them to further selection (Lebot, 1999; Whistler, 2009). The coconut (*Cocos nucifera*) was naturally established on the islands prior to human arrival (Harries *et al.*, 2002). It contributed to Lapita colonization, being a precious source of water, calories, lipids, fibres and an exceptional multi-purpose tree.

Nowadays, the most important staple crops in the Pacific correspond to ancient Melanesian domesticates: aroids, bananas, island cabbage (or aibika, *Abelmoschus manihot*), breadfruit (*Artocarpus altilis*) and yams. The arrival of American introductions more recently, that is cassava (*Manihot esculenta*), cocoyam (*Xanthosoma sagittifolium*) and sweet potato (*Ipomoea batatas*) were rapidly adopted and are now major staple food crops. All these species are cross-pollinating and highly heterozygous, but always asexually propagated in farmers' fields.

The traditional cropping systems combine vegeculture and local fruit and nut tree species. Depending on the cultural background of the communities and their environment (wet or dry), the first and dominant crop is often taro (*C. esculenta*), the greater yam (*D. alata*) or sweet potato followed by other root and tuber crops (cassava and cocoyam), bananas and island cabbage. Variation exists depending on the location, but the story is the same: shifting cultivation, successive clonal generations of different species, often with intercropping of minor species and trees, and progressive and unmanaged fallow to restore soil fertility.

One of the major objectives of the cropping systems is to manage environmental risks. Fruit and nut trees are included with food plants that lack tap roots and, therefore, are highly susceptible to sudden drought. The trees also provide windbreaks to damaging tropical depressions and cyclones. Oceania is often cited as one of the most vulnerable places on Earth due to the instability of the El Niño Southern Oscillation (ENSO). ENSO events vary greatly in rainfall, cyclonic strength and occurrence, and are likely to become even harder to predict as global warming gathers pace, with its warmer air temperatures and increased temperature extremes. Since 1960, all PICTs have exhibited warming trends with the last decade being the warmest. These trends are projected to continue and become more erratic and unpredictable (Lough *et al.*, 2016).

Considering the cultural significance of the Pacific staple food crops, but also the socio-economic projections of the region due to its vulnerability to climate uncertainties, plant breeding would seem an approach with the potential to strengthen smallholders' capacity to adapt to uncertain circumstances. The aim of this review is to present, for each of the most important crops, the achievements to date.

Methodology

We searched the following databases: CAB Abstracts, CAB Heritage, Agricola, Agritrop, Medline (PubMed) and used Google for information on 12 different staple crops. We also did searches on coconut, an important cash crop in the PICTs, in order to have a benchmark to compare with. The following keywords were entered in each database to obtain lists of references from 1950 to 2022: 'species name', 'breeding', 'genetic resources', 'Oceania', 'Pacific'. Further, we used references from the articles obtained to check for additional material. We also discussed with colleagues and checked for any upcoming studies not yet published or from consultants' reports and the grey literature. References were then classified into two groups, those dealing with 'germplasm characterization' and those dealing with 'breeding' *per se*, that is involving manual controlled crosses and/or documenting the reproductive biology of the species. For each species, all references were read and analysed and the most important points are reported in this review. The complete list of references is presented in Supplementary Material 1. As there are 338 references, they cannot be all cited in the following review and only the most recent or significant ones are cited for each species.

Results

The number of studies retrieved per species from 1950 to 2022, for plant genetic resource characterization and breeding, is presented in Table 11.1.

Aroids (*Colocasia esculenta, Alocasia macrorrhizos, Cyrtosperma merkusii, X. sagittifolium*)

Most studies deal with taro (*C. esculenta*) and the use of morphological and molecular markers (isozymes, AFLPs, RAPDs, SSRs, DaRTs, ITS) to characterize the genetic diversity of the Pacific and to compare it with the diversity of Asia. Where breeding has occurred, it has been to

Table 11.1. The number of references retrieved from all databases for each species (from 1950 to 2022) regarding germplasm characterization and breeding and genetic improvement activities in the PICTs.

Crop	Species	Ploidy	Germplasm characterization	Breeding	Total
Taro	*Colocasia esculenta*	2×, 3×	39	33	72
Giant taro	*Alocasia macrorrhizos*	2×	4	1	5
Swamp taro	*Cyrtosperma merkusii*	2×	1	–	1
Cocoyam	*Xanthosoma sagittifolium*	2×, 3×	-	–	–
Greater yam	*Dioscorea alata*	2×, 3×, 4×	16	2	18
Lesser yam	*Dioscorea esculenta*	2×	1	–	1
Pacific yam	*Dioscorea nummularia*	4×, 6×, 8×	1	–	1
Bananas/ plantains	*Musa* spp.	2×, 3×, 4×	50	–	50
Sweet potato	*Ipomoea batatas*	6×	30	10	40
Cassava	*Manihot esculenta*	2×, 3×	9	1	10
Island cabbage	*Abelmoschus manihot*	2×	6	1	7
Breadfruit	*Artocarpus altilis*	2×, 3×	37	–	37
Coconut	*Cocos nucifera*	2×	52	44	96
Total			246	92	338

deal with TLB, taro leaf blight (*Phytophthora colocasiae*) (Lebot and Ivancic, 2022). However, the first attempt at breeding aroids in the Pacific was with giant swamp taro (*Cyrtosperma merkusii*). In the mid-1970s, at the request of the Government of Kiribati, the University of the South Pacific (USP), Alafua Campus, Samoa, was asked to breed *C. merkusii* with tolerance to salinity. Saltwater was said to be seeping into the pits traditionally used to raise giant swamp taro in that part of Micronesia. As there was no local experience in plant breeding, and it was not possible to do the work at a distance, the seed was produced in Samoa and sent to Kiribati for testing. Unfortunately, there was no successful outcome.

The focus then turned to taro and in 1976, UNDP/FAO and SPC established a regional root crop programme, which continued until 1987 (Guarino and Jackson, 1986). *Ex situ* field collections were established in almost all countries and morpho-agronomic characterization was conducted with standardized IPGRI descriptors. However, the work did not result in an understanding of the true diversity, which exists in the region; this had to wait for a thorough genetic analysis, first using isozymes and then more instructive molecular markers. After 1987, the work lapsed and failure to support the collections resulted in many losses.

Later, in the mid-1990s, the work was resurrected with the establishment of two regional networks: TaroGen (*Taro Genetic Resources: Conservation and Utilisation*), funded by AusAid, and TANSAO (*Taro Network for Southeast Asia and Oceania*), funded by the EU. These programmes had slightly different objectives, but combined, they collected and described, morphologically and by molecular means, the taro of the Pacific Islands, including that of Papua New Guinea. They also made core collections representing the genetic diversity of these regions and ensured their safety by storing them in a purpose-built facility, opened in 1998, that became known as the Centre for Pacific Crops and Trees (CePaCT).

In time, this attracted support from the Global Crop Diversity Trust (Berlin, Germany), with an annual grant in perpetuity. CePaCT, based at SPC, Fiji, maintains *in vitro* more than one thousand accessions of taro, mostly from the Pacific region (Ebert and Waiqanabete, 2018).

The results from the collecting and characterization showed high morpho-agronomic variation among cultivars (vegetative and corm flesh colours, plant height, leaf shape, number of suckers, flowering and spathe colour) in comparison with wild taro, but allelic diversity is poor in comparison. Some cultivars are somatic mutants selected clonally, but most result from spontaneous cross-pollinations in farmers' fields, and the selection of volunteer seedlings by farmers seeking morphological variants, which they test and keep if found to have acceptable characteristics.

Compared to South-east Asia, there is very limited allelic diversity in the Pacific Islands (Kreike *et al.*, 2004). All taro cultivars from the Pacific region are diploid of dasheen types with a single, well-developed central corm. Over the last decades, a few triploids (of eddoe-types with small lateral cormels) have been introduced from temperate Asia to attempt to develop an export crop, but, unfortunately, these triploids are not adapted to wet and warm climates.

Compared to the efforts put into conserving germplasm, utilization of the diversity has not been as great. There have been attempts at breeding over the past 50 years, but they have not all fulfilled their initial promise. Breeding programmes were initiated in Samoa and Solomon Islands in the late 1970s, in Hawaii, Papua New Guinea, and Fiji in the 1980s, and Vanuatu in the 1990s. At first, the technical bottleneck was the absence of flowering in most cultivars, but when a protocol was reported from the Caribbean using gibberellic acid in 1975, this changed the situation (Lebot and Ivancic, 2022). In the spate of a few years, several programmes were established, and new lines were released.

In the Solomon Islands in the late 1970s, the programme produced *Agriculture Taro*, a cultivar with tolerance to TLB from a cross between wild taro introduced from Bangkok, Thailand and local cultivars. In Samoa, *Alafua Sunrise* was released in 1988, a product from crossing between local cultivars, and, in Fiji, similar work saw *Samoa Hybrid* released in 1984, followed by three others within the next decade. All have continued to be popular in local markets. In Papua New Guinea, breeding was for tolerance to TLB which had been present since World War II and was widespread. The programme began in 1993 with the transfer of the breeder and seed from the Solomon Islands. Initially, breeding made slow progress. Wild traits, such as profuse stolon production, corm acridity and low yields, were common in progenies from controlled backcrosses between the wild variety from Bangkok and local cultivars (Lebot and Ivancic, 2022). Later, under TaroGen, recurrent phenotypic selection was invoked to speed up the chances of successful outcomes. A similar strategy was used in Vanuatu under TANSAO aiming at introducing good quality cultivars from South-east Asia with tolerance to TLB.

In parallel with breeding in Papua New Guinea, and also under TaroGen, breeding in Samoa aimed to overcome TLB that had been introduced in 1993 and which had annihilated a multi-million-dollar export industry within 6 months. Here, in contrast to Papua New Guinea, a participatory approach was used with farmers making selections within populations derived from crosses between local cultivars and those from Micronesia and the Philippines, which had been identified by DNA fingerprinting to be unlike taro from other Pacific island countries. In this way, acceptable progeny for domestic and export markets was produced within 5 years.

Samoa has benefitted from being able to grow taro once again. The cultivars now grown may not be as favoured as those grown before TLB, but they are acceptable to households, domestic and export markets. Since 1993, Fiji has taken over the export market but, unfortunately, has increased its vulnerability to TLB. It is now growing in large areas with the same export cultivar, *Niue*, grown previously by Samoa. In Fiji, it is known as *Tausala ni Samoa*. This unfortunate situation concerns the innate features of root and tuber crops.

Distributing clonally propagated crops is unlike distributing seeds of crops that produce sexually. It is expensive. If TLB is not present, there may be no incentive to distribute a few plants to each grower as insurance against an outbreak, and to convince farmers to keep them.

Fiji realises the problem, and some of the new TLB-resistant Samoa cultivars have been distributed through official schemes, but it remains to be seen whether farmers keep them.

Today, taro breeding rarely occurs due to the absence of financial support and the lack of pressing demand from the industry and growers. Where it is maintained, it is to cross the new TLB-tolerant hybrids with local cultivars. In most countries, farmers still work with traditional cultivars. There are no seed industries in PICTs, so the wide distribution of improved taro hybrids remains a slow process between and within countries (Lebot and Ivancic, 2022).

The giant taro (*Alocasia macrorrhizos*) is another important aroid in central Polynesia (Tonga, Samoa, Wallis and Futuna) but is poorly documented. Only five references were retrieved, four dealt with taxonomy and cladistics and only one with breeding. The swamp taro (*Cyrtosperma merkusii*) is important only on atolls and only one reference was retrieved from databases. No breeding programme was developed on *C. merkusii*. No references were retrieved for elephant foot yam (*Amorphophallus paeoniifolius*) and cocoyam (*Xanthosoma sagittifolium*) although cocoyam hybrids have been produced through controlled hand pollination by VARTC (the Vanuatu Agricultural Research and Technical Center) and were evaluated prior to their distribution to growers.

Yams (*Dioscorea* Spp.)

The greater yam (*D. alata*) has very high cultural significance, especially in Melanesia, where hundreds of cultivars exist. Most of the work conducted over recent decades has aimed at characterizing genetic resources with morpho-agronomic descriptors, molecular markers, major compounds and secondary metabolites. The results have shown the region to be one of the great diversities (Chaïr *et al.*, 2016). It did not immediately lead to breeding, however.

Like all *Dioscorea* species, the greater yam is dioecious (male and female flowers on different plants) with variable ploidy levels (di-, tri- or tetraploids). Before controlled hand crosses can be made, male and female plants with compatible ploidy levels need to be identified. Of the hundreds of accessions collected in the 1980s very few produced flowers, and those that did were mostly male. Consequently, ploidy levels were not determined as breeding was not considered a viable option. The focus was only on selecting accessions against the major constraint of yam production, the anthracnose fungus, *Colletotrichum gloeosporioides*.

However, in 2012, a few fertile tetraploid cultivars were identified in Vanuatu, creating an opportunity for polyploidy breeding (Abraham *et al.*, 2012). Unfortunately, when crosses were made with local male plants, the local female plants were so susceptible to anthracnose that they were unable to develop to the stage of producing mature capsules with fertile seeds. The answer was to test female plants from another source.

In collaboration with the CTCRI (Central Tuber Crops Research Institute, Trivandrum, Kerala, India), hundreds of botanical seeds collected from open-pollinated plants were introduced to Vanuatu. Most seedlings (more than 96%) produced long, deep-forming tubers with oxidizing flesh, which are unacceptable traits to farmers. Few produced compact tubers and female flowers.

However, 20 were selected as parents and crossed to Vanuatu male plants. Numerous hybrids were produced but, again, most were unsuitable because of oxidizing tuber flesh. Nevertheless, after four cycles of controlled crosses, acceptable hybrids (resistant to anthracnose, with compact tuber shape and non-oxidizing tuber flesh) were produced, and these were propagated and released to farmers.

Controlled crosses were also conducted between *D. alata* and *D. nummularia* (the closest related species) and vigorous hybrids were produced (Lebot *et al.*, 2019). Unfortunately, these too showed oxidizing tuber flesh.

The Vanuatu yam programme is no longer active, and no country in the region has benefited from this ground-breaking research. Movement of yam germplasm from one country to another is problematic due to the presence of different strains of common viruses, requiring transfer through internationally recognized quarantine centres where accessions can be checked to mitigate risk. No breeding activities on the minor yams (*D.*

bulbifera, D. esculenta, D. pentaphylla) have been conducted.

Bananas and Plantains (*Musa* Spp.)

Bananas and plantains are major foods in the Pacific and their genetic resources have been well documented because of their global importance. Wild bananas (*Musa acuminata* subspecies) are endemic in New Guinea. There are many fertile diploid cultivars with sexual seed formation and parthenocarpy, and these genotypes are of great interest to banana breeders worldwide. Of particular interest is *M. acuminata* subsp. *banksii*. It occurs on the northern *Sahul* plate and in the oceanic islands of Samoa. *M. schizocarpa* and *M. acuminata* subsp. *errans* are also of interest; they are endemic to New Guinea, and hybridization, genetic recombinations and introgressions between them are frequent (Perrier *et al.*, 2011).

Also studied have been the Pacific plantains, known by their Hawaiian names of *Iholena*, *Maoli* and *Popo'ulu*; these are triploids (AAB) in the international classification. They are dispersed throughout Melanesia, Polynesia and Micronesia and each genotype presents numerous cultivars known under different local names and selected clonally as somatic mutants. These plantains were likely first domesticated in Melanesia, as molecular markers reveal that the *M. acuminata* contribution to their AAB genome comes from subsp. *banksii* (Lebot *et al.*, 1993). Another important banana is *M. troglodytarum* (syn. *M. fehi*). It is unique to the Pacific, has erect bunches and shows significant genetic diversity in Melanesia and Eastern Polynesia (especially in Tahiti and the Marquesas).

It is apparent that most research has been directed at germplasm collecting, characterization, classification and preservation of the Pacific's unique banana genetic resources by international centres. Within the PICTs, Pacific plantains and *Fehi* types are being replaced by more vigorous exotic varieties (mostly FHIA hybrids), and local cultivars are disappearing. What has been collected has not been retained locally, and there has been no attempt at breeding.

Sweet Potato (*Ipomoea batatas*)

Most references deal with germplasm characterization and only a quarter with breeding (Table 11.1). The way of introduction of sweet potatoes to the Pacific has attracted the attention of archaeologists and anthropologists for many years, and an impressive volume of research has been produced. In this region, the sweet potato is an ancient introduction from South America, but how it arrived is an enigma. The current theory is that Polynesians reached South America and brought back the sweet potato from Peru, with its name, *Kumara*. Archaeological evidence from Easter Island and Hawaii suggests that its introduction pre-dates western contact and that sweet potato reached Eastern Polynesia and New Zealand, where it was the staple of the Maoris, about 1000 years ago. When Captain Cook arrived in Tahiti and Hawaii, it was already widely cultivated. How far the sweet potato travelled westwards is still debated, although it is known from etymological studies that the name *kumala* is used in local languages in Vanuatu and as *kumara* in Solomon Islands, the same or similar to its name in Polynesia.

In the West Pacific, introductions arrived by a different route, possibly from Portuguese colonies to New Guinea, which was also the destination of sweet potato from the Philippines, brought there by Spanish galleons from their base in Mexico. This is supported by DNA studies that have shown that Papua New Guinea cultivars are principally derived from the Northern neotropical gene pool of Central America rather than from Southern Peru, the region where the Pacific cultivars are thought to have originated (Roullier *et al.*, 2012).

However, it is likely that introductions from other areas eventually reached Papua New Guinea where sexual recombination between them played a prominent role in the country becoming a second centre of diversity of sweet potato. Natural cross-pollination between different genotypes is frequent in farmers' fields, and seedlings exhibiting morphological variation are readily cloned and evaluated by farmers for palatability and other important traits (Roullier *et al.*, 2013). It is not uncommon for a farmer to cultivate a dozen or more cultivars and exchange them with others to diversify varietal portfolios.

Breeding to improve sweet potato production only began within the PICTs in the early 1980s. At that time, the crop in Tonga was devastated by the scab fungus, *Elsinoe batatas*, and in 1981, a programme by USP, Samoa, began to breed for resistance. Success was achieved rapidly by selecting local susceptible cultivars and crossing them with plants derived from seedlings of open-pollinated seed introduced from the Solomon Islands.

Breeding in Vanuatu in the next decade was investigated using selections and breeders' lines from IITA, Nigeria, in hybridizations with local cultivars. Breeders' lines TIB2, TIB9 and TIB10 performed well under local conditions and were used as parents. Further exploration of broadening the genetic base occurred in the mid-1990s with several hundred seeds introduced from the CIP Regional Office, Bogor, Java, Indonesia. Genotypes tolerant to scabs were selected and used as parents. Polycross blocks, composed of carefully selected parents, were established; plants were staked to induce flowering and pollination and the half-sib progenies evaluated. In this way, major traits were easily captured through recurrent phenotypic selection. They are scab tolerant and possibly tolerant to weevils (*Cylas formicarius*). It appears that purple-red-skinned lines, rich in chlorogenic acids and coumarins, present a degree of protection. Improved hybrids have been released to farmers in Vanuatu (Lebot *et al.*, 2021).

Cassava (*Manihot esculenta*)

Cassava was introduced to the PICTs in the early 1800s from different geographical sources by missionaries and/or European settlers. It gained importance rapidly because of its ease of cultivation and, in some islands, it has already replaced traditional root crops, especially yam in lowlands areas exposed to droughts. All references deal with the characterization of germplasm and none with breeding. Of the three subspecies of cassava – *M. esculenta* ssp. *esculenta* (which corresponds to cultivated cassava) and two wild subspecies, *flabellifolia* and *peruviana* – only ssp. *esculenta* has been widely distributed in the Pacific Islands. The wild *M. glaziovii* has been introduced as an ornamental providing well-appreciated shade in most villages, but hybrids with *M. esculenta* are unknown.

Most sources report that only sweet cassava varieties have been introduced to the Pacific, except for a very few bitter varieties (i.e. those presumed to be high in cyanogenic glycosides) that have been introduced to parts of northern Micronesia. Commonly, farmers eliminate seedlings with a bitter taste when they chew the fresh roots of newly identified variants.

Studies conducted with molecular markers have revealed significant allelic diversity that is surprising for such a recently introduced species (Sardos *et al.*, 2008). Cassava flowers spontaneously and is predominantly allogamous with cross-pollination occurring frequently. Flowers and fruits are known to farmers who use their formation as a sign of plant maturity and time for harvest. Naturally occurring cassava triploids are high yielding but are more difficult to peel than diploids (Sardos *et al.*, 2009). Their very high dry matter content makes them suitable for flour production but they are prone to rapid postharvest physiological decay (PPD), like most landraces high in dry matter. Cassava is valued as a foodbank stored in the ground and some landraces can stay for up to 3 years.

Farmers report that new cultivars appear on a regular basis. It seems that the high diversity observed is a fairly recent event. This is confirmed by the limited number of accessions reported in the early 1980s by UNDP/FAO-funded projects and the high diversity observed nowadays in farmers' fields. Farmers frequently collect seedlings from gardens planted to different genotypes. The most vigorous or attractive plants are cloned and evaluated, especially for taste, sticky texture, early maturity, ability of roots to remain in the soil without deteriorating up to 2–3 years, high yields, compact clusters of roots around the basal stem and ease of harvest. All cultivars are therefore clones of hybrids, and these hybrids are often exchanged between farmers as stem cuttings. This dynamic traditional system of exchanging cultivars leads to an efficient distribution of new genotypes.

Fortunately, cassava is virtually free of diseases in all PICTs, so breeding is not needed to overcome viruses and phytoplasmas that affect cassava elsewhere. However, in a region prone to cyclones, plant architecture is important and breeding in Vanuatu is to evaluate half-sib progenies produced in polycross blocks for low stature.

Island Cabbage, Bélé, Aibika (*Abemoschus manihot*)

This green leafy vegetable is a necessary complement to starch-based traditional foods in the Pacific. It is rich in proteins and minerals and is highly appreciated and consumed daily, especially in Melanesian countries. Most consumers prefer green cultivars with no anthocyanin pigmentation as they are thought to be responsible for bitterness. Genetic resources have been documented and Melanesia appears as an area of diversity for this species with more morpho-agronomic variation documented than in any other geographical region (Bruerre, 2010). In Vanuatu, about 225 accessions were collected, photographed and characterized morphologically using standardized descriptors. Cultivars flower annually and produce capsules typical of the Malvaceae family with fertile seeds. It is thought that most cultivars are diploids with 66 chromosomes. Not much is known of the allelic diversity found in the region. Breeding has not been carried out in the Pacific.

Breadfruit (*Artocarpus altilis*)

Breadfruit was first identified in Tahiti by Sir Joseph Banks, a botanist travelling with Captain Cook. The yield of the Tahitian seedless (triploids) trees was so impressive that it was soon considered as a practical solution to feed the slaves of the British colonies in the Caribbean. This led to the famous, but ill-fated, Bounty breadfruit-collecting expedition organized by the British Admiralty.

The origins of breadfruit are complex, with both *A. camansi* and *A. mariannensis* involved to different degrees in different parts of the Pacific. It appears that cultivars of breadfruit have arisen in two ways. In Melanesia and Polynesia, they are derived from a single species, *A. camansi*, while in Micronesia, breadfruit originated by hybridization of seeded *A. altilis* cultivars derived from *A. camansi* with *A. mariannensis* as the other parent. In both regions, human selection over many thousands of years led to the morphological variation observed today, which includes both sterile and fertile diploids and sterile triploids. Triploids resulted from unreduced gametes of one of the two progenitors (Ragone, 2001).

Cultivated seeded types grow spontaneously throughout the Melanesian Islands of Solomon Islands and Vanuatu where selected trees are often cloned using root cuttings. Some of these trees present very small, aborted seeds but it is unknown if they are seedless triploids. The selection of triploids seems to have occurred in central Polynesia where clones with variable morphological traits but with a narrow genetic base were selected and dispersed clonally (Zerega *et al.*, 2004). These triploids are mostly cultivated in Samoa, the Society and the Marquesas Islands. So far, no breeding activities have been attempted.

Coconut (*Cocos nucifera*)

As breeding material, the coconut varieties are generally grouped according to their growth habit. Tall varieties are mostly cross pollinating and flower 6–10 years after planting. Dwarf varieties have a shorter stature, are normally self-pollinating and start bearing about 3 years after planting. Tall varieties dominate in the landscapes of the Pacific Islands and as cash crops in the plantations, while Dwarf varieties and some rare 'intermediate types' are confined to gardens and backyards.

The coconut has attracted significant research and extension efforts from donors (Table 11.1), as copra has been an export cash crop since the early 20th century. The first known attempt to improve the coconut palm in the Pacific dates back to 1926 in Fiji with Maréchal who produced a hybrid by crossing the Niu Leka Dwarf (NLAD), originally present in Fiji, Tonga and Samoa, and a Malayan Red Dwarf (MRD).

Ambitious research programmes started at the beginning of the 1960s with the establishment of large germplasm collections in experiment stations dedicated to coconut in Solomon Islands (Yandina), Vanuatu (Saraoutou), Papua New Guinea (Madang) and Fiji (Taveuni). Coconut breeding programmes mainly focused on improving copra yield by crossing Dwarf and Tall varieties of different origins. The most widespread hybrid was the MRD ×Rennell Island Tall (RIT) produced in the seed gardens of several PICTs. However, seed garden management and hybrid seeds and seedling diffusion are particularly costly, and these programmes had limited impact on overall production.

The Saraoutou research station, now part of VARTC, was established in 1962 in Vanuatu by IRHO (*Institut de Recherches pour les Huiles et Oléagineux*), a French research institute. Breeding the local Tall (Vanuatu Tall or VTT), by mass selection and controlled crossing of the best genitors, resulted in an elite Vanuatu Tall population with a higher yield (+56%) compared to the unselected VTT. Nuts of elite VTT are easily produced in open pollination and have been widely released in the country since 2018 with the support of *Vanuatu Value Chains program* (VaVaC) financed under the 11th European Development Fund. It has been suggested that decentralized production, at the farmer level, of this elite population would result in significant improvement in productivity (Labouisse *et al.*, 2011).

Coconut foliar decay (CFD), a lethal disease endemic to Vanuatu, has seriously hampered the coconut breeding programme of this country. The causal agent is a DNA single-strand virus, transmitted by *Myndus taffini*. Although all populations of Vanuatu Tall are fully tolerant, all exotic varieties and hybrids were found susceptible to different degrees, except two hybrids. The first one, the Vanuatu Red Dwarf ×Vanuatu Tall (VRD × VTT), has a high-yielding potential, but its dissemination was discontinued in 1996 due to several drawbacks (low germination rate, low copra content and frequent dropping of still immature bunches). The second one, the Vanuatu Tall x Rennell Island Tall (VTT × RIT) yields 34% more than elite Vanuatu Tall and shows a very high level of tolerance to CFD tolerance. This result was obtained thanks to a conventional breeding programme conducted from 1967 to 2008 (Labouisse *et al.*, 2011). VARTC is still running a seed garden for producing VTT × RIT hybrid on a very small scale.

Saraoutou station has also been a driving force for agronomic research in the Pacific by training technical and research staff of the PICTs on coconut germplasm conservation and hybrid production. From 1989 to 2000, with the implementation of the project *Production and dissemination of improved coconut cultivars* in the framework of the *Pacific regional agricultural programme* (PRAP) funded by the European Union, 36 new hybrids were created by crossing Dwarf varieties and local Tall varieties originated from eight PICTs. About 57 ha of comparative trials was established at VARTC and the observations of 9900 hybrid trees recorded over several years constitute a database of regional interest.

Conventional coconut breeding is a time-consuming and expensive process due to the plant's biology and the need to conserve and test cultivars over large field areas. Frequent cyclones in the Pacific also make it difficult to record correct data throughout the experiment. It is expected that modern technologies based on DNA sequencing will accelerate breeding programmes by identifying markers linked to characters of interest at an early stage.

Conclusions

This short review has shown the many attempts that have been made to improve staple food crops of PICTs over the last 70 years by conventional breeding. The attempts have unlocked an understanding of the abundant genetic resources that exist and their potential. Many lessons have been learned. Most importantly, it is now known that the crops have a narrow genetic base and because they are vegetatively propagated, they have no means of adaptation in farmers' fields except for rare somatic mutations. But perhaps the greatest understanding is that when major challenges occurred, conventional plant breeding provided acceptable solutions, and solutions came quickly. An outbreak of TLB, for instance, was solved within 4–5 years, scab of sweet potato within a year or two and anthracnose resistance of yams under a decade. Each of these was a significant achievement, requiring intimate knowledge of the crops, their genetic resources within the region and worldwide, breeding strategies likely to achieve success and international collaboration.

However, a paradox exists. If there have been successes from plant breeding, why is it no longer a priority, especially with its potential to mitigate against forthcoming climate change which is likely to be detrimental to food and nutritional security? It is well known that the Pacific is more vulnerable than any other region in the world. Apart from increased temperatures and ocean rise, ENSO events will occur in varying intensity every 3 to 8 years and Pacific island farmers will face extremes of climate, excessive and prolonged rainfall and flooding, extended droughts, severe cyclones and exceptional tides (McGregor, 2011).

Furthermore, the paradox is surprising when we consider that most people in Pacific island countries are growing staple food crops, mostly roots and tubers, for household use and/or for local and occasionally export markets. In some countries, production involves more than 75% of the population, with considerable contributions to gross domestic product (GDP).

We provide four possible reasons for the paradox: (i) a serious under-investment in agricultural research by PICTs and regional organizations, perhaps because gains are always uncertain and often long term, and priorities lie elsewhere; (ii) sharing genetic resources involves difficult-to-manage biosecurity concerns and high cost of in-country distributions; (iii) short-term, project-based funding by development partners with high impact expectations and (iv) a precedence for alternate crop improvement strategies that are considered to be more appropriate.

Having made a case for plant breeding based on several successes, and having established that a paradox exists because plant breeding is not being used in this time of climate crisis, we look at the alternate strategies applied presently to crop improvement and ask if plant breeding could make a difference. We select sweet potato, by far the most important food crop staple of the region.

One strategy for sweet potato improvement is to distribute pathogen-tested (PT) accessions from collections at SPC's CePaCT, collections of cultivars from the Pacific Islands (including breeders' lines) and those from the international agricultural centres (CIP, CIAT and IITA) assembled to create a 'climate ready' resource. Although there is undeniable value in conserving the region's crop genetic resources, it is uncertain whether these distributions have created the desired effects due to a lack of studies evaluating their adoption (Iese *et al.*, 2018).

An alternate strategy, which also uses PT, is underway in several South Pacific countries. It is based on farmers in Melanesian countries realizing the benefits of collecting seedlings found growing naturally in their fields. They propagate them, and those found acceptable are distributed through family networks and markets. However, after a few years, yields decline, creating a cycle of discovery, evaluation and sharing. The modern-day assumption is that the initial boost in performance is due to freedom from viruses. Therefore, to capitalize on these traditional practices – reliant on natural seed production and virus filtering – PT schemes have been established, whereby a few cultivars grown commercially for urban markets are collected, treated for viruses by modern technologies and given back to growers. So far, the results look promising, and yields are impressive as they are in other countries where PT schemes have been established, for instance, in Australia, China, South Africa and the USA.

But could plant breeding be used to improve the schemes now operating in the Pacific? Our concern is that PT may narrow crop diversity over the long term by tying growers to cultivars that no longer have the flexibility to change in response to changing climates: they deny the potential represented in this region's diverse gene pool – Papua New Guinea, for instance, is recognized as one of the world's major centres of sweet potato diversity (Roullier *et al.*, 2013). Additionally, we are concerned that clonal selections made this way will not be suited to the highly varied environments of the countries where GxE interactions preclude the widespread use of a few varieties.

We consider that another way would see sweet potato breeding within a core sample representative of the genetic diversity of the region, and for PT to be confined to the distribution of breeders' lines to growers, in ways that it is used by CePaCT for the distribution of TLB-resistant taro selections from breeding cycles in both Papua New Guinea and Samoa.

We are also keen to see an entirely new approach developed for sweet potato and other staple root and tuber crops that involves breeding to meet current needs. It is one that does not deny the potential of fixed clonal variation, but also considers allelic diversity to broaden the genetic base of crops, and sexual recombination using true or botanical seed (Lebot, 2013). The use of botanical seeds is an important and novel part of root and tuber crop improvement. To be successful, it requires the exploitation of crop gene pools, an understanding of species' floral biology and participatory evaluation, where farmers have a major role.

This type of breeding programme is comparatively easy to operate in PICTs compared to conventional centralized breeding programmes. Molecular studies have shown that, over time, introduced genotypes do recombine naturally with local materials and volunteer seedlings are captured clonally by farmers, strengthening adaptation and sustainability (Sardos *et al.*, 2012). The cassava and sweet potato diversity found in the PICTs are examples of such processes. Participatory selection can thus bypass a lengthy on-station process, and farmers can select genotypes adapted to local conditions and consumer preferences.

Major diseases (anthracnose, scab, TLB) and pests (*Cylas, Oryctes, Papuana* spp.) are spreading rapidly across the region. In the absence of genetically improved planting materials, farmers will face increasing difficulties, while pressure on the land continues to intensify. Plant breeding of staple food crops is a strategy worth pursuing, to attenuate the impact of impending challenges.

Recommendations

There are several critical issues that need to be addressed in order to strengthen the capacity of smallholders in PICTs to adapt to forthcoming events, of which the impact of the climate crisis is the most immediate:

- Create a collaborative decentralized network of government agencies and universities, using conventional breeding to improve food crop staples.
- Prioritize crops of concern, applying the proven models of TANSAO and TaroGen to bring about the crop improvements required.
- Broaden the narrow genetic base of food crop staples by introducing allelic diversity from core samples.
- Use botanical seed, supported by *in vitro* methods, to overcome biosecurity concerns.
- Foster farmers' participation (acknowledging the role of women and youth for their dominant roles in food production), allowing for rapid distribution, evaluation and adoption of new lines and their integration with locally adapted genotypes.
- Support universities to train scientists and technicians to integrate genetic improvement and breeding techniques into present curricula.

Australian Agency for International Development

Acknowledgement

Numerous development agencies have contributed to germplasm characterization and conservation activities in the PICTs, including AusAid, CIRAD, Global Crop Trust, EU, NZAid, UNDP-FAO, the World Bank and many others. Their contributions are highly appreciated.

Conflicts of Interests

The authors have no conflicts of interest to declare.

List of Acronyms

AusAID	Australian Agency for International Development
CePaCT	Centre for Pacific Crops and Trees
CIAT	Centro Internacional de Agricultura Tropical (Cali, Colombia)
CIP	International Potato Center (Lima, Peru)
CIRAD	Centre de Coopération Internationale en Recherche Agronomique pour le Développement
CTCRI	Central Tuber Crops Research Institute (Trivandrum, Kerala, India)
ENSO	El Niño Southern Oscillation
EU	European Union
FAO	Food and Agriculture Organization
FHIA	Fundación Hondureña de Investigación Agrícola
GDP	Gross Domestic Product
IITA	International Institute for Tropical Agriculture (Ibadan, Nigeria)
INEA	International Network for Edible Aroids
IPGRI	International Plant Genetic Resources Institute
NARI	National Agricultural Research Institute (Papua New Guinea)
NZAid	New Zealand Aid Programme
PICTs	Pacific Island Countries and Territories
PPD	Postharvest Physiological Deterioration
PT	Pathogen Tested
SPC	Secretariat of the Pacific Community
TLB	Taro Leaf Blight (*Phytophthora colocasiae*)
UNDP	United Nations Development Programme
USP	University of the South Pacific
TANSAO	Taro Network for Southeast Asia and Oceania
TaroGen	Taro Genetic Resources: Conservation and Utilization
VARTC	Vanuatu Agricultural Research and Technical Centre

References

Abraham, K., Nemorin, A., Lebot, V. and Arnau, G. (2012) Meiosis and sexual fertility of autotetraploid clones of greater yam *Dioscorea alata* L. *Genetic Resources and Crop Evolution* 60(3), 819–823.

Bruerre, M. (2010) The Vanuatu Island Cabbage Project. SPC-VARTC Report, Saraoutou, Santo.

Chaïr, H., Sardos, J., Supply, A., Mournet, P., Malapa, R. *et al.* (2016) Plastid phylogenetics of Oceania Yams (*Dioscorea* spp., Dioscoreaceae) reveals natural interspecific hybridization of the greater yam (*D. alata*). *Botanical Journal of the Linnean Society* 180(3), 319–333. DOI: 10.1111/boj.12374.

Choin, J., Mendoza-Revilla, J., Arauna, L.R., Cuadros-Espinoza, S., Cassar, O. *et al.* (2021) Genomic insights into population history and biological adaptation in Oceania. *Nature* 592, 583–589. DOI: 10.1038/s41586-021-03236-5.

Ebert, A.W. and Waiqanabete, L.M. (2018) Conserving and sharing taro genetic resources for the benefit of global taro cultivation: A core contribution of the centre for Pacific crops and trees. *Biopreservation and Biobanking* 16, 361–367. DOI: 10.1089/bio.2018.0017.

Georgeou, N., Hawksley, C., Wali, N., Lountain, S., Rowe, E. *et al.* (2022) Food security and smallholder farming in Pacific Island countries and territories: A scoping review. *PLOS Sustainability and Transformation* 1(4), e0000009. DOI: 10.1371/journal.pstr.0000009.

Guarino, L. and Jackson, G.V.H. (1986) Describing and Documenting Root Crops in the South Pacific. UNDP FAO Suva, Fiji, RAS/83/001 Field document 12, p. 141.

Harries, H., Baudoin, L. and Cardeña, R. (2002) Floating, boating and introgression: Molecular techniques and the ancestry of the coconut palm populations on Pacific Islands. *Ethnobotany Research and Applications* 2, 37–53.

Iese, V., Holland, E., Wairiu, M., Havea, R., Patolo, M.N. *et al.* (2018) Facing food security risks: The rise and rise of the sweet potato in the Pacific Islands. *Global Food Security* 18, 48–56. DOI: 10.1016/j.gfs.2018.07.004.

Kirch, P.V. (2017) *On the Road of the Winds: An Archaeological History of the Pacific Islands before European Contact*. 2nd edn, University of California Press, Oakland, CA, p. 408.

Kreike, C.M., van Eck, H.J. and Lebot, V. (2004) Genetic diversity of taro, *Colocasia esculenta* (L.) Schott, in South-East Asia and the Pacific. *Theoretical and Applied Genetics* 109, 761–768. DOI: 10.1007/s00122-004-1691-z.

Labouisse, J.P., Siléyé, T., Bonnot, F. and Baudoin, L. (2011) Achievements in breeding coconut hybrids for tolerance to coconut foliar decay disease in Vanuatu, South Pacific. *Euphytica* 177, 1–13.

Lebot, V. (1999) Biomolecular evidence for plant domestication in Sahul. *Genetic Resources and Crop Evolution* 46(6), 619–628.

Lebot, V. (2013) Coping with insularity: the need for crop genetic improvement to strengthen adaptation to climatic change and food security in the Pacific. *Environment, Development and Sustainability* 15, 1405– 1423. DOI: 10.1007/s10668-013-9445-1.

Lebot, V. and Ivancic, A. (2022) Taro (*Colocasia esculenta* (L.) Schott), breeding history, objectives, methods and strategies: A review of fifty years of sporadic efforts. *Euphytica* 218, 166. DOI: 10.1007/s10681-022-03118-5.

Lebot, V., Aradhya, M., Manshardt, R. and Meilleur, B. (1993) Genetic relationships among cultivated bananas and plantains from Asia and the Pacific. *Euphytica* 67(3), 163–175.

Lebot, V., Abraham, K., Kaoh, J., Rogers, C. and Molisalé, T. (2019) Development of anthracnose resistant hybrids of the Greater Yam (*Dioscorea alata* L.) and interspecific hybrids with *D. nummularia* Lam. *Genetic Resources and Crop Evolution* 66(4), 871–883.

Lebot, V., Leo, P. and Legendre, L. (2021) Phenotyping chlorogenic acids and coumarins in sweet potato [*Ipomoea batatas* (L.) lam.] breeding lines for enhanced tolerance to periderm pathogens. *Euphytica* 217, 59.

Lough, J., Gupta, A.S., Power, S.B., Grose, M.R. and McGree, S. (2016) Observed and projected changes in surface climate of tropical islands. In: Taylor, M., McGregor, A. and Dawson, B. (eds) *Chap 2. Vulnerability of Pacific Island Agriculture and Forestry to Climate Change*. Pacific Community, Nouméa, New Caledonia, p. 554.

Loy, T.H., Spriggs, M. and Wickler, S. (1992) Direct evidence for human use of plants 28,000 years ago: Starch residues on stone artefacts from the Northern Solomon Islands. *Antiquity* 66, 898–912. DOI: 10.1017/S0003598X00044811.

McGregor, A. (2011) Assessing the social and economic value of germplasm and crop improvement as a climate change adaptation strategy: Samoa and Vanuatu case studies. A background case study prepared for IUCN's report, Climate Change Adaptation in the Pacific: Making Informed Choices, prepared for the Australian Department of Climate Change and Energy Efficiency (DCCEE), IUCN, Suva, Fiji.

Pacific Community (2020) Pacific Islands Populations. Available at: https://sdd.spc.int/news /2020/09/23 /2020-pacific-islands-population-poster (accessed 8 March 2023).

Perrier, X., De Langhe, E., Donohue, M., Lentfer, C.J., Vrydaghs, L. *et al.* (2011) Multidisciplinary perspectives on banana (*Musa* spp.) domestication. *Proceedings of the National Academy of Sciences (PNAS)* 108(28), 11311–11318.

Ragone, D. (2001) Chromosome numbers and pollen stainability of three species of Pacific Island breadfruit (*Artocarpus*, Moraceae). *American Journal of Botany* 88, 693–696.

Roullier, C., Benoit, L., McKey, D. and Lebot, V. (2012) Historical collections reveal patterns of diffusion of sweet potato in Oceania obscured by modern plant movements and recombination. *PNAS, Proceedings of the National Academy of Sciences (US)* 110(6), 2205–2210.

Roullier, C., Kambuou, R., Paofa, J., McKey, D. and Lebot, V. (2013) On the origin of sweet potato (*Ipomoea batatas* (L.) Lam) genetic diversity in New Guinea, a secondary centre of diversity. *Heredity* 110(6), 594–604. DOI: 10.1038/hdy.2013.14.

Sardos, J., Malapa, R., McKey, D., Noyer, J.L., Duval, M.F. *et al.* (2008) Genetic diversity of cassava (*Manihot esculenta* Crantz) introduced in the oceanic islands of Vanuatu, Melanesia, South Pacific. *Genome* 51, 912–921.

Sardos, J., Rodier-Goud, M., Dambier, D., Malapa, R., Noyer, J.L. *et al.* (2009) Evidence for spontaneous polyploidization in cassava *Manihot esculenta* Crantz. *Plant Systematics and Evolution* 283(3–4), 203–209.

Sardos, J., Noyer, J.L., Malapa, R., Bouchet, S. and Lebot, V. (2012) Genetic diversity of taro (*Colocasia esculenta* (L.) Schott) in Vanuatu (Oceania): An appraisal of the Distribution of Allelic Diversity (DAD) with SSR markers. *Genetic Resources and Crop Evolution* 59(5), 805–820.

Summerhayes, G., Leavesley, M., Fairbain, A., Mandui, H., Field, J. *et al.* (2010) Human adaptation and plant use in highland new guinea 49,000 to 44,000 years ago. *Nature Reports* 330, 78–81. DOI: 10.1126/science.1193130.

Whistler, W.A. (2009) Plants of the canoe people: An ethnobotanical voyage through Polynesia. In: *National Tropical Botanical Garden*. Kaua'i, Hawaii, p. 241.

Zerega, N.J.C., Ragone, D. and Motley, T.J. (2004) Complex origins of breadfruit (Artocarpus altilis, Moraceae): Implications for human migrations in Oceania. *American Journal of Botany* 91, 760–766.

12 The path to climate neutrality for California dairies

C.J. McCabe, H.M. El Mashad and F.M. Mitloehner*
Department of Animal Science, University of California, Davis, One Shields Ave., Davis, CA 95616-8521, USA

Abstract

In recent years, dairy farms have come under scrutiny with pressure to curb their environmental impacts. Since 1950, the California dairy industry has made strides in reducing greenhouse gas (GHG) emissions per kilogram of milk produced. However, total GHG emissions have remained near constant over the past 15 years. Most on-farm dairy production emissions are in the form of methane (CH_4) emissions produced via enteric fermentation, where CH_4 is produced as a byproduct of feed digestion, and manure management. Methane is a powerful but short-lived GHG. Historically, GWP100 has been utilized to describe a GHG emission's warming impact over a 100-year time span. To better characterize the impact of CH_4 on atmospheric warming, a relatively new accounting system named global warming potential star (GWP*) has been proposed to consider the production and degradation of this short-living GHG. Characterizing greenhouse gases by how they warm our atmosphere instead of the number of emissions produced is a better metric for the true impact of the emissions on atmospheric warming.

The goal of this research is to use GWP* to analyze the impact of potential GHG emissions scenarios from California dairy and the impact of those scenarios on atmospheric warming. Utilizing GWP* can help discern when an industry or sector has achieved climate neutrality or no annual warming contributions from industry. This paper also investigates the necessary amount of CH_4 reduction needed and the time point at which the dairy sector can achieve climate neutrality. The scenarios are business-as-usual (BAU), 40% reduction in manure CH_4 emissions (40 MAN) by applying anaerobic digestion and its alternative technologies, and 40 MAN along with a 10.6% reduction in enteric fermentation CH_4 emissions via 1/3 of California's cows fed the feed additive 3-nitrooxypropanol (40 MAN+EF). Under GWP100 in 2030, carbon dioxide equivalents (CO_2e) for the 40 MAN and the 40 MAN + EF scenarios were reduced by 18 and 22%, respectively compared to the BAU. For all three scenarios, the relative warming impact of the industry decreased over time due to constant herd sizes and total annual emissions. By aggressively decreasing CH_4 emissions under the 40 MAN and 40 MAN+EF scenarios, there is the possibility for the California dairy industry to reach climate neutrality by the year 2027. These scenarios have more CH_4 naturally removed in the atmosphere than is emitted, thus lowering atmospheric contributions from the industry. These scenarios could be adopted by dairies in other states and countries to help the global dairy industry to achieve climate goals through persistent CH_4 mitigation.

Keywords: GWP*, methane, atmospheric warming, sustainable agriculture

*Corresponding author: fmmitloehner@ucdavis.edu

Introduction

The California dairy industry is the state's leading commodity, producing over 19 billion kilograms of milk per year which generates over $7.4 billion of the nearly $50 billion California agriculture industry (CDFA, 2021). Dairy farms also make up 45% of California's annual methane (CH_4) budget with 25% resulting from manure management and 20% from enteric fermentation (California Air Resources Board (CARB), 2022b). This potent greenhouse gas (GHG) has come under scrutiny in recent years with the passage of legislation and pressure to reduce the environmental impact of milk production. In March 2017, the California Air Resources Board (CARB) approved the Short-Lived Climate Pollutant (SLCP) Reduction Strategy to reduce CH_4 emissions based on Senate Bill 1383 (State of California, 2016). This mandates the entire dairy industry to reduce the 2013 manure management CH_4 emissions by 40% by 2030. The California legislature does not currently mandate enteric fermentation CH_4 reductions. However, regulations are likely to be created in the future to meet California's climate goals (State of California, 2006).

SLCPs are a category of pollutants that have lifetimes much shorter than CO_2 from days to a few decades. Along with CH_4, other pollutants in this category include black carbon, tropospheric ozone, and CH_4, which have lifetimes of days, weeks, and 12 years, respectively (Haines *et al.*, 2017). Once CH_4 is produced and released to the atmosphere, the majority will be oxidized to CO_2 via the process of hydroxyl oxidation in about 12 years (Cantrell *et al.*, 1990; Kirschke *et al.*, 2013). The produced CO_2 can then remain in the atmosphere for up to 1000 years or be taken up by various carbon dioxide sinks (Solomon *et al.*, 2009; Muñoz and Schmidt, 2016). Although this CO_2 is long living in the atmosphere, these emissions are part of the biogenic carbon cycle whereby the carbon is taken up by the feedstocks that are utilized as a source of dairy feed (Muñoz and Schmidt, 2016). As a result, no new carbon is added to the atmosphere, but simply recycled. Reducing SLCPs has been proposed as a major opportunity to slow projected global warming by 0.5°C over the next 25 years (Shindell *et al.*, 2017).

Conventionally, the 100-year global warming potential (GWP100) metric has been used to determine the 100-year time-integrated climate impact of a GHG. A GHG's GWP100 calculation is determined by comparing the emissions of a one-ton pulse of the GHG of interest over a 100-year period compared to a one ton pulse emission of CO_2 (Reay *et al.*, 2007). The application of the conventional GWP100, for converting SLCPs into CO_2 equivalents, often misrepresents the impacts of the SLCPs on global temperature (Allen *et al.*, 2016). Therefore, GWP* was proposed to relate the rate of SLCP emissions to CO_2 emissions and their subsequent impact on warming (Allen *et al.*, 2018).

GWP* can be used to determine the effects of mitigation efforts on future annual emissions and warming impacts (Smith *et al.*, 2021). Under GWP100, CH_4 emissions in a scenario in which emissions are increasing, are undervalued as more CH_4 is produced than removed from the atmosphere, whereas, in a constant or falling scenario, CH_4 emissions are consistently overvalued as an equal or greater amount is removed than produced respectively (Allen *et al.*, 2018). GWP* can predict the cooling effects that result from reductions in SLCP emission rates (Cain *et al.*, 2019). Climate neutrality is achieving no additional warming impacts from the activities of a firm or an entity at regional, subnational, and national scales (Pineda and Faria, 2019). While net zero GHG emissions are achieving the balance between the metric-weighted anthropogenic GHG emissions from an entity with the metric-weighted anthropogenic GHG removal over a specific period (Pineda and Faria, 2019; Matthews *et al.*, 2021). Thereby under net zero emissions, everything produced in a given year must be taken out of the atmosphere or offset, whereas climate neutrality focuses on the annual warming contribution for those gasses produced and removed in a given year.

Applying GWP* to predict the warming impact of the U.S. dairy industry's CH_4 emissions on climate showed that the CH_4 emissions alone, not accounting for other greenhouse gasses of production, have not contributed to additional warming from 1986 to 2017 (Liu *et al.*, 2021). Furthermore, the industry will approach climate neutrality by 2031 for CH_4 emissions alone if they are reduced by 1% per year (Liu *et al.*, 2021).

Dairy anaerobic digesters and other manure management technologies, and feed additives have been employed across dairy farms to reduce CH_4 emissions from manure and enteric fermentation, respectively (Rotz, 2018). The California Department of Food and Agriculture (CDFA) has established the Dairy Digester Research and Development Program (DDRDP) and the Alternative Manure Management Practice (AMMP) program to help dairy farms install anaerobic digesters and other manure management technologies to reduce CH_4 emissions. Research has demonstrated 3-nitrooxypropanol (3-NOP) to be an effective CH_4 reducer; however, its application is still limited on farms without regulatory approval (Melgar *et al.*, 2020).

To understand the impact that California dairy can have on future atmospheric warming, there is a need to determine the effect of emission reductions resulted from the increase of manure management technologies and feed additives. Therefore, the objectives of this study were to determine: (1) the atmospheric warming impact of the California dairy industry; and (2) the time frame by which the California dairy industry could achieve climate neutrality via the use of anaerobic digestion and alternative manure management and feed additive technologies from 2019 to 2030.

Review Methodology

The present review comprises GHG emission data and associated literature for the California dairy industry and its respective impact on atmospheric warming. Data was obtained from the California Air Resources Board (CARB) and the United States Department of Agriculture (USDA) National Agriculture Statistics Service (NASS) to assess emissions and cattle population, respectively. Relevant literature was obtained from searching databases for scientific manuscripts relevant to dairy's impact on climate. Discussions were conducted with experts to determine relevant material for this review.

Materials and Methods

California dairy industry data collection

California dairy cattle production and California dairy GHG emissions data were collected from the United States Department of Agriculture (USDA) and California Air Resources Board (CARB), respectively (Naranjo *et al.*, 2020; California Air Resources Board (CARB), 2022b). Dairy cattle production data included lactating cow and heifer populations as of January 1st of each year and annual milk production from 1990 to 2020. Emissions data from CARB were collected from the state's scoping plan over the years 2000–2019 (California Air Resources Board, 2021). The data included enteric fermentation CH_4, manure management CH_4, and manure management nitrous oxide (N_2O). The available data from 2000 to 2019 was used to predict the enteric fermentation and manure management emissions of CH_4 and N_2O for the years 1990–1999. Multiple linear regression models were employed, using R Software (version 4.1.3), to predict the emissions of CH_4 and N_2O as a function of both annual milk yield per cow and lactating herd number as follows:

$$\text{Enteric CH}_4\ (\text{Million metric tons; [MMT]}) = -0.23 + 1.45 \times 10^{-7}X_1 + 2.75 \times 10^{-5}X_2 \qquad R^2 = 0.87$$

$$\text{Manure Management CH}_4\ (\text{MMT}) = -0.504 + 2.34 \times 10^{-7}X_1 + 4.63 \times 10^{-5}X_2 \qquad R^2 = 0.87$$

$$\text{Manure Management N}_2\text{O}\ (\text{MMT}) = -4.11 \times 10^{-4} + 7.51 \times 10^{-10}X_1 + 4.26 \times 10^{-8}X_2 \qquad R^2 = 0.87$$

Where:
X_1 = Lactating cow population;
X_2 = Annual milk yield, kg/cow

Indirect emissions from dairy farm operations (feed production, energy use, etc.) were assumed at 0.24 kg carbon dioxide equivalents (CO_2e)/kg milk (Naranjo *et al.*, 2020).

GWP vs. GWP* accounting calculations

The emission data of CH_4 and N_2O were converted to CO_2e to calculate the industry's GWP100 under AR5 scoping values (IPCC, 2014):

$$GWP100\ (CO_2e) = CO_2 + 28 \times CH_4 + 265 \times N_2O \quad (1)$$

To transition from the production of the industry's emissions to the warming impact, GWP* was utilized from 2010 onward due to the 20-year timespan needed for metric calculations (Smith *et al.*, 2021). Equation 2 was developed based on the concepts developed by Smith *et al.* (2021) to convert GWP100 to GWP*:

$$GWP * (CO_2we) = CO_2 + 4.53\ CH_4\ (t) - 4.25\ CH_4\ (t - 20) + 265 \times N_2O \quad (2)$$

Where t is equal to the current year being analyzed.

Scenarios to Reduce Atmospheric Warming Impact of the California Dairy Industry

Three scenarios were proposed and analyzed to determine the industry's projected warming between the present year (2019) and the target year (2030). The three scenarios included (1) a business-as-usual (BAU) where all industry emissions were assumed to stay constant; (2) a 40% reduction in manure management CH_4 emissions (40 MAN) in line with SB 1383; and (3) the 40 MAN emissions along with an assumed 10.6% reduction in total enteric fermentation CH_4 emissions (40 MAN +EF). The 40 MAN reduction was assumed to be met via the construction of anaerobic digesters and alternative manure management techniques such as manure solid separators and weeping walls. The 10.6% reduction in the total enteric fermentation CH_4 emissions was estimated by assuming one-third of the California dairy industry's cows would be receiving 3-NOP, which has an estimated 31.7% reduction in CH_4 emissions when fed at a rate of 127 mg/kg of dry matter (Feng and Kebreab, 2020). For the industry's CH_4 emission scenario projections, consistent emissions reductions were assumed to meet the stated goals between the present year (2019) and the goal year (2030). The year 2019 is assumed to be the present year because that is the year with the most recently available scoping plan (California Air Resources Board (CARB), 2022b).

Additionally, milk production was assumed to continue the similar trend observed from 1990 to 2020 with a 0.8% increase in annual milk produced per cow through 2030 (Fig. 12.1). The proposed milk yield increase of 0.8% annually is a reasonable assumption that is justified by another study that projected USA milk yield will continue to increase within the next 50 years (Britt *et al.*, 2018). They mentioned that the milk yield will climb at an accelerated rate and will be doubled by 2067 based on current production levels due to genetic selection that will positively affect yield and health traits, and improvements in feed quality and management. Although cows are sensitive to changes in temperature, as will be increasingly likely in the future, additional management in cooling cows and genetic merit gains will help offset the increasingly warmer environment (Gunn *et al.*, 2019).

Lactating cow numbers were expected to reduce by 0.5% annually in line with CARB's scoping goals and suggestions on how SB 1383 will be achieved (California Air Resources Board (CARB), 2022b). The California lactating herd reached its absolute maximum in 2009 and has begun to slightly decrease in recent years due to the migration of dairy cows to other states and resource challenges within California (Fig. 12.1). Along with this, manure management N_2O emissions were assumed to be held consistent from current emissions levels between 2019 and 2030.

Under all three evaluated scenarios, GWP100 and GWP* were calculated to determine the associated GHG emissions and the warming impact of the respective emissions. The California dairy industry was determined to achieve climate neutrality when the warming impact of the industry reached zero or went below zero annual carbon dioxide warming equivalents (CO_2we).

Results

Historic dairy population and GHG emissions

Methane emissions represent approximately 75–80% of the California dairy industry's GHG emissions from farm to gate, based on GWP100

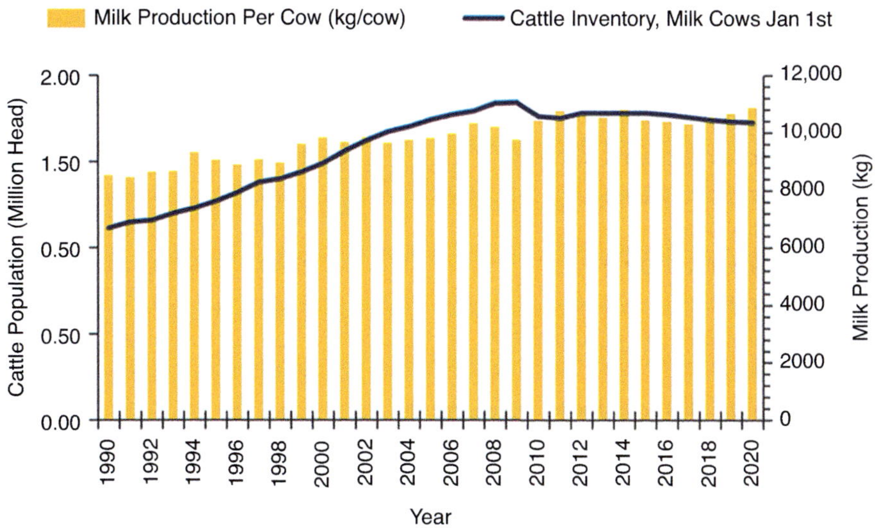

Fig. 12.1. California dairy industry lactating cow population and milk yield per cow from 1990 to 2020 (Naranjo *et al.*, 2020). Lactating cattle population is referenced from the United States Department of Agriculture National Agricultural Statistics Service based on estimates of animal numbers based on January 1st of each year.

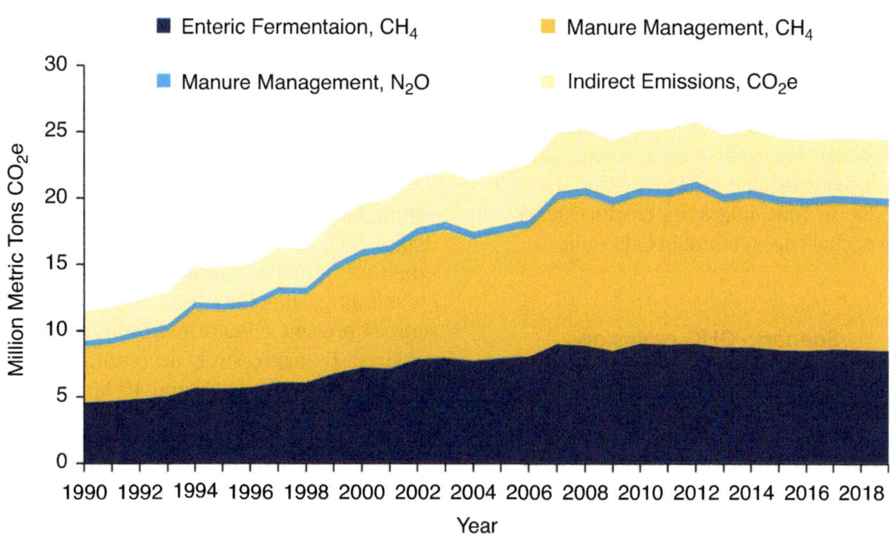

Fig. 12.2. California dairy industry's greenhouse gas emissions profile by dairy farm source from 2000 to 2019 (California Air Resources Board (CARB), 2022a). Indirect emissions include everything that is necessary for dairy production including emissions associated with feed production and farm management. Data here is reported in carbon dioxide equivalents and includes carbon dioxide and nitrous oxide emissions (IPCC, 2014).

Table 12.1. California cow population, production parameters, and greenhouse gas emissions under Business-as-Usual scenario between 2019 and 2030.

	2019	2025	2030
Number of Lactating Cows (Million)	1.73	1.69	1.64
Milk Production Per Cow (thousand kgs/year)	10.64	11.33	11.82
Total Milk Production (Billion kg/year)	18.41	19.09	19.43
Enteric Emissions (MMT CO_2e)	8.51	8.51	8.51
Manure CH_4 Emissions (MMT CO_2e)	10.91	10.91	10.91
Total CO_2e Emission (MMT[a])	24.41	24.57	24.65
Cradle to Farmgate Footprint (kg CO_2e/kg milk[b])	1.37	1.32	1.29
CO_2we Emissions (MMT[c])	31.65	19.69	7.54
Cradle to Farmgate Warming Footprint (kg CO_2we/kg milk)	1.72	1.03	0.39

[a] CO_2e emission calculated based on using global warming potential 100 (GWP 100) factors from AR 5 of 1 for CO_2, 28 for CH_4, and 265 for N_2O (Smith *et al.*, 2021).
[b] Includes all emissions divided by the number of milk production (kg) per year.
[c] Calculated using GWP* (Cain *et al.*, 2019).

(Fig. 12.2). With a similar number of lactating cows since 2007, the industry has stayed stable at 24–25 million metric tons of CO_2e annually. The California herd achieved its absolute maximum lactating-cow population in 2009 with 1.84 million animals (Fig. 12.1) and has maintained consistent levels with a lactating herd population of 1.75 million lactating cows from approximately 2007–2019 (USDA, 2022). In recent years, the lactating herd population decreased to 1.72 million cows due to cows migrating to other states and resource challenges present in California.

Scenario GHG emissions

Beginning with the BAU scenario, the industry continues its historic environmental gains by producing more milk (18.41 vs. 19.43 billion kilograms) with a fewer number of cows in 2019 compared to 2030 (1.73 vs. 1.64 million) as shown in Table 12.1. This results in a decrease in GHG emission intensities or GHG emissions per kg of milk (1.37 vs. 1.29 kg CO_2e per kg of milk). However, there is no pronounced difference in the total GHG emissions contributed by the industry (24.41 vs. 24.65 MMT CO_2e) due to gains in milk production per cow without a change in overall emissions. From 2019 to 2030, the 40 MAN scenario reduces industry manure management CH_4 emissions from 0.37 to 0.24 MMT CH_4 and the 40 MAN +EF scenario reduces enteric fermentation CH_4 from 0.30 to 0.27 MMT CH_4 through implementing 3-NOP. By implementing the 40 MAN and 40 MAN +EF scenarios, the industry's total GHG emissions decreases by 2030 from 24.41 to 20.15 and 19.16 MMT CO_2e, respectively (Fig. 12.3).

Regardless of the scenario studied, based on the GWP* approach, the warming impact of the GHG profile decreases over time due to consistent herd populations and emission factors (Fig. 12.4). However, even though CH_4 emissions are consistent under the BAU scenario, outstanding emissions from indirect and N_2O sources prevent climate neutrality from being achieved. By aggressively decreasing CH_4 emissions under the 40 MAN and 40 MAN +EF scenarios, there is the possibility for the California dairy industry to reach climate neutrality by the year 2027. By this target year, the industry has the potential to reach climate-negative milk production, whereby warming equivalents are going below zero and the industry is offsetting historical emissions contributions.

Atmospheric warming implications for the proposed scenarios

Analyzing the cumulative warming emissions of the scenarios from 2019 to 2030 creates a

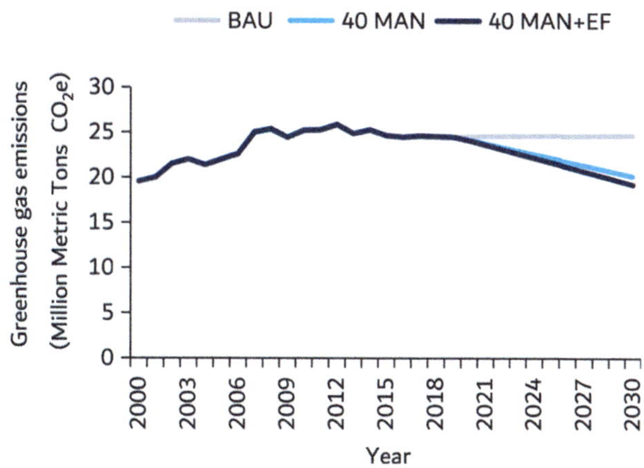

Fig. 12.3. California dairy industry's annual greenhouse gas emissions between the various scenarios analyzed from 2000 to 2030. Business as Usual (BAU) assumes a continuation of California dairy industry trends of a slight decrease in herd population. 40 Manure (40 MAN) assumes a 40% reduction in manure management emissions scenarios in line with SB 1383 legislation. 40 Manure +enteric fermentation (40 MAN +EF) assumes a reduction in manure management emissions as well as a 10.6% decrease in enteric fermentation emissions.

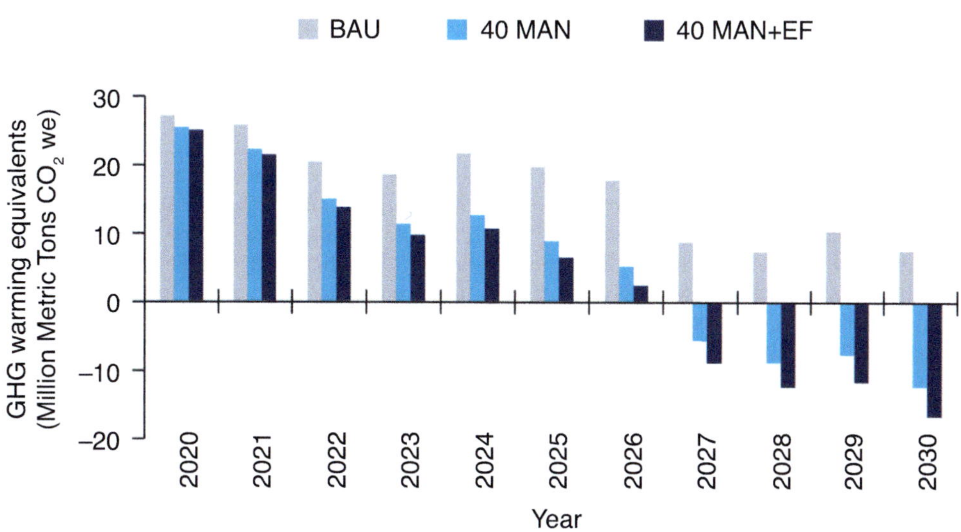

Fig. 12.4. California dairy industry annual warming impact of greenhouse gas emissions based on the three analyzed scenarios from 2020 to 2030. Business as Usual (BAU) assumes a continuation of California dairy industry trends of a slight decrease in herd population. 40 Manure (40 MAN) assumes a 40% reduction in manure management emissions scenarios in line with SB 1383 legislation. 40 Manure +enteric fermentation (40 MAN +EF) assumes the reduction in manure management emissions as well as a 10.6% decrease in enteric fermentation emissions.

divergence in the impact on atmospheric warming between scenarios. By using the assumption that one Tera ton of CO_2 raises the global atmospheric temperature by 1.8°C (Smith *et al.*, 2021), then under BAU the atmosphere would see an increase by 3.33×10^{-4} °C through the remainder of the decade (Cain *et al.*, 2019). Likewise, the 40 MAN and 40 MAN +EF scenarios would result in an increase in warming of 1.21×10^{-4} and 7.71×10^{-5} °C, respectively. Although climate neutrality can be achieved under these scenarios, this metric is only achieved on an annual basis starting in 2027. The industry will still contribute positive GHG emissions and warming from 2019 until climate neutrality is realized, which causes the projected warming to occur when measuring over the entire scenario timeline. Once climate neutrality has been accomplished, the CH_4 reductions are not great enough to overcome historical industry emissions in the form of CO_2 and N_2O throughout the scenario period. Additional reductions in CH_4 emissions are necessary to continue to decrease the warming impact of the industry in the decades to come (Smith *et al.*, 2021). Further modeling techniques utilizing either the FaIR model or other climate modeling software are needed to understand the sensitivity of these changes in emission metrics to their impact on global atmospheric temperatures (Allen *et al.*, 2018).

Discussion

The studied scenarios are the first to look at how policy and implementation of CH_4 reduction strategies will reduce California dairy's atmospheric warming contributions. By implementing manure management strategies via the creation of manure anaerobic digesters and feed additives, the industry can achieve climate neutrality as early as 2027. The industry's warming impact will decrease over time, with constant herd populations and aggressive reductions in CH_4 by applying manure management technologies and enteric fermentation inhibitors. Not to mention that CH_4 is a precursor for forming ozone molecules, which have detrimental impacts on both human health and crop performance (Van Dingenen *et al.*, 2018). Thereby aggressive on-farm reductions of CH_4 can have multiple benefits for agriculture production, ecosystems, and the planet.

California dairy's progress toward climate goals

Since the implementation of SB 1383, the California Department of Food and Agriculture (CDFA) has provided grants in combination with private investment to help to construct or begin construction of approximately 113 projects throughout the state (CDFA 2022b). The biogas produced from these digesters has been used for electricity generation, fuel cells, or upgraded into renewable natural gas. Additionally, CDFA's Alternative Manure Management Program (AMMP) grants have constructed 114 projects, which include mechanical lagoon solid separators, compost-bedded pack barns, and weeping walls (CDFA, 2022a). These technologies help to separate the organic matter, which would traditionally undergo anaerobic biochemical processes during manure storage and produce CH_4 and other GHGs (Montes *et al.*, 2013).

To meet the State's climate goals, one of the pathways the dairy industry can pursue is to continue employing manure management technologies and feed additives to reduce manure and enteric fermentation CH_4, respectively. One of the limitations to our scenario is that 3-NOP is not currently available for sale in the U.S., but it is expected to achieve regulatory approval within the next few years given its prior approval in Brazil, Chile, and the European Union (Feng and Kebreab, 2020; Mitloehner *et al.*, 2020). It is currently going through regulatory approval as a veterinary drug in the United States, with anticipation that it could be first approved in 2024 (Mitloehner *et al.*, 2020). Over the next few decades, there will likely be other feed additive candidates that can be utilized to reduce enteric CH_4. Another candidate to reduce enteric CH_4 is red seaweed (*A. taxiformis*), but further work such as sourcing large enough quantities and ensuring palatability is needed before it is widely adopted across farms (Vijn *et al.*, 2020; Stefenoni *et al.*, 2021). Furthermore, dairy digesters will continue to be a crucial tool to reduce manure CH_4 emissions. California currently has 236 dairy digesters, which collect and harvest CH_4 from 254 dairy farms throughout the state (Dairy Cares, 2022). However, with CDFA-funded dairy digester projects through 2023, the industry will be 42.6% of the way toward the SB 1383 goal (CDFA, 2022b). To meet the remaining GHG reduction stated under SB 1383, an estimated 420 additional projects (50% AMMP technologies and 50% digesters) will be necessary to reduce 4.4 MMT CO_2e (California Air Resources Board (CARB),

2022b). Given the success observed to this point, there is the potential for the industry to achieve its 2030 climate goals and these scenarios to be accomplished. For additional consideration, it is important to note that approximately one-fifth of all California dairies have either constructed or are planning to build a privately financed or public-private partnership digester. Our case study is not able to perform sensitivity analysis on the likelihood of those other approximately 80% of the state's farms adopting dairy digesters. Although government and private market incentives have helped facilitate dairy digester adoption, this study would need additional data collection into farm behavior to understand these motivators for anaerobic digester adoption (Cowley and Wade Brorsen, 2018; California Air Resources Board (CARB), 2022a). Beyond 2030, further emission reductions are needed through farms adopting feed additives and digesters to help the industry offset its historical GHG emissions and continue to maintain climate neutrality.

Climate neutrality for the U.S. and global dairy industry

Although California dairy can play a big role in producing climate-neutral milk, the majority of GHG emissions fall outside the scope of the state's dairy industry with >80% of the national herd located in other states of the U.S. (Macdonald *et al.*, 2020). A recent publication looked at the U.S. dairy industry's ability to achieve climate neutrality given an assumed annual 1–2% reduction in industry-wide emissions (Place and Mitloehner, 2021). This scenario predicted U.S. dairy could achieve climate neutrality by 2041. While California will play a big role in helping meet these scenarios, other states are likely to follow suit in mandating dairy industry emission reductions and providing incentives to help producers adopt these practices (Liu *et al.*, 2021). States whereby the dairy cow population has matured over the last couple of decades (e.g. New York and Pennsylvania) could more easily achieve climate neutrality under GWP* using the CH_4 reduction technologies compared to states with expanding dairy cow herds (e.g. Idaho and Texas) (USDA, 2022). To achieve state-specific climate neutrality, the latter will need greater reductions in CH_4 emissions.

As animal-source protein consumer demand continues to increase worldwide and especially in developing nations, this supply will likely be met with expanding herd sizes (Alexandratos and Bruinsma, 2012). This will result in increased CH_4 emissions and an increase in the global dairy industry's warming impact under GWP* (Allen *et al.*, 2018; Liu *et al.*, 2021; Smith *et al.*, 2021). Although the global dairy industry has greatly reduced country-specific CH_4 intensities of milk production (kg milk/CH_4 emissions), mitigation technologies will need to be adopted to reduce total CH_4 emissions (FAO, 2018). Anaerobic digesters have already been shown to reduce CH_4 emissions, offset fossil fuel use, and provide a consistent and reliable form of energy. For example, the Indian dairy sector has rapidly implemented anaerobic digesters that reduced manure management CH_4 emissions by 19.12 MMT CO_2e (84%) from 1990 to 2022 (Bosoli *et al.*, 2019). Anaerobic digesters have also been susceptible to CH_4 leakage and have increased the CH_4 footprint per kg of milk on farms where they have been implemented in the developing world (Bruun *et al.*, 2014; York *et al.*, 2017). Although the European Union has been a rapid adopter of anaerobic digesters, most of the substrates utilized are biofuel crops and the manure management CH_4 emission reductions have been outpaced by the herd size reduction observed from 1990 to 2020 (European Environment Agency, 2022). By implementing strong, aggressive reductions in CH_4 emissions as outlined in these scenarios, the California dairy sector can serve as a model to reduce the global dairy industry's impact on atmospheric warming. To make this happen, policymakers in California and around the world will be increasingly required to balance the need to meet the world's growing demand for milk and positively impact rural communities in a climate smart way (Rojas-Downing *et al.*, 2017; Lane *et al.*, 2019).

Conclusion

California dairy's GHG footprint is highly biased toward CH_4 emissions from enteric and manure management sources. By aggressively decreasing CH_4 emissions and utilizing GWP* to account for the industry's true warming impact, the present study demonstrates that the California dairy industry has the potential to contribute no additional annual warming and achieve climate neutrality before the end of the decade. Achieving the 40% reduction in manure management CH_4 emissions, set out under SB 1383, is enough to help the industry achieve climate neutrality. Climate-negative warming (or cooling) continues onwards through 2030 and helps the industry offset historical emissions. Continued reductions in CH_4 emissions will be necessary for climate neutrality to be achieved by California dairy to offset annual CO_2 and N_2O emission contributions. For California and the global dairy industry, there is the opportunity to achieve climate neutrality through a range of policies and technology programs to create a climate-compatible dairy sector.

Acknowledgements

The authors would like to thank the California Dairy Research Foundation (CDRF) for their support of this research.

Conflicts of Interest

The authors have no conflicts of interest to declare.

References

Alexandratos, N. and Bruinsma, J. (2012) World Agriculture towards 2030/2050: The 2012 Revision. WORLD Agriculture. Available at: www.fao.org/economic/esa (accessed 4 October 2021).

Allen, M.R., Fuglestvedt, J.S., Shine, K.P., Reisinger, A., Pierrehumbert, R.T. *et al.* (2016) New use of global warming potentials to compare cumulative and short-lived climate pollutants. *Nature Climate Change* 6(8), 773–776.

Allen, M.R., Shine, K.P., Fuglestvedt, J.S., Millar, R.J., Cain, M. *et al.* (2018) A solution to the misrepresentations of CO2-equivalent emissions of short-lived climate pollutants under ambitious mitigation. *npj Climate and Atmospheric Science* 1(1), 1–8. Available at: https://www.nature.com/articles/s41612-018-0026-8.

Bosoli, D., Blumenthal, C., Andrews, S. and Marks, J. (2019) Global Non-CO2 Greenhouse Gas Emission Projections & Mitigation 2015–2050. Available at: https://www.epa.gov/sites/default/files/2020-05/documents/epa_non-co2_greenhouse_gases_rpt-epa430r19010.pdf

Britt, J.H., Cushman, R.A., Dechow, C.D., Dobson, H., Humblot, P. *et al.* (2018) Invited review: Learning from the future—A vision for dairy farms and cows in 2067. *Journal of Dairy Science* 101(5), 3722–3741.

Bruun, S., Jensen, L.S., Khanh Vu, V.T. and Sommer, S. (2014) Small-scale household biogas digesters: An option for global warming mitigation or a potential climate bomb? *Renewable and Sustainable Energy Reviews* 33, 736–741.

California Air Resources Board (2021) California Greenhouse Gas Emissions for 2000 to 2019 – Trends of Emissions and Other Indicators. Vol. 20. Sacramento; Available at: https://ww3.arb.ca.gov/cc/inventory/pubs/reports/2000_2019/ghg_inventory_trends_00-19.pdf.

Cain, M., Lynch, J., Allen, M.R. and Fuglestvedt, J.S. (2019) Improved calculation of warming-equivalent emissions for short-lived climate pollutants. *npj Climate and Atmospheric Science* 2(1), 29. DOI: 10.1038/s41612-019-0086-4.

California Air Resources Board (CARB) (2022a) Analysis of Progress toward Achieving the 2030 Dairy and Livestock Sector Methane Emissions Target, California Air Resources Board. Sacramento. Available at: https://ww2.arb.ca.gov/resources/documents/dairy-livestock-sb1383-analysis

California Air Resources Board (CARB) (2022b) Draft 2022 State Scoping Plan Update. Available at: https://ww2.arb.ca.gov/our-work/programs/ab-32-climate-change-scoping-plan/2022-scoping-plan-documents

California Department of Food & Agriculture (CDFA) (2021) California Agricultural Statistics Review 2020–2021. Available at: https://www.cdfa. ca.gov/Statistics/PDFs/2021_Ag_Stats_Review.pdf

California Department of Food & Agriculture (CDFA) (2022a) Alternative Manure Management Program-Project Level Data. Sacramento. Available at: https://www.cdfa.ca.gov/oefi/AMMP/

California Department of Food & Agriculture (CDFA) (2022b) Report of Funded Projects (2015–2022) . Available at: https://www.cdfa.ca.gov/oefi/ddrdp/docs/2022_DDRDP _Legislative_Report.pdf

Cantrell, C.A., Shetter, R.E., Mcdanml, A.H., Calvert, J.G., Davidson, A. *et al.* (1990) Carbon kinetic Isotope effect in the oxidation of methane by the hydroxyl radical. *Journal of Geophysical Research: Atmospheres* 95(22), 455–462.

Cowley, C. and Wade Brorsen, B. (2018) The hurdles to greater adoption of anaerobic digesters. *Agricultural and Resource Economics Review* 47(1), 132–157.

Dairy Cares (2022) Dairy Digesters | Dairy Digesters in California | Creating Clean Energy. Available at: https://www.dairycares.com/dairy-digesters (accessed 8 December 2022).

European Environment Agency (2022) Methane Emissions in the EU: The Key to Immediate Action on Climate Change. Available at: https://www.eea.europa.eu/publications/methane-emissions-in-the-eu (accessed 30 March 2023).

FAO (2018) Climate Change and the Global Dairy Cattle Sector-The role of the dairy sector in a low-carbon future. Rome. Available at: https://www.fao.org/3/CA2929EN/ca2929en.pdf

Feng, X. and Kebreab, E. (2020) Net reductions in greenhouse gas emissions from feed additive use in California dairy cattle. *PLoS One* 15(Sep), 1–13. DOI: 10.1371/journal.pone.0234289.

Gunn, K.M., Holly, M.A., Veith, T.L., Buda, A.R., Prasad, R, *et al.* (2019) Projected heat stress challenges and abatement opportunities for U.S. milk production. *PLoS One* 14(3), e0214665. DOI: 10.1371/journal.pone.0214665.

Haines, A., Amann, M., Borgford-Parnell, N., Leonard, S., Kuylenstierna, J. *et al.* (2017) Short-lived climate pollutant mitigation and the sustainable development goals. *Nature Climate Change* 7(12), 863–869. DOI: 10.1038/s41558-017-0012-x.

IPCC (2014) Climate Change 2014 Synthesis Report Summary for Policymakers. IPCC 5th Assess Rep 1–31.

Kirschke, S., Bousquet, P., Ciais, P., Saunois, M., Canadell, J. *et al.* (2013) Three decades of global methane sources and sinks. *Nature Geoscience* 6, 813–823.

Lane, D., Murdock, E., Genskow, K., Betz, C.R. and Chatrchyan, A. (2019) Climate change and dairy in New York and Wisconsin: Risk perceptions, vulnerability, and adaptation among farmers and advisors. *Sustainability* 11(13), 3599. Available at: https://www.mdpi.com/2071-1050/11/13/3599/htm

Liu, S., Proudman, J. and Mitloehner, F.M. (2021) Rethinking methane from animal agriculture. *CABI Agriculture and Bioscience* 2(1), 1–13. DOI: 10.1186/s43170-021-00041-y.

Macdonald, J.M., Law, J. and Mosheim, R. (2020) Consolidation in US. dairy farming. *ERS* 274, 54. Available at: www.ers.usda.gov

Matthews, J.B.R., Fuglestvedt, J.S., Masson-Delmotte, V., Möller, V., Méndez, C. *et al.* (2021) Annex VII: Glossary. Pages 2215–2256 in Climate Change 2021: The Physical Science Basis. Contribution of Working Group I to the Sixth Assessment Report of the Intergovernmental Panel on Climate Change 2021, 2215–2256.

Melgar, A., Welter, K.C., Nedelkov, K., Martins, C.M.M.R., Harper, M.T. *et al.* (2020) Dose-response effect of 3-nitrooxypropanol on enteric methane emissions in dairy cows. *Journal of Dairy Science* 103(7), 6145–6156.

Mitloehner, F.M., Kebreab, E., Tricarico, J.M., Wallace, J., Gooch, C. *et al.* (2020) Dairy feed additives to reduce enteric methane emissions. *Newtrient* 2020(Feb), 1–8. Available at. Available at: https://hoards.com/file-384-newtrient-update-additivespdf

Montes, F., Meinen, R., Dell, C., Rotz, A., Hristov, A.N. *et al.* (2013) Special Topics-Mitigation of methane and nitrous oxide emissions from animal operations: II. A review of manure management mitigation options. *Journal of Animal Science* 91(11), 5070–5094.

Muñoz, I. and Schmidt, J.H. (2016) Methane oxidation, biogenic carbon, and the IPCC's emission metrics. Proposal for a consistent greenhouse-gas accounting. *The International Journal of Life Cycle Assessment* 21(8), 1069–1075. DOI: 10.1007/s11367-016-1091-z.

Naranjo, A., Johnson, A., Rossow, H. and Kebreab, E. (2020) Greenhouse gas, water, and land footprint per unit of production of the California dairy industry over 50 years 103(4), 3760–3773. Available at: http://www.journalofdairyscience.org/article/S0022030220300746/fulltext (accessed 5 January 2022).

Pineda, A.C. and Faria, P. (2019) Towards a science-based approach to climate neutrality in the corporate sector – Discussion paper. In: *SBTi*. In, pp. 1–29. Available at: https://sciencebasedtargets.org/resources/files/Towards-a-science-based-approach-to-climate-neutrality-in-the-corporate-sector-Draft-for-comments.pdf

Place, S. and Mitloehner, F.M. (2021) Pathwayto Climate Neutrality for U.S. Beef and Dairy Cattle Production. Davis. Available at: https://clear. ucdavis.edu/sites/g/files/dgvnsk7876/files/inline-files/White-paper-climate-neutrality-beef-dairy.pdf

Reay, D., Sabine, C., Smith, P. and Hymus, G. (2007) IntergovernmentalPanel on Climate Change. Fourth Assessment Report. Geneva, Switzerland:Inter-governmental Panel on Climate Change. Cambridge University Press, Cambridge, UK. Available at: https://www.ipcc.ch/site/assets/uploads/2018/02/ar4_syr_full_report.pdf

Rojas-Downing, M.M., Nejadhashemi, A.P., Harrigan, T. and Woznicki, S.A. (2017) Climate change and livestock: Impacts, adaptation, and mitigation. *Climate Risk Management* 16, 145–163.

Rotz, C.A. (2018) Modeling greenhouse gas emissions from dairy farms. *Journal of Dairy Science* 101(7), 6675–6690. DOI: 10.3168/jds.2017-13272.

Shindell, D., Borggord-Parnell, N., Brauer, M., Haines, A., Kuylenstierna, J.C.I. *et al.* (2017) A climate policy pathway for near-and long-term benefits. *Science (80-)* 356(6337), 493–495.

Smith, M.A., Cain, M. and Allen, M.R. (2021) Further improvement of warming-equivalent emissions calculation. *npj Climate and Atmospheric Science* 4(1), 1–3. Available at: https://www.nature.com/articles/s41612-021-00169-8

Solomon, S., Plattner, G.K., Knutti, R. and Friedlingstein, P. (2009) Irreversible climate change due to carbon dioxide emissions. *Proceedings of the National Academy of Sciences USA* 106(6), 1704–1709.

State of California (2006) Assembly Bill No. 32. Sacramento. pp. 1–13. Available at: http://www.leginfo.ca.gov/pub/05-06/bill/asm/ab_0001-0050/ab_32_bill_20060927_chaptered.pdf

State of California (2016) Senate Bill No.1383. Available at: https://leginfo. legislature.ca.gov /faces/billNavClient.xhtml?bill_id=201520160SB1383

Stefenoni, H.A., Räisänen, S.E., Cueva, S.F., Wasson, D.E., Lage, C.F.A. *et al.* (2021) Effects of the macroalga *Asparagopsis taxiformis* and oregano leaves on methane emission, rumen fermentation, and lactational performance of dairy cows. *Journal of Dairy Science* 104(4), 4157–4173.

USDA (2022) Quick Stats. National Agricultural Statistics Service (NASS). Available at: https://quickstats.nass.usda.gov/ (accessed 10 January 2022).

Van Dingenen, R., Crippa, M.J., Anssens-Maenhout, G., Guizzardi, D. and Dentener, F. (2018) Global trends of methane emissions and their impacts on ozone concentrations. *Vol. EUR29394EN, JRC Science for Policy Report* 1–90.

Vijn, S., Compart, D.P., Dutta, N., Foukis, A., Hess, M. *et al.* (2020) Key considerations for the use of seaweed to reduce enteric methane emissions from cattle. *Frontiers in Veterinary Science* 7(Dec), 1–9.

York, L., Heffernan, C. and Rymer, C. (2017) A comparison of policies to reduce the methane emission intensity of smallholder dairy production in India. *Agriculture, Ecosystems & Environment* 246, 78–85.

13 Recent advances in breeding maize for drought, heat and combined heat and drought stress tolerance in Sub-Saharan Africa

Baffour Badu-Apraku[1]*, Morakinyo A.B. Fakorede[2], Charles Nelimor[3], Alimatu S. Osuman[1,4,5], Tégawendé O. Bonkoungou[1], Oyekunle Muhyideen[6] and Richard O. Akinwale[2]

[1]International Institute of Tropical Agriculture (IITA), PMB 5320, Ibadan, Nigeria; [2]Department of Crop Production and Protection, Obafemi Awolowo University, P.O. Box 2586, Ile-Ife, Nigeria; [3]Savanna Agricultural Research Institute (CSIR-SARI), P.O. Box TL 52, Tamale, Ghana; [4]West Africa Center for Crop Improvement (WACCI), University of Ghana, PMB 30 Legon, Accra, Ghana; [5]Crops Research Institute, Ghana. P.O. Box 3785, 039-5028 Kumasi, Ghana; [6]Department of Plant Science, Institute for Agricultural Research/Ahmadu Bello University, Zaria, Nigeria

Abstract

Drought, heat, and combined drought and heat are important abiotic stresses constraining the production and productivity of maize (*Zea mays* L.) in sub-Saharan Africa (SSA). In the face of climate change, these stresses are likely to occur simultaneously and put at risk food and economic security in SSA. This review describes maize breeding activities conducted by the International Institute of Tropical Agriculture (IITA) in partnership with national scientists under the Drought Tolerant Maize for Africa (DTMA) and Stress Tolerant Maize for Africa (STMA) projects, which together sought to develop and deploy multiple stress tolerant hybrids, and open-pollinated varieties. Emphasis was on (i) developing a reliable methodology for screening maize for tolerance to drought stress (DS), heat stress (HS), and combined drought and heat stress (CDHS) using key secondary traits and grain yield, (ii) use of appropriate breeding techniques for tailoring maize for tolerance to DS, HS and CDHS, (iii) exploring diverse sources of germplasm for genetic enhancement of maize, (iv) extensive multilocational evaluation to identify genotypes with stable performance under the stresses, and (v) application of genomic tools to accelerate genetic gains in maize breeding at IITA. At IITA, the performance of maize hybrids under stresses of DS, HS and CDHS have been improved using conventional breeding techniques/procedures. These techniques/procedures have led to accelerated genetic gains in yield that were 26–49% higher than the best commercial hybrid checks under CDHS and DS. Additive gene action has been consistently found to be more important than the non-additive among early maize under DS and CDHS while both the additive and non-additive have been reported to be important for the extra-early maize. The most reliable secondary traits for selecting for improved grain yield under the stresses include anthesis-silking interval, ears per plant, and plant and ear aspects. Several early and extra-early landraces have been identified as potential sources of tolerance to DS, HS, and CDHS. Several quantitative trait loci (QTLs) associated with grain yield and key secondary traits have been identified *via*

*Corresponding author: b.badu-apraku@cgiar.org

genome-wide association studies in landraces and inbred lines. Those desirable QTLs, upon validation, could be invaluable for genomics-enabled breeding.

Keywords: maize, drought, heat, combined drought and heat stress, conventional breeding, molecular breeding, genetic gain, West and Central Africa

Review Methodology

Reviewed are research papers dealing with drought, heat, and drought-plus-heat tolerance, particularly in West and Central Africa (WCA). First, we dealt with the physiology of drought and combined drought and heat stress (CDHS) tolerance and the effects on maize morphology and grain yield. Then we reviewed the research resulting from the mandate of the Semi-Arid Food Grain Research and Development (SAFGRAD) Project, covering more than 40 years in sub-Saharan Africa (SSA) from which substantial information has been accumulated by different researchers, including the most recent efforts by IITA and International Maize and Wheat Improvement Centre (CIMMYT). Our attention was focused on the synthesis of how the research was done, the research conclusions, and the missing gaps that new research may fill. Our emphasis was on the contribution to the genetic improvement of drought and CDHS tolerance in maize, grouping together similar findings and pointing out dissimilar results with a possible explanation of why such results were observed. Evidence of new research methods pertaining to drought and drought-plus-heat tolerance in maize was thoroughly and critically scrutinized and we identified unanswered research questions on DS and CDHS tolerance in maize. We searched through the existing literature from about 1980 to date; therefore, the review covered conventional and the ongoing molecular aspects of DS, HS, and CDHS tolerance research in maize.

General Overview

Beginning from 1950, scientists in WCA started noticing drought during the crop-growing seasons. Initially, it occurred as dry spells and plant breeders easily overcame this by utilizing the information on genotype ×environment interaction (GEI) present within crop species. As time went on, the dry spells were longer, covering wider areas of land, and leading to serious crop yield losses in some cases and, in others, complete crop failure. The situation called for attention; therefore, in 1977, the Ministers of Foreign Affairs of the WCA states met to deliberate on the serious crop yield-limiting environmental factor termed drought. One important outcome of the meeting was the establishment of the Semi-Arid Food Grain Research and Development (SAFGRAD) Project. The IITA, Ibadan, Nigeria teamed up with SAFGRAD as the executing agency for the maize (*Zea mays* L.) component, which later became autonomous and was named West and Central Africa Collaborative Maize Research Network (WECAMAN). SAFGRAD covered the National Agricultural Research Systems (NARS) of the 17 countries in West Africa, while WECAMAN covered only eight (i.e. Benin, Burkina Faso, Cameroon, Cote d'Ivoire, Ghana, Mali, Nigeria, and Senegal) of the 17 but later expanded to 11 countries (Guinea, Tchad, and Togo). The work of SAFGRAD and the first WECAMAN was restricted to the savannas, but the expanded WECAMAN covered all agro-ecologies of the 11 countries. Research and development activities of SAFGRAD and WECAMAN were funded primarily by United States Agency for International Development (USAID) but were later joined by the United Nations Development program (UNDP), International Fund for Agricultural Development (IFAD), and some other donors, especially the Bill and Melinda Gates Foundation. This collaboration of IARC-NARS-Donor has resulted in the development and production of drought and heat-tolerant maize varieties that are presently revolutionizing maize production in the savannas of WCA.

Drought and Combined Drought and Heat Stress in Maize in Sub-Saharan Africa

Over 300 million people in SSA depend on maize as their main food source, making it the most important staple food crop in the region. Maize has the potential to address the challenges of food insecurity and malnutrition presently facing the region. Maize is cultivated on about 100 million hectares of land across the developing world where the majority of people depend on it for at least 30% of their dietary calorie needs (Shiferaw *et al.*, 2011). The demand for maize in the developing world is anticipated to double by the middle of this century as a result of increased acceptance and rising population growth (Rosegrant *et al.*, 2009). Because of the high incoming solar radiation, prevailing low humidity, and low night temperature, the savannas of WCA have less diseases and insect pest pressures, which have made this part of Africa the potential maize production belt of Africa (Badu-Apraku *et al.*, 2017). Studies have shown that *Striga hermonthica* parasitism and recurrent drought are important biotic and abiotic constraints to maize production in the SSA (Badu-Apraku *et al.*, 2010; Amegbor *et al.*, 2017; Badu-Apraku *et al.*, 2020). This is partly due to the frequency and unpredictability of the occurrence of drought in the region, which often coincides with *Striga* parasitism on farmer's field, resulting in high yield losses (Badu-Apraku *et al.*, 2018). Damage to the maize crop due to DS is unpredictable as it is determined by different factors and is not limited to rainfall patterns, moisture and water holding capacity of the soil and water losses through evapotranspiration. There are usually differences in plant responses to drought depending on plant growth stage and other environmental factors (Demirevska *et al.*, 2009). The key yield-limiting factors under an inadequate supply of soil moisture include reduced absorption of photosynthetically active radiations, impaired radiation use efficiency, and decreased harvest index (Earl and Davis, 2003).

The early (90–95 days to physiological maturity) and extra-early (80–85 days to physiological maturity) maturing varieties play a vital role in reducing hunger in the SSA and more importantly in WCA. This is due to their better response to fertilizer application, ability to escape extreme climatic conditions, and their inherent ability to attain physiological maturity in a relatively shorter period (Pswarayi and Vivek, 2008; Badu-Apraku *et al.*, 2013a). Due to the inherent ability of extra-early maturing varieties to both tolerate and escape drought, maize is now grown in the drier areas of the southern guinea savannas (SGS) and some parts of the northern fringes of the guinea savanna (NGS) of WCA, which is the home of other cereals such as sorghum (*Sorghum bicolor* L.) and millet (*Pennisetum glaucum* [L.] R. Br.) (Badu-Apraku and Fakorede, 2017; Badu-Apraku *et al.*, 2020). Early and extra-early maize varieties are also in high demand by farmers in the rainforest agroecology of WCA where they are compatible in intercrop with other arable crops such as cassava (*Manihot esculenta* Crantz) (Adeniyan *et al.*, 2014; Badu-Apraku and Fakorede, 2017), pigeon pea (*Cajanus cajan* [L.] Millsp) (Dania *et al.*, 2014), and cowpea (*Vigna unguiculata* [L.] Walp.) (Takim, 2012; Badu-Apraku and Fakorede, 2017).

Despite being an important food security crop in SSA, the production and average grain yield (GY) of maize per hectare are low on farmers' fields because of recurring droughts during the cropping season. It has been estimated that about 15% of the annual yield loss in SSA is caused by DS (Edmeades, 2008). Edmeades *et al.* (1999) and Lafitte and Edmeades (1995) reported that about 17% of yield loss was caused by DS, while in southern Africa DS causes as much as 60% yield loss (Rosen and Scott, 1992; Monneveux *et al.*, 2006). The significant reduction in crop yields caused by drought is through the detrimental effect on plant growth, physiology, and reproduction (Barnabás *et al.*, 2008). Maize is more sensitive to DS than most other cereal crops. During early vegetative growth, water is critical to both the morphological and physiological development of the maize plant. The size and number of leaves (structural components) are reduced in maize because of DS. Under DS, the relative water content, involved in maintaining the plant-water relationship and determining the metabolic status of cells and tissue is reduced, resulting in high leaf temperature due to reduced transpiration cooling (Siddique *et al.*, 2001). Additionally, transpiration is affected; photosynthesis is impaired due to stomata and non-stomata limitations, and light interception is affected through distorted leaf structure (Aslam *et al.*, 2015). Translocation of assimilates to the reproductive parts for their elongation rather than roots is the most probable cause of the severe effects of drought at the reproductive stage (Taiz and Zeiger, 2006). Reproductive-stage drought (RSD) interferes with plant growth, nutrient, and water relations, thereby affecting photosynthesis and assimilate partitioning, causing a reduction in kernel size, abnormal differentiation of the spikelet, inhibiting ear elongation, and ultimately reducing GY (Farooq *et al.*, 2009; Praba *et al.*, 2009; Wang *et al.*, 2016). Reduction in GY due to RSD is consequent upon limiting invertase activities in the ovaries, causing a reduced flux of hexose sugar accumulation, altered hormonal balance, leading to ovary abortion (Chourey *et al.*, 2010). Photosynthesis is reduced when drought occurs just before the flowering stage leading to delayed silk extrusion, thereby increasing the period between anthesis and silking, and ultimately causing a reduction in yield due to a lack of synchrony between anthesis and silking (Bolaños and Edmeades, 1996). The timing of pollen shed in maize is not significantly affected by drought; however, pollen mortality, specific gravity, settling speed, pollen shape, and dispersal are affected due to pollen dehydration (Aylor, 2002). In addition to delayed silk extrusion which could be up to 5–9 days, DS causes pre-fertilization drying of silk, which results in fertilization failure and barrenness (Dass *et al.*, 2001). Reproductive growth stages such as pollination, embryo development, endosperm development, and kernel development, are also threatened by drought (Aslam *et al.*, 2015). Daryanto *et al.* (2016) analysed the data of studies from 1980 to 2015 and reported that DS accounted for about 40% reduction in GY in maize. Globally, an annual loss of 15–20% in maize yield was reported between 1980 and 2008 resulting from recurrent drought (Lobell *et al.*, 2011). Moreover, maize production in SSA is rain-fed, thus exposing the region to high yield loss concomitant to drought occurrence. Drought has become a regular occurrence in SSA, due to irregular rainfall patterns in most agro-ecologies in the sub-region (Badu-Apraku *et al.*, 2020). Reduction in maize GY as a result of DS depends on the growth stage of the crop at which the stress occurs, and on the intensity and duration of the drought (Betran *et al.*, 2003; Badu-Apraku and Fakorede, 2017). During the plant developmental stages, water requirements for maize production are high, with a total of 250 litres of water required per plant during the growing season (Du Plessis, 2003; Warburton *et al.*, 2017). The peak period of water demand during the growth cycle of the maize plant is from 2 weeks before, to 2 weeks after pollination (Warburton *et al.*, 2017; Nielsen, 2018). Grain yield reduction in maize could be as high as 90% when drought occurrence coincides with the reproductive and grain-filling stages of development (NeSmith and Ritchie, 1992).

In addition to the recurrent drought in SSA, excessive radiation and high temperatures are the main factors that determine plant growth and development. HS speeds up the rate of respiration in the seed thereby reducing the water content of the seed, thus reducing the germination potential of the seeds resulting in poor germination and stand establishment. The undesirable effects of high temperatures on cereal crops differ depending on the timing, duration, and harshness of the HS (Fahad *et al.*, 2017). High temperatures affect the quantity and viability of pollen produced resulting in reduced fertilization of ovules and thus reduced sink capacity (Schoper *et al.*, 1986). At the early reproductive stage of maize, elevated temperatures also delay the emergence of silks (female reproductive tissues) leading to the asynchrony of flowers which further reduces fertilization (Cicchino *et al.*, 2011). Thompson (Thompson, 1975) reported GY reduction of maize of about 10% when there was a temperature increase in the US Corn Belt from 22 to 28°C during the grain-filling period. Badu-Apraku *et al.* (1983) also reported a yield loss of 42% of temperate maize when average daily temperatures increased by 6°C.

In SSA, the growth cycle of maize is greatly affected by the reduction and irregular rainfall caused by HS and DS because of global warming through climate variability (Fakorede *et al.*, 2003). Under field conditions, DS and HS occur simultaneously but the latter is more destructive to the maize plant during flowering as extreme heat denatures the pollen grains and prevents the accessibility of the silks (Badu-Apraku *et al.*, 2017). The combined effects of HS and DS on maize have detrimental effects on the growth and development of the crop and lead to total yield losses as compared to the effects of the individual stresses on the crop (Savage and Jacobson, 1935; Jagtap *et al.*, 1998; Jiang and Huang, 2001; Heyne and Brunson, 1940). Several researchers have reported that the interactions among the individual stresses have an appreciably greater negative impact on crop yield and productivity than the stress components applied individually (Mittler and Blumwald, 2010; Suzuki *et al.*, 2014). CDHS elicits varying effects on crop growth, development, and yield processes of plants (Prasad *et al.*, 2015). Globally, crop yield losses caused by CDHS alone equated to approximately US \$200 billion annually between 1980 and 2012 (Prasad *et al.*, 2015), while within the same period, DS only caused \$50 billion damage to agricultural production (NOAA National Centers for Environmental Information (NCEI) U.S. Billion-Dollar Weather and Climate Disasters, 2023). These growing losses represent a major impediment to global food security, with a significant impact on the livelihood of smallholder farmers in SSA. The combination of DS and HS perhaps requires a hostile response of the plant (Prasad *et al.*, 2008) such that it would increase the stomatal conductance when the temperature increased beyond the optimal for the leaves to be cooled through increased transpiration. Conversely, if the temperature increased beyond the optimal coincided with drought, the plant is unable to open the stomata and the leaf temperature could be 2–5°C higher (Rizhsky *et al.*, 2002; Rizhsky *et al.*, 2004). At the whole plant level, activities of ribulose 1,5-bisphosphate carboxylase/oxygenase (Rubisco) is highly reduced at temperatures higher than 35°C thereby restricting photosynthesis (Crafts-Brandner and Law, 2000; Griffin *et al.*, 2004).

Review of DTMA and STMA Projects-Stress Resilient Maize Varieties for African Farmers

For the past three decades, maize has emerged as an important food and economic security crop for millions of people in SSA. Despite its importance, the average GY of maize is still low in farmers' fields in SSA due to constant exposure of the crop to recurrent drought, parasitic weeds (primarily *Striga hermonthica*), diseases, insect pests, inappropriate crop management practices, inadequate input use, and institutional factors. The Drought Tolerant Maize for Africa (DTMA) project was thus launched in 2006 to develop and promote the commercialization of maize hybrids and varieties with stable yields and enhanced tolerance to drought, resistance to *Striga*, and major diseases prevalent in Eastern, Southern, and Western Africa. The DTMA Project, financed by the Bill and Melinda Gates Foundation (BMGF) and the Howard G. Buffet Foundation (HGBF) was run in three phases (2006–2015) and jointly implemented by CIMMYT and IITA in close collaboration with the National Agricultural Research Systems (NARES), private seed companies, community-based organizations (CBOs), Non-governmental Organizations (NGOs), and universities in 13 target countries across Africa, including Ethiopia, Kenya, Tanzania, Uganda, Angola, Malawi, Mozambique, Zambia, Zimbabwe, Benin, Ghana, Mali, and Nigeria. The first two phases of this project focused primarily on product development, testing, registration, and release of maize varieties through the building of the required partnerships with both public and private sector institutions while strengthening their capacity to promote and deploy stress-resilient products to farmers. The third phase had a strong emphasis not only on the development of better stress-resilient open-pollinated varieties (OPVs) and hybrids but also it facilitated the production of quality seeds, promotion, and delivery of these OPVs and hybrids to African farmers. The Stress Tolerant Maize for Africa (STMA) project followed the DTMA project in 2016 with continued funding from the BMGF and USAID to develop maize varieties with combined multiple stress tolerance and enhanced productivity that could effectively respond to emerging and future production challenges for out scaling.

The two projects catalyzed the release of 254 new stress-resilient maize hybrids and OPVs in target countries across Africa between 2007 and 2017. Out of these, IITA in collaboration with its national and private sector partners released 44 hybrids and 54 OPVs of extra-early-, early-, intermediate-, and late-maturity combining tolerance to drought with resistance to *Striga* in West Africa. These OPVs and hybrids have high levels of resistance to the major diseases in the savannas and have shown average yield advantages of 20–30% over the commercial checks across stressful and favourable field conditions in on-farm trials conducted in West Africa. These efforts have contributed to the release of new maize varieties in the target countries. The DTMA project also worked with public sector partners and private seed companies and produced approximately 52,000 metric tonnes (MT) of the seed of the stress-resilient maize OPVs and hybrids in 2014 alone. The STMA project planned, produced, and disseminated an additional 50,000 MT seeds of multiple stress-tolerant OPV and hybrid maize in eight target countries in SSA at the end of the project in 2019. Under the DTMA and STMA projects, IITA's national partners and seed companies produced a total of 115,816 MT seed of stress-resilient maize OPVs and hybrids from 2007 to 2017, which covered nearly 5.8 million ha of land and benefited millions of farmers in West Africa alone. This was driven by IITA's engagement in the constant supply of breeder seeds of the released OPVs and parents of hybrids to the NARS and private seed companies under the two projects. Significant investments were also made under these projects in training techniques and management staff of the national partners and seed companies to strengthen their human capacity to produce and market good quality seeds of stress-resilient OPVs and hybrids to a greater number of farmers in West Africa. The DTMA and STMA projects have established effective product pipelines with several promising stress-resilient OPVs and hybrids in the pipeline for release and commercialization. Consequently, increased investment in strengthening the public and private sectors will still be needed to speed up the uptake and commercialization of new multiple stress resilient OPVs and hybrids that can replace the old ones in farmers' fields to help achieve higher yield gains.

In conclusion, maize is now cultivated by the majority of farmers in SSA Africa due to its adaptability to various farming systems and potential to increase output in comparison to other grain crops. The new maize varieties available to farmers have had a considerable impact largely due to the better response of maize to fertilizer and the availability of high-yielding, multi-stress-tolerant OPVs and hybrids, which have had a major impact in stimulating production, especially in the savanna areas since the 1980s. The production of stress-resilient maize varieties that can be consumed as both green maize and grain has helped in addressing seasonal and transitory food insecurities, has increased maize production and helped to overcome chronic food insecurity, increasing the availability of food for the most vulnerable groups in West Africa. However, a vibrant and efficient functioning seed sector is required to ensure wide distribution of the new multiple stress-tolerant OPVs and hybrids required to increase farmers' incomes, reduce food prices for consumers, and improve the profitability and sustainability of the seed industry. Results of an impact assessment study in rural Nigeria by Abdoulaye *et al.* (2018) showed that the adoption of improved maize varieties on productivity and household welfare outcomes have increased maize productivity and per capita expenditure by 32 and 22%, respectively. The authors found that without the adoption of improved multiple stress-tolerant OPVs and hybrids, the prevalence of food insecurity would have been higher by 6 and 50%, respectively.

Drought and Heat Response Mechanisms in Maize

Drought tolerance in maize denotes varieties that can adapt to sub-optimal water conditions with minimal yield reduction. The consequences of DS are difficult to identify because they not only affect yield but also affect other yield potentials. However, minimal differences in yield under drought and optimum water conditions are often used as the basis for selection under this condition (Mitra, 2001). Maize response to drought is broadly classified into four categories: drought escape, drought avoidance, tolerance, and recovery (Badu-Apraku and Fakorede, 2017). Drought escape in maize involves shortening of the life cycle of the plant such that the onset of drought does not coincide with critical periods of development. Drought avoidance involves physiological processes such as the regulation of the stomata, leaf wax, and deep root system (Van Oosten *et al.*, 2016).

Drought recovery refers to the inherent ability of the plant to produce appreciable yield after a period of exposure to DS, which causes dehydration (Lawlor, 2013). Due to climate change and other variables, drought avoidance and tolerance are the most reliable response mechanisms (Heschel and Riginos, 2005). Drought tolerance is the ability of a plant to maintain its growth, development, and to have a relatively high yield under DS (Aslam *et al.*, 2015). There is difficulty in identifying target traits in improving maize for tolerance to DS, due to its diverse effects at different stages of development. Plant phenology, normal growth and development, good grain filling, and translocation of assimilates are important traits for targeting the improvement of maize for tolerance to drought (Edmeades, 2013). One important lesson learned from the drought-tolerant studies spanning over three decades of breeding was that the drought-tolerant traits had differential responses to the levels of stress severity. Therefore, before initiating a breeding program for drought tolerance in maize, it is imperative for breeders to determine the level of stress tolerance of the most appropriate germplasm and traits of interest.

Breeding Maize for Drought Tolerance

It has been well established that the genetic improvement of maize for tolerance to drought could result in genetic gains (Edmeades *et al.*, 1999; Badu-Apraku *et al.*, 2011a). One important way to close the gap between potential and realized yield under DS is to adopt agronomic practices that effectively maximize water availability to the plant (Campos *et al.*, 2004). Therefore, drought-tolerant maize genotypes could be valuable germplasm resources in environments with the erratic occurrence of varying intensities of drought (Akaogu *et al.*, 2017). Drought-tolerant maize varieties offer the most economic and sustainable opportunity to stabilize maize yields (Badu-Apraku *et al.*, 2017). An important strategy to improve maize production and productivity in SSA is to develop drought-tolerant genotypes for resource-poor farmers.

Over the years, significant gains have been recorded in improving maize yield under DS. However, progress has been slow since drought is a complex quantitative trait controlled by multiple genes and signalling pathways (Bartels and Sunkar, 2005; Fleury *et al.*, 2010), highly influenced by genotype ×environment [G×E] effects, and characterized by low yield heritability (Ziyomo and Bernardo, 2013; Maazou *et al.*, 2016). To achieve accelerated genetic gains for improvement for GY under drought and to better understand the complex physiological nature of the responses of maize to drought at the different stages of growth, the use of marker-assisted selection (MAS) was proposed (Jacobsen and Schouten, 2007; Ribaut and Ragot, 2007). Since major QTL are not influenced by environmental effects, they can be verified and introgressed into target genotypes without linkages (Gupta *et al.*, 2010). However, progress from MAS is limited by challenges such as the number of QTL regulating the traits of interest, distribution of the QTL alleles, and epistasis. To overcome these challenges, the use of genomic selection (GS) was proposed. The use of GS increases the precision of molecular markers in the identification of genotypes with genes for drought tolerance (Shikha *et al.*, 2017); it also identifies the genomic regions with alleles associated with each specific trait in a population (Li *et al.*, 2018). Even though both MAS and GS rely on the marker-trait association, GS identifies the genomic regions associated with the specific drought-tolerant trait in a wider population as against MAS which considers recombinant inbred lines (RIL) from a bi-parental cross (the application of MAS and GS for stress tolerance maize breeding is given under the next section of this review).

The relative performance of genotypes under DS and non-stress (NS) conditions has played a significant role in selecting promising genotypes under limiting water conditions. Maize improvement through conventional means for drought tolerance in recent years has been achieved through multi-locational, multi-year evaluations with an emphasis on tolerance to drought at the flowering stage (Badu-Apraku *et al.*, 2003; Aslam *et al.*, 2015). Identification and use of secondary traits for indirect selection have, under drought and well-watered conditions effectively improved the yield of tropical maize for tolerance to DS (Ziyomo and Bernardo, 2013; Badu-Apraku *et al.*, 2015; Badu-Apraku *et al.*, 2016b). However, under different moisture conditions, high variation in heritability estimates, and a decrease in heritability as drought intensifies occurred (Bolaños and Edmeades, 1996). Nevertheless, some traits such as the ASI and the number of ears per plant have displayed constant or increased genetic variances and heritability and high genetic correlation for GY under drought. Indirect selection through secondary traits has led to gains in GY under drought. Badu-Apraku *et al.* (2018) reported 423 kg/ha/cycle was achieved in GY using secondary traits under drought. It has been established over decades through the characterization of germplasm in breeding programs that cultivated plants, in general, have relatively lower levels of tolerance to biotic and abiotic stresses when compared to crop wild relatives (Mammadov *et al.*, 2018). Novel and useful genetic variation has been identified in teosinte (*Zea diploperennis* Iltis, Doebley, and Guzman) and incorporated into maize gene pools *via* hybridization, backcrossing, and/or selection (Warburton *et al.*, 2017). This has provided evidence that useful genetic variation can be tapped from teosinte for the improvement of domesticated maize. Several tropical maize populations have been developed by introgressing various traits for *Striga* resistance and drought tolerance from teosinte.

Breeding maize for heat and combined drought and heat stress tolerance

The use of conventional plant breeding methods and skills has been very useful in selecting stress tolerance in maize. Substantial progress has been made by the international and national research institutes in SSA in developing maize cultivars with improved DS and/or HS tolerance. For example, the maize breeding programs of CIMMYT and ITA in collaboration with national agricultural research institutes (NARS) in West Africa have successfully developed drought-tolerant inbred lines, hybrids, and open-pollinated varieties. The CDHS-tolerant maize cultivars are usually developed by screening maize germplasm under induced DS with elevated temperatures and identifying plants with superior performance (Ehlers and Hall, 1998). Phenotypic characters associated with HS-tolerance in maize have been identified by maize researchers and these include canopy temperature, stomatal conductance (Reynolds *et al.*, 1994), thylakoid membrane stability (Kumar *et al.*, 2012), stay green characteristic (Kumar *et al.*, 2010), waxy leaves (Richards, 1996), chlorophyll content, grain-filling duration, and GY (Yang *et al.*, 2002). Even though conventional breeding for tolerance to CDHS in maize has, to a large extent, been successful, maize scientists have been faced with several challenges due to the complexity of the traits associated with CDHS, coupled with the fact that the genes associated with HS tolerance are quantitatively inherited (Parmar *et al.*, 2017). Genetic gains from selection for tolerance to CDHS using conventional breeding approaches have been very limited due to a lack of adequate progress partly because of the long time required for screening the available germplasm before identifying the appropriate germplasm to initiate breeding programs targeting tolerance to CDHS.

The development of maize germplasm with tolerance to abiotic stresses is very complex due to the multiple traits involved. Drought tolerance in maize is controlled by many minor genes located on the chromosomes known as Quantitative Trait Loci (QTL) (Mohammadi *et al.*, 2005). The development of stress-tolerant cultivars is one of the most sustainable strategies to cope with climate change and the predicted reduced maize grain yield (Badu-Apraku and Fakorede, 2017). Accelerated gains in improvement for tolerance to CDHS in maize could be achieved by selecting under the prevailing harsh environments to identify germplasm that performs better in the target environments. Alternatively, QTL mapping could be adopted, followed by marker-assisted selection approaches (Ashraf *et al.*, 2008). QTL mapping helps in obtaining information on the total number of genes, the locations, and action patterns (Fahad *et al.*, 2017). In selecting drought tolerant materials, the major challenge generally encountered is the interaction between QTL and the environment (Tuberosa and Salvi, 2006). Therefore, to minimize the effects of the environment on the QTLs, a mandatory isomerization is done to ensure appropriate characterization of the QTLs once they are identified (Salvi and Tuberosa, 2005).

Modern breeding approaches such as QTLs, genome sequencing, and genome-wide association mapping have not been extensively utilized in breeding maize for CDHS. Aside from breeding maize for tolerance to DS, HS tolerance has emerged as an important objective of the maize breeding programs in SSA and a high priority for the IITA-MIP (Badu-Apraku and Fakorede, 2017) and the CIMMYT Global Maize Program (Cairns and Prasanna, 2018). The development and deployment of tropical maize OPVs and hybrids with enhanced tolerance to CDHS are considered the most economical and environmentally sustainable way to address the challenges of climate change on maize production (Rezende *et al.*, 2020). The genetic resources of maize are crucial to achieving this goal by providing the genetic diversity necessary for the development of varieties capable of meeting projected increases in food demands and environmental challenges. With their evolutionary history and adaptation to low-input agricultural systems, the landraces of maize in particular harbour abundant genetic diversity that could be necessary for breeding maize for climate resilience (Dwivedi *et al.*, 2016). However, despite their great potential, earlier genetic diversity studies involving landrace accessions of maize focused on the analysis of variation in genetic parameters to the detriment of traits conferring tolerance to abiotic stresses. Besides high yield potential and tolerance to DS and HS, appropriate cultivars for large areas of WCA including Sahel West Africa and Eastern and Southern Africa where growing cycles are short, must be extra-early (80–85 days) or early (90–95 days) maturing (Gedil and Menkir, 2019). The success of a tropical maize improvement program for drought tolerance is well reflected in genetic gains under DS (Cairns *et al.*, 2013). However, these gains are not adequate to alleviate concerns over increasing global change and rising food insecurity issues. New drought-tolerant source populations are urgently required to ensure increased genetic gains from selection to sustain food security (Cairns *et al.*, 2013). Moreover, compared to DS, research on tolerance to HS in tropical maize is still at the maiden stage (Cairns *et al.*, 2013; Alam *et al.*, 2017), and more studies are urgently needed to complement HS maize breeding efforts. Additionally, the knowledge available on the genetic mechanisms controlling the response of maize to DS and/or HS was very scarce (Yuan *et al.*, 2019). It was believed that detailed genetic characterization of early and extra-early maize accessions under DS, HS, and CDHS would contribute significantly to the identification of superior lines as sources of novel genetic variation. Such studies could be extremely invaluable for further genetic mapping and help in elucidating the genetic basis of abiotic stress tolerance and identifying genomic regions modulating the inheritance of adaptive traits. The loci, when identified and functionally validated would be extremely useful to uncover the underlying candidate genes for use in marker-assisted selection (MAS) and genomic selection (GS). This, in turn, would fast-track the genetic enhancement of maize for adaptation to climate change.

Strategy for screening for drought tolerance and combined heat and drought stress at IITA

At IITA, screening of maize germplasm for drought tolerance is carried out under both naturally occurring and managed DS. Locations that are prone to drought are exploited for screening for the ability of maize plants to withstand drought. The occurrence of such natural drought opens an avenue to assess the response of different genotypes to moisture deficit. Screening under natural DS is limited by the non-uniformity of DS in the field and lack of control over the timing as well as the severity of DS. Therefore, to confirm performance, promising genotypes selected under such conditions are subsequently evaluated in multi-dry-land locations and/or under managed drought conditions. Also, screening for reactions to drought is carried out during the dry season at a site with irrigation facilities. The normal practice is to withdraw irrigation water at the growth stage of interest to determine the degree of genetic variation to moisture deficit. Breeding success is moderate to high for moisture stress induced during flowering and or grain filling period. To simulate moisture stress at flowering, irrigation is withdrawn at this time, although irrigation water is provided about 18 days after pollen shed to prevent complete yield loss. An alternative method adopted in the IITA Maize Program is to irrigate the maize plants up to 21, 25, and 28 days after planting for the extra-early-, early and intermediate-maturing genotypes, respectively. Thereafter, irrigation water is withdrawn for the rest of the growth cycle. Effective screening for drought tolerance depends not only on the removal of the source of water supply but also on the aridity of the site and the soil type, texture, and depth. The managed drought site at Ikenne in the forest agro-ecology of Nigeria is well drained and has a high water holding capacity and the screening is done during the dry season when the drought probability is highest, the relative humidity is lowest, and the harmattan is most severe during the year. The situation is even more severe at Kadawa in the Sudan savanna where natural terminal drought occurs.

At Kadawa, the trials for CDHS are irrigated every 4 days for the first 21 and 28 days after planting for the extra-early and early maturing maize, respectively. The plants are then subjected to severe DS at elevated temperatures for 3 weeks during the month of April when the day temperatures usually vary from 35 to 45°C and night temperature ranges from 22 to 27°C. Irrigation water is supplied only once a week during grain-filling for CDHS trials and from planting till harvesting of the crop for HS trials with day temperatures ranging from 33 to 40°C and night temperatures from 24 to 28°C. Managed DS at Ikenne is achieved by planting at the end of November and supplying water to the plants through a sprinkler irrigation system that provides 17 mm of water per week as described earlier. The extra-early and early trials are irrigated for the first 3 and 4 weeks after planting (WAP), respectively, after which irrigation water is withdrawn till physiological maturity so that the plants rely on stored soil moisture for growth and development. Data for each trial are recorded on days to 50% anthesis (AD) and silking (DS), and anthesis–silking interval (ASI) is computed as the differences between DS and AD. Other data recorded are plant and ear heights (PHT; EHT), root and stalk lodging (RL; SL), plant aspect, ear aspect (PASP; EASP), husk cover (HC), and ears per plant (EPP) at harvest. Additional data obtained for the managed drought and CDHS experiments at about 70 DAP include stay-green characteristics also termed as leaf death score (LDS). Leaf-firing (LF) and tassel blast (TB) are recorded under HS and CDHS conditions during flowering. Leaf firing (drying of new/immature leaves) and TB (drying of the complete tassel or most of it without pollen extrusion) are recorded as the number of plants in a plot with the symptoms and calculated as a percentage based on the total number of plants in the plot. In field trials, TB often may not seem to be directly related to GY, as genotypes with severe TB may still produce good yields, as they may get pollen from other genotypes in the trials. However, when a genotype with this trait is planted in isolation, where no other source of pollen is available, it will end up with severe yield losses. Therefore, all genotypes with TB symptoms under HS are rejected, irrespective of their yields. Grain yield (kg ha^{-1}) under managed drought and CDHS is computed from shelled grain weight per plot and adjusted to 15% moisture content. However, under optimal conditions, GY (kg ha^{-1}) is estimated from ear weight per plot, based on 80% shelling percentage, also adjusted to 15% moisture content.

Below are some important considerations in field screening for drought tolerance under managed DS and CDHS:

i. A level (non-sloping) field is preferred. This will prevent slope-induced moisture stress differences in the field and ensure greater uniformity of the stress on genotypes under screening.
ii. The experimental field is divided into blocks to prevent border effects.
iii. Laying of pipes and sprinklers is done in such a way as to ensure uniformity of irrigation water and, in effect, DS, HS, and CDHS.
iv. Planting time is chosen carefully, so that the flowering and early grain-filling stages are exposed to natural DS, HS, and CDHS.
v. Germplasm to be screened is grouped based on maturity so that irrigation water regime is timed to similar growth stage for all genotypes screened.
vi. Each block is irrigated to field capacity whenever irrigation is desired.
vii. The pipes, sprinklers, and sprinkler heads are kept clean and ensured to deliver

uniform amount of water as desired by the breeder.

viii. An appropriate experimental design must be employed for replicated experiments. For augmented designs (unreplicated experiments), checks are planted in a systematic manner in each block.

Genetic analysis of drought and combined drought and heat stress tolerance in maize

The ability or capacity of an individual genotype to transfer its superior performance to its progeny in a cross is referred to as combining ability. This is achieved by intermating inbred lines and evaluating their performance in hybrid combinations. There are two types of combining abilities, the specific combining ability (SCA), and the general combining ability (GCA). Specific combining ability refers to the performance of an inbred line in hybrid combinations with a specific inbred line relative to its hybrid performance with other inbred parents. General combining ability is the average performance of an inbred parent in hybrid combinations compared to other inbred parents in the same hybrid combinations (Kearsey and Pooni, 1996). General combining ability is an indication of the additive gene action or breeding value, while the SCA provides information on the non-additive gene action. When additive gene action predominates, the implication is that genetic gains from selection could be maximized through population improvement methods such as the S_1, full-sib, and half-sib family selection methods while the importance of the non-additive gene action indicates that hybridization of inbred lines would be the best approach (Acquaah, 2012; Badu-Apraku *et al.*, 2021b). The knowledge of the mode of gene action controlling the inheritance of DS is, therefore, crucial to the adoption of appropriate breeding strategies and identifying target genes in developing drought-tolerant varieties. There have been conflicting reports on the mode of gene action controlling the inheritance of GY under drought. For instance, Derera *et al.* (2007) reported that GY, ASI, and EASP were predominantly controlled by additive gene action under drought conditions, while both gene actions-controlled yield under the optimal conditions in the 72 testcrosses evaluated. Similarly, Badu-Apraku *et al.* (2004) reported the predominance of additive to non-additive gene action for yield and other traits in early maturing full-sib families evaluated under drought conditions for improved GY. A diallel study among nine early white inbred lines revealed significant additive and non-additive gene actions but additive gene action was larger than non-additive gene action for the inheritance of GY and most traits under drought (Badu-Apraku *et al.*, 2011b). This result implied that population improvement through appropriate selection methods as well as hybridization could be employed for the improvement of GY and other traits among the set of parents studied. In a study involving crosses of 12 IITA and five CIMMYT early yellow inbred lines evaluated under DS, significant additive and non-additive gene actions were observed but with a greater preponderance of additive gene action over non-additive (Badu-Apraku *et al.*, 2016a). Similar results have been reported for crosses involving 17 early maturing white maize inbreds from IITA and CIMMYT (Badu-Apraku *et al.*, 2016b) and from a North Carolina Design II study of 150 early white and yellow maize hybrids generated from 30 inbred parents (Oyekunle and Badu-Apraku, 2013). Thus, additive gene action was consistently more important than non-additive gene action for early maturing maize under DS regardless of the genetic background of the parents involved in the study, kernel colour of the inbred parents, or the type of mating design used for the analysis.

The scenario is not the same with the extra-early germplasm. For instance, a diallel study involving 17 extra-early white maize inbreds revealed significant additive and non-additive gene actions but with a greater preponderance of non-additive gene action over additive gene action for GY and most agronomic traits under DS (Badu-Apraku *et al.*, 2013b). Furthermore, in a Design II study involving 30 white QPM inbreds used to generate 150 hybrids, both additive and non-additive gene actions were responsible for the inheritance of grain yield and other traits across environments, but there was a preponderance of non-additive gene action over additive gene action for most measured traits across environments (Badu-Apraku *et al.*, 2016b). Akaogu *et al.* (2017) reported a preponderance of additive gene action over non-additive among 120 extra-early testcross maize hybrids. A combining ability study of 20 extra-early yellow inbred lines under drought stress revealed significant additive and non-additive gene actions but with a greater preponderance of additive gene effects over the non-additive effects across test environments (Badu-Apraku *et al.*, 2013c). Similar results have been reported for a study of 20 newly developed extra-early maturing Provitamin A (PVA) maize inbred lines (Badu-Apraku *et al.*, 2020). Additionally, Annor and Badu-Apraku (2016) concluded that additive gene action had a higher effect on the inheritance of grain yield and most other measured traits under drought when 96 extra-early QPM hybrids derived from 24 inbred parents were evaluated under contrasting drought environmental conditions. Annor *et al.* (2019) reported that both GCA and SCA effects showed significance for grain yield and most measured traits under drought. However, a higher component of SCA sum of squares to GCA showed that the GY was largely regulated by non-additive gene action in the studied crosses under drought conditions. Using line-by-tester design, Amegbor *et al.* (2017) crossed 63 extra-early maturing maize inbred lines to four testers to generate 252 hybrids. The hybrids were subsequently evaluated along with four commercial checks under drought and rainfed conditions with the objective to determine the combining ability of the extra-early inbred lines, compute the heritability of measured traits, assess the performance of the inbreds in hybrid combinations, and investigate the associations among traits under drought and optimal conditions. Results revealed that the GCA and SCA for the traits were significant. The genotypic sum of squares due to GCA effects for GY were 57.1 and 53.4% under drought and rainfed conditions, respectively. Hybrid TZdEEI 91 × TZEEI 21 out-yielded the best check by 49.13% under drought while TZdEEI 55 × TZEEI 13 out-yielded the best check by 39.05% under optimal conditions. The most promising hybrids with consistently high GY under drought and rainfed conditions were TZdEEI 54 × TZEEI 13, TZdEEI 91 × TZEEI 21, and TZdEEI 55 × TZEEI 21 and should be further evaluated for possible commercial production in SSA. In another study, Obisesan (2021) evaluated GY performance and stability of 210 maize hybrids across stress and NS environments, determined the mode of gene action controlling the inheritance of yield and other traits under *Striga* infestation, DS, NS conditions, and across stress environments; assessed the combining abilities of 44 inbred lines and classified the lines to heterotic groups based on their combining ability; determined the efficiency of inbred testers in classifying the lines into heterotic groups, and examined the inter-trait relationships among the inbred lines and hybrids. Results revealed that additive and non-additive gene actions controlled the inheritance of yield under *Striga* infestation and drought conditions. Across the stresses, additive gene action controlled yield and most other measured traits.

Because of the emerging threats of extreme temperatures to maize in SSA, plant breeders of the IITA-MIP, using S_1 recurrent selection and other breeding schemes, have successfully improved an extra-early white population (TZEE-W Pop DT C_0 STR C_5) for tolerance to drought and CDHS and resistance to *Striga hermonthica* parasitism, resulting in TZEE-W Pop HDT C_1 STR C_5 (Badu-Apraku and Fakorede, 2017). The improved population (TZEE-W Pop HDT C_1 STR C_5) was subjected to pedigree selection and advanced to the S_6 stage of inbreeding. Several outstanding inbred lines with tolerance to multiple stresses have been developed in the program and the performance of the inbred lines in hybrid combinations has clearly established that several outstanding hybrids with tolerance to drought, CDHS, and resistance to *Striga hermonthica* parasitism have been identified for further testing on-farm and commercialization in SSA. Genetic studies have been carried out to maximize the potential usefulness of the inbred lines in the development of productive hybrids and synthetics for DS and/or HS-prone environments. For instance, Nasser *et al.* (2020) studied the GCA effects of 20 inbred lines and SCA of the crosses under CDHS and NS environments to assess the yield and stability of the derived hybrids under contrasting environments, and to investigate inter-trait relationships among traits of the hybrids under contrasting environments in Niger. They found that both GCA and SCA effects for GY and other measured traits were significant, with the preponderance of GCA effects over SCA effects under both the CDHS and NS environments. Inbred ENT 13 showed positive and significant GCA effects for GY under both CDHS and NS environments indicating that the inbred line could be used for developing outstanding hybrids with good levels of tolerance to CDHS. This also implied that the inbred line could serve as an invaluable genetic resource for the introgression of favourable alleles into tropical early-maturing maize populations for accelerated genetic gains from selection for improved GY. Hybrids ENT 13×TZEI 167 and TZEI 157×ENT 13 were among the highest yielding and stable hybrids across research environments. It was thus recommended that these outstanding hybrids should be further tested extensively for possible commercialization to help improve food security in SSA. Furthermore, it was also concluded that plant and ear aspects, and days to 50% silking could be combined with GY in a selection index for improvement of yield under CDHS without yield penalties under NS environments.

From the reviews, there are contradictions in the mode of gene action for GY and other important traits under contrasting drought conditions. Additive gene action has been consistently found to be more important than the non-additive among early maize under DS and CDHS while both the additive and non-additive have been reported to be important for the extra-early maize. This may imply greater genetic complexity among extra-early maize. It may, therefore, be concluded that the mode of gene action of newly developed maize inbred lines should be determined to select the most appropriate breeding strategies to be adopted for newly developed inbred lines in a hybrid development program.

Selecting for stress tolerance using secondary traits

According to Badu-Apraku and Lum (2010) and Bolaños and Edmeades (1996) selection for stress tolerance solely on the basis of yield performance alone without taking into account the expression of other suitable secondary traits is not effective in accelerating genetic gains from the selection. The implication is that adequate knowledge of secondary traits that contribute significantly to variations in GY under DS are required by breeders for index selection to guarantee accelerated genetic gains from the selection. However, employing adaptive traits which are derivatives of primary traits such as grain yield used in selecting for improved drought tolerance in maize under DS are inadequately quantified (Bolaños and Edmeades, 1996). Therefore, before a trait is considered as a secondary trait, it must be genetically different and associated with GY in the target environment (Bolaños and Edmeades, 1996). The preference is that the correlation between the trait and yield should be close to 1.0, so that it is simpler and cheaper to measure the trait alone because it is highly heritable (Bolaños and Edmeades, 1996). Falconer (1981) documented that when the product of the square root of heritability of a secondary trait and the genetic correlation between the secondary trait and GY exceed the square root of the heritability of GY, then it would be cheaper and faster to select for that secondary trait alone if the priority trait is GY. This makes it cheaper and easier to evaluate because it is stable within the measurement period, and it is not associated with yield loss under non-stressed conditions. Additionally, the trait is observed at or before flowering and therefore it is a reliable estimator of yield potential before the final harvest (Bolaños and Edmeades, 1996).

Secondary traits such as leaf and stem elongation rate and ASI have been used to select for high GY under drought while leaf rolling and erectness scores (LRS; LES), canopy temperature (CT), leaf chlorophyll concentration (CHL), stay green characteristic (LDS), tassel size (TS), leaf osmotic concentration (OSM), and lodging (LOD) under drought conditions have been identified as key secondary traits for selecting for high GY in stress environments (Bolaños and Edmeades, 1996). Contrarily, secondary traits such as TB, LF, and tassel sterility have been identified as secondary traits for selecting tolerance to HS (Zaidi *et al.*, 2016). Bolaños and Edmeades (1996) reported that the correlation between some secondary traits and GY under DS conditions is large, ranging from 0.5 to 0.9 for EPP, kernels per ear, tassel size, PHT, and ASI while some traits have low heritability, ranging from 0.1 to 0.4 for weight per kernel, canopy temperature, LDS, chlorophyll concentration, leaf osmotic concentration; and other traits such as lodging, leaf rolling and erectness scores, leaf and stem elongation rate and tassel branch number were found to be less than one. But the authors recommended that to achieve an accelerated and improved yield and yield stability under DS conditions, it is always better to select for a combination of these traits as well as GY as compared to selecting for GY alone. It is important to note that all these traits are vital in selecting maize genotypes that are tolerant to drought and heat stress. However, where resources for collecting data are limited, one could measure shelled grain weight, anthesis date and EPP as the minimum required data set for efficient selection under drought conditions. Also, different researchers have proposed different selection criteria for drought tolerance e.g. drought susceptibility index (DSI) (Fischer and Maurer, 1978), mean productivity (MP) and stress tolerance (TOL) (Rosielle and Hamblin, 1981), drought resistance index (DI), and stress tolerance index (STI) (Fernandez, 1992). Bänzinger *et al.* (2000) also developed a selection index based on both primary traits (yield) and secondary traits which was later modified by Hao *et al.* (2011). To identify productive single-cross hybrids under CDHS conditions, a base index (I) that integrated superior GY under CDHS with PASP, EASP, and the LDS at 70 days after planting, ASI, and increased EPP was adopted (Badu-Apraku *et al.*, 2015). To minimize the effects of different scales of measurement, each trait was standardized using a mean of zero and standard deviation of 1; a positive base index value was considered as tolerance to CDHS while a negative value indicated susceptibility.

Genetic gains from selection for drought tolerance in maize under the Drought Tolerant Maize for Africa Project

The demand for maize as food, feed, and industrial raw material is continuously increasing in SSA. This increasing demand is driven by expanding populations and rising incomes in all countries of the sub-region. The importance of maize in the food basket in SSA has been increasing rapidly during the last few decades (Badu-Apraku and Fakorede, 2017), and several dependable improved technologies (high-yielding and stable cultivars with accompanying agronomic practices) are available to farmers (Badu-Apraku and Fakorede, 2017). IITA had developed extra-early, early, intermediate/late maturing maize with tolerance/resistance to *Striga*, drought, and low soil N under the Drought Tolerant Maize for Africa Project. Tolerance/resistance to the stresses had been increased in the germplasm through population improvement and effective screening methods. The breeding program had developed several quality protein maize (QPM), provitamin A and QPM/provitamin A as well as normal endosperm white and yellow endosperm early and extra-early maturing open-pollinated and hybrids with resistance to multiple stresses. Several stress tolerant/resistant open-pollinated and hybrid cultivars developed in this program had been released and adopted by farmers in WCA. Even though information on the genetic enhancement of the open-pollinated varieties has been published (Badu-Apraku and Fakorede, 2017), there is limited information to justify the huge investments that had been made by the IITA-MIP breeding program during the past three decades in developing extra-early and early maturing hybrids under drought and well-watered conditions. Therefore, during the past couple of years, the effort has been devoted to the assessment of the genetic gains in GY and associated changes in other agronomic traits of the early and extra-early multiple stress-tolerant maize hybrids during the last three decades. The objectives of these studies have been to assess progress made in breeding for improved maize hybrids and to determine traits of potential value for accelerating genetic gains in future breeding as well as to identify the most stable and high-yielding hybrids. The first study was conducted at 17 locations in West Africa from 2017 to 2019 to [i] determine the rate of genetic improvement in the GY, and [ii] identify the most stable and high *per se* performance in GY of 54 early maize hybrids developed during three breeding periods, 2008–2010 (period 1), 2011–2013 (period 2), and 2014–2016 (period 3) under DS and NS conditions. Under DS, mean GY ranged from 658 kg ha^{-1} for the period 1 hybrids to 2766 kg ha^{-1} for period 3 hybrids with a corresponding annual genetic gain of 4.14% and average annual rate of increase of 59.94 kg ha^{-1} year^{-1}. Under NS conditions, GY ranged from 3407 kg ha^{-1} for the period 1 hybrids to 5539 kg ha^{-1} for the period 3 hybrids with an annual genetic gain of 1.43% and an average annual rate increase of 61.14 kg ha^{-1} year^{-1}. The increase in GY was associated with a significant decrease in LDS, increased EPP, improvement in plant and ear aspects, and root and stalk lodging. Hybrids 18 (TZEI 18 × TZdEI 352), 17 (TZdEI 352 × TZdEI 315), and 30 (TZE-Y Pop DT C5 STR C5 ×ENT 13) were the most outstanding across drought and NS environments. It was concluded that substantial progress had been achieved in SSA by breeding for high-yielding and drought-tolerant hybrids during the past three decades. A second study involving the extra-early maturing hybrids sought to (i) estimate the gains in GY under drought and rainfed conditions, and (ii) identify high-yielding and stable hybrids developed during the past two decades under drought and rainfed conditions. Sixty extra-early maturing hybrids developed in IITA during three breeding periods: (2008–2010, 2011–2013, and 2014–2016) were assessed under managed drought and rainfed conditions across 14 environments in Nigeria to estimate genetic gains in GY using linear regression analysis. Considerably high genetic gains (4.1%) per year for GY were obtained under managed drought environments. Six of the top ten hybrids identified in the study were from period 3, an indication of the high genetic gains from selection in period 3. In addition, two of the top six period 3 hybrids—(TZEEI 29 × TZEEI 13) and (TZdEEI 64 × TZEEI 54) were found to be the most stable and high-yielding across the 14 test environments. These hybrids have been suggested for thorough assessments in preparation for commercialization in SSA.

Genotype × environment interaction in drought and heat stress environments

Analyses of several multi-environment trials (METs) data of early and extra-early maize in the WCA sub-region have shown the tremendous contribution of G × E interaction (GEI) to the total variation (Badu-Apraku *et al.*, 2007; Badu-Apraku *et al.*, 2008; Badu-Apraku *et al.*, 2009), which justified extensive testing of cultivars in multiple environments over several years to aid decisions on cultivar recommendations. Significant GEI leads to the identification of genotypes with specific and broad adaptation to evaluation environments, classification of sites into mega-environments and identification core testing sites. Mega-environment can be defined as a portion of a crop species' growing region with a homogenous environment in which some genotypes perform similarly (Gauch and Zobel, 1996; Gauch and Zobel, 1997). Gauch and Zobel (1996) used the maize MET dataset for the identification of maize mega-environments. Menkir (2003) used Geographic Information System (GIS) to analyse long-term data collected on seven climatic variables relevant to maize production to refine the stratification of the sub-region into well-defined agro-ecological zones. GIS separated the locations into four distinct zones, namely, mid-altitude, rainforest, moist savanna, and dry savanna. Similarly, Setimela *et al.* (2007) used the maize MET data set of CIMMYT and agro-climatic data to identify maize mega-environments for sub-Saharan Africa including WCA. The WCA subregion was stratified into four distinct groups viz. dry savanna, wet savanna, dry mid-altitude, and wet mid-altitude. Findings from a MET study involving 20 cultivars evaluated under four induced drought environments revealed significant genotypic and GEI for GY and other drought adaptive traits (Badu-Apraku *et al.*, 2010). More importantly, it was reported in the same study that the proportion of total variation explained by the GEI was consistently larger than that of genotype for GY and other agronomic traits except husk cover and LDS scores.

Classification of mega-environments

To characterize test environments used for drought trials in the sub-region, a study was conducted using 12 early maturing maize cultivars, over 3 years at 16 locations in West Africa using genotype main effect plus genotype by environment interaction (GGE) biplot analysis. Results revealed that Zaria (Nigeria), Nyankpala (Ghana), and Ejura (Ghana) displayed the highest discriminating ability. Two mega-environments were identified. Bagou, Nyankpala, Bagauda, Ikenne, and Mokwa constituted the first (ME1) while Ejura, Ina, and Sotuba represented the second mega-environment (ME2). Furthermore, ME1 was more useful for evaluating early maize genotypes for tolerance to drought than ME2. Zaria, a location in the northern Guinea savanna of Nigeria that was identified in another study as closest to the ideal environment under NS (63), was the least suitable location for evaluating genotypes for drought tolerance. Nyankpala in Ghana and Ikenne in Nigeria were identified as the core testing sites for drought tolerance in early maturing maize cultivars.

Inter-trait relationships among tropical maize under drought and heat stress

Adequate knowledge of the association among traits of crops is crucial in determining secondary traits that can be used in indirect selection for a target trait. Genetic variability among genotypes and environmental conditions greatly influences the relationship among traits. Badu-Apraku *et al.* (2011a), using GGE biplot analysis, found that EASP, PASP, ASI, and EPP, apart from being significantly correlated with yield, were the most reliable traits for indirect selection for improved yield among early maize inbred lines under DS. In a Design I study of 250 full-sib progenies from an early yellow maize population evaluated under managed DS, stepwise multiple regression analysis identified EASP, EPP, PASP, LDS, and ASI as traits with significant direct influences on GY, accounting for about 72% of the total variation in GY (Talabi *et al.*, 2017). In the same study, EASP had the highest direct effect (−0.509) on yield among the five traits. Similarly, sequential path analysis of the relationship between yield and other measured traits of early-maturing S_1 lines developed from cycle C_0 of TZE-Y Pop DT STR evaluated under DS revealed that only ASI, EASP, and EPP had a significant direct effect on yield, accounting for approximately 66% of the total variation in yield under induced DS (Badu-Apraku *et al.*, 2018). Of these three, EPP had the largest positive direct effect (0.470).

Based on the relationship among traits, a base index involving ASI, PASP, EASP, EPP, and LDS was formulated for selecting for improved yield of maize under DS. However, controversies on the reliability of LDS in the base index ensued. Thus, a study was conducted involving 90 extra-early inbred lines evaluated under drought to test the hypothesis that LDS was not effective for selection under drought. While PASP, PHT, and EHT were identified as the most reliable traits for simultaneous selection for yield, the LDS was found unreliable for selecting drought-tolerant genotypes. Path–coefficient and genotype main effect plus genotype ×environment interaction (GGE) biplot analyses identified EASP, PHT, and ASI as reliable secondary traits for selecting genotypes for drought tolerance (Badu-Apraku *et al.*, 2012). A contrary finding was, however, reported in a study by Talabi *et al.* (2017) on 250 early maturing full-sib progenies where EASP, EPP, PASP as well as LDS were identified as contributing significantly to GY.

In another study involving maize landraces under DS, HS, and CDHS, path coefficient analysis showed that EASP, SL, and PASP had significant direct effects on GY under DS while EASP, LF, LDS, EPP, and SL had a direct effect on yield under HS (Nelimor *et al.*, 2020a). In the same study, yield under the NS environment had a significant association with yield under DS ($r = 0.66^{**}$), HS ($r = 0.75^{**}$), and CDHS ($r = 0.62^{**}$). This suggests that the yield performance of maize under DS, HS, and CDHS can be predicted from yield performance under a stress-free environment. The relationship between yield under DS and HS was moderately positive ($r = 0.48^{*}$) and this corroborated the finding of Cairns *et al.* (2013) who reported a moderate positive correlation ($r = 0.49^{*}$) between yield under the two stresses. Other drought adaptive traits such as ASI under NS condition, had significant association only with ASI under CDHS. Similarly, LDS, EPP, and EASP under DS and HS showed no relationship, suggesting that the mechanism of tolerance under the two stresses may be different. Osuman *et al.* (2020) also reported that EPP, PASP, EASP, LF, and TB were the most reliable traits for selecting genotypes for tolerance to CDHS.

From the foregoing, it is clear that ASI, EPP, EASP, and PASP were the most reliable secondary traits for selecting for improved GY in both early and extra-early maize germplasm under DS and HS. The traits were consistently associated with yield in several independent studies under the two stresses. In addition to these traits, plant and ear heights were particularly important for extra-early maize. The reliability of LDS is still in question as it was found to be contributing significantly to yield in some studies but not in others.

Heterosis and heterotic groupings under drought environment

Information on heterosis in germplasm is essential to determine the genetic response of a set of genotypes, organize a large number of parental materials into groups for the purpose of hybrid development, and identify testers that can be used in further improvement programs. The bases of heterosis are both genetic and environmental variation. Heterosis is greatly reduced especially under stressful environments just as genetic variability. Oyekunle and Badu-Apraku (2013) reported a reduction in the values of mid-parent and high-parent heterosis from 45.3 and 18.4% under DS to 111.9 and 102.6% under well-watered conditions. Thus, heterotic pattern and groups vary in terms of number and composition under drought vis-a-vis NS environments. Important genetic factors that determine heterotic grouping and pattern are the number of parental genotypes involved, extent of variability among genotypes evaluated, mating design used for the genetic analysis, proportion of additive to non-additive gene action, and the heterotic grouping method used.

Although there is a dearth of information on the heterotic response of early and extra-early maize germplasm under HS and CDHS, several studies have shown diverse heterotic patterns and groups under DS. Despite the great influence of environment on morphological traits of inbreds, useful information on genetic relatedness among parental materials was obtained using morphological characteristics or *per se* performance. For instance, based on the *per se* performance of 35 early maturing maize landraces, agronomic data under moisture deficit classified genotypes into two major groups by cluster analysis (Menkir *et al.*, 2009). In the study, group I contained 24 farmers' varieties that were characterized by lower GY, early flowering, shorter heights, lower ear placement, longer ASI, and higher leaf death scores compared to those in group II. All improved varieties were classified in group II, with higher yields, later flowering characters, taller and higher ear placement, shorter ASI, slightly higher number of ears per plant, and lower leaf death scores, as opposed to those in group I. The grouping was quite different from that under well-watered environments. Furthermore, grouping based on combining ability has been explored for tropical maize of these maturity classes and the results obtained have been interesting. Where testers are available in a breeding program, the use of line ×tester mating design becomes a novel idea in grouping the inbreds. However, at the beginning or at the early stage of a breeding program, proven testers are usually not available and, in such situation, diallel and North Carolina Design II become effective, not only to classify lines into groups but to identify inbred testers that can be used for improving population and for line ×tester study. Badu-Apraku *et al.* (2011a) reported two heterotic groups among nine white early-maturing inbred lines in a diallel study under drought. Results from another diallel study involving 91 crosses derived from 14 early maturing yellow-endosperm QPM inbred lines evaluated under drought classified the inbreds into three groups, each based on SCA effects of GY and the heterotic group's SCA and GCA (HSGCA) methods. The heterotic grouping based on the GCA of multiple traits (HGCAMT) method identified only two groups. Close correspondence was observed in the pattern of grouping of both grouping methods in terms of placement of inbreds into the same heterotic groups. Another diallel study of 17 white early maturing maize inbreds (11 IITA inbreds and six CIMMYT inbreds) reported that HSGCA, SCA effects of GY and HGCAMT grouping methods classified the lines into three groups under drought and HSGCA was the most efficient method (Badu-Apraku *et al.*, 2016b). Inbred TZEI 19 was identified as the best tester across research environments based on HSGCA method. Another similar diallel study involved crosses of 12 IITA and five CIMMYT early yellow inbreds under drought. The HGCAMT method revealed four heterotic groups while the HSGCA method identified three heterotic groups (Badu-Apraku *et al.*, 2016a).

In summary, from the two diallel studies involving crosses between IITA and CIMMYT lines, it is observed that the content of the groups established by the two grouping methods was very consistent. It is noteworthy that the grouping of the inbreds was not completely mutually exclusive as there were groups containing lines from both IITA and CIMMYT programs. This suggests the lines from the different programs might share sources of genes for tolerance to drought. In an NCD II study among 30 early maturing inbreds, three testers (TZEEI 79, TZEEI 76, and TZEEI 63) and three opposing heterotic groups (TZEEI 95, TZEEI 78, TZEEI 79), (TZEEI76, TZEEI 82), and (TZEEI 63, TZEEI 58) were identified across drought environments (Badu-Apraku and Oyekunle, 2012). TZEEI 79 was identified as a proven inbred tester while the highest positive GCA effects for GY and TZEEI 95 × TZEEI 79 was identified as the potential single-cross tester for the development of three-way and double-cross hybrids under drought environments. Akaogu (Akaogu *et al.*, 2012) in a line ×tester study used three testers, TZEEI 79, TZEEI 63, and TZEEI 95 to classify 39 extra-early yellow-grained inbreds into three heterotic groups under DS, where the three testers were able to classify only 20% of the 39 lines into complementary heterotic groups. In a recent study, 12 extra-early maize inbreds newly developed by introgressing novel genes from *Zea diploperennis* were classified into three heterotic groups, while TZEEI 79 and TZdEEI 7 were identified as inbred testers and TZdEEI 7 × TZEEI 12 as a single-cross tester across drought environments (Shaibu *et al.*, 2021).

Assessment of early and extra-early maturing maize landraces for tolerance to drought, heat and combined drought and heat stress

One important prerequisite for developing climate-resilient maize cultivars is the effective identification of diverse sources of germplasm with good levels of tolerance to abiotic stresses. Two separate studies were conducted with the primary objective of identifying landraces that combine high grain yield potential with desirable secondary traits under DS, HS, and CDHS for use in maize breeding programs as donors of tolerance to these stresses. In the first study, 33 landraces from Burkina Faso (6), Ghana (6) and Togo (21), and three drought-tolerant populations/varieties were evaluated under optimal NS, MDS, HS, and CDHS for 2 years. Results underscored weak phenotypic and genetic correlations between GY under NS and CDHS, DS and HS, as well as DS and CDHS, which suggested the existence of independent genetic control of yield in these stresses. Grain yield under HS was, however, highly and positively correlated with GY under CDHS, suggesting that HS-tolerant genotypes would most likely tolerate CDHS conditions. Yield reduction averaged 46, 55, and 66%, under DS, HS, and CDHS, respectively. Accession GH-3505 was highly tolerant to DS, while GH-4859 and TG-1353 performed well under the three stresses. The second study was conducted using 66 extra-early maturing maize landraces and six DS and/or HS-tolerant populations. Results showed that performance under DS was predictive of yield potential under CDHS, but tolerance to HS was independent of tolerance to DS and CDHS. Yield reduction was lower under HS (23%) compared to DS (49%) and CDHS (58%). Accessions TZm-1162, TZm-1167, TZm-1472, and TZm-1508 showed particularly good performance under the three stresses. These desirable landraces could be invaluable sources of genes/alleles for adapting maize to climate change (Nelimor *et al.*, 2019; Nelimor *et al.*, 2020a).

Molecular approaches in improving maize for adaptation to drought and combined drought and heat

The recent breakthroughs in DNA-based molecular protocols, to include genomics-based methods have provided huge opportunities for plant breeders to precisely and accurately dissect complex quantitative traits (targeted manipulations in crop plants) to achieve rapid genetic gains at a reduced cost. Within the IITA-MIP, molecular strategies are employed routinely for multiple purposes; from characterization and heterotic grouping of untapped germplasm (e.g.

landraces and newly developed breeding lines), through monitoring of line purity, to trait/gene mapping, marker-assisted selection, and/or genomic selection (Badu-Apraku and Fakorede, 2017).

Assessment of genetic diversity in landraces and inbred lines

To sustain genetic gains from breeding innovations, it is important for maize breeders to have continued access to novel/useful genetic variability. The most widespread application of molecular marker-based technologies within the IITA early and extra-early maize breeding program, thus far, centres on genetic characterization, diversity analysis, and heterotic grouping of germplasm to facilitate the planning of crosses for the development of high-yielding cultivars/ideotypes for target environments. In this direction, secluded gene pools (e.g. landraces) from dry and warm environments, and modern breeding lines are being characterized for tolerance to abiotic stresses such as DS, HS, and CDHS, respectively using state-of-the-art genotyping technologies.

A simple sequence repeat (SSR)/microsatellite marker-based diversity analysis revealed considerable genetic diversity among drought and *Striga* resilient extra-early maturing white inbred lines, which facilitated their classification into distinct heterotic groups consistent with pedigree rather than their reaction to the stresses (Badu-Apraku *et al.*, 2013b). Thirty-one (Dass *et al.*, 2001) unique/private alleles were obtained from SSR maker data of 21 multiple stress tolerant (e.g. drought) inbred lines, which could be crucial for breeding and diagnostic purposes (Oyekunle *et al.*, 2015). The findings of single nucleotide polymorphism (SNP) markers-based genetic diversity assessment of 128 inbred lines that can tolerate drought and *Striga* enabled the identification of suitable lines for hybrid combinations (Mengesha *et al.*, 2017). Similarly, Adu *et al.* (2019) examined the genetic diversity and population structure of 94 early maturing inbred lines that combined drought tolerance with resistance against *Striga* and maize streak virus (MSV) using SNP markers. The genetic distance of the pairs of inbred lines ranged from 0.06 to 0.65, with ~68% of the inbred pairs falling between 0.42 and 0.46. Only 3% of the pairs of inbreds lines showed a genetic distance of <0.30, clearly demonstrating the wide genetic variability among multiple stress-tolerant early and extra-early maize inbred lines available in the IITA-MIP, which should be harnessed for hybrid development. In the case of landraces, Nelimor *et al.* (2020b) used 5794 DNA markers from the DArTseq genotyping-by-sequencing (GBS) platform to assess genetic diversity, population structure, and gene flows amongst a panel of 208 accessions comprising gene pools from Burkina Faso (BFP), Ghana (GHP) and Togo (TGP), and a set of DS- and/ or HS-tolerant (DHT) open-pollinated populations (dubbed, the reference population) developed by IITA-MIP. Results showed that; (i) accessions clustered by adaptation; the Sahel landraces (BFP) formed a unique group, whereas the Coastal landraces (GHP and TGP) were mixed (Fig. 13.1), (ii) genetic divergence, estimated as Wright's fixation index (F_{ST}), between DHT and BFP was twice as large as the divergence between DHT and other pools (Table 13.1) and, (iii) based on gene flow estimates, relatively little of the genetic diversity available in the BFP has been used in modern breeding (Table 13.1). Thus, it was anticipated that landraces of maize from Burkina Faso would be invaluable for pre-breeding for allele mining of candidate genes conferring DS and/or HS adaptation. Indeed, under field conditions of DS and CDHS, the resilience of maize landraces from Burkina Faso against the stressors was unrivalled by their Ghanaian and Togolese counterparts (Nelimor *et al.*, 2019; Nelimor *et al.*, 2020a). In a recent study, analyses of DArTseq-based SNP data from 162 DS and CDHS-tolerant early maturing inbred lines within the IITA-MIP underscored enormous genetic variability stratified into six distinct heterotic groups that opened a fertile ground for their exploration and use in climate-ready hybrid maize development (Osuman *et al.*, 2020). For the determination of line purity, Obeng-Bio *et al.* (2020) analysed 70 early provitamin

A quality protein maize germplasm with varying levels of tolerance to DS from the IITA-MIP. Results showed that there were relatively few heterozygous individuals in the set of inbred lines, with the majority of the inbreds being

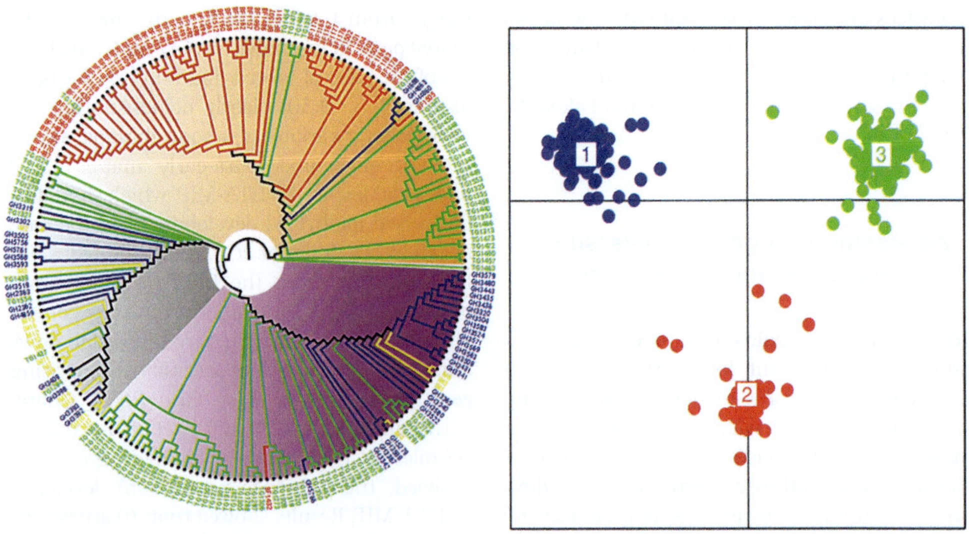

Fig. 13.1. Diversity analysis of maize accessions from Sahel and Coastal West Africa (from Nelimor *et al.*, 2020b). Phylogenetic tree estimated through neighbor-joining method for 208 maize accessions (left). Green, red, blue, and yellow clades, and taxa represent Togolese, Burkinabe, Ghanaian, and modern drought and/or combined drought and heat tolerant populations, respectively. The orange, violet, and ash highlights represent subpopulations 1, 2, and 3, respectively. Discriminant analysis of principal components (DAPC) using 5974 DArTseq markers (right). The axes represent the first two Linear Discriminants (LD). Each colour represents a cluster, while each dot represents an individual. Cluster 1 and 2: Mixture of Ghana, Togo and improved populations, cluster 3: Burkina Faso landraces.

Table 13.1. Measure of genetic population differentiation (F_{ST}) (lower diagonal) and estimation of gene flow (Nm) (upper diagonal) within and among four gene pools of maize accessions.

	Burkina Faso	Ghana	Togo	Reference population
Burkina Faso	–	1.18	1.31	0.98
Ghana	0.27	–	2.63	2.82
Togo	0.24	0.14	–	2.03
Reference population	0.31	0.14	0.17	–
Average F_{ST}	0.28	0.18	0.18	0.21

approximately 94% homozygous and having their loci fixed with minimal segregation. A more comprehensive study by Badu-Apraku *et al.* (2021a) assessed the genetic diversity and population structure of a maize collection of 439 early and extra-early inbred lines and testers with varying levels of tolerance to multiple stresses (e.g. DS and CDHS) using 9642 DArTseq SNP markers. The findings showed that DArTseq SNP-based genetic distance (GD) methods were effective in categorizing inbred lines into heterotic groups similar to defined heterotic groups previously identified in combining ability studies where mating designs/classical grouping methods such as SCA of GY, HSGCA, and HGCAMT were deployed. However, the authors anticipate that continuous refinement of defined heterotic groups would be helpful to sustain adequate genetic gains from selection in the IITA early and extra-early maize breeding program.

Identification of QTLs for drought and combined heat and drought tolerance

Several studies have elucidated the genetic basis of drought-adaptive traits in tropical maize; with emphasis on GY and its components, including key secondary traits such as ASI, and LDS. Under different levels of DS, several QTLs with small-to-moderate effects, but generally unstable across different DS environments, were found to be associated with GY (Ribaut and Ragot, 2007). In a bi-parental double haploid population, Cerrudo *et al.* (2018) reported QTL regions associated with GY and secondary traits in bins 1.02 and 1.03. Apart from individual population-based QTL mapping studies, few meta-QTL analyses of drought tolerance have been conducted in multiple bi-parental maize populations. A meta-QTL analysis of drought tolerance using three bi-parental populations identified 83 and 62 QTLs for GY and ASI, respectively; six constitutively expressed QTLs for GY were mapped onto chromosomes 1, 4, 5, and 10. In a similar study involving three populations, 203 QTLs were identified for different secondary/morpho-physiological traits (Almeida *et al.*, 2014). The 8 Mb region in bin 3.06 was delimited as a hotspot of two interesting candidate genes, namely *zmm16* (MADS-domain transcription factor) and *psbs1* (photosystem II unit) controlling reproductive organ development and photosynthate accumulation, respectively. In another meta-analysis study, Semagn *et al.* (2013) evaluated 18 bi-parental maize under DS and optimal environments. Fifteen (15) out of the 68 'consensus' QTLs identified for GY and associated traits were not stable across environments and genetic backgrounds. Genetic analysis of two $BC_1F_{2:3}$ backcross populations showed that QTL associated with early vigor and SG confer drought tolerance in tropical maize (Trachsel *et al.*, 2016). Trachsel *et al.* (2016) further speculated that seven chromosomal segments located in bins 2.02, 2.05, 2.06, 2.07, 5.02, 5.04, and 8.06 could serve as good candidates in MAS to improve maize adaptation to DS. Unfortunately, the use of QTL-based MAS in drought-tolerance maize breeding is hampered by various factors, including a lack of consistent and major phenotypic effects of QTLs in diverse recipient genetic backgrounds.

Genome-wide association mapping studies (GWAS) offer huge opportunities for the identification and cloning of genes at target QTLs to facilitate MAS. As a result, several GWAS for DS tolerance have been reported in maize during the last few years. GWAS analyses led to the identification of 83 genetic loci harbouring 42 candidate genes regulating seedling stage drought tolerance (Wang *et al.*, 2016). It is noteworthy that the peak SNP (chr9.S_94178074) observed in the same study overlapped with the previously reported QTL9.3 (Semagn *et al.*, 2013) and is directly located in the *ZmVPP1*, which encodes a vacuolar-type H + pyrophosphatase (H+ -PPase). For reproductive phase drought tolerance, GWAS revealed 42 genomic regions harbouring 33 stress-related genes; *GRMZM2G125777* was strongly associated with ear relative position, hundred-kernel weight, and timing of male and female flowering (Xue *et al.*, 2013).

In the case of HS tolerance, a maize panel of 662 DH lines of tropical and sub-tropical origin was investigated using GWAS (Longmei *et al.*, 2021). Twelve (12) loci, accounting for 18.14–18.65% of the phenotypic variation in GY were found localized on chromosomes 1, 3, 6, 7, and 10. Further, the authors uncovered significant haplotype trait associations on chromosomes 1, 3, and 7, which could be invaluable for the genomic-assisted breeding of heat-tolerant maize cultivars. Seetharam *et al.* (2021) constituted an association-mapping panel, dubbed the heat-tolerant association mapping (HTAM) panel by involving 543 tropical maize inbred lines from diverse genetic backgrounds. The HTAM panel was test crossed and phenotyped across nine locations under managed HS at reproductive stages before being subjected to individual location GWAS analysis using 955,690 SNPs. Results showed 269 novel significant SNPs for GY, out of which, 175 were within 140 unique gene models underlying various biological pathway responses to different abiotic stresses. Notably, SNP S1_283602921, S2_184967723, mapped on chromosomes 1, 2, and S5_182091381 located on chromosome five were in the gene model *GRMZM2G065374/Zm00001d034298*, and *GRMZM2G137426/Zm00001d017138* coding for *bHLH47* and *bHLH128* genes, respectively. Further, the authors performed a

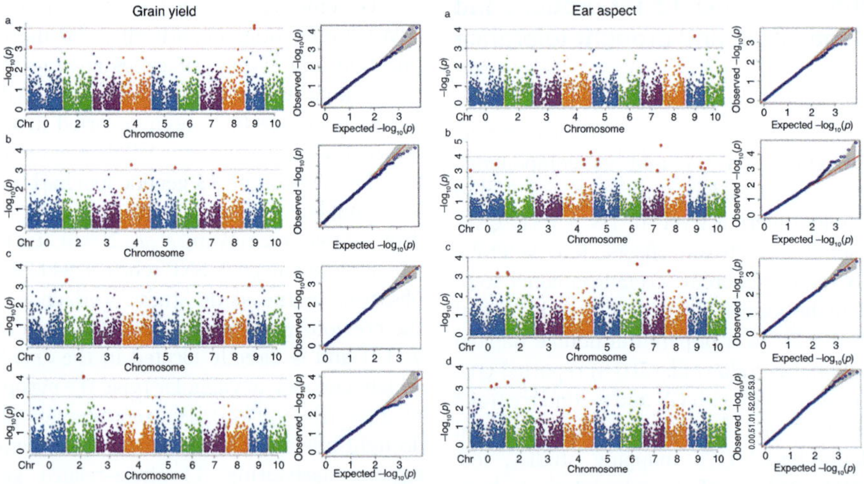

Fig. 13.2. Manhattan plots and Q-Q plots showing markers associated with grain yield (left) and ear aspect (right) in a maize panel of 108 landraces from Sahel and coastal West Africa. Non-stress (a), drought (b), heat stress (c), and combined drought and heat stress (d) conditions (From Nelimor *et al.* unpublished data, 2021).

haplotype trend regression (HTR) analysis of the significant SNPs and found 13 variants (five haplotype blocks and eight single SNP variants) to be significant for GY under HS. In particular, variants SNP_3 (S5_24337729), and Hap_6 (S6_164097936, S6_164097938) were found within the intervals of two previously reported meta-QTLs, MQTL5.3 and MQTL6.2, for GY and ASI under DS and NS conditions (Semagn *et al.*, 2013). Yuan *et al.* (2019) detected 549 SNPs that were significantly associated with 12 traits under one or combined environments (under NS, DS, and HS conditions], out of these SNPs: S1_140960558 on chromosome one was associated with days to 50% anthesis under the three conditions and 76 SNPs were linked to GY under both DS and HS, confirming the complex nature of these traits. Preliminary results of the marker-trait association study in the IITA early and extra-early maize program showed five significant peaks on chromosomes 1, 2, 4, 5, and 8 (Fig. 13.2; Supplementary Table T13.1).

They represented candidate genomic regions associated with GY under NS, DS, HS, and CDHS conditions (Nelimor *et al.*, 2020a). Thus, the results suggested that genes controlling GY in maize might be positioned on the above-listed chromosomes, which is consistent with the findings of Yuan *et al.* (2019). In this ongoing study, hotspot regions for EASP, a trait strongly associated with GY, were distributed on chromosomes 1, 2, 4, 5, and 9 (Nelimor *et al.*, 2020b). In a similar study, 162 tropical maize inbred lines developed by IITA were genotyped by Osuman *et al.* (2020) using GBS-derived DArTseq markers. The lines were evaluated under CHDS and terminal DS (TDS) conditions to: (i) map the genomic regions associated with GY and its related traits under CHDS conditions, (ii) identify markers linked to the various traits under TDS conditions, and (iii) predict putative candidate genes underlying the stable genomic regions for the studied traits under stressful conditions. Results of the study revealed that under CHDS conditions, 66 SNPs were associated with 15 traits, i.e. AD, ASI, ear rot (EARO), EASP, EPP, HC, LDS, LF, PASP, PHT, RL, days to 50% silking (SD), TB and GY (at Bonferroni-correction $-\log_{10}[P] \geq 3.89$) across the 10 chromosomes in maize genome with the SNPs causing 8.37–49.23% variation in phenotype. These indicated that the 15 traits are complex in nature with numerous loci affecting the phenotypes. In the same study, hotspot for GY was on Chr04 (SNP_42991074, SNP_49826316, SNP_168609387, SNP_213750918 and SNP_231684692), which was not surprising as Yuan *et al.* (2019) earlier reported 4 SNPs for GY under HS conditions on Chr04. Similarly,

Chr02 and Chr07 were hot regions for SL and Chr06 for RL thus, pinpointing the important roles of these chromosomes in regulating maize response under CHDS conditions. In this same study, four repetitive SNPs: SNP_161703060, SNP_196800695, SNP_195454836, and SNP_51772182 on chromosomes 1, 2, 5, and 7, respectively, showed pleiotropic effects on both AD and SD. This implied that the four SNPs regulated the two traits simultaneously; hence, they could be targeted for MAS to improve these traits concurrently. This may be the basis for the strong correlation coefficient (r=0.97) that they observed between AD and SD. Additionally, the researchers (Osuman *et al.*, 2022) identified markers linked to ten traits of agronomic importance: that is, AD, EHT, EARO, EASP, EPP, HC, RL, SD, SL, and GY under TDS. Twenty-seven SNPs were associated with these ten traits at the peak of 3.90–24.11 with 12.53–64.79% phenotypic variation on all chromosomes except Chr10. The most prominent SNP [SNP_269120178] on Chr01 with −logP (24.11) and R^2 (64.11) plus other two SNPs (SNP_104170418 and SNP_162101608) on Chr02 were linked to GY. Two SNPs (SNPs: SNP_51109816 and SNP_99059711) linked to EPP (Supplementary Table T13.2). These five SNPs could be targeted for MAS to improve GY and ear-related traits under TDS conditions. The stability of QTLs/SNPs across multiple environments and genetic background is a requisite for their utilization for practical plant breeding (Tuberosa *et al.*, 2002; Awata *et al.*, 2020). Hence, the researchers (Osuman *et al.*, 2022) further used means of traits (AD, ASI, EARO, EASP, EPP, LDS, PASP, RL, SD, SL, and GY) under CHDS and TDS for mapping. In all, 24 SNPs were associated with these traits across nine chromosomes. Three SNPs on Chr01 (SNP_203396231, SNP_265278865 & SNP_269120178) and two SNPs on Chr03 [SNP_112491022 and SNP_99059711] were linked to GY and EPP, respectively (Fig. 13.3; Supplementary Table T13.3). These five SNPs could be targeted for allele pyramiding aimed at improving these traits simultaneously. Also, the results of the mapping with the means across CHDS and TDS conditions revealed that some SNPs were detected across either CHDS or TDS and combined conditions (means) (Table 13.2). The authors considered these SNPs as stable

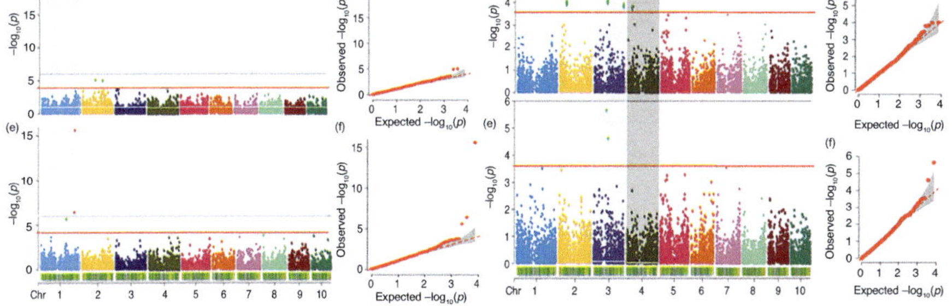

Fig. 13.3. Manhattan plots and Q-Q plots showing markers associated with grain yield (left) and ears per plant (right) in a maize panel of 162 inbred lines developed by IITA-MIP. Under combined drought and heat stress conditions (a-b); under terminal drought (c-d). Across combined drought and heat stress and terminal drought conditions (e-f) (From Osuman *et al.* unpublished data, 2021). The threshold of 3.89 (Bonferroni correction) was adopted with the red line in the Manhattan plots. The *X*-axis represents chromosome number while the *Y*-axis represents $-log_{10}[P]$. The *X* and *Y* axis in the QQ plots represent the expected and observed $-log_{10}[P]$, respectively. Red line in the QQ-plots with the shaded regions indicates a 95% confidence interval.

Table 13.2. Common SNPs detected across either CHD or TD and combined effects [means] in a maize panel of 162 inbred lines.

		Conditions[a]		
SNPs[b]	Chr.[c]	CHD	TDS	Combined effects
SNP_269120178	1	–	GY	GY
SNP_11473743	2	AD	AD & SD	–
SNP_24521844	2	–	EARO	EARO
SNP_27502516	2	–	EARO	EARO
SNP_99059711	3	–	EPP	EPP
SNP_82875264	5	–	RL	RL
SNP_164264714	5	–	RL	RL
SNP_125871560	6	–	SL	SL
SNP_51772182	7	EASP & LASP	PASP	–
SNP_43475091	8	–	RL	RL
SNP_89731392	9	–	SL	RL
SNP_158056460	9	LD	LD	–

[a]CHD = combined heat and drought conditions, TD = terminal drought stress and Combined with means are averages across CHD and TDS conditions. Days to 50% anthesis (AD), ear rot (EARO), ear aspect (EASP), leaf death (LD), plant aspect (PASP), root lodging (RL), stalk lodging (SL), and grain yield (GY).
[b]Single nucleotide polymorphisms.
[c]chromosomes.

and recommended them for further validation and use in maize breeding programs. In all, 105 SNPs were found to be associated with the traits measured under the three scenarios. These SNPs could be useful in future genomic selection breeding programs targeted at all the traits at a go. If validated, those significant genetic markers/loci (Supplementary Tables 1–3) could be targets for MAS to improve maize for tolerance of HS and CDHS. Also, the predicted candidate genes may serve as the foundation for future functional validations to unravel the molecular mechanisms involved in maize response to CHDS or TDS.

Marker-aided selection/breeding maize for tolerance to drought and heat combined with drought stress

While QTLs/loci conferring DS and/or HS tolerance in maize have been identified, this information has not been adequately deployed for the development of climate-resilient maize cultivars. Nonetheless, few QTLs associated with climate-adaptive traits have been introgressed into maize breeding pipelines using indirect marker-based selection strategies such as marker-assisted backcross (MABC), marker-assisted recurrent selection (MARS), and genomic selection (GS). For example, using MABC, favourable alleles at five QTLs, responsible for 38% of the total phenotypic variance in ASI, were introgressed from a drought-tolerant donor to CML287 an elite, drought-susceptible recurrent parent (Wang *et al*., 2016). The best hybrids under DS outyielded the control by 50%, while no yield penalty was observed under WW conditions.

At IITA-MIP, a maize population developed through crosses between two elite *Striga*-resistant inbred lines (Acr.Syn-W-S2-173-B*4) and (TZLComp.1C4-S1-37-5-B*3) that were also tolerant to drought and resistant to the major lowland foliar diseases were subjected to MARS (Abdulmalik *et al*., 2017; Bankole *et al*., 2017). Subsequently, 60 S_1 lines extracted from each of four cycles C_0, C_1, C_2, and C_3 were crossed to a tester, TZISTR1138, to generate testcrosses, which were evaluated along with testcrosses of each of the parental lines of the bi-parental cross to the same tester (P_1 ×T, P_2 ×T), a cross of the two parental lines ($P_1 \times P_2$) and standard hybrid checks 9022–13 and 8338–1 under two each of

DS and WW environments. Results highlighted relative genetic gain in GY for at least one of the cycles under each evaluation condition. Bankole *et al.* (2017) evaluated 50 S_1 lines from each of three selection cycles (C_0, C_1, C_2) subjected to MARS under DS and WW conditions. Results showed that testcrosses derived from the second selection cycle recorded significantly higher yields than the cycle zero testcrosses under DS, resulting in an average genetic gain in GY of 7% per year. Further analyses of changes in the frequency of favourable alleles revealed increments of favourable alleles in the S_1 lines from 114 in C_0 to 124 in C_3 (Abdulmalik *et al.*, 2017), and from 0.51 at C_0 to 0.515 at C_2 (Bankole *et al.*, 2017).

In the CIMMYT eastern and southern Africa MIP, evaluations of testcrosses of 10 populations generated using a slightly modified MARS technique revealed genetic gains in GY of 105 kg ha^{-1} yr^{-1} under WW and 51 kg ha^{-1} yr^{-1} under DS, and up to 16% across DS and WW conditions (Beyene *et al.*, 2015a). A follow-up study was conducted to compare the performance and GY stability of the maize populations developed using MARS and pedigree selection procedures (Beyene *et al.*, 2016). Results showed that the top 10 hybrids derived through MARS produced 0.5–46.3 and 3.4–13.3% higher mean GY than hybrids developed using conventional pedigree selection under DS and WW conditions, respectively. However, mean days to anthesis of the best 10 MARS-derived hybrids were comparable to those of hybrids developed using the pedigree method under both conditions. These results demonstrated the effectiveness of MARS in improving GY (Beyene *et al.*, 2015a, 2015b; Abdulmalik *et al.*, 2017; Bankole *et al.*, 2017) without compromising other desirable traits (Abdulmalik *et al.*, 2017; Bankole *et al.*, 2017).

In the case of GS, Beyene *et al.* evaluated eight CIMMYT tropical bi-parental maize populations derived through GS and reported that (i) the average genetic gain in GY per cycle was 0.086 t/ha under DS, (ii) cycle-3-GS derived hybrids significantly out-yielded those derived from C_0, and (iii) it is possible to achieve three GS cycles in a year (Beyene *et al.*, 2015b). Following the study, Badu-Apraku *et al.* (2019) evaluated 200 testcrosses of an early yellow biparental maize population generated through GS and obtained no yield reward under DS but observed negative effects on flowering traits and EHT (Badu-Apraku *et al.*, 2019). A recent empirical study within the CIMMYT eastern and southern Africa MIP, compared the performance of tropical maize hybrids selected through GS and phenotypic selection (PS) under DS and WW conditions (Beyene *et al.*, 2019). Cost-benefit analyses showed that GS reduced the cost by 32% over the PS though similar selection gains were observed. These results (Beyene *et al.*, 2019; Badu-Apraku *et al.*, 2019) demonstrated that the application of GS with a focus on GY under DS and WW conditions can lead to the selection of maize genotypes that exhibit similar performance as those generated from conventional PS but at a lower cost. Therefore, GS should be integrated into mainstream breeding pipelines to increase the efficiency of maize breeding programs in SSA.

In summary, this section of the review discussed recent innovations in molecular and/or genomics-based methods that empower maize improvement for adaptation to climate stressors, particularly DS, HS, and CDHS. Compared to DS, progress in breeding maize specifically for HS and CDHS adaptation has been slow in SSA, though genetic diversity for tolerance to these stressors appears to be enormous in African maize germplasm. Apart from the shortened breeding cycle, the deployment of genomic approaches (including MABC, MARS, and GS), does offer yield reward and has proven to be relatively cheaper and effective in tailoring maize for DS tolerance in SSA. Breeding works on HS and CDHS are expected to rely on the same pre-breeding and genomics tools that have benefited the breeding efforts on drought. As a step to accelerate the development of HS and CDHS-tolerant maize cultivars, QTLs and candidate genes regulating HS and CDHS adaptive traits are being identified in maize breeding programs including IITA-MIP. Once validated, the desirable QTLs and/or genes may usher in a new era of genomics-aided adaptation breeding of maize to HS and CDHS in SSA.

Conflicts of Interest

The authors have no conflicts of interest to declare.

References

Abdoulaye, T., Wossen, T. and Awotide, B. (2018) Impacts of improved maize varieties in Nigeria: ex-post assessment of productivity and welfare outcomes. *Food Security* 10, 369–379.

Abdulmalik, R.O., Menkir, A., Meseka, S.K., Unachukwu, N., Ado, S.G. *et al.* (2017) Genetic gains in grain yield of a maize population improved through marker assisted recurrent selection under stress and non-stress conditions in West Africa. *Frontiers in Plant Science* 8, 1–11.

Acquaah, G. (2012) *Principles of Plant Genetics and Breeding*. 2nd edn. Wiley-Blackwell, Oxford.

Adeniyan, O.N., Aluko, O.A., Olanipekun, S.O., Olasoji, J.O. and Aduramigba-Modupe, V.O. (2014) Growth and yield performance of cassava/maize intercrop under different plant population density of maize. *Journal of Agricultural Science* 6(8), 35–40.

Adu, G.B., Badu-Apraku, B., Akromah, R., Garcia-Oliveira, A.L., Awuku, F.J. *et al.* (2019) Genetic diversity and population structure of early-maturing tropical maize inbred lines using SNP markers. *PLoS ONE* 14(4), 1–12.

Akaogu, I.C., Badu-Apraku, B., Adetimirin, V.O., Vroh-Bi, I., Oyekunle, M. *et al.* (2012) Genetic diversity assessment of extra-early maturing yellow maize inbreds and hybrid performance in Striga-infested and Striga-free environments. *Journal of Agricultural Science* 151(4), 519–537.

Akaogu, I.C., Badu-Apraku, B. and Adetimirin, V.O. (2017) Combining ability and performance of extra-early maturing yellow maize inbreds in hybrid combinations under drought and rain-fed conditions. *Journal of Agricultural Science* 155(10), 1520–1540.

Alam, M.A., Seetharam, K., Zaidi, P.H., Dinesh, A., Vinayan, M.T. *et al.* (2017) Dissecting heat stress tolerance in tropical maize (Zea mays L). *Field Crops Research* 204, 110–119.

Almeida, G.D., Nair, S., Borém, A., Cairns, J., Trachsel, S. *et al.* (2014) Molecular mapping across three populations reveals a QTL hotspot region on chromosome 3 for secondary traits associated with drought tolerance in tropical maize. *Molecular Breeding* 34, 701–715.

Amegbor, I.K., Badu-Apraku, B. and Annor, B. (2017) Combining ability and heterotic patterns of extra-early maturing white maize inbreds with genes from *Zea diploperennis* under multiple environments. *Euphytica* 213, 1–16.

Annor, B. and Badu-Apraku, B. (2016) Gene action controlling grain yield and other agronomic traits of extra-early quality protein maize under stress and non-stress conditions. *Euphytica* 212, 213–228.

Annor, B., Badu-Apraku, B., Nyadanu, D., Akromah, R. and Fakorede, M.A.B. (2019) Testcross performance and combining ability of early maturing maize inbreds under multiple-stress environments. *Scientific Reports* 9(1), 1–11.

Ashraf, M., Athar, H.R., Harris, P.J.C. and Kwon, T.R. (2008) Some prospective strategies for improving crop salt tolerance. *Advances in Agronomy* 97, 45–110.

Aslam, M., Maqbool, M.A. and Cengiz, R. (2015) *Drought Stress in Maize (Zea mays L.): Effects, Resistance Mechanism, Global Achievements and Biological Strategies for Improvement*. Springer, Cham, Switzerland, p. 74. DOI: 10.1007/978-3-319-25442-5.

Awata, L.A.O., Beyene, Y., Gowda, M., Suresh, L.M., Jumbo, M.B. *et al.* (2020) Genetic analysis of QTL for resistance to maize lethal necrosis in multiple mapping populations. *Genes* 11(1), 1–22.

Aylor, D.E. (2002) Settling speed of corn (*Zea mays*) pollen. *Journal of Aerosol Science* 33(11), 1599–1605.

Badu-Apraku, B. and Fakorede, M.A.B. (2017) *Advances in Genetic Enhancement of Early and Extra-Early Maize for Sub-Saharan Africa*. Springer, Cham, Switzerland.

Badu-Apraku, B. and Lum, A.F. (2010) The pattern of grain yield response of normal and quality protein maize cultivars in stress and non-stress environments. *Agronomy Journal* 102(2), 381–394.

Badu-Apraku, B. and Oyekunle, M. (2012) Genetic analysis of grain yield and other traits of extra-early yellow maize inbreds and hybrid performance under contrasting environments. *Field Crops Research* 129, 99–110.

Badu-Apraku, B., Hunter, R.B. and Tollenaar, M. (1983) Effect of temperature during grain filling on whole plant and grain yield in maize (*Zea mays* L). *Canadian Journal of Plant Science* 63(2), 357–363.

Badu-Apraku, B., Abamu, F.J., Menkir, A., Fakorede, M.A.B., Obeng-Antwi, K. *et al.* (2003) Genotype by environment interactions in the regional early maize variety trials in West and Central Africa. *Maydica* 48(2), 93–104.

Badu-Apraku, B., Fakorede, M.A.B., Menkir, A., Kamara, A.Y., Akanvou, L. *et al.* (2004) Response of early maturing maize to multiple stresses in the Guinea savanna of west and central Africa. *Journal of Genetics and Breeding* 58, 119–130.

Badu-Apraku, B., Fakorede, M.A.B. and Lum, A.F. (2007) Evaluation of experimental varieties from recurrent selection for Striga resistance in two extra-early maize populations in the savannas of West and Central Africa. *Experimental Agriculture* 43(2), 183–200.

Badu-Apraku, B., Fakorede, M.A.B. and Lum, A.F. (2008) S1 family selection in early maturing maize population in Striga-infested and Striga-free environments. *Crop Science* 48(5), 1984–1994.

Badu-Apraku, B., Fakorede, M.A.B., Lum, A.F. and Akinwale, R.O. (2009) Improvement of yield and other traits of extra-early maize under stress and non-stress environments. *Agronomy Journal* 101(2), 381–389.

Badu-Apraku, B., Akinwale, R.O. and Fakorede, M.A.B. (2010) Selection of early maturing maize inbred lines for hybrid production using multiple traits under striga-infested and striga-free environments. *Maydica* 55(3), 261–274.

Badu-Apraku, B., Fakorede, M.A.B., Oyekunle, M. and Akinwale, R.O. (2011a) Selection of extra-early maize inbreds under low-N and drought at flowering and grain-filling for hybrid production. *Maydica* 56(2), 1721–1735.

Badu-Apraku, B., Oyekunle, M., Akinwale, R.O. and Lum, A.F. (2011b) Combining ability of early-maturing white maize inbreds under stress and nonstress environments. *Agronomy Journal* 103(2), 544–547.

Badu-Apraku, B., Oyekunle, M., Menkir, A. and Haruna, A. (2013a) Comparative performance of early-maturing maize cultivars developed in three eras under drought stress and well-watered environments in West Africa. *Crop Science* 53(4), 1298–1311.

Badu-Apraku, B., Akinwale, R.O., Franco, J. and Oyekunle, M. (2012) Assessment of reliability of secondary traits in selecting for improved grain yield in drought and low-nitrogen environments. *Crop Science* 52(5), 2050–2062.

Badu-Apraku, B., Oyekunle, M., Akinwale, R.O. and Aderounmu, M. (2013b) Combining ability and genetic diversity of extra-early white maize inbreds under stress and non-stress environments. *Crop Science* 53(1), 9–26.

Badu-Apraku, B., Oyekunle, M., Fakorede, M.A., Vroh, I., Akinwale, R.O. *et al.* (2013c) Combining ability, heterotic patterns, and genetic diversity of extra-early yellow inbreds under contrasting environments. *Euphytica* 192, 413–433.

Badu-Apraku, B., Fakorede, M.A.B., Gedil, M., Talabi, A.O., Annor, B. *et al.* (2015) Heterotic responses among crosses of IITA and CIMMYT early white maize inbred lines under multiple stress environments. *Euphytica* 206, 245–262.

Badu-Apraku, B., Fakorede, M.A., Gedil, M., Annor, B., Talabi, A.O. *et al.* (2016a) Heterotic patterns of IITA and CIMMYT early maturing yellow maize inbreds under contrasting environments. *Agronomy Journal* 108(4), 1321–1336.

Badu-Apraku, B., Fakorede, M.A.B., Talabi, A.O., Oyekunle, M., Akaogu, I.C. *et al.* (2016b) Gene action and heterotic groups of early white quality protein maize inbreds under multiple stress environments. *Crop Science* 56(1), 183–199.

Badu-Apraku, B., Ifie, B.E., Talabi, A.O., Obeng-Bio, E. and Asiedu, R. (2018) Genetic variances and heritabilities of traits of an early yellow maize population after cycles of improvement for Striga resistance and drought tolerance. *Crop Science* 58(6), 2261–2273.

Badu-Apraku, B., Talabi, A.O., Fakorede, M.A.B., Fasanmade, Y., Gedil, M. *et al.* (2019) Yield gains and associated changes in an early yellow bi-parental maize population following genomic selection for Striga resistance and drought tolerance. *BMC Plant Biology* 19(1), 1–17.

Badu-Apraku, B., Fakorede, M.A.B., Talabi, A.O., Oyekunle, M., Aderounmu, M. *et al.* (2020) Genetic studies of extra-early provitamin-A maize inbred lines and their hybrids in multiple environments. *Crop Science* 60(3), 1325–1345.

Badu-Apraku, B., Garcia-Oliveira, A.L., Petroli, C.D., Hearne, S., Adewale, S.A. *et al.* (2021a) Genetic diversity and population structure of early and extra-early maturing maize germplasm adapted to sub-Saharan Africa. *BMC Plant Biology* 21, 1–15.

Badu-Apraku, B., Obisesan, O., Olumide, O.B. and Toyinbo, J. (2021b) Gene action, heterotic patterns, and inter-trait relationships of early maturing pro-vitamin a maize inbred lines and performance of testcrosses under contrasting environments. *Agronomy* 11(7), 1371.

Badu-Apraku, B., Yallou, C.G., Obeng-Antwi, K., Alidu, H., Talabi, A.O. *et al.* (2017) Yield gains in extra-early maize cultivars of three breeding eras under multiple environments. *Agronomy Journal* 109(2), 418–431.

Bankole, F., Menkir, A., Olaoye, G., Crossa, J., Hearne, S. *et al.* (2017) Genetic gains in yield and yield related traits under drought stress and favorable environments in a maize population improved using marker assisted recurrent selection. *Frontiers in Plant Science*, 8, 808.

Bänzinger, M., Edmeades, G.O., Beck, D. and Bellon, M. (2000) Breeding for drought and nitrogen stress tolerance in maize: from theory to practice. *Mexico DF CIMMYT* 44–46.

Barnabás, B., Jäger, K. and Fehér, A. (2008) The effect of drought and heat stress on reproductive processes in cereals. *Plant Cell and Environment* 31(1), 11–38.

Bartels, D. and Sunkar, R. (2005) Drought and salt tolerance in plants. *Critical Reviews in Plant Sciences* 24(1), 23–58.

Betran, F.J., Beck, D., Bänziger, M. and Edmeades, G.O. (2003) Genetic analysis of inbred and hybrid grain yield under stress and nonstress environments in tropical maize. *Crop Science* 43(3), 807–817.

Beyene, Y., Semagn, F.K., Mugo, S.N., Tarekegne, A.T., Babu, R. *et al.* (2015a) Genetic gains in grain yield through genomic selection in eight bi-parental maize populations under drought stress. *Crop Science* 55(1), 154–163.

Beyene, Y., Semagn, K., Crossa, J., Mugo, S., Atlin, G.N. *et al.* (2015b) Improving maize grain yield under drought stress and non-stress environments in sub-Saharan Africa using marker-assisted recurrent selection. *Crop Science* 56(1), 344–353.

Beyene, Y., Semagn, K., Mugo, S., Prasanna, B.M., Tarekegne, A. *et al.* (2016) Performance and grain yield stability of maize populations developed using marker-assisted recurrent selection and pedigree selection procedures. *Euphytica* 208, 285–297.

Beyene, Y., Gowda, M., Olsen, M., Robbins, K.R., Pérez-Rodríguez, P. *et al.* (2019) Empirical comparison of tropical maize hybrids selected through genomic and phenotypic selections. *Frontiers in Plant Science* 10, 1502.

Bolaños, J. and Edmeades, G.O. (1996) The importance of the anthesis-silking interval in breeding for drought tolerance in tropical maize. *Field Crops Research* 48(1), 65–80.

Cairns, J.E. and Prasanna, B.M. (2018) Developing and deploying climate-resilient maize varieties in the developing world. *Current Opinion in Plant Biology* 45, 226–230.

Cairns, J.E., Crossa, J., Zaidi, P.H., Grudloyma, P., Sanchez, C. *et al.* (2013) Identification of drought, heat, and combined drought and heat tolerant donors in maize. *Crop Science* 53(4), 1335–1346.

Campos, H., Cooper, M., Habben, J.E., Edmeades, G.O. and Schussler, J.R. (2004) Improving drought tolerance in maize: A view from industry. *Field Crops Research* 90(1), 19–34.

Cerrudo, D., Cao, S., Yuan, Y., Martinez, C., Suarez, E.A. *et al.* (2018) Genomic selection outperforms marker assisted selection for grain yield and physiological traits in a maize doubled haploid population across water treatments. *Front Plant Science* 9, 366.

Chourey, K., Jansson, J., Verberkmoes, N., Shah, M., Chavarria, K.L. *et al.* (2010) Direct cellular lysis/protein extraction protocol for soil metaproteomics. *Journal of Proteome Research* 9(12), 6615–6622.

Cicchino, M., Rattalino-Edreria, J.I., Uribelarrea, M. and Otegui, M.E. (2011) Heat stress in field-grown maize: Response of physiological determinants of grain yield. *Crop Science* 50(4), 1438–1448.

Crafts-Brandner, S.J. and Law, R.D. (2000) Effect of heat stress on the inhibition and recovery of the ribulose-1,5-bisphosphate carboxylase/oxygenase activation state. *Planta* 212, 67–74.

Dania, S.O., Fagbola, O. and Alabi, R.A. (2014) The yield and farmer's economic advantage of maize-pigeon pea intercrop as affected by the application of organic base fertilizer. *European Scientific Journal* 10(25), 357–374.

Daryanto, S., Wang, L. and Jacinthe, P.A. (2016) Global synthesis of drought effects on maize and wheat production. *PLoS ONE* 11(5), 1–15.

Dass, S., Arora, P., Kumari, M. and Pal, D. (2001) Morphological traits determining drought tolerance in maize (*Zea mays* L). *Indian Journal of Agricultural Research* 35(3), 190–193.

Demirevska, K., Zasheva, D., Dimitrov, R., Simova-Stoilova, L., Stamenova, M. *et al.* (2009) Drought stress effects on Rubisco in wheat: Changes in the Rubisco large subunit. *Acta Physiologiae Plantarum* 31, 1129–1138.

Derera, J., Tongoona, P., Vivek, B.S., Laing, M.D. (2007) Gene action controlling grain yield and secondary traits in southern African maize hybrids under drought and non-drought environments. *Euphytica* 162, 411–422.

Du Plessis, J. (2003) *Maize Production*, Department of Agriculture, Pretoria, South Africa 1-38.

Dwivedi, S.L., Ceccarelli, S., Blair, M.W., Upadhyaya, H.D., Are, A.K. *et al*. (2016) Landrace germplasm for improving yield and abiotic stress adaptation. *Trends in Plant Science* 21(1), 31–42.

Earl, H.J. and Davis, R.F. (2003) Effect of drought stress on leaf and whole canopy radiation use efficiency and yield of maize. *Agronomy Journal* 95(3), 688–696.

Edmeades, G.O. (2008) Drought tolerance in maize: an emerging reality. In: James, C. (ed.) *ISAAA Brief 39 Global Status of Commercialized Biotech/GM Crops*. ISAAA, pp. 197–217.

Edmeades, G.O. (2013) *Progress in Achieving and Delivering Drought Tolerance in Maize—An Update*. ISAAA, Ithaca, NY, p. 130.

Edmeades, G.O., Bolaños, J., Chapman, S.C., Lafitte, H.R. and Bänziger, M. (1999) Selection improves drought tolerance in tropical maize populations: I Gains in biomass, grain yield, harvest index. *Crop Science* 39(5), 1306–1315.

Ehlers, J.D. and Hall, A.E. (1998) Heat tolerance of contrasting cowpea lines in short and long days. *Field Crops Research* 55(1–2), 11–21.

Fahad, S., Bajwa, A.A., Nazir, U., Anjum, S.A., Farooq, A. *et al*. (2017) Crop production under drought and heat stress: Plant responses and management options. *Frontiers in Plant Science* 8, 1–16.

Fakorede, M.A.B., Badu-Apraku, B., Kamara, A.Y., Menkir, A. and Ajala, S.O. (2003) Maize revolution in West and Central Africa: An overview. *Proceedings of a Regional Maize Workshop* 14–18.

Falconer, D.A. (1981) *Introduction to Quantitative Genetics*. Longman Scientific, New York.

Farooq, M., Wahid, A., Kobayashi, N.S., Fujita, D.B. and Basra, S.M. (2009) Plant drought stress: Effects, mechanisms and management. In: Lichtfouse, E., Navarrete, M., Debaeke, P., Véronique, S. and Alberola, C. (eds) *Sustainable Agriculture*. Springer, Dordrecht. DOI: 10.1007/978-90-481-2666-8_12.

Fernandez, G.C. (1992) Effective selection criteria for assessing plant stress tolerance In: *Proceeding of the International Symposium on Adaptation of Vegetables and other Food Crops in Temperature and Water Stress*, Shanhua, Taiwan, pp. 257–270.

Fischer, R.A. and Maurer, R. (1978) Drought resistance in spring wheat cultivars' grain yield responses. *Australian Journal of Agricultural Research* 29(5), 897–912.

Fleury, D., Jefferies, S., Kuchel, H. and Langridge, P. (2010) Genetic and genomic tools to improve drought tolerance in wheat. *Journal of Experimental Botany* 61(12), 3211–3222.

Gauch, H.G. and Zobel, R.W. (1996) Predictive and postdictive success of statistical analyses of yield trials. *Theoretical and Applied Genetics* 76, 1–10.

Gauch, H.G. and Zobel, R.W. (1997) Identifying mega-environments and targeting genotypes. *Crop Science* 37(2), 311–326.

Gedil, M. and Menkir, A. (2019) An integrated molecular and conventional breeding scheme for enhancing genetic gain in maize in Africa. *Frontiers in Plant Science* 10, 1–17.

Griffin, J.J., Ranney, T.G. and Pharr, D.M. (2004) Heat and drought influence photosynthesis and water relations, and soluble carbohydrates of two ecotypes of redbud (*Cercis canadensis*). *Journal of American Society for Horticultural Science* 129(4), 497–502.

Gupta, P.K., Langridge, P. and Mir, R.R. (2010) Marker-assisted wheat breeding: Present status and future possibilities. *Molecular Breeding* 26, 145–161.

Hao, Z., Li, X., Xie, C., Weng, J., Li, M. *et al*. (2011) Identification of functional genetic variations underlying drought tolerance in maize using SNP markers. *Journal of Integrative Plant Biology* 53(8), 641–652.

Heschel, M.S. and Riginos, C. (2005) Mechanisms of selection for drought stress tolerance and avoidance in *Impatiens capensis* (Balsaminaceae). *American Journal of Botany* 92(1), 37–44.

Heyne, E.G. and Brunson, A.M. (1940) Genetic Studies of heat and drought tolerance in maize. *Agronomy Journal* 32, 803–814.

Jacobsen, E. and Schouten, H.J. (2007) Cisgenesis strongly improves introgression breeding and induced translocation breeding of plants. *Trends in Biotechnology* 25(5), 219–223.

Jagtap, S.S., Alabi, R.T. and Adeleye, O. (1998) The influence of maize density on resource use and productivity; An experimental and simulation study. *African Crop Science Journal*. 6(3), 259–272.

Jiang, Y. and Huang, B. (2001) Drought and heat stress injury to two cool-season turfgrasses in relation to antioxidant metabolism and lipid peroxidation. *Crop Science* 41(2), 436–442.

Kearsey, M.J. and Pooni, H.S. (1996) *The Genetical Analysis of Quantitative Traits*. Chapman and Hall, Stanley Thornes Publishers, London. DOI: 10.1007/978-1-4899-4441-2.

Kumar, U., Joshi, A.K., Kumari, M., Paliwal, R., Kumar, S. *et al.* (2010) Identification of QTLs for stay green trait in wheat (Triticum aestivum L) in the "Chirya 3" × "Sonalika" population. *Euphytica* 174, 437–445.

Kumar, S., Gupta, D. and Nayyar, H. (2012) Comparative response of maize and rice genotypes to heat stress: Status of oxidative stress and antioxidants. *Acta Physiologiae Plantarum* 34, 75–86.

Lafitte, H.R. and Edmeades, G.O. (1995) Association between traits in tropical maize inbred lines and their hybrids under high and low soil nitrogen. *Maydica* 419, 259–267.

Lawlor, D.W. (2013) Genetic engineering to improve plant performance under drought: Physiological evaluation of achievements, limitations, and possibilities. *Journal of Experimental Botany* 64(1), 695–709.

Li, Y., Ruperao, P., Batley, J., Edwards, D., Khan, T. *et al.* (2018) Investigating drought tolerance in chickpea using genome-wide association mapping and genomic selection based on whole-genome resequencing data. *Frontiers in Plant Science* 9, 190.

Lobell, D.B., Bänziger, M., Magorokosho, C. and Vivek, B. (2011) Nonlinear heat effects on African maize as evidenced by historical yield trials. *Nature Climate Change* 1(1), 42–45.

Longmei, N., Gill, G.K., Zaidi, P.H., Kumar, R., Nair, S.K. *et al.* (2021) Genome wide association mapping for heat tolerance in sub-tropical maize. *BMC Genomics* 22(1), 1–14.

Maazou, A.R.S., Tu, J. and Liu, Z. (2016) Breeding for drought tolerance in maize (*Zea mays* L). *American Journal of Plant Sciences* 7(14), 1858–1870.

Mammadov, J., Buyyarapu, R., Guttikonda, S.K., Parliament, K., Abdurakhmonov, I.Y. *et al.* (2018) Wild relatives of maize, rice, cotton, and soybean: Treasure troves for tolerance to biotic and abiotic stresses. *Frontiers in Plant Science* 9, 886.

Mengesha, W.A., Menkir, A., Unakchukwu, N., Meseka, S., Farinola, A. *et al.* (2017) Genetic diversity of tropical maize inbred lines combining resistance to *Striga hermonthica* with drought tolerance using SNP markers. *Plant Breeding* 136(3), 338–343.

Menkir, A. (2003) The role of GIS in the development and targetting of maize germplasm to farmers' needs in West and Central Africa. *Maize Revolution in West and Central Africa*, 16–30.

Menkir, A., Badu-Apraku, B., Ajala, S., Kamara, A. and Ndiaye, A. (2009) Response of early maturing maize landraces and improved varieties to moisture deficit and sufficient water supply. *Plant Genetic Resources: Characterization and Utilization* 7(3), 205–215.

Mitra, J. (2001) Genetics and genetic improvement of drought tolerance in crop plants. *Current Science* 80(6), 758–762.

Mittler, R. and Blumwald, E. (2010) Genetic engineering for modern agriculture: Challenges and perspectives. *Annual Review of Plant Biology* 61, 443–462.

Mohammadi, M., Taleei, A., Zeinali, H. and Naghavi, M. (2005) QTL analysis for phenologic traits in doubled haploid population of barley. *International Journal of Agriculture* 7(5), 820–823.

Monneveux, P., Sánchez, C., Beck, D. and Edmeades, G.O. (2006) Drought tolerance improvement in tropical maize source populations: Evidence of progress. *Crop Science* 46(1), 180–191.

Nasser, L.M., Badu-Apraku, B., Gracen, V.E. and Mafouasson, H.N.A. (2020) Combining ability of early-maturing yellow maize inbreds under combined drought and heat stress and well-watered environments. *Agronomy* 10(10), 1585.

Nelimor, C., Badu-Apraku, B., Tetteh, A. and N'guetta, A.S.P. (2019) Assessment of genetic diversity for drought, heat and combined drought and heat stress tolerance in early maturing maize landraces. *Plants* 8(11), 518.

Nelimor, C., Badu-Apraku, B., Tetteh, A.Y., Garcia-Oliveira, A.L. and N'guetta, A.S.P. (2020a) Assessing the potential of extra-early maturing landraces for improving tolerance to drought, heat, and both combined stresses in maize. *Agronomy* 10(3), 318.

Nelimor, C., Badu-Apraku, B., Garcia-Oliveira, A.L., Tetteh, A., Agre, P. *et al.* (2020b) Genomic analysis of selected maize landraces from Sahel and Coastal West Africa reveals their variability and potential for genetic enhancement. *Genes* 11(9), 1054.

NeSmith, D.S. and Ritchie, J.T. (1992) Effects of soil water-deficits during tassel emergence on development and yield component of maize (*Zea mays*). *Field Crops Research* 28(3), 251–256.

Nielsen, R.L. (2018) Effects of Stress during Grain Filling in Corn; Corny News Network. Purdue University. Available at: https://wwwagrypurdueedu/ext/corn/news/timeless/grainfillstresshtml (accessed 30 December 2021).

NOAA National Centers for Environmental Information (NCEI) U.S. Billion-Dollar Weather and Climate Disasters (2023). Available at: https://www. ncei.noaa.gov/access/billions/events. DOI:10.25921/stkw-7w73 (accessed 27 March 2023).

Obeng-Bio, E., Badu-Apraku, B., Ifie, B.E., Danquah, A., Blay, E.T. *et al.* (2020) Genetic diversity among early provitamin A quality protein maize inbred lines and the performance of derived hybrids under contrasting nitrogen environments. *BMC Genetics* 21, 1–13.

Obisesan, I.O. (2021) Genetic analysis of multiple-stress tolerant extra-early maize (*Zea mays* L.) inbred lines under contrasting environments [Unpublished PhD thesis]. Ile-Ife. Nigeria: Department of Crop Production and Protection, Obafemi Awolowo University.

Osuman, A.S., Badu-Apraku, B., Ifie, B.E., Tongoona, P., Obeng-Bio, E. *et al.* (2020) Genetic diversity, population structure and inter-trait relationships of combined heat and drought tolerant early-maturing maize inbred lines from west and central Africa. *Agronomy* 10(9), 1324.

Osuman, A.S., Badu-Apraku, B., Karikari, B., Ifie, B.E., Tongoona, P. *et al.* (2022) Genome-wide association study reveals genetic architecture and candidate genes for yield and related traits under terminal drought, combined heat and drought in tropical maize germplasm. *Genes* 13(2), 349.

Oyekunle, M. and Badu-Apraku, B. (2013) Genetic analysis of grain yield and other traits of early-maturing maize inbreds under drought and well-watered conditions. *Journal of Agronomy and Crop Science* 200(2), 92–107.

Oyekunle, M., Badu-Apraku, B., Hearne, S. and Franco, J. (2015) Genetic diversity of tropical early-maturing maize inbreds and their performance in hybrid combinations under drought and optimum growing conditions. *Field Crops Research* 170, 55–65.

Parmar, N., Singh, K.H., Sharma, D., Singh, L., Kumar, P. *et al.* (2017) Genetic engineering strategies for biotic and abiotic stress tolerance and quality enhancement in horticultural crops: A comprehensive review. *3 Biotech* 7, 1–35.

Praba, M.L., Cairns, J.E., Babu, R.C. and Lafitte, H.R. (2009) Identification of physiological traits underlying cultivar differences in drought tolerance in rice and wheat. *Journal of Agronomy and Crop Science* 195(1), 30–46.

Prasad, P.V., Staggenborg, S.A. and Ristic, Z. (2008) Impacts of drought and/or heat stress on physiological, developmental, growth, and yield processes of crop plants. *Response of Crops to Limited Water: Understanding and Modeling Water Stress Effects on Plant Growth Processes* 1, 301–355.

Prasad, P.V.V., Djanaguiraman, M., Perumal, R. and Ciampitti, I.A. (2015) Impact of high temperature stress on floret fertility and individual grain weight of grain sorghum: Sensitive stages and thresholds for temperature and duration. *Frontiers in Plant Science* 6, 1–11.

Pswarayi, A. and Vivek, B.S. (2008) Combining ability amongst CIMMYT's early maturing maize (*Zea mays* L) germplasm under stress and non-stress conditions and identification of testers. *Euphytica* 162(3), 353–362.

Reynolds, M.P., Balota, M., Delgado, M.I.B., Amani, I. and Fischer, R.A. (1994) Physiological and morphological traits associated with spring wheat yield under hot, irrigated conditions. *Functional Plant Biology* 21(6), 717–730.

Rezende, W.S., Beyene, Y., Mugo, S., Ndou, E., Gowda, M. *et al.* (2020) Performance and yield stability of maize hybrids in stress-prone environments in eastern Africa. *Crop Journal* 8(1), 107–118.

Ribaut, J.M. and Ragot, M. (2007) Marker-assisted selection to improve drought adaptation in maize: The backcross approach, perspectives, limitations, and alternatives. *Journal of Experimental Botany* 58(2), 351–360.

Richards, R.A. (1996) Defining selection criteria to improve yield under drought. *Plant Growth Regulation* 20, 157–166.

Rizhsky, L., Liang, H. and Mittler, R. (2002) The combined effect of drought stress and heat shock on gene expression in tobacco. *Plant Physiology* 130(3), 1143–1151.

Rizhsky, L., Liang, H., Shuman, J., Shulaev, V., Davletova, S. *et al.* (2004) When defense pathways collide the response of arabidopsis to a combination of drought and heat stress. *Plant Physiology* 134(4), 1683–1696.

Rosegrant, M.R., Ringler, C., Sulser, T.B., Ewing, M., Palazzo, A. *et al.* (2009) *Agriculture and Food Security under Global Change: Prospects for 2025/2050*. International Food Policy Research Institute, Washington, DC, p. 89.

Rosen, S. and Scott, L. (1992) Famine grips sub-Saharan Africa. *Agricultural Outlook AO-US Department of Agriculture, Economic Research Service (USA).*

Rosielle, A.A. and Hamblin, J. (1981) Theoretical aspects of selection for yield in stress and non-stress environment. *Crop Science* 21(6), 943–946.

Salvi, S. and Tuberosa, R. (2005) To clone or not to clone plant QTLs: Present and future challenges. *Trends in Plant Science* 10(6), 297–304.

Savage, D.A. and Jacobson, L.A. (1935) The killing effect of heat and drought on buffalo grass and blue grama grass at Hays, Kansas. *Agronomy Journal* 27(7), 566–582.

Schoper, J.B., Lambert, R.J. and Vasilas, B.L. (1986) Maize pollen viability and ear receptivity under water and high temperature stress. *Crop Science* 26(5), 1029–1033.

Seetharam, K., Kuchanur, P.H., Koirala, K.B., Tripathi, M.P., Patil, A. *et al.* (2021) Genomic regions associated with heat stress tolerance in tropical maize (*Zea mays* L). *Scientific Reports* 11(1), 1–11.

Semagn, K., Beyene, Y., Warburton, M.L., Tarekegne, A., Mugo, S. *et al.* (2013) Meta-analyses of QTL for grain yield and anthesis silking interval in 18 maize populations evaluated under water-stressed and well-watered environments. *BMC Genomics* 14(1), 1–6.

Setimela, P.S., Vivek, B., Banziger, M., Crossa, J. and Maideni, F. (2007) Evaluation of early to medium maturing open pollinated maize varieties in SADC region using GGE giplot based on the SREG model. *Field Crops Research* 103(3), 161–169.

Shaibu, A.S., Badu-Apraku, B. and Ayo-Vaughan, M.A. (2021) Enhancing drought tolerance and *Striga hermonthica* resistance in maize using newly derived inbred lines from the wild maize relative, *Zea diploperennis*. *Agronomy* 11(1), 177.

Shiferaw, B., Prasanna, B.M., Hellin, J. and Bänziger, M. (2011) Crops that feed the world 6. Past successes and future challenges to the role played by maize in global food security. *Food Security* 3, 307–327.

Shikha, M., Kanika, A., Rao, A.R., Mallikarjuna, M.G., Gupta, H.S. *et al.* (2017) Genomic selection for drought tolerance using genome-wide SNPs in maize. *Frontiers in Plant Science* 8, 550.

Siddique, K.H.M., Regan, K.L., Tennant, D. and Thomson, B.D. (2001) Water use and water use efficiency of cool season grain legumes in low rainfall mediterranean-type environments. *European Journal of Agronomy* 15(4), 267–280.

Suzuki, N., Rivero, R.M., Shulaev, V., Blumwald, E. and Mittler, R. (2014) Abiotic and biotic stress combinations. *New Phytologist* 203(1), 32–43.

Taiz, L. and Zeiger, E. (2006) *Plant Physiology Sinauer*. Associates Press, Sunderland, MA, USA.

Takim, F.O. (2012) Advantages of maize-cowpea intercropping over sole cropping through competition indices. *Journal of Agriculture and Biodiversity Research* 1(4), 53–59.

Talabi, A.O., Badu-Apraku, B. and Fakorede, M.A.B. (2017) Genetic variances and relationship among traits of an early-maturing maize population under drought-stress and low-N environments. *Crop Science* 57(2), 1–12. DOI: 102135/cropsci2016030177.

Thompson, L.M. (1975) Weather variability, climatic change, and grain production. *Journal of Soil and Water Conservation* 40(4), 386–389.

Tuberosa, R. and Salvi, S. (2006) Genomics-based approaches to improve drought tolerance of crops. *Trends in Plant Science* 11(8), 405–412.

Trachsel, S., Sun, D., SanVicente, F.M., Zheng, H., Atlin, G.N. *et al.* (2016) Identification of QTL for early vigor and stay-green conferring tolerance to drought in two connected advanced backcross populations in tropical maize (*Zea mays* L). *PloS One* 11(3), e0149636.

Tuberosa, R., Salvi, S., Sanguineti, M.C., Landi, P., Maccaferri, M. *et al.* (2002) Mapping QTLS regulating morpho-physiological traits and yield: Case studies, shortcomings and perspectives in drought-stressed maize. *Annals of Botany* 89(7), 941–963.

Van Oosten, M.J., Costa, A., Punzo, P., Landi, S., Ruggiero, A. *et al.* (2016) Genetics of drought stress tolerance in crop plants. *Drought Stress Tolerance in Plants* 2, 39–70.

Wang, X., Wang, H., Liu, S., Ferjani, A., Li, J. *et al.* (2016) Genetic variation in ZmVPP1 contributes to drought tolerance in maize seedlings. *Nature Genetics* 48(10), 1233–1241.

Warburton, M.L., Rauf, S., Marek, L., Hussain, M., Ogunola, O. *et al.* (2017) The use of crop wild relatives in maize and sunflower breeding. *Crop Science* 57(3), 1227–1240.

Xue, Y., Warburton, M.L., Sawkins, M., Zhang, X., Setter, T. *et al.* (2013) Genome-wide association analysis for nine agronomic traits in maize under well-watered and water-stressed conditions. *Theoretical and Applied Genetics* 126(10), 2587–2596.

Yang, J., Sears, R., Gill, B. and Paulsen, G. (2002) Quantitative and molecular characterization of heat tolerance in hexaploid wheat. *Euphytica* 126, 275–282.

Yuan, Y., Cairns, J.E., Babu, R., Gowda, M., Makumbi, D. *et al.* (2019) Genome-wide association mapping and genomic prediction analyses reveal the genetic architecture of grain yield and flowering time under drought and heat stress conditions in maize. *Frontiers in Plant Science* 9, 1919.
Zaidi, P.H., Zaman-Allah, M., Trachsel, S., Seetharam, K., Cairns, J.E. *et al.* (2016) Phenotyping for Abiotic Stress Tolerance in Maize: Heat Stress Field Manual CIMMYT, Hyderabad, India.
Ziyomo, C. and Bernardo, R. (2013) Drought tolerance in maize: Indirect selection through secondary traits versus genome wide selection. *Crop Science* 53(4), 1269–1275.

14 Statistical testing of range-size change as affected by the extremes of seasonal maximum temperatures

M.S. Dhanoa[1], R. Sanderson[2], J.L. Ellis[1], C.D. Powell[1], S. López[3], A. Shepherd[4], L.M. Cardenas[5] and J. France[1*]

[1]Centre for Nutrition Modelling, Department of Animal Biosciences, University of Guelph, Guelph ON, N1G 2W1, Canada; [2]Institute of Biological, Environmental and Rural Sciences, Aberystwyth University, Gogerddan, Aberystwyth, Ceredigion, SY23 3EB, UK; [3]Instituto de Ganadería de Montaña (IGM), CSIC-Universidad de León, Departamento de Producción Animal, Universidad de León, E-24007, León, Spain; [4]Institute of Biological and Environmental Sciences, University of Aberdeen, Aberdeen, Scotland, UK; [5]Rothamsted Research, North Wyke, Okehampton, Devon, UK

Abstract

Any geo-position on Earth will have its own seasonal pattern of maximum day temperatures. Climate change driven extreme temperature episodes may disturb such patterns and create temperature range changes. The significance of such changes can be tested statistically by maximum temperature range data analysis using the distribution of the ratio of range to standard deviation. Structural pattern trends, percentiles and kurtosis summaries bring out more details of seasonal temperatures.

Keywords: extreme value distribution, kurtosis, maximum temperature, range to standard deviation ratio distribution, split-line model, structural patterns

Introduction

Extreme weather effects (e.g. heatwaves, fires, floods, storms, cyclones) of climate change are causing difficulties globally. The records of maximum day temperatures show variable length episodes of very high temperatures. As a consequence of such episodes, at any given geo-position, a long maximum temperature series will contain variable temperature ranges. Statistical analysis of such a feature in maximum temperature series can be tested with reference to the distribution of the ratio of range (w) over standard deviation (SD). David *et al.* (1954) described the distribution of w/SD, and included a table of significance testing upper and lower limit ratio values. For further details about such ratios of variables, the reader is referred to Geary (1930) and Fieller (1932), and for more recent consideration, to Shanmugalingam (1982) and

*Corresponding author: jfrance@uoguelph.ca

Dhanoa *et al.* (2018). In Australia, maximum temperatures tend to vary considerably due to extreme weather events such as bushfires. Such events will impact the seasonal pattern of temperatures and may modify yearly and monthly temperature distribution by causing extra skewness or even thick tails (Dhanoa *et al.*, 2020). Inter-year shifting of seasonal temperature patterns makes modelling a challenge. The objective of the present work, therefore, is to review statistical testing of range-size change as affected by the extremes of seasonal maximum temperatures (polar vortex ice storms in North America create similar effects on day-minimum and day-maximum temperature values and ranges).

Review Methodology

A systematic literature review of various electronic bibliographic databases (e.g. Web of Science, Google Scholar) was conducted to search and retrieve articles. The primary search terms were Extreme Value Distribution and Maximum Temperature Range and no limits were placed on the publication date. In addition to electronic resources, bibliographies of book chapters and technical reports were queried. The most relevant published articles, book chapters and technical reports were considered in producing this article.

Materials and Methods

Lindley and Miller (1953) provided multiplicative factors for the conversion of data range to standard deviation up to normal sample size 13. Davies (1967) reported multiplicative factors for calculating SD confidence intervals or the confidence interval for the ratio of two standard deviations. Upper and lower limits for *w*/SD distribution are available in David *et al.* (1954). An example data set (subset years 1971 to 2009) for the application of the *w*/SD distribution was provided courtesy of the Australian government's Bureau of Meteorology in Melbourne (37.6709° S, 144.8433° E) [BoM Climate and Ocean Data Services, Ticket K8P2636828 (SEC=UNCLASSIFIED), BoM Meteorology Station # 86282, http://www.bom.gov.au/climate/data/index.shtml]. UK data were obtained from the website of the Meteorological Office, the UK's national weather service, for the Hurley site (51.5279° N, 0.8226° W) of the former Institute of Grassland and Environmental Research (https://www.metoffice.gov.uk, search for Hurley site). All analyses and calculations were performed using the statistical software Genstat (VSN International, 2022).

Results and Discussion

Attributes of the northern and southern hemispheres are different in relation to land mass, sea surfaces and human population. To serve the needs of the latter, the emergence of intensive agriculture, deforestation, all kinds of industries and other such activities have created collateral effects on our weather and climate. As a result of greenhouse gases (CO_2, CH_4, N_2O) and pollution, extremes of seasonal weather, including the inflation of maximum temperature range (range being the difference between highest and lowest daily maximum temperatures), are evident all over the planet. Herein, changes in the range of maximum temperatures are analysed, with reference to the distribution of *w*/SD. Temperature summaries for the Australian site compared to the British site show large differences as illustrated below using Windrose diagrams (Goedhart and Payne, 2021). The British site runs from 1953 to 1991 and the Australian site from 1971 to 2019 with 21 common years (1971–1991). Fig. 14.1a and 14.b shows that during the UK winter (some sub-zero temperatures), Australian maximum summer temperatures are way above structural levels of seasonal patterns, whilst the Australian winter temperatures (during the UK summer) stay relatively consistent and mild.

The Earth's orbital position around the Sun (just over eight light minutes away) determines seasonal weather patterns (maximum temperatures) in the northern and southern hemispheres, which are usually bell-shaped. This leads to contrasting weather types such as (when viewed from the northern hemisphere) summer in the northern hemisphere but winter in the southern hemisphere and vice versa.

(a)

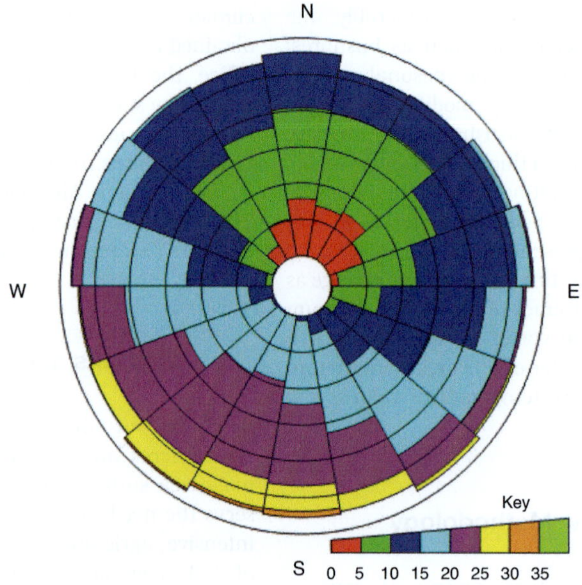

(b)

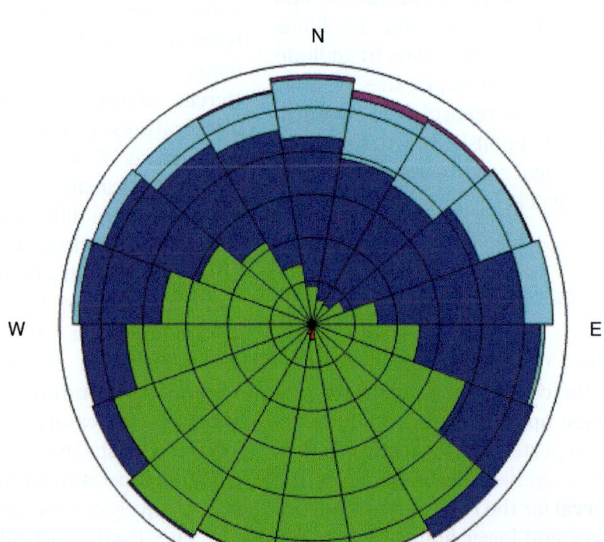

Fig. 14.1. Windrose diagrams of daily maximum temperature (°C) for (a) the UK site and (b) the Australian site. UK temperatures cover the period 1953–1991 and Australian temperatures the period 1971–2019 (rings drawn at 1% intervals). For each year, angles are defined as 1/365 or 1/366 of 360 degrees.

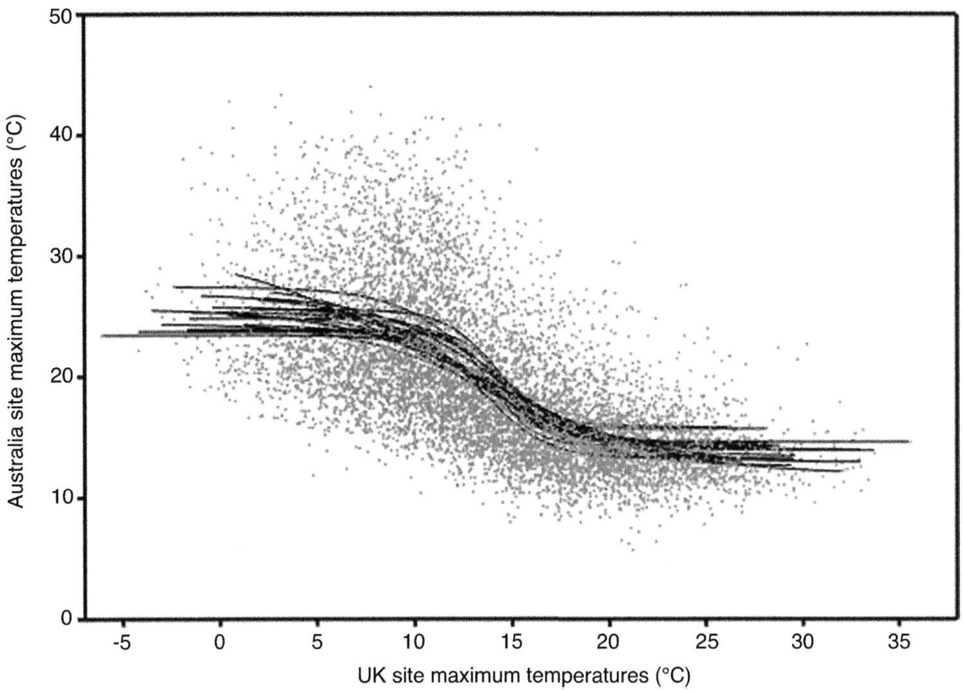

Fig. 14.2. Paired Australian site vs UK site fitted logistic curves [$A + C/(1 + \exp(-B^*(X-M)))$] for each common year (1971–1991).

However, within any calendar day, maximum temperature observations are paired because they are both within the time period of one Earth rotation around its own axis regardless of its orbital position (the Moon continues in its own orbit around planet Earth). Therefore, a plot of the Australian maximum temperature vs any northern hemisphere site (e.g. Hurley) should show an inverse relationship with a switch from high to low values of the Australian site as depicted in Fig. 14.2. To highlight the underlying structural trend (structural because the Sun is heating both hemispheres), a logistic equation $Y = A + C/(1 + e^{-B^*(X-M)})$ can be used to approximate this trend (this equation has both upper and lower asymptotes). The parameters of the logistic obtained by fitting the Australian site vs the UK site maximum temperatures are given in Table 14.1. Position shifting of seasonal patterns, usually seen when comparing multiple years, is reflected by structural trend shifts, as shown in Fig. 14.2.

Due to episodes of extreme high temperature, the maximum temperature range shows an upwards shift as the fit of a split-line model (Vieth, 1989) illustrates in Fig. 14.3. From Table 14.6 of David *et al.* (1954), lower and upper significant limits for a sample size of 39 are 5%, lower = 3.64, upper = 5.12; 2.5%, lower = 3.55, upper = 5.31; and 1.0%, lower = 3.44, upper = 5.51. The ratio *w*/SD and these significance limits are presented in Fig. 14.4, which shows that 15 years of ratios breach the 5% upper limit, with one even beyond the 1.0% limit.

Range changing relatively extreme maximum temperature episodes are quite common across all continents, including the Polar Regions. Example data from both sites used in our analysis show incidents of relatively extreme maximum temperature. Such incidents affect yearly data distribution, causing some shifts of data percentiles. The highest maximum temperatures recorded over the 21 common years are shown in Table 14.2 for both sites. As a result of these high observations, there are positional changes to the 95th and 98th percentiles as shown in Fig. 14.5. Of course,

Table 14.1. Logistic curve parameters obtained fitting the Australian site vs the UK site maximum temperatures over the common years (1971–1991).

Logistic curve fitted as:
Response variate *Y*: Maximum temperature (°C) [←Australian site data]
Explanatory variate *X*: Maximum temperature (°C) [←British site data]
Fitted curve[a]: $Y=A + C/(1+\exp(-B^*(X-M))$

Parameter estimates of curve as fitted in Genstat

Year	*B*	*M*	*C*	*A*	*A*+*C*
1971	–0.9707	14.70	9.60	14.31	23.91
1972	–0.5346	12.60	9.48	15.83	25.31
1973	–0.3297	13.08	14.07	12.63	26.70
1974	–0.3416	12.06	14.47	13.09	27.56
1975	–0.3657	13.56	11.16	13.98	25.14
1976	–0.4069	13.19	11.90	13.90	25.80
1977	–0.4210	13.60	12.35	13.03	25.38
1978	–0.2778	14.12	11.69	12.77	24.46
1979	–0.3264	12.68	13.40	13.45	26.85
1980	–0.9289	14.11	8.19	15.71	23.90
1981	–0.6288	14.36	10.51	14.32	24.84
1982	–0.4412	13.57	12.97	14.45	27.42
1983	–0.1924	10.82	19.01	11.87	30.88
1984	–0.3443	14.88	11.39	12.99	24.38
1985	–0.9858	14.09	9.44	14.32	23.76
1986	–0.6337	13.63	10.50	13.25	23.74
1987	–0.5552	14.03	9.20	14.26	23.46
1988	–0.4365	14.42	11.47	14.17	25.64
1989	–0.3218	14.16	13.84	12.94	26.78
1990	–0.7278	14.46	10.95	14.61	25.56
1991	–0.2458	14.26	13.02	12.63	25.66

[a]*A* = lower asymptote, *A*+*C*=upper asymptote, *M* = mid-point of the sigmoid (inflexion point), *B* = shape parameter.

these results are from a single site in both hemispheres, so upscaling should give a better view of such climate change driven features in seasonal weather patterns over various size geo-dimensions.

Some of the maximum observations in Fig. 14.5 may be extreme observations and, if so, they can alter the estimates of the data mean, median and range. To demonstrate this effect, summaries are calculated from the whole dataset and then repeated using data up to the 98th, 96th, 94th, 92nd and 90th percentiles to obtain estimates as shown in Table 14.3. Weather forecasters tend to cite monthly means when commenting on really high day temperatures. A more appropriate and better comparison may be the mean + half range rather than just the mean. There may be a case to calculate the monthly average maximum and monthly average minimum over the same time period that gave the monthly average.

Range, standard deviation and kurtosis are important features for the study of data distribution. Observation outliers and extreme values can cause skewness and heavy tails. Heavy tails can lead to long or thick tail distributions such as the extreme or generalised extreme value distributions (Eastoe, 2017; Dhanoa *et al.*, 2020).

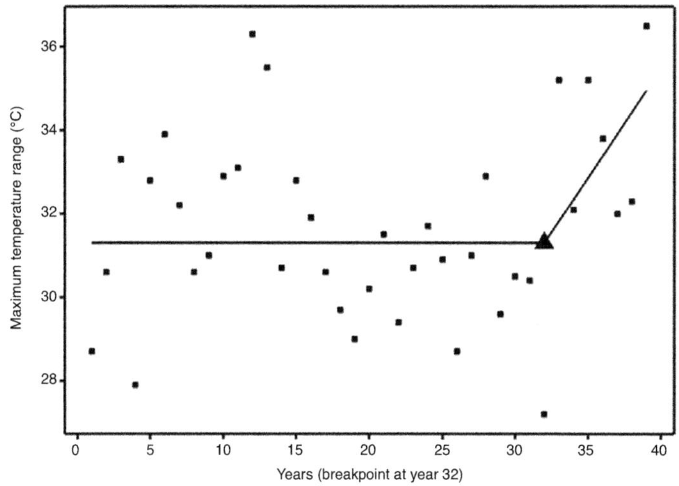

Fig. 14.3. Empirical split-line model fit to the Australian data to detect the trend change in maximum temperature range over the years 1–39 (1971–2009). Left of the break-point the trend line is a constant plateau shape.

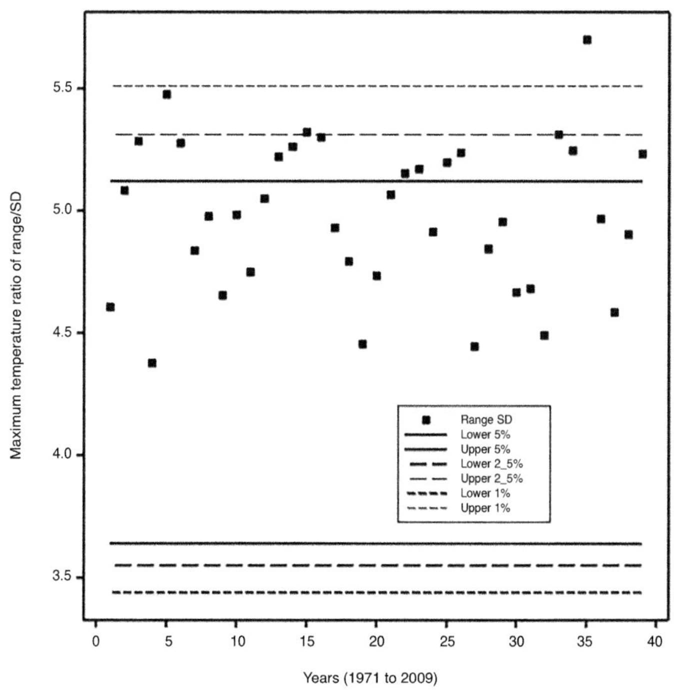

Fig. 14.4. Maximum temperature range size test for the Australian site using the range/SD ratio distribution. Upper and lower significance limits were taken from David *et al*. (1954).

Table 14.2. Highest daily maximum temperature (°C) and range[a] recorded for each of the 21 common years.

	British site		Australian site	
year	maximum	range	maximum	range
1971	28.8	30.50	38.0	28.70
1972	26.1	27.60	39.0	30.60
1973	29.4	27.40	39.8	33.30
1974	24.8	22.20	36.4	27.90
1975	33.0	32.50	39.0	32.80
1976	33.7	34.10	41.4	33.90
1977	26.7	26.90	40.2	32.20
1978	26.9	29.90	38.8	30.60
1979	29.6	30.60	41.0	31.00
1980	28.1	27.90	41.4	32.90
1981	28.4	30.00	41.7	33.10
1982	28.2	30.60	44.0	36.30
1983	32.0	31.20	43.3	35.50
1984	29.9	28.70	36.4	30.70
1985	28.2	32.40	42.8	32.80
1986	29.5	31.20	38.3	31.90
1987	28.4	34.50	39.6	30.60
1988	28.1	26.50	39.4	29.70
1989	33.0	30.70	38.6	29.00
1990	35.6	33.60	39.0	30.20
1991	27.7	31.20	40.8	31.50

[a]Range = highest minus lowest daily maximum temperatures.

The kurtosis value of the data distribution with respect to the normal distribution kurtosis value of 3 can be indicative of relatively longer tails causing data range changes. The density plot (Baird, 2021) of kurtosis values for our example Australian data (Fig. 14.6) shows, they do exceed a value of 3 for many years because of the effect of extreme maximum temperatures on the data range.

Concluding Remarks

The study of seasonal maximum temperature series with respect to the *w*/SD ratio can statistically establish distortion of the data range by extreme temperature episodes. Analysis of such data series, whether contemporary or from past archives, can help to identify the onset of extreme temperature events. Earth temperatures are affected by many influences in our solar system including human life and its support activities adding to climatic changes and leading to extreme weather consequences. A feature of weather data analysis is that it can provide evidence about trigger factors leading to other weather events or weather data information related to the observed changes. Given the complexity of Earth's temperature patterns, various analytical tools are needed. Statistical options to summarise data, as presented in this communication, do bring out data-based information. However, features may not be noticeable if monthly average data are used. This is because the distribution of a large number of sample means (something like 30 plus) from any

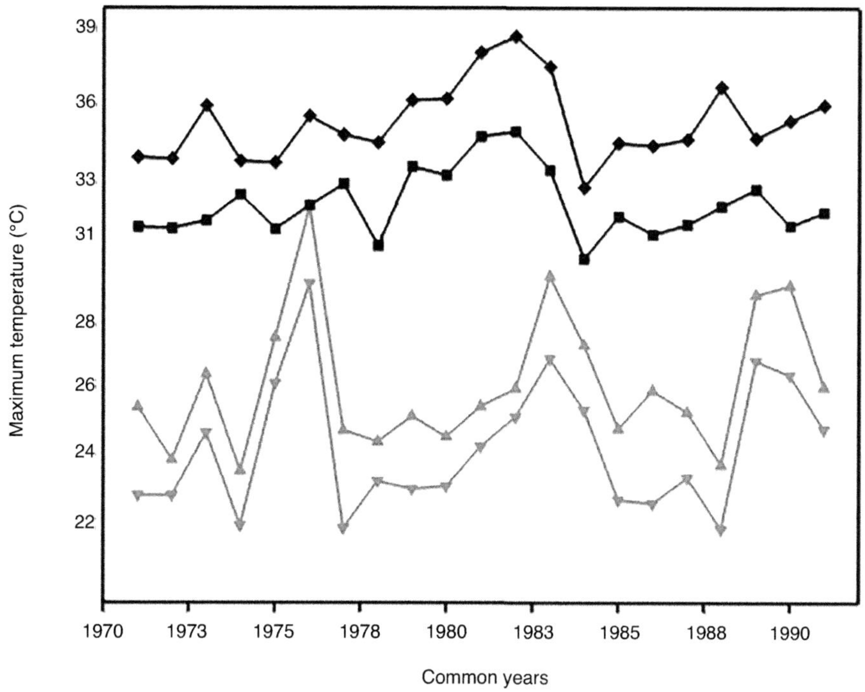

Fig. 14.5. Percentiles from each common year. British site (grey): 95th percentile indicated by an inverted triangle and 98th percentile by an upright triangle. Australian site (black): 95th percentile a square and 98th percentile a diamond.

Table 14.3. Shifting effect of upper tail (extreme) data on the estimates of mean and median (Australian site data 1971–1991).

Summary	All data	<98th percentile	<96th percentile	<94th percentile	<92nd percentile	<90th percentile
Observations	7669	7514	7359	7198	7046	6890
Missing values	a	0	0	0	0	0
Mean	19.53	19.15	18.84	18.54	18.28	18.02
Median	18.4	18.2	18.0	17.8	17.6	17.5
Minimum	5.7	5.7	5.7	5.7	5.7	5.7
Maximum	44.0	35.4	33.0	31.3	30.0	28.8
Range	38.3	29.7	27.3	25.6	24.3	23.1
SD	6.395	5.889	5.527	5.214	4.953	4.711
Range/SD	5.989	5.043	4.939	4.909	4.906	4.903
Lower quartile	14.5	14.4	14.4	14.4	14.3	14.2
Upper quartile	23.1	22.7	22.4	22.1	21.7	21.4
Mean+(range/2)	38.68	34.0	32.49	31.34	30.43	29.57

[a]Standard deviation.

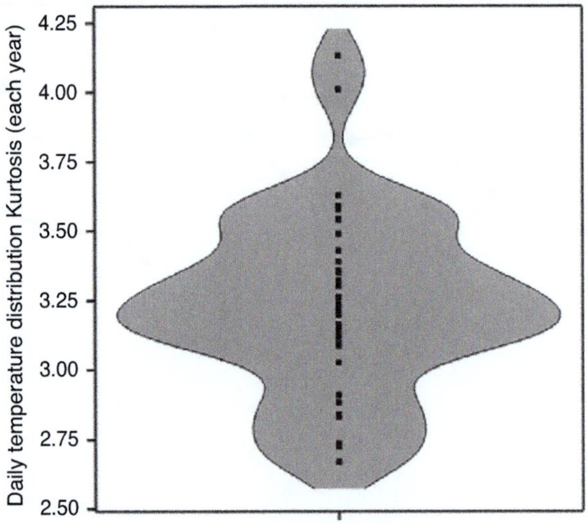

Fig. 14.6. Density plot of kurtosis estimates for the Australian site data over the years 1971–2009. Kurtosis estimate points (filled squares) are plotted along a vertical line, with kernel density smoothing on either side indicating the density of points along the line.

parent population, according to the central limit theorem, will be approximately normal. As a result, kurtosis will regress towards the normal distribution value of 3. This means shrinkage of that parent population tails and one loses sight of relatively extreme observations, which in the context of climate change driven weather extremes is of heightened interest. The basic time unit of Earth temperatures is a day, which is the time period for a single Earth rotation around its own axis regardless of its orbital position. Day temperature observations are taken in tandem in both hemispheres as a consequence of Earth planet shape and rotation. Therefore, daily temperature series analysis can furnish more information about various features, as we have illustrated in this communication. There is apparently little use made of temperature range although it is a valuable attribute of seasonal temperature patterns (e.g. temperature average ±range/2 is better than just the average when commenting on high or low day temperatures). Monthly averages of minimum and maximum temperatures should also be included in addition to monthly average summaries. Anomalies based on day-minimum and day-maximum temperatures can also be useful weather features of interest.

Funding

This research was funded in part by The Canada Research Chairs program, grant number 045867 (Natural Sciences and Engineering Research Council of Canada, Ottawa).

Conflict of Interest

The authors declare no conflict of interest.

References

Baird, D.B. (2021) DXDENSITY procedure: Produces one-dimensional density (or violin) plots. Available at: https://genstat.kb.vsni.co.uk/knowledge-base/dxdensity/

David, H.A., Hartley, H.O. and Pearson, E.S. (1954) The distribution of the ratio, in a single normal sample, of range to standard deviation. *Biometrika* 41, 482–493.

Davies, O.L. (1967) *Statistical Methods in Research and Production*. Oliver & Boyd, Edinburgh and London.

Dhanoa, M.S., Sanderson, R., Shanmugalingam, S., Lopez, S., Murray, J.M.D. *et al*. (2018) The distribution of the ratio of two correlated measured variables may not always be normal: Case studies related to meat quality and animal nutrition. *E-planet* 16(1), 43–50.

Dhanoa, M.S., Louro, A., Cardenas, L.M., Shepherd, A., Sanderson, R. *et al*. (2020) A strategy for modelling heavy-tailed greenhouse gases (GHG) data using the generalised extreme value distribution: Are we overestimating GHG flux using the sample mean? *Atmospheric Environment* 237, 117500.

Eastoe, E. (2017) Extreme value distributions. *Significance* 14, 12–13.

Fieller, E.C. (1932) The distribution of the index in a normal bivariate population. *Biometrika* 24, 428–440.

Geary, R.C. (1930) The frequency distribution of the quotient of two normal variables. *Journal of the Royal Statistical Society (Series B)* 93, 442–446.

Goedhart, P.W. and Payne, R.W. (2021) WINDROSE procedure: Plots rose diagrams of circular data like wind speeds. Available at: https://genstat.kb.vsni.co.uk/knowledge-base/windrose/

Lindley, D.V. and Miller, J.C.P. (1953) *Elementary Statistical Tables*. Cambridge University Press, Cambridge, UK.

Shanmugalingam, S. (1982) On the analysis of the ratio of two correlated normal variables. *The Statistician* 31, 251–258.

Vieth, E. (1989) Fitting piecewise linear regression functions to biological responses. *Journal of Applied Physiology* 67(1), 390–396.

VSN International (2022) *Genstat Reference Manual (Release 22)*, VSN International, Hemel Hempstead, UK.

15 Genome editing in rice: New paths for an old crop

Valeria Oliveira Nizolli, Victoria Freitas de Oliveira, Luciano Carlos da Maia, Camila Pegoraro and Antonio Costa de Oliveira*
Faculdade de Agronomia Eliseu Maciel, Departamento de Fitotecnia, Universidade Federal de Pelotas, Campus Capão do Leão, 96010-610, Pelotas, RS, Brazil

Abstract
Rice is one of the most consumed cereals worldwide and feeds about 3 billion people. However, the occurrence of climate changes has influenced abiotic and biotic stresses and consequently affected the production and quality of rice grains. The development of new, higher yielding cultivars is necessary to ensure global food security. Although great progress has already been achieved by conventional breeding, biotechnological tools, such as transgenics and genome editing, can aid in meeting future demands. Gene editing is characterized by cutting and modifying target genes. Among the genome editing techniques, CRISPR/Cas (clustered regularly interspaced short palindromic repeats/Cas), has been proposed because of its ease of manipulation. Variants such as multiple Cas proteins, base editing and prime editing, which aim to increase editing efficiency have also been proposed. Edited plants are more accepted because they are transgene-free. Thus, the objective of this review is to describe the different gene editing techniques and their respective applications in rice breeding. It was observed that the CRISPR/Cas tool was efficient in gene editing in studies related to yield, tolerance to biotic and abiotic stresses, and rice grain quality. However, the impact of this approach in breeding programs depends on the cultivation of edited plants on a large scale in the field. Based on the progress made so far, it is believed that rice breeders can bet on CRISPR/Cas and its variants to help face the current and future challenges of climate change and high food demand.

Keywords: CRISPR-Cas, MNs, ZFNs

Introduction

Rice (*Oryza sativa* L.) production is ca. 750 million tons and it is consumed by approximately 3 billion people in the world. A total of 50% increase in the current consumption is projected for 2050, which could mean a demand as high as 1.125 billion tonnes. The occurrence of biotic stresses (diseases – viruses, bacteria, fungi, nematodes; pests; weeds) and abiotic stresses (drought, submergence, salinity, heat, cold and heavy metals) is a limiting factor for rice production. Also, climate change influences the frequency, intensity and duration of these stresses (Dar *et al.*, 2021; Tabassum *et al.*, 2021). This means that it is necessary to develop new rice cultivars with tolerance to stress and higher yield potential since the expansion of the cultivated area is limited (Makino *et al.*, 2020).

In addition to increasing yield, rice quality must also be considered, as it directly

*Corresponding author: acostol@gmail.com

influences consumer acceptance as well as market price. Quality includes physical (head rice recovery, grain size and shape, and grain color), biochemical (amylose content, gelatinization temperature, viscosity and aroma) and nutritional (micronutrient, digestibility, protein and lipid content) properties (Sultana *et al.*, 2022). Additionally, reducing toxic elements such as arsenic and cadmium in rice grains are important attributes for obtaining safe food and for maintaining human health (Zhao and Wang, 2020).

Yield, tolerance to biotic and abiotic stresses and grain quality in rice have been successfully improved using conventional breeding (crosses and induced random mutations). However, further efforts by breeders are needed to ensure global food safety, health and consumer acceptance in the future. Despite all the success of conventional breeding so far, it is important to highlight that it is a time-consuming process and prone to human bias (Tabassum *et al.*, 2021). Therefore, biotechnology tools such as transgenics and genome editing, with greater precision and speed, can be used to help conventional rice breeding to meet current and future demands.

It has been shown that the use of transgenic crops has increased yields by 22%, in addition to increasing farmer profits. Additionally, there was a reduction in the number of insecticides used in *Bt* crops and the possibility of using herbicides with less impact on the environment when using herbicide-tolerant crops (Lucht, 2015). Despite the economic importance of transgenic crops, there is concern on the part of consumers regarding the adverse effect of transgenes on the environment and human health. Therefore, the cultivation of transgenics is not authorized in some countries, and even in countries where it is authorized, this concern is common among consumers (Ishii and Araki, 2016).

Unlike transgenics, genome editing does not involve inserting genes and sequences from external organisms, only target genes are identified, cut and modified (Son and Lim, 2021). Genome editing is a technique based on site-directed nucleases, including meganucleases (MN), zinc-finger nucleases (ZFN), transcription activator-like effector nucleases (TALENs) and clustered regularly interspaced short palindromic repeats (CRISPR)/CRISPR-associated (Cas) system. CRISPR/Cas9 is the most used tool for editing due to its simplicity and easy manipulation. CRISPR/Cas variants, such as multiple Cas proteins, base editing and prime editing, were developed to improve and increase the efficiency of the editing process. This tool allows some strategies to be used to remove or prevent the integration of transgenes into the genome (Gu *et al.*, 2021). Thus, genome-edited crops are transgene-free and may have better acceptance than transgenic crops.

Taken together, the objective of this review is to address genome editing techniques and the application of this tool in improving yield, tolerance to biotic and abiotic stresses and grain quality in rice.

Review Methodology

We searched databases including CAB Abstracts, Google Scholar, PubMed and ScienceDirect (Keyword search terms used: CRISPR-Cas; MNs; ZFNs; TALENs; Oryza sativa L.; plant breeding, etc.). We used the references from the articles obtained by this method to check for additional relevant material. We also spoke to colleagues for any upcoming studies not yet published.

Gene editing techniques in rice: Current

Gene editing techniques normally consist of three steps, the first being when an exogenously engineered enzyme has nuclease activity and the ability to identify the target DNA sequence. Therefore, the manipulated nuclease binds to the sequence of interest and leads to a double strand break (DSB) in regions close to the target gene. Then, endogenous non-homologous end joining (NHEJ) or homology-directed repair (HDR) pathways promote DSB repair. NHEJ is an error-prone repair mechanism and often provides insertions or deletions (Indel), while HDR tends to be a more accurate repair of DSBs. These techniques have shown promising results in several studies using organisms and plants (Osakabe and Osakabe, 2015; Wang *et al.*, 2016; Peng *et al.*, 2018; Wada *et al.*, 2020).

Different genes that control traits associated with rice yield were edited. These include grain size (GS), grain number (GN), grain width and weight (GW), dense and erect panicle (DEP). Regarding quality, the *Waxy* (WX) gene has already been edited, encoding the granule-bound starch synthase I (GBSSI), which determines the amylose content in the endosperm. This is the main contributor to eating and cooking quality. Genes encoding other enzymes involved in starch synthesis (isoamylase – ISA and starch branching enzymes – SBEs) were also edited. In addition, genes associated with fragrance (*BADH2*) and mineral composition (*NRAMP*) were edited (Table 15.1).

Genes responsible for tolerance to drought and salinity were the most sought after in editing studies for tolerance to abiotic stresses in rice. When biotic stresses are considered, the editing of genes involved in resistance to the main rice diseases is observed, such as *SWEET* genes, associated with rice susceptibility to bacterial blight. Genes aimed at herbicide resistance were also edited, mainly the enzyme coding for the acetolactate synthase (ALS), aiming at resistance to the herbicides of the chemical group of imidazolines, fundamental in the control of the invasive red rice (Table 15.1).

Meganuclease

In 1985, meganucleases were discovered in yeast (Jacquier and Dujon, 1985). They are homing endonucleases found in mitochondria and belong to the LAGLIDADG site-specific endonuclease family. These enzymes identify genes that are intron-free, and that are capable of combining with long DNA containing 14–40 base pairs. They present two different forms that are used in gene editing known as I-SCeI and I-CreI (Maeder and Gersbach, 2016). Each meganuclease has one or two conserved LAGLIDADG motifs in the protein chains, in order to transform into a homodimer protein, and thus be able to cut the palindromic DNA. These enzymes naturally function as mobile genetic elements and promote DSB to introduce introns at sites of interest (Maeder and Gersbach, 2016). I-SCeI has the ability to cleave sequences located at the bottom and top strands of DNA, resulting in lower toxicity and developing homologous recombination (Metzger *et al.*, 2011; Zheng *et al.*, 2020). Artificial meganucleases induce DSBs and consequently recruit the endogenous repair machinery, which is performed by NHEJ, being able to promote random insertions and deletions (*Indel*) or, if a donor DNA is provided as a repair template, it is performed by HDR (Tröder and Zevnik, 2022).

However, even though there are numerous naturally occurring meganucleases, due to their fixed recognition sequence, it represents a major limitation when it is necessary to choose different sites of interest (Tröder and Zevnik, 2022). Furthermore, it takes a long time and a high cost to construct an artificial meganuclease, further restricting the frequent and widespread use in gene editing (Zheng *et al.*, 2020).

The technique is considered restricted, few works have been developed using meganuclease. A meganuclease was designed for specific cleavage of a target sequence, and combined with repair mediated by homologous recombination, enabling the precise and targeted insertion of herbicide resistance genes (*hppd, epsps*) into cotton (D'Halluin *et al.*, 2008). Meganucleases can also be used to make gene silencing mutations (Iqbal *et al.*, 2020). In maize, meganuclease I-CreI was used at two loci of interest, *liguleless1* and *MS26* (Gao *et al.*, 2010; Lyznik *et al.*, 2012; Djukanovic *et al.*, 2013). Gene silencing studies were verified in immature corn embryos, and it was observed that meganucleases promoted the induction of mutations (Indel) in the target loci, with a frequency of 5.8 and 3%, respectively. However, in rice, no study has been performed using meganucleases (Daboussi *et al.*, 2015; Zhu *et al.*, 2017; Viana *et al.*, 2019; Nizolli *et al.*, 2021).

Zinc finger nucleases

Zinc fingers, which are protein motifs, were identified in 1985. They are part of a transcription factor, identified in frog oocytes, and are found in several species, including humans (Miller *et al.*, 1985; Klug, 2010). The motif name comes from the structure of a zinc ion binding to the target region in the DNA. Compared with other DNA binding motifs, its difference is that a cascade

Table 15.1. Application of CRISPR/Cas9 system, prime editing, base editing and Cas proteins variants for improving rice.

Trait	Gene	Author	
Yield-related traits	*GW2, GW5, TGW6*	(Xu *et al.*, 2016)	CRISPR-Cas9
	GS9	(Zhao *et al.*, 2018)	CRISPR-Cas9
	Gn1a, DEP1, GS3, IPA1	(Li *et al.*, 2016)	CRISPR-Cas9
	Gn1a, DEP1	(Huang *et al.*, 2018)	CRISPR-Cas9
	GRF4	(Li *et al.*, 2016))	CRISPR-Cas9
	PYLs	(Miao *et al.*, 2018; Usman *et al.*, 2020a)	CRISPR-Cas9
	SPLs	(Hua *et al.*, 2019)	Base editing
	Gn1a, GS3	(Shen *et al.*, 2018)	CRISPR-Cas9
	SD1	(Hu *et al.*, 2019)	CRISPR-Cas9
	GA20ox2	(Han *et al.*, 2019)	CRISPR-Cas9
	GS3, GW2, Gn1a	(Zhou *et al.*, 2019)	CRISPR-Cas9
	SCM1/SD1, SCM3/ OsTB1/FC1, SCM2/ APO1	(Cui *et al.*, 2020)	CRISPR-Cas9
	FWL4	(Gao *et al.*, 2020)	CRISPR-Cas9
	PIN5b, GS3, MYB30	(Zeng *et al.*, 2020)	CRISPR-Cas9
	SPL16	(Usman *et al.*, 2020b)	CRISPR-Cas9
	GW7, ER1	(Xiong *et al.*, 2021)	dCas9
	Os03g01240, Os04g39780, Os11g35410	(Lowder *et al.*, 2018)	dCas9
	SLR1, SPLs	(Hua *et al.*, 2020)	Base editing
	Hd2, Hd4, Hd5	(Li *et al.*, 2017)	CRISPR-Cas9
	GS3	(Huang *et al.*, 2020)	CRISPR-Cas9
Quality	*ISA1*	(Chao *et al.*, 2019)	CRISPR-Cas9
	Waxy	(Zhang *et al.*, 2018; Huang *et al.*, 2020)	CRISPR-Cas9
	Waxy	(Xu *et al.*, 2021)	Base editing
	SBEI, SBEIIb	(Chao *et al.*, 2019)	CRISPR-Cas9
	SBEIIb	(Sun *et al.*, 2017)	Base editing
	AAP6, AAP10	(Li *et al.*, 2017)	CRISPR-Cas9
Nutrition	*Nramp5*	(Tang *et al.*, 2017; Wang *et al.*, 2020)	CRISPR-Cas9
	HAK1	(Chang *et al.*, 2020)	CRISPR-Cas9
	CYP97A4, DSM2, CCD4a, CCD4b, CCD7	(Nieves-Cordones *et al.*, 2017)	CRISPR-Cas9
	FAD2-1	(Yang *et al.*, 2017)	CRISPR-Cas9

Continued

Table 15.1. Continued

Trait	Gene	Author	
	Or	(Abe *et al.*, 2018)	CRISPR-Cas9
	SSU-crtl, ZmPsy	(Dong *et al.*, 2020)	CRISPR-Cas9
Fragrance	*BADH2*	(Ashokkumar *et al.*, 2020; Shao *et al.*, 2017; Usman *et al.*, 2020)	CRISPR-Cas9
Male Sterile	*TMS5*	(Zhou *et al.*, 2016; Barman *et al.*, 2019)	CRISPR-Cas9
	TMS10	(Yu *et al.*, 2017)	CRISPR-Cas9
	CSA	(Li *et al.*, 2016)	CRISPR-Cas9
Biotic stress tolerance	*ERF922*	(Zhou *et al.*, 2018)	CRISPR-Cas9
	Bsrk1	(Zhou *et al.*, 2015)	CRISPR-Cas9
	SWEET13	(Jiang *et al.*, 2013)	CRISPR-Cas9
	SWEET11, SWET14	(Zhou *et al.*, 2014; Xu *et al.*, 2019)	CRISPR-Cas9
	SWEETs	(Kim *et al.*, 2019)	CRISPR-Cas9
	SWEET11 (8N3)	(Zeng *et al.*, 2020)	CRISPR-Cas9
	SWEET11, SWEET13, SWEET14	(Oliva *et al.*, 2019)	CRISPR-Cas9
	SWEET14	(Liu *et al.*, 2020; Zafar *et al.*, 2020)	CRISPR-Cas9
	Xa13	(Macovei *et al.*, 2018)	CRISPR-Cas9
	eIF4G	(Xie and Yang, 2013)	CRISPR-Cas9
	MPK5	(Li *et al.*, 2019)	CRISPR-Cas9
	TMS5, Pi21, Xa13	(Ma *et al.*, 2018)	CRISPR-Cas9
	Pi21	(Nawaz *et al.*, 2020)	CRISPR-Cas9
	SEC3A	(Shah *et al.*, 2019)	CRISPR-Cas9
	PTF1	(Shen *et al.*, 2017)	CRISPR-Cas9
Abiotic stress tolerance	*Ann3*	(Bo *et al.*, 2019)	CRISPR-Cas9
	NAC041	(Mao *et al.*, 2018)	CRISPR-Cas9
	PRX2	(Lou *et al.*, 2017)	CRISPR-Cas9
	SAPK2	(Huang *et al.*, 2017)	CRISPR-Cas9
	TIFY1a, TIFY1b	(Liao *et al.*, 2019)	CRISPR-Cas9
	SRL1, SRL2	(Zhang *et al.*, 2019)	CRISPR-Cas9
	RR22	(Santosh Kumar *et al.*, 2020)	CRISPR-Cas9
	DST	(Ogata *et al.*, 2020)	CRISPR-Cas9
	ERA1	(Yue *et al.*, 2020)	CRISPR-Cas9
	miR535	(Xu *et al.*, 2014)	CRISPR-Cas9
	SAP	(Park *et al.*, 2022)	CRISPR-Cas9

Continued

Table 15.1. Continued

Trait	Gene	Author	
Herbicide resistant	*BEL*	(Li *et al.*, 2016)	CRISPR-Cas9
	EPSPS	(Sun *et al.*, 2016)	CRISPR-Cas9
	ALS	(Shimatani *et al.*, 2017; Li *et al.*, 2019)	CRISPR-Cas9
	ALS	(Liu *et al.*, 2020; Zhang *et al.*, 2021)	Base editing
	ACC	(Endo *et al.*, 2016)	Base editing
Other traits	*DL, ALS, NCED1, AO1*	(Endo *et al.*, 2016)	Cpf1
	BADH2, MPK2, Os02g23823, PDS	(Shan *et al.*, 2013)	CRISPR-Cas9
	Os02g23823, PDS, MPK2	(Wang *et al.*, 2015)	CRISPR-Cas9
	DL, PDS, YSA	(Mikami *et al.*, 2015a)	CRISPR-Cas9
	PDS, DEP1	(Shan *et al.*, 2014)	CRISPR-Cas9
	Lazy1, CAO1	(Miao *et al.*, 2013)	CRISPR-Cas9
	ROC5, SPP, YSA	(Feng *et al.*, 2013)	CRISPR-Cas9
	ROC5, YSA	(Lowder *et al.*, 2015)	CRISPR-Cas9
	PDS, PMS3, EPSPS, DERF1, MSH1, MYB5,	(Zhang *et al.*, 2014)	CRISPR-Cas9
	MYB1, ROC5, SPP, YSA		
	ROC5, DEP1	(Zheng *et al.*, 2016)	CRISPR-Cas9
	MYB1	(Mao *et al.*, 2013)	CRISPR-Cas9
	AOX1a, AOX1b, AOX1c, BEL	(Xu *et al.*, 2015)	CRISPR-Cas9
	EPFL9	(Yin *et al.*, 2017)	CRISPR-Cas9
	CDKA2, CDKB1, CDKB2	(Endo *et al.*, 2015)	CRISPR-Cas9
	YSA, CDKB2	(Mikami *et al.*, 2015b)	CRISPR-Cas9
	DMC1A, DMC1B	(Mikami *et al.*, 2016)	CRISPR-Cas9
	FTL11, Os07g0261200, Os02g0700600	(Ma *et al.*, 2015)	CRISPR-Cas9
	CDC48, NRT1.1B, SPL14	(Zong *et al.*, 2017)	Base editing
	BBM1	(Khanday *et al.*, 2019)	CRISPR-Cas9
	REC8, PAIR, OSD1, MTL	(Wang *et al.*, 2019)	CRISPR-Cas9
	SF3B1	(Butt *et al.*, 2019)	CRISPR-Cas9
	AAT, NRT1.1B, CDC48	(Hua *et al.*, 2020)	Cas9/APOBEC

Continued

Table 15.1. Continued

Trait	Gene	Author	
	DEP1, EPFL9	(Zhong *et al.*, 2018)	Cpf1
	BEL, PDS, EPSPS, BEL, RLK	(Wang *et al.*, 2017)	Cpf1
	ALS, APO1	(Hua *et al.*, 2020)	Prime editing
	PDS, ACC1, Wx	(Xu *et al.*, 2020)	Prime editing
	ALS, ACC, DEP1	(Xu *et al.*, 2020)	Prime editing
	ALS, KO2, DEP1, PDS	(Tang *et al.*, 2020)	Prime editing
	NRT1.1B	(Lu and Zhu, 2017)	Base editing

of modules is used where a small number of key residues present in the protein determine a recognition. Each module carries a unique DNA binding motif for three subsequent base pairs of the target sequence. The modular conformation of these elements makes them enormously modifiable in linking target DNA sequences of many lengths (Klug, 2010). In 1996, researchers joined these zinc finger modules with the DNA sequence cleavage domain of the restriction enzyme *Fok*I to form a nuclease that can be programmed, determined as zinc finger nucleases (ZFN) (Kim *et al.*, 1996; Tröder and Zevnik, 2022).

A restriction endonuclease known as *Fok*I, from *Flavobacterium okeanokoites*, is composed of a DNA-binding domain at N-terminal residues and a cleavage domain at C-terminal residues. To form a ZFN, the C-terminal region is used together with zinc fingers. It is known that it takes two C-terminal *Fok*I with 96 amino acid residues to form the DNA cleavage domain of ZFNs, and they act as dimers when two binding sites are located in close proximity and separated by only six base pairs (Paschon *et al.*, 2019; Zheng *et al.*, 2020). The target sites recognized by the ZFNs can be extended within a space with a coherent variation through the binding of three or more zinc fingers, where each one constitutes around 30 amino acids, identifying three to four base pairs (Paschon *et al.*, 2019).

ZFN technology is used by inserting DSBs into specific regions in the DNA genome. Thus, genes of interest can be accessed through mutagenesis caused by NHEJ repair or altered by HDR if a homologous DNA template is available (Guo *et al.*, 2010; Viana *et al.*, 2019). However, some limitations are observed in the ZFNs system. For example, it is required to produce a zinc finger array for each chosen target gene (Chen *et al.*, 2019). In addition, the number of target genes is restricted and there is a likelihood of an overlap between the catalytic domain and the DNA binding domain, being capable of impairing the specificity of the protein (Lopes Filho *et al.*, 2020).

The ZFN technique has been successfully applied to rice. ZFNs were used to efficiently induce targeted mutation at the *SSIVa* locus to alter plant height, grain filling and starch content (Jung *et al.*, 2018). In addition, ZFNs were applied for the targeted alteration of the endogenous *malate dehydrogenase* (*MDH*) gene, which resulted in an increase in yield (Shukla *et al.*, 2013; Tiwari and Lata, 2019).

Transcription activator-like effector nucleases

It has been found that the DNA binding domain of the ZFNs technique (zinc fingers) can be replaced by more versatile and more affordable DNA binding modules to construct. In 2010, a great leap forward in the area of nuclease designs took place with the creation of TALENs (Christian *et al.*, 2010; Miller *et al.*, 2011). The modules come from the TALE protein, identified in the phytopathogenic bacteria *Xanthomonas*. These bacteria have the ability to promote a specific binding of TALEs in sequences in order to control the gene expression response of the infected plant cell in their favour. Designed

TALENs that contain a DNA cleavage domain of *Fok*I and another DNA binding domain – TALE (Zheng *et al.*, 2020) show greater efficacy and promote fewer off-target errors when compared to the ZFNs technique (Feng *et al.*, 2014; Zheng *et al.*, 2020).

In the TALENs technique, the DNA binding domain is composed of N-terminal residues, a nuclear localization signal (NLS) that is essential for recognizing DNA regions of interest, a transcriptional activation domain and the central repetitive portion containing 13–29 tandem repeat units, known as TAL effectors. Each TAL has 34 amino acids and in two of them, there are variations at position 12 and 13, also known as repeat variable di-residues (RVDs), and indicate the specificity of the repeat that concerns a single nucleotide, defining an easy code 1:1 for protein-DNA interaction (i.e. 1 repeat unit for one nucleotide) (Mussolino and Cathomen, 2012). The amino acid located in position 13 acts to identify specific nucleotides, while the one located in the 12 regions establishes the di-residues that vary in repetition (Boch and Bonas, 2010; Zheng *et al.*, 2020).

The most common RVDs used by many experimental validations are known as amino acids NN, NG, HD and NI that show a binding affinity with a unique preference for G/A, T, C and A, respectively, attributing specificity to the target (Mussolino and Cathomen, 2012). As we align the repeater modules (RVDs) in a specific structure, it is able to produce TALENs with the precise target sequence. However, there is a limitation for alternative destination locations for TALEN. A thymine located at position 0, that is precursor of the TALE repeat-linked sequences, is unalterably indispensable (Mak Nga-Sze *et al.*, 2012). Other more up-to-date versions of TALEs have also been observed in nature in which thymine is replaced by cytosine at position 0, with no effect on its activity (Bhardwaj and Nain, 2021).

The TALEN RVDs are manipulated in each module to recognize the sequence to be edited. The TALEN clipping domain generates a double-strand break in the target sequence, which will be repaired by NHEJ or HDR if a donor DNA is provided (Christian *et al.*, 2010; Joung and Sander, 2013; Chen *et al.*, 2014; Daboussi *et al.*, 2015; Zhu *et al.*, 2017).

TALENs have been widely used to improve traits such as industrial quality, resistance to biotic stress and tolerance to abiotic stress. Regarding resistance to the pathogen *Xanthomonas oryzae* in rice, TALEN was used to disrupt the *Os11N3* gene of susceptibility to bacterial growth as a method of preventing the virulence procedure of the bacteria (Li *et al.*, 2012). TALEN was also used for LOX3 lipoxygenase mutagenesis in order to increase the longevity of rice grains during storage (Ma and Liu, 2015). The TALEN technique was used to edit the *OsSWEET11* and *OsSWEET15* genes, which act in seed filling (Yang *et al.*, 2018; Tiwari and Lata, 2019). TALENs have also been used in work related to herbicide resistance, through double-point mutations in the rice acetolactate synthase gene (*OsALS*) (Li *et al.*, 2016).

Clustered Regularly Interspaced Short Palindromic Repeats (CRISPR)

When studying bacterial immune systems, researchers Jennifer Doudna and Emmanuelle Charpentier described a mechanism, which they called CRISPR/Cas, that would revolutionize gene editing technology (Jinek *et al.*, 2012). The CRISPR/Cas technology works as a genomic memory against invading pathogens (Wiedenheft *et al.*, 2012; Bortesi and Fischer, 2015; Anzalone *et al.*, 2020).

Soon after its discovery, it was realized that the bacterium's CRISPR/Cas system could be reprogrammed to target a wide range of DNA or RNA sequences, just by altering the spacer sequences within a guide RNA molecule (Bortesi and Fischer, 2015; Chen *et al.*, 2019). When compared with other available genome editing techniques, such as those mentioned above, the success of the technology is evident, both in the scientific community and in plant breeding programs (Gao, 2021). This is because the CRISPR system does not depend on the modification of proteins to determine the targets to be edited, as in MG, ZFNs and TALEN, which simplifies the process and significantly reduces its cost, allowing the widespread use of the technology (Chen *et al.*, 2019; Anzalone *et al.*, 2020; Li *et al.*, 2020).

CRISPR/Cas gene editing tools are flexible, highly efficient and have been widely applied to improve multiple traits in plants. To date, gene editing is being used to generate a variety of heritable genetic modifications in plants, including small random insertions/ deletions (Indels); point mutations or nucleotide substitutions; insertions or deletions of DNA fragments; and targeted chromosomal rearrangements (Chen *et al.*, 2019; Li *et al.*, 2020; Tabassum *et al.*, 2021). However, the versatility of the CRISPR/Cas system and its constant optimization, in addition to increasing editing accuracy, has expanded the range of possibilities, such as manipulating gene regulation and even producing epigenetic changes (Liu *et al.*, 2022).

Recently, several orthologues and variants of Cas with additional properties have been identified and leveraged for use in gene editing (Chen *et al.*, 2019; Anzalone *et al.*, 2020; Chen and Gao, 2020). Among the Cas proteins, Cas8, Cas9, Cas12a and their variants can be highlighted, the three most used for genome editing being those derived from *Francisella novicida* (*Fn*Cas12a), *Lachnospiraceae bacterium* ND2006 (*Lb*/Cas12a) and *Acidaminococcus* sp. BV3L6 (*As*Cas12a) (Moreno-Mateos *et al.*, 2017; Malzahn *et al.*, 2019; Chen and Gao, 2020; Schindele and Puchta, 2020). Recently, applications of Cas12b proteins and their variations of *Alicyclobacillus acidoterrestris* (*Aac*Cas12b), *Alicyclobacillus acidiphilus* (*Aa*Cas12b), *Bacillus thermoamylovorans* (*Bth*Cas12b) and *Bacillus hisashii* (*Bh*Cas12b-v4) for rice genome editing have been reported (Ming *et al.*, 2020). In the Cas13 proteins, we have Cas13a, Cas13b, Cas13d, Cas13X, Cas13Y, Cas13bt, and yet, among its various orthologues, Cas13a from *Leptotrichia wadei* (*Lwa*Cas13a) was the most efficient so far (Abudayyeh *et al.*, 2017; Cox *et al.*, 2017; Konermann *et al.*, 2018; Chen and Gao, 2020; Xu *et al.*, 2021; Kannan *et al.*, 2022). Cas9 itself has orthologues, such as *Streptococcus pyogenes* Cas9 (*Sp*Cas9), *Staphylococcus aureus* Cas9 (*Sa*Cas9), *Streptococcus thermophilus* Cas9 (*St*Cas9), *Neisseria meningitides* Cas9 (*Nm*Cas9), *Campylobacter jejuni* Cas9 (*Cj*Cas9), among others (Tabassum *et al.*, 2021). Recently, it is already possible to find the Cas9 protein in modified forms, such as the catalytically dead or endonuclease deficient Cas9 (dCas9) and nickase Cas9 (nCas9). New tools for precise genetic modification are also available, such as Base editing (BEs) and Prime editing (PEs), which have greatly expanded the possibilities of genome editing (Chen and Gao, 2020; Morgante *et al.*, 2020; Zhu *et al.*, 2020).

Different studies with CRISPR/Cas and its variants in the coding and non-coding genes have been carried out in rice. These studies aim to improve agronomic traits such as yield, tolerance to biotic and abiotic stresses, grain quality, among other traits of agronomic importance (Table 15.1). Some studies are exploratory, while others are targeted and bring extraordinary results for application in plant improvement. In fact, the studies developed to date point to the potential of genome editing, specifically CRISPR-Cas and its variants, to help improve agronomic traits in rice.

CRISPR/Cas9

CRISPR/Cas9 are the most studied and used to date (Anzalone *et al.*, 2020). Among these, CRISPR/Cas9 has been adopted to improve yield, stress resilience, herbicide resistance and industrial quality in several cultivated plant species (Table 15.1) (Tabassum *et al.*, 2021). In CRISPR/Cas9 (Fig. 15.1), Cas9 nucleases are guided by a designed sgRNA (single guide RNA), complementary to the target sequence, which in prokaryotes would result from the fusion of crRNA and tracrRNA. The Cas9 enzyme recognizes sgRNA through its recognition domain (Rec) (Chen *et al.*, 2019; Anzalone *et al.*, 2020; Schindele and Puchta, 2020). The complex formed by Cas9 and sgRNA travels along the double strand of the target DNA until it finds a PAM site, which is recognized by the Rec domain. Then, Cas9 performs a DSB on the two DNA strands three base pairs upstream of the PAM, within the protospacer sequence, using two distinct nuclease domains (RuvC and HNH) (Chen *et al.*, 2019; Anzalone *et al.*, 2020; Schindele and Puchta, 2020). The RuvC and HNH domains cleave the complementary and non-complementary strands of DNA, respectively, and specifically at the third nucleotide upstream of the PAM site, allowing the pairing of 20–24 nucleotides between the complementary target DNA and the sgRNA (Chen *et al.*, 2019; Anzalone *et al.*, 2020; Schindele and Puchta, 2020).

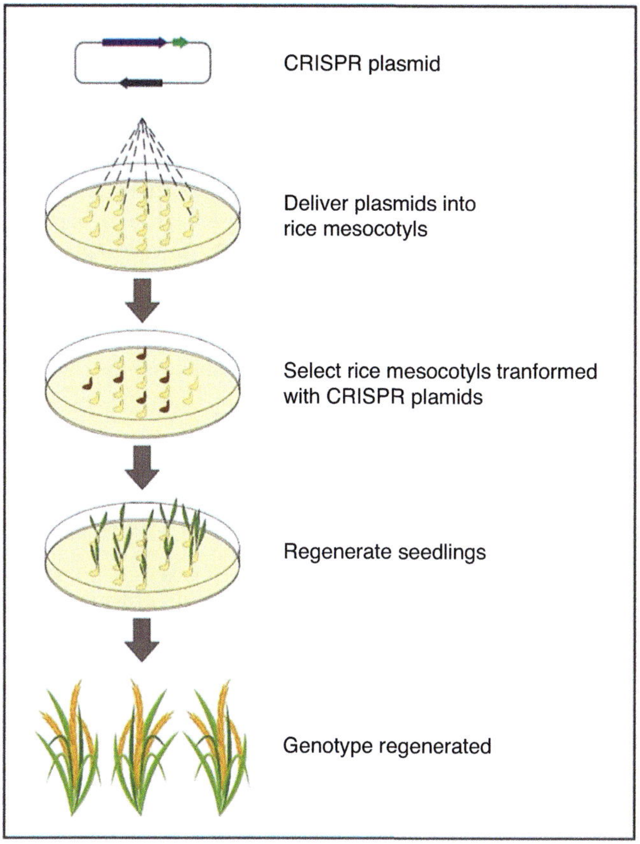

Fig. 15.1. CRISPR technique application.

The CRISPR/Cas9 system was used to edit several genes involved in the expression of characters associated with rice yield (Table 15.1). In one of the most recent works (Huang *et al.*, 2022), a description of how sgRNA/Cas9 mediated the editing of a region related to grain size 3 (GS3), a QTL identified in the regulation of rice grain length and weight. For this purpose, an allele was introduced by editing, whose mutation in the organ size regulation domain of the GS3 protein was able to produce longer grains, resulting in a significant genetic improvement of the hybrid rice grain.

For quality improvement, CRISPR/Cas9-mediated editing of the S7 site of the *Waxy* (*Wx*) gene promoter, resulted in a change in gene expression (Huang *et al.*, 2020). This result agrees with a proposed strategy to create new *Wx* alleles with a wider range of high amylose grains (Zhang *et al.*, 2018).

The high potential of CRISRP/Cas9 has also been successfully exploited to improve tolerance to biotic and abiotic stresses in rice. In breeding for plant disease resistance, CRISPR/Cas9 editing has aided in the development of broad-spectrum resistance against various bacteria, fungi and viruses through silencing susceptibility genes (S) and insertion of resistance genes (R) (Table 15.1). In this sense, CRISPR/Cas9-mediated genome editing made it possible to introduce mutations in three promoters of the *SWEET* gene (SWEET11, *SWEET13*, *SWEET14*) that resulted in rice lines with broad-spectrum resistance to *Xanthomonas oryzae* pv. Oryzae (Oliva *et al.*, 2019). Another example of the high potential of CRISPR/Cas9 technology is presented (Nawaz *et al.*, 2020). Mutant lines were generated to investigate the role of the *Pi21* gene against resistance to *Magnaporthe oryzae*, increasing the range of breeding options for resistance to such an important pathogen. In breeding for abiotic stresses, genome editing via CRISPR/Cas9 can be used to develop genotypes tolerant to drought, salinity, cold and osmotic stress (Table 15.1). Aiming at drought tolerance, the senescence-associated protein (*OsSAP*) gene was edited via CRISPR/Cas9 (Park *et al.*, 2022).

CRISPR/Cas12a (Cpf1)

Still aiming to overcome the limitations of Cas9, CRISPR/Cas12a systems were introduced, which act on DNA and RNA, and have great potential due to their reliability (Tabassum *et al.*, 2021). Cas12a nucleases (Fig. 15.2), originally described as Cpf1, are guided by sgRNAs that are nearly half the size needed for Cas9 (Cas12a has 43 nucleotides, while Cas9 has 80 nucleotides) (Zetsche *et al.*, 2015; Anzalone *et al.*, 2020). This is because Cas12a, in bacteria, does not require tracrRNA, only crRNAs are needed to cleave both strands of DNA, using a single RuvC-type nuclease domain. In addition, the complex formed by Cas12a and sgRNA recognizes T-rich PAM sequences instead of G (5'-TTTN-3'),

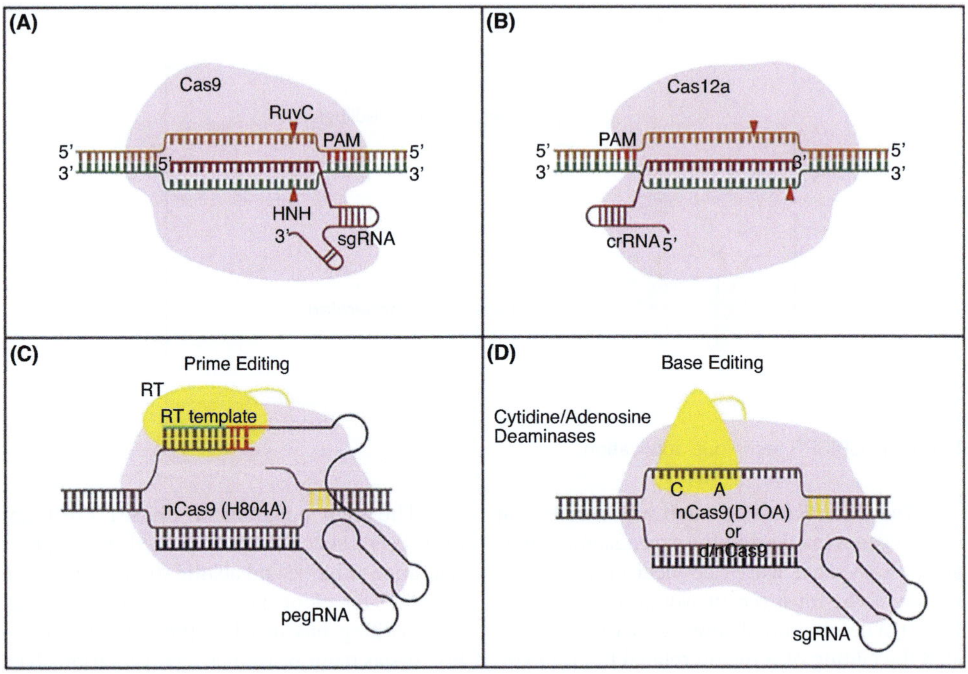

Fig. 15.2. A. Cas9 – Cas9 Nucleases have two splicing domains (RuvC and HNH) and are guided by single guide RNAs (sgRNAs) to generate a DSB (double-strand break). DSBs occur three base pairs upstream of PAM (protospacer adjacent motif). B. Cas12a – Cas12a Nucleases bind to crRNAs (CRISPR RNAs) forming a complex that recognizes T-rich PAM (protospacer adjacent motif) sequences and cleaves both strands of DNA generating a DSB. Cas12a has a single RuvC-like nuclease domain. C. Prime Editing – Fusion of an nCas9 (nickase) to a projected RT (reverse transcriptase). nCas9-RT will fuse to a pegRNA (prime editing guide RNA), which will specify the target site in its sequence and encode the desired edit. nCas9 has one of the mutated cleavage sites, being known as H840A when the mutation inactivates HNH. RT uses the newly released 3' end at the cleavage site of the target DNA to initiate reverse transcription using the specified pegRNA sequence as a template. Reverse transcription generates a 3' DNA patch that contains the edited sequence that will be incorporated into the genome. D. Base editing – dCas9 (dead) with both mutated splice sites or nCas9 (nickase) with only one of the mutated splice sites, is fused to the cytosine domain deaminase, or the adenosine deaminase domain. The modified Cas9 binds to DNA, guided by the sgRNA molecule (single guide RNA), directing the complex so that cytosine or adenosine deaminase converts C>T or A>G, respectively, at the target site.

cleaving at staggered and distant PAM sites, which may facilitate the reduction of off-targets in genomes rich in GC content (Guo *et al.*, 2010; Zetsche *et al.*, 2015; Fonfara *et al.*, 2016; Chen *et al.*, 2019). Several variants of Cas12, including Cas12a, cut DNA in a staggered fashion within the distal PAM regions of the protospacer results in cohesive ends on the DNA double-strand, which may increase the efficiency of HR repairs (Zaidi *et al.*, 2017; Chen *et al.*, 2019; Liu *et al.*, 2022). Some Cas12 variants, such as Cas12f, robustly cleave single-stranded DNA targets. However, other variants of Cas12, including Cas12b and Cas12i, predominantly cut double-stranded DNA (Chen *et al.*, 2019; Anzalone *et al.*, 2020). Other Cas enzymes, such as Cms1 and AaCas12b, also have a smaller size than Cas9 and Cas12a and an AT-rich PAM site (Begemann *et al.*, 2017; Chen *et al.*, 2019; Anzalone *et al.*, 2020).

Among the applications in plants, the designed vector of *Francisella novicida* (FnCas12a), carried out targeted mutagenesis in rice, where it was possible to observe deletions in the target region of the DL, ALS, NCED1 and AO1 genes (Endo *et al.*, 2016). Similar multiplex mutations were performed in rice plants using the CRISPR/Cas12 system (Wang *et al.*, 2017; Zhong *et al.*, 2018).

dCas9 (base editing e prime editing)

In addition to the modifications already discussed so far, new knowledge and fusions with other biotechnological techniques are bringing surprising advances to the gene editing system (Anzalone *et al.*, 2020; Morgante *et al.*, 2020). As a strategy to increase the specificity of the CRISPR/Cas9 system, it is possible to avoid performing double-stranded DNA cutting by modifying the cleavage domains of Cas9 so that it is only cleaving one strand of DNA. This is because, when the non-target DNA strand is displaced by the sgRNA spacer, it forms a single-strand DNA (R-loop) that is exposed and accessible to other molecules. After the formation of the R-loop, Cas9 undergoes conformational changes that result in the activation of its nuclease domains (RuvC and HNH) (Soda *et al.*, 2018; Yang *et al.*, 2018; Morgante *et al.*, 2020; Zhu *et al.*, 2020). However, these conformational changes can be prevented by mismatches between the target DNA and the sgRNA spacer, which ends up restricting nuclease activation to sequences that are highly complementary to sgRNA. Thus, with the inactivation of any nuclease domain, a Cas9 nickase (nCas9) is created, an enzyme that cleaves only one strand of DNA (Soda *et al.*, 2018; Morgante *et al.*, 2020; Zhu *et al.*, 2020). Inactivation of both nuclease domains generates the catalytically dead or endonuclease deficient Cas9 (dCas9), which is able to properly recognize when bound to the sgRNA complex and serves as a bridge to recruit proximal effectors to specific genomic sites (Soda *et al.*, 2018; Morgante *et al.*, 2020; Zhu *et al.*, 2020).

In rice, the dCas9/TV gene activator was able to upregulate *OsGW7* and *OsER1* up to 3,738-fold, with one sgRNA targeting each promoter (Xiong *et al.*, 2021). On the other hand, successfully tested multiplex strategies for dCas9-based transcriptional activation of three rice genes (*Os03g01240, Os04g39780, Os11g35410*) were performed in protoplasts (Lowder *et al.*, 2018).

From the development of nickases and dCas9, new technologies began to use the ability of these enzymes to recognize the specific editing site, but without performing the cut in the double strand of DNA (Bikard *et al.*, 2013; Morgante *et al.*, 2020; Zhu *et al.*, 2020). For example, the fusion of dCas9 to transcription factors can lead to transcriptional activation and repression of target genes, without promoting gene editing per se (Bikard *et al.*, 2013; Morgante *et al.*, 2020; Zhu *et al.*, 2020). Above all, two editing techniques based on nickases and dCas9: base editing and prime editing were also developed.

Prime editing

Prime editing (Fig. 15.2) is a versatile gene editing approach capable of minimizing off-target effects, being able to introduce all 12 types of possible point mutations, small insertions and deletions in a precise and targeted manner without requiring DSBs or donor DNA templates (Anzalone *et al.*, 2019; Anzalone *et al.*, 2020; Morgante *et al.*, 2020; Zhu *et al.*, 2020). This technology requires the fusion of a nCas9 to an engineered reverse transcriptase

(RT). nCas9 RT will merge with a prime editing guide RNA (pegRNA), which will specify the target site in its sequence and encode the desired cut. nCas9 has one of the mutated cutting sites, being known as D10A when the mutation inactivates RuvC and H840A when the mutation inactivates HNH. nCas9 D10A performs the cut on the strand complementary to pegRNA, while the nCas9 H840A performs the cut on the strand not complementary to pegRNA (Zhu *et al.*, 2020). Prime editing uses the newly released 3' end at the target DNA site to initiate reverse transcription using the specified pegRNA sequence as a template. This is because pegRNA contains a template for RT encoding the desired edits for hybridizing the 3' end of the cut DNA strand to initiate reverse transcription, encoding whatever is desired to edit (Morgante *et al.*, 2020; Lin *et al.*, 2020; Zhu *et al.*, 2020). Reverse transcription then generates a 3' DNA patch that contains the edited sequence and ultimately leads to the incorporation of that sequence into the genome. The H840A system differs from the D10A by using an additional sgRNA to cut the unedited strand at a location distant from the target indicated by the pegRNA, increasing editing efficiency (Zhu *et al.*, 2020). However, cutting the unedited strand also increases the frequency of unwanted insertions and deletions (Indels) at the target location (Chen *et al.*, 2021).

This powerful technology was tested in rice and wheat, where base substitutions, multiple base substitutions, insertions and deletions were performed simultaneously (Guo *et al.*, 2010; Chen *et al.*, 2021; Tabassum *et al.*, 2021). Prime editing was applied on a regenerated transgenic rice line in order to successfully repair an artificially mutated EGFP gene containing two successive stop codons (TAG TGA), converting them into TAC GGA (tyrosine-glycine) (Hua *et al.*, 2020). However, despite its versatility and efficiency, prime editing can vary between editing classes, target loci and cell types (Anzalone *et al.*, 2019; Zhu *et al.*, 2020). In plants, the technology remains limited to using orthogonal strategies, such as using reverse transcriptase orthologues with different activities, using ribozymes to produce precise pegRNAs, and increasing the temperature of the culture to increase catalytic activities (Zhu *et al.*, 2020). Still, it is a robust technology to be exploited in plant breeding programs.

Base editing

Editing genomes using Base editing technology (Fig. 15.2) allows the generation of base conversions, programmable and irreversibly, without requiring DSBs. Base editing consists of a dCas9 that has three versions, varying according to the altered bases (Chen *et al.*, 2019; Zhu *et al.*, 2020; Liu *et al.*, 2022). dCas9 can be fused to the cytosine deaminase domain, called cytosine base editors (CBEs), and, when fused to the adenosine deaminase domain, called adenine base editors (ABEs), which were developed to produce base transitions C:G-to—T:A and A:T-to-G:C, respectively (Chen *et al.*, 2019; Zhu *et al.*, 2020; Liu *et al.*, 2022). dCas9 can also be fused to uracil glycosylase inhibitor (UGI), being able to prevent U excision during repair, increasing the efficiency of CBEs (Chen *et al.*, 2019; Zhu *et al.*, 2020; Liu *et al.*, 2022). This is because, in the case of CBEs, cytosine deaminase converts cytosines (C) to uridines (U) in the released single-stranded DNA (ssDNA), and the U:G mismatch can be repaired to U:A, resulting in a base pair T:A. Fusions with uracil glycosylase inhibitor (UGI) are able to prevent U excision (which would return the original C:G sequence) and significantly increase CBE editing efficiency (Chen *et al.*, 2019; Zhu *et al.*, 2020; Liu *et al.*, 2022). For ABEs, adenine deaminase converts adenines (A) to inosines (I), which can be recognized as guanines (G). Thus, the function of the modified Cas9 is only to bind to DNA, guided by the sgRNA molecule, thus, directing the complex so that cytosine (or adenosine) deaminase converts a single base at the target site (Chen *et al.*, 2019; Anzalone *et al.*, 2020; Mishra *et al.*, 2020). The UGI inhibitor, on the other hand, alters the cellular repair pathway of uracil removal (Zhu *et al.*, 2020).

Genome editing work using base editing technology in rice (Table 15.1) demonstrates that it is possible to obtain a range of mutations by substituting many individual amino acids in the domains of genes that control economically important traits, as demonstrated with the editing of *Wx* alleles aiming at improving rice quality (Xu *et al.*, 2021).

Gene editing techniques in rice improvement: Challenge and development of transgene-free rice varieties

DNA sequencing, quantitative trait loci (QTLs), genome-wide association (GWAS) and transcriptomics (RNASeq and microarray) studies provide information about numerous genes involved in the control of agronomic traits such as yield (Ren *et al.*, 2020; Beerelli *et al.*, 2022), tolerance to biotic (Zhang *et al.*, 2016; Jiang *et al.*, 2020) and abiotic (Hsu and Tung, 2017; Song *et al.*, 2018; Bhandari *et al.*, 2020; Kulkarni *et al.*, 2020; Nayyeripasand *et al.*, 2021) stresses, and grain quality (Du *et al.*, 2013; Gao *et al.*, 2016; Roda *et al.*, 2020; Zhao *et al.*, 2021). These candidate genes can be edited to improve these agronomic traits in rice. Taking as example, the listed events in Table 15.1, one can observe that most events involve yield improvement. Although some single genes/transcription factors can have effects that impact yield increases, most likely these increases will remain in their majority coming from hybridization followed by selection strategies. On the other hand, climate changes have affected the environment and have facilitated the life of some pathogens. Therefore, those strategies focusing on single biotic stress resistance events are more likely to thrive commercially. In a similar way, abiotic stress tolerance due to single genes is also more likely to reach the farmer. Herbicide tolerance is also a trait likely to spread, since we need to count on different molecules to control weeds and perform proper management to avoid. Another set of traits that should have a major role in the near future is those related to symbiosis with nitrogen-fixing microorganisms. Mineral fertilizers are becoming a matter of national security to many countries and several soil management techniques, such as crop rotation, no-tillage and plant growth-promoting bacteria are on the agenda. Last, but not least, water use efficiency needs to be tackled. Although water deficit is a complex trait, recent advances in studies relating lower stomata number and drought tolerance point to that direction.

One of the greatest contributions of CRISPR/Cas technology in the agricultural area is perhaps the possibility of improving multiple traits simultaneously, directly in elite lines, streamlining the development of commercial cultivars (Gao *et al.*, 2020). In addition to this are the benefits of mutations carried out with the technology being classified as non-transgenic genetically modified organisms (GMOs) (Jansing *et al.*, 2019; Anzalone *et al.*, 2020). This is because, unlike transgenic GMOs, in genomic editing by CRISPR the insertion of exogenous DNA can be removed after performing crosses and selections in crop improvement (Khatodia *et al.*, 2016).

Classical gene editing results in indel mutations in specific sequences, and when they occur in exons result in knockout of genes encoding proteins. Exon editing can also result in partial knockout or gain-of-function mutations. This tool allows for other possibilities, such as the regulation of the transcription of a target gene. Thus, cis-elements present in promoters can be edited, preventing the binding of transcription factors (TF) and RNA polymerase; and knocking out or changing TF transcription, making it impossible for it to activate the target gene. These modifications can also be directed towards the induction of the target gene. Additionally, you can edit UTRs, splicing sites and miRNA precursors (non-coding genes), modulating post-transcriptional regulation. Finally, one can edit sequences that encode sites of post-translational modification and sites of interaction between proteins, directing protein expression (Tan *et al.*, 2020). Thus, it appears that editing allows accessing and modifying any sequence of interest in the genome, opening up countless possibilities for improvement. Therefore, genome editing via CRISPR-Cas can be used to improve any trait of interest, as long as the genes involved in controlling that trait are known (Tan *et al.*, 2020).

However, one of the problems observed in this technology is the emergence of off-target mutations. "Off-target" refers to nuclease cleavage at unintended genomic sites caused by the nuclease tolerating some incompatibilities between gRNA and the target nucleotide sequence (Liu *et al.*, 2022). To reduce the occurrence of off-target mutations, the system has been improved to increase efficiency, new applications and, since then, in addition to modifications that we will see later, many Cas9 orthologues and variants have been characterized

(Gao, 2021; Tabassum *et al.*, 2021). Thus, some nucleases may have applications that are slightly different from those seen above, as is the case with the Cas13 effector proteins and their variants. These have been modified to recognize and edit RNA targets with potential application in viral RNA interference in plants (Aman *et al.*, 2018; Mahas *et al.*, 2019). Processing of pre-crRNA into Cas13a has been validated in plants for application against various invading viruses (Aman *et al.*, 2018). However, comprehensive screening studies are still required to validate and compare the various CRISPR/Cas13 orthologues, including Cas13a, Cas13b and Cas13d (Mahas *et al.*, 2019).

Taken together, the genetic improvement of rice can rely on the CRISPR/Cas tool and its variants to face current and future challenges, and fulfil the task of feeding a constantly growing population.

Conclusion

Although conventional breeding has been decisive so far, there is still a long way to go in order to meet future demands and face the challenges of rice cultivation. Genome editing technology, specifically the CRISPR/Cas tool and its variants, has already shown its potential to help breeders on this journey and revolutionize agriculture.

The research developments so far show that the CRISPR/Cas technology and its variants are more accessible and easier to manipulate for genome editing than other tools (MNs, ZFNs and TALENs). In this sense, a large number of studies using CRISPR/Cas and its variants have been developed in rice. Among these, we can highlight the success of this tool in the editing of genes involved in yield, tolerance to biotic and abiotic stresses, and grain quality. On top of these, novel traits from wild-relatives can be brought to cultivated crops or wild-relatives can be tamed by editing adaptation traits. Exciting results are expected as such actions become routine, originating neo-domestication examples. However, the potential application of genome editing in breeding still depends on the cultivation response of large-scale edited plants in the field.

Finally, it is worth noting that different strategies were developed to eliminate transgenes from the edited genome. In this way, plants improved from genome editing and conventional breeding are similar in terms of risks to the environment and human health, which are practically non-existent. Therefore, it is expected that edited plants, in addition to meeting the demands, have good acceptance among consumers.

Authors' Contributions

VNO and VFO wrote the first draft. CP, LCM and ACO read and contributed to the following versions of the manuscript.

Acknowledgement

The authors are thankful to the Conselho Nacional de Desenvolvimento Científico e Tecnológico (CNPq), Coordenação de Aperfeiçoamento de Pessoal de Nível Superior (CAPES) grant no. 001 and Fundação de Amparo à Pesquisa do Estado do Rio Grande do Sul (FAPERGS) for grants and fellowships.

References

Abe, K., Araki, E., Suzuki, Y., Toki, S. and Saika, H. (2018) Production of high oleic/low linoleic rice by genome editing. *Plant Physiology Biochemistry* 131, 58–62.

Abudayyeh, O.O., Gootenberg, J.S., Essletzbichler, P., Han, S., Joung, J. *et al.* (2017) RNA targeting with CRISPR–Cas13. *Nature* 550(7675), 280–284.

Aman, R., Ali, Z., Butt, H., Mahas, A., Aljedaani, F. *et al.* (2018) RNA virus interference via CRISPR/Cas13a system in plants. *Genome Biology* 19(1), 1–9.

Anzalone, A.V., Koblan, L.W. and Liu, D.R. (2020) Genome editing with CRISPR–Cas nucleases, base editors, transposases and prime editors. *Nature Biotechnology* 38, 824–844.

Anzalone, A.V., Randolph, P.B., Davis, J.R., Sousa, A.A., Koblan, L.W. *et al.* (2019) Search-and-replace genome editing without double-strand breaks or donor DNA. *Nature* 576(7785), 149–157.

Ashokkumar, S., Jaganathan, D., Ramanathan, V., Rahman, H., Palaniswamy, R. *et al.* (2020) Creation of novel alleles of fragrance gene *OsBADH2* in rice through CRISPR/Cas9 mediated gene editing. *PLoS ONE* 15, e0237018.

Barman, H.N., Sheng, Z., Fiaz, S., Zhong, M., Wu, Y. *et al.* (2019) Generation of a new thermo-sensitive genic male sterile rice line by targeted mutagenesis of *TMS5* gene through CRISPR/Cas9 system. *BMC Plant Biology* 19, 1–9.

Beerelli, K., Balakrishnan, D., Addanki, K.R., Surapaneni, M., Rao Yadavalli, V. *et al.* (2022) Mapping of QTLs for yield traits using F2:3:4 populations derived from two alien introgression lines reveals qTGW8.1 as a consistent QTL for grain weight from *Oryza nivara*. *Front Plant Science* 13, 790221–790221.

Begemann, M.B., Gray, B.N., January, E., Singer, A., Kesler, D.C. *et al.* (2017) Characterization and validation of a novel group of type V, class 2 nucleases for in vivo genome editing. *bioRxiv* 2017, 192799.

Bhandari, A., Sandhu, N., Bartholome, J., Cao-Hamadoun, T.V., Ahmadi, N. *et al.* (2020) Genome-wide association study for yield and yield related traits under reproductive stage drought in a diverse indica-aus rice panel. *Rice* 13, 1–22.

Bhardwaj, A. and Nain, V. (2021) TALENs—An indispensable tool in the era of CRISPR: A mini review. *Journal of Genetic Engineering and Biotechnology* 19(125), 1–10.

Bikard, D., Jiang, W., Samai, P., Hochschild, A., Zhang, F. *et al.* (2013) Programmable repression and activation of bacterial gene expression using an engineered CRISPR-Cas system. *Nucleic Acids Research* 41(15), 7429–7437.

Bo, W., Zhaohui, Z., Huanhuan, Z., Xia, W., Binglin, L. *et al.* (2019) Targeted mutagenesis of NAC transcription factor gene, *OsNAC041*, leading to salt sensitivity in rice. *Rice Science* 26, 98–108.

Boch, J. and Bonas, U. (2010) *Xanthomonas AvrBs3* family-type III effectors: Discovery and function. *Annual Review of Phytopathology* 48, 419–436.

Bortesi, L. and Fischer, R. (2015) The CRISPR/Cas9 system for plant genome editing and beyond. *Biotechnology Advances* 33, 41–52.

Butt, H., Eid, A., Momin, A.A., Bazin, J., Crespi, M. *et al.* (2019) CRISPR directed evolution of the spliceosome for resistance to splicing inhibitors. *Genome Biology* 20, 1–9.

Chang, J.D., Huang, S., Konishi, N., Wang, P., Chen, J. *et al.* (2020) Overexpression of the manganese/cadmium transporter *OsNRAMP5* reduces cadmium accumulation in rice grain. *Journal of Experimental Botany* 71, 5705–5715.

Chao, S., Cai, Y., Feng, B., Jiao, G., Sheng, Z. *et al.* (2019) Editing of the rice isoamylase gene *ISA1* provides insights into its function in starch formation. *Rice Science* 26, 77–87.

Chen, K. and Gao, C. (2020) Genome-edited crops: How to move them from laboratory to market. *Frontiers of Agricultural Science and Engineering* 7, 181–187.

Chen, P.J., Hussmann, J.A., Yan, J., Knipping, F., Ravisankar, P. *et al.* (2021) Enhanced prime editing systems by manipulating cellular determinants of editing outcomes. *Cell* 184, 5635–5652.

Chen, K., Shan, Q. and Gao, C. (2014) An efficient TALEN mutagenesis system in rice. *Methods* 69, 2–8.

Chen, K., Wang, Y., Zhang, R., Zhang, H. and Gao, C. (2019) CRISPR/Cas genome editing and precision plant breeding in agriculture. *Annual Review of Plant Biology* 70, 667–697.

Christian, M., Cermak, T., Doyle, E.L., Schmidt, C., Zhang, F. *et al.* (2010) Targeting DNA double-strand breaks with TAL effector nucleases. *Genetics* 186, 757–761.

Cox, D.B., Gootenberg, J.S., Abudayyeh, O.O., Franklin, B., Kellner, M.J. *et al.* (2017) RNA editing with CRISPR-Cas13. *Science* 358(6366), 1019–1027.

Cui, Y., Hu, X., Liang, G., Feng, A., Wang, F., *et al.* (2020) Production of novel beneficial alleles of a rice yield-related QTL by CRISPR/Cas9. *Plant Biotechnology Journal* 18, 1987–1989.

D'Halluin, K., Vanderstraeten, C., Stals, E., Cornelissen, M. and Ruiter, R. (2008) Homologous recombination: A basis for targeted genome optimization in crop species such as maize. *Plant Biotechnology Journal* 6, 93–102.

Daboussi, F., Stoddard, T.J. and Zhang, F. (2015) Engineering meganuclease for precise plant genome modification. In: *Advances in New Technology for Targeted Modification of Plant Genomes*. Springer, New York, pp. 21–38.

Dar, M.H., Bano, D.A., Waza, S.A., Zaidi, N.W., Majid, A. *et al.* (2021) Abiotic stress tolerance-progress and pathways of sustainable rice production. *Sustainability* 13, 2078.

Djukanovic, V., Smith, J., Lowe, K., Yang, M., Gao, H. *et al.* (2013) Male-sterile maize plants produced by targeted mutagenesis of the cytochrome P450-like gene (MS26) using a re-designedI–CreI homing endonuclease. *The Plant Journal* 76, 888–899.

Dong, O.X., Yu, S., Jain, R., Zhang, N., Duong, P.Q. *et al.* (2020) Marker-free carotenoid-enriched rice generated through targeted gene insertion using CRISPR-Cas9. *Nature Communications* 11, 1–10.

Du, J., Zeng, D., Wang, B., Qian, Q., Zheng, S. *et al.* (2013) Environmental effects on mineral accumulation in rice grains and identification of ecological specific QTLs. *Environmental Geochemistry and Health* 35, 161–170.

Endo, A., Masafumi, M., Kaya, H. and Toki, S. (2016) Efficient targeted mutagenesis of rice and tobacco genomes using Cpf1 from *Francisella novicida*. *Scientific Reports* 6, 1–9.

Endo, M., Mikami, M. and Toki, S. (2015) Multigene knockout utilizing off-target mutations of the CRISPR/Cas9 system in rice. *Plant and Cell Physiology* 56, 41–47.

Feng, Z., Zhang, B., Ding, W., Liu, X., Yang, D.L. *et al.* (2013) Efficient genome editing in plants using a CRISPR/Cas system. *Cell Research* 23, 1229–1232.

Feng, Y., Zhang, S. and Huang, X. (2014) A robust TALENs system for highly efficient mammalian genome editing. *Scientific Reports* 4, 1–7.

Fonfara, I., Richter, H., Bratovič, M., Le Rhun, A. and Charpentier, E. (2016) The CRISPR-associated DNA-cleaving enzyme Cpf1 also processes precursor CRISPR RNA. *Nature* 532(7600), 517–521.

Gao, C. (2021) Genome engineering for crop improvement and future agriculture. *Cell* 184(6), 1621–1635.

Gao, Q., Li, G., Sun, H., Xu, M., Wang, H. *et al.* (2020) Targeted mutagenesis of the rice FW 2.2-like gene family using the CRISPR/Cas9 system reveals OsFWL4 as a regulator of tiller number and plant yield in rice. *International Journal of Molecular Sciences* 21, 809.

Gao, Y., Liu, C., Li, Y., Zhang, A., Dong, G. *et al.* (2016) QTL analysis for chalkiness of rice and fine mapping of a candidate gene for *qACE9*. *Rice* 9, 1–10.

Gao, H., Smith, J., Yang, M., Jones, S., Djukanovic, V., Nicholson, M.G. *et al.* (2010) Heritable targeted mutagenesis in maize using a designed endonuclease. *The Plant Journal* 61, 176–187.

Gu, X., Liu, L. and Zhang, H. (2021) Transgene-free genome editing in plants. *Frontiers in Genome Editing* 3, 805317.

Guo, J., Gaj, T. and Barbas III, C.F. (2010) Directed evolution of an enhanced and highly efficient *FokI* cleavage domain for zinc finger nucleases. *Journal of Molecular Biology* 400, 96–107.

Han, Y., Teng, K., Nawaz, G., Feng, X., Usman, B. *et al.* (2019) Generation of semi-dwarf rice (*Oryza sativa* L.) lines by CRISPR/Cas9-directed mutagenesis of *OsGA20ox2* and proteomic analysis of unveiled changes caused by mutations 3. *Biotech* 11, 1–7.

Hsu, S.K. and Tung, C.W. (2017) RNA-Seq analysis of diverse rice genotypes to identify the genes controlling coleoptile growth during submerged germination. *Front Plant Science* 8, 762.

Hu, X., Cui, Y., Dong, G., Feng, A., Wang, D. *et al.* (2019) Using CRISPRCas9 to generate semi-dwarf rice lines in elite landraces. *Scientific Reports* 9, 1–7.

Hua, K., Jiang, Y., Tao, X. and Zhu, J.K. (2020) Precision genome engineering in rice using prime editing system. *Plant Biotechnology Journal* 18, 2167–2169.

Hua, K., Tao, X., Liang, W., Zhang, Z., Gou, R. *et al.* (2020) Simplified adenine base editors improve adenine base editing efficiency in rice. *Plant Biotechnology Journal* 18, 770–778.

Hua, K., Tao, X. and Zhu, J.K. (2019) Expanding the base editing scope in rice by using Cas9 variants. *Plant Biotechnology Journal* 17, 499–504.

Huang, J., Gao, L., Luo, S., Liu, K., Qing, D. *et al.* (2022) The genetic editing of GS3 via CRISPR/Cas9 accelerates the breeding of three-line hybrid rice with superior yield and grain quality. *Molecular Breeding* 42, 1–4.

Huang, L., Li, Q., Zhang, C., Chu, R., Gu, Z. *et al.* (2020) Creating novel Wx alleles with fine-tuned amylose levels and improved grain quality in rice by promoter editing using CRISPR/Cas9 system. *Plant Biotechnology Journal* 18, 2164–2166.

Huang, X., Zeng, X., Li, J. and Zhao, D. (2017) Construction and analysis of tify1a and tify1b mutants in rice (*Oryza sativa*) based on CRISPR/Cas9 technology. *Journal of Agricultural Biotechnology* 25, 1003–1012.

Huang, L., Zhang, R., Huang, G., Li, Y., Melaku, G. *et al.* (2018) Developing superior alleles of yield genes in rice by artificial mutagenesis using the CRISPR/Cas9 system. *The Crop Journal* 6, 475–481.

Iqbal, Z., Iqbal, M.S., Ahmad, A., Memon, A.G. and Ansari, M.I. (2020) New prospects on the horizon: Genome editing to engineer plants for desirable traits. *Current Plant Biology* 24, 100171.

Ishii, T. and Araki, M. (2016) Consumer acceptance of food crops developed by genome editing. *Plant Cell Reports* 35, 1507–1518.

Jacquier, A. and Dujon, B. (1985) An intron-encoded protein is active in a gene conversion process that spreads an intron into a mitochondrial gene. *Cell* 41, 383–394.

Jansing, J., Schiermeyer, A., Schillberg, S., Fischer, R. and Bortesi, L. (2019) Genome editing in agriculture: Technical and practical considerations. *International Journal of Molecular Sciences* 20(12), 2888.

Jiang, H., Feng, Y., Qiu, L., Gao, G., Zhang, Q. *et al.* (2020) Identification of blast resistance QTLs based on two advanced backcross populations in rice. *Rice* 13, 1–2.

Jiang, W., Zhou, H., Bi, H., Fromm, M., Yang, B. *et al.* (2013) Demonstration of CRISPR/Cas9/sgRNA-mediated targeted gene modification in Arabidopsis, tobacco, sorghum and rice. *Nucleic Acids Research* 41, e188.

Jinek, M., Chylinski, K., Fonfara, I., Hauer, M., Doudna, J.A. *et al.* (2012) A programmable dual-RNA–guided DNA endonuclease in adaptive bacterial immunity. *Science* 337(6096), 816–821.

Joung, J.K. and Sander, J.D. (2013) TALENs: A widely applicable technology for targeted genome editing. *Nature Reviews Molecular Cell Biology* 14, 49–55.

Jung, Y.J., Nogoy, F.M., Lee, S.K., Cho, Y.G. and Kang, K.K. (2018) Application of ZFN for site directed mutagenesis of rice *SSIVa* gene. *Biotechnol and Bioprocess Engineering* 23(1), 108–115.

Kannan, S., Altae-Tran, H., Jin, X., Madigan, V.J., Oshiro, R. *et al.* (2022) Compact RNA editors with small Cas13 proteins. *Nature Biotechnology* 40(2), 194–197.

Khanday, I., Skinner, D., Yang, B., Mercier, R. and Sundaresan, V. (2019) A male-expressed rice embryogenic trigger redirected for asexual propagation through seeds. *Nature* 565, 91–95.

Khatodia, S., Bhatotia, K., Passricha, N., Khurana, S.M. and Tuteja, N. (2016) The CRISPR/Cas genome-editing tool: Application in improvement of crops. *Frontiers in Plant Science* 7, 506.

Kim, Y.G., Cha, J. and Chandrasegaran, S. (1996) Hybrid restriction enzymes: Zinc finger fusions to *Fok I* cleavage domain. *Proceedings of the National Academy of Sciences* 93, 1156–1160.

Kim, Y.A., Moon, H. and Park, C.J. (2019) CRISPR/Cas9-targeted mutagenesis of *Os8N3* in rice to confer resistance to *Xanthomonas oryzae* pv. oryzae. *Rice* 12, 67.

Klug, A. (2010) The discovery of zinc fingers and their development for practical applications in gene regulation and genome manipulation. *Quarterly Reviews of Biophysics* 43, 1–21.

Konermann, S., Lotfy, P., Brideau, N.J., Oki, J., Shokhirev, M.N. and Hsu, P.D. (2018) Transcriptome engineering with RNA-targeting type VI-D CRISPR effectors. *Cell* 173(3), 665–676.

Kuang, Y., Li, S., Ren, B., Yan, F., Spetz, C. *et al.* (2020) Base-editing-mediated artificial evolution of *OsALS1* in planta to develop novel herbicide-tolerant rice germplasms. *Molecular Plant* 13, 565–572.

Kulkarni, S.R., Balachandran, S.M., Ulaganathan, K., Balakrishnan, D., Praveen, M. *et al.* (2020) Molecular mapping of QTLs for yield related traits in recombinant inbred line (RIL) population derived from the popular rice hybrid KRH-2 and their validation through SNP genotyping. *Scientific Reports* 10, 1–21.

Li, S., Gao, F., Xie, K., Zeng, X., Cao, Y. *et al.* (2016) *The OsmiR396c-OsGRF4-OsGIF1* regulatory module determines grain size and yield in rice. *Plant Biotechnology Journal* 14, 2134–2146.

Li, J., Li, H., Chen, J., Yan, L. and Xia, L. (2020) Toward precision genome editing in crop plants. *Molecular Plant* 13(6), 811–813.

Li, S., Li, J., He, Y., Xu, M., Zhang, J. *et al.* (2019) Precise gene replacement in rice by RNA transcript-templated homologous recombination. *Nature Biotechnology* 37, 445–450.

Li, C., Li, W., Zhou, Z., Chen, H., Xie, C. *et al.* (2020) A new rice breeding method: CRISPR/Cas9 system editing of the *Xa13* promoter to cultivate transgene-free bacterial blight-resistant rice. *Plant Biotechnology Journal* 18, 313.

Li, M., Li, X., Zhou, Z., Wu, P., Fang, M. *et al.* (2016) Reassessment of the four yield-related genes *Gn1a*, *DEP1*, *GS3*, and *IPA1* in rice using a CRISPR/Cas9 system. *Frontiers in Plant Science* 7, 377.

Li, T., Liu, B., Chen, C.Y. and Yang, B. (2016) TALEN-mediated homologous recombination produces site-directed DNA base change and herbicide resistant rice. *Journal Genetics and Genomics* 43, 297–305.

Li, T., Liu, B., Spalding, M.H., Weeks, D.P. and Yang, B. (2012) High-efficiency TALEN-based gene editing produces disease-resistant rice. *Nature Biotechnology* 30, 390–392.

Li, J., Meng, X., Zong, Y., Chen, K., Zhang, H. *et al.* (2016) Gene replacements and insertions in rice by intron targeting using CRISPR–Cas9. *Nature Plants* 2, 1–6.

Li, S., Shen, L., Hu, P., Liu, Q., Zhu, X. *et al.* (2019) Developing disease-resistant thermosensitive male sterile rice by multiplex gene editing. *Journal of Integrative Plant Biology* 61, 1201–1205.

Li, J., Sun, Y., Du, J., Zhao, Y. and Xia, L. (2017) Generation of targeted point mutations in rice by a modified CRISPR/Cas9 system. *Molecular Plant* 10, 526–529.

Li, Q., Zhang, D., Chen, M., Liang, W., Wei, J. *et al.* (2016) Development of japonica photo-sensitive genic male sterile rice lines by editing carbon starved anther using CRISPR/Cas9. *Journal of Genetics and Genomics* 43, 415–419.

Li, X., Zhou, W., Ren, Y., Tian, X., Lv, T. *et al.* (2017) High-efficiency breeding of early-maturing rice cultivars via CRISPR/Cas9-mediated genome editing. *Journal of Genetics and Genomics* 44, 175–178.

Liao, S., Qin, X., Luo, L., Han, Y., Wang, X. *et al.* (2019) CRISPR/Cas9-induced mutagenesis of semi-rolled leaf1, 2 confers curled leaf phenotype and drought tolerance by influencing protein expression patterns and ROS scavenging in rice (*Oryza sativa* L.).). *Agronomy* 9, 728.

Lin, Q., Zong, Y., Xue, C., Wang, S., Jin, S. *et al.* (2020) Prime genome editing in rice and wheat. *Nature Biotechnology* 38(5), 582–585.

Liu, G., Lin, Q., Jin, S. and Gao, C. (2022) The CRISPR-Cas toolbox and gene editing technologies. *Molecular Cell* 82, 333–347.

Liu, X., Qin, R., Li, J., Liao, S., Shan, T. *et al.* (2020) A CRISPR-Cas9-mediated domain-specific base-editing screen enables functional assessment of ACCase variants in rice. *Plant Biotechnology Journal* 18, 1845–1847.

Lopes Filho, J.H., da Silva, V.C., dos Santos, J.C., Dante, R., Gerhardt, I. *et al.* (2020) Introduction to genomic editing in plants: Challenges of modern agriculture for the present and the future. In: Molinari, H.B., Vieira, L.R., Silva, N., Prado, G.S. and Lopes Filho, J.H. (eds) *CRISPR Technology in Plant Genomic Editing: Biotechnology Applied to Agriculture*. Embrapa Agroenergy-Scientific Book (ALICE), Brasília.

Lou, D., Wang, H., Liang, G. and Yu, D. (2017) *OsSAPK2* confers abscisic acid sensitivity and tolerance to drought stress in rice. *Frontiers in Plant Science* 8, 993.

Lowder, L.G., Zhang, D., Baltes, N.J., Paul, J.W., Tang, X. *et al.* (2015) A CRISPR/Cas9 toolbox for multiplexed plant genome editing and transcriptional regulation. *Plant Physiology* 169, 971–985.

Lowder, L.G., Zhou, J., Zhang, Y., Malzahn, A., Zhong, Z. *et al.* (2018) Robust transcriptional activation in plants using multiplexed CRISPR-Act2. 0 and mTALE-Act systems. *Molecular Plant* 11, 245–256.

Lu, Y. and Zhu, J.K. (2017) Precise editing of a target base in the rice genome using a modified CRISPR/Cas9 system. *Molecular Plant* 10, 523–525.

Lucht, J.M. (2015) Public acceptance of plant biotechnology and GM crops. *Viruses* 7, 4254–4281.

Lyznik, L.A., Djukanovic, V., Yang, M. and Jones, S. (2012) Double-strand break-induced targeted mutagenesis in plants. In: Dunwell, J. and Wetten, A. (eds) *Transgenic Plants. Methods in Molecular Biology*, Vol. 847. pp. 399–416.

Ma, J., Chen, J., Wang, M., Ren, Y., Wang, S. *et al.* (2018) Disruption of *OsSEC3A* increases the content of salicylic acid and induces plant defense responses in rice. *Journal of Experimental Botany* 69, 1051–1064.

Ma, D. and Liu, F. (2015) Genome editing and its applications in model organisms Genom. *Proteomics and Bioinformatics Research* 13, 336–344.

Ma, X., Zhang, Q., Zhu, Q., Liu, W., Chen, Y. *et al.* (2015) A robust CRISPR/Cas9 system for convenient, high-efficiency multiplex genome editing in monocot and dicot plants. *Molecular Plant* 8, 1274–1284.

Macovei, A., Sevilla, N.R., Cantos, C., Jonson, G.B., Slamet-Loedin, I. *et al.* (2018) Novel alleles of rice *eIF4G* generated by CRISPR/Cas9-targeted mutagenesis confer resistance to Rice tungro spherical virus. *Plant Biotechnology Journal* 16, 1918–1927.

Maeder, M.L. and Gersbach, C.A. (2016) Genome-editing technologies for gene and cell therapy. *Molecular Therapy* 24(3), 430.

Mahas, A., Aman, R. and Mahfouz, M. (2019) CRISPR-Cas13d mediates robust RNA virus interference in plants. *Genome Biology* 20(1), 1–16.

Mak Nga-Sze, A., Bradley, P., Cernadas, R.A., Bogdanove, A.J. and Stoddard, B.L. (2012) The crystal structure of TAL effector PthXo1 bound tolts DNA target. *Science* 335(6069), 716–720.

Makino, A., Kaneta, Y., Obara, M., Ishiyama, K., Kanno, K. *et al.* (2020) High yielding ability of a large-grain rice cultivar, Akita 63. *Scientific Reports* 10, 1–8.

Malzahn, A.A., Tang, X., Lee, K., Ren, Q., Sretenovic, S. *et al.* (2019) Application of CRISPR-Cas12a temperature sensitivity for improved genome editing in rice, maize, and Arabidopsis. *BMC Biology* 17(1), 1–14.

Mao, Y.F., Zhang, H., Xu, N.F., Zhang, B.T., Gou, F. *et al.* (2013) Application of the CRISPR–Cas system for efficient genome engineering in plants. *Molecular Plant* 6, 2008–2011.

Mao, X., Zheng, Y., Xiao, K., Wei, Y., Zhu, Y. *et al.* (2018) *OsPRX2* contributes to stomatal closure and improves potassium deficiency tolerance in rice. *Biochemical and Biophysical Research Communications* 495, 461–467.

Metzger, M.J., McConnell-Smith, A., Stoddard, B.L. and Miller, A.D. (2011) Single-strand nicks induce homologous recombination with less toxicity than double-strand breaks using an AAV vector template. *Nucleic Acids Research* 39, 926–935.

Miao, J., Guo, D., Zhang, J., Huang, Q., Qin, G. *et al.* (2013) Targeted mutagenesis in rice using CRISPR-Cas system. *Cell Research* 23, 1233–1236.

Miao, C., Xiao, L., Hua, K., Zou, C., Zhao, Y. *et al.* (2018) Mutations in a subfamily of abscisic acid receptor genes promote rice growth and productivity. *Proceedings of the National Academy of Sciences of USA* 115, 6058–6063.

Mikami, M., Toki, S. and Endo, M. (2015a) Comparison of CRISPR/Cas9 expression constructs for efficient, targeted mutagenesis in rice. *Plant Molecular Biology* 88, 561–572.

Mikami, M., Toki, S. and Endo, M. (2015b) Parameters affecting frequency of CRISPR/Cas9 mediated targeted mutagenesis in rice. *Plant Cell Reports* 34, 1807–1815.

Mikami, M., Toki, S. and Endo, M. (2016) Precision targeted mutagenesis via Cas9 paired nickases in rice. *Plant and Cell Physiology* 57, 1058–1068.

Miller, J., McLachlan, A.D. and Klug, A. (1985) Repetitive zinc-binding domains in the protein transcription factor IIIA from Xenopus oocytes. *The EMBO Journal* 4(6), 1609–1614.

Miller, J.C., Tan, S., Qiao, G., Barlow, K.A., Wang, J. *et al.* (2011) A TALE nuclease architecture for efficient genome editing. *Nature Biotechnology* 29(2), 143–150.

Ming, M., Ren, Q., Pan, C., He, Y., Zhang, Y. *et al.* (2020) CRISPR–Cas12b enables efficient plant genome engineering. *Nature Plants* 6(3), 202–208.

Mishra, R., Joshi, R.K. and Zhao, K. (2020) Base editing in crops: Current advances, limitations and future implications. *Plant Biotechnology Journal* 18(1), 20–31.

Moreno-Mateos, M.A., Fernandez, J.P., Rouet, R., Vejnar, C.E., Lane, M.A. *et al.* (2017) CRISPR-Cpf1 mediates efficient homology-directed repair and temperature-controlled genome editing. *Nature Communications* 8(1), 1–9.

Morgante, C., Arraes, F., Pinto, C., de Melo, B.P. and Grossi-de-Sa, M.F. (2020) Modulation of gene expression in plants via CRISPR/dCas9 technology. In: Molinari, H.B., Vieira, L.R., Silva, N., Prado, G.S. and Lopes Filho, J.H. (eds) *CRISPR Technology in Plant Genomic Editing: Biotechnology Applied to Agriculture*. Embrapa Agroenergy-Scientific Book (ALICE), Brasília.

Mussolino, C. and Cathomen, T. (2012) TALE nucleases: Tailored genome engineering made easy. *Current Opinion in Biotechnology* 23, 644–650.

Nawaz, G., Usman, B., Peng, H., Zhao, N., Yuan, R. *et al.* (2020) Knockout of *Pi21* by CRISPR/Cas9 and iTRAQ-based proteomic analysis of mutants revealed new insights into *M. oryzae* resistance in elite rice line. *Genes* 11, 735.

Nayyeripasand, L., Garoosi, G.A. and Ahmadikhah, A. (2021) Genome-wide association study (GWAS) to identify salt-tolerance QTLs carrying novel candidate genes in rice during early vegetative stage. *Rice* 14, 9.

Nieves-Cordones, M., Mohamed, S., Tanoi, K., Kobayashi, N.I., Takagi, K. *et al.* (2017) Production of low- Cs^+ rice plants by inactivation of the K+ transporter *OsHAK1* with the CRISPR-Cas system. *The Plant Journal* 92, 43–56.

Nizolli, V.O., Pegoraro, C. and Oliveira, A.C. (2021) Rice blast: Strategies and challenges for improving genetic resistance. *Crop Breeding and Applied Biotechnology* 21, e387721S9.

Ogata, T., Ishizaki, T., Fujita, M. and Fujita, Y. (2020) CRISPR/Cas9-targeted mutagenesis of *OsERA1* confers enhanced responses to abscisic acid and drought stress and increased primary root growth under nonstressed conditions in rice. *PLOS ONE* 15, e0243376.

Oliva, R., Ji, C., Atienza-Grande, G., Huguet-Tapia, J.C., Perez-Quintero, A. *et al.* (2019) Broad-spectrum resistance to bacterial blight in rice using genome editing. *Nature Biotechnology* 37, 1344–1350.
Osakabe, Y. and Osakabe, K. (2015) Genome editing with engineered nucleases in plants. *Plant and Cell Physiology* 56, 389–400.
Park, J.R., Kim, E.G., Jang, Y.H., Jan, R., Farooq, M. *et al.* (2022) Applications of CRISPR/Cas9 as new strategies for short breeding to drought gene in rice. *Frontiers in Plant Science* 13, 850441.
Paschon, D.E., Lussier, S., Wangzor, T., Xia, D.F., Li, P.W. *et al.* (2019) Diversifying the structure of zinc finger nucleases for high-precision genome editing. *Nature Communications* 10, 1–2.
Peng, J., Botha-Oberholster, A.M., Jaganathan, D., Venkataraman, G., Ramasamy, K. *et al.* (2018) CRISPR for crop improvement: An update review. *Frontiers in Plant Science* 9(985), 1–17.
Ren, J., Zhang, F., Gao, F., Zeng, L., Lu, X. *et al.* (2020) Transcriptome and genome sequencing elucidates the molecular basis for the high yield and good quality of the hybrid rice variety Chuanyou6203. *Scientific Reports* 10, 1–5.
Roda, F.A., Marques, I., Batista-Santos, P., Esquível, M.G., Ndayiragije, A. *et al.* (2020) Rice biofortification with zinc and selenium: A transcriptomic approach to understand mineral accumulation in flag leaves. *Frontiers in Genetics* 11, 543.
Santosh Kumar, V.V., Verma, R.K., Yadav, S.K., Yadav, P., Watts, A. *et al.* (2020) CRISPR-Cas9 mediated genome editing of drought and salt tolerance (*OsDST*) gene in indica mega rice cultivar MTU1010. *Physiology and Molecular Biology of Plants* 26, 1099–1110.
Schindele, P. and Puchta, H. (2020) Engineering CRISPR/LbCas12a for highly efficient, temperature-tolerant plant gene editing. *Plant Biotechnology Journal* 18(5), 1118.
Shah, P.R., Varanavasiappan, S., Kokiladevi, E., Ramanathan, A. and Kumar, K.K. (2019) Genome editing of rice *PFT1* gene to study its role in rice sheath blight disease resistance. *International Journal of Current Microbiology and Applied Sciences* 8, 2356–2364.
Shan, Q., Wang, Y., Li, J. and Gao, C. (2014) Genome editing in rice and wheat using the CRISPR/Cas system. *Nature Protocols* 9, 2395–2410.
Shan, Q., Wang, Y., Li, J., Zhang, Y., Chen, K. *et al.* (2013) Targeted genome modification of crop plants using a CRISPR-Cas system. *Nature Biotechnology* 31, 686–688.
Shao, G., Xie, L., Jiao, G., Wei, X., Sheng, Z. *et al.* (2017) CRISPR/CAS9-mediated editing of the fragrant gene *Badh2* in rice. *Chinese Journal of Rice Science* 31, 216–222.
Shen, C., Que, Z., Xia, Y., Tang, N., Li, D. *et al.* (2017) Knock out of the annexin gene *OsAnn3* via CRISPR/Cas9-mediated genome editing decreased cold tolerance in rice. *Journal of Plant Biotechnology* 60, 539–547.
Shen, L., Wang, C., Fu, Y., Wang, J., Liu, Q. *et al.* (2018) QTL editing confers opposing yield performance in different rice varieties. *Journal of Integrative Plant Biology* 60, 89–93.
Shimatani, Z., Kashojiya, S., Takayama, M., Terada, R., Arazoe, T. *et al.* (2017) Targeted base editing in rice and tomato using a CRISPR-Cas9 cytidine deaminase fusion. *Nature Biotechnology* 35, 441–443.
Shukla, V., Gupta, M., Urnov, F., Guschin, D., Jan, M. *et al.* (2013) Targeted modification of malate dehydrogenase. WO Patent Publication Number: WO 2013166315 A1.
Soda, N., Verma, L. and Giri, J. (2018) CRISPR-Cas9 based plant genome editing: Significance, opportunities and recent advances. *Plant Physiology and Biochemistry* 131, 2–11.
Son, E. and Lim, S.S. (2021) Consumer acceptance of gene-edited versus genetically modified foods in Korea. *International Journal of Environmental Research and Public Health* 18, 3805.
Song, J., Li, J., Sun, J., Hu, T., Wu, A. *et al.* (2018) Genome-wide association mapping for cold tolerance in a core collection of rice (*Oryza sativa* L.) landraces by using high-density single nucleotide polymorphism markers from specific-locus amplified fragment sequencing. *Frontiers in Plant Science* 9, 875.
Sultana, S., Faruque, M. and Islam, M.R. (2022) Rice grain quality parameters and determination tools: A review on the current developments and future prospects. *International Journal of Food Properties* 25, 1063–1078.
Sun, Y., Jiao, G., Liu, Z., Zhang, X., Li, J. *et al.* (2017) Generation of high-amylose rice through CRISPR/Cas9-mediated targeted mutagenesis of starch branching enzymes. *Frontiers in Plant Science* 8, 298.
Sun, Y., Zhang, X., Wu, C., He, Y., Ma, Y. *et al.* (2016) Engineering herbicide-resistant rice plants through CRISPR/Cas9-mediated homologous recombination of acetolactate synthase. *Molecular Plant* 9, 628–631.
Tabassum, J., Ahmad, S., Hussain, B., Mawia, A.M., Zeb, A. *et al.* (2021) Applications and potential of genome-editing systems in rice improvement: Current and future perspectives. *Agronomy* 11(7), 1359.

Tan, Y.Y., Du, H., Wu, X., Liu, Y.H., Jiang, M. *et al.* (2020) Gene editing: An instrument for practical application of gene biology to plant breeding. *The Journal of Zhejiang University Science B: Biomedicine & Biotechnology* 21, 460–473.

Tang, L., Mao, B., Li, Y., Lv, Q., Zhang, L. *et al.* (2017) Knockout of OsNramp5 using the CRISPR/Cas9 system produces low Cd-accumulating indica rice without compromising yield. *Scientific Reports* 7, 1–2.

Tang, X., Sretenovic, S., Ren, Q., Jia, X., Li, M. *et al.* (2020) Plant prime editors enable precise gene editing in rice cells. *Molecular Plant* 13, 667–670.

Tiwari, S. and Lata, C. (2019) Genome engineering in rice: Applications, advancements and future perspectives. In: Singh, S.P., Upadhyay, S.K., Pandey, A. and Kumar, S. (eds) *Molecular Approaches in Plant Biology and Environmental Challenges. Energy, Environment, and Sustainability*. Springer Nature, Singapore 15, 323–327

Tröder, S.E. and Zevnik, B. (2022) History of genome editing: From meganucleases to CRISPR. *Laboratory Animals* 56(1), 60–68.

Usman, B., Nawaz, G., Zhao, N., Liao, S., Liu, Y. *et al.* (2020a) Precise editing of the *OsPYL9* gene by RNA-guided Cas9 nuclease confers enhanced drought tolerance and grain yield in rice (*Oryza sativa* L.) by regulating circadian rhythm and abiotic stress responsive proteins. *International Journal of Molecular Sciences* 21, 7854.

Usman, B., Nawaz, G., Zhao, N., Liao, S., Qin, B. *et al.* (2020b) Programmed editing of rice (*Oryza sativa* L.) *OsSPL16* Gene Using CRISPR/Cas9 improves grain yield by modulating the expression of pyruvate enzymes and cell cycle proteins. *International Journal of Molecular Sciences* 22, 249.

Usman, B., Nawaz, G., Zhao, N., Liu, Y. and Li, R. (2020) Generation of high yielding and fragrant rice (*Oryza sativa* L.) lines by CRISPR/ Cas9 targeted mutagenesis of three homoeologs of cytochrome *P450* gene family and *OsBADH2* and transcriptome and proteome profiling of revealed changes triggered by mutations. *Plants* 9, 788.

Viana, V.E., Pegoraro, C., Busanello, C. and Costa de, A. (2019) Mutagenesis in rice: The basis for breeding a new super plant. *Frontiers in Plant Science* 10, 1326.

Wada, N., Ueta, R., Osakabe, Y. and Osakabe, K. (2020) Precision genome editing in plants: State-of-the-art in CRISPR/Cas9-based genome engineering. *BMC Plant Biology* 20, 1–2.

Wang, C., Liu, Q., Shen, Y., Hua, Y., Wang, J. *et al.* (2019) Clonal seeds from hybrid rice by simultaneous genome engineering of meiosis and fertilization genes. *Nature Biotechnology* 37, 283–286.

Wang, M., Mao, Y., Lu, Y., Tao, X. and Zhu, J. (2017) Multiplex gene editing in rice using the CRISPR-Cpf1 system. *Molecular Plant* 10, 1011–1013.

Wang, C., Shen, L., Fu, Y., Yan, C. and Wang, K. (2015) A simple CRISPR/Cas9 system for multiplex genome editing in rice. *Journal of Genetics and Genomics* 42, 703–706.

Wang, F., Wang, C., Liu, P., Lei, C., Hao, W. *et al.* (2016) Enhanced rice blast resistance by CRISPR/ Cas9-targeted mutagenesis of the ERF transcription factor gene *OsERF922*. *PloS One* 11, e0154027.

Wang, S., Yang, Y., Guo, M., Zhong, C., Yan, C. *et al.* (2020) Targeted mutagenesis of amino acid transporter genes for rice quality improvement using the CRISPR/Cas9 system. *The Crop Journal* 8, 457–464.

Wiedenheft, B., Sternberg, S.H. and Doudna, J.A. (2012) RNA-guided genetic silencing systems in bacteria and archaea. *Nature* 482(7385), 331–338.

Xie, K. and Yang, Y. (2013) RNA-guided genome editing in plants using a CRISPR–Cas system. *Molecular Plant* 6, 1975–1983.

Xiong, X., Liang, J., Li. Z., Gong, B.Q. and Li, J.F. (2021) Multiplex and optimization of dCas9-TV-mediated gene activation in plants. *Journal of Integrative Plant Biology* 63, 634–645.

Xu, R., Li, J., Liu, X., Shan, T., Qin, R. *et al.* (2020) Development of plant prime-editing systems for precise genome editing. *Plant Communications* 1, 100043.

Xu, R.F., Li, H., Qin, R.Y., Li, J., Qiu, C.H. *et al.* (2015) Generation of inheritable and "transgene clean" targeted genome-modified rice in later generations using the CRISPR/Cas9 system. *Scientific Reports* 5, 1–10.

Xu, R., Li, H., Qin, R., Wang, L., Li, L. *et al.* (2014) Gene targeting using the *Agrobacterium tumefaciens*-mediated CRISPR-Cas system in rice. *Rice* 7, 5.

Xu, Y., Lin, Q., Li, X., Wang, F., Chen, Z. *et al.* (2021) Fine-tuning the amylose content of rice by precise base editing of the *Wx* gene. *Plant Biotechnology Journal* 19, 11–13.

Xu, Z., Xu, X., Gong, Q., Li, Z., Li, Y. *et al.* (2019) Engineering broad-spectrum bacterial blight resistance by simultaneously disrupting variable TALE-binding elements of multiple susceptibility genes in rice. *Molecular Plant* 12, 1434–1446.

Xu, R., Yang, Y., Qin, R., Li, H., Qiu, C. *et al.* (2016) Rapid improvement of grain weight via highly efficient CRISPR/Cas9-mediated multiplex genome editing in rice. *Journal of Genetics and Genomics* 43, 529–532.

Xu, W., Zhang, C., Yang, Y., Zhao, S., Kang, G. *et al.* (2020) Versatile nucleotides substitution in plant using an improved prime editing system. *Molecular Plant* 13, 675–678.

Xu, C., Zhou, Y., Xiao, Q., He, B., Geng, G. *et al.* (2021) Programmable RNA editing with compact CRISPR–Cas13 systems from uncultivated microbes. *Nature Methods* 18(5), 499–506.

Yang, X., Chen, L., He, J. and Yu, W. (2017) Knocking out of carotenoid catabolic genes in rice fails to boost carotenoid accumulation, but reveals a mutation in strigolactone biosynthesis. *Plant Cell Reports* 36, 1533–1545.

Yang, J., Luo, D., Yang, B., Frommer, W.B. and Eom, J.S. (2018) *SWEET 11* and 15 as key players in seed filling in rice. *New Phytologist* 218(2), 604–615.

Yin, X., Biswal, A.K., Dionora, J., Perdigon, K.M., Balahadia, C.P. *et al.* (2017) CRISPR-Cas9 and CRISPR-Cpf1 mediated targeting of a stomatal developmental gene *EPFL9* in rice. *Plant Cell Reports* 36, 745–757.

Yu, J., Han, J., Kim, Y.J., Song, M., Yang, Z. *et al.* (2017) Two rice receptor-like kinases maintain male fertility under changing temperatures. *Proceedings of the National Academy of Sciences* 114, 12327–12332.

Yue, E., Cao, H. and Liu, B. (2020) *OsmiR535:* A potential genetic editing target for drought and salinity stress tolerance in *Oryza sativa*. *Plants* 9, 1337.

Zafar, K., Khan, M.Z., Amin, I., Mukhtar, Z., Yasmin, S. *et al.* (2020) Precise CRISPR-Cas9 mediated genome editing in super basmati rice for resistance against bacterial blight by targeting the major susceptibility gene. *Frontiers in Plant Science* 11, 575.

Zaidi, S.S.E.A., Mahfouz, M.M. and Mansoor, S. (2017) CRISPR-Cpf1: A new tool for plant genome editing. *Trends in Plant Science* 22(7), 550–553.

Zeng, X., Luo, Y., Vu, N.T., Shen, S., Xia, K. *et al.* (2020) CRISPR/ Cas9-mediated mutation of *OsSWEET14* in rice cv. Zhonghua11 confers resistance to *Xanthomonas oryzae* pv. *oryzae* without yield penalty. *BMC Plant Biology* 20, 1–11.

Zeng, Y., Wen, J., Zhao, W., Wang, Q. and Huang, W. (2020) Rational improvement of rice yield and cold tolerance by editing the three genes *OsPIN5b*, *GS3*, and *OsMYB30* with the CRISPR–Cas9 system. *Frontiers in Plant Science* 10, 1663.

Zetsche, B., Gootenberg, J.S., Abudayyeh, O.O., Slaymaker, I.M., Makarova K.S. *et al.* (2015) Cpf1 is a single RNA-guided endonuclease of a class 2 CRISPR-Cas system. *Cell* 163(3), 759–771.

Zhang, J., Zhang, H., Botella, J.R. and Zhu, J.K. (2018) Generation of new glutinous rice by CRISPR/ Cas9-targeted mutagenesis of the Waxy gene in elite rice varieties. *Journal of Integrative Plant Biology* 60, 369–375.

Zhang, A., Liu, Y., Wang, F., Li, T., Chen, Z. *et al.* (2019) Enhanced rice salinity tolerance via CRISPR/ Cas9-targeted mutagenesis of the *OsRR22* gene. *Molecular Breeding* 39, 1–10.

Zhang, R., Chen, S., Meng, X., Chai, Z., Wang, D. *et al.* (2021) Generating broad-spectrum tolerance to ALS-inhibiting herbicides in rice by base editing. *Science China Life Sciences* 64, 1624–1633.

Zhang, H., Zhang, J., Wei, P., Zhang, B., Gou, F. *et al.* (2014) The CRISPR/Cas9 system produces specific and homozygous targeted gene editing in rice in one generation. *Plant Biotechnology Journal* 12, 797–807.

Zhang, Y., Zhao, J., Li, Y., Yuan, Z., He, H. *et al.* (2016) Transcriptome analysis highlights defense and signaling pathways mediated by rice *pi21* gene with partial resistance to *Magnaporthe oryzae*. *Frontiers in Plant Science* 7, 1834.

Zhao, D.S., Li, Q.F., Zhang, C.Q., Zhang, C., Yang, Q.Q. *et al.* (2018) *GS9* acts as a transcriptional activator to regulate rice grain shape and appearance quality. *Nature Communications* 9, 1–4.

Zhao, F.J. and Wang, P. (2020) Arsenic and cadmium accumulation in rice and mitigation strategies. *Plant Soil* 446, 1–21.

Zhao, L., Zhao, C., Zhou, L., Zhao, Q., Zhu, Z. *et al.* (2021) QTL mapping for starch paste viscosity of rice (*Oryza sativa* L.) using chromosome segment substitution lines derived from two sequenced cultivars with the same Wx allele. *BMC Genomics* 22, 1–2.

Zheng, N., Li, L. and Wang, X. (2020) Molecular mechanisms, off-target activities, and clinical potentials of genome editing systems. *Clinical and Translational Medicine* 10, 412–426.

Zheng, X., Yang, S., Zhang, D., Zhong, Z., Tang, X. *et al.* (2016) Effective screen of CRISPR/Cas9-induced mutants in rice by single-strand conformation polymorphism. *Plant Cell Reports* 35, 1545–1554.

Zhong, Z., Zhang, Y., You, Q., Tang, X., Ren, Q. *et al.* (2018) Plant genome editing using FnCpf1 and LbCpf1 nucleases at redefined and altered PAM sites. *Molecular Plant* 11, 999–1002.

Zhou, H., He, M., Li, J., Chen, L., Huang, Z. *et al.* (2016) Development of commercial thermo-sensitive genic male sterile rice accelerates hybrid rice breeding using the CRISPR/Cas9-mediated *TMS5* editing system. *Scientific Reports* 6, 1–2.

Zhou, X., Liao, H., Chern, M., Yin, J., Chen, Y. *et al.* (2018) Loss of function of a rice TPR-domain RNA-binding protein confers broad-spectrum disease resistance. *Proceedings of the National Academy of Sciences* 115, 3174–3179.

Zhou, H., Liu, B., Weeks, D.P., Spalding, M.H. and Yang, B. (2014) Large chromosomal deletions and heritable small genetic changes induced by CRISPR/Cas9 in rice. *Nucleic Acids Research* 42, 10903–10914.

Zhou, J., Peng, Z., Long, J., Sosso, D., Liu, B.O. *et al.* (2015) Gene targeting by the TAL effector *PthXo2* reveals cryptic resistance gene for bacterial blight of rice. *The Plant Journal* 82, 632–643.

Zhou, J., Xin, X., He, Y., Chen, H., Li, Q. *et al.* (2019) Multiplex QTL editing of grain-related genes improves yield in elite rice varieties. *Plant Cell Reports* 38, 475–485.

Zhu, C., Bortesi, L., Baysal, C., Twyman, R.M., Fischer, R. *et al.* (2017) Characteristics of genome editing mutations in cereal crops. *Trends in Plant Science* 22, 38–52.

Zhu, H., Li, C. and Gao, C. (2020) Applications of CRISPR–Cas in agriculture and plant biotechnology. *Nature Reviews Molecular Cell Biology* 21(11), 661–677.

Zong, Y., Wang, Y., Li, C., Zhang, R., Chen, K. *et al.* (2017) Precise base editing in rice, wheat and maize with a Cas9-cytidine deaminase fusion. *Nature Biotechnology* 35, 438–440.

16 Plant adaptability to climate change and drought stress for crop growth and production

Guillermo Gomer Cotrina Cabello[1], Alfonso Ruiz Rodriguez[2], Aqarab Husnain Gondal[3]*, Franklin Ore Areche[2], Denis Dante Corilla Flores[2], Jhon Adolfo Quincho Astete[4], Becquer Frauberth Camayo-Lapa[5], Rafael Julian Malpartida Yapias[4], Abdul Jabbar[6], José Yovera Saldarriaga[7], William Herminio Salas-Contreras[2] and Dante Daniel Cruz Nieto[8]

[1]Daniel Alcides Carrión National University, Cerro de Pasco, Peru; [2]National University of Huancavelica, Huancavelica, Peru; [3]Institute of Soil and Environmental Sciences, University of Agriculture Faisalabad, Punjab, Pakistan; [4]Altoandina National Autonomous University of Tarma, Tarma, Peru; [5]National University of Central Peru, Huancayo, Peru; [6]Department of Citrus Research Institute Sargodha, Sargodha, Punjab, Pakistan; [7]Santiago Antúnez of Mayolo National University, Peru; [8]José Faustino Sánchez Carrión National University, Huacho, Peru

Abstract

Abiotic factors pose a significant constraint for food security and agricultural production worldwide, and the issue has been exacerbated by extreme and rapid climate change. Heat and drought are the most important limiting factors that have a significant influence on crop growth and production. For better management, it is critical to understand the biochemical, ecological and physiological responses to these stresses. Plant responses to these challenges may be divided into three categories: phonological, physiological and biochemical. This review gives a thorough description of plant adaptations towards drought and heat stress, with a particular emphasis on identifying similarities and variations. As a result of physical damage, biological disruption and biochemical abnormalities, suboptimal water supplies and unusual temperatures negatively impact crop development and yields. However, both of these stressors have a wide range of impacts and are thus complex to explain in terms of mechanics. More profound knowledge of how plants respond to various challenges can lead to more practical solutions and management. A distinctive aspect of the phenomenon is comparing fundamental behaviour with abiotic stresses.

Keywords: heat, drought, climate change

*Corresponding author: aqarabhusnain944@gmail.com

Introduction

Climate changing is adversely affecting agricultural crop productivity, which is a significant constraint on world food security on one side and acceleration of human population on other side (Raza *et al.*, 2019). Furthermore, according to an FAO global assessment conducted in connection with population growth, an additional 70% food will be required by 2050 (Food and Agricultural Organization (FAO), 2009). It is highly challenging for agriculturists to overcome future food uncertainty in this most populous world. Global warming and potential climatic irregularities mean crops often have an increasing number of biogenic and abiogenic stresses that significantly influence crop productivity. Heat and drought are the most severe climate change-related impacts for agriculture output and food security.

All plants have specific range of optimum temperature to grow. Temperature above this range have negative impacts on growth stages and development. The gradual rise in mean annual temperature or heat is one of the most threatening abiotic stresses. Elevated temperatures may alter the developmental and biological plant processes, which affect yield and quality of produce (Calleja-Cabrera *et al.*, 2020). Although higher temperature is beneficial for some crops of cooler regions, overall, elevated temperature has negative impact on crop production throughout the world and causes yield decline to a large extent.

Extended intervals without rain that are linked to rising temperatures are projected to cause more frequent drought. Drought is increasing worldwide, due to lower precipitation and changing rainfall pattern. At reproductive or growth phases, drought stress has the most destructive influence on crop output. Droughts impair plant growth, reproduction and physiology, which significantly impact agricultural outputs (Fahad *et al.*, 2017). They produce embolism at the xylem, which hinders water movement from the ground to the leaves, and can potentially lead to plant mortality via hydraulic collapse if uncontrolled (Sevanto, 2014). According to a study that examined data from research published between 1980 and 2015, drought has reduced wheat and maize yields by up to 40% over the world (Daryanto *et al.*, 2016). Lobell *et al.* (2011) report that the production of wheat dropped by 5.5% and the output of maize decreased by 3.8% globally between 1980 and 2011. In addition, between 1880 and 2012, the average combined ocean and terrestrial temperature rose by 0.85°C (Pachauri *et al.*, 2014). Every decade is expected from now on an average increase of 0.2°C. Increased accumulation of greenhouse gases (CO_2, CH_4, etc.) is the major contributor to global climate change. A 30–150% increase in CO_2 and methane concentrations has been seen over the last 250 years (Prentice *et al.*, 2001). More than any other environmental factor, these stresses limit plant development and output. For instance, it was projected that every 1°C rise in temperature would result in a 6% loss in global wheat yield. Different models predict a yield loss of 6–10% for every 1°C increase in the average temperature of the growing season, because of these two and other stressors (Guarino and Lobell, 2011). Even though rising temperatures are favourable for crop productivity in certain colder parts globally, generally, effects on world nutrition safety remain unfavourable. Furthermore, climate change poses a major effect on the stability of agricultural production. The top ten global crops, which include sugarcane, palm, soybean, wheat, maize and rice, have seen a significant impact on yields in many places throughout the world (Ray *et al.*, 2019).

Drought vs. Climatic Stress

Drought, excessive salinity and cold may all cause a cellular dehydration that manifests as symptoms of physiological dehydration. Drought stress occurs when soil and atmospheric humidity are low and the ambient air temperature is above normal. In this situation, the evapotranspiration flux is out of balance with the soil's absorption of water (Suzuki *et al.*, 2016). When the temperature of the soil and air rises over a certain threshold level for a certain period of time, it is referred to as heat stress. Temperature affects yields in a comprehensive multi-location research. Plant growth and agricultural yield of important crops are being adversely affected by these unfavourable circumstances, which cause the development of drought-prone regions (Suzuki *et al.*, 2016; Ray *et al.*, 2019). Both drought and heat must be evaluated together since their combined impact is greater than that of each of them alone. Numerous genes influence the body's response to abiotic stress, such as heat or drought, whereas the underlying processes are more complicated than those for biotic stress, which tend to be characterized by monogenic resistance. Heat and drought responses may be complicated further by various environmental factors, both biotic and abiotic, which make research more difficult. Due to oxidative, osmotic and thermal stressors caused by water scarcity and soil salinity, these issues are indisputably important obstacles to production in agriculture (Barnabás *et al.*, 2008). It has also been noted that the lower stomatal conductance and transpiration under these circumstances may cause heat stress when leaf temperatures increase. Growth and performance of plants deteriorate significantly in tropical and subtropical areas when there is a lack of water and high temperatures (Suzuki *et al.*, 2016).

Direct and Indirect Impacts of Heat and Drought on Plant Parts

Droughts affect plants when root water supplies are insufficient or water loss is excessive due to transpiration. Drought damage is frequently unanticipated due to a number of factors such as evapotranspiration losses, rainfall patterns and water storage capacity. Drought alters the connections between nutrients, growth and water, assimilates partitioning and photosynthesis, and considerably reduces crop yield. According to Demirevska *et al.* (2009), reaction of plants to water scarcity differs across plant species, depending on the plant's stage of growth and development and other environmental variables. Furthermore, it also disrupts the hormonal balance of the plants. Excessive reactive oxygen production, which results in oxidative stress, is one of the primary consequences of heat stress (Hasanuzzaman *et al.*, 2012). The primary variables that diminish plant yield are a decrease in the soil humidity level and reduction in PAR absorption, lower efficiency of radiation and a low harvest index (Earl and Davis, 2003). To deal with the severe consequences of drought stress, plants show adaptation in their physiological processes and growth (Wang *et al.*, 2020). Climate change and drought alter plant function in number of ways as shown in Fig. 16.1.

The biochemical, hormonal, physiological and morphological changes caused by heat stress significantly affect plant growth and production. Increases in atmospheric temperature, referred to as heat shocks, have become the most significant limiting factors in agricultural output on a global scale. The changing seasons and distribution of crops may be affected by increased temperature (Porter, 2005). Proteins may be severely damaged, their synthesis disrupted, necessary enzymes inactivated and membranes damaged due to heat stress. Heat stress may affect cell division considerably as well (Smertenko *et al.*, 1997). All of these variables may impair plant growth and possibly contribute to oxidative damage. Additionally, short-term heat exposure during seed filling may result in rapid seed filling, resulting in low quality and decreased yield. We looked at the primary responses of crop plants to drought and climate change or heat stress. Aquaporins, which are integral membrane proteins that serve as channels for the transport of specific tiny solutes and water, regulate all of these parameters. Furthermore, heat and drought reduced the internal CO_2 contents of the plants, vegetative growth, and early maturity and inhibits the enzyme activities including Rubisco involved in ATP synthesis. Severe heat and drought stop the movement of water resulting in less nutrient uptake and almost all the plant parts damaged (Orians *et al.*, 2019). Indirect effects included weeds that are already responsible for 34% of crop losses, with insects accounting for 18% and disease for 16%. Weeds, insects and disease already have a huge detrimental influence on

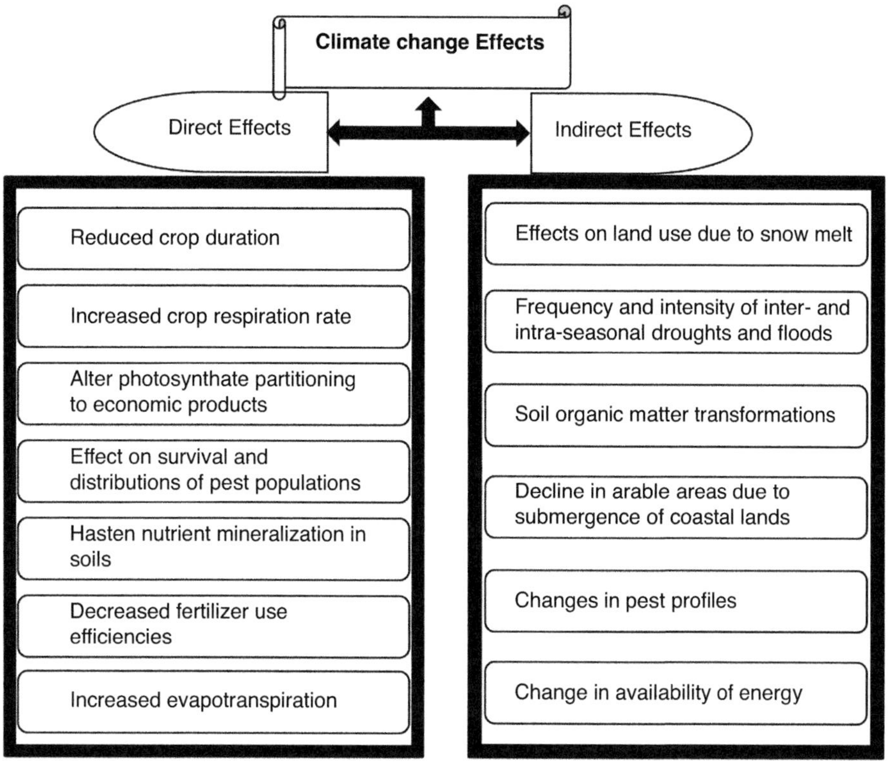

Fig. 16.1. Effects of climate change on plant parts or processes.

our agricultural production system because of climate change (Skendžić *et al.*, 2021). Insect pests thrive in warmer climates because they may complete their life cycles faster, reducing the amount of time they spend in susceptible phases (Ding *et al.*, 2011). Insect, weed and disease populations are expanding northward as a result of the warming climate. Crops are more susceptible to disease because of the increasing climate effects on temperature and moisture stress. Fig. 16.2 represents the indirect heat and drought effects on plants and soil system.

Direct and Indirect Impacts of Heat and Drought on Soil

A healthy soil ecosystem is critical to a farm or ranch's natural productivity. Extremes of temperature and precipitation may harm soils as a result of changing climatic conditions. Temperature rises may disrupt ecosystem services supplied by soil, such as water holding capacity, carbon sequestration and nutrient uptake by plants and animals. The frequency and severity of severe precipitation events has already increased and is likely to continue to rise, which will lead to increased soil erosion if conservation strategies are not implemented (Kiem *et al.*, 2016). At times when rainfall exceeds the soil capacity to infiltrate, erosion occurs. Topsoil is carried away by water that cannot penetrate the soil; thus, it flows off the land as a result. Crops can no longer rely on the water and soil that discharge after high rainstorm events. Climate change-induced shifts in rainfall patterns are expected to lead to more frequent and powerful rainstorms. In Iowa, for example, even though the state's yearly precipitation has remained flat, the number of days with significant rainfall has grown dramatically (Gelybó *et al.*, 2018). Sediment contamination is further exacerbated by

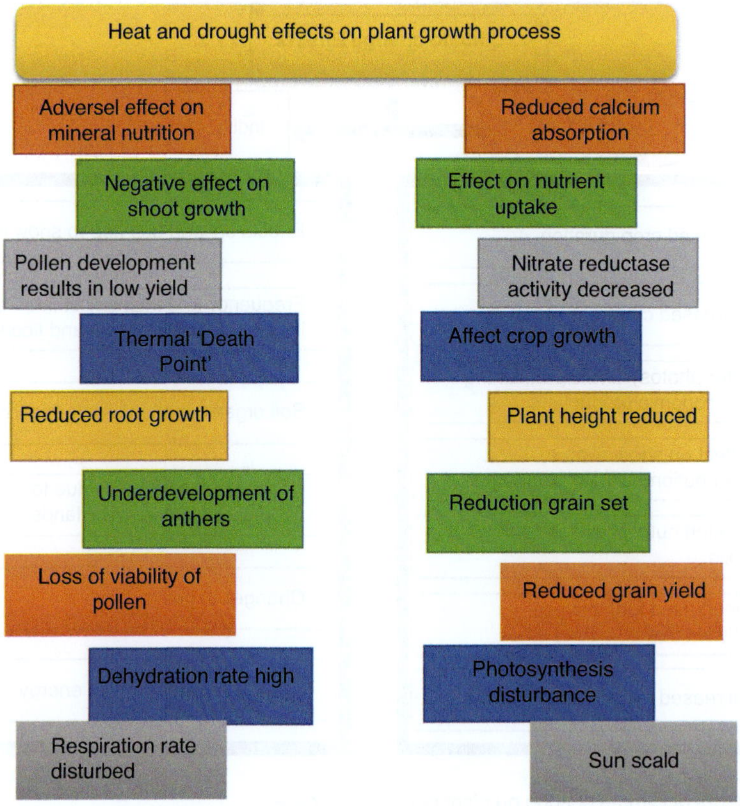

Fig. 16.2. Effects of drought and heat on plant parts and soil processes.

severe precipitation events, which causes soil erosion. Soil erosion may be reduced by using agricultural leftovers, mulch, or cover crops to keep soil covered. A rise in the frequency and severity of intense precipitation will need better soil management strategies.

Phonological Adaptation of Plants Towards Abiotic Stresses

Effects of drought on crop growth

Drought causes poor germination and seedling establishment, which are the primary causes of less seedling development from increased area. Several studies have indicated that drought stress negatively impacts planting germination and growth (Jaleel *et al.*, 2009). Drought stress has been linked to a reduced seed germination potential, hypocotyl length, early seedling, dry mass of the root, and vegetative development in various main field crops such as rice, lucerne and pea. Cell division, expansion and differentiation are the primary mechanisms by which plants grow. Drought inhibits cell elongation and mitosis, resulting in slowed growth (Hussain *et al.*, 2008). Depletion slows cell growth, owing to the reduction in turgor. Aridity inhibits cell elongation because of the restricted water transport from the xylem to neighbouring cells. Drought conditions can reduce the size and quantity of individual leaves. The amount of assimilates available and the turgor pressure determines how much the leaf expands. Water deficit inhibits leaf expansion due to a poor rate of photosynthesis and reduced turgor pressure. Fresh and dry biomasses are also severely decreased during periods of water scarcity. The height, stem, leaf size and breadth of maize plants were significantly reduced under water-stressed circumstances (Kamara *et al.*, 2003). Kamara *et al.* (2003) observed in a further investigation that the accumulation of maize biomass was significantly decreased during moisture deficient circumstances applied at numerous phases of growth.

Effects of temperature on crop growth

The additional chief restrictive factor for plant growth and improvement in hot climates is excessive radiation and high temperatures. High temperatures can cause growth inhibition, twigs and leaves to scorch, leaf senescence, and decoloration of leaves and fruits, as well as sunburn-like symptoms. High temperatures can decrease the germination potential of seeds, leading to poor germination and standing. The time, duration and degree of thermal stress affect cereal crops more strongly. Under comparable conditions, high-temperature stress lowered the spike number and florets in rice per plant, and the seed in sorghum under the same environmental circumstances was also considerably affected. Anthers and pollens in interior florets were more vulnerable to high temperatures than ovules. The sterility of the floret has been associated with inadequate pollen shedding, poor germination of pollen grains, dehiscence, reduced elongation of the pollen tube and decline in *in vivo* pollen germination at high temperatures (>30°C). In maize and sugarcane under high-temperature stress, a significant decrease was noted in growth and net assimilation rate. The deposit of biomass, internodal length and early sugarcane leaf senescence were reported to decrease significantly because of high temperatures (Ebrahim *et al.*, 1998). Sunburn of leaves, leaf abscission and senescence, scorching, fruit damage and discoloration, twigs, root, stem and shoot growth inhibition are the most common phonological indications of extreme heat stress (Rodríguez *et al.*, 2005). Abnormal plants, plant emergence, lower germination rates, poor showing strength and decreased radicles are the main effects of heat stress in numerous plant species that are grown (Kumar *et al.*, 2011). The germination rate of wheat was explicitly prohibited at extremely high temperatures (45°C) and caused cell death, where the seedling rate was lowered (Cheng *et al.*, 2009). The number of tillers, plant height and biomass of rice crop reduced significantly due to heat stress (Mitra and Bhatia, 2008). Other issues include reducing the net assimilation rate that has been established in sugarcane and millet, as the major cause of the poor growth rate (Srivastava *et al.*, 2012).

Effects of drought on crop production

Yield is the result of a combination of various physiological and phonological processes. Drought stress has a deleterious impact commonly on these processes. Drought stress resulted in significant output losses in major field crops. Pre-anthesis drought lowers anthesis time, but after anthesis, drought lessened the grain filling period for cereals. Taiz and Zeiger (2006) described four major enzymes that control the grain filling process in cereals, namely adenosine diphosphate, sucrose synthase, starch synthase, starch branching enzyme and glucose pyrophosphorylase. Drought effects have been observed to reduce the activity of these enzymes, which has a negative influence on primary cereal crop yield (Ahmadi and Baker, 2001). Drought stress during flowering can cause total sterility in pearl millet caused by a disruption in assimilating transport for the developing ear (Yadav *et al.*, 2004).

A drought-related decline in yields may be due to various factors, including disturbed assimilation, reduced photosynthesis and poor flag leaf formation. Drought circumstances during the tasseling stage of maize resulted in significant yield losses (Anjum *et al.*, 2011). Drought stress in cotton caused an impressive decrease in boll production and abortion of produced bolls, affecting the fibre yield (Pettigrew, 2004). Drought conditions resulted in a significant reduction in barley grain yield, owing to a lower number of viable tillers and grains and a lower 1000-grain weight. Drought stress during the flowering stage of the pigeon pea resulted in a 50% drop in seed output (Nam *et al.*, 2001).

Effects of temperature on crop production

Almost all plant tissues are vulnerable to high heat throughout all developmental and growth phases; the multiplicative tissues are particularly sensitive, and unusual temperature fluctuations during the reproductive phase can result in the loss of the entire grain production cycle (Lobell *et al.*, 2011). Higher temperature shock waves during the propagative stage of major cereals can cause significant yield reductions in temperate sites. Heat stress has a detrimental impact on the value of the end product in oilseed and cereal crops because it reduces the oil, protein contents and starch significantly. Ferris *et al.* (1998) found that under high-temperature conditions, the grain weight (1000-grain weight) and total grain number of wheat crop decreased significantly. Heat stress affected the performance of several rice growth and yield aspects, resulting in lower rice output.

Heat stress for very short periods, during the reproductive phase, may lead to significant reduction in flowers and floral buds; however, a great disparity in sensitivity among and within plant varieties and species exists (Sato *et al.*, 2006). Lower production of flowers or flower production without seed or fruit may result from a short spell of heat during propagative stage (Maheswari *et al.*, 2012). Fahad *et al.* (2017) described that the rice tillering phase was highly responsive to higher night temperatures. Heat induction caused ample yield losses in groundnut and common beans. Heat stress has a remarkable impact on fertilization, growth of fertilized embryos and meiosis in tomato crop, resulting in significant yield losses. Heat, drought and stress reduce the growth and output of various major crops; however, the damage varies depending on the intensity of the stress and the crop's growth phase. Unfertilized embryo and pro-embryo, anomaly in style and stigmatic positions, compromised meiosis in female and male tissues, impaired pollen tube and pollen germination growth, distressed fertilization procedures, unfertilized embryo, reduced ovule viability, hindrance in growth of endosperm and condensed pollen grains engaged by the stigma are all problems of abiotic stresses including excessive heat (Cao *et al.*, 2008). Generally, the propagative stage is more susceptible to this stress, resulting in a significant loss in yield. The rise in the seasonal average temperature at 1°C has shown a drop in grain yields of cereals of 4.1–10.0% (Wang *et al.*, 2012). In relation to tolerant varieties, sensitive crop variants are more seriously damaged by heat stress. Heat stress changes the early dough, and maturity affects the kernel drying process and causes the loss of wheat grain yield (Saitoh, 2008). Heat also decreases the weight of the single kernel and contributes mostly to the loss in return (Kutcher *et al.*, 2010).

Physiological Responses

Effect of drought on nutrient and water relations

The leaf water potential, leaf stomatal conductance, transpiration rate, substomatal conductance and canopy temperature influence water relations. Drought stress affects all of these parameters in plants, although stomatal and substomatal conductance is the most valuable in this regard (Farooq *et al.*, 2009). Under drought conditions, there was a considerable decrease in transpiration rate and water potential, increased canopy and leaf temperature. The water-efficiency ratio of dry matter generated by water consumed is also a critical component in plant physiological control. This increase in water efficiency is primarily due to the accretion of dry matter by overwhelming minimum water through the closure of stomata and a reduced evaporation rate. Potato showed lower water use efficiency when faced with a water shortage early in the season, resulting in poor biomass development and yield. Drought stress has a substantial influence on the interaction between plant nutrients. Plant roots absorb several vital minerals, such as calcium, silicon, magnesium and nitrogen along with water; however, dry circumstances block the transfer of these elements by diffusion and mass flow, limiting the plant development (Barber, 1995).

Plants increase the surface area and length of their roots and change their architecture to catch less transportable nutrients. A shortage in moisture level might hamper the root development and reduce the consumption of more minor moving minerals such as phosphorous. In a plant's nutritional linkages, root-microbe interactions are equally crucial. Reduced oxygen and carbon flow to nodules, as well as nitrogen build-up owing to drought stress, decreased the capacity of some legumes to fix nitrogen. A shortage of water in the soil affects the structure and activity of soil microbial colonies, disrupting plant feeding interactions. The responses of different crop species to nutrient uptake under water stress differ. Under drought conditions, N uptake rises, P uptake decreases and potassium uptake stays unaffected. However, nutritional connections are more complex as several nutrients interactively affect one another and overall plant physiology. This element requires a thorough molecular investigation.

Heat stress

The status of water in plants is of significant relevance under changing climatic circumstances. Usually, when adequate water is available, plants tend to balance their tissue water content independently of temperature fluctuations, but the increase in temperature becomes disastrous with less water. Unfortunately, heat stress is often connected with water shortages on the soil, especially in tropical and subtropical regions. Although there was adequate amount of water in the soil, sugarcane demonstrated a dramatic loss of water content in leaf tissue when exposed to hot temperatures. It recommends that heat stress might have a detrimental impact on root conductivity. Tomato showed a comparable decrease in root conductivity and water content during heat stress. When plants are exposed to heat, water loss is often higher throughout the day owing to a higher rate of transpiration, which eventually impairs various critical plant physiological systems. Heat stress affects the bulk, volume and growth of roots, restricting nutrients and water for above-ground plant components (Huang *et al.*, 2012). There is less evidence on the direct impact of heat stress on crop nutrients.

The activity of essential food metabolic enzymes, such as nitrate reductase, can be drastically decreased under high-temperature stress. Because of several factors, including a nutrient intake per unit surface of the root, decreased root mass and nutrition intake may decrease under thermal stress (Basirirad, 2000). Overall, heat and drought stress harm the plant nutritional cycle, availability and absorption by interacting with various physiological processes. Photosynthesis in plants affected by drought and excessive heat is one of the principal physiological processes (Farooq *et al.*, 2009). Defective photosynthetic machinery functioning, reduction in leaf expansion and leaf ageing are the primary variables affecting it (Wahid *et al.*, 2007). Under drought conditions, stomatal closure limits the availability of CO_2, making plants more vulnerable to photodamage. The reduction of humidity causes unfavourable changes in photosynthetic machinery, impacts photosynthetic pigments negatively and impairs significant enzyme functions, leading to large losses in yield and growth (Zandalinas *et al.*, 2016). Heat stress would also affect photosynthetic processes by disrupting pigment photosynthesis, reducing photosystem II activity and decreasing the ability of RuBP for regeneration (Sharkey, 2005). Some of the significant impacts on photosynthesis are described below of drought and heat stress.

Effect of drought on photosynthetic pigments

Drought has various impacts on thylakoid membranes and photosynthetic pigments (Kumar *et al.*, 2011). Drought conditions have also been linked to decreased chlorophyll concentration. Several researchers have found that when cereals are subjected to moisture stress, their chlorophyll content decreases. It differs with cultivar and crops, e.g. during water stress, chlorophyll concentrations in specific cultivars of black gram raised while they dropped in other crops. The difference in behaviour was related to changes in enzymatic activities involved in chlorophyll synthesis. The quantity of chlorophyll a in the dry plant was shown to be higher than the concentration of chlorophyll b. Chlorophyll a concentration under drought-stressed plants was greater than that of chlorophyll b. The chlorophyll a/b ratio in *Brassica* species was revealed to be lower under drought conditions.

Effect of temperature on photosynthetic pigments

High-temperature exposure generally results in a decrease in chlorophyll production. Lower chlorophyll accumulation in plants may be caused by the increased breakdown and reduced chlorophyll production or a combination of both during extreme heat stress (Allakhverdiev *et al.*, 2008). Inhibition in chlorophyll synthesis under extreme heat is strictly linked to the inactivation of several enzymes. In wheat, enzymatic activities, i.e. 5-aminolevulinate dehydratase needed for pyrrol production, reduced dramatically and the chlorophyll contents in cucumber was decreased 60% at temperature 42°C due to the inhibition of 5-aminolevulinate synthesis (Ashraf and Karim, 1991). Under higher temperatures, the output of the protochlorophyllide was reduced by 70%. Heat stress enhanced the breakdown of chlorophyll a, as well as b in mature leaves (Maroco *et al.*, 1997). These effects on chlorophyll pigments and other photosynthetic machinery are thought to be related to oxidative damage. In heat-tolerant tomato and sugarcane varieties, the chlorophyll a/b ratio increased, but the chlorophyll to carotenoid ratio decreased significantly. This demonstrates that the alteration in the balance of the photosynthetic pigment also has a function in heat shock tolerance.

Effect of drought on the photosynthetic process

To minimize water loss through transpiration, practically all the plants' first and predominant response to water stress is stomatal closure. Stomatal closure may occur in response to a drop in leaf water potential or a decline in atmospheric moisture level (Fiscus *et al.*, 2005). The stomatal closure reduces CO_2 uptake, resulting in no assimilation due to oxidative stress, and also improves heat dissipation in leaves (Vu *et al.*, 1999). Interestingly, soil moisture status has a more significant impact on stomatal control than leaf water content, which might explain why stomata respond to ABA generated by roots under dry circumstances. However, under drought conditions, stomatal responses vary greatly among plant species. Reduced stomatal conductance limits photosynthesis in light drought, but reduced Rubisco activity becomes the most important factor affecting photosynthesis (Porter, 2005; Jaleh, 2017). Water scarcity induces cell shrinkage, resulting in reduced cell volume; thus, the cellular substance becomes extra sticky, leading to protein denaturation.

Higher solute concentration in the cytoplasm may also lead to ionic toxicity, which negatively influences enzymatic functions that can take part in various plant functions, including photosynthesis. The rate at which the Rubisco enzyme is created and eliminated determines its absorption in leaves. Because it has a half-life of many days, it is very stable even in times of severe water scarcity. Rubisco's diminished synthesis, on the other hand, causes severe damage due to a reduction in its small subunits (Parry *et al.*, 2003). Under drought stress, the binding of inhibitors such as 2-carboxyaribinitol 1-phosphate to the catalytic site of Rubisco is also prevalent, affecting enzyme activity (Lawlor and Cornic, 2002).

Similarly, drought and heat stress harm other essential photosynthetic enzymes. Under mild drought, reduced phosphorylation and inadequate ATP generation have been identified as the fundamental causes limiting photosynthesis (Marchand *et al.*, 2005). Under drought circumstances, the lower generation of nicotinamide adenine dinucleotide phosphate decreases ATP synthesis due to depressed regulation of the non-cyclic electron transport chain.

Effect of temperature on the photosynthetic process

Higher temperature affects the light-dependent chemical processes in the carbon metabolism and thylakoid in the stroma, which are significant sources of damage (Haag, 2019). The membrane of the thylakoid is extremely vulnerable to heat stress and in chloroplasts, major changes occur, such as changes in thylakoids' structural organization, loss of granum stacking and swelling of grana (Rivas and Barber, 1997). The PSII's thermo-tolerance adjustment is controlled by enhanced photon flux density and leaf temperature. The PSII is very sensitive to temperature, and its activity is considerably involved even in partial heat stress. Under higher temperatures, the oxygen-evolving complex is also severely damaged, leading to an uneven distribution of electrons to the PSII acceptor site. Denaturation of the D_1 and D_2 proteins occurs at higher temperatures (Toth *et al.*, 2005). Heat stress caused damage to several PSII components in barley and wheat (Yang *et al.*, 2006). Cotton's photosynthetic activity was further hampered by a halt in RuBP regeneration and a reduction in the electron transport chain. The PSII stromal enzymes and chloroplast are mostly stable at high temperatures, and the PSII electron transmission system is functioning. In recent research, Fahad *et al.* (Ray *et al.*, 2019) have discovered that the photosynthetic activity of two rice cultivars has declined dramatically at high day and night temperatures. High temperature has more impact on the photosynthetic efficiency of C_3 plants than that of C_4 plants (Rogiers *et al.*, 2012).

Photosynthesis was reduced due to chlorophyll pigment degradation, a blockage of the PSII reaction center, a drop in quantum efficiency, a drop in leaf nitrogen contents, electron transport and down-regulation of PSII photochemistry. High-temperature stress has a significant impact on starch and sucrose synthesis because key enzymes such as sucrose phosphate adenosine, diphosphateglucose pyrophosphorylase, invertase and synthase have lower activity. Due to reducing the active state of the CO_2 binder enzyme Rubisco, net photosynthesis is hampered for some plant species. At the same time, the catalytic activity of Rubisco increases with temperature, and its weak affinity to CO_2 and its ability to bind to O_2 limit the increase in its net photosynthetic rate. Despite all of the harmful consequences of higher temperatures on photosynthesis, the ideal temperature needs for photosynthesis are anticipated to rise as CO_2 levels increase in the atmosphere. Rogiers *et al.* (2012) noticed a decline of 60% with an increase from 25 to 45 °C temperature in average rates of photosynthesis of *Vitis vinifera* leaves. This photosynthetic decrease has been related to stomach closure of 15–30%.

Effect of drought on assimilate partitioning

Drought disturbs the balance of assimilated products, as most are moved into the roots to improve water consumption. The rate of photosynthesis and the concentration of sucrose in the leaves are used to transport assimilates from the source to the sink. Dryness impacts photosynthesis and reduces the sugar content, reducing the export rate between sources. Drought also reduces the sink's ability to digest incoming agents appropriately. The invertase acid activity that impacts phloem loading and unloading is also decreased. As a result, moisture stress affects dry matter partitioning (Dinar and Rudich, 1985).

Effect of temperature on assimilate partitioning

The heat stress reduces the sources and sinks, which have substantial impacts on growth and, in the end, economic production. In assimilating heat stress partitioning, however, considerable variation in distinct genotypes of wheat was reported. In two heat-sensitive tomato cultivars, Dinar and Rudich (1985) found carbon transportation to the apex was drastically reduced.

Wardlaw (1974) examined the reactions of the source of wheat, the sink and the transportation track to high-temperature stress and found the photosynthetic rate to be ideal between 20 and 30°C, but a rapid drop over 30°C was observed. The flags were loaded according to the same manner (Tietjen *et al.*, 2017). The motion inside the trunk, however, showed that the temperature was 1 to 50°Cindependent. It was discovered that, in addition to a lower photosynthetic rate, the impact of temperature stress on the wheat division was associated with the questionable source and sink activity. From these results, it may be concluded that improving the mobility efficiency of leaf and other plant components of assimilates might constitute an essential strategy to enhance grain filling in cereals.

Oxidative Damage: A Communal Response

Most abiotic stresses in plants are usually in a sequential stage of oxidative damage. Drying stress causes oxidative damage to the plants by allowing reactive oxygen species (ROS) to grow. ROS represents a significant danger that proteins and lipids will be destroyed for cell functioning. According to Moran *et al.* (1994), during drought conditions, the pea plant's protein and lipid peroxidation increased fourfold compared to normal conditions. The ROS is primarily produced in chloroplasts; however, oxygen interactions with electron transport chain (ETC) elements in mitochondria also produce ROS. ROS-generating processes can be either non-enzymatic or enzymatic. Under high-temperature stress, ROS generation has also been documented. Plants often rely on antioxidant defence, which can be non-enzymatic or enzymatic, to deal with oxidative stress. Enzymatic protection is often regarded as the most effective. The POD, GR, CAT and SOD are the major enzymes involved in this system. Aside from these enzymes, non-enzymatic components such as glutathione and carotenoid can also contribute to the antioxidant system. Enzymes such as SOD, CAT and POD either directly scavenge ROS or indirectly protect plants by controlling non-enzymatic defence (Pettigrew, 2004). An increased level of malondialdehyde has been seen in response to ROS, which is a strong indicator of drought-induced oxidative damage. As a result, maintaining more significant levels of antioxidants can be a practical approach for plants to combat the detrimental effects of ROS. Plants use phytohormones, which are natural defensive compounds that help them sustain more significant amounts of antioxidants under stress conditions. They aid plant adaptation towards the varying environment by modulating growth, source or sink transitions, nutrient allocation and development (Fahad *et al.*, 2017).

Conclusion

Abiotic stresses are reducing crop yield all across the world. Drought and heat stress cause plants to respond in a variety of ways, the most notable of which is by altering their development and morphology. Despite the fact that drought and heat stress may have a negative impact on the plant's growth and development, reproductive growth is the most affected. Anthesis or grain filling stress may have a major impact on crop production if it is mild. Damage to the photosynthetic machinery, oxidative stress and membrane instability are also caused by these forces. The capacity of plants to withstand these pressures differs significantly across species. Recent advances have been achieved in limiting the adverse consequences of these abiotic stressors, either through the use of genetic methods or by the induction of stress tolerance.

Acknowledgements

The authors are highly thankful to Institute of Soil and Environmental Sciences for their moral support.

References

Ahmadi, A. and Baker, D.A. (2001) The effect of water stress on the activities of key regulatory enzymes of the sucrose to starch pathway in wheat. *Plant Growth Regulation* 35, 81–91.

Allakhverdiev, S.I., Kreslavski, V.D., Klimov, V.V., Los, D.A., Carpentier, R. *et al.* (2008) Heat stress: An overview of molecular responses in photosynthesis. *Photosynthesis Research* 98, 541–550.

Anjum, S.A., Xie, X.Y., Wang, L.C., Saleem, M.F., Man, C. *et al.* (2011) Morphological, physiological and biochemical responses of plants to drought stress. *African Journal of Agricultural Research* 6(9), 2026–2032.

Ashraf, M. and Karim, F. (1991) Screening of some cultivars/lines of black gram (*Vigna mungo* L. Hepper) for resistance to water stress. *Tropical Agriculture* 68, 57–62.

Barber, S.A. (1995) *Soil Nutrient Bioavailability: A Mechanistic Approach*, 2nd edn. Wiley, New York, NY.

Barnabás, B., Jäger, K. and Fehér, A. (2008) The effect of drought and heat stress on reproductive processes in cereals. *Plant, Cell & Environment* 31(1), 11–38.

Basirirad, H. (2000) Kinetics of nutrient uptake by roots: Responses to global change. *New Phytologist* 147, 155–169. DOI: 10.1046/j.1469-8137.2000.00682.x.

Calleja-Cabrera, J., Boter, M., Oñate-Sánchez, L. and Pernas, M. (2020) Root growth adaptation to climate change in crops. *Frontiers in Plant Science* 11, 544.

Cao, Y.Y., Duan, H., Yang, L.N., Wang, Z.Q., Zhou, S.C. *et al.* (2008) Effect of heat stress during meiosis on grain yield of rice cultivars differing in heat tolerance and its physiological mechanism. *Acta Agronomica Sinica* 34, 2134–2142.

Cheng, L., Zou, Y., Ding, S., Zhang, J., Yu, X. *et al.* (2009) Polyamine accumulation in transgenic tomato enhances the tolerance to high temperature stress. *Journal of Integrative Plant Biology* 51(5), 489–499.

Daryanto, S., Wang, L. and Jacinthe, P.A, P.A. (2016) Global synthesis of drought effects on maize and wheat production. *PLoS One* 11(5), e0156362.

Demirevska, K., Zasheva, D., Dimitrov, R., Simova-Stoilova, L., Stamenova, M. *et al.* (2009) Drought stress effects on Rubisco in wheat: Changes in the Rubisco large subunit. *Acta Physiologiae Plantarum* 31(6), 1129.

Dinar, M. and Rudich, J. (1985) Effect of heat stress on assimilate partitioning in tomato. *Annals of Botany* 56, 239–248.

Ding, Y., Hayes, M.J. and Widhalm, M. (2011) Measuring economic impacts of drought: A review and discussion. *Disaster Prevention and Management: An International Journal* 20(4), 434–446.

Earl, H.J. and Davis, R.F. (2003) Effect of drought stress on leaf and whole canopy radiation use efficiency and yield of maize. *Agronomy Journal* 95(3), 688–696.

Ebrahim, M.K., Zingsheim, O., El-Shourbagy, M.N., Moore, P.H. and Komor, E. (1998) Growth and sugar storage in sugarcane grown at temperature below and above optimum. *Journal of Plant Physiology* 153, 593–602.

Fahad, S., Bajwa, A.A., Nazir, U., Anjum, S.A., Farooq, A. *et al.* (2017) Crop production under drought and heat stress: Plant responses and management options. *Frontiers in Plant Science* 8, 1147.

Farooq, M., Wahid, A., Kobayashi, N., Fujita, D. and Basra, S. (2009) Plant drought stress: Effects, mechanisms and management. *Sustainable Agriculture* 29(1), 153–188.

Ferris, R., Ellis, R.H., Wheeler, T.R. and Hadley, P. (1998) Effect of high temperature stress at anthesis on grain yield and biomass of field grown crops of wheat. *Annals of Botany* 82, 631–639.

Fiscus, E.L., Booker, F.L. and Burkey, K.O. (2005) Crop responses to ozone: Uptake, modes of action, carbon assimilation and partitioning. *Plant, Cell & Environment* 28, 997–1011.

Food and Agricultural Organization (FAO) (2009). Available at: http://www.fao.org/news/archive/news-by-date/2009

Gelybó, G., Tóth, E., Farkas, C., Horel, Á., Kása, I. *et al.* (2018) Potential impacts of climate change on soil properties. *Agrokémia És Talajtan* 67(1), 121–141.

Guarino, L. and Lobell, D.B. (2011) A walk on the wild side. *Nature Climate Change* 1, 374–375. DOI: 10.1038/nclimate1272.

Haag, J. (2019) *Molecular and Biochemical Enhancement of Chlorophyll in Sports Turf*. Lulu.com.

Hasanuzzaman, M., Hossain, M.A., Silva, J.A.T. and Fujita, M. (2012) Plant responses and tolerance to abiotic oxidative stress: Antioxidant defenses is a key factor. In: Bandi, V., Shanker, A.K., Shanker, C. and Mandapaka, M. (eds) *Crop Stress and Its Management: Perspectives and Strategies*. Springer, Berlin, Germany, pp. 261–316.

Huang, B., Rachmilevitch, S. and Xu, J. (2012) Root carbon and protein metabolism associated with heat tolerance. *Journal of Experimental Botany* 63, 3455–3465.

Hussain, M., Malik, M.A., Farooq, M., Ashraf, M.Y. and Cheema, M.A. (2008) Improving drought tolerance by exogenous application of glycinebetaine and salicylic acid in sunflower. *Journal of Agronomy and Crop Science* 194(3), 193–199.

Jaleel, C.A., Manivannan, P.A., Wahid, A., Farooq, M., Al-Juburi, H.J. *et al.* (2009) Drought stress in plants: A review on morphological characteristics and pigments composition. *International Journal of Agriculture and Biology* 11, 100–105.

Jaleh, D. (2017) Metabolic adjustments, assimilate partitioning, and alterations in source-sink relations in drought-stressed plants. In: *Photoassimilate Distribution in Plants And Crops*. Routledge, pp. 407–420.

Kamara, A.Y., Menkir, A., Badu-Apraku, B. and Ibikunle, O. (2003) The influence of drought stress on growth, yield and yield components of selected maize genotypes. *The Journal of Agricultural Science* 141, 43–50.

Kiem, A.S., Johnson, F., Westra, S., van Dijk, A., Evans, J.P. *et al.* (2016) Natural hazards in Australia: Droughts. *Climatic Change* 139(1), 37–54.

Kumar, S., Kaur, R., Kaur, N., Bhandhari, K., Kaushal, N. *et al.* (2011) Heat-stress induced inhibition in growth and chlorosis in mungbean (*Phaseolus aureus* roxb.) is partly mitigated by ascorbic acid application and is related to reduction in oxidative stress. *Acta Physiologiae Plantarum* 33, 2091–2101.

Kutcher, H.R., Warland, J.S. and Brandt, S.A. (2010) Temperature and precipitation effects on canola yields in Saskatchewan, Canada. *Agricultural and Forest Meteorology* 150, 161–165.

Lawlor, D.W. and Cornic, G. (2002) Photosynthetic carbon assimilation and associated metabolism in relation to water deficits in higher plants. *Plant Cell Environment* 25, 275–294.

Lobell, D.B., Schlenker, W. and Costa-Roberts, J. (2011) Climate trends and global crop production since 1980. *Science* 333(6042), 616–620.

Maheswari, M., Yadav, S.K., Shanker, A.K., Kumar, M.A. and Venkateswarlu, B. (2012) Overview of plant stresses: Mechanisms, adaptations and research pursuit. In: *Crop Stress and Its Management: Perspectives and Strategies*. Springer, Dordrecht, The Netherlands, pp. 1–18.

Marchand, F.L., Mertens, S., Kockelbergh, F., Beyens, L. and Nijs, I. (2005) Performance of high Arctic tundra plants improved during but deteriorated after exposure to a simulated extreme temperature event. *Global Change Biology* 11, 2078–2089.

Maroco, J.P., Pereira, J.S. and Chaves, M.M. (1997) Stomatal responses to leaf-to-air vapour pressure deficit in *Sahelian* species. *Australian Journal of Plant Physiology* 24, 381–387.

Mitra, R. and Bhatia, C.R. (2008) Bioenergetic cost of heat tolerance in wheat crop. *Current Science* 94, 1049–1053.

Moran, J.F., Becana, M., Iturbe-Ormaetxe, I., Frechilla, S., Klucas, R.V. *et al.* (1994) Drought induces oxidative stress in pea plants. *Planta* 194, 346–352.

Nam, N.H., Chauhan, Y.S. and Johansen, C. (2001) Effect of timing of drought stress on growth and grain yield of extra-short-duration pigeonpea lines. *The Journal of Agricultural Science* 136, 179–189.

Orians, C.M., Schweiger, R., Dukes, J.S., Scott, E.R. and Müller, C. (2019) Combined impacts of prolonged drought and warming on plant size and foliar chemistry. *Annals of Botany* 124(1), 41–52.

Pachauri, R.K., Allen, M.R., Barros, V.R., Broome, J., Cramer, W. *et al.* (2014) *Climate change 2014: Synthesis report*. Contribution of Working Groups I, II and III to the Fifth Assessment Report of the Intergovernmental Panel on Climate Change. IPCC, p. 151.

Parry, M., Andralojc, P., Mitchell, R.A., Madgwick, P. and Keys, A. (2003) Manipulation of rubisco: The amount, activity, function and regulation. *Journal of Experimental Botany* 54, 1321–1333.

Pettigrew, W.T. (2004) Physiological consequences of moisture deficit stress in cotton. *Crop Science* 44, 1265–1272. DOI: 10.2135/cropsci2004.1265.

Porter, J.R. (2005) Rising temperatures are likely to reduce crop yields. *Nature* 436(7048), 174–174.

Prentice, I.C., Farquhar, G.D., Fasham, M.J.R., Goulden, M.L., Heimann, M. *et al.* (2001) *The Science of Climate Change*. Contributions of Working Group I to the Third Assessment Report of the Intergovernmental Panel on Climate Change. Cambridge University Press, pp. 183–237. Available at: https://www. ipcc.ch/site/assets/uploads/2018/03/WGI_TAR

Ray, D.K., West, P.C., Clark, M., Gerber, J.S., Prishchepov, A.V. *et al.* (2019) Climate change has likely already affected global food production. *PLoS ONE* 14(5), e0217148.

Raza, A., Razzaq, A., Mehmood, S.S., Zou, X., Zhang, X. *et al.* (2019) Impact of climate change on crops adaptation and strategies to tackle its outcome: A review. *Plants* 8(2), 34.

Rivas, J.D.L. and Barber, J. (1997) Structure and thermal stability of photosystem II reaction centers studied by infrared spectroscopy. *Biochemistry* 36, 8897–8903.

Rodríguez, M., Canales, E. and Borrás-Hidalgo, O. (2005) Molecular aspects of abiotic stress in plants. *Biotecnología Aplicada* 22, 1–10.

Rogiers, S.Y., Greer, D.H., Hatfield, J.M., Hutton, R.J., Clarke, S.J. *et al.* (2012) Stomatal response of an anisohydric grapevine cultivar to evaporative demand, available soil moisture and abscisic acid. *Tree Physiology* 32, 249–261.

Saitoh, H. (2008) *Ecological and Physiology of Vegetable*. Nousangyoson Bunka Kyoukai, Tokyo, Japan.

Sato, S., Kamiyama, M., Iwata, T., Makita, N., Furukawa, H. *et al.* (2006) Moderate increase of mean daily temperature adversely affects fruit set of *Lycopersicon esculentum* by disrupting specific physiological processes in male reproductive development. *Annals of Botany* 97, 731–738.

Sevanto, S. (2014) Phloem transport and drought. *Journal of Experimental Botany* 65(7), 1751–1759.

Sharkey, T.D. (2005) Effects of moderate heat stress on photosynthesis: Importance of thylakoid reactions, rubisco deactivation, reactive oxygen species, and thermotolerance provided by isoprene. *Plant, Cell & Environment* 28, 269–277.

Skendžić, S., Zovko, M., Živković, I.P., Lešić, V. and Lemić, D. (2021) The impact of climate change on agricultural insect pests. *Insects* 12(5), 440.

Smertenko, A., Dráber, P., Viklický, V. and Opatrný, Z. (1997) Heat stress affects the organization of microtubules and cell division in *Nicotiana tabacum* cells. *Plant, Cell & Environment* 20(12), 1534–1542.

Srivastava, S., Pathak, A.D., Gupta, P.S., Shrivastava, A.K. and Srivastava, A.K. (2012) Hydrogen peroxide-scavenging enzymes impart tolerance to high temperature induced oxidative stress in sugarcane. *Journal of Environmental Biology* 33, 657–661.

Suzuki, N., Bassil, E., Hamilton, J.S., Inupakutika, M.A., Zandalinas, S.I. *et al.* (2016) ABA is required for plant acclimation to a combination of salt and heat stress. *PLoS One* 11(1), e0147625.

Taiz, L. and Zeiger, E. (2006) *Plant Physiology*, 4th edn. Sinauer Associates Inc Publishers, Sunderland, MA.

Tietjen, B., Schlaepfer, D.R., Bradford, J.B., Lauenroth, W.K., Hall, S.A. *et al.* (2017) Climate change-induced vegetation shifts lead to more ecological droughts despite projected rainfall increases in many global temperate drylands. *Global Change Biology* 23(7), 2743–2754.

Toth, S.Z., Schansker, G., Kissimon, J., Kovacs, L., Garab, G. *et al.* (2005) Biophysical studies of photosystem II-related recovery processes after a heat pulse in barley seedlings (*Hordeum vulgare* L.). *Journal of Plant Physiology* 162, 181–194.

Vu, J.C.V., Gesch, R.W., Allen, L.H., Boote, K.J. and Bowes, G. (1999) CO_2 enrichment delays a rapid, drought-induced decrease in rubisco small subunit transcript abundance. *Journal of Plant Physiology* 155, 139–142.

Wahid, A., Gelani, S., Ashraf, M. and Foolad, M.R. (2007) Heat tolerance in plants: An overview. *Environmental and Experimental Botany* 61, 199–223.

Wang, J.Z., Cui, L.J., Wang, Y. and Li, J.L. (2012) Growth, lipid peroxidation and photosynthesis in two tall fescue cultivars differing in heat tolerance. *Biologia Plantarum* 53, 247–242.

Wang, Y., Duan, L., Liu, T., Li, J. and Feng, P. (2020) A non-stationary standardized streamflow index for hydrological drought using climate and human-induced indices as covariates. *Science of the Total Environment* 699, 134278.

Wardlaw, I.F. (1974) Temperature control of translocation. In: Bielske, R.L., Ferguson, A.R. and Cresswell, M.M. (eds) *Mechanism of Regulation of Plant Growth*. Bull Royal Society, Wellington, pp. 533–538.

Yadav, R.S., Hash, C.T., Bidinger, F.R., Devos, K.M. and Howarth, C.J. (2004) Genomic regions associated with grain yield and aspects of post flowering drought tolerance in pearl millet across environments and tester background. *Euphytica* 136, 265–277.

Yang, X., Chen, X., Ge, Q., Li, B., Tong, Y. *et al.* (2006) Tolerance of photosynthesis to photoinhibition, high temperature and drought stress in flag leaves of wheat: A comparison between A hybridization line and its parents grown under field conditions. *Plant Science* 171, 389–397.

Zandalinas, S.I., Rivero, R.M., Martínez, V., Gómez-Cadenas, A. and Arbona, V. (2016) Tolerance of citrus plants to the combination of high temperatures and drought is associated to the increase in transpiration modulated by a reduction in abscisic acid levels. *BMC Plant Biology* 16, 1–16.

17 Potential of plant growth-promoting rhizobacteria to improve crop productivity and adaptation to a changing climate

Bekele Gelena Kelbessa[1], Mukesh Dubey[2], Vittoria Catara[3], Farideh Ghadamgahi[1], Rodomiro Ortiz[1] and Ramesh Raju Vetukuri[1*]

[1]Department of Plant Breeding, Swedish University of Agricultural Sciences, Lomma, SE-234 22, Sweden; [2]Department of Forest Mycology and Plant Pathology, Uppsala Biocenter, Swedish University of Agricultural Sciences, Uppsala, 75651, Sweden; [3]Department of Agriculture, Food and Environment, University of Catania, Catania, 95124, Italy

Abstract

Abiotic and biotic stresses negatively affect the growth and yield of crops worldwide. It is estimated that abiotic and biotic stresses account for global food crop losses of 50% and 30%, respectively. With the rapidly growing world population expected to reach 10 billion by 2050, the issue of food security is becoming a global concern. Climate change, which has its roots in industrialization, urbanization, and agricultural activities, further exacerbates food crop losses. Various innovative agricultural technologies, such as stress-tolerant varieties, smart irrigation systems, and shifting cropping calendars, have been used to both increase crop productivity and mitigate the effects of various stresses, but most of these strategies are time-intensive and costly. Therefore, strategies that are both sustainable and environmentally friendly, as well as economical, are required to face the challenge of providing the world's human population with adequate food. In this context, plant growth-promoting rhizobacteria (PGPR) have the potential to increase crop productivity by mitigating the effects of various abiotic and biotic stresses associated with climate change. This approach can also improve the sustainability of agroecosystems by reducing the use of unsustainable agrochemicals, which contribute to greenhouse gas emissions. This review presents current information on how PGPR can mitigate the impacts of abiotic and biotic stresses, associated with climate change, food security, as well as the mechanisms underlying PGPR-induced tolerance to these stresses.

Keywords: plant growth-promoting rhizobacteria, climate change, crop productivity, abiotic stress, biotic stress, stress-tolerance mechanisms, sustainable agriculture

Introduction

Increasing agricultural production without damaging the natural environment is one of the greatest challenges of our time and will be crucial to meeting the food needs of the world's rapidly growing population (Hannah *et al.*, 2020; Naamala and Smith, 2020). In the coming three decades, global food production should be increased by about 60–100% to meet the future

*Corresponding author: ramesh.vetukuri@slu.se

demand for food (Inbaraj, 2021; Boukhatem *et al.*, 2022). However, abiotic and biotic stresses have negatively impacted the quantity and quality of major agricultural crops such as cereals, pulses, and vegetables, resulting in decreased global food security (Syed Ab Rahman *et al.*, 2018; Majeed *et al.*, 2018; Qiao *et al.*, 2022). A recent report indicated that abiotic stresses (drought, salt, heat, nutrient deficiency, and other weather events) cause approximately 50–70% of global food losses (Kumar and Verma, 2018; Francini and Sebastiani, 2019). Most weather-dependent crops, including maize, rice, wheat, barley, millet, and sorghum, which provide about 50% of the protein and 56% of the total food energy for most of the world, are highly affected by climate change (Toulotte *et al.*, 2022). Furthermore, the increase in extreme weather disasters caused by climate change has reduced total agricultural productivity by about 21% since 1961, with tropical areas such as Africa (−34%) experiencing a greater impact than temperate regions such as Europe and Central Asia (−7.1%) (Ortiz-Bobea *et al.*, 2021).

On the other hand, plant diseases caused by various biotic agents, for example, fungi, oomycetes, bacteria, viruses, and nematodes, are a key component of the agroecosystem and can directly affect plant health to reduce crop productivity (Wang *et al.*, 2022). Recent estimates of main food crop losses caused by biotic stress have reached 20–40% (Velasquez *et al.*, 2018). Progressive climate change not only directly affects resistance mechanisms, but also indirectly influences the occurrence, biology, and speed of disease development, along with the frequency of suitable infection conditions and survival of pathogens (Sangiorgio *et al.*, 2020). However, it is important to note that climate change may have positive, neutral, and negative effects on crop disease development, as each pathogen responds differently to these changes. Together with indirect losses as a consequence of climate change, biotic stresses have emerged as a main limiting factor to achieving sustainable food production. In other words, both abiotic and biotic stress factors severely limit crop production, with climate change further exacerbating these stress factors (Yang *et al.*, 2021).

Consistent food production with minimal environmental impacts is necessary to sufficiently feed the ever-growing human population. Various techniques are being used to meet this need, including the development of stress-tolerant varieties through modern breeding and biotechnological approaches (e.g. CRISPR/Cas), smart irrigation systems, shifting cropping calendars, and agrochemicals (Fatima *et al.*, 2020; Olenska *et al.*, 2020). However, most of these techniques are time-consuming, costly, increase greenhouse gas emissions, and result in residual crop contamination that poses a risk to human and animal health (Caulier *et al.*, 2018; Fiodor *et al.*, 2021). In this regard, the use of plant growth-promoting rhizobacteria (PGPR) and the compounds they produce are considered a cost-effective and efficient biotechnological tool to improve crop productivity in various adverse environmental conditions. Although the use of PGPR is an ancient practice, scientists have given it considerable attention in the last two decades.

Review Methodology

In this review, different databases, such as the Web of Science (https://www.webofscience.com/), Scopus (https://www.scopus.com/), PubMed (https://pubmed.ncbi.nlm.nih.gov/), and Google Scholar (https://scholar.google.com/), were searched for both qualitative and quantitative data published in the last 7 years. The following search terms were used: 'plant beneficial bacteria'; 'crop productivity'; 'climate change'; 'abiotic stress'; 'biotic stress'; 'microbiome', and 'microbe assisted plant breeding'. Combinations of these terms were also used. Thousands of papers related to plant growth-promoting bacteria were found, of which 158 were included in this review. Manual searches that applied different search terms were also performed on different journal websites, e.g. Frontiers (https://www.frontiersin.org/), Elsevier (https://journalfinder.elsevier.com/), Scientific Reports (https://www.nature. com/), PLOS ONE (https://journals.plos.org/plosone/), and MDPI (https://www.mdpi.com/), in order to identify additional relevant articles that may have been missed in the first search.

Impact of Climate Change on Crop Productivity

Crop health and yield depend on both endogenous (e.g. genotype, hormone levels, and photosynthetic efficiency, among others) and exogenous (e.g. light, water, temperature, and mineral nutrients) factors (Berni *et al.*, 2022). Changes in these factors can negatively affect the physiology of plants, along with the related biochemical and molecular mechanisms, resulting in low crop production (Backer *et al.*, 2018). Drought is one of the most destructive abiotic factors, affecting 64% of the global land area and ultimately threatening agricultural productivity all over the world (Ma *et al.*, 2020; Tyagi *et al.*, 2022). Drought conditions adversely affect crop water potential, which impairs the normal physiological and biochemical (photosynthesis, respiration, and nutrient uptake) functioning of crops (Khataar *et al.*, 2018; Naseem *et al.*, 2018; Ma *et al.*, 2022; Vishnupradeep *et al.*, 2022). Furthermore, drought stress affects the photosynthetic efficiency of crops by altering the number of photosynthetic pigments and reducing the activity of various photosynthetic enzymes, for example, ribulose-1,5-biphosphate carboxylase/oxygenase (RuBisCo), fructose-1,6-biphosphatase (FBPase), and phosphoenolpyruvate carboxylase (PEPCase), along with enzymes related to nutrient uptake and transport (Jochum *et al.*, 2019; Chain *et al.*, 2020; Gupta *et al.*, 2022; Mukhtar *et al.*, 2022).

The reduction in stomatal aperture (stomatal closure) due to drought restricts the loss of water to the atmosphere and enhances plant tolerance to the onset of drought conditions. However, stomatal closure disrupts the CO_2 influx rate through the leaf mesophyll as well as the chloroplast electron transport, which impairs the photosynthetic activity of the plant (Driesen *et al.* 2020). Moreover, drought stress leads to secondary stress, called oxidative stress, during which the excessive accumulation of reactive oxygen species (ROS), such as superoxide (O_2^-), hydrogen peroxide (H_2O_2), and hydroxyl radicals (OH^-), damages membrane permeability, initiates lipid peroxidation, and degrades both nucleic acids and proteins (Abdelaal *et al.*, 2021; Singh *et al.*, 2022). Drought also negatively affects the composition of the soil microbiome due to limited nutrient availability and increased soil heterogeneity; this exacerbates plant stress (Naylor and Coleman-Derr, 2018). Taken together, drought leads to a 9–10% decline in global cereal production (Lesk *et al.*, 2016).

Soil salinity is another destructive stress factor, affecting 20% of total cultivated and 33% of irrigated land; this causes an estimated $31 billion in yield losses worldwide every year [www.fao.org/global-soil-partnership, accessed 1 July 2022] (Shilev, 2020; Zheng *et al.*, 2022). Soil salinity is caused by the excessive accumulation of water-soluble salts, such as sodium chloride (NaCl), sodium nitrate ($NaNO_3$), sodium carbonate (Na_2CO_3), and magnesium chloride ($MgCl_2$), among others; this soil salinity impacts plants by causing nutrient deficiency, imbalanced cellular homeostasis, and inhibited protein synthesis (Gupta *et al.*, 2022). According to an estimate, around 1.5 billion hectares of agricultural land worldwide are affected by soil salinity [www.fao.org/global-soil-partnership, accessed 1 July 2022].

Furthermore, soil salinity severely damages the physiology and metabolism of plants by simultaneously causing two stresses, namely, osmotic potential change and ionic toxicity. The accumulation of Na^+ and Cl^- ions outside the roots influences plant-water potential by decreasing the uptake of water and mineral nutrients (Eida *et al.*, 2019). Moreover, the accumulation of water-soluble Na^+ and Cl^- ions within plant cells (ionic toxicity) results in an imbalance of ions in the cytosol, causing increased competition for other essential nutrients such as Ca^{2+} and K^+, which are important cofactors for enzymes and protein synthesis (Singh *et al.*, 2022). As is the case in drought stress, prolonged exposure to salt stress induces oxidative stress, which leads to the excessive accumulation of ROS in plant tissues. The main downstream effect of this increase in ROS is damage to nucleic acids, membrane lipids, and proteins (Heydarian *et al.*, 2021; Ouertani *et al.*, 2022). Soil salinization not only causes global production losses for most crops, but also affects the soil microbiome and the physicochemical properties of soil (Kumawat *et al.*, 2022).

On the other hand, increasing global average temperatures, a direct result of anthropogenic greenhouse gas emission into the atmosphere, have become a major constraint to crop production. An increase in ambient temperature negatively affects the morphological, biochemical, and physiological functioning of plants, namely, plasma membrane permeability, water content, and photosynthetic activity (Mukhtar *et al.*, 2020, 2022). Moreover, heat-induced inhibition of light absorption and light distribution disrupt the electron transport chain and the photosystem complexes (PSI and PSII) of plants (Heydarian *et al.*, 2021; Mukhtar *et al.*, 2022). Heat stress has also been reported to decrease the accumulation of the protein proline, which plays a vital role in the growth and flowering of plants (Khan *et al.*, 2020). Although the effect of temperature on yields varies based on the crop and geographical area, a 1°C increase in the global average temperature has reduced the global yields of wheat, maize, and rice by 2.9 ± 2.3%, 7.1 ± 2.8%, and 5.2 ± 2%, respectively (Wang *et al.*, 2020a).

When considering environmental pollutants, the accumulation of heavy metals such as arsenic (As), cadmium (Cd), chromium (Cr), copper (Cu), mercury (Hg), lead (Pb), and cobalt (Co) through both urbanization and industrial and agricultural activities (e.g. chemical fertilizers and pesticides) has become a societal concern (Jebara *et al.*, 2021). Elevated concentrations of heavy metals in plants damage cytoplasmic enzymes, disrupt the biochemical activities of certain compounds, contaminate the food chain, and pose a serious threat to human health and the environment (Mandal *et al.*, 2021; Sepehri and Khatabi, 2021). In addition, heavy metal pollution affects soil structure, decreases microbial diversity, impairs root growth, and leads to water imbalance and nutrient deficiency, all of which reduce plant productivity (Barman *et al.*, 2022).

The impact of the above-mentioned abiotic stresses on food crop yields is expected to increase due to changing climate conditions (Saijo and Loo, 2020). For example, climate change directly or indirectly affects rainfall distribution, evapotranspiration rates, global average temperature, and soil nutrient status, all of which are key determinants of agricultural crop productivity (Fig. 17.1).

Projections of the severity of abiotic stresses under climate change estimate that more than 50% of arable land will be affected by 2050 (Bokhari *et al.*, 2019). Similarly, the high winds and frequent rainstorms associated with climate change will lead to severe soil degradation and nutrient stress, which will ultimately limit food production (Amouzou *et al.*, 2018).

Biotic Stress

Plants are targeted by various biotic agents, including fungi, oomycetes, bacteria, viruses, and nematodes; these organisms cause biotic stress by disrupting the normal physiological and metabolic mechanisms of host plants, thereby reducing the productivity of agricultural crops (Hashem *et al.*, 2019). The potato, for example, is the third most popular crop in terms of human consumption after rice and wheat. However, it is susceptible to various pathogens such as late blight, stem canker, early blight, powdery scab, and potato virus, all of which cause severe losses in quality and yield (Vilvert *et al.*, 2022). Potato late blight, which is caused by *Phytophthora infestans*, accounts for several billion dollars in yield losses worldwide each year (Yao *et al.*, 2016). Wheat, which accounts for almost one-fifth of total human dietary protein, is also significantly affected by various pathogens such as head blight, blotches, and rusts, resulting in annual yield losses of 15–20% (Figueroa *et al.*, 2018). Rice is one of the most important crops on Earth as it is consumed by half of the world's population (Sanya *et al.*, 2022). However, rice production is threatened by pathogens such as bacteria panicle blight and rice blasts, among many others (Suárez-Moreno *et al.*, 2019). The negative impact of plant diseases on food crops is expected to worsen due to climate change, mainly due to the increased emergence and virulence of pathogens as well as the reduced fitness of hosts (Burdon and Zhan, 2020). Concerning a solution to these biotic challenges, the role of PGPR in controlling pathogens in maize (Zhang *et al.*, 2021), rice (Ngalimat *et al.*, 2021), wheat (Kiani *et al.*, 2021; Li *et al.*, 2022a), potato (Vrieze *et al.*, 2018; Ghadamgahi *et al.*, 2022), and tomato

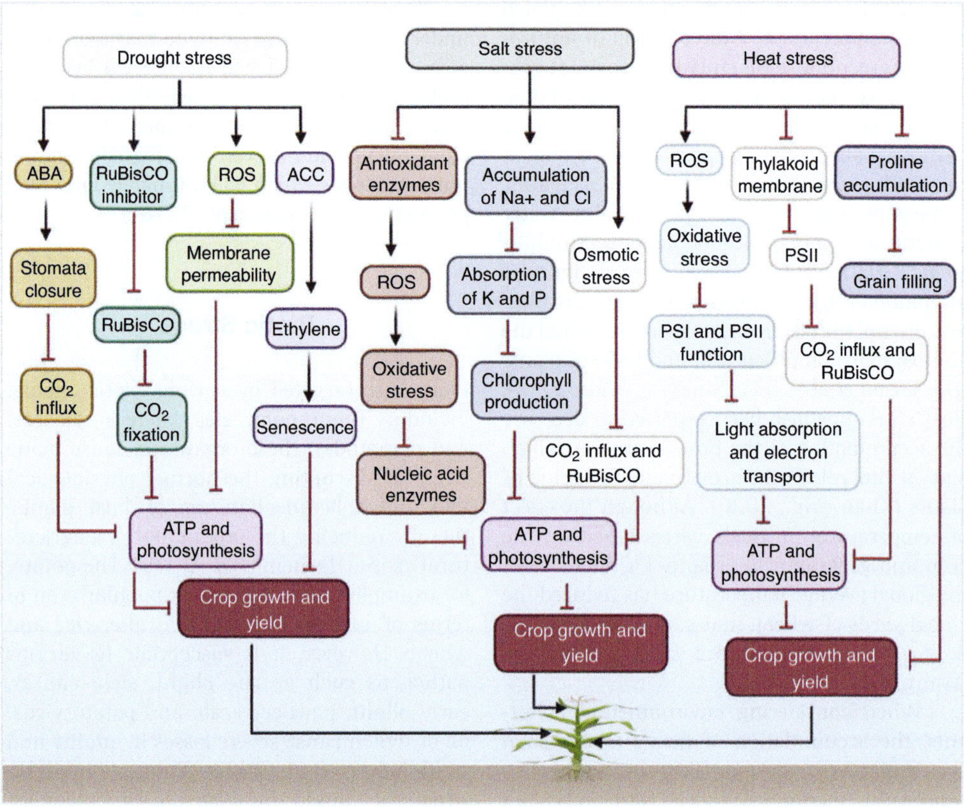

Fig. 17.1. Illustration of selected abiotic stress factors (drought, salt, and temperature) on crop growth performance and yield. Positive inhibition of plant physiological processes by the respective stress is indicated by black arrows (↑), whereas negative inhibition is represented by red, blunt-ending lines (T). The figure was created with Bio render (https://app.biorender.com/template). Regulation of physiological processes is indicated by black arrows and signaling pathways.

(Djebaili *et al.*, 2021; Minchev *et al.*, 2021) has been studied extensively.

Role of PGPR in Mitigating Climate Change Effects on Crop Productivity

All of the above-mentioned stresses affect overall crop performance; unfortunately, the current management approaches are usually cost-intensive and some stresses, such as high temperature, cannot be easily managed under field conditions. Nevertheless, the use of PGPR has been reported to enhance crop productivity by mitigating these stresses either through direct or indirect mechanisms (Batool *et al.*, 2020). In general, PGPR directly supports the well-being of crop plants by regulating growth-stimulating hormones, increasing the availability of essential nutrients (e.g. nitrogen, phosphorus), and enhancing the bioavailability of iron via the secretion of siderophores (Kudoyarova *et al.*, 2019; Wang *et al.*, 2019). The microbes described as PGPR also indirectly influence the intrinsic tolerance of crop plants to biotic and abiotic stresses by producing osmolytic substances, maintaining ion homeostasis, synthesizing extracellular polymeric substances, secreting hydrolytic enzymes, activating various antioxidant enzymes, contributing to antibiosis, inducing systemic resistance, and synthesizing pathogenesis-related proteins in plants (Ayub *et al.*, 2020; Pandey and Gupta, 2020).

Although PGPR usually utilizes multiple mechanisms to help plants tolerate biotic stresses and a changing climate, many of their activities and modes of regulation are interrelated. So far, the potential role of PGPR strains from *Bacillus*, *Burkholderia*, *Azotobactoer*, *Enterobacter*, *Pseudomonas*, *Serratia*, and *Streptomyces* genera in the management of plant stresses have been widely reported, and in some cases, the underlying mechanisms have been clarified (Dhar Purkayastha *et al.*, 2018; Bandyopadhyay *et al.*, 2022). The inoculation of cereal crops such as maize, wheat, rice, finger millet, and sorghum with single or combined bacterial strains in the form of consortia has been reported to increase crop productivity and resistance to various abiotic and biotic stresses (Albdaiwi *et al.*, 2019; Liu *et al.*, 2022). Horticultural crops such as potatoes and tomatoes also contribute to global food security, and a wide body of research indicates that the use of PGPR increases both yield performance and how well these plants tolerate various stresses (Trdan *et al.*, 2019; Batool *et al.*, 2020). However, how PGPR promotes plant growth and mitigates the negative effects of various stresses may differ across the host range as well as from strain to strain (Kohl *et al.*, 2019).

Mechanisms Employed by PGPR

Production of phytohormones

PGPR synthesize various phytohormones, namely, indole-3-acetic acid (IAA), cytokinins, gibberellins (GA), abscisic acid (ABA), 1-amino cyclopropane-1-carboxylate (ACC) deaminase, and jasmonates; these hormones are intricately involved in regulating plant growth or the response to various biotic and environmental stresses (Jochum *et al.*, 2019; Sangiorgio *et al.*, 2020; Liu *et al.*, 2022). Bacteria-synthesized IAA regulates various developmental processes in plants to enhance tolerance to various abiotic stresses (Ciftci *et al.*, 2021; Zheng *et al.*, 2022). The inoculation of plants with IAA-producing bacteria has been shown to promote lateral root growth, increase root surface area, and alter root structure in a way that benefits the uptake of nutrients and water; this could be expected to enhance crop productivity under adverse environmental conditions (Wang *et al.*, 2020b). Another study reported that inoculation with three *Bacillus* strains (viz: *Bacillus megaterium* DD-2, *Bacillus aryabhattai* DD-3, and *Bacillus subtilis* DD-4) promoted the growth of shoot and underground parts of rice via the synthesis of IAA, gibberellins, and siderophores along with the dissolution of potassium and phosphorus (Liu *et al.*, 2022).

Bacteria produced-cytokinin is another hormone that regulates many aspects of plant growth and development, e.g. cell division, vascular development, maintenance of shoot and root meristem activity, apical dominance in response to environmental stimuli, and adventitious and lateral root formation (Heydarian *et al.*, 2021; Mekureyaw *et al.*, 2022). This hormone also mediates signal exchange between plant roots and shoots when a plant is under stress (Boukhatem *et al.*, 2022). Park *et al.* (2017) reported that the cytokinin-producing *B. aryabhattai* SRB02 strain enhances the growth and tolerance capacity of soybean to oxidative, nitrosative, and temperature gradients. Similarly, the inoculation of tomato plants with the cytokinin-synthesizing *Pseudomonas fluorescens* G20-18 strain improved chlorophyll content, increased the accumulation of ABA in the roots, and increased the activity of enzymes involved in carbohydrate metabolism and antioxidant capacity; this resulted in improved tomato plant growth and tolerance to drought stress (Mekureyaw *et al.*, 2022). According to the report by Akhtar *et al.* (2020), cytokinin-producing bacteria increase crop productivity by optimizing nutrient supply and modulating host immunity when plants are exposed to selected stresses (drought, salt and heat).

Recently, Yang *et al.* (2021) showed that inoculation with a four-species bacterial consortium (viz. *Paenibacillus amylolyticus*, *Xanthomonas retroflexus*, *Microbacterium oxydans*, and *Stenotrophomonas rhizophila*) synergistically increased the drought tolerance of *Arabidopsis* by improving biofilm production, chlorophyll content, and upregulating endogenous ABA signaling. A similar result was reported when a bacterial consortium composed of *B. megaterium* CAM12 and *Pantoea agglomerans* CAH6 was used to inoculate mung bean plants; the results indicated enhanced crop drought tolerance and growth compared to single inoculation with these strains (Silambarasan *et al.*, 2019). Moreover, Vega-Celedon *et al.* (2021) presented results that a bacterial consortium consisting of *Pseudomonas* sp. TmR5a and *Curtobacterium* sp. BmP22c promoted the growth of tomato plants under normal and cold stress conditions via the synthesis of ACC deaminase and inhibition of ice recrystallization.

PGPR upregulates the biosynthesis of ABA, a hormone that enhances the tolerance ability of plants to various environmental stresses such as drought, salt, and heat by inducing ABA-responsive transcriptional factors such as basic leucine zipper transcription factor (bZIP1) and 9-cis-epoxycarotenoid dioxygenase (NCED1); both of these genes are involved in the regulation of stomatal activity (Kim *et al.*, 2022). Akhtyamova *et al.* (2021) reported that the inoculation of barley roots with *B. subtilis* IB22 resulted in ABA accumulation in the roots, which subsequently improved crop productivity. Similarly, the application of *B. subtilis* IB-22 and *Pseudomonas mandelii* IB-Ki14 to wheat plants upregulated ABA accumulation in roots, but decreased ABA accumulation in leaves, increased leaf area, reduced the loss of chlorophyll, and decreased the leaf transpiration rate (Arkhipova *et al.*, 2020).

Gibberellins are another class of phytohormones that are involved in key plant processes, including seed dormancy, seed germination, root development, flowering and fruit ripening (Binenbaum *et al.*, 2018). The production of this hormone is not limited to plants but is also a common feature of soil microbes, including PGPR. The results of numerous studies have confirmed that bacteria-produced GA_3 has a potentially important role in promoting plant growth and improving tolerance to various abiotic stresses (Kenneth *et al.*, 2019; Vanissa *et al.*, 2020). Ethylene is a gaseous hormone that at low concentrations is involved in the regulation of different plant growth stages including seed germination, formation of adventitious roots, flowering, senescence, as well as stress tolerance (Batool *et al.*, 2020; Olenska *et al.*, 2020). However, the overproduction of ethylene due to various abiotic stresses negatively affects root/shoot proliferation and accelerates senescence (Fiodor *et al.*, 2021). Many researchers have reported that PGPR participates in ethylene homeostasis under various environmental stresses through the production of the enzyme ACC deaminase (ACCD) (Pandey and Gupta, 2020; Zarei *et al.*, 2020; Singh *et al.*, 2022). Bacteria-mediated ACCD production inhibits ethylene accumulation by degrading ACC to ammonia and α-ketobutyrate; this reaction reduces the availability of substrates for ethylene formation (Jha and Mohamed, 2022). According to Tiwari *et al.* (2021), *Bacillus amyloliquefaciens*-SN13 upregulates the expression of stress-related genes (e.g. AP2/ERF, GST, ERD4, and ARF2) and alters phytohormone levels (e.g. IAA, ABA, and GA) in OsNAM-overexpressed *Arabidopsis* plants; these results suggest that *B. amyloliquefaciens*-SN13 can influence the ability of transgenic plants to mitigate salt stress.

Nutrient acquisition

Primary macronutrients, such as nitrogen (N), phosphorus (P), and potassium (K), are the main limiting factors that determine the quantity and quality of agricultural production, especially under stressful conditions (Wang *et al.*, 2020b). All plants require nitrogen for the synthesis of amino acids, nucleotides, phospholipids, and chlorophyll (Olenska *et al.*, 2020). However, the major source of nitrogen in nature, atmospheric nitrogen (N_2), is not easily accessible to plants (Rilling *et al.*, 2018). This challenge could be exacerbated by global climate change, which may decrease the biogeochemical transformation of N (Naamala and Smith, 2020). Nitrogen fixation by PGPR has emerged as a promising biological alternative to the application of N fertilizer and could lead to environmentally sustainable crop production (Mpanga *et al.*, 2019). PGPR utilizes the nitrogenase enzyme complex to reduce atmospheric N_2 to ammonia (NH_3), which can mitigate the negative effects of abiotic stress by increasing nutrient availability to plants (Aloo *et al.*, 2020). PGPR strains, especially those representing the genera *Bacillus*, *Pseudomonas*, *Serratia*, *Agrobacterium*, *Azotobacter*, *Erwinia*, *Azospirillum*, *Klebsiella*, and *Burkholderia*, have been reported to colonize various niches, for instance, the rhizosphere of plants, and contribute to meeting the nitrogen requirements of non-legumes (Martinez *et al.*, 2018; Backes *et al.*, 2021).

However, the efficiency and effectiveness of PGPR-mediated N fixation depend strongly on plant genotype, soil properties, and environmental stresses. For example, plant age and genotype had a strong influence on the structure and composition of diazotrophs colonizing plant roots. According to a study by Rodriguez-Blanco *et al.* (2015), the nitrogen-fixing bacterial community in different maize varieties varied depending on the age and genotype of the plant. Similarly, Carvalho *et al.* (2022) tested the association of nitrogen-fixing bacterial strains in different sugarcane genotypes in germinated roots and found that the nitrogen fixation efficiency of the tested bacterial strains was genotype-dependent. Furthermore, in this study, the authors also reported that genes were differentially expressed in different genotypes of sugarcane in response to association with diazotrophic bacteria. A recent report also demonstrated that the distribution of nitrogen-fixing bacterial communities and the nitrogenase-encoding gene (*nifH* gene) were strongly influenced by the different sweet potato genotypes grown in the field (Marques *et al.*, 2019). On the other hand, soil and weather factors such as fluctuations in soil pH, moisture status, nutrient availability, and temperature directly affect the growth, survival, and metabolic activity of nitrogen-fixing bacterial strains or indirectly by altering plant growth and root architecture (Lindström and Mousavi, 2020; Abdul Rahman *et al.*, 2021). However, the mechanisms and degree of nitrogen fixation efficiency affected by each stress may vary.

Phosphorus is another essential macronutrient that affects crop production in many agroecosystems (Lee *et al.*, 2019; Adnan *et al.*, 2020). Plants require the appropriate amount of phosphorus to perform various metabolic processes such as photosynthesis, biosynthesis, signal transduction, respiration, and energy transfer (Alori *et al.*, 2017). Although phosphorus plays an important role in plant metabolism, it is not readily available to plants due to poor assimilation of the dominant forms (Adnan *et al.*, 2017; Billah *et al.*, 2019). Phosphate solubilization and/or mobilization is a valuable mechanism through which PGPR enhances the bioavailability of P by secreting organic chelators such as citric, gluconic, and malic acids following the metabolization of sugars in root exudates (Lucke *et al.*, 2020). PGPR-produced organic acids are also important in increasing the bioavailability of $HPO4^{-2}$, Fe, Mn, and Zn by lowering the soil pH (Etesami and Maheshwari, 2018). Several studies have shown that bacteria from the genera *Bacillus*, *Pseudomonas*, *Serratia*, *Arthrobacter*, *Enterobacter*, and *Erwinia*, among others, increase the bioavailability of $HPO4^{-2}$ or dihydrogen phosphate ($H2PO4^{-1}$) in soil, which plants can then uptake under stress conditions (Billah *et al.*, 2019; Singha *et al.*, 2021). Adnan *et al.* (2020) reported that the inoculation of maize plants with phosphate-solubilizing bacteria improved phosphorus utilization in a salt-stressed environment. Similar results were also reported when lucerne (El Moukhtari *et al.*, 2022) and maize (Vanissa *et al.*, 2020) were inoculated with *Pseudomonas* sp., *Anthrobacter* sp., and *Bacillus* sp. in a low-phosphorus and salt-stressed environment. The inoculation of greenhouse tomato plants with a microbial consortium comprising *Achromobacter* sp., *Bacillus* sp., *Delftia* sp., and *Enterobacter* sp. improved mineral uptake and alleviated salt stress relative to the application of a single strain (Kapadia *et al.*, 2021).

Siderophore production

Iron is an essential micronutrient that has recently been in the research spotlight for its role in promoting plant growth and mitigating various environmental stresses (Patel *et al.*, 2018). Although iron regulates vital plant processes such as chlorophyll biosynthesis, photosynthesis, and respiration, the soluble form of iron (Fe^{+3}) is not readily available to plants (Kenneth *et al.*, 2019; Ojuederie *et al.*, 2019). As is the case with other essential nutrients, the lack of bioavailable iron in the soil impairs plant growth and adaptability to various stresses, especially under iron-deficient conditions. PGPR sequesters iron by producing low molecular mass compounds (200–2000 Da) called siderophores, such as carboxylates, catecholate, and hydroxamates in iron-deficient soils (Patel *et al.*, 2018; Sun *et al.*, 2022). Due to the high affinity for iron, the presence of siderophores can promote plant growth and increase the

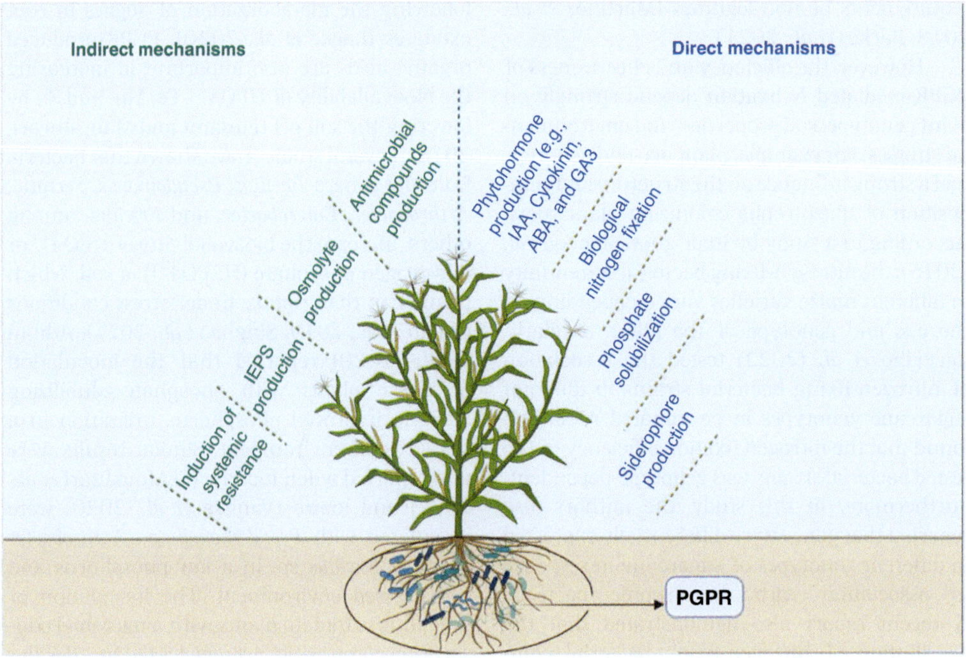

Fig. 17.2. A schematic representation of how PGPR directly and indirectly promotes sustainable crop production. The figure was created with Bio render (https://app.biorender.com/template).

survival rate of plants under climate change by sequestering iron from the surrounding plant rhizosphere (Backer *et al.*, 2018) (Fig. 17.2).

Osmotic balance

In response to abiotic stresses, PGPR can synthesize certain osmoprotectant substances, such as betaine, proline, glycine, and flavonoids (Etesami and Maheshwari, 2018; Fiodor *et al.*, 2021; Jha and Mohamed, 2022), which synergistically work with the osmolytes secreted by plants (e.g. proline, trehalose, and polyamines) to improve the survival rates of plants under fluctuating abiotic stresses. The osmolytes secreted by plants act as signaling molecules that are directly involved in the osmoregulation of plant cells, especially during extreme events such as drought, high salinity, and elevated temperatures. For instance, prior research has shown that the inoculation of plants with certain PGPR strains can increase plant tolerance to drought stress through the synthesis of osmolytes and ACCD, with the latter involved in the inhibition of ethylene signaling pathways in plant cells (Din *et al.*, 2019; Sagar *et al.*, 2020).

Furthermore, PGPR has also been reported to produce phenolic compounds, for example salicylic acid (SA), which is a signaling molecule that is activated during drought stress to upregulate various genes that regulate the synthesis of antioxidants, chaperones, and other secondary metabolites (Ojuederie *et al.*, 2019). According to Akhtar *et al.* (2020), the inoculation of maize with *Bacillus licheniforms* FMCH001 enhances the tolerance ability to drought *via* the upregulation of catalase (CAT) antioxidant activity in the roots; this dynamic will ultimately enhance crop productivity. Similarly, the treatment of quinoa plants with *Burkholderia phytofirmans* PsJN under salt stress conditions was found to increase the activity of antioxidant enzymes such as catalase and dehydroascobate reductase (Yang *et al.*, 2020). Likewise, the cultivation of chickpea plants in conjuction with *Bacillus megaterium*, *B. subtilis*, and *B. thuringiensis* PGPR strains improved plant growth and drought tolerance by increasing chlorophyll content, sugar content, root and shoot weight, yield, and nutrient content (Khan *et al.*, 2018). However, it is important to state that the response of PGPR to stress conditions is influenced by several factors, including the intensity and duration of the stress, plant age, and bacterial strain (Rahman *et al.*, 2018). PGPR has been reported to regulate water potential and stomatal opening in plants by increasing the accumulation of osmolytic compounds through hydraulic conductivity (Arkhipova *et al.*, 2020).

Ion homeostasis

PGPR maintains ionic balance through different pathways. For example, bacteria that colonize the roots help plants with salt stress in the soil via chemotaxis and the production of various compounds, such as ESP, IAA, and ACCD (Gupta *et al.*, 2022). According to Gupta and Pandey (2019), the treatment of French beans with a bacterial consortium containing *Aneurinibacillus aneurinilyticus* ACC02 and *Paenibacillus* sp. ACC06 mitigated the adverse effects of salt stress by improving ACCD activity, maintaining the K^+/Na^+ ratio and preserving the chlorophyll content. In another study, inoculation of rice with *Achromobacter xylosoxidase* NBH1, *Microbacterium esteraromaticum* JVBH1, and *Lysinibacillus* sp. BBH2 increased the productivity of rice plants under salt stress by enhancing total chlorophyll content, proline accumulation, and phenolic content (Prittesh *et al.*, 2020). Moreover, Eida *et al.* (2019) reported that the inoculation of *Arabidopsis* with five bacterial strains (viz. *Arthrobacter* sp. JZ12, *Bacillus* sp.JZ34, *Microbacterium barkeri* JZ37, *Pantoea stewartii* JZ2, and *P. stewartii* JZ29) enhanced the shoot and root growth of the plant under salt stress by reducing the Na^+/K^+ shoot ratio and inducing tissue-specific transcriptional changes in ion transporters. In addition, the inoculation of maize with *Bacillus mycoides* PM35 alleviated salt stress by increasing photosynthetic pigments, radical scavenging capacity, carotenoid levels, protein content, and soluble sugar contents (Ali *et al.*, 2022).

The potential of certain bacterial strains in improving crop productivity under salt stress has also been highlighted for barley (Azadikhah *et al.*, 2019) and sweet potato (Li *et al.*, 2022b). Nutrient imbalances are common in plant cells due to the accumulation of Na^+ ions, which compete with the K^+ ions that are important for protein synthesis (Eida *et al.*, 2019). PGPR is known to maintain ion homeostasis which is necessary for plant growth and tolerance by upregulating the K^+ protein transporter to reduce Na^+ accumulation in plant leaves (Ayub *et al.*, 2020). Bacteria-mediated stimulation of K^+ influx into plant leaves has been empirically shown to increase chlorophyll content and alleviate the ionic toxicity effect (Etesami and Maheshwari, 2018; Shilev, 2020). Thus, plants that have been inoculated with PGPR can avoid the ionic toxicity effect by maintaining a stable Na^+/K^+ ratio and increasing ion exclusion *via* the roots and regulation of ion transporter proteins.

Furthermore, numerous studies have revealed that inoculation with PGPR strains minimizes the negative effects of heat stress on plant growth and productivity through the upregulation of various antioxidant enzymes, activation of HSPs, increased production of extracellular polysaccharide (EPS), and synthesis of ACCD, which is involved in the regulation of ethylene degradation (Mukhtar *et al.*, 2020; Choi *et al.*, 2022). Some other bacterial strains were also found to markedly alleviate the adverse effect of heat stress by enhancing the accumulation of HSFs in plant tissues that regulate stress-responsive genes.

PGPR utilizes various mechanisms, e.g. metal reduction and oxidation, formation of insoluble complexes, chelation *via* siderophore production, precipitation, methylation, bioaccumulation and intracellular sequestration, to extract and immobilize heavy metals from the soil (Jebara *et al.*, 2021; Vishnupradeep *et al.*, 2022). Some PGPR strains (e.g. *Pseudomonas putida* and *Arthrobacter viscosus*) also undergo structural changes to the cell envelope, cell wall, and plasma membrane in response to heavy metal toxicity (Mandal *et al.*, 2021). These bacterial strains synthesize numerous components, such as EPS, carbohydrates, extracellular proteins, and humic acids, that interact with heavy metals and prevent them from entering the cell (Mandal *et al.*, 2021). Wei *et al.* (2022) reported that the inoculation of tomato plants with two bacterial strains (*Serratia* sp. D23 and *Sphingomonas* sp. D36) reduces Cd toxicity and concentration by modulating subcellular Cd localization, reducing oxidative stress, and upregulating stress-related gene expression.

Under situations of stress, PGPR can also reprogram plant cells by regulating metabolism and molecular signaling pathways. It has been reported that many PGPR strains increase the accumulation of various compounds, such as proteins, phenols, soluble sugars, flavonoids, and ABA, which alter the metabolic changes triggered by stress to reduce the overall effects (Bandyopadhyay *et al.*, 2022). For instance, Abd El-daim *et al.* (2019) reported that the inoculation of wheat with *Bacillus velezensis* 5113 induced metabolic and molecular reprogramming, which subsequently translated to increased plant growth and stress tolerance to heat, cold, and drought. In addition to biochemical processes, PGPR also modulate molecular mechanisms, such as the synthesis of late embryogenesis proteins (LEA) and lipochitooligosaccharides (LCOs), which leads to the production of nodulation factors, regulation of microbe-associated molecular patterns (MAMPs), as well as regulation of various stress-responsive genes (Du *et al.*, 2018; Heydarian *et al.*, 2021).

Production of extracellular polymeric substances (EPS)

To establish relationships with plant roots and the soil, PGPR releases various EPS, such as polysaccharides, polyamides, and polyesters, into the rhizosphere (Etesami and Maheshwari, 2018; Singh *et al.*, 2022). Some PGPR strains secretes EPS into the soil as mucilages ('sticky matrix'), which bind to plant roots and form biofilms to increase the surface area for plant water and nutrient uptake from the surrounding rhizosphere; these biofilms can also serve as a protective layer against adverse effects (Dhar Purkayastha *et al.*, 2018; Haque *et al.*, 2020). The EPS released by PGPR also improves soil porosity and aeration, as well as restores moisture by increasing soil particle adhesion and creating macropores (Naseem *et al.*, 2018). The bacterial-mediated production of EPS plays a crucial role in the protection of plants from ionic toxicity potentially a result of salt stress by binding cations such as Na^+ (Bhagat *et al.*, 2021). EPS can also function as signaling molecules for communication between PGPR and plant roots, and occasionally initiate defence against infection by pathogenic microbes (Saha *et al.*, 2020). Thus, bacterial inoculation has the potential to improve soil structure and reduce the initial effects of osmotic stress. Many bacterial strains have been reported to produce EPS in response to various stresses, including pathogens, drought, soil salinity, heat stress, and metal contamination (Marasco *et al.*, 2022). For example, Fatima and Arora (2021) reported that the inoculation of sunflowers with *Pseudomonas entomophila* PE3 enhances plant growth, yield, and salt-tolerance response (Fig. 17.3).

Production of antimicrobial compounds

PGPR not only mitigates plant responses to climatic stress, but also indirectly improves plant health by suppressing existing and new pathogens, which are partly as a consequence of climate change, through the production of various antimicrobial compounds such as antibiotics (e.g. bacteriocins, cyclic lipopeptides, and oomycin), hydrolytic enzymes (e.g. protease, lipase, chitinase, amylase, and beta-1,3-glucanases), and siderophore compounds (Soenens and Imperial, 2020; Narasimha Murthy *et al.*, 2021). The potential role of bacteria-produced antibiotics, hydrolytic enzymes, and siderophores in controlling plant pathogens has been studied in potatoes (Caulier *et al.*, 2018; Vrieze *et al.*, 2018), maize (Bjelić *et al.*, 2018) and wheat (Besset-Manzoni *et al.*, 2019), along with tomato (Rasool *et al.*, 2021), red pepper (Kwon *et al.*, 2022), pigeon pea (Dukare and Paul, 2021), and strawberry plants (Hong *et al.*, 2022). It has also been reported that certain halophytic bacterial strains suppress economically important pathogens, such as *Ralstonia solanacearum*, *Verticillium dahlia*, and *Clavibacter michiganensis*, under high salt stress (Hong *et al.*, 2022). It is also important to note that PGPR must compete with pathogenic microbes for C and N containing compounds to grow when colonizing the surface of roots (Kenneth *et al.*, 2019; Narasimha Murthy *et al.*,

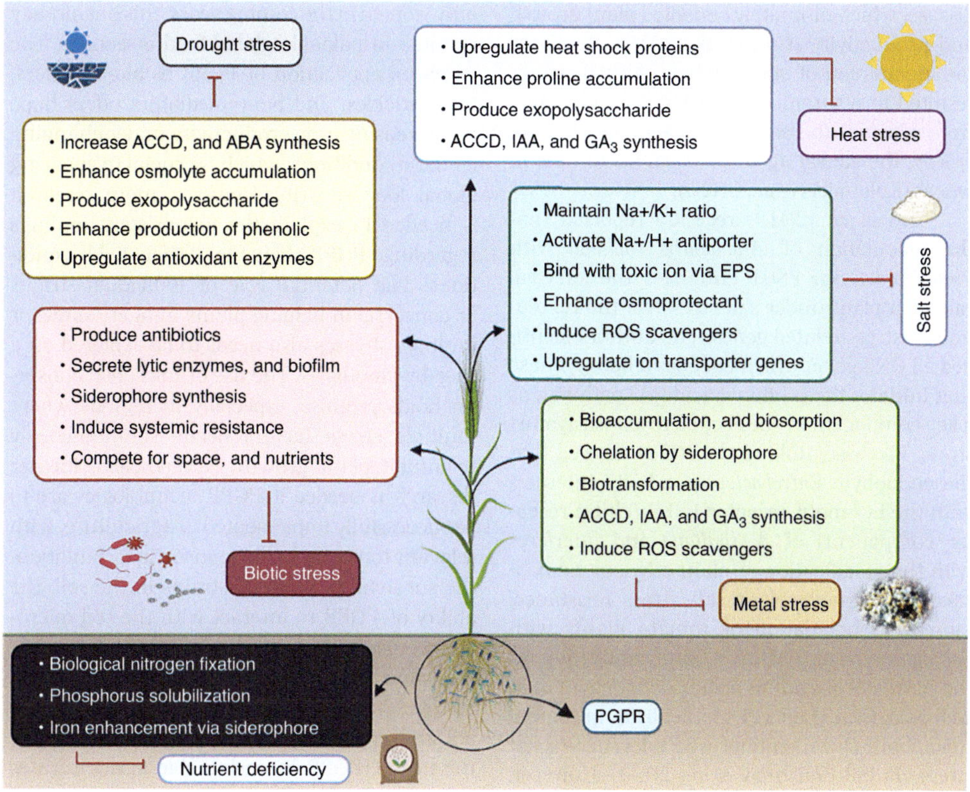

Fig. 17.3. Mechanisms by which PGPR improves plant resistance to abiotic and biotic stresses. Production of antibiotics, polysaccharides, osmolytes, different phytohormones, ROS scavengers, antioxidant enzymes, siderophores, and nutrient enhancement by PGPR is indicated by black arrows (↑), whereas negative inhibition of both abiotic and biotic stresses by PGPR-mediated compounds is indicated by red blunt-ended lines (T). The figure was created with Biorender (https://app.biorender.com/template).

2021). When competing for nutrients, PGPR secretes secondary metabolites that suppress the activity of pathogenic microbes (Lucke *et al.*, 2020). Some PGPR strains efficiently control plant diseases when used alone, while others demonstrate better performance when applied as a bacterial consortium.

Induction of stress resistance

PGPR-mediated induced systemic resistance (ISR) may lead to physiological and metabolic changes in the host plant, which will assist in responding to abiotic and biotic stresses (Kohl *et al.*, 2019). Prior results have shown that PGPR plays a substantial role in protecting plants from a wide range of stresses via ISR (Maksimov *et al.*, 2019; Baek *et al.*, 2020; Laborda *et al.*, 2020; Samaras *et al.*, 2021; Zhang *et al.*, 2021). PGPR-mediated activation of induced resistance has been found to be regulated by interconnected signaling pathways in which phytohormones, such as jasmonate and ethylene, play important regulatory roles (Backer *et al.*, 2018; Romera *et al.*, 2019). Many PGPR species representing the genera *Bacillus*, *Pseudomonas*, and *Serratia* have been reported to have roles in ISR (Romera *et al.*, 2019; Olenska *et al.*, 2020). For instance, the inoculation of tomato plants with *Bacillus amyloliquefaciens* Ba13 was found to induce systemic resistance to tomato yellow leaf virus disease, which ultimately benefited plant growth and productivity (Guo *et al.*, 2019). Similarly, the inoculation of spring wheat with *B subtilis* resulted in a situation in which 25% of plants were resistant to *Pseudomonas syringae* pv. *atrofaciens*, the causal agent of basal bacteriosis in wheat (Kolomiiets *et al.*, 2020).

Chu *et al.* (2019) recently reported that the inoculation of *Arabidopsis thaliana* with *Pseudomonas* sp. PS01 enhances the survival rate of a plant under salt stress via the regulation of stress-related genes. The authors identified a LOX2 gene, which encodes a lipoxygenase that initiates the jasmonic acid (JA) pathway, as a key component of the elicited response against stress. Likewise, Rolli *et al.* (2022) showed that the endophyte *Entrobacter* sp. SA187 interacts with the Lys-motif receptor LYK4 (chitin receptor component) of *A. thaliana* and interferes with the modification of plant ethylene homeostasis to improve plant salt stress resistance. Moreover, inoculation of tomato plants with *Sphingobacterium* BHU-AV3 significantly induced the synthesis of various antioxidant enzymes, as well as activated energy metabolism, in tomato roots, both of which improved tolerance to salt stress via ISR (Vaishnav *et al.*, 2020). However, the effectiveness of ISR against abiotic and biotic stresses depends on the bacterial strains, the ethylene and jasmonate pathway, and which stresses a plant is exposed to. There is evidence that combining several bacterial strains as a consortium may lead to stronger ISR than when a single strain is used to protect against abiotic and biotic stresses (Table 17.1); nevertheless, further investigation is needed to quantify the effect of using bacterial consortia to improve plant tolerance to biotic and abiotic stresses.

Conclusions and Prospects

The growing human population, abiotic and biotic stresses associated with climate change, and a clear reduction in available arable land are exerting significantly negative impacts on agricultural food production, which are expected to only worsen in the coming years; this has already resulted in palpable global food insecurity. The potential application of PGPR as bio-fertilizers, bio-pesticides, and bio-remediators offers hope for increasing crop productivity under changing climatic conditions, which is crucial to ensuring global food security. However, more research is needed to explain the inconsistent findings regarding PGPR benefits under field conditions. The potential role of individual strains or consortia in helping plants mitigate single or multiple stresses also needs to be explored on a case-by-case basis. The use of microbial consortia holds promise, especially in regions where multiple stress factors occur simultaneously to inhibit plant growth. Nevertheless, further research is needed if PGPR technologies are to be successfully implemented in agriculture, with relevant topics including consortia formulation, the survival of bacterial strains in the soil, the ability of PGPR to interact with the soil microbiome, appropriate compatibility with the target crop, and adaptation to different environmental factors. This knowledge would enable researchers and practitioners to optimize and increase the use of PGPR in sustainable agriculture as an integral part of strategies focused on crop production and productivity under changing climate conditions. Future research should also aim to identify novel stress-tolerant PGPR strains, as well as which genes and metabolites are involved in abiotic and biotic stress tolerance in the context of global climate change. This could be approached by applying metagenomics, molecular breeding tools such as CRISPR, and metabolomics approaches. Recently, the environmentally friendly nature of plant microbiome engineering in boosting crop productivity under changing climate conditions has also gained considerable attention from scientists. However, the microbiome of plants, including its interrelations with other microbes, should be researched in more depth so that researchers have a better scientific basis from which to apply new, innovative strategies for boosting crop productivity.

Table 17.1. Examples of how different PGPR strains guard against various abiotic stress in plants

Crop	PGPR strain	Stress type	Mechanisms	Reference
Maize Drought	*Bacillus licheniformis*	Drought	✓ Enhances activity of the antioxidant enzyme catalase in roots	(Akhtar *et al*., 2020)
	(*FMCH001*)		✓ Enhances water use efficiency	
Pepper	*Bacillus butanolivorans*	Drought	✓ Activates peroxidase and glutathione peroxidase	(Kim *et al*., 2022)
	KJ40		✓ Increases expression of the sHSP, Cadhn, bZIP1, and NCED1 genes,	
			✓ Improves polyphenol content	
Tomato	*Pseudomonas fluorescens G20–18*	Drought	✓ Improve chlorophyll content of leaves,	(Mekureyaw *et al*., 2022)
			✓ Increases the accumulation of ABA in the roots	
			✓ Increases the activity of various enzymes involved in	
			carbohydrate metabolism	
			✓ improves antioxidant capacity	
Arabidopsis	*Bacillus amyloliquefaciens SN13*	*Drought*	✓ Increases the concentrations of various osmolytes	(Tiwari *et al*., 2021)
			✓ Upregulates expression of the AP2/ERF, GST, ERD4, and ARF2 genes, which are related to phytohormone production	
Tomato	*Bacillus safensis* SCAL1 Heat		✓ Increase the activity of ACC deaminase, IAA, GA3, and Kinetin	(Mukhtar *et al*., 2022)
			✓ Increase exopolysaccharides	
			✓ Stimulate the synthesis of antioxidant enzymes	
			✓ Improves chlorophyll content	
Rice	*Brevibacterium linens* RS16	Heat	✓ Enhance the activity of ACC deaminase	(Choi *et al*., 2022)
			✓ Upregulates heat-shock proteins (HSP16 and HSP26)	
			✓ Stimulate the synthesis of antioxidant enzymes	

Continued

Table 17.1. Continued

Crop	PGPR strain	Stress type	Mechanisms	Reference
Wheat	*Bacillus safensis*	Heat	✓ Increases expression of HSPs and small HSPs	(Sarkar *et al.*, 2021)
	Pseudomonas brassicacearum		✓ Increases the synthesis of redox and antioxidant enzymes	(Ashraf *et al.*, 2019)
			✓ Upregulates the ADC1 and ADC2 genes, both of which are related to putrescine synthesis	
			✓ Improves phenol contents	
			✓ Enhances catalase and oxidase production	
			✓ Improves phosphate solubilization	
Arabidopsis	*Streptomyces sp.*	Salt	✓ Improves antioxidant activity	(Manullang and Chuang, 2020)
			✓ Enhances proline and protein contents	
			✓ Produces IAA hormone	
			✓ Improves phosphate solubilization	
			✓ Increases catalase enzyme activity	
			✓ Upregulates nitrogen assimilation (GS1) proteins	
			✓ Increases expression of a gene (cFBPase) related to carbohydrate metabolism	
			✓ Upregulates light-harvesting chlorophyll protein (Lhcb1)	

Funding

This research work was supported by FORMAS (2019–01316) and the Swedish Research Council (2019–04270), Novo Nordisk Fonden (0074727) and Partneskap Alnarp.

Conflicts of Interest

The authors have no conflicts of interest to declare.

References

Abdelaal, K., AlKahtani, M., Attia, K., Hafez, Y., Kiraly, L. *et al.* (2021) The role of plant growth-promoting bacteria in alleviating the adverse effects of drought on plants. *Biology-Basel* 10(6).

Abd El-Daim, I.A., Bejai, S. and Meijer, J. (2019) *Bacillus velezensis* 5113 induced metabolic and molecular reprogramming during abiotic stress tolerance in wheat. *Scientific Reports* 9(1), 16282. DOI: 10.1038/s41598-019-52567-x.

Abdul Rahman, N.S.N., Abdul Hamid, N.W. and Nadarajah, K. (2021) Effects of abiotic stress on soil microbiome. *International Journal of Molecular Sciences* 22(16), 520. DOI: 10.3390/biology10060520.

Adnan, M., Shah, Z., Fahad, S., Arif, M., Alam, M. *et al.* (2017) Phosphate-solubilizing bacteria nullify the antagonistic effect of soil calcification on bioavailability of phosphorus in alkaline soils. *Scientific Reports* 7(1), 16131. DOI: 10.1038/s41598-017-16537-5.

Adnan, M., Fahad, S., Zamin, M., Shah, S., Mian, I.A. *et al.* (2020) Coupling phosphate-solubilizing bacteria with phosphorus supplements improve maize phosphorus acquisition and growth under lime induced salinity stress. *Plants-Basel* 9(7), 900. DOI: 10.3390/plants9070900.

Akhtar, S.S., Amby, D.B., Hegelund, J.N., Fimognari, L., Grosskinsky, D.K. *et al.* (2020) *Bacillus licheniformis* FMCH001 increases water use efficiency via growth stimulation in both normal and drought conditions. *Frontiers in Plant Science* 11, 297. DOI: 10.3389/fpls.2020.00297.

Akhtyamova, Z., Arkhipova, T., Martynenko, E., Nuzhnaya, T., Kuzmina, L. *et al.* (2021) Growth-promoting effect of rhizobacterium (*Bacillus subtilis* IB22) in salt-stressed barley depends on abscisic acid accumulation in the roots. *International Journal of Molecular Sciences* 22(19), 10680. DOI: 10.3390/ijms221910680.

Albdaiwi, R.N., Khyami-Horani, H., Ayad, J.Y., Alananbeh, K.M. and Al-Sayaydeh, R. (2019) Isolation and characterization of halotolerant plant growth promoting rhizobacteria from durum wheat (*Triticum turgidum* subsp. durum) cultivated in saline areas of the dead sea region. *Frontiers in Microbiology* 10, 1639. DOI: 10.3389/fmicb.2019.01639.

Ali, B., Wang, X., Saleem, M.H., Azeem, M.A., Afridi, M.S. *et al.* (2022) *Bacillus mycoides* PM35 reinforces photosynthetic efficiency, antioxidant defense, expression of stress-responsive genes, and ameliorates the effects of salinity stress in maize. *Life* 12(2), 219. DOI: 10.3390/life12020219.

Aloo, B.N., Makumba, B.A. and Mbega, E.R. (2020) Plant growth promoting rhizobacterial biofertilizers for sustainable crop production: The past, present, and future. *Frontiers in Plant Science* 13, 1002448. DOI: 10.3389/fpls.2022.1002448.

Alori, E.T., Glick, B.R. and Babalola, O.O. (2017) Microbial phosphorus solubilization and its potential for use in sustainable agriculture. *Frontiers in Microbiology* 8, 971. DOI: 10.3389/fmicb.2017.00971.

Amna,Ud Din, B., Sarfraz, S., Xia, Y. and Kamran, M.A. (2019) Mechanistic elucidation of germination potential and growth of wheat inoculated with exopolysaccharide and ACC-deaminase producing bacillus strains under induced salinity stress. *Ecotoxicology and Environmental Safety* 183, 109466. DOI: 10.1016/j.ecoenv.2019.109466.

Amouzou, K.A., Naab, J.B., Lamers, J.P.A. and Becker, M. (2018) Productivity and nutrient use efficiency of maize, sorghum, and cotton in the West African Dry Savanna. *Journal of Plant Nutrition and Soil Science* 181(2), 261–274. DOI: 10.1002/jpln.201700139.

Arkhipova, T., Martynenko, E., Sharipova, G., Kuzmina, L., Ivanov, I. *et al.* (2020) Effects of plant growth promoting rhizobacteria on the content of abscisic acid and salt resistance of wheat plants. *Plants-Basel* 9(11), 1429. DOI: 10.3390/plants9111429.

Ashraf, A., Bano, A. and Ali, S.A. (2019) Characterisation of plant growth-promoting rhizobacteria from rhizosphere soil of heat-stressed and unstressed wheat and their use as bio-inoculant. *Plant Biology* 21(4), 762–769. DOI: 10.1111/plb.12972.

Ayub, M.A., Ahmad, H.R., Ali, M., Rizwan, M., Ali, S. *et al.* (2020) Chapter 3—salinity and its tolerance strategies in plants. In: Tripathi, D.K., Pratap Singh, V., Chauhan, D.K., Sharma, S., Prasad, S.M. *et al.* (eds) *Plant Life Under Changing Environment*. Academic Press, pp. 47–76. DOI: 10.1016/B978-0-12-818204-8.00003-5.

Azadikhah, M., Jamali, F., Nooryazdan, H.R. and Bayat, F. (2019) Growth promotion and yield enhancement of barley cultivars using ACC deaminase producing *Pseudomonas fluorescens* strains under salt stress. *Spanish Journal of Agricultural Research* 17(1), 801. DOI: 10.5424/sjar/2019171-13828.

Backer, R., Rokem, J.S., Ilangumaran, G., Lamont, J., Praslickova, D. *et al.* (2018) Plant growth-promoting rhizobacteria: Context, mechanisms of action, and roadmap to commercialization of biostimulants for sustainable agriculture. *Frontiers in Plant Science* 9, 1473. DOI: 10.3389/fpls.2018.01473.

Backes, A., Vaillant-Gaveau, N., Esmaeel, Q., Barka, E.A. and Jacquard, C. (2021) A biological agent modulates the physiology of barley infected with *Drechslera teres*. *Scientific Reports* 11(1), 8330. DOI: 10.1038/s41598-021-87853-0.

Baek, D., Rokibuzzaman, M., Khan, A., Kim, M.C., Park, H.J. *et al.* (2020) Plant-growth promoting *Bacillus oryzicola* YC7007 modulates stress-response gene expression and provides protection from salt stress. *Frontiers in Plant Science* 10, 1646. DOI: 10.3389/fpls.2019.01646.

Bandyopadhyay, P., Yadav, B.G., Kumar, S.G., Kumar, R., Kogel, K.H. *et al.* (2022) *Piriformospora indica* and *Azotobacter chroococcum* consortium facilitates higher acquisition of N, P with improved carbon allocation and enhanced plant growth in *Oryza sativa*. *Journal of Fungi* 8(5), 453. DOI: 10.3390/jof8050453.

Barman, D., Dutta, I. and Jha, D.K. (2022) Heavy metal resistant bacteria from coal dumping site with plant growth promoting potentials. *Biologia* 77(2), 533–545. DOI: 10.1007/s11756-021-00963-y.

Batool, T., Ali, S., Seleiman, M.F., Naveed, N.H., Ali, A. *et al.* (2020) Plant growth promoting rhizobacteria alleviates drought stress in potato in response to suppressive oxidative stress and antioxidant enzymes activities. *Scientific Reports* 10(1), 16975. DOI: 10.1038/s41598-020-73489-z.

Berni, R., Hausman, J.F., Villas-Boas, S. and Guerriero, G. (2022) Impact of pseudomonas sp. SVB-B33 on stress-and cell wall-related genes in roots and leaves of hemp under salinity. *Horticulturae* 8(4), 336. DOI: 10.3390/horticulturae8040336.

Besset-Manzoni, Y., Joly, P., Brutel, A., Gerin, F., Soudiere, O. *et al.* (2019) Does *in vitro* selection of biocontrol agents guarantee success in planta? A study case of wheat protection against Fusarium seedling blight by soil bacteria. *PLoS One* 14(12), e0225655. DOI: 10.1371/journal.pone.0225655.

Bhagat, N., Raghav, M., Dubey, S. and Bedi, N. (2021) Bacterial exopolysaccharides: Insight into their role in plant abiotic stress tolerance. *Journal of Microbiology and Biotechnology* 31(8), 1045–1059. DOI: 10.4014/jmb.2105.05009.

Billah, M., Khan, M., Bano, A., Ul Hassan, T., Munir, A. *et al.* (2019) Phosphorus and phosphate solubilizing bacteria: Keys for sustainable agriculture. *Geomicrobiology Journal* 36(10), 904–916. DOI: 10.1080/01490451.2019.1654043.

Binenbaum, J., Weinstain, R. and Shani, E. (2018) Gibberellin localization and transport in plants. *Trends in Plant Science* 23(5), 410–421. DOI: 10.1016/j.tplants.2018.02.005.

Bjelić, D., Marinković, J., Tintor, B. and Mrkovački, N. (2018) Antifungal and plant growth promoting activities of indigenous rhizobacteria isolated from maize (*Zea mays* L.) rhizosphere. *Communications in Soil Science and Plant Analysis* 49(1), 88–98. DOI: 10.1080/00103624.2017.1421650.

Bokhari, A., Essack, M., Lafi, F.F., Andres-Barrao, C., Jalal, R. *et al.* (2019) Bioprospecting desert plant Bacillus endophytic strains for their potential to enhance plant stress tolerance. *Scientific Reports* 9(1), 18154. DOI: 10.1038/s41598-019-54685-y.

Boukhatem, Z.F., Merabet, C. and Tsaki, H. (2022) Plant growth promoting actinobacteria, the most promising candidates as bioinoculants? *Frontiers in Agronomy* 4, 849911. DOI: 10.3389/fagro.2022.849911.

Burdon, J.J. and Zhan, J. (2020) Climate change and disease in plant communities. *PLoS Biology* 18(11), e3000949. DOI: 10.1371/journal.pbio.3000949.

Carvalho, T.L.G., Rosman, A.C., Grativol, C., de, M.N.E., Baldani, J.I. *et al.* (2022) Sugarcane genotypes with contrasting biological nitrogen fixation efficiencies differentially modulate nitrogen metabolism, auxin signaling, and microorganism perception pathways. *Plants (Basel)* 11, 15. DOI: 10.3390/plants11151971.

Caulier, S., Gillis, A., Colau, G., Licciardi, F., Liepin, M. *et al.* (2018) Versatile antagonistic activities of soil-borne *Bacillus* spp. and *Pseudomonas* spp. against *Phytophthora infestans* and other potato pathogens. *Frontiers in Microbiology* 9, 143. DOI: 10.3389/fmicb.2018.00143.

Chain, J.M., Tubert, E., Graciano, C., Castagno, L.N., Recchi, M. *et al.* (2020) Growth promotion and protection from drought in *Eucalyptus grandis* seedlings inoculated with beneficial bacteria embedded in a superabsorbent polymer. *Scientific Reports* 10(1), 18221. DOI: 10.1038/s41598-020-75212-4.

Choi, J., Roy Choudhury, A., Walitang, D.I., Lee, Y. and Sa, T. (2022) ACC deaminase-producing *Brevibacterium linens* RS16 enhances heat-stress tolerance of rice (*Oryza sativa* L.). *Physiologia Plantarum* 174(1), e13584. DOI: 10.1111/ppl.13584.

Chu, T.N., Tran, B.T.H., Van Bui, L. and Hoang, M.T.T. (2019) Plant growth-promoting rhizobacterium Pseudomonas PS01 induces salt tolerance in *Arabidopsis thaliana*. *BMC Research Notes* 12(1), 11. DOI: 10.1186/s13104-019-4046-1.

Ciftci, C., Tekdal, D. and Cetiner, S. (2021) 3—the importance of plant growth–promoting rhizobacteria for plant productivity. In: Singh, V.P., Singh, S., Tripathi, D.K., Prasad, S.M., Bhardwaj, R. et al. (eds) *Abiotic Stress and Legumes*. Academic Press, pp. 69–80. DOI: 10.1016/B978-0-12-815355-0.00004-7.

Dhar Purkayastha, G., Mangar, P., Saha, A. and Saha, D. (2018) Evaluation of the biocontrol efficacy of a *Serratia marcescens* strain indigenous to tea rhizosphere for the management of root rot disease in tea. *PLoS One* 13(2), e0191761. DOI: 10.1371/journal.pone.0191761.

Djebaili, R., Pellegrini, M., Ercole, C., Farda, B., Kitouni, M. *et al.* (2021) Biocontrol of soil-borne pathogens of *Solanum lycopersicum* L. and *Daucus carota* L. by plant growth-promoting actinomycetes: *In vitro* and *in planta* antagonistic activity. *Pathogens* 10(10), 1305. DOI: 10.3390/pathogens10101305.

Driesen, E., Van de Ende, W., De Proft, M. and Saeys, W. (2020) Influence of environmental factors light, CO_2, temperature, and relative humidity on stomatal opening and development: A review. *Agronomy* 10, 12. DOI: 10.3390/agronomy10121975.

Dukare, A. and Paul, S. (2021) Biological control of fusarium wilt and growth promotion in pigeon pea (*Cajanus cajan*) by antagonistic rhizobacteria, displaying multiple modes of pathogen inhibition. *Rhizosphere* 17, 100278. DOI: 10.1016/j.rhisph.2020.100278.

Du, Y.T., Zhao, M.J., Wang, C.T., Gao, Y., Wang, Y.X. *et al.* (2018) Identification and characterization of GmMYB118 responses to drought and salt stress. *BMC Plant Biology* 18(1), 320. DOI: 10.1186/s12870-018-1551-7.

Eida, A.A., Alzubaidy, H.S., de Zelicourt, A., Synek, L., Alsharif, W. *et al.* (2019) Phylogenetically diverse endophytic bacteria from desert plants induce transcriptional changes of tissue-specific ion transporters and salinity stress in *Arabidopsis thaliana*. *Plant Science* 280, 228–240. DOI: 10.1016/j.plantsci.2018.12.002.

El Moukhtari, A., Lamsaadi, N., Farssi, O., Oubenali, A., El Bzar, I. *et al.* (2022) Silicon-and phosphate-solubilizing pseudomonas alkylphenolica PF9 alleviate low phosphorus availability stress in alfalfa (*Medicago sativa* L.). *Frontiers in Agronomy* 4, 823396. DOI: 10.3389/fagro.2022.823396.

Etesami, H. and Maheshwari, D.K. (2018) Use of plant growth promoting rhizobacteria (pgprs) with multiple plant growth promoting traits in stress agriculture: Action mechanisms and future prospects. *Ecotoxicology and Environmental Safety* 156, 225–246. DOI: 10.1016/j.ecoenv.2018.03.013.

Fatima, T. and Arora, N.K. (2021) *Pseudomonas entomophila* PE3 and its exopolysaccharides as biostimulants for enhancing growth, yield and tolerance responses of sunflower under saline conditions. *Microbiological Research* 244, 126671. DOI: 10.1016/j.micres.2020.126671.

Fatima, Z., Ahmed, M., Hussain, M., Abbas, G., Ul-Allah, S. *et al.* (2020) The fingerprints of climate warming on cereal crops phenology and adaptation options. *Scientific Reports* 10(1), 18013. DOI: 10.1038/s41598-020-74740-3.

Figueroa, M., Hammond-Kosack, K.E. and Solomon, P.S. (2018) A review of wheat diseases-A field perspective. *Molecular Plant Pathology* 19(6), 1523–1536. DOI: 10.1111/mpp.12618.

Fiodor, A., Singh, S. and Pranaw, K. (2021) The contrivance of plant growth promoting microbes to mitigate climate change impact in agriculture. *Microorganisms* 9(9), 1841. DOI: 10.3390/microorganisms9091841.

Francini, A. and Sebastiani, L. (2019) Abiotic stress effects on performance of horticultural crops. *Horticulturae* 5(4), 67. DOI: 10.3390/horticulturae5040067.

Ghadamgahi, F., Tarighi, S., Taheri, P., Saripella, G.V., Anzalone, A. *et al.* (2022) Plant growth-promoting activity of *Pseudomonas aeruginosa* FG106 and its ability to act as a biocontrol agent against potato, tomato and taro pathogens. *Biology* 11(1), 140. DOI: 10.3390/biology11010140.

Guo, Q., Li, Y., Lou, Y., Shi, M., Jiang, Y. *et al.* (2019) Bacillus amyloliquefaciens Ba13 induces plant systemic resistance and improves rhizosphere microecology against tomato yellow leaf curl virus disease. *Applied Soil Ecology* 137, 154–166. DOI: 10.1016/j.apsoil.2019.01.015.

Gupta, A., Mishra, R., Rai, S., Bano, A., Pathak, N. *et al.* (2022) Mechanistic insights of plant growth promoting bacteria mediated drought and salt stress tolerance in plants for sustainable agriculture. *International Journal of Molecular Sciences* 23(7), 3741. DOI: 10.3390/ijms23073741.

Gupta, S. and Pandey, S. (2019) ACC deaminase producing bacteria with multifarious plant growth promoting traits alleviates salinity stress in French bean (*Phaseolus vulgaris*) plants. *Frontiers in Microbiology* 10, 1506. DOI: 10.3389/fmicb.2019.01506.

Hannah, L., Roehrdanz, P.R., Bahadur, K.C.K., Fraser, E.D.G., Donatti, C.I. *et al.* (2020) The environmental consequences of climate-driven agricultural frontiers. *PLoS One* 15(2), e0236028. DOI: 10.1371/journal.pone.0236028.

Haque, M.M., Mosharaf, M.K., Khatun, M., Haque, M.A., Biswas, M.S. *et al.* (2020) Biofilm producing rhizobacteria with multiple plant growth-promoting traits promote growth of tomato under water-deficit stress. *Frontiers in Microbiology* 11, 542053. DOI: 10.3389/fmicb.2020.542053.

Hashem, A., Tabassum, B. and Abd_Allah, E.F. (2019) *Bacillus subtilis*: A plant-growth promoting rhizobacterium that also impacts biotic stress. *Saudi Journal of Biological Sciences* 26(6), 1291–1297. DOI: 10.1016/j.sjbs.2019.05.004.

Heydarian, Z., Gruber, M., Coutu, C., Glick, B.R. and Hegedus, D.D. (2021) Gene expression patterns in shoots of *Camelina sativa* with enhanced salinity tolerance provided by plant growth promoting bacteria producing 1-aminocyclopropane-1-carboxylate deaminase or expression of the corresponding acds gene. *Scientific Reports* 11(1), 4260. DOI: 10.1038/s41598-021-83629-8.

Hong, S.R., Kim, T.Y., Won, S.J., Moon, J.H., Ajuna, H.B. *et al.* (2022) Control of fungal diseases and fruit yield improvement of strawberry using *Bacillus velezensis* CE 100. *Microorganisms* 10(2), 365. DOI: 10.3390/microorganisms10020365.

Inbaraj, M.P. (2021) Plant-microbe interactions in alleviating abiotic stress—A mini review. *Frontiers in Agronomy* 3, 667903. DOI: 10.3389/fagro.2021.667903.

Jebara, S.H., Abdelkrim, S., Saadani, O., Fatnassi, I.C., Mannai, K. *et al.* (2021) Chapter 18—potential use of efficient resistant plant growth promoting rhizobacteria in biofertilization and phytoremediation of heavy metal contaminated soil. In: Hasanuzzaman, M. and Prasad, M.N.V. (eds) *Handbook of Bioremediation*. Academic Press, pp. 285–293. DOI: 1016/B978-0-12-819382-2.00018-1.

Jha, Y. and Mohamed, H.I. (2022) Inoculation with *Lysinibacillus fusiformis* strain YJ4 and *Lysinibacillus sphaericus* strain YJ5 alleviates the effects of cold stress in maize plants. *Gesunde Pflanzen* 1–19. DOI: 10.1007/s10343-022-00666-7.

Jochum, M.D., McWilliams, K.L., Borrego, E.J., Kolomiets, M.V., Niu, G.H. *et al.* (2019) Bioprospecting plant growth-promoting rhizobacteria that mitigate drought stress in grasses. *Frontiers in Microbiology* 10, 2106. DOI: 10.3389/fmicb.2019.02106.

Kapadia, C., Sayyed, R.Z., Enshasy, H.A.E., Vaidya, H., Sharma, D. *et al.* (2021) Halotolerant microbial consortia for sustainable mitigation of salinity stress, growth promotion, and mineral uptake in tomato plants and soil nutrient enrichment. *Sustainability (Switzerland)* 13(15), 8369. DOI: 10.3390/su13158369.

Kenneth, O.C., Nwadibe, E.C., Kalu, A.U. and Unah, U.V. (2019) Plant growth promoting rhizobacteria (PGPR): A novel agent for sustainable food production. *American Journal of Agricultural and Biological Sciences* 14, 35–54. DOI: 10.3844/ajabssp.2019.35.54.

Khan, M.A., Asaf, S., Khan, A.L., Jan, R., Kang, S.M, *et al.* (2020) Thermotolerance effect of plant growth-promoting *Bacillus cereus* SA1 on soybean during heat stress. *BMC Microbiology* 20(1), 175. DOI: 10.1186/s12866-020-01822-7.

Khan, N., Bano, A. and Zandi, P. (2018) Effects of exogenously applied plant growth regulators in combination with PGPR on the physiology and root growth of chickpea (Cicer arietinum) and their role in drought tolerance. *Journal of Plant Interactions* 13(1), 239–247. DOI: 10.1080/17429145.2018.1471527.

Khataar, M., Mohhamadi, M.H. and Shabani, F. (2018) Soil salinity and matric potential interaction on water use, water use efficiency and yield response factor of bean and wheat. *Scientific Reports* 8(1), 2679. DOI: 10.1038/s41598-018-20968-z.

Kiani, T., Mehboob, F., Hyder, M.Z., Zainy, Z., Xu, L.S. *et al.* (2021) Control of stripe rust of wheat using indigenous endophytic bacteria at seedling and adult plant stage. *Scientific Reports* 11(1), 14473. DOI: 10.1038/s41598-021-93939-6.

Kim, S.T., Yoo, S.J., Weon, H.Y., Song, J. and Sang, M.K. (2022) *Bacillus butanolivorans* KJ40 contributes alleviation of drought stress in pepper plants by modulating antioxidant and polyphenolic compounds. *Scientia Horticulturae* 301, 111111. DOI: 10.1016/j.scienta.2022.111111.

Kohl, J., Kolnaar, R. and Ravensberg, W.J. (2019) Mode of action of microbial biological control agents against plant diseases: Relevance beyond efficacy. *Frontiers in Plant Science* 10, 845. DOI: 10.3389/fpls.2019.00845.

Kolomiiets, Y.V., Grigoryuk, I.P., Likhanov, A.F., Butsenko, L.M., Pasichnyk, L.A. *et al.* (2020) Induction of wheat resistance against the causative agent of basal bacteriosis with growth-promoting bacteria. *Cytology and Genetics* 54(6), 514–521. DOI: 10.3103/s0095452720060067.

Kudoyarova, G., Arkhipova, T., Korshunova, T., Bakaeva, M., Loginov, O. *et al.* (2019) Phytohormone mediation of interactions between plants and non-symbiotic growth promoting bacteria under edaphic stresses. *Frontiers in Plant Science* 10, 1368. DOI: 10.3389/fpls.2019.01368.

Kumar, A. and Verma, J.P. (2018) Does plant-microbe interaction confer stress tolerance in plants: A review? *Microbiological Research* 207, 41–52. DOI: 10.1016/j.micres.2017.11.004.

Kumawat, K.C., Nagpal, S. and Sharma, P. (2022) Potential of plant growth-promoting rhizobacteria-plant interactions in mitigating salt stress for sustainable agriculture: A review. *Pedosphere* 32(2), 223–245. DOI: 10.1016/s1002-0160(21)60070-x.

Kwon, H.T., Lee, Y., Kim, J., Balaraju, K., Kim, H.T. *et al.* (2022) Identification and characterization of *Bacillus tequilensis* GYUN-300: An antagonistic bacterium against red pepper anthracnose caused by *Colletotrichum acutatum* in Korea. *Frontiers in Microbiology* 13, 826827. DOI: 10.3389/fmicb.2022.826827.

Laborda, P., Chen, X., Wu, G., Wang, S., Lu, X. *et al.* (2020) *Lysobacter gummosus* OH17 induces systemic resistance in *Oryza sativa* 'Nipponbare. *Plant Pathology* 69(5), 838–848. DOI: 10.1111/ppa.13167.

Lee, K.E., Adhikari, A., Kang, S.M., You, Y.H., Joo, G.J, *et al.* (2019) Isolation and characterization of the high silicate and phosphate solubilizing novel strain *Enterobacter ludwigii* GAK2 that promotes growth in rice plants. *Agronomy-Basel* 9(3), 144. DOI: 10.3390/agronomy9030144.

Lesk, C., Rowhani, P. and Ramankutty, N. (2016) Influence of extreme weather disasters on global crop production. *Nature* 529(7584), 84–87. DOI: 10.1038/nature16467.

Lindström, K. and Mousavi, S.A. (2020) Effectiveness of nitrogen fixation in rhizobia. *Microbial Biotechnology* 13(5), 1314–1335. DOI: 10.1111/1751-7915.13517.

Li, Q., Huang, Z., Deng, C., Lin, K.-H., Hua, S. *et al.* (2022b) *Endophytic klebsiella* sp. San01 promotes growth performance and induces salinity and drought tolerance in sweet potato (*Ipomoea batatas* L.). *Journal of Plant Interactions* 17(1), 608–619. DOI: 10.1080/17429145.2022.2077464.

Li, S., Xu, J.Q., Fu, L.M., Xu, G.H., Lin, X.M. *et al.* (2022a) Biocontrol of wheat crown rot using *Bacillus halotolerans* QTH8. *Pathogens* 11(5), 595. DOI: 10.3390/pathogens11050595.

Liu, Z.P., Zhang, X.L., Li, L.B., Xu, N., Hu, Y. *et al.* (2022) Isolation and characterization of three plant growth-promoting rhizobacteria for growth enhancement of rice seedling. *Journal of Plant Growth Regulation* 41(3), 1382–1393. DOI: 10.1007/s00344-021-10393-4.

Lucke, M., Correa, M.G. and Levy, A. (2020) The role of secretion systems, effectors, and secondary metabolites of beneficial rhizobacteria in interactions with plants and microbes. *Frontiers in Plant Science* 11, 589416. DOI: 10.3389/fpls.2020.589416.

Ma, H., Li, P., Liu, X., Li, C., Zhang, S. *et al.* (2022) Poly-γ-glutamic acid enhanced the drought resistance of maize by improving photosynthesis and affecting the rhizosphere microbial community. *BMC Plant Biology* 22(1), 11. DOI: 10.1186/s12870-021-03392-w.

Majeed, A., Muhammad, Z. and Ahmad, H. (2018) Plant growth promoting bacteria: Role in soil improvement, abiotic and biotic stress management of crops. *Plant Cell Reports* 37(12), 1599–1609. DOI: 10.1007/s00299-018-2341-2.

Maksimov, I.V., Sorokan, A.V., Burkhanova, G.F., Veselova, S.V., Alekseev, V.Y. *et al.* (2019) Mechanisms of plant tolerance to RNA viruses induced by plant-growth-promoting microorganisms. *Plants* 8(12), 575. DOI: 10.3390/plants8120575.

Mandal, S., Saha, K.K. and Mandal, N.C. (2021) Molecular insight into key eco-physiological process in bioremediating and plant-growth-promoting bacteria. *Frontiers in Agronomy* 3, 664126. DOI: 10.3389/fagro.2021.664126.

Manullang, W. and Chuang, H.W. (2020) Streptomyces sp. mitigates abiotic stress response and promotes plant growth. *Journal of Plant Protection Research* 60(3), 263–274. DOI: 10.24425/jppr.2020.133955.

Marasco, R., Fusi, M., Mosqueira, M., Booth, J.M., Rossi, F. *et al.* (2022) Rhizosheath-root system changes exopolysaccharide content but stabilizes bacterial community across contrasting seasons in a desert environment. *Environmental Microbiome* 17(1), 14. DOI: 10.1186/s40793-022-00407-3.

Marques, J.M., Mateus, J.R., Silva, T.F., Couto, C.R.D., Blank, A.F. *et al.* (2019) Nitrogen fixing and phosphate mineralizing bacterial communities in sweet potato rhizosphere show a genotype-dependent distribution. *Diversity-Basel* 11(12), 231. DOI: 10.3390/d11120231.

Martinez, O.A., Encina, C., Tomckowiack, C., Droppelmann, F., Jara, R. *et al.* (2018) Serratia strains isolated from the rhizosphere of rauli (*Nothofagus alpina*) in volcanic soils harbour PGPR mechanisms and promote rauli plantlet growth. *Journal of Soil Science and Plant Nutrition* 18(3), 804–819. DOI: 10.4067/S0718-95162018005002302.

Ma, Y., Dias, M.C. and Freitas, H. (2020) Drought and salinity stress responses and microbe-induced tolerance in plants. *Frontiers in Plant Science* 11, 591911. DOI: 10.3389/fpls.2020.591911.

Mekureyaw, M.F., Pandey, C., Hennessy, R.C., Nicolaisen, M.H., Liu, F.L. *et al.* (2022) The cytokinin-producing plant beneficial bacterium *Pseudomonas fluorescens* G20-18 primes tomato (*Solanum lycopersicum*) for enhanced drought stress responses. *Journal of Plant Physiology* 270, 153629. DOI: 10.1016/j.jplph.2022.153629.

Minchev, Z., Kostenko, O., Soler, R. and Pozo, M.J. (2021) Microbial consortia for effective biocontrol of root and foliar diseases in tomato. *Frontiers in Plant Science* 12, 756368. DOI: 10.3389/fpls.2021.756368.

Mpanga, I.K., Gomez-Genao, N., Moradtalab, N., Wanke, D., Chrobaczek, V. *et al.* (2019) The role of N form supply for PGPM-host plant interactions in maize. *Journal of Plant Nutrition and Soil Science* 182(6), 908–920. DOI: 10.1002/jpln.201900133.

Mukhtar, T., Ur Rehman, S., Smith, D., Sultan, T., Seleiman, M.F. *et al.* (2020) Mitigation of heat stress in *Solanum lycopersicum* l. by ACCdeaminase and exopolysaccharide producing *Bacillus cereus*: Effects on biochemical profiling. *Sustainability (Switzerland)* 12(6), 2159. DOI: 10.3390/su12062159.

Mukhtar, T., Ali, F., Rafique, M., Ali, J., Afridi, M.S. *et al.* (2022) Biochemical characterization and potential of *Bacillus safensis* strain SCAL1 to mitigate heat stress in *Solanum lycopersicum* L. *Journal of Plant Growth Regulation*. DOI: 10.1007/s00344-021-10571-4.

Naamala, J. and Smith, D.L. (2020) Relevance of plant growth promoting microorganisms and their derived compounds, in the face of climate change. *Agronomy-Basel* 10(8), 1179. DOI: 10.3390/agronomy10081179.

Narasimha Murthy, K., Soumya, K., Udayashankar, A.C., Srinivas, C. and Jogaiah, S. (2021) 7—biocontrol potential of plant growth-promoting rhizobacteria (PGPR) against *Ralstonia solanacearum*: Current and future prospects. In: Jogaiah, S. (ed.) *Biocontrol Agents and Secondary Metabolites*. Woodhead Publishing, pp. 153–180. DOI: 10.1016/B978-0-12-822919-4.00007-7.

Naseem, H., Ahsan, M., Shahid, M.A. and Khan, N. (2018) Exopolysaccharides producing rhizobacteria and their role in plant growth and drought tolerance. *Journal of Basic Microbiology* 58(12), 1009–1022. DOI: 10.1002/jobm.201800309.

Naylor, D. and Coleman-Derr, D. (2018) Drought stress and root-associated bacterial communities. *Frontiers in Plant Science* 8, 2223. DOI: 10.3389/fpls.2017.02223.

Ngalimat, M.S., Hata, E.M., Zulperi, D., Ismail, S.I., Ismail, M.R. *et al.* (2021) Plant growth-promoting bacteria as an emerging tool to manage bacterial rice pathogens. *Microorganisms* 9(4), 682. DOI: 10.3390/microorganisms9040682.

Ojuederie, O.B., Olanrewaju, O.S. and Babalola, O.O. (2019) Plant growth promoting rhizobacterial mitigation of drought stress in crop plants: Implications for sustainable agriculture. *Agronomy* 9(11), 712. DOI: 10.3390/agronomy9110712.

Olenska, E., Malek, W., Wojcik, I., Swiecicka, I., Thijs, S. *et al.* (2020) Beneficial features of plant growth-promoting rhizobacteria for improving plant growth and health in challenging conditions: A methodical review. *Science of the Total Environment* 743, 140682. DOI: 10.1016/j.scitotenv.2020.140682.

Ortiz-Bobea, A., Ault, T.R., Carrillo, C.M., Chambers, R.G. and Lobel, D.B. (2021) Anthropogenic climate change has slowed global agricultural productivity growth. *Nature Climate Change* 11(4), 306–328. DOI: 10.1038/s41558-021-01000-1.

Ouertani, R.N., Abid, G., Ben Chikha, M., Boudaya, O., Mejri, S. *et al.* (2022) Physiological and biochemical analysis of barley (*Hordeum vulgare*) genotypes with contrasting salt tolerance. *Acta Physiologiae Plantarum* 44(5), 51. DOI: 10.1007/s11738-022-03388-5.

Pandey, S. and Gupta, S. (2020) Evaluation of pseudomonas sp. for its multifarious plant growth promoting potential and its ability to alleviate biotic and abiotic stress in tomato (*Solanum lycopersicum*) plants. *Scientific Reports* 10(1), 20951. DOI: 10.1038/s41598-020-77850-0.

Park, Y.G., Mun, B.G., Kang, S.M., Hussain, A., Shahzad, R. *et al.* (2017) *Bacillus aryabhattai* SRB02 tolerates oxidative and nitrosative stress and promotes the growth of soybean by modulating the production of phytohormones. *PLoS One* 12(3), eo173203. DOI: 10.1371/journal.pone.0173203.

Patel, P., Trivedi, G. and Saraf, M. (2018) Iron biofortification in mungbean using siderophore producing plant growth promoting bacteria. *Environmental Sustainability* 1(4), 357–365. DOI: 10.1007/s42398-018-00031-3.

Prittesh, P., Avnika, P., Kinjal, P., Jinal, H.N., Sakthivel, K. *et al.* (2020) Amelioration effect of salt-tolerant plant growth-promoting bacteria on growth and physiological properties of rice (*Oryza sativa*) under salt-stressed conditions. *Archives of Microbiology* 202(9), 2419–2428. DOI: 10.1007/s00203-020-01962-4.

Qiao, L., Wang, X.H., Smith, P., Fan, J.L., Lu, Y.L. *et al.* (2022) Soil quality both increases crop production and improves resilience to climate change. *Nature Climate Change* 12(6), 574. DOI: 10.1038/s41558-022-01376-8.

Rahman, M., Rahman, M., Sabir, A.A., Mukta, J.A., Khan, M.M.A. *et al.* (2018) Plant probiotic bacteria Bacillus and Paraburkholderia improve growth, yield and content of antioxidants in strawberry fruit. *Scientific Reports* 8(1), 2504. DOI: 10.1038/s41598-018-20235-1.

Rasool, M., Akhter, A., Soja, G. and Haider, M.S. (2021) Role of biochar, compost and plant growth promoting rhizobacteria in the management of tomato early blight disease. *Scientific Reports* 11(1), 6092. DOI: 10.1038/s41598-021-85633-4.

Rilling, J.I., Acuña, J.J., Sadowsky, M.J. and Jorquera, M.A. (2018) Putative nitrogen-fixing bacteria associated with the rhizosphere and root endosphere of wheat plants grown in an andisol from southern Chile. *Frontiers in Microbiology* 9(Nov), 2710. DOI: 10.3389/fmicb.2018.02710.

Rodriguez-Blanco, A., Sicardi, M. and Frioni, L. (2015) Plant genotype and nitrogen fertilization effects on abundance and diversity of diazotrophic bacteria associated with maize (*Zea mays* L.). *Biology and Fertility of Soils* 51(3), 391–402. DOI: 10.1007/s00374-014-0986-8.

Rolli, E., de Zélicourt, A., Alzubaidy, H., Karampelias, M., Parween, S. *et al.* (2022) The Lys-motif receptor LYK4 mediates Enterobacter sp. SA187 triggered salt tolerance in *Arabidopsis thaliana*. *Environmental Microbiology* 24(1), 223–239. DOI: 10.1111/1462-2920.15839.

Romera, F.J., García, M.J., Lucena, C., Martínez-Medina, A., Aparicio, M.A. *et al.* (2019) Induced systemic resistance (ISR) and Fe deficiency responses in dicot plants. *Frontiers in Plant Science* 10, 287. DOI: 10.3389/fpls.2019.00287..

Sagar, A., Sayyed, R.Z., Ramteke, P.W., Sharma, S., Marraiki, N. *et al.* (2020) ACC deaminase and antioxidant enzymes producing halophilic enterobacter sp. PR14 promotes the growth of rice and millets under salinity stress. *Physiology and Molecular Biology of Plants* 26(9), 1847–1854. DOI: 10.1007/s12298-020-00852-9.

Saha, I., Datta, S. and Biswas, D. (2020) Exploring the role of bacterial extracellular polymeric substances for sustainable development in agriculture. *Current Microbiology* 77(11), 3224–3239. DOI: 10.1007/s00284-020-02169-y.

Saijo, Y. and Loo, E.P. (2020) Plant immunity in signal integration between biotic and abiotic stress responses. *New Phytologist* 225(1), 87–104. DOI: 10.1111/nph.15989.

Samaras, A., Roumeliotis, E., Ntasiou, P. and Karaoglanidis, G. (2021) *Bacillus subtilis* MBI600 promotes growth of tomato plants and induces systemic resistance contributing to the control of soilborne pathogens. *Plants* 10(6), 1113. DOI: 10.3390/plants10061113.

Sangiorgio, D., Cellini, A., Donati, I., Pastore, C., Onofrietti, C. *et al.* (2020) Facing climate change: Application of microbial biostimulants to mitigate stress in horticultural crops. *Agronomy-Basel* 10(6), 794. DOI: 10.3390/agronomy10060794.

Sanya, D.R.A., Syed-Ab-Rahman, S.F., Jia, A.Q., Onesime, D., Kim, K.M. *et al.* (2022) A review of approaches to control bacterial leaf blight in rice. *World Journal of Microbiology & Biotechnology* 38(7), 113. DOI: 10.1007/s11274-022-03298-1.

Sarkar, J., Chakraborty, U. and Chakraborty, B. (2021) High-temperature resilience in *Bacillus safensis* primed wheat plants: A study of dynamic response associated with modulation of antioxidant machinery, differential expression of hsps and osmolyte biosynthesis. *Environmental and Experimental Botany* 182, 104315. DOI: 10.1016/j.envexpbot.2020.104315.

Sepehri, M. and Khatabi, B. (2021) Combination of siderophore-producing bacteria and piriformospora indica provides an efficient approach to improve cadmium tolerance in Alfalfa. *Microbial Ecology* 81(3), 717–730. DOI: 10.1007/s00248-020-01629-z.

Shilev, S. (2020) Plant-growth-promoting bacteria mitigating soil salinity stress in plants. *Applied Sciences-Basel* 10(20), 7326. DOI: 10.3390/app10207326.

Silambarasan, S., Logeswari, P., Cornejo, P. and Kannan, V.R. (2019) Role of plant growth-promoting rhizobacterial consortium in improving the *Vigna radiata* growth and alleviation of aluminum and drought stresses. *Environmental Science and Pollution Research* 26(27), 27647–27659. DOI: 10.1007/s11356-019-05939-9.

Singha, K.M., Singh, B. and Pandey, P. (2021) Host specific endophytic microbiome diversity and associated functions in three varieties of scented black rice are dependent on growth stage. *Scientific Reports* 11(1), 12259. DOI: 10.1038/s41598-021-91452-4.

Singh, R.P., Pandey, D.M., Jha, P.N. and Ma, Y. (2022) ACC deaminase producing rhizobacterium enterobacter cloacae ZNP-4 enhance abiotic stress tolerance in wheat plant. *PLoS ONE* 17(5), e0267127. DOI: 10.1371/journal.pone.0267127.

Soenens, A. and Imperial, J. (2020) Biocontrol capabilities of the genus *Serratia*. *Phytochemistry Reviews* 19(3), 577–587. DOI: 10.1007/s11101-019-09657-5.

Suárez-Moreno, Z.R., Vinchira-Villarraga, D.M., Vergara-Morales, D.I., Castellanos, L., Ramos, F.A. *et al.* (2019) Plant-growth promotion and biocontrol properties of three *Streptomyces* spp. isolates to control bacterial rice pathogens. *Frontiers in Microbiology* 10, 290. DOI: 10.3389/fmicb.2019.00290.

Sun, Y.X., Wu, J.L., Shang, X.Y., Xue, L.G., Ji, G.Y. *et al.* (2022) Screening of siderophore-producing bacteria and their effects on promoting the growth of plants. *Current Microbiology* 79(5), 150. DOI: 10.1007/s00284-022-02777-w.

Syed Ab Rahman, S.F., Singh, E., Pieterse, C.M.J. and Schenk, P.M. (2018) Emerging microbial biocontrol strategies for plant pathogens. *Plant Science* 267, 102–111. DOI: 10.1016/j.plantsci.2017.11.012.

Tiwari, S., Gupta, S.C., Chauhan, P.S. and Lata, C. (2021) An osnam gene plays important role in root rhizobacteria interaction in transgenic arabidopsis through abiotic stress and phytohormone crosstalk. *Plant Cell Reports* 40(1), 143–155. DOI: 10.1007/s00299-020-02620-1.

Toulotte, J.M., Pantazopoulou, C.K., Sanclemente, M.A., Voesenek, L. and Sasidharan, R. (2022) Water stress resilient cereal crops: Lessons from wild relatives. *Journal of Integrative Plant Biology* 64(2), 412–430. DOI: 10.1111/jipb.13222.

Trdan, S., Vučajnk, F., Bohinc, T. and Vidrih, M. (2019) The effect of a mixture of two plant growth-promoting bacteria from Argentina on the yield of potato, and occurrence of primary potato diseases and pest–short communication. *Acta Agriculturae Scandinavica, Section B—Soil & Plant Science* 69(1), 89–94. DOI: 10.1080/09064710.2018.1492628.

Tyagi, R., Pradhan, S., Bhattacharjee, A., Dubey, S. and Sharma, S. (2022) Management of abiotic stresses by microbiome-based engineering of the rhizosphere. *Journal of Applied Microbiology* 133(2), 254–272. DOI: 10.1111/jam.15552.

Vaishnav, A., Singh, J., Singh, P., Rajput, R.S., Singh, H.B, *et al.* (2020) Sphingobacterium sp. BHU-AV3 induces salt tolerance in tomato by enhancing antioxidant activities and energy metabolism. *Frontiers in Microbiology* 11, 443. DOI: 10.3389/fmicb.2020.00443.

Vanissa, T.T.G., Berger, B., Patz, S., Becker, M., Turečková, V. *et al.* (2020) The response of maize to inoculation with *Arthrobacter* sp. and *Bacillus* sp. in phosphorus-deficient, salinity-affected soil. *Microorganisms* 8(7), 1005. DOI: 10.3390/microorganisms8071005.

Vega-Celedon, P., Bravo, G., Velasquez, A., Cid, F.P., Valenzuela, M. *et al.* (2021) Microbial diversity of psychrotolerant bacteria isolated from wild flora of Andes mountains and Patagonia of Chile towards the selection of plant growth-promoting bacterial consortia to alleviate cold stress in plants. *Microorganisms* 9(3), 538. DOI: 10.3390/microorganisms9030538.

Velasquez, A.C., Castroverde, C.D.M. and He, S.Y. (2018) Plant-pathogen warfare under changing climate conditions. *Current Biology* 28(10), R619–R634. DOI: 10.1016/j.cub.2018.03.054.

Vilvert, E., Stridh, L., Andersson, B., Olson, Å., Aldén, L. *et al.* (2022) Evidence based disease control methods in potato production: A systematic map protocol. *Environmental Evidence* 11(1), 6. DOI: 10.1186/s13750-022-00259-x.

Vishnupradeep, R., Bruno, L.B., Taj, Z., Karthik, C., Challabathula, D. *et al.* (2022) Plant growth promoting bacteria improve growth and phytostabilization potential of *Zea mays* under chromium and drought stress by altering photosynthetic and antioxidant responses. *Environmental Technology and Innovation* 25, 102154. DOI: 10.1016/j.eti.2021.102154..

Vrieze, M., Germanier, F., Vuille, N. and Weisskopf, L. (2018) Combining different potato-associated pseudomonas strains for improved biocontrol of *Phytophthora infestans*. *Frontiers in Microbiology* 9, 2573. DOI: 10.3389/fmicb.2018.02573.

Wang, B.X., Hof, A.R. and Ma, C.S. (2022) Impacts of climate change on crop production, pests and pathogens of wheat and rice. *Frontiers of Agricultural Science and Engineering* 9(1), 4–18. DOI: 10.15302/j-fase-2021432.

Wang, J., Li, R., Zhang, H., Wei, G. and Li, Z. (2020b) Beneficial bacteria activate nutrients and promote wheat growth under conditions of reduced fertilizer application. *BMC Microbiology* 20(1), 38. DOI: 10.1186/s12866-020-1708-z.

Wang, X.H., Li, Q., Sui, J.K., Zhang, J.M., Liu, Z.Y. *et al.* (2019) Isolation and characterization of antagonistic bacteria *Paenibacillus jamilae* HS-26 and their effects on plant growth. *Biomed Research International* 3638926. DOI: 10.1155/2019/3638926.

Wang, X.H., Zhao, C., Muller, C., Wang, C.Z., Ciais, P. *et al.* (2020a) Emergent constraint on crop yield response to warmer temperature from field experiments. *Nature Sustainability* 3(11), 908–916. DOI: 10.1038/s41893-020-0569-7..

Wei, T., Sun, Y., Yashir, N., Li, X., Guo, J. *et al.* (2022) Inoculation with rhizobacteria enhanced tolerance of tomato (*Solanum lycopersicum* L.) plants in response to cadmium stress. *Journal of Plant Growth Regulation* 41(1), 445–460. DOI: 10.1007/s00344-021-10315-4.

Yang, A., Akhtar, S.S., Fu, Q., Naveed, M., Iqbal, S. *et al.* (2020) *Burkholderia phytofirmans* PsJN stimulate growth and yield of quinoa under salinity stress. *Plants* 9(6), 672.

Yang, N., Nesme, J., Roder, H.L., Li, X.J., Zuo, Z.L. *et al.* (2021) Emergent bacterial community properties induce enhanced drought tolerance in Arabidopsis. *NPJ Biofilms and Microbiomes* 7(1), 82. DOI: 10.1038/s41522-021-00253-0.

Yao, Y., Li, Y., Chen, Z., Zheng, B., Zhang, L. *et al.* (2016) Biological control of potato late blight using isolates of trichoderma. *American Journal of Potato Research* 93(1), 33–42. DOI: 10.1007/s12230-015-9475-3.

Zarei, T., Moradi, A., Kazemeini, S.A., Akhgar, A. and Rahi, A.A. (2020) The role of ACC deaminase producing bacteria in improving sweet corn (*Zea mays* L. var saccharata) productivity under limited availability of irrigation water. *Scientific Reports* 10(1), 20361. DOI: 10.1038/s41598-020-77305-6.

Zhang, M., Zhang, C., Zhang, S.S., Yu, H.L., Pan, H.Y. *et al.* (2021) *Klebsiella jilinsis* 2N3 promotes maize growth and induces resistance to northern corn leaf blight. *Biological Control* 156, 104554. DOI: 10.1016/j.biocontrol.2021.104554.

Zheng, L.H., Ma, X., Lang, D.Y., Zhang, X.J., Zhou, L. *et al.* (2022) Encapsulation of *Bacillus pumilus* G5 from polyvinyl alcohol-sodium alginate (PVA-SA) and its implications in improving plant growth and soil fertility under drought and salt soil conditions. *International Journal of Biological Macromolecules* 209, 231–243. DOI: 10.1016/j.ijbiomac.2022.04.017.

18 Climate change and allergy-related impact on humans

Jae-Won Oh*

Department of Pediatrics, College of Medicine, Hanyang University, Hanyang University Guri Hospital, 153 Gyeongchun-Ro, Guri Gyeonggi-Do Korea, Seoul, Korea

Abstract

Climate change represents a massive threat to global health that could affect many disease factors in the twenty-first century. The influence of a changing climate on allergic disease has generated interest among the public and the scientific community. Climate change is a continuing process that affects allergy, and there are multiple potential consequences of climatic change on plant phenology. Impacts on pollen may be one of the most important consequences of climate change on human health. A growing number of people are contracting allergic diseases caused by pollen. Airborne pollen is one of the common causatives and triggering agents of respiratory allergy in a changing climate. Moreover, allergenic plants grow in soil disturbed by human activities and deforestation with air pollution, and it is impossible to avoid plants that cause allergies because pollen can travel many kilometers on the wind. Weather conditions may alter pollen concentrations. A number of studies have shown that increased CO_2 concentration and atmospheric temperature increase pollen concentration. Hence, most works on the impact of climate change on aeroallergens include its impact on the amount and allergenicity of pollen. It is as yet unknown what complex interactions between pollens, meteorological variables and air pollutants occur as a result of climate change. Considering the effect of climate change on long-term trends in pollen levels and emerging viral infections, it is crucial to report and forecast the associated risk for future human health.

Keywords: pollen, allergy, Climate Change, allergy sensitization, pollen season

Introduction

It is now widely accepted that the temperature of the earth is increasing, as confirmed by the warming of the oceans, rising sea levels, glaciers melting, sea ice retreating in the Arctic, and diminished snow cover in the Northern Hemisphere. Moreover, changes are also occurring in the amount, intensity, frequency, and type of precipitation and the occurrence of extreme events such as heat waves, droughts, floods and hurricanes. As stated in the recent Working Group I Report of the Intergovernmental Panel on Climate Change, human-induced warming reached approximately 1°C (likely between 0.8°C and 1.2°C) above pre-industrial levels in 2017, increasing at 0.2°C (likely between 0.1°C and 0.3°C) per decade (high confidence) (Mitchell *et al.*, 2018). Major changes involving the atmosphere and climate have a major impact on the biosphere and human environment. Many

*jaewonoh@hanyang.ac.kr

prevalent human diseases are linked to climate fluctuations, from cardiovascular mortality and respiratory illnesses due to heatwaves to altered transmission of infectious diseases and malnutrition from crop failures (Epstein, 2005; Mitchell *et al.*, 2018). Therefore, climate change represents a threat to global health that could affect many disease factors in the twenty-first century.

Among previous reports, climate change influences the development of asthma and allergic diseases and influences pollen production that induces seasonal allergic manifestations. The influence of a changing climate on allergic disease has generated interest among the public and the scientific community. The climate is constantly changing and variable over time and is dependent on incoming solar radiation, outgoing thermal radiation and the composition of the atmosphere on earth. Climate change is a constant process that affects allergies. Impacts on pollen may be one of the important consequences of climate change on human health. There are multiple interrelated, potential consequences of climatic change on plant phenology. Therefore, climate change can subsequently influence the occurrence of pollen allergy.

The rapid proliferation of trees and weeds, as well as air pollution by greenhouse gases caused by the increase in traffic and the construction of modern buildings and factories, constitute a harmful effect conducive to allergies. Nobody can be protected from allergies associated with plants by staying in the cities as distinct from rural areas. Plants producing allergenic pollen grow abundantly in areas where humans are disturbing the natural environment to build houses, roadways and agricultural activities. Plants producing allergenic pollen such as ragweed grow best in soil disturbed by human activities and deforestation in a changing world. It is almost impossible to avoid pollen from these plants which causes allergies because pollen can travel many kilometers in the wind (Wopfter *et al.*, 2005). However, the ability to avoid large doses is critical for sensitive people. The intensity of the allergic reaction depends on the amount of exposure. Weather conditions, including rainfall, atmospheric temperature, humidity, wind speed and direction, may alter the concentrations of plant pollens and other allergens, which can subsequently influence the occurrence of allergic diseases (Kim *et al.*, 2021).

A growing number of people are suffering from allergic diseases caused by pollen because of climate change. Moreover, the seasonal and regional variations of pollen have changed globally. As the sensitization rate to pollen has increased in children recently, rapidly proliferating allergenic plants have emerged as a dangerous element to allergic children (Lee *et al.*, 2021). Most of the work on the impact of climate change on aeroallergens can be divided into a number of distinct areas, including the impact on the amount, allergenicity, season of pollen, as well as distribution of allergenic plants (Ziska and Beggs, 2014; Schmidt, 2016). Studies of the effect of climate change on the distribution of allergenic pollens have focused typically on the analysis of observed pollen counts and their regression relationships with local meteorological and climatic factors (Epstein, 2005; Haines *et al.*, 2006; Reid and Gamble, 2009). The onset, duration and intensity of the pollen season vary annually. Weather variables and CO_2 are among the main factors affecting phenology and pollen production by plants (Garcıá-Mozo *et al.*, 2002; Singer *et al.*, 2005). In addition, weather patterns influence the movement and dispersion of all aeroallergens in the atmosphere through the action of winds and rainfall, depending on atmospheric stability (Zhang *et al.*, 2014; Rojo *et al.*, 2021).

Review Methodology

I focused on recent works regarding research on the relationship between pollen allergy and climate change such as CO_2 level and allergic sensitization rate to pollen recently and also searched databases including Google Scholar, and ScienceDirect (Keyword search terms for thunderstorm-induced asthma and the other respiratory diseases).

I also used the references from the articles obtained by this method to check for additional relevant material and the recent book on pollen allergy.

Climate Change and Respiratory Allergy

The effects of climate changes on respiratory allergy are still unclear and current knowledge is provided by epidemiological and experimental studies on the relationship between asthma and environmental factors, such as meteorological variables, and airborne allergens such as pollen and air pollution. Data about the influence of weather on asthma are poor and inconclusive. Weather affects asthma directly, acting on airways, or indirectly, influencing airborne allergens and pollutant levels. The complexity of the aerosol reaching the airways and the several compounds that play a role in this relationship might explain the controversial results of studies conducted so far.

Numerous studies have shown that air pollution is consistently associated with adverse health effects, and it has a quantifiable impact on respiratory diseases, cardiovascular diseases and so on. Air pollution and aeroallergens seem critical in evaluating the possible effects of climate change on allergic respiratory diseases. An increasing body of evidence suggests that climate change might affect both environmental factors. Data linking changes in environmental variables and changes in the incidence and prevalence of asthma are still lacking, even if an increasing body of evidence shows the adverse effects of ambient air pollution on allergic respiratory diseases (Rossi *et al.*, 1993; Eccles, 2002; Prather *et al.*, 2003; Haines *et al.*, 2006; Confalonieri *et al.*, 2007). Possible impacts of climate change on allergy and asthma prevalence are beyond the scope of this work, but the role of predictable changes in air pollution and aeroallergen exposure will be considered.

Pollen as an Aeroallergen

Though representing only a small proportion of the airborne particles present in the atmosphere, pollen grains can be causative agents of allergic respiratory responses in pollen-allergic subjects, and pollen allergy, because of its elevated prevalence and associated costs, is now a public health problem (D'Amato *et al.*, 2007; D'Amato *et al.*, 2011).

Given the size and weight of pollen particles, as aeroallergens they fall in the 20–60 μm range of most particulate matter and their allergenic constituents are proteins with a molecular size between 10,000 and 40,000 Da. Protective mechanisms in the nasal mucosa and upper tracheobronchial passages remove most of the larger particles, so only those of 5 μm or smaller reach the alveoli of the lungs. These are considerations in the pathogenesis of allergic rhinitis, bronchial asthma and hypersensitivity pneumonitis, as well as the irritant effects of chemical and particulate pollutants (Stewart, 2000).

Climate change may affect air pollutant levels in several ways: the influence on regional weather may have an effect on the severity and frequency of air pollution episodes and anthropogenic emissions; the enhancement of the urban heat island effect may increase some secondary pollutants such as ozone, and it can indirectly increase natural sources of air pollutant emissions such as decomposition of vegetation, soil erosion and wildfires (Gangoiti *et al.*, 2001; Expert Group on Climate Change and Health in the UK, 2001; Mickley *et al.*, 2004; D'Amato, 2004; Beggs and Bambrick, 2005; Moore *et al.*, 2006). Tropospheric ozone (O_3) is formed in the presence of bright sunshine and high temperatures by the reaction between volatile organic compounds (VOC) and nitrogen oxides (NOx), both emitted from natural and anthropogenic sources. O_3 concentrations are increasing in most regions, and this trend is expected to continue over the next 50 years.

Climate change appears to induce an increased concentration of all health-related air pollutants. Of particular concern are potential changes in tropospheric O_3. However, the consequence of higher temperature on O_3 concentration might be partially counterbalanced by a decrease in demand for heating systems during mild winters due to global warming. Prediction about the effects of climate change on health-related air pollution is hampered by several limits: future emissions depend on numerous factors, such as population growth, economic development, energy use and production; current knowledge about weather effects on air pollution is still unsatisfactory; there is still a need for better emission inventories and observational datasets, and long-term effects and actual enforcement of international agreements to reduce air pollution and greenhouse gases emissions are unpredictable (Beck *et al.*, 2013).

Effects of Weather on Allergen Load and Pollen Potency

Plant species require a certain amount of heat to complete their development; then air temperature plays a key role, together with other factors, such as day length, water and nutrient availability, and soil type. Increasing worldwide evidence demonstrates that the timing of life cycle events of a large number of species has responded to the increase in Earth's temperature. Changes involve also plants producing allergenic pollen, with expected consequences on atopic populations (Frenguelli, 2002; Beggs, 2004; Teranishi *et al.*, 2006; Cecchi *et al.*, 2007). Data provided by 30 years of observations within the International Phenological Gardens Network showed that spring events advanced by 6 days, the highest rate of phenological changes observed in Europe (Cleland *et al.*, 1972).

Both climate and weather play a key role in the production, release and bioavailability of pollen-derived allergens. Plants discriminate day from night by means of photoreceptors, that is, pigments that capture different wavelengths that may promote or inhibit flowering. These pigments synchronize biological activities with the day and night cycle. Plants may respond to day lengths exceeding a critical threshold in late spring or early summer, securing time for seed maturation. Increasing light intensity during springtime and early summer is likely to influence flowering in summer-flowering plants and modify the effect of the photoperiod.

The variation in pollen potency can be attributed to the synthesis of allergens in pollen during the ripening of the pollen. In the week before pollination, no allergens are found in pollen, which otherwise is ready to release. Subsequently, a rapid increase in pollen allergen content occurs along with the development of anthers. Anthers react to weather conditions, and when the conditions are optimal for pollination, they open and release pollen. If the weather conditions are favorable for early pollination, ripe anthers will release less ripe pollen with less allergen. Otherwise, when the weather conditions delay the opening of the anthers, riper pollen with more allergen is released (Oh, 2018).

Pollen season is defined as the period during which pollen is present in the air. The term is used either collectively, referring to pollen from any plant taxon or from each taxon separately. The pollen season in a certain area is related to the local flowering season, as for pollen to be present in the air, it must be produced and released by mature flowers. However, pollen seasons and flowering seasons usually do not fully coincide because of the effects of mid- and/or long-range transport. In temperate regions, the main pollination period covers approximately half the year, from spring to autumn. The effect of temperature on pollination is greater in spring and early summer-flowering plants, whereas pollination of plants that flower in late summer and fall is generally correlated with their photoperiod. The complex relationship between weather and pollen concentration in the atmosphere plays a key role in modulating allergen levels (Oh, 2018).

Impact of Climate Change

The impact on pollen may be one of the most important consequences of climate change on human health. There are multiple interrelated potential consequences of climatic change on plant phenology. Because the influence of climate change is complex, there is no predictable quantitative assessment of how climate change may affect the levels of pollen allergy in humans in the future (Reid and Gamble, 2009).

An altered climate will affect the range of allergenic plant species and the length of the pollen season, and CO_2 is one of the most important greenhouse gases. Its atmospheric concentration has increased from a pre-industrial value of about 280 to 380 ppm recently. Elevated atmospheric CO_2 levels may increase plant productivity and pollen production affecting plant ecology. As CO_2 absorbs infrared radiation (heat), an increase in atmospheric concentrations will elevate surface temperature. Since 1970, the average surface air temperature has increased by approximately 1.6 °F (up to 1.0°C). A number

Fig. 18.1. (a) Open top chambers at the National Institute of Forest Science in Suwon, Korea. Each chamber has different CO_2 concentration for 1 year and 400 ppm of CO_2 express recent concentration, 600 ppm and 800 ppm of CO_2 express future air concentration, and (b) its internal view showing a CO_2 injection nozzle and white oak plants. Permitted from Kim *et al*. (2018).

of studies reported a significant increase in allergenic proteins with a comparison of atmospheric CO_2 concentration among the current, and future levels based on the climate chamber experiments (Fig. 18.1). Under future higher CO_2 concentrations, people will be exposed to higher allergen levels from oak pollen. However, the actual level of elevated allergy risk is difficult to predict due to the underlying plant physiology and ecology resulting from complex impacts of planetary environment change, including air temperature, rainfall, soil nutrition and changing flowering periods (Kim *et al*., 2018; Choi *et al*., 2018).

Change of pollen season

Weather conditions, including rainfall, humidity, wind speed and air temperature, may alter pollen concentrations. Increased temperatures may lead to earlier pollination of many plants and a longer duration of pollination. Additionally, climate change is thought to increase the quantity of allergenic protein present in individual pollen grains.

The duration of the pollen season is also extended, especially in summer and in late flowering species. Moreover, there is some evidence of significantly stronger allergenicity in pollen from trees growing at increased temperatures. Changes also involve plants producing allergenic pollen, with expected consequences on allergic individuals (Beggs, 2004; Zhang and Steiner, 2022; Teranishi *et al*., 2006). Pollen production started earlier and ended later with the year and extended the duration of pollen season over the last 20 years in Korea. This trend might be suggesting that longer pollen seasons increase the duration of exposure, potentially resulting in more sensitization (Fig. 18.2) (Lee *et al*., 2021; Kim *et al*., 2012). Longer pollen seasons also lead to longer symptomatic periods in those with allergic disease, and higher pollen concentrations may lead to more severe symptoms. An earlier start of the pollen season was confirmed in studies focused on allergenic plants. Due to the earlier onset, the pollen seasons are altered more often by weather conditions in late autumn or early spring (Emberlin *et al*., 1999; Ziska *et al*., 2003). Ziska *et al*. recently demonstrated a clear positive correlation between recent global warming and an increase in the seasonal duration and pollen concentrations for multiple allergenic plant species on a decadal basis for the northern hemisphere. This study demonstrated a significant increase in pollen season duration over time, increasing by 0.9 days per year on

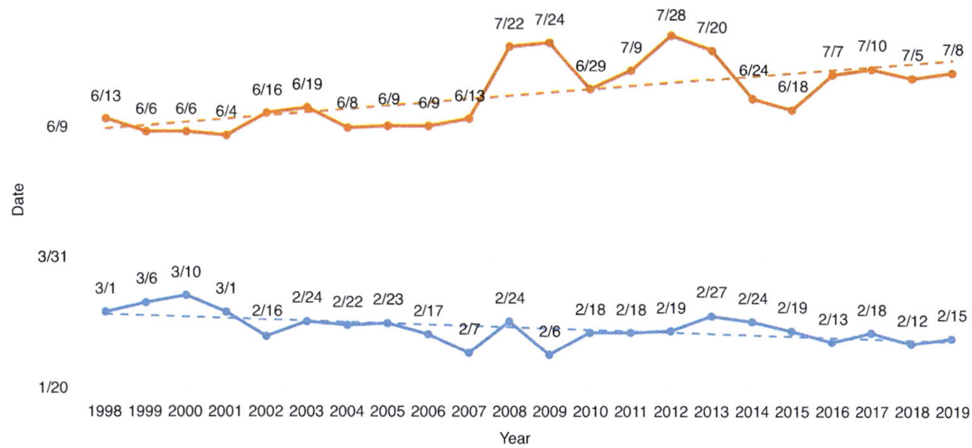

Fig. 18.2. The start and end dates of the pollen season for the major allergenic tree pollens in the Seoul metropolitan area during the past 22 years. Pollen production has started earlier and ended later with year and extended the duration of pollen season for last 22 years in Korea. This trend might be suggested that longer pollen seasons increase the duration of exposure, potentially resulting in more sensitization. Permitted from Lee *et al.* (2021).

average. These data indicate that recent climatic changes, through temperature, are affecting pollen concentrations as well as season duration and timing in the northern hemisphere (Ziska *et al.*, 2019).

Increased pollen sensitization rates with changing pollen season

Sensitization to pollen plays a crucial role in the development of allergic disorders. However, the effect is less pronounced in individuals sensitized to only one type of pollen than in those sensitized to many types of pollen. Since different pollen types are released at different periods of the pollen season, atopic individuals sensitized to more than one type of pollen have a longer period of exposure. The duration of the pollen season is also influenced by factors affecting pollen dispersal. A longer pollen season and larger amounts of atmospheric pollen may enhance human exposure to allergenic pollens, potentially leading to an increase in allergic sensitization (Schoos *et al.*, 2019).

A recent study demonstrated a positive correlation between changes in the pollen season over 20 years and sensitization rates to major tree pollens among children. With climate change, a continuous increase in the concentrations of tree pollens and thus in sensitization rates, furthermore, the incidence of pollen allergy which has been more common in adults over teenage, especially in younger children, can be expected and pollen-sensitization rates increased in children less than 10 years of age relatively compared with those over 10 years of age for over 20 years (Fig. 18.3) (Lee *et al.*, 2021).

Thunderstorm-related asthma

It is not clear if an increase in thunderstorm events has been caused directly by climate change. However, modeling each extreme event is neither feasible nor necessary. All weather events should now be considered in the context of climate change. There is evidence that, during pollen season, thunderstorms can be associated with allergic asthma outbreaks in patients suffering from pollen allergy (Suphioglu *et al.*, 1992; Celenza *et al.*, 1996; Antò and Sunyer, 1997; Newson *et al.*, 1998; Hew *et al.*, 2019). Pollen grains can be carried by thunderstorms at ground level, where, after rupturing by osmotic shock, they may release into the atmosphere part of their content, including inhalable

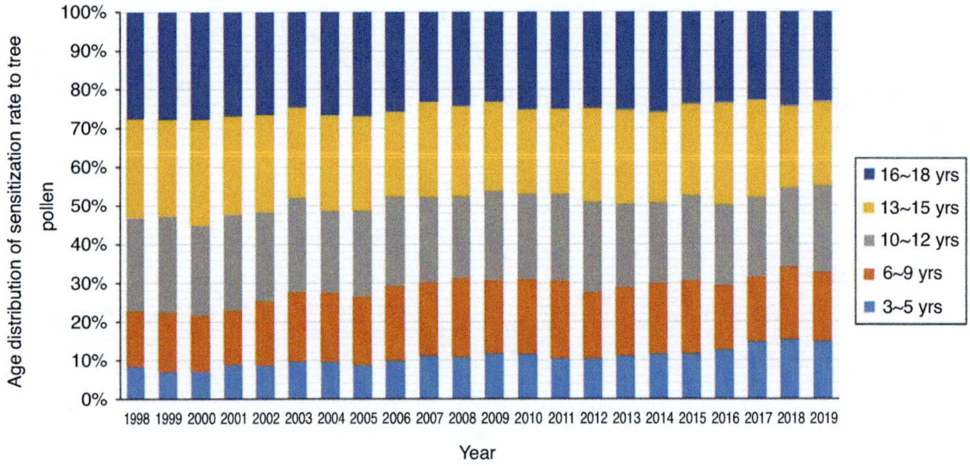

Fig. 18.3. Age distribution of sensitization rates to the major pollens such as oak, birch, alder and elm, and so on in the Seoul metropolitan area during the study period. Between 1998 and 2019, pollen-sensitization rates increased in children less than 10 years of age relatively compared with the other age range. Permitted from Lee *et al*. (2021).

allergen-carrying cytoplasmic starch granules (<5μ) or other paucimicronic components. Such small particles may be inhaled into the distal airways and induce asthma in subjects with pollen allergies (Suphioglu *et al.*, 1992; Hew *et al.*, 2019). Demonstrations have shown that changes in the weather such as rain or humidity may induce hydration of pollen grains and sometimes also their fragmentation which generates atmospheric biological aerosols carrying allergens (Celenza *et al.*, 1996; Antò and Sunyer, 1997). During the first phase of a thunderstorm, patients suffering from pollen allergy may inhale a high concentration of the allergenic material, like biological antigenic aerosols dispersed in the atmosphere, which can induce asthmatic reactions in some cases (Celenza *et al.*, 1996; Antò and Sunyer, 1997; Newson *et al.*, 1998; Girgis *et al.*, 2000). Thunderstorms can induce attacks of severe asthma (D'Amato *et al.*, 1994; Rosas *et al.*, 1998; Marks *et al.*, 2001; Wardman *et al.*, 2002) and, in some circumstances, are a common cause of epidemics of exacerbations of asthma requiring attendance at an Emergency Department. A study in New South Wales, Australia found that thunderstorm outflows were detected in 33% of epidemics, compared with 3% of non-epidemic days (Marks *et al.*, 2001).

Pollen and respiratory virus

A recent, large cohort study from South Korea reported that asthma exacerbations in school-aged children are associated with co-exposure to multiple seasonal environmental factors, that is, ozone, rhinovirus and tree pollen (Lee *et al.*, 2019; Park *et al.*, 2016). A number of recent studies reported pollen exposure weakens the immunity against certain seasonal respiratory viruses by diminishing the antiviral interferon response. A study recently reported that pollen grains of various plant taxa release as yet unidentified compounds that down-modulate the production of antiviral λ-interferons in respiratory epithelial cells, and provided evidence from human and mouse models that pollen exposure leads to enhanced susceptibility to infection with two different respiratory viruses: human rhinovirus and respiratory syncytial virus (Gilles *et al.*, 2020). Damialis *et al.* recently demonstrated 44% of the infection rate variability. Infection rates increased after higher pollen concentrations most frequently during the four previous days. Without lockdown, an increase of pollen abundance by 100 pollen/m^3 resulted in a 4% average increase in infection rates. Their dataset is the most comprehensive: the dataset is the largest possible, worldwide from 130 stations, across 31 countries including Korea and five continents. This study suggested two mechanisms of the innate immune response, inflammasome activation and antiviral interferon response of pollen, which appear to be modulated towards the same direction by pollen and SARS-CoV-2 (Damialis *et al.*, 2021).

Forecast for Pollen

To reduce the exposure to pollen and improve the measures listed earlier, it is crucial to provide a reliable aerobiological forecast to patients with pollen allergies. Although the quality of the weather forecast has improved substantially in the last decades, the effects of weather variables on aeroallergen load are not completely understood. For instance, rainfall clears the pollen from the air, thus reducing the risk of exacerbation of allergic rhinitis and asthma (Rossi *et al.*, 1993; Jamason *et al.*, 1997).

Weather conditions, including rainfall, atmospheric temperature, humidity, wind speed and wind direction, may alter the concentrations of plant pollens and other allergens, which can subsequently influence the occurrence of allergic diseases such as asthma, allergic rhinitis, allergic conjunctivitis and even atopic dermatitis (Lee *et al.*, 2012; Kim *et al.*, 2017). Many studies have demonstrated that increased CO_2 concentration and atmospheric temperature increase pollen concentration (Kim *et al.*, 2018; Choi *et al.*, 2018). Most work on the impacts of climate change on aeroallergens can be divided into a number of distinct areas, including the impacts on pollen amount, pollen allergenicity, pollen season and plant and pollen distribution. Meteorological factors such as mean temperature, wind speed, humidity, amount of sunlight and degree days can directly affect the biological and chemical components of this interaction (Reid and Gamble, 2009; Beggs, 2010). The accumulation of daily sunshine, and other influencing meteorological factors such as humidity and precipitation, should be considered to further improve the accuracy of the modeled start dates and season lengths of birch and oak pollen and develop regression models for pollen prediction. Daily fluctuations in pollen concentration are linked to a variety of meteorological factors such as temperature, rainfall and the duration of sunshine. Temperature and rainfall are especially important in determining pollen concentrations, although the relationship is complex and influenced by other variables (Seo *et al.*, 2020). The Korea Meteorological Administration developed monthly pollen concentration models and has forecasted trees/weeds pollen risks in spring and autumn. The probability density function regression model was developed using daily weather variables and the daily pollen concentration during the past 20 years.

Conclusion

Looking to the future, it is as yet unknown whether complex interactions with pollens, meteorological variables and air pollutants may also play a role in a changing planetary environment. Even though there is published evidence on the effects of various environmental parameters, these usually refer to preliminary results and investigations of only a single factor. If one takes into account the huge effect of climate change and globalization on the long-term trends in airborne pollen levels, as well as emerging viral infections, it must be of utmost importance to forecast and warn of the associated risk for human health in future pandemics and take appropriate measures to reduce it as much as possible. However, many things are changing and we do not know what will happen, therefore we should monitor and develop an effective prediction model.

Conflicts of Interest

The authors have no conflicts of interest to declare.

References

Antò, J.M. and Sunyer, J. (1997) Thunderstorms: A risk factor for asthma attacks. *Thorax* 52, 669–670.

Beck, I., Jochner, S., Gilles, S., McIntyre, M., Buters, J.T. *et al.* (2013) High environmental ozone levels lead to enhanced allergenicity of birch pollen. *PLOS One* 20, 8–e80147.

Beggs, P.G. and Bambrick, H.J. (2005) Is the global rise of asthma an early impact of anthropogenic climate change? *Environmental Health Perspectives* 113, 915–919.

Beggs, P.J. (2004) Impacts of climate change on aeroallergens: Past and future. *Clinical and Experimental Allergy* 34, 1507–1513.

Beggs, PJ. (2010) Adaptation to impacts of climate change on aeroallergens and allergic respiratory diseases. *International Journal of Environmental Research and Public Health* 7, 3006–3021.

Cecchi, L., Torrigiani Malaspina, T., Albertini, R., Zanca, M., Ridolo, E. *et al.* (2007) The contribution of long-distance transport to the presence of ambrosia pollen in central northern Italy. *Aerobiologia* 23, 145–151.

Celenza, A., Fothergill, J., Kupek, E. and Shaw, RJ. (1996) Thunderstorm associated asthma: A detailed analysis of environmental factors. *BMJ* 312, 604–607.

Choi, Y.J., Oh, H.R., Oh, J.W., Kim, K.R., Kim, B.J. *et al.* (2018) Chamber and field studies demonstrate differential amb a 1 contents in common ragweed depending on CO2 levels. *Allergy, Asthma and Immunology Research* 10, 278–282.

Cleland, E.E., Chuine, I., Menzel, A., Mooney, H.A. and Schwartz, M.D. (1972) Allergen carriage by atmospheric aerosol. I. Ragweed pollen determinants in smaller micronic fractions. *The Journal of Allergy and Clinical Immunology* 50, 289–293.

Confalonieri, U., Menne, B., Akhtar, R., Ebi, K.L., Hauengue, M. *et al.* Parry, M.L., Canziani, O.F., Palutikof, J.P., van der Linden, P.J. and Hanson, C.E. (eds). (2007) *Impacts, adaptation and vulnerability*. Contribution of working group II to the fourth assessment report of the intergovernmental panel on climate change, Cambridge University Press, Cambridge, UK, pp. 391–431.

D'Amato, G., Liccardi, G., Gilder, J.A., Baldacci, S. and Viegi, G. (1994) Thunderstorm-associated asthma in pollinosis patients. *BMJ* 309, 131–133.

D'Amato, G. (2004) Air pollution, reactive oxygen species, and allergic bronchial asthma: the therapeutic role of antioxidants. In: Hughes, D.A., Darlington, L.G. and Bendich, A. (eds) *Diet and Human Immune Function*. Humana Press, Totowa, New Jersey, USA, pp. 397–409.

D'Amato, G., Cecchi, L., Bonini, S., Nunes, C., Annesi-Maesano, I. *et al.* (2007) Allergenic pollen and pollen allergy in Europe. *Allergy* 62, 976–990.

D'Amato, G., Rottem, M., Dahl, R., Blaiss, M., Ridolo, E. *et al.* (2011) Annesi-maesano and the WAO special committee on climate change and allergy. Climate change, migration and allergic respiratory diseases: An update for the allergist. *World Allergy Organization Journal* 4, 120–125.

Damialis, A., Gilles, S., Sofiev, M., Sofieva, V., Kolek, F. *et al.* (2021) COVID-19/POLLEN study group, Traidl-Hoffmann C. Higher airborne POLLEN concentrations correlated with increased SARS-cov-2 infection rates, as evidenced from 31 countries across the globe. *Proceedings of the National Academy of Sciences of the United States of America* 118, e2019034118.

Eccles, R. (2002) An explanation for the seasonality of acute upper respiratory tract viral infections. *Acta Oto-Laryngologica* 122, 183–191.

Emberlin, J., Mullins, J., Corden, J., Jones, S., Millington, W. *et al.* (1999) Regional variations in grass pollen seasons in the UK, long-term trends and forecast models. *Clinical and Experimental Allergy* 29, 347–356.

Epstein, PR. (2005) Climate change and human health. *The New England Journal of Medicine* 353, 1433–1436.

Expert Group on Climate Change and Health in the UK (2001) Health effects of climate change in the UK. Department of Health, London, pp. 199–214.

Frenguelli, G. (2002) Interactions between climate changes and allergenic plants. *Monaldi Archives for Chest Disease* 57, 141–143.

Gangoiti, G., Millan, M.M., Salvador, R. and Mantilla, E. (2001) Long-range transport and re-circulation of pollutants in the western Mediterranean during the project regional cycles of air pollution in the west-central Mediterranean area. *Atmospheric Environment* 35, 6267–6276.

Garcıá-Mozo, H., Galán, C., Belmonte, J., Fernández, D. and Rodriguez, FJ. (2002) Modelling start of oak pollen season in different climatic zones in Spain. *Agricultural and Forest Meteorology* 110, 247–257.

Gilles, S., Beck, I., Lange, S., Ring, J., Behrendt, H. *et al.* (2020) Pollen exposure weakens innate defense against respiratory viruses. *Allergy* 75, 576–587.

Girgis, S.T., Marks, G.B., Downs, S.H., Kolbe, A., Car, G.N. *et al.* (2000) Thunderstorm associated asthma in an inland town in southeastern Australia. Who Is at Risk?. *European Respiratory Journal* 16, 3–8.

Haines, A., Kovats, R.S., Campbell-Lendrum, D. and Corvalan, C. (2006) Climate change and human health: Impacts, vulnerability, and mitigation. *Lancet* 367, 2101–2109.

Hew, M., Lee, J., Susanto, N.H., Prasad, S., Bardin, P.G. *et al.* (2019) The 2016 Melbourne thunderstorm asthma epidemic: Risk factors for severe attacks requiring hospital admission. *Allergy* 74, 122–130.

Jamason, P.F., Kalkstein, L.S. and Gergen, PJ. (1997) A synoptic evaluation of asthma hospital admissions in New York city. *American Journal of Respiratory and Critical Care Medicine* 156, 1781–1788.

Kim, J.H., Oh, J.W., Lee, H.B., Kim, S.W., Kang, I.J. *et al.* (2012) Changes in sensitization rate to weed allergens in children with increased weeds pollen counts in Seoul metropolitan area. *Journal of Korean Medical Science* 27, 350–355.

Kim, J.R., Han, M.J. and Oh, J.W. (2021) Forecast for pollen allergy: A review from field observation to modeling and services in Korea. *Immunology and Allergy Clinics of North America* 41, 127–141.

Kim, K.R., Kim, M., Choe, H.S., Han, M.J., Lee, H.R. *et al.* (2017) A biology-driven receptor model for daily pollen allergy risk in Korea based on weibull probability density function. *International Journal of Biometeorology* 61, 259–272.

Kim, K.R., Oh, J.W., Woo, S.Y., Seo, Y.A., Choi, Y.J. *et al.* (2018) Does the increase in ambient CO2 concentration elevate allergy risks posed by oak pollen? *International Journal of Biometeorology* 62, 1587–1594.

Lee, H.R., Kim, K.R., Choi, Y.J. and Oh, J.W. (2012) Meteorological impact on daily concentration of pollen in Korea. *Korean Journal of Agricultural and Forest Meteorology* 14, 99–107.

Lee, S.K., Kim, J.H., Choi, Y.J., Yang, S., Kim, C.R. *et al.* (2021) Increased sensitization rates to tree pollens in allergic children and adolescents and a change in the pollen season in the metropolitan area of Seoul, Korea. *Pediatric Allergy and Immunology* 32, 872–879.

Lee, S.W., Yon, D.K., James, C.C., Lee, S., Koh, H.Y. *et al.* (2019) Short-term effects of multiple outdoor environmental factors on risk of asthma exacerbations: Age-stratified time-series analysis. *The Journal of Allergy and Clinical Immunology2019;144:1542–1550.*

Marks, G.B., Colquhoun, J.R., Girgis, S.T., Koski, M.H., Treloar, A.B. *et al.* (2001) Thunderstorm outflows preceding epidemics of asthma during spring and summer. *Thorax* 56, 468–471.

Mickley, L.J., Jacob, D.J. and Field, B.D. (2004) Effects of future climate change on regional air pollution episodes in the United States. *Geophysical Research Letters* 31, L24103.

Mitchell, D., Allen, M.R., Hall, J.W., Muller, B., Rajamani, L. *et al.* (2018) The myriad challenges of the Paris agreement. *Philosophical Transactions of the Royal Society A: Mathematical, Physical and Engineering Sciences* 376(2119), 20180066. DOI: 10.1098/rsta.2018.0066.

Moore, D., Copes, R., Fisk, R., Joy, R., Chan, K. *et al.* (2006) Population health effects of air quality changes due to forest fires in British Columbia in 2003: Estimates from physician-visit billing data. *Canadian Journal of Public Health* 97, 105–108.

Newson, R., Strachan, D., Archibald, E., Emberlin, J., Hardaker, P. *et al.* (1998) Acute asthma epidemics, weather and pollen in England, 1987–1994. *European Respiratory Journal* 11, 694–701.

Oh, J.W. (2018) The formation of pollen. In: Oh, J.W. (ed.) *Pollen Allergy in A Changing World: A Guide to Scientific Understanding and Clinical Practice*. Springer Nature, New York, pp. 9–16.

Park, H.J., Lee, J.H., Park, K.H., Kim, K.R., Han, M.J. *et al.* (2016) A six-year study on the changes in airborne pollen counts and skin positivity rates in Korea: 2008-2013 57, 714–720.

Prather, M., Gauss, M., Berntsen, T., Isaksen, I., Sundet, J. *et al.* (2003) Fresh air in the 21st century? *Geophysical Research Letters* 30, 1100–1103.

Reid, C.E. and Gamble, J.L. (2009) Aeroallergens, allergic disease, and climate change: Impacts and adaptation. *EcoHealth* 6, 458–470.

Rojo, J., Oteros, J., Picornell, A., Maya-Manzano, J.M., Damialis, A. *et al.* (2021) Effects of future climate change on birch abundance and their pollen load. *Global Change Biology* 27, 5934–5949.

Rosas, I., McCartney, H.A., Payne, R.W., Calderón, C., Lacey, J. *et al.* (1998) Analysis of the relationship between environmental factors (aeroallergens, air pollution and weather) and asthma emergency admissions to a hospital Mexico city. *Allergy* 53, 394–401.

Rossi, O.V.J., Kinnula, V.L., Tienari, J. and Huhti, E. (1993) Association of severe asthma attacks with weather, pollen, and air pollutants. *Thorax* 48, 244–248.

Schmidt, CW. (2016) Pollen overload seasonal allergies in a changing climate. *Environmental Health Perspectives* 124, A71–A75.

Schoos, A.M., Jelding-Dannemand, E., Stokholm, J., Bonnelykke, K., Bisgaard, H. *et al*. (2019) Single and multiple time-point allergic sensitization during childhood and risk of asthma by age 13. *Pediatric Allergy and Immunology* 30, 716–723.

Seo, Y.A., Kim, K.R., Cho, C., Oh, J.W. and Kim, TH. (2020) Deep neural network-based concentration model or oak pollen allergy warning in South Korea. *Allergy, Asthma Immunology Research* 12, 149–163.

Singer, B.D., Ziska, L.H., Frenz, D.A., Gebhard, D.E. and Straka, JG. (2005) Increasing Amb a 1 content in common ragweed (*Ambrosia artemisiifolia*) pollen as a function of rising atmospheric CO2 concentration. *Functional Plant Biology* 32, 667–670.

Stewart, G.A. (2000) The molecular biology of allergens. In: Busse, W.W. and Holgate, S.T. (eds) *Asthma and Rhinitis*. Blackwell Scientific, Boston, pp. 1107–1142.

Suphioglu, C., Singh, M.B., Taylor, P., Knox, R.B., Bellomo, R. *et al*. (1992) Mechanism of grass-pollen-induced asthma. *Lancet* 339, 569–572.

Teranishi, H., Katoh, T., Kenda, K. and Hayashi, S. (2006) Global warming and the earlier start of the Japanese-cedar (*Cryptomeria japonica*) pollen season in Toyama, Japan. *Aerobiologia* 22, 90–94.

Wardman, A.E., Stefani, D. and MacDonald, J.C. (2002) Thunderstorm-associated asthma or shortness of breath epidemic: A Canadian case report. *Canadian Respiratory Journal* 9, 267–270.

Wopfter, N., Gadermaier, G., Egger, M., Asero, R., Ebner, C. *et al*. (2005) The spectrum of allergens in ragweed and mugwort pollen. *International Archives of Allergy and Immunology* 138, 337–346.

Zhang, Y. and Steiner, AI. (2022) Projected climate-driven changes in pollen emission season length and magnitude over the continental United States. *Nature Communications*. DOI: 10.1038/s41467-022-28764-0.

Zhang, Y., Bielory, L. and Georgopoulos, PG. (2014) Climate change effect on *Betula* (birch) and *Quercus* (oak) pollen seasons in the United States. *International Journal of Biometeorology* 58, 909–919.

Ziska, L.H. and Beggs, P.J. (2014) Anthropogenic climate change and allergen exposure: The role of plant biology. *International Journal of Biometeorology* 58, 1–6.

Ziska, L.H., Gebhard, D.E., Frenz, D.A., Faulkner, S., Singer, B.D. *et al*. (2003) Cities as harbingers of climate change: Common ragweed, urbanization, and public health. *The Journal of Allergy and Clinical Immunology* 111, 290–295.

Ziska, L.H., Makra, L., Harry, S.K., Bruffaerts, N., Hendrickx, M. *et al*. (2019) Temperature-related changes in airborne allergenic pollen abundance and seasonality across the Northern hemisphere: A retrospective data analysis. *Lancet Planet Health* 3, e124–31.

19 Grazing intensity in grassland ecosystems: implications for carbon storage and functional properties

Deb Raj Aryal[1,2*]

[1]Consejo Nacional de Ciencia y Tecnología (CONACyT), Av. Insurgentes Sur 1582, C.P. 03940, Mexico City, Mexico; [2]Facultad de Ciencias Agronómicas, Universidad Autónoma de Chiapas, Villaflores, Carr. Ocozocoautla—Villaflores km 84.5, C.P. 30470, Chiapas, Mexico

Abstract

Grazing lands cover more than one-third of the terrestrial land surface and are still expanding. The intensity of grazing is also growing with time due to the increase in the number of animals per unit of land or the duration of herbivory. Changes in grazing intensity exert strong pressure on ecosystem functioning, including carbon (C) sequestration. Grazing response to ecosystem C retention can be explained by the changes in biogeochemical processes that regulate C fluxes. Here, we summarized the ecosystem functional properties that affect C storage in grasslands as a result of changing grazing intensities. We synthesized 151 published studies globally and discussed the changes in different ecosystem functional property responses to grazing. High-intensity grazing mostly decreased C storage while light- to moderate-intensity grazing optimized the trade-offs between C sequestration and animal production. The synthesis showed that 86% of the studies reported either an increase or no change in soil organic carbon (SOC) stocks under light grazing while under heavy grazing, 74% of the studies reported a significant decrease in SOC stocks. We also discussed the changes in ecosystem functional properties whereby grazing affects C sequestration. Some of them include: the changes in net primary productivity, soil organisms, enzymatic activity, soil respiration, and the complex interactions of these processes with climate change. Grassland management practices should be ecosystem-specific to ensure optimum production and enhance C sequestration. It is also important to understand the time factor (hysteresis), especially on how long it takes to sequester a certain amount of C from grassland management practices.

Keywords: pasturelands, grazing intensity, rangeland management, mechanisms of carbon sequestration, soil organic matter, belowground processes, net primary productivity

Introduction

To meet the increasing demand for animal food products, land used for livestock grazing is increasing (Song *et al.*, 2018). Globally, the area of grazing land grew from 668 million hectares (Mha) to 3284 Mha between 1700 and 2015 (Klein Goldewijk *et al.*, 2017). Currently, 3312 Mha (about 40%) of the global land surface is used for animal grazing (Houghton and Nassikas, 2017). These are natural rangelands or induced/cultivated pasturelands with grasses

*drajar@conacyt.mx

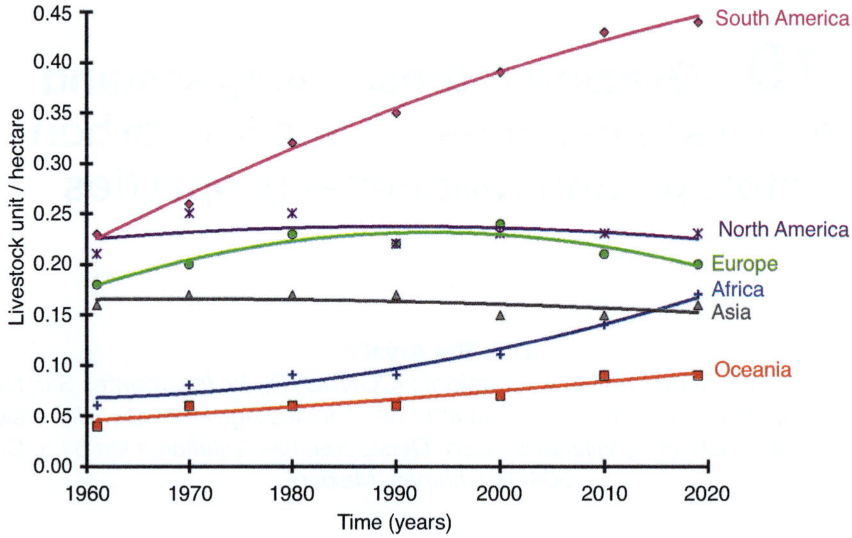

Fig. 19.1. The regional trend of livestock land-use intensity expressed in terms of livestock unit per hectare of agricultural land (data source: FAOSTAT, 2021, accessed on 21/01/2022).

and forbs as the dominant vegetation. It is estimated that grazing lands store about 34% of the global terrestrial carbon (C) (Eze *et al.*, 2018). Land-use change for the expansion of grazing lands, increasing grazing intensity on the same land or producing and transporting feedstuffs for intensive animal farming, all cause C emissions to the atmosphere (Fetzel *et al.*, 2017; Winkler *et al.*, 2021). The sustainability of global grassland ecosystems depends on proper grazing management that focuses on multiple ecosystem services and the conservation of their biodiversity (Zhang *et al.*, 2021). Along with land expansion for animal production, the intensity of grazing has also increased with time in many regions. Between 1961 and 2019, the world-average number of cattle per unit of agricultural land (livestock unit per hectare) changed from 0.14 to 0.20 (FAOSTAT, 2021). However, the regional trends were quite different, with South America showing the highest increase in animal units per hectare (Fig. 19.1).

Grazing intensity can be defined as the relationship between the intensity of herbivory (biomass demand by herbivores) and the net primary production of a land unit during a certain period. The number of animals per unit of land during a period of herbivory, type of animals, the number of days that the animals occupy the paddocks, frequency of herbivory, and fallow events, all determine the intensity of grazing (Petz *et al.*, 2014; Fetzel *et al.*, 2017).

On one hand, C stocks in grassland ecosystems depend on the delicate balance between input and output. On the other hand, animal grazing exerts a strong pressure on the C allocation patterns between the above- and belowground pools because aboveground plant biomass is directly removed during herbivory (Jin *et al.*, 2022). In the meantime, animal urine and fecal deposition may affect belowground processes related to C and nitrogen (N) cycling in the grasslands (Garcia *et al.*, 2021). The ecological responses of belowground components can be different according to the herbivory pressure and trampling (Morales Ruiz *et al.*, 2021). Grazing intensity can have different effects on ecosystem C storage depending on climatic conditions, geography, soil characteristics, plant species composition, and land management practices (Aryal *et al.*, 2018b; Wade *et al.*, 2022). It is also important to note that the discrepancies in trends and patterns of grazing intensity-related C storage variations can partly be attributed to the methodological differences in defining intensity gradients, duration of grazing manipulations, sample size, lack of baseline data, and spatiotemporal coverage (Derner *et al.*, 2019).

A better understanding of the effects of grazing intensity on ecosystem C storage could contribute to the sustainable management of grazing lands, avoid their degradation, mitigate C emissions to the atmosphere, and reduce the greenhouse gas footprint of the animal products (Schuman *et al.*, 2002). Many studies have demonstrated that high-intensity grazing often limits aboveground C storage in grasslands because of herbivory but the effects on belowground C storage are diverse (Wigley *et al.*, 2020; Aryal *et al.*, 2018a). Herbivory from high-intensity grazing removed 77% of annual aboveground biomass production in the grasslands of North Dakota (Shariff *et al.*, 1994). In alpine shrubland ecosystems on the Qinghai-Tibetan Plateau, the belowground biomass of the understory herbaceous plants greatly declined with increasing grazing intensity, contributing to the decrease in soil organic carbon (SOC) stocks (Wang *et al.*, 2022).

Some studies also report that light- to moderate-intensity grazing enhanced C sequestration. A study in US Midwest showed that proper grazing management and feedlot intensification have the potential to offset GHG emissions through soil C sequestration (Stanley *et al.*, 2018). A study in a meadow steppe ecosystem reported a significant decline in both above- and belowground C in assimilation in the heavily grazed grasslands, while belowground C allocation increased in lightly grazed lands (Jin *et al.*, 2022). In a global meta-analysis, a 15% reduction in grassland SOC was attributed to grazing management, with a higher impact in the tropical region (Eze *et al.*, 2018).

In the mixed-grass prairie of Saskatchewan, Canada, SOC stocks increased from 24.7 to 57.4 Mg ha^{-1} with the increase in cattle stocking rates and were attributed to the abundance of introduced vegetation, particularly the rhizomatous grass *Poa pratensis* (Bork *et al.*, 2020).

Prolonged grazing also affects SOC storage, a study found that long-term grazing lowered SOC stocks following the conversion of Hawaiian dry tropical forests to pasturelands (Elmore and Asner, 2006). In contrast, a study in northern Australia mentioned that soil organic matter (SOM) increased under heavy stocking rates compared with moderate stocking (Segoli *et al.*, 2015). In the Basque Country of Northern Spain, regenerative rotational grazing (shorter herbivory and longer fallow period) achieved 30% higher grass production and 3.6% higher SOC storage than conventional grazing (de Otálora *et al.*, 2021). A meta-analysis of 122 studies revealed that heavy grazing decreased SOC content in topsoil and productivity of aboveground vegetation but an increase in belowground biomass and SOC contents in deeper soil horizons (Jiang *et al.*, 2020). High-intensity grazing can also reduce soil water holding capacity and nutrient cycling potential, increase soil erosion, and ultimately reduce C sequestration (Fan *et al.*, 2019). In Inner Mongolian grasslands, long-term grazing exclusion (21–34 years) and rotational grazing significantly increased labile SOC stocks and potential N mineralization rate compared with continuous grazing (Dong *et al.*, 2021). Different grazing regimes also affect elemental coupling and SOM turnover due to the differences in organic matter input and rhizosphere activity (Rumpel *et al.*, 2015). A study synthesized the three main ways that grazing affects SOC storage: (a) reduction in net primary productivity due to direct foraging; (b) differential C allocation between above- and belowground pools; and (c) grazing can alter soil properties through trampling and animal dung/urine addition (Jiang *et al.*, 2020). Grazing responses to grassland ecosystem functional properties are diverse, and there is no consensus on the effect of different grazing intensities on C storage (Fig. 19.2). In this paper, we systematically reviewed how grazing intensity management affects fundamental processes and functional properties related to above-and belowground C sequestration (Fig. 19.3).

Review Methodology

We reviewed 151 published research articles collected from Web of Science, Google Scholar, Scopus, CABI Reviews, and Consorcio Nacional de Recursos de Información Cientifica y Tecnológica (CONRICYT) Mexico database to synthesize the changes in carbon storage and ecosystem functional properties as a result of differences in animal grazing intensity. To search the articles, we used these keywords: grazing

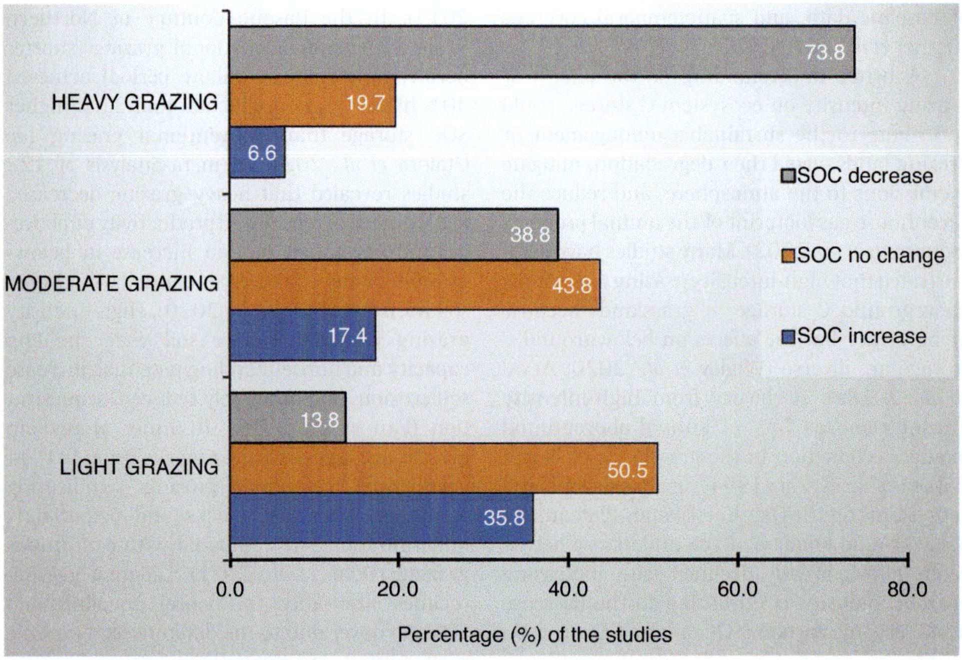

Fig. 19.2. Percentage of the studies (out of 151) reporting increase, decrease, or no change in soil organic carbon stocks at different grazing intensities.

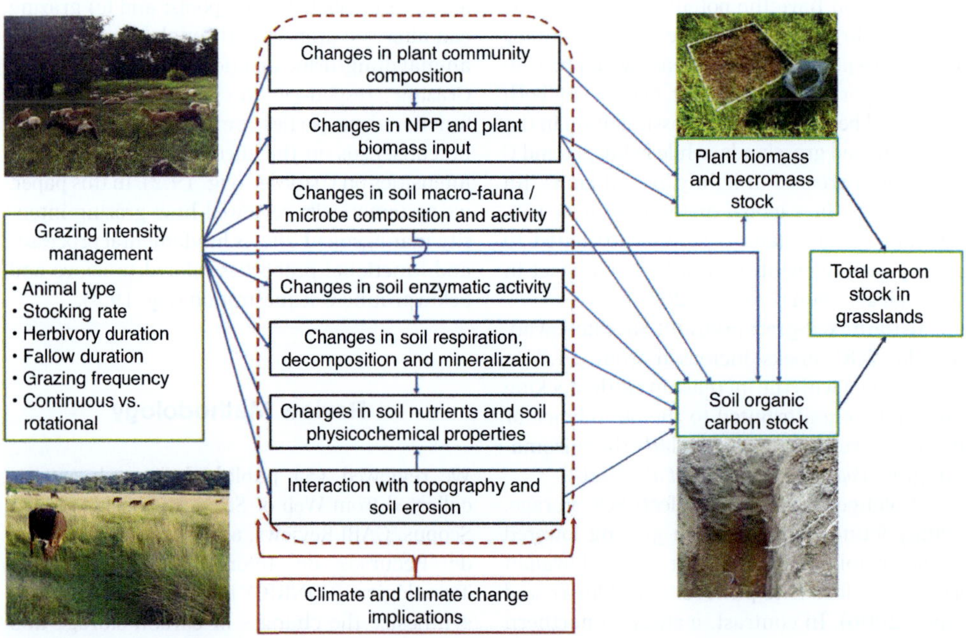

Fig. 19.3. Conceptual diagram linking grazing intensity with possible changes in ecosystem functional properties and carbon storage in grasslands. NPP, net primary production.

intensity, carbon sequestration, carbon storage, grazing management, soil organic carbon, soil respiration, soil micro-organisms, soil enzymatic activity, soil macrofauna, soil nutrients, soil properties, grasslands, rangelands, and pasturelands in different combinations. We used the Boolean operators "AND" and "OR" to connect search terms.

Both meta-analysis and original research papers were included in the review process. Papers published between 1993 and May 28, 2022 are included in the review. Only articles that explain at least one of the mechanisms or ecosystem functional properties related to the grassland C dynamics were considered for the review. We compiled the studies in an excel database and calculated the percentage of the total number of studies reporting either an increase, no change, or decrease in SOC storage as result of light, moderate, or heavy grazing. Grazing regimes (light, moderate, and heavy) were directly extracted from their respective studies without normalization because of the differences in climate, soil, animal species, and land-use history.

Changes in Carbon Storage and Functional Properties

Changes in soil organic carbon stocks

Global soils store about 1505 Pg of C of 1m depth, which is greater than that stored in plant biomass (550 Pg) and atmosphere (860 Pg) (Friedlingstein *et al.*, 2022; Lal, 2018; Scharlemann *et al.*, 2014). In grassland ecosystems, about 89% of the total C is stored in their soils (Eze *et al.*, 2018). The SOC stocks are determined by various interconnected mechanisms and biogeochemical processes that control C fluxes and stabilize or destabilize soil organic matter (SOM) (Dignac *et al.*, 2017). Some of them include the C input from above- and belowground primary productivity, C fluxes from heterotrophic respiration, organic matter decomposition, mineralization, and stabilization (Wiesmeier *et al.*, 2019; Basile-Doelsch *et al.*, 2020). Many biotic and abiotic factors and their interactions affect SOC storage in grasslands, including vegetation type and plant diversity, soil micro-organisms, soil macro-fauna, climate, topography, and soil properties (Basile-Doelsch *et al.*, 2020; Fan *et al.*, 2021; Rattan, 2016; Wiesmeier *et al.*, 2019). A study in North American shortgrass steppe showed that grazing exclusion for more than seven decades altered plant community composition but did not affect total soil C, nitrogen (N), and their turnover rates (Derner *et al.*, 2019). A global meta-analysis demonstrated that continuous grazing significantly reduced SOC and N, while increasing soil compaction compared with rotational or no grazing (Byrnes *et al.*, 2018). Similar results were shown in another study where the authors mentioned that SOC stocks increase at light to moderate grazing, but decrease under heavy or overgrazing (Zhan *et al.*, 2020). In this synthesis, we observed that out of 151 studies reviewed, 35.8% reported an increase in SOC stocks under light grazing, while 50.5% reported no change and 13.8% mentioned a decrease. Under heavy grazing, 73.8% of the studies reported a significant decrease in SOC stocks, while 19.7% reported no change and only 6.6% reported an increase. Under moderate grazing, the majority of the studies reported either no change or an increase in SOC stocks (Fig. 19.2).

A meta-analysis of 129 studies reported that heavy grazing intensity significantly altered the C:N:P stoichiometry of plant biomass and soils in grasslands (He *et al.*, 2020). Another meta-analysis of 115 studies indicated that low-intensity grazing enhances soil C, but moderate- and high-intensity grazing leads to SOC losses (Zhou *et al.*, 2017). Another global meta-analysis of 83 studies showed that the impact of grazing intensity on SOC stocks varied with the climatic region, and grazing intensity below the carrying capacity of the grasslands resulted in a decrease in SOC storage (Abdalla *et al.*, 2018). In contrast, seasonal and continuous grazing increased SOC storage compared with grass-cutting and removal sites in southwestern China (Balasubramanian *et al.*, 2020). In north-eastern Brazil, heavy grazing with cattle and goats significantly depleted SOC stocks in the upper 5 cm, which did not affect SOC storage to deeper soils (Schulz *et al.*, 2016). Another study on the crop-livestock system in Brazil showed that low- to moderate-intensity cattle grazing provided a better synergy between C sequestration and animal productivity (da

Silva *et al.*, 2014). In the Colombian Amazon, 20 years after forest-to-pasture conversion, SOC stocks in heavily grazed sites decreased by 20% compared with the original forest SOC level, whereas in lightly grazed sites SOC increased by 41% (Navarrete *et al.*, 2016). A study in the Himalayan alpine grasslands of India reported an average SOC stock of 90.6 Mg ha^{-1} in non-grazed grasslands and 50.4 Mg ha^{-1} in highly grazed areas, where an increase in herd size resulted in a decrease in SOC stocks (Khan *et al.*, 2019). However, a study in the Inner Mongolian grasslands showed that grazing intensity explained less than 9.5% of the total SOC variations, while climate change explained 15.3–34.9% of it (Xin *et al.*, 2020). Grazing intensity may strongly influence aboveground vegetation and topsoil C (litter, fine roots, and root exudation), while climatic change could be more responsible for the changes in deeper soil C. In many regions, historical management of grasslands may be more important than the current grazing pressure for C accumulation (Xin *et al.*, 2020).

One of the mechanisms of SOC stabilization or destabilization in grasslands is the priming effect from the incorporation of animal manure during grazing (Dignac *et al.*, 2017). A global meta-analysis stated that animal dung and urine addition could increase SOC storage in grasslands due to the priming effect of animal manure incorporation on organic matter turnover (Li *et al.*, 2021). However, it is also demonstrated that increasing grazing intensity could uncouple C and N cycles and increase the risk of N leaching (Soussana and Lemaire, 2014). Since soil organic matter with adequate C:N stoichiometry acts as a buffer to provide nutrients for micro-organisms, such uncoupling can have a negative priming effect on SOC storage (Soussana and Lemaire, 2014).

Changes in NPP and plant biomass input

Grazing continuously removes aboveground plant biomass (AGB) and changes its capacity to regenerate (Fig. 19.3). Doubling of long-term (>30 years) grazing intensity reduced aboveground net primary production by 24% (1.5 ± 0.2 vs. 1.2 ± 0.2 Mg ha^{-1} yr^{-1}) in North American shortgrass steppe and mixed-grass prairie (Irisarri *et al.*, 2016). The authors reported that increased grazing intensity reduced C3 graminoid biomass and increased C4 grass biomass. Differential responses of grazing intensity on forage productivity were reported in the prairie of North Dakota, where light-grazed treatments (1.3 animal unit month per hectare, AUM ha^{-1}) produced significantly higher aboveground biomass (4.2 Mg ha^{-1} yr^{-1}) compared with non-grazed (3.6 Mg ha^{-1} yr^{-1}) and extreme grazed (6.9 AUM ha^{-1}) treatments (2.8 Mg ha^{-1} yr^{-1}) (Patton *et al.*, 2007). The authors mentioned that low-to-moderate grazing increases biomass production more than grazing exclusion, but the grazing effect depends upon the growing conditions during the year. In a northern mixed grass prairie, heavy grazing led to declines in standing dead biomass, litter, and root biomass but an increase in root decomposition (40% vs. 60% mass loss yr^{-1}) (Biondini *et al.*, 1998). Grazing continuously disturbs the functional balance between the nutrient-absorbing root system and the transpiring leaf canopy. In the grasslands of Inner Mongolia, heavy grazing significantly decreased live root biomass and belowground NPP compared with reference non-grazed sites (Gao *et al.*, 2008). However, in the natural mountain grassland of central Argentina, fine root production and turnover rates were twofold higher at grazed sites compared with ungrazed sites (Pucheta *et al.*, 2004). In the semi-arid agro-pastoral transition zone of northern China, the plant species richness, plant height, canopy cover, and AGB stocks decreased significantly with high-intensity grazing (Yan *et al.*, 2013). High intensity continuously stocked grazing of Marandu palisadegrass pastures at a 15 cm canopy height caused a long-term decrease in plant persistence and animal production in Brazilian Savannah, particularly when soil phosphorus availability is low (Euclides *et al.*, 2019). In contrast, a study in Wyoming, USA, reported that grazing enhanced root growth, aboveground biomass turnover, and redistribution of C within the plant-soil system resulting in an 11%–24% increase in SOC, where grazing exclusion caused an immobilization of C in excessive aboveground plant litter (Reeder and Schuman, 2002). Other studies also reported a significant reduction in above- and belowground plant biomass, plant litter, and productivity as a consequence of increased grazing intensity (Yan *et al.*, 2013; Chen *et al.*, 2015; del Pozo *et al.*, 2006; Ji *et al.*, 2020; Mwendera *et al.*, 1997; Ritchie, 2014; Smith *et al.*, 2014; Sun *et al.*, 2011; Zhang *et al.*, 2018). A study in the temperate mixed grass-shrub steppe of Patagonia stated that moderate grazing increased grass productivity compared with grazing exclusion or heavy grazing while standing dead grass biomass stock only in heavy grazing (Oñatibia and Aguiar, 2016). A study in Northern Patagonia, Argentina, indicated that long-term overgrazing reduced, on average, 35% of the total ecosystem C pool with a significant reduction in NPP (419 vs. 128g m^{-2} $year^{-1}$) (Enriquez *et al.*, 2015). Shrub biomass was not modified by grazing management (Oñatibia and Aguiar, 2016). In a semi-arid savanna of Laikipia, Kenya, herbivore exclusion resulted in a 22% increase in grass cover, 31% woody canopy cover, and 54% increase in SOC storage (Wigley *et al.*, 2020). A meta-analysis mentioned that grazing reduced biomass productivity, where AGB correlated positively with SOC ($r=0.59$, $p<0.01$), available phosphorus ($r=0.52$, $p<0.01$), and bulk density (Hao and He, 2019).

Canopy C exchange rate and plant biomass accumulation were less in high-intensity grazing compared with low-intensity grazing in Piracicaba, Brazil (Silva *et al.*, 2016). In Rio Grande de Sul, Brazil, high-intensity grazing significantly reduced soil C input leading to negative C balance and lower-particulate organic C stocks, compared with low-intensity grazing (da Silva *et al.*, 2014). In southern Brazil, moderate grazing offsets 34%–98% of GHG emissions mainly due to SOC enhancement and significantly reduced net global warming potential compared with intensive and light grazing or grazing exclusion (Ribeiro *et al.*, 2020). A study in the Australian rangelands reported that controlled grazing enhanced SOC and N compared with uncontrolled grazing, but the effect of grazing varied between soil types (Waters *et al.*, 2017). In the Mongolian Plateau, grazing reduced the rates of soil C and N mineralization due to changes in plant biomass and soil environment (Wang *et al.*, 2020). Grazing increased decomposition rates and C transfer from roots to particulate fractions of organic matter in Theix, France (Damien *et al.*, 2015). Similarly, grazing increased total soil C contents in the northeastern Tibetan Plateau due to increased belowground productivity (Yang *et al.*, 2016). Some studies mentioned that grazing did not affect biomass productivity above- and belowground, while others reported an increase in belowground biomass (Bork *et al.*, 2019; Wang and Wesche, 2016). For example, in the C4 grasslands of Panamá, aboveground productivity increased with grazing intensity (Wilsey *et al.*, 2002).

Changes in CO_2 exchange and soil respiration

The loss of C through soil respiration directly affects SOC storage and can be influenced by the intensity of grazing (Fig. 19.3). On the short-grass steppe of northeastern Colorado, USA, ecosystem CO_2 exchange and soil respiration rates did not differ significantly between heavy grazed (60% biomass removal) and grazing exclusion (LeCain *et al.*, 2002). In an Eddy covariance study in Belgium, the variations in gross primary production were negatively correlated ($R = -0.63$) to grazing intensity, defined as the product of the stocking rate and the grazing duration (Jérôme *et al.*, 2014). In the tropical grasslands of Sao Paulo, Brazil, heavy grazing significantly decreased annual CO_2 emissions from soils (64.3 Mg CO_2 ha^{-1} yr^{-1}) compared with light grazing (97.8 Mg CO_2 ha^{-1} yr^{-1}) (Brito *et al.*, 2015; da Silva Cardoso *et al.*, 2017). In the lowland Savanna of southern Kenya, heavy-to-light grazing showed a 15%–25% higher soil respiration compared with non-graze and over-grazed sites (Wachiye *et al.*, 2022).

In the Loess Plateau of China, a study with Li-COR-automated soil CO_2 flux measurements showed that light grazing and grazing exclusion increased soil CO_2 emission by 15% and 33% relative to adjacent croplands (Yu *et al.*, 2019). In the steppe of Inner Mongolia, average soil respiration was the highest under rotational grazing (1.26 μmol m^{-2} s^{-1}), followed by grazing exclusion (0.98 μmol m^{-2} s^{-1}) and continuous grazing (0.94 μmol m^{-2} s^{-1}) (Nie *et al.*, 2019). The variations in soil respiration were attributed to soil temperature, soil moisture, potential substrate, root biomass, and soil microbial activity

(Nie *et al.*, 2019). In a C4 grassland of Panamá, grazed and ungrazed plots had no significant differences in CO_2 uptake (24.2 vs. 20.3 μmol/m^2/s), while ecosystem respiration rates were significantly lower in grazed plots compared with ungrazed ones (LeCain *et al.*, 2002). In adaptive multi-paddock grazing in Canada, the seasonal mean CO_2 emissions increased with higher cattle stocking rates but did not differ from conventional grazing (Ma *et al.*, 2021). In the Tibetan plateau, China, the effect of grazing intensity on soil CO_2 emissions differed between the seasons of the year and was affected by soil moisture and temperature (Chen *et al.*, 2016). In Puy de Dome, France, a reduction in grazing intensity did not increase the C sink of managed temperate grasslands over time if not supplied with the additional N fertilizer (Allard *et al.*, 2007).

Changes in composition and activity of soil micro-organisms

Soil microorganisms regulate the key functions in terrestrial C cycling through organic matter decomposition or stabilization. The persistence of SOC should, therefore, be viewed through the lens of decomposers and their functional complexity (Lehmann *et al.*, 2020). Different intensities of grazing can change the quantity and quality of organic compounds entering the soils (Fig. 19.3) through aboveground litter production, fine root turnover, and root exudates and eventually can alter the structure, diversity, and activity of soil microorganisms (Faghihinia *et al.*, 2020; Rong *et al.*, 2022; Santiago *et al.*, 2021). High-intensity grazing decreased soil microbial biomass C by 21.6%, according to a meta-analysis (Zhou *et al.*, 2017). A study in desert Steppe at Siziwang Banner in Inner Mongolia, China, indicated that 15 year grazing acted as environmental filtering of soil microbial rare taxa with the SOC reduction of both labile and recalcitrant SOC but the loss was not statistically significant (Fan *et al.*, 2021). Another study in the desert Steppe indicated that grazing practice altered the soil prokaryotic community and increased SOC contents in the large macro-aggregates and silt clay fractions (Fan *et al.*, 2021). A 13 year grazing-manipulated experiment in northern China stated that soil arbuscular mycorrhizal hyphal length and density decreased with grazing intensity where root colonization was not affected (Faghihinia *et al.*, 2020). Moderate grazing enhanced the ratio of external to internal fungal structures (Faghihinia *et al.*, 2020). In the semi-arid grasslands of New Mexico, cattle grazing increased microbial community size by 42%, especially gram-positive bacteria and actinobacteria (Ghimire *et al.*, 2019). But in the semi-arid grasslands of Southern Patagonia, high-intensity grazing significantly decreased the microbial biomass C and microbial coefficient (Santiago *et al.*, 2021). Another study in the semi-arid steppe of Patagonia reported that high-intensity grazing resulted in a significant decrease in total arbuscular mycorrhizal fungi spore abundance, diversity, and evenness but moderate grazing did not show any significant effect (Dudinszky *et al.*, 2019).

In the Inner Mongolian grasslands, light grazing increased the microbial α-diversity and shifted the microbial composition from fungal to bacterial dominance (Xun *et al.*, 2018). This change can lead to the variation from the decomposition and turnover of recalcitrant to labile organic C. In a temperate grassland steppe in northern China, grazing increased the ratio of bacterial-to-fungal biomass, where soil pH and SOC content affected the bacterial community structure (Rong *et al.*, 2022). In temperate grasslands of Central France, 14 years of grazing intensification triggered the proliferation of Gram (+) bacteria and subsequent faster decomposition of organic matter (Klumpp *et al.*, 2009). In the semi-arid steppe of Inner Mongolia, heavy grazing reduced soil microbial biomass, their activity, and protozoan abundance, but increased nematode abundance (Qi *et al.*, 2011). Grazing-mediated changes in plant community composition, SOC, and N contents may contribute to the changes in microbial biomass, their activity, and soil CO_2 emissions (Ford *et al.*, 2013; Ingram *et al.*, 2008). A meta-analysis showed that light-to-moderate grazing did not affect the microbial community composition, but heavy grazing significantly reduced the soil's total microbial community size to 28.12% (Zhao *et al.*, 2017).

Changes in activities of soil macro-fauna

The composition and diversity of soil macro-fauna have a key role in the fragmentation of organic detritus and humification, the process by which organic detritus is transformed into soil organic matter (Lavelle *et al.*, 1993). By interacting with many other biotic and abiotic factors, they contribute significantly to soil hydraulic properties, soil structure, pedogenesis, and other soil processes that regulate C sequestration (Morriën *et al.*, 2017). Changing grazing regimes can alter the food webs and the consortium of soil macro-fauna (Fig. 19.3), leading to changes in soil fundamental processes (Schon *et al.*, 2008). The abundance of macrofauna, especially earthworms and beetles, was thought to have reduced soil bulk density and augmented water retention in low-intensity rotational grazing sites compared with conventional high-intensity grazing in the plains of eastern Colombia (Teutscherova *et al.*, 2021). These organisms played a crucial role in improving soil structure through bioturbation (Teutscherova *et al.*, 2021). In Indre, France, litter-dwelling (arthropods and mollusks) and soil-dwelling (earthworms) macroinvertebrates decreased with higher intensity grazing, but such trends differed between study sites (Ponge *et al.*, 2015). In the semi-arid Horqin steppe of northern China, grazing significantly reduced the diversity of soil macro-fauna (e.g. *Thomisidae*, *Reduviidae*, *Staphylinidae*, and *Curculionidae*) as measured by Shannon's index (Tao *et al.*, 2013).

In a semi-arid continental meadow steppe of Inner Mongolia, China, heavy grazing reduced soil nematode diversity *via* changes in grass species composition, soil moisture, aboveground biomass, and belowground biomass (Pan *et al.*, 2021). In the Brazilian Amazon, the richness of dung beetle (coleopterans) and the rates of dung pats disintegration did not differ between conventional open pasture and silvopasture with *Eucalyptus urograndis* tree and *Urochloa brizantha* cv. Marandu grass (Lopes *et al.*, 2020). Such homogeneity between different habitat complexities can be linked to the similarity in the intensity of cattle grazing.

Changes in soil enzymatic activity

Soil enzymes play a fundamental role in catalyzing important reactions related to organic matter decomposition and mineralization. Such microbially produced extracellular enzymes are necessary for the breakdown and depolymerization of senescent organic matter of plant, animal, or microbial origin (Chen *et al.*, 2018). Grazing intensity can alter the presence and concentrations of different soil enzymes that accelerate or decelerate the processes related to the degradation of recalcitrant or labile organic substrates (Fig. 19.3). In Dehesa grassland of Central Spain, grazing abandonment reduced soil microbial enzyme activity related to litter decomposition (Oggioni *et al.*, 2020). But in the Basque Country of northern Spain, increasing grazing intensity did not affect the activity of six enzymes (*β-glucosidase*, *β-glucosaminidase*, *sulfatase*, *acid phosphatase*, *L-alanine aminopeptidase*, and *L-leucine aminopeptidase*) (de Otálora *et al.*, 2021). In the *Stipa kirschnii*-dominated grasslands of Inner Mongolia, heavy grazing reduced soil urease and sucrase activities compared with light grazing (Zhang *et al.*, 2022). In the Hongyuan Alpine Meadow of China, heavy grazing reduced the *β−1,4-glucosidase* activity and root exudation with a significant effect on rhizosphere priming due to the decrease in labile carbon. In the same study, moderate grazing improved *β-N-acetylglucosaminidase* activity compared with the non-grazed grasslands (Sun *et al.*, 2017). In the Mediterranean grasslands of northwestern Greece, heavy grazing significantly reduced soil *β-glucosidase* activity and SOC but not *phosphatase* and *urease* activity (Panayiotou *et al.*, 2017). In the semi-arid grasslands of New Mexico, cattle grazing increased *β-glucosaminidase* activity by 95% compared with ungrazed croplands (Dudinszky *et al.*, 2019).

Decomposition and mineralization

Decomposition of the dead biomass and mineralization of SOM are the critical determinants of ecosystem C storage (Wei *et al.*, 2021; Aryal *et al.*, 2022). Changing grazing regimes can alter the plant community composition of the grasslands and the quality of litter and fine roots to decompose (Kooch *et al.*, 2020). Changes in lignin, N, or other elemental compositions of plant tissues can delay or stimulate litter decomposition above- and belowground. Grazing exclusion for 7 years reduced ammonium mineralization rates and soil microbial activity in northern England (Medina-Roldán *et al.*, 2012). In the grasslands of North Dakota, moderate grazing resulted in higher litter decomposition and N mineralization rates than non-grazed and heavily grazed grasslands (Shariff *et al.*, 1994). A meta-analysis of 115 studies showed that grazing significantly increased soil respiration, net N mineralization, and soil N nitrification by 4.3%, 34.7%, and 25.9%, respectively, in grazed grasslands compared with non-grazed ones (Zhou *et al.*, 2017). In the rangelands of Central Iran, 17 years of grazing significantly decreased soil respiration and microbial biomass C but did not affect litter decomposition rates (Raiesi and Asadi, 2006). In the grasslands of Inner Mongolia, China, grazing increased soil C and N mineralization (Wu *et al.*, 2012). Moderate grazing stimulated N mineralization, nitrification, and ammonification compared with heavy and light grazing in the Inner Mongolian steppe, (Xu *et al.*, 2007). In the same region, grazing-related soil trampling increased litter decomposition rates and litter carbon concentrations, and promoted microbial utilization of litter C and its physiochemical stabilization (Wei *et al.*, 2021). Adequate stocking density fostered humification indices and SOC storage in the tropical pasturelands of Sao Paulo, Brazil (Segnini *et al.*, 2019). In the alpine meadow of Haibei, China, the effect of grazing on litter decomposition rates and the resulting release of organic C, N, and phosphorus per unit of land were mediated by warming (Li *et al.*, 2022). Thus, global climate change can have a large impact on decomposition and mineralization processes and hence soil C sequestration in the grassland ecosystems (Fig. 19.3).

Soil nutrients and soil physicochemical properties

Changes in grazing management influence the dynamics of soil N, P, and other nutrients. Grazing can also change soil bulk density, porosity, and other physicochemical properties (Fig. 19.3). A synthesis of 53 published studies indicated that grazing significantly decreased SOC, total N, total P, and soil moisture content (Liu *et al.*, 2021). The study reported that grazing significantly increased soil bulk density and soil temperature, all affecting C storage.

In northern Australia, grazing intensity with 80% pasture utilization for 26 years caused a significant loss of soil total N from the upper 10 cm soil (Pringle *et al.*, 2014). In the Karoo biome, South Africa, long-term (>75 years) of high-intensity grazing depleted soil N storage by 27.5% compared to non-grazing. Grazing also increased soil compaction, reduced infiltration rates, and consequently reduced SOC storage (Talore *et al.*, 2016). In Central Germany, grazing pressure did not affect soil N, and microbial biomass, where oxalate soluble iron contents explained the variations of SOC storage (Nüsse *et al.*, 2017). In a 5 year grazing management study in northern China, soil N storage decreased linearly with increasing animal stocking rates (He *et al.*, 2011). Similar finding was reported in the Inner Mongolian meadow steppe, China, where high-intensity grazing for 30 years caused a significant loss of soil N contents (Han *et al.*, 2008). A study across the Inner Mongolian arid and semi-arid grasslands in northern China, showed that grazing exclusion enhanced C accumulation in aboveground biomass, litter, and live roots in more fertile sites than less fertile ones (Yu *et al.*, 2021). In another study in Inner Mongolia, China, soil total C, N, and available P concentrations significantly increased with grazing intensity but the concentrations of Mg, Cu, Fe, Mn, Zn did not change with grazing intensity (Hou *et al.*, 2020). In the eastern Qinghai–Tibetan Plateau, China, Yak grazing increased soil bulk density and total potassium stocks but decreased SOC, N, and P stocks (Ma *et al.*, 2016). In a meadow steppe of northeast China, soil N mineralization was affected by cattle grazing through two major mechanisms, that is, reduced quantity of plant litter input into the soil due to the reduction in aboveground net primary productivity and the changes in soil temperature and moisture regime (Zhang *et al.*, 2022). In the Savanna grasslands of Kenya, dry season soil water content decreased with high-intensity grazing, while soil temperature showed a reverse trend, which directly influenced soil respiration rates (Wachiye *et al.*, 2022).

Grazing mediated changes in plant community composition

Plant species composition and diversity determine ecosystem structure, functions, and services, including C sequestration (Zhang *et al.*, 2021; Sarai *et al.*, 2022). Some studies stated that higher plant diversity increases rhizosphere C inputs and fosters the activity of soil macro- and micro-organisms (Lange *et al.*, 2015; Steinbeiss *et al.*, 2008; Sullivan *et al.*, 2017). A meta-analysis of 116 published studies showed that the effect of grazing intensity followed the intermediate disturbance hypothesis where species richness and diversity increased with light and moderate grazing intensities and decreased at high-intensity grazing (Wang and Tang, 2019). Heavy grazing modified grass-specific biomass composition, thus diminishing biomass quality in intensively grazed areas in arid rangelands of Patagonia (Oñatibia and Aguiar, 2016). Plant species richness and abundance of each species were higher in the light- and moderate-grazed grasslands than in heavy-grazed ones in Southern Ethiopia (Angassa, 2014). Increased grazing enhanced C stocks for C4-dominated grasslands compared with C3-dominated or mixed grasslands (Abdalla *et al.*, 2018). In Qinghai-Tibetan Plateau, high-intensity continuous grazing decreased the plant biomass and caused a shift in plant community composition and dominance from palatable to undesirable plant species (Zhang *et al.*, 2018). Such a negative relationship between total aboveground biomass and animal stocking rate indicates the reduction in total plant biomass (Fig. 19.3). In South African grasslands, high-intensity continuous grazing had a negative impact on plant biodiversity (O'Connor *et al.*, 2010). In the semi-arid grasslands of Inner Mongolia, China, livestock grazing shifted plant community composition (changing C3:C4 proportions) and reduced aboveground biomass storage (Liang *et al.*, 2021). It is also reported that increasing grazing intensity increased SOC on C4-dominated grasslands, but decreased in C3-dominated grasslands. This change in species composition also affected soil C:N ratio and soil respiration (Liang *et al.*, 2021). Intensive grazing led to the loss of readily decomposable C4-derived C components in the Bermudagrass pasturelands of Florida (Silveira *et al.*, 2014). But in the shortgrass Great plains of North America, grazing increased the dominance of C4 plants relative to C3 plants, affecting the proportion of root-derived C input to the soil (Derner *et al.*, 2006). In the sub-humid Espinal grasslands of Chile, higher stocking rates increased the abundance of composites, decreased the grasses, and did not change the legumes. Such changes led to a decline in overall aboveground biomass storage (del Pozo *et al.*, 2006). In contrast, moderate-to-heavy grazing for more than 75 years did not change the original plant species composition, while light or grazing exclusion increased the presence of invasive species in the shortgrass Steppe of north-central Colorado (Porensky *et al.*, 2017). A study in Northern China showed that aridity and grazing pressure significantly reduced plant species richness of the grasslands (Liu *et al.*, 2022).

The effect of topography and soil erosion

Topography is an important variable to consider in grazing intensity management for SOC storage in the grasslands. High-intensity grazing can exacerbate soil erosion in the slopy grasslands and hence deplete C storage (Fig. 19.3). In the foothills of the Rocky Mountains in Alberta, heavy grazing reduced SOC, total N, and soil protein contents in the lower position of the slope, while the differences were not significant in the upper part of the slope (Zhang *et al.*, 2018). A study in the highlands of Ethiopia reported an interaction between grazing intensity and slope, where moderate grazing was sufficient to reduce vegetation cover and biomass yield on a 4–8% slope but not on a 0–4% slope (Mwendera *et al.*, 1997). In the Loess Plateau of China, organic C loss from soil erosion was 56% lower in the light grazed and 29% lower in the ungrazed grasslands compared with croplands, indicating that light grazing had higher potential than non-grazing practices in reducing SOC loss from erosion and CO_2 emission (Yu *et al.*, 2019).

Implications of Climate Change on Grazing

In addition to grazing management, the fate of SOC persistence in grassland ecosystems depends upon global climate change and its interaction with soil biochemical processes (Fig. 19.3). A meta-analysis of 223 published studies showed that climate change factors had additive effects on grazing intensity in reducing above- and below ground C storage (Zhou *et al.*, 2019). The authors reported that the combined effects of grazing intensity with global climate change factors decreased aboveground C pool by 4.0%, soil C pool by 4.7%, and soil respiration by 2.7%. In eastern Australia, increasing grazing intensity reduced the link between the soil nutrient cycling index and decomposition enzymes, labile C, and inorganic phosphorus contents in the soil. However, the environmental factors related to aridity had no significant effect on this relationship (Eldridge and Delgado-Baquerizo, 2018). A global synthesis of the 30 year publications stated that the negative effects of 2°C warming on aboveground C sequestration were greater in grazed grasslands compared with ungrazed ones. They also mentioned that high-intensity grazing had higher effects on net primary productivity and species diversity than the effects of climate change (Li *et al.*, 2018). In the mixed-grass prairie of North America, ecosystem CO_2 exchange rates strongly reduced with the reduction in summer precipitation, where light grazing improved ecosystem C sequestration compared with heavy grazing (Chimner and Welker, 2011). In northern Australia, interannual variability in rainfall had a large effect on soil N contents, where higher SOM was found under heavy grazing (Segoli *et al.*, 2015). In southern Patagonia, the effect of grazing intensity on soil microbial communities (fungi and bacteria) varied largely with the environmental characteristics (humid vs dry climate) of the sites (Santiago *et al.*, 2021). The authors reported a lower microbial biomass carbon value under heavy grazing (412 μg C g^{-1} dry soil) and a higher value in the sites with moderate grazing (512 μg C g^{-1} dry soil). Furthermore, the variation in evapotranspiration rates due to climate-driven changes in plant community structure can affect SOC stocks and turnover rates (Frank *et al.*, 2012). A meta-analysis showed that SOC stocks reduced by 16% in dry climates (<600 mm) compared with 8% in wet climates (>1000 mm) due to grassland degradation resulting principally from overgrazing (Dlamini *et al.*, 2016).

Final Considerations and Outlook

Animal species, stocking rates, duration of herbivory/fallow, continuous vs rotational grazing, frequency of grazing during the rotation are some variables that determine the intensity of grazing (Eldridge *et al.*, 2017; Ji *et al.*, 2020). However, the variables related to primary productivity of the grasslands, seasonality of the production, plant species, and their capacity to regenerate (resprout) after grazing disturbance should also be considered in determining grazing intensity. Most studies reported that high-intensity grazing significantly reduces C storage in grassland ecosystems but light-to-moderate grazing could balance the trade-offs between C sequestration and animal production. Changes in ecosystem CO_2 exchange, above and belowground productivity, organic matter decomposition, and soil respiration are some mechanisms that directly affect carbon sequestration. Grazing can change the structure and composition of plant species, microbial composition and their activities, soil enzymatic activities, soil nutrient concentrations, and its physicochemical properties. Climate, topography, and climate change-related interactions are the important factors that can alter C sequestration potential with a change in grazing intensity.

A study stated that rehabilitation of degraded grasslands globally can sequester approximately 45 Tg C yr^{-1}, which can be achieved by cessation of overgrazing and reducing intensity from heavy-to-moderate or light grazing (Conant and Paustian, 2002). In addition to regulating animal stocking density to optimize herbivory, incorporation of trees and shrubs on the grazing lands, diversification of grazing animals, and regenerative multi-paddock grazing have been practiced to avoid the loss of SOC and other nutrients from overgrazing (Silveira *et al.*, 2014; Badgery *et al.*, 2020; Mosier *et al.*, 2021; Wang *et al.*, 2019). Furthermore, grazing management should consider the uncertainties related to global climate change and its feedback on ecosystem C turnover (Van Groenigen *et al.*, 2014). The effect of interaction between climatic change and grazing intensity on SOC storage can be more pronounced in one region than others, which is one of the most important uncertainties in the efforts to promote carbon sequestration. To optimize the trade-offs between grassland C sequestration and animal production, land managers should also understand the importance of the time factor, especially on how long it takes to lose or sequester a certain amount of C from grassland management practices.

Acknowledgments

We extend our special recognition to all the authors of the published research reviewed in this paper. We are thankful to Consejo Nacional de Ciencia y Tecnologia (CONACYT), Mexico for financing the proyect "382/2015 Estrategias de mitigación de impactos ambientales del sector agropecuario en Chiapas." We acknowledge the anonymous reviewers for their valuable comments and suggestions on the earlier versions of the manuscript, which greatly improved the quality of this review.

References

Abdalla, M., Hastings, A., Chadwick, D., Jones, D.L., Evans, C. *et al.* (2018) Critical review of the impacts of grazing intensity on soil organic carbon storage and other soil quality indicators in extensively managed grasslands. *Agriculture, Ecosystems & Environment* 253, 62–81.

Allard, V., Soussana, J.F., Falcimagne, R., Berbigier, P., Bonnefond, J.M. *et al.* (2007) The role of grazing management for the net biome productivity and greenhouse gas budget (CO_2, N_2O and CH_4) of semi-natural grassland. *Agriculture, Ecosystems & Environment* 121(1–2), 47–58.

Angassa, A. (2014) Effects of grazing intensity and bush encroachment on herbaceous species and rangeland condition in southern Ethiopia. *Land Degradation & Development* 25(5), 438–451.

Aryal, D.R., Corzo, R.R., Cruz, A.L., Sanabria, C.V., Gómez, H. *et al.* (2018a) Biomass accumulation in forests with high pressure of fuelwood extraction in Chiapas, Mexico. *Revista Árvore* 42(3), e420307.

Aryal, D.R., Morales Ruiz, D.E., Tondopó Marroquín, C.N., Pinto Ruiz, R., Guevara Hernández, F. *et al.* (2018b) Soil organic carbon depletion from forests to grasslands conversion in Mexico: A review. *Agriculture* 8(11), 181.

Aryal, D.R., De Jong, B.H., Gaona, S.O., Vega, J.M., Olguín, L.E. *et al.* (2022) Fine wood decomposition rates decline with the age of tropical successional forests in Southern Mexico: Implications to ecosystem carbon storage. *Ecosystems* 25, 661–677.

Badgery, W.B., Kemp, D., Yingjun, Z., Zhongwu, W., Guodong, H. *et al.* (2020) Optimising grazing for livestock production and environmental benefits in Chinese grasslands. *The Rangeland Journal* 42(5), 347–358.

Balasubramanian, D., Zhou, W.J., Ji, H.L., Grace, J., Bai, X.L. *et al.* (2020) Environmental and management controls of soil carbon storage in grasslands of Southwestern China. *Journal of Environmental Management* 254, 109810.

Basile-Doelsch, I., Balesdent, J. and Pellerin, S. (2020) Reviews and syntheses: The mechanisms underlying carbon storage in soil. *Biogeosciences* 17(21), 5223–5242.

Biondini, M.E., Patton, B.D. and Nyren, PE. (1998) Grazing intensity and ecosystem processes in a northern mixed-grass prairie, USA. *Ecological Applications* 8(2), 469–479.

Bork, E.W., Lyseng, M.P., Hewins, D.B., Carlyle, C., Chang, S.X. *et al.* (2019) Herbage biomass and its relationship to soil carbon under long-term grazing in Northern temperate grasslands. *Canadian Journal of Plant Science* 99(6), 905–916.

Bork, E.W., Raatz, L.L., Carlyle, C.N., Hewins, D.B. and Thompson, K.A. (2020) Soil carbon increases with long-term cattle stocking in northern temperate grasslands. *Soil Use and Management* 36(3), 387–399.

Brito, L.F., Azenha, M.V., Janusckiewicz, E.R., Cardoso, A.S., Morgado, E.S. *et al.* (2015) Seasonal fluctuation of soil carbon dioxide emission in differently managed pastures. *Agronomy Journal* 107(3), 957–962.

Byrnes, R.C., Eastburn, D.J., Tate, K.W. and Roche, LM. (2018) A global meta-analysis of grazing impacts on soil health indicators. *Journal of Environmental Quality* 47(4), 758–765.

Chen, J., Zhou, X., Wang, J., Hruska, T., Shi, W. *et al.* (2016) Grazing exclusion reduced soil respiration but increased its temperature sensitivity in a Meadow grassland on the *Tibetan plateau*. *Ecology and Evolution* 6(3), 675–687.

Chen, J., Luo, Y., Van, K.J., Hungate, B.A., Cao, J. *et al.* (2018) A keystone microbial enzyme for nitrogen control of soil carbon storage. *Science Advances* 4(8), eaaq1689.

Chen, W., Huang, D., Liu, N., Zhang, Y., Badgery, W.B. *et al.* (2015) Improved grazing management may increase soil carbon sequestration in temperate steppe. *Scientific Reports* 5(1), 1–13.

Chimner, R.A. and Welker, JM. (2011) Influence of grazing and precipitation on ecosystem carbon cycling in a mixed-grass prairie. *Pastoralism: Research, Policy and Practice* 1(1), 1–15.

Conant, R.T. and Paustian, K. (2002) Potential soil carbon sequestration in overgrazed grassland ecosystems. *Global Biogeochemical Cycles* 16(4), 90–91.

Damien, H., Nathalie, V., Frédérique, L., Gael, A., Julien, P. *et al.* (2015) How does soil particulate organic carbon respond to grazing intensity in permanent grasslands? *Plant and Soil* 394(1), 239–255.

da Silva Cardoso, A., de Figueiredo Brito, L., Janusckiewicz, E.R., da Silva Morgado, E., Barbero, R.P. *et al.* (2017) Impact of grazing intensity and seasons on greenhouse gas emissions in tropical grassland. *Ecosystems* 20(4), 845–859.

da Silva, F.D., Amado, T.J.C., Ferreira, A.O., Assmann, J.M., Anghinoni, I. *et al.* (2014) Soil carbon indices as affected by 10 years of integrated crop–livestock production with different pasture grazing intensities in southern Brazil. *Agriculture, Ecosystems & Environment* 190, 60–69.

del Pozo, A., Ovalle, C., Casado, M.A., Acosta, B. and de Miguel, J.M. (2006) Effects of grazing intensity in grasslands of the espinal of central Chile. *Journal of Vegetation Science* 17(6), 791–798.

de Otálora, X., Epelde, L., Arranz, J., Garbisu, C., Ruiz, R, *et al.* (2021) Regenerative rotational grazing management of dairy sheep increases springtime grass production and topsoil carbon storage. *Ecological Indicators* 125, 107484.

Derner, J.D., Boutton, T.W. and Briske, DD. (2006) Grazing and ecosystem carbon storage in the north American great plains. *Plant and Soil* 280(1), 77–90.

Derner, J.D., Augustine, D.J. and Frank, DA. (2019) Does grazing matter for soil organic carbon sequestration in the western North American great plains? *Ecosystems* 22(5), 1088–1094.

Dignac, M.F., Derrien, D., Barre, P., Barot, S., Cécillon, L. *et al.* (2017) Increasing soil carbon storage: Mechanisms, effects of agricultural practices and proxies. *A Review. Agronomy for Sustainable Development* 37(2), 1–27.

Dlamini, P., Chivenge, P. and Chaplot, V. (2016) Overgrazing decreases soil organic carbon stocks the most under dry climates and low soil pH: A meta-analysis shows. *Agriculture, Ecosystems & Environment* 221, 258–269.

Dong, L., Martinsen, V., Wu, Y., Zheng, Y., Liang, C. *et al.* (2021) Effect of grazing exclusion and rotational grazing on labile soil organic carbon in north China. *European Journal of Soil Science* 72(1), 372–384.

Dudinszky, N., Cabello, M.N., Grimoldi, A.A., Schalamuk, S. and Golluscio, RA. (2019) Role of grazing intensity on shaping arbuscular mycorrhizal fungi communities in Patagonian semiarid steppes. *Rangeland Ecology & Management* 72(4), 692–699.

Eldridge, D.J. and Delgado-Baquerizo, M. (2018) Grazing reduces the capacity of landscape function analysis to predict regional-scale nutrient availability or decomposition, but not total nutrient pools. *Ecological Indicators* 90, 494–501.

Eldridge, D.J., Delgado-Baquerizo, M., Travers, S.K., Val, J. and Oliver, I. (2017) Do grazing intensity and herbivore type affect soil health? insights from a semi-arid productivity gradient. *Journal of Applied Ecology* 54(3), 976–985.

Elmore, A.J. and Asner, GP. (2006) Effects of grazing intensity on soil carbon stocks following deforestation of a Hawaiian dry tropical forest. *Global Change Biology* 12(9), 1761–1772.

Enriquez, A.S., Chimner, R.A., Cremona, M.V., Diehl, P. and Bonvissuto, GL. (2015) Grazing intensity levels influence C reservoirs of wet and mesic meadows along a precipitation gradient in Northern Patagonia. *Wetlands Ecology and Management* 23(3), 439–451.

Euclides, V.P., Montagner, D.B., Macedo, M.C.M., de Araújo, A.R., Difante, G.S. *et al.* (2019) Grazing intensity affects forage accumulation and persistence of *Marandu palisadegrass* in the *Brazilian savannah*. *Grass and Forage Science* 74(3), 450–462.

Eze, S., Palmer, S.M. and Chapman, PJ. (2018) Soil organic carbon stock in grasslands: Effects of inorganic fertilizers, liming and grazing in different climate settings. *Journal of Environmental Management* 223, 74–84.

Faghihinia, M., Zou, Y., Chen, Z., Bai, Y., Li, W. *et al.* (2020) The response of grassland mycorrhizal fungal abundance to a range of long-term grazing intensities. *Rhizosphere* 13, 100178.

Fan, F., Liang, C., Tang, Y., Harker-Schuch, I. and Porter, JR. (2019) Effects and relationships of grazing intensity on multiple ecosystem services in the Inner Mongolian steppe. *Science of the Total Environment* 675, 642–650.

Fan, J., Jin, H., Zhang, C., Zheng, J., Zhang, J. *et al.* (2021) Grazing intensity induced alternations of soil microbial community composition in aggregates drive soil organic carbon turnover in a desert steppe. *Agriculture, Ecosystems & Environment* 313, 107387.

Fan, J., Zhang, C., Jin, H., Zhang, J. and Han, G. (2021) Grazing accelerates labile and recalcitrant soil carbon loss driving by rare microbial taxa in a desert steppe. *Land Degradation & Development* 32(15), 4241–4253.

FAOSTAT (2021) Data: Land, input and sustainability. Food and Agriculture Organization (FAO) of the United Nations. Available at: https://www.fao.org/faostat/en/#data (accessed 21 January 2022).

Fetzel, T., Havlik, P., Herrero, M. and Erb, K. (2017) Seasonality constraints to livestock grazing intensity. *Global Change Biology* 23(4), 1636–1647.

Ford, H., Rousk, J., Garbutt, A., Jones, L. and Jones, DL. (2013) Grazing effects on microbial community composition, growth and nutrient cycling in salt marsh and sand dune grasslands. *Biology and Fertility of Soils* 49(1), 89–98.

Frank, D.A., Pontes, A.W. and McFarlane, KJ. (2012) Controls on soil organic carbon stocks and turnover among North American ecosystems. *Ecosystems* 15(4), 604–615.

Friedlingstein, P., Jones, M.W., O'Sullivan, M., Andrew, R.M., Bakker, D.C. *et al.* (2022) Global carbon budget 2021. *Earth System Science Data* 14(4), 1917–2005.

Gao, Y.Z., Giese, M., Lin, S., Sattelmacher, B., Zhao, Y. *et al.* (2008) Belowground net primary productivity and biomass allocation of a grassland in Inner Mongolia is affected by grazing intensity. *Plant and Soil* 307(1), 41–50.

Garcia, L., Dubeux, J.J., Sollenberger, L.E., Vendramini, J.M., DiLorenzo, N. *et al.* (2021) Nutrient excretion from cattle grazing nitrogen-fertilized grass or grass–legume pastures. *Agronomy Journal* 113(4), 3110–3123.

Ghimire, R., Thapa, V.R., Cano, A. and Acosta-Martinez, V. (2019) Soil organic matter and microbial community responses to semiarid croplands and grasslands management. *Applied Soil Ecology* 141, 30–37.

Han, G., Hao, X., Zhao, M., Wang, M., Ellert, B.H. *et al.* (2008) Effect of grazing intensity on carbon and nitrogen in soil and vegetation in a meadow steppe in Inner Mongolia. *Agriculture, Ecosystems & Environment* 125(1–4), 21–32.

Hao, Y. and He, Z. (2019) Effects of grazing patterns on grassland biomass and soil environments in China: A meta-analysis. *PLoS One* 14(4), e0215223.

He, M., Zhou, G., Yuan, T., Groenigen, K.J., Shao, J. *et al.* (2020) Grazing intensity significantly changes the C: N: P stoichiometry in grassland ecosystems. *Global Ecology and Biogeography* 29(2), 355–369.

He, N., Zhang, Y., Yu, Q., Chen, Q., Pan, Q. *et al.* (2011) Grazing intensity impacts soil carbon and nitrogen storage of continental steppe. *Ecosphere* 2(1), 1–10.

Hou, D., Guo, K. and Liu, C. (2020) Asymmetric effects of grazing intensity on macroelements and microelements in grassland soil and plants in Inner Mongolia grazing alters nutrient dynamics of grasslands. *Ecology and Evolution* 10(16), 8916–8926.

Houghton, R.A. and Nassikas, AA. (2017) Global and regional fluxes of carbon from land use and land cover change 1850–2015. *Global Biogeochemical Cycles* 31(3), 456–472.

Ingram, L., Stahl, P., Schuman, G., Buyer, J., Vance, G. *et al.* (2008) Grazing impacts on soil carbon and microbial communities in a mixed-grass ecosystem. *Soil Science Society of America Journal* 72(4), 939–948.

Jérôme, E., Beckers, Y., Bodson, B., Heinesch, B., Moureaux, C. *et al.* (2014) Impact of grazing on carbon dioxide exchanges in an intensively managed Belgian grassland. *Agriculture, Ecosystems & Environment* 194, 7–16.

Jiang, Z., Hu, Z., Lai, D.Y., Han, D., Wang, M. *et al.* (2020) Light grazing facilitates carbon accumulation in subsoil in chinese grasslands: A meta-analysis. *Global Change Biology* 26(12), 7186–7197.

Irisarri, J.G.N., Derner, J.D., Porensky, L.M., Augustine, D.J., Reeves, J.L. *et al.* (2016) Grazing intensity differentially regulates ANPP response to precipitation in North American semiarid grasslands. *Ecological Applications* 26(5), 1370–1380.

Ji, L., Qin, Y., Jimoh, S.O., Hou, X., Zhang, N. *et al.* (2020) Impacts of livestock grazing on vegetation characteristics and soil chemical properties of alpine meadows in the eastern Qinghai-Tibetan Plateau. *Ecoscience* 27(2), 107–118.

Jin, D., Yan, R., Li, L., Qi, J., Chen, J. *et al.* (2022) Stocking rate changed the magnitude of carbon sequestration and flow within the plant-soil system of a meadow steppe ecosystem. *Plant and Soil* 473(1), 33–47.

Khan, R.W.A., Shaheen, H., Mehmood, A. and Naseer Awan, S. (2019) Grazing intensity impacts on soil carbon stocks of western Himalayan alpine paddocks. *Carbon Management* 10(6), 533–540.

Klein Goldewijk, K., Beusen, A., Doelman, J. and Stehfest, E. (2017) Anthropogenic land use estimates for the holocene – HYDE 3.2. *Earth System Science Data* 9(2), 927–953.

Klumpp, K., Fontaine, S., Attard, E., Le Roux, X., Gleixner, G. *et al.* (2009) Grazing triggers soil carbon loss by altering plant roots and their control on soil microbial community. *Journal of Ecology* 97(5), 876–885.

Kooch, Y., Moghimian, N., Wirth, S. and Noghre, N. (2020) Effects of grazing management on leaf litter decomposition and soil microbial activities in Northern Iranian Rangeland. *Geoderma* 361, 114100.

Lal, R. (2018) Digging deeper: A holistic perspective of factors affecting soil organic carbon sequestration in agroecosystems. *Global Change Biology* 24(8), 3285–3301.

Lange, M., Eisenhauer, N., Sierra, C.A., Bessler, H., Engels, C. *et al.* (2015) Plant diversity increases soil microbial activity and soil carbon storage. *Nature Communications* 6(1), 1–8.

Lavelle, P., Blanchart, E., Martin, A., Martin, S. and Spain, A. (1993) A hierarchical model for decomposition in terrestrial ecosystems: application to soils of the humid tropics. *Biotropica* 130–150.

LeCain, D.R., Morgan, J.A., Schuman, G.E., Reeder, J.D. and Hart, RH. (2002) Carbon exchange and species composition of grazed pastures and exclosures in the shortgrass steppe of Colorado. *Agriculture, Ecosystems & Environment* 93(1–3), 421–435.

Lehmann, J., Hansel, C.M., Kaiser, C., Kleber, M., Maher, K. *et al.* (2020) Persistence of soil organic carbon caused by functional complexity. *Nature Geoscience* 13(8), 529–534.

Liang, M., Smith, N.G., Chen, J., Wu, Y., Guo, Z. *et al.* (2021) Shifts in plant composition mediate grazing effects on carbon cycling in grasslands. *Journal of Applied Ecology* 58(3), 518–527.

Li, B., Song, H., Cao, W., Wang, Y., Chen, J. *et al.* (2021) Responses of soil organic carbon stock to animal manure application: A new global synthesis integrating the impacts of agricultural managements and environmental conditions. *Global Change Biology* 27(20), 5356–5367.

Li, B., Lv, W., Sun, J., Zhang, L., Jiang, L. *et al.* (2022) Warming and grazing enhance litter decomposition and nutrient release independent of litter quality in an alpine meadow. *Journal of Plant Ecology* rtac009.

Liu, C., Li, W., Xu, J., Wei, W., Xue, P. *et al.* (2021) Response of soil nutrients and stoichiometry to grazing management in alpine grassland on the Qinghai-Tibet plateau. *Soil and Tillage Research* 206, 104822.

Liu, J., Isbell, F., Ma, Q., Chen, Y., Xing, F. *et al.* (2022) Overgrazing, not haying, decreases grassland topsoil organic carbon by decreasing plant species richness along an aridity gradient in Northern China. *Agriculture, Ecosystems & Environment* 332, 107935.

Li, W., Li, X., Zhao, Y., Zheng, S. and Bai, Y. (2018) Ecosystem structure, functioning and stability under climate change and grazing in grasslands: current status and future prospects. *Current Opinion in Environmental Sustainability* 33, 124–135.

Lopes, L.B., Pitta, R.M., Eckstein, C., Grossi, P.C., Sindeaux, E. *et al.* (2020) Diversity of coleopterans associated with cattle dung in open pastures and silvopastoral systems in the Brazilian amazon. *Agroforestry Systems* 94(6), 2277–2287.

Ma, W., Ding, K. and Li, Z. (2016) Comparison of soil carbon and nitrogen stocks at grazing-excluded and yak grazed alpine meadow sites in Qinghai–Tibetan plateau, China. *Ecological Engineering* 87, 203–211.

Ma, Z., Shrestha, B.M., Bork, E.W., Chang, S.X., Carlyle, C.N. *et al.* (2021) Soil greenhouse gas emissions and grazing management in northern temperate grasslands. *Science of The Total Environment* 796, 148975.

Medina-Roldán, E., Paz-Ferreiro, J. and Bardgett, RD. (2012) Grazing exclusion affects soil and plant communities, but has no impact on soil carbon storage in an upland grassland. *Agriculture, Ecosystems & Environment* 149, 118–123.

Morales Ruiz, D.E., Aryal, D.R., Pinto Ruiz, R., Guevara Hernández, F., Casanova Lugo, F. *et al.* (2021) Carbon contents and fine root production in tropical silvopastoral systems. *Land Degradation & Development* 32(2), 738–756.

Morriën, E., Hannula, S.E., Snoek, L.B., Helmsing, N.R., Zweers, H. *et al.* (2017) Soil networks become more connected and take up more carbon as nature restoration progresses. *Nature Communications* 8(1), 1–10.

Mosier, S., Apfelbaum, S., Byck, P., Calderon, F., Teague, R. *et al.* (2021) Adaptive multi-paddock grazing enhances soil carbon and nitrogen stocks and stabilization through mineral association in southeastern US grazing lands. *Journal of Environmental Management* 288, 112409.

Mwendera, E., Saleem, M.M. and Woldu, Z. (1997) Vegetation response to cattle grazing in the Ethiopian highlands. *Agriculture, Ecosystems & Environment* 64(1), 43–51.

Navarrete, D., Sitch, S., Aragão, L.E. and Pedroni, L. (2016) Conversion from forests to pastures in the Colombian amazon leads to contrasting soil carbon dynamics depending on land management practices. *Global Change Biology* 22(10), 3503–3517.

Nie, C., Li, Y., Niu, L., Liu, Y., Shao, R. *et al.* (2019) Soil respiration and its Q10 response to various grazing systems of a typical steppe in Inner Mongolia, China. *PeerJ* 7, e7112.

Nüsse, A., Linsler, D., Kaiser, M., Ebeling, D., Tonn, B. *et al.* (2017) Effect of grazing intensity and soil characteristics on soil organic carbon and nitrogen stocks in a temperate long-term grassland. *Archives of Agronomy and Soil Science* 63(12), 1776–1783.

Oggioni, S.D., Ochoa-Hueso, R. and Peco, B. (2020) Livestock grazing abandonment reduces soil microbial activity and carbon storage in a Mediterranean dehesa. *Applied Soil Ecology* 153, 103588.

Oñatibia, G.R. and Aguiar, MR. (2016) Continuous moderate grazing management promotes biomass production in patagonian arid rangelands. *Journal of Arid Environments* 125, 73–79.

O'Connor, T., Kuyler, P., Kirkman, K. and Corcoran, B. (2010) Which grazing management practices are most appropriate for maintaining biodiversity in South African grassland? *African Journal of Range & Forage Science* 27(2), 67–76.

Panayiotou, E., Dimou, M. and Monokrousos, N. (2017) The effects of grazing intensity on soil processes in a Mediterranean protected area. *Environmental Monitoring and Assessment* 189(9), 1–11.

Pan, F., Yan, R., Zhao, J., Li, L., Hu, Y. *et al.* (2021) Effects of grazing intensity on soil nematode community structure and function in different soil layers in a meadow steppe. *Plant and Soil* 1–14.

Patton, B.D., Dong, X., Nyren, P.E. and Nyren, A. (2007) Effects of grazing intensity, precipitation, and temperature on forage production. *Rangeland Ecology & Management* 60(6), 656–665.

Petz, K., Alkemade, R., Bakkenes, M., Schulp, C.J., Velde, M. *et al.* (2014) Mapping and modelling trade-offs and synergies between grazing intensity and ecosystem services in rangelands using global-scale datasets and models. *Global Environmental Change* 29, 223–234.

Ponge, J.F., Salmon, S., Benoist, A. and Geoffroy, JJ. (2015) Soil macrofaunal communities are heterogeneous in heathlands with different grazing intensity. *Pedosphere* 25(4), 524–533.

Porensky, L.M., Derner, J.D., Augustine, D.J. and Milchunas, DG. (2017) Plant community composition after 75 yr of sustained grazing intensity treatments in shortgrass steppe. *Rangeland Ecology & Management* 70(4), 456–464.

Pringle, M., Allen, D., Phelps, D., Bray, S., Orton, T. *et al.* (2014) The effect of pasture utilization rate on stocks of soil organic carbon and total nitrogen in a semi-arid tropical grassland. *Agriculture, Ecosystems & Environment* 195, 83–90.

Pucheta, E., Bonamici, I., Cabido, M. and Díaz, S. (2004) Below-ground biomass and productivity of a grazed site and a neighbouring ungrazed exclosure in a grassland in central Argentina. *Austral Ecology* 29(2), 201–208.

Qi, S., Zheng, H., Lin, Q., Li, G., Xi, Z. *et al.* (2011) Effects of livestock grazing intensity on soil biota in a semiarid steppe of Inner Mongolia. *Plant and Soil* 340(1), 117–126.

Raiesi, F. and Asadi, E. (2006) Soil microbial activity and litter turnover in native grazed and ungrazed rangelands in a semiarid ecosystem. *Biology and Fertility of Soils* 43(1), 76–82.

Rattan, L. (2016) Potential and challenges of conservation agriculture in sequestration of atmospheric CO_2 for enhancing climate-resilience and improving productivity of soil of small landholder farms. *CAB Reviews* 11(9), 1–16.

Reeder, J.D. and Schuman, GE. (2002) Influence of livestock grazing on C sequestration in semi-arid mixed-grass and short-grass rangelands. *Environmental Pollution* 116(3), 457–463.

Ribeiro, R.H., Ibarr, M.A., Besen, M.R., Bayer, C. and Piva, JT. (2020) Managing grazing intensity to reduce the global warming potential in integrated crop–livestock systems under no-till agriculture. *European Journal of Soil Science* 71(6), 1120–1131.

Ritchie, ME. (2014) Plant compensation to grazing and soil carbon dynamics in a tropical grassland. *PeerJ* 2, e233.

Rong, Y., Monaco, T.A., Liu, Z., Zhao, M. and Han, G. (2022) Soil microbial community structure is unaltered by grazing intensity and plant species richness in a temperate grassland steppe in northern China. *European Journal of Soil Biology* 110, 103404.

Rumpel, C., Crème, A., Ngo, P., Velásquez, G., Mora, M. *et al.* (2015) The impact of grassland management on biogeochemical cycles involving carbon, nitrogen and phosphorus. *Journal of Soil Science and Plant Nutrition* 15(2), 353–371.

Santiago, T., Veronica, G. and Marina, GP. (2021) Soil microbial communities respond to an environmental gradient of grazing intensity in south Patagonia Argentina. *Journal of Arid Environments* 184, 104300.

Sarai, S.S., De, B.H., Esperanza, H.L., Jorge, M.V., Danilo, M.R. *et al.* (2022) Fine root biomass stocks but not the production and turnover rates vary with the age of tropical successional forests in southern Mexico. *Rhizosphere* 100474.

Scharlemann, J.P., Tanner, E.V., Hiederer, R. and Kapos, V. (2014) Global soil carbon: Understanding and managing the largest terrestrial carbon pool. *Carbon Management* 5(1), 81–91.

Schon, N., Mackay, A., Minor, M., Yeates, G. and Hedley, M. (2008) Soil fauna in grazed New Zealand hill country pastures at two management intensities. *Applied Soil Ecology* 40(2), 218–228.

Schulz, K., Voigt, K., Beusch, C., Almeida-Cortez, J.S., Kowarik, I. *et al.* (2016) Grazing deteriorates the soil carbon stocks of caatinga forest ecosystems in Brazil. *Forest Ecology and Management* 367, 62–70.

Schuman, G., Janzen, H. and Herrick, J. (2002) Soil carbon dynamics and potential carbon sequestration by rangelands. *Environmental Pollution* 116(3), 391–396.

Segnini, A., Xavier, A.A.P., Otaviani-Junior, P.L., Oliveira, P.P.A., Pedroso A de F, F. *et al.* (2019) Soil carbon stock and humification in pastures under different levels of intensification in Brazil. *Scientia Agricola* 76, 33–40.

Segoli, M., Bray, S., Allen, D., Dalal, R., Watson, I. *et al.* (2015) Managing cattle grazing intensity: Effects on soil organic matter and soil nitrogen. *Soil Research* 53(6), 677–682.

Shariff, A.R., Biondini, M.E. and Grygiel, CE. (1994) Grazing intensity effects on litter decomposition and soil nitrogen mineralization. *Rangeland Ecology & Management/Journal of Range Management Archives* 47(6), 444–449.

Silva, V.J., Pedreira, C.G., Sollenberger, L.E., Silva, L.S., Yasuoka, J.I. *et al.* (2016) Carbon assimilation, herbage plant-part accumulation, and organic reserves of grazed 'mulato II'brachiariagrass pastures. *Crop Science* 56(5), 2853–2860.

Silveira, M.L., Xu, S., Adewopo, J., Franzluebbers, A.J. and Buonadio, G. (2014) Grazing land intensification effects on soil C dynamics in aggregate size fractions of a spodosol. *Geoderma* 230, 185–193.

Smith, S.W., Vandenberghe, C., Hastings, A., Johnson, D., Pakeman, R.J. *et al.* (2014) Optimizing carbon storage within a spatially heterogeneous upland grassland through sheep grazing management. *Ecosystems* 17(3), 418–429.

Song, X.P., Hansen, M.C., Stehman, S.V., Potapov, P.V., Tyukavina, A. *et al.* (2018) Global land change from 1982 to 2016. *Nature* 560(7720), 639–643.

Soussana, J.F. and Lemaire, G. (2014) Coupling carbon and nitrogen cycles for environmentally sustainable intensification of grasslands and crop-livestock systems. *Agriculture, Ecosystems & Environment* 190, 9–17.

Stanley, P.L., Rowntree, J.E., Beede, D.K., DeLonge, M.S. and Hamm, MW. (2018) Impacts of soil carbon sequestration on life cycle greenhouse gas emissions in Midwestern USA beef finishing systems. *Agricultural Systems* 162, 249–258.

Steinbeiss, S., Beßler, H., Engels, C., Temperton, V.M., Buchmann, N. *et al.* (2008) Plant diversity positively affects short-term soil carbon storage in experimental grasslands. *Global Change Biology* 14(12), 2937–2949.

Sullivan, M.J., Talbot, J., Lewis, S.L., Phillips, O.L., Qie, L. *et al.* (2017) Diversity and carbon storage across the tropical forest biome. *Scientific Reports* 7(1), 1–12.

Sun, D., Wesche, K., Chen, D., Zhang, S., Wu, G. *et al.* (2011) Grazing depresses soil carbon storage through changing plant biomass and composition in a Tibetan alpine meadow. *Plant, Soil and Environment* 57(6), 271–278.

Sun, G., Zhu-Barker, X., Chen, D., Liu, L., Zhang, N. *et al.* (2017) Responses of root exudation and nutrient cycling to grazing intensities and recovery practices in an alpine meadow: an implication for pasture management. *Plant and Soil* 416(1), 515–525.

Talore, D.G., Tesfamariam, E.H., Hassen, A., Du, J., Klampp, K. *et al.* (2016) Long-term impacts of grazing intensity on soil carbon sequestration and selected soil properties in the arid eastern cape, South Africa. *Journal of the Science of Food and Agriculture* 96(6), 1945–1952.

Tao, L.R., Lin, Z.H., Yong, Z.X. and Zhu, F. (2013) Effects of cultivation and grazing exclusion on the soil macro-faunal community of semiarid sandy grasslands in Northern China. *Arid Land Research and Management* 27(4), 377–393.

Teutscherova, N., Vázquez, E., Sotelo, M., Villegas, D., Velásquez, N. *et al.* (2021) Intensive short-duration rotational grazing is associated with improved soil quality within one year after establishment in Colombia. *Applied Soil Ecology* 159, 103835.

Van Groenigen, J.K., Qi, X., Osenberg, C.W., Luo, Y. and Hungate, B.A. (2014) Faster decomposition under increased atmospheric CO_2 limits soil carbon storage. *Science* 344(6183), 508–509.

Wachiye, S., Pellikka, P., Rinne, J., Heiskanen, J., Abwanda, S. *et al.* (2022) Effects of livestock and wildlife grazing intensity on soil carbon dioxide flux in the savanna grassland of Kenya. *Agriculture, Ecosystems & Environment* 325, 107713.

Wade, C., Sonnier, G. and Boughton, EH. (2022) Does grazing affect soil carbon in subtropical humid seminatural grasslands? *Rangeland Ecology & Management* 80, 10–17.

Wang, B., Wu, L., Chen, D., Wu, Y., Hu, S. *et al.* (2020) Grazing simplifies soil micro-food webs and decouples their relationships with ecosystem functions in grasslands. *Global Change Biology* 26(2), 960–970.

Wang, C. and Tang, Y. (2019) A global meta-analyses of the response of multi-taxa diversity to grazing intensity in grasslands. *Environmental Research Letters* 14(11), 114003.

Wang, J., Li, W., Cao, W. and Wang, S. (2022) Effects of different intensities of long-term grazing on plant diversity, biomass and carbon stock in alpine shrubland on the Qinghai-Tibetan plateau. *PeerJ* 10, e12771.

Wang, L., Delgado-Baquerizo, M., Wang, D., Isbell, F., Liu, J. *et al.* (2019) Diversifying livestock promotes multidiversity and multifunctionality in managed grasslands. *Proceedings of the National Academy of Sciences* 116(13), 6187–6192.

Wang, Y. and Wesche, K. (2016) Vegetation and soil responses to livestock grazing in central Asian grasslands: A review of Chinese literature. *Biodiversity and Conservation* 25(12), 2401–2420.

Waters, C.M., Orgill, S.E., Melville, G.J., Toole, I.D. and Smith, WJ. (2017) Management of grazing intensity in the semi-arid rangelands of southern Australia: Effects on soil and biodiversity. *Land Degradation & Development* 28(4), 1363–1375.

Wei, Y., Zhang, Y., Wilson, G.W., Guo, Y., Bi, Y. *et al.* (2021) Transformation of litter carbon to stable soil organic matter is facilitated by ungulate trampling. *Geoderma* 385, 114828.

Wiesmeier, M., Urbanski, L., Hobley, E., Lang, B., Lützow, M. *et al.* (2019) Soil organic carbon storage as a key function of soils—A review of drivers and indicators at various scales. *Geoderma* 333, 149–162.

Wigley, B.J., Augustine, D.J., Coetsee, C., Ratnam, J. and Sankaran, M. (2020) Grasses continue to trump trees at soil carbon sequestration following herbivore exclusion in a semiarid African savanna. *Ecology* 101(5), e03008.

Wilsey, B.J., Parent, G., Roulet, N.T., Moore, T.R. and Potvin, C. (2002) Tropical pasture carbon cycling: relationships between C source/sink strength, aboveground biomass and grazing. *Ecology Letters* 5(3), 367–376.

Winkler, K., Fuchs, R., Rounsevell, M. and Herold, M. (2021) Global land use changes are four times greater than previously estimated. *Nature Communications* 12(1), 1–10.

Wu, H., Wiesmeier, M., Yu, Q., Steffens, M., Han, X. *et al.* (2012) Labile organic C and N mineralization of soil aggregate size classes in semiarid grasslands as affected by grazing management. *Biology and Fertility of Soils* 48(3), 305–313.

Xin, X., Jin, D., Ge, Y., Wang, J., Chen, J. *et al.* (2020) Climate change dominated long-term soil carbon losses of Inner Mongolian grasslands. *Global Biogeochemical Cycles* 34(10), e2020GB006559.

Xun, W., Yan, R., Ren, Y., Jin, D., Xiong, W. *et al.* (2018) Grazing-induced microbiome alterations drive soil organic carbon turnover and productivity in meadow steppe. *Microbiome* 6(1), 1–13.
Xu, Y., Li, L., Wang, Q., Chen, Q. and Cheng, W. (2007) The pattern between nitrogen mineralization and grazing intensities in an Inner Mongolian typical steppe. *Plant and Soil* 300(1), 289–300.
Yang, Z., Xiong, W., Xu, Y., Jiang, L., Zhu, E. *et al.* (2016) Soil properties and species composition under different grazing intensity in an alpine meadow on the eastern Tibetan plateau, China. *Environmental Monitoring and Assessment* 188(12), 1–12.
Yan, L., Zhou, G. and Zhang, F. (2013) Effects of different grazing intensities on grassland production in China: A meta-analysis. *PLoS One* 8(12), e81466.
Yu, H., Li, Y., Oshunsanya, S.O., Are, K.S., Geng, Y. *et al.* (2019) Re-introduction of light grazing reduces soil erosion and soil respiration in a converted grassland on the loess plateau, China. *Agriculture, Ecosystems & Environment* 280, 43–52.
Yu, L., Sun, W. and Huang, Y. (2021) Grazing exclusion enhances plant and topsoil carbon stocks in arid and semiarid grasslands. *Agriculture, Ecosystems & Environment* 320, 107605.
Zhang, B., Thomas, B.W., Beck, R., Liu, K., Zhao, M. *et al.* (2018) Labile soil organic matter in response to long-term cattle grazing on sloped rough fescue grassland in the foothills of the rocky mountains, Alberta. *Geoderma* 318, 9–15.
Zhang, C., Dong, Q., Chu, H., Shi, J., Li, S. *et al.* (2018) Grassland community composition response to grazing intensity under different grazing regimes. *Rangeland Ecology & Management* 71(2), 196–204.
Zhang, R., Wang, J. and Niu, S. (2021) Toward a sustainable grazing management based on biodiversity and ecosystem multifunctionality in drylands. *Current Opinion in Environmental Sustainability* 48, 36–43.
Zhang, X., Zhang, W., Sai, X., Chun, F., Li, X. *et al.* (2022) Grazing altered soil aggregates, nutrients and enzyme activities in a *Stipa kirschnii* steppe of Inner Mongolia. *Soil and Tillage Research* 219, 105327.
Zhang, Y., Zhao, J., Xin, X., Wang, M., Pan, F. *et al.* (2022) Effects of stocking rate on the interannual patterns of ecosystem biomass and soil nitrogen mineralization in a meadow steppe of Northeast China. *Plant and Soil* 473, 9–31.
Zhan, T., Zhang, Z., Sun, J., Liu, M., Zhang, X. *et al.* (2020) Meta-analysis demonstrating that moderate grazing can improve the soil quality across China's grassland ecosystems. *Applied Soil Ecology* 147, 103438.
Zhao, F., Ren, C., Shelton, S., Wang, Z., Pang, G. *et al.* (2017) Grazing intensity influence soil microbial communities and their implications for soil respiration. *Agriculture, Ecosystems & Environment* 249, 50–56.
Zhou, G., Zhou, X., He, Y., Shao, J., Hu, Z. *et al.* (2017) Grazing intensity significantly affects belowground carbon and nitrogen cycling in grassland ecosystems: A meta-analysis. *Global Change Biology* 23(3), 1167–1179.
Zhou, G., Luo, Q., Chen, Y., He, M., Zhou, L. *et al.* (2019) Effects of livestock grazing on grassland carbon storage and release override impacts associated with global climate change. *Global Change Biology* 25(3), 1119–1132.

20 The importance of conserving crop wild relatives in preparing agriculture for climate change

Christopher P. Cockel[1]*, Filippo Guzzon[2], Maraeva Gianella[1] and Jonas V. Müller[1]

[1]Royal Botanic Gardens, Kew, Millennium Seed Bank, Ardingly, RH17 6TN, UK;
[2]Centre for Pacific Crops and Trees (CePaCT), Land Resources Division (LRD), Pacific Community (SPC), Private Mail Bag, Suva, Fiji

---

Abstract

Climate change is widely acknowledged to have severe implications for global food production and therefore food security. Utilising crop wild relatives (CWR) to help build resilience in domesticated crops is seen as part of the solution assuming that important genetic traits can be transferred to domesticated crops and that the resulting improved crop varieties can be farmed on a sufficiently large scale. CWR can be exploited as a valuable source of alleles related to adaptive traits to counter abiotic and biotic stresses resulting from climate change, and to improve yield and nutrition. This review aims to critically analyse the degree to which CWR have been utilised by crop breeders and researchers and will draw conclusions about the importance of CWR in preparing agriculture for climate change. Examples for CWR-derived improved varieties of nine major crops and at different plant development stages are presented. However, working with CWR is often seen as 'marginal'. It is problematic due to the complex processes involved and length of time needed for traditional crop breeding, shortage of funds, the limited availability of CWR germplasm, overcoming risk aversion among farmers, and a reluctance by the wider agricultural community to appreciate the scale of the problem and to embrace the potential benefits that could derive from using CWR in crop breeding. Given the observed lack of availability of CWR in accessible germplasm collections, additional resources must be directed towards ensuring CWR are protected *in situ* and conserved *ex situ* so they can be made available to researchers.

Keywords: crop wild relatives, food security, climate change, plant genetic resources, pre-breeding, germplasm, characterisation, machine learning

Introduction

With the recent conclusion of the COP26 UN Climate Change Conference (UNCC, 2021), more attention than ever before is being directed to the urgent need to find solutions to the potentially catastrophic impacts of unabated climate change, such as increased global food insecurity resulting from a combination of reduced crop yields and higher food prices brought on by extreme weather (especially more severe heat and drought stresses), something

---

*Corresponding author: c.cockel@kew.org

that had already been acknowledged for more than a decade (Met Office & WFP's Office for Climate Change, Environment and Disaster Risk Reduction, 2012).

Domestication and plant breeding programmes have led global systems of food production to depend on a small set of cultivated plants that, in turn, have been, and still are, subjected to genetic erosion (Dempewolf *et al.*, 2017). This narrow range of available plant genetic resources, in terms of both inter- and intra-specific genetic diversity, is incapable of responding adequately to the potentially devastating consequences of global warming and human population growth, such as malnutrition (i.e. hunger, obesity, micronutrient deficits), increased carbon dioxide emissions (Cohen, 2010) and insufficient food production (Ulian *et al.*, 2020; Cortés and López-Hernández, 2021).

A partial solution for mitigating and adapting future food production to climate change can be found by exploiting the genetics of resilience to be found in crop wild relatives (CWR). CWR are defined as wild plant ancestors that are closely related to domesticated varieties and species (Maxted *et al.*, 2006). CWR persist in nature without human intervention, but many of them are threatened by human activity (Meilleur and Hodgkin, 2004).

Conserving CWR alone is not sufficient, it is their conservation together with their utilisation by crop breeders to harness natural resilience traits contained in the plants' genetics that make them potentially valuable. If specific traits (such as tolerance to heat or drought stress, or resistance to certain pests and diseases) can be isolated and transferred into domesticated crops, the crops that humankind relies upon may be better able to feed the generations yet to come.

Here, some context is needed to understand why it is likely that CWR can be useful under future climate change scenarios, with predicted increased prevalence of certain pests and diseases (Dempewolf *et al.*, 2017). Unlike farmed elite crop cultivars/varieties that receive inputs of water, fertilisers, and pesticides, which are protected from competition from weeds, CWR lead a harsher existence in often much less hospitable conditions. They do not receive additional irrigation water or nutrients, nor do they benefit from extra protection from pests and diseases, herbivores, or competition from other plants. As such, to survive, CWR have evolved to be resilient. For example, wild relatives of tomato and potato in the Solanaceae family (e.g. *Solanum incanum* L.) are often toxic if eaten and are armed with thorns to deter herbivores (War *et al.*, 2012). Moreover, the bulk of genetic diversity in crop gene pools is found in CWR, especially when compared with the eroded genetic basis of modern agricultural cultivars that needs to be enlarged to select for the productive and resilient cultivars of the future (Dempewolf *et al.*, 2017).

The feasibility of the genetic transfer from wild material to crops depends on the degree of relatedness (i.e. gene pool level) and therefore the existing crossing barriers, low hybrid fertility or sterility. Within the Primary Gene Pool (GP-1), a crop and its close wild forms can cross producing fertile hybrids. The Secondary Gene Pool (GP-2) includes less closely related species, and gene flow to crops, even if difficult, is still possible using conventional breeding techniques. The Tertiary Gene Pool (GP-3) includes species from which gene transfer to crops requires embryo rescue, chromosome doubling, or bridging species to obtain hybrids (Harlan and de Wet, 1971; Singh *et al.*, 2021).

Traditional crop breeding takes time, due to the complex process of crossing, selection and testing needed to produce a new elite cultivar (Voss-Fels and Stahl, 2019). It also relies on an efficient selection procedure and a degree of luck (Zamir, 2001).

The so-called 'pre-breeding' (= identifying and isolating desirable traits and/or genes from wild relatives or landraces that cannot be used directly in breeding populations and transferring them into breeding lines) (Nass and Paterniani, 2000; Sharma *et al.*, 2013) can be used as a way of speeding up this process and preparing material ready to be incorporated into crop breeding programmes. Pre-breeding using wild wheat progenitors began at ICARDA in Aleppo in 1994 (Valkoun, 2001). Maximising the potential of pre-breeding was one of the central aims of the recently concluded 'Adapting Agriculture to Climate Change' project, which contained the strap line of 'Collecting, protecting and *preparing* crop wild relatives' (Dempewolf *et al.*, 2014). Indeed, pre-breeding data for 14 crops associated with this project have been made available *via* the Germinate platform (Raubach *et al.*, 2021).

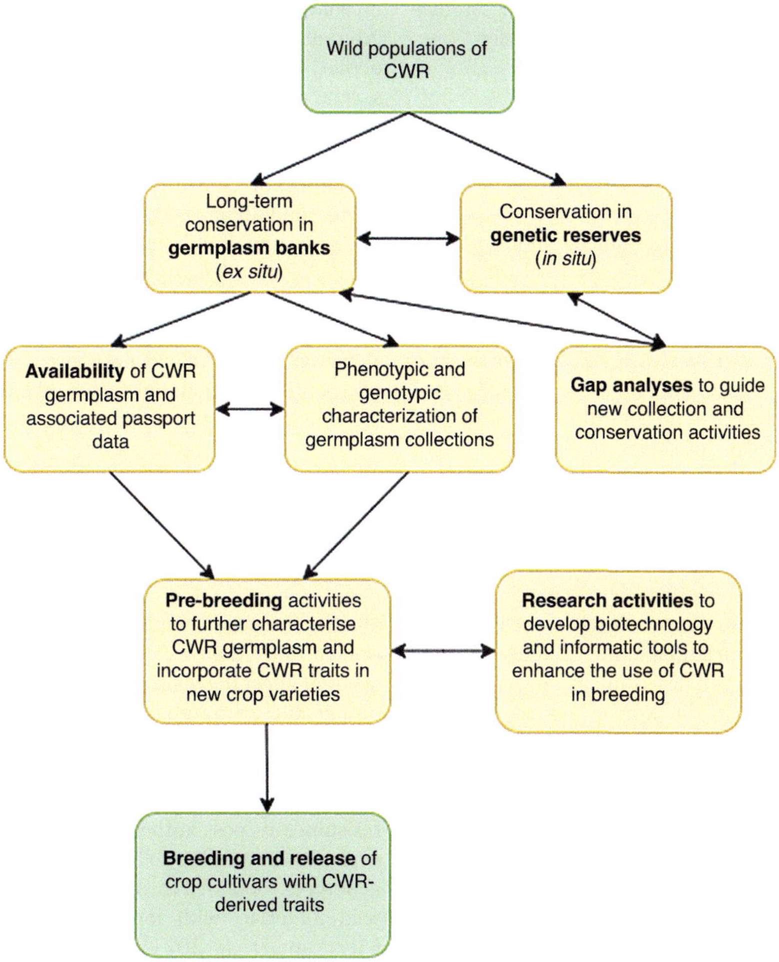

Fig. 20.1. A summary of the conservation and breeding pipeline for crop wild relatives.

Following a conservation and breeding pipeline for CWR (Fig. 20.1), this review will summarise current research in seeking to answer the question of how important *ex situ* and *in situ* conservation of CWR is for crop breeding and in preparing and adapting agriculture and future food production to climate change.

Review Methodology

We produced a narrative review with the aim of understanding the use of CWR in breeding as well as current gaps in the conservation, accessibility and use of these genetic resources. We searched Google Scholar (https://scholar.google.com/) and PubMed (https://pubmed.ncbi.nlm.nih.gov/) resources using search terms such as the importance of crop wild relatives for breeding, CWR and biotic stress, abiotic stress, drought stress, salt stress, heat stress, germination, phenological state, CWR conservation (*in situ*, *ex situ* genetic reserves). The results of this literature search were then analysed based on the publication date, using the most recent (2017 and following) articles to give examples where possible, otherwise going back to the most representative cases. Studies were analysed and included in the review if they included successful transfer of traits from CWR to crops, excluding screening/characterisation studies with no mention of effective trait transfer.

Importance of the Crop Wild Relatives for Breeding

CWR, together with landraces and neglected and underutilised crop species, hold the potential to improve food security, through crop improvement and/or standing as novel crops by themselves (Ulian *et al.*, 2020; Cortés and López-Hernández, 2021). The impact measured or predicted of CWR on agricultural productivity implies massive monetary value contained in these genetic resources. Now outdated figures from 1997 to 98 estimate that introgression of genetic traits from CWR to crops contributed US$20 billion to the US economy per year, and US$115 billion worldwide (Pimentel *et al.*, 1997), and that the wholesale farm values of CWR alone contributed to the US economy with US$9 billion annually (Phillips and Meelleur, 1998). In 2013, it was estimated that the potential value of benefits from CWR traits (Millennium Seed Bank, accessions belonging to 29 priority crops) was US$120 billion (PwC, PricewaterhouseCoopers, 2013). Overall, the value of the varieties developed with the use of wild relatives in breeding of crops has no doubt increased since these estimates were generated, due to improved breeding techniques/knowledge of CWR useful traits (Benlioğlu and Adak, 2019). Based on their socio-economic importance (actual and potential for variety improvement, considering also their conservation status and threat level), CWR can be prioritised for conservation actions (Kell *et al.*, 2017). Our literature review indicates that CWR have been extensively studied as sources of valuable alleles for adaptive traits to abiotic and biotic stresses imposed by climate change, but also to improve yield in terms of quantity and nutritional value, and to optimise the use of plant resources (Dempewolf *et al.*, 2017; Kilian *et al.*, 2021). Already, CWR are playing an important role in maintaining food security (Zair *et al.*, 2021), such as in the case of wild relative *Aegilops tauschii*, a close relative of bread wheat (*Triticum aestivum*) that was shown to contain useful drought, heat, salinity and disease resistance traits valuable for wheat breeding (Kishii, 2019). Indeed, most modern crop varieties will already contain some elements derived from breeding with CWR. Although several studies highlight the useful traits carried by CWR (Dempewolf *et al.*, 2017; Warburton *et al.*, 2017; Mammadov *et al.*, 2018; López-Hernández and Cortés, 2019; Cortés *et al.*, 2021; Cortés and Blair, 2021; Buitrago-Bitar *et al.*, 2021; Pratap *et al.*, 2021; Kilian *et al.*, 2021), it emerges that few modern CWR-derived varieties are currently cultivated or in an advanced state of selection (see Table 20.1 for examples). However, several studies have already identified alleles and wild material that could be used in breeding programmes (Kilian *et al.*, 2021).

Overall, crop adaptation to climate change can target different plant development stages, that is, germination, seedling establishment, vegetative growth, and fruiting. At the germination stage, cultivated materials normally perform better when seeds are subjected to abiotic stresses, compared with wild species (Fuller and Allaby, 2009). Water stress and drought affect the plant physiology at the germination stage, through its vegetative growth and photosynthesis, to yield and seed/fruit quality (Zia *et al.*, 2021). Two successful uses of CWR to improve crop tolerance to drought are improved resistance in post-anthesis in barley (Kalladan *et al.*, 2013) and the selection of durum wheat genotypes (obtained through backcrossing with wild material) with long grain-filling period (Aberkane *et al.*, 2021). An introgression line of wild Dongxiang/cultivate *japonica* Ningjing rice showed significantly enhanced salt stress resistance at all the stages – germination, establishment, vegetative growth and seed yield (Quan *et al.*, 2018). A recent phenotypic evaluation of 200 CWR-derived lines of *Oryza sativa* L., for agronomic performance over two seasons, obtained a promising cluster for salt tolerance (Tin *et al.*, 2021). An introgression line carrying two quantitative treat loci (QTL) for heat stress tolerance was identified while transferring genes from the wild rice *Oryza rufipogon* Griff. (Lei *et al.*, 2013). Wheat-*Leymus racemosus* chromosomes introgression lines showed increased heat tolerance at the seedling and reproductive stages, with one line also showing high yield potential (Mohammed *et al.*, 2014).

Tolerance to biotic stress is often monogenic and has been more extensively transferred from

Table 20.1. Examples of improved varieties with CWR-derived traits.

Crop	CWR-derived improved variety: trait	Citation
Wheat	Increased protein content Waterlogging tolerance	Hodgkin and Hajjar (2008)
Cassava	Increased protein content	Hodgkin and Hajjar (2008)
Barley	Drought tolerance	Hajjar and Hodgkin (2007)
Rice	Salt tolerance Yield increase	Tin *et al*. (2021) Hodgkin and Hajjar (2008)
Pea	Resistance to pea bruchid weevil	Coyne *et al*. (2020)
Chickpea	Resistance to *Ascochyta* blight	Coyne *et al*. (2020)
Tomato	Higher soluble solids content	Rick (1974)
Peanut	Resistance to root knot nematode	Simpson and Starr (2001)
Potato	Resistance to potato blight	Hodgkin and Hajjar (2008)
Common bean	Yield increase Antrachnose resistance Seed mineral accumulation	Blair *et al*. (2006) Garzón *et al*. (2008) Blair and Izquierdo (Blair and Izquierdo, 2012)

CWR to crops, when compared with abiotic stress tolerance. Successful examples are resistance to the root knot nematode in peanuts (*Arachis batizocoi* Krapov. & W.C. Gregory, *A. cardenasii* Krapov & W.C. Gregory, *A. diogoi* Hoehne), resistance to corn blight (*Tripsacum dactyloides* L.), resistance to potato blight (*Solanum demissum* Lindl.) and bruchid resistance in pea (*Pisum fulvum* Sibth. & Sm.) (Coyne *et al.*, 2020).

A review by Ebert (Ebert, 2020) shows that wild relatives of Solanaceae have been also extensively used to improve the nutritional profiles of aubergine/eggplant and tomato. Regarding cereals, using *Triticum dicoccoides* (Körn. ex Asch. & Graebn.) Schweinf in durum and bread wheat improvement led to an increased protein content (Levy and Feldman, 1987). While Hajjar and Hodgkin (2007) reported the use in 2004 of six barley cultivars with drought tolerance derived from *Hordeum spontaneum* K. Koch being released for use in Syria by the International Center for Agricultural Research in the Dry Areas (ICARDA).

Some researchers point to the difficulty involved in incorporating CWR into crop breeding programmes as a barrier to their utilisation, such as the risk of incorporating undesirable traits/alleles, known as 'linkage drag' (Ford-Lloyd *et al.*, 2011). Unwanted traits such as delayed flowering and lower yields (Cortés and López-Hernández, 2021), adaptations that confer resilience to wild plants, are some of those traits that were originally bred out during domestication ('domestication syndrome' (Allaby, 2014)). This problem can be mitigated by using molecular marker-assisted backcrossing using genome-specific markers (Warburton *et al.*, 2017).

Nevertheless, the use of CWR in crop breeding presents different difficulties imposed by the intrinsic traits of the wild plants. The pre-breeding phase *per se* can be limited by the costs in terms of time and resources, that is, multiple crossings or time-consuming trait evaluations (Dempewolf *et al.*, 2017). Moreover, CWR genetic material conserved in seed banks often lacks characterisation and evaluation data, which complicates its use in pre-breeding (Maxted *et al.*, 2010). The already available data are often scattered, non-standardised and distributed across multiple databases; hence, ready-to-use, comprehensive and accessible information is needed, for example, the Germinate platform (Raubach *et al.*, 2021). Field trials are often inefficient for trait selection due to the intrinsic variability in growth and phenological states of the wild material, which also shows high variation in the phenotypic manifestation of the desired traits among populations and years of harvest (Orsenigo *et al.*, 2017; Cortés and López-Hernández, 2021). In addition, the polygenic nature of several traits, especially the ones related to abiotic stress tolerance, complicates their transfer in the form of alleles with the traditional breeding techniques (Zsögön *et al.*, 2016). These traits are often lost with the domestication syndrome and because the adaptation to abiotic stresses relies on unwanted traits ('linkage drag') such as long vegetative stages, delayed flowering and reduced yield. On the other hand, monogenic traits have been more extensively exploited and successfully transferred to crops without carrying other undesirable traits (Prohens *et al.*, 2017; Cortés and López-Hernández, 2021).

Finally, there is also the potential issue of cross incompatibility between the crop and its wild relative caused by karyotype differences or by genomic rearrangement (Coyne *et al.*, 2020), which complicates efforts of breeders to fully exploit desired resilience traits from CWR.

Fortunately, advances in biotechnology and informatic tools allow breeders to partially overcome the obstacles to a broader use of CWR in crop breeding. A recent review by Cortés and López-Hernández (2021) reports that major 'big data' developments from both empirical and theoretical genomics can significantly improve the production of new, improved and resilient crops. Summarising their extensive review, they note that the heritability of adaptive traits and their underlying genes can be studied through genome-wide selection scans (GWSS), genome-environment associations (GEA) and genome-wide association studies (GWAS). Then, clustering and machine learning (ML) can speed up gene bank characterisation and germplasm collection activities through the distinction between abiotic stress tolerant and susceptible genotypes. Useful genetic variation can then be transferred to elite lines in genomic-assisted breeding programmes such as genomic prediction (GP) and inter-specific marker- and genomic-assisted backcrossing (MAB and GABC) schemes, or by exploiting multi-trait gene editing strategies (e.g. CRISPR-Cas9). Indeed, genomic-enabled prediction can be applied for the selection adaptive heritable traits also including genomic x environment interactions (G x E), for example, in drought and heat environments (Crossa *et al.*, 2016). Esposito *et al.* (2019) also point to the utility of employing a range of ML approaches in identifying possible genes and MicroRNAs (miRNA) responsible for cold tolerance in *Solanum commersonii*, a wild relative of the cultivated potato (*Solanum tuberosum*).

New gene editing techniques also open the possibility of realising *de novo* domestication of wild species. Instead of transferring the abiotic stress-related polygenic traits from CWR to elite crops, which are difficult to manipulate without the aforementioned drawbacks, *de novo* domestication theory aims to target the known monogenic traits of domestication (e.g. yield) in wild material, thereby conserving the already present adaptive traits (Zsögön *et al.*, 2016; Gasparini *et al.*, 2021).

Maxted *et al.* (2010) highlight the need for the characterisation and evaluation processes to be speeded up and improved to maintain the pace of the rapidly evolving biotechnological and bioinformatic tools. Indeed, the conservation of plant genetic resources (PGRs) must remain a tool and not become a barrier to the advancement in crop breeding.

Conservation of Crop Wild Relatives

Conventionally, two main strategies are employed for the conservation of plant diversity, *in situ* and *ex situ* conservation. *In situ* conservation is the conservation of the diversity in its natural habitat. It involves the designation, management and monitoring of identified target populations at the location where they are currently found. In contrast, *ex situ* conservation is the conservation of plant diversity outside its natural habitat (Maxted, 2013). This involves the sampling, transfer and storage of a sample of a plant population away from the original location where it was collected (Veteläinen *et al.*, 2009). Depending on the biological characteristics of different species, different conservation methodologies are employed for different plant species. A well-established pathway is the long-term conservation of seeds in seed banks (Walters and Pence, 2021). Conserving seeds in seed banks benefits from the fact that the majority of plant species produce orthodox seeds, that is, seeds that can tolerate drying to low moisture content and subsequent storage in sub-zero temperatures (De Vitis *et al.*, 2020). Conserving orthodox seeds in seed banks is considered a very efficient conservation strategy. It allows the storage of many seed accessions in relatively small spaces; moreover, seed collections are economically viable and can provide a good sample of the genetic diversity within species, usually remaining viable for the long-term (Li and Pritchard, 2009). For other species with recalcitrant seeds (= short-lived seeds that cannot withstand drying to low moisture conditions) and/or vegetative reproduction, other methodologies tend to be employed, such as *in vitro* storage, field gene banks and cryopreservation.

Castañeda-Álvarez *et al.* (2016) carried out an extensive analysis of the representativeness of CWR samples conserved in germplasm banks compared with natural populations, considering 1076 plant taxa related to 81 crops. From their analysis, it clearly emerged that CWR were poorly represented in gene banks. For 29.1% CWR taxa, no germplasm accessions existed, and further 23.9% of CWR taxa were represented by fewer than 10 accessions. Moreover, over 95% of the taxa were considered as underrepresented in gene banks when taking into account their full range distribution, with collecting gaps emerging especially in the Mediterranean and the Near East, western and southern Europe, Southeast and East Asia, and South America.

To address this gap in availability of CWR material for breeding, the 'Adapting Agriculture to Climate Change' project was conceived, and to-date has been the most ambitious and concerted initiative to collect, conserve and use CWR to prepare agriculture to meet the challenges posed by a changing climate. Yet, the ultimate measure of the project's success rests on the ability of crop breeders to effectively harness resilience traits from these CWR to develop new crop varieties.

The 10 year project was initiated in 2011 (Dempewolf *et al.*, 2014) and focused on collecting the seed of the wild relatives of crops of major importance for food security. The seed collecting phase of the project started in 2013 and was concluded in 2019, and it was carried out in 25 countries in Africa, Asia, the Americas and Europe, involving CWR taxa of 28 crop gene pools. A total of 4587 seed accessions were collected, corresponding to 28 crops and 27 genera. A total of 3667 unique samples of 27 target genera, covering 253 species (284 taxa), were duplicated at the Millennium Seed Bank (UK) by 21 countries. A total of 3190 seed accessions have been shared by the MSB with 10 international gene banks within the CGIAR system for multiplication and use, and passport data for 3544 accessions are available *via* the Genesys PGR website (https://www.genesyspgr.org/). Despite the overall success of the project, several issues were encountered (e.g. low collecting success for some important crop gene pools and in some centres of plant biodiversity) (Müller *et al.*, 2021). The accessions collected in the project are already being used in several pre-breeding activities worldwide (Kilian *et al.*, 2021).

Filling the gaps in CWR conservation is important not only to increase their availability as plant breeding resources but also to protect CWR from extinction, bearing in mind that several CWR taxa are considered as threatened with extinction at global, continental and national levels (Bilz *et al.*, 2011; IUCN, n.d.; Rossi *et al.*, 2013). Additionally, many CWR populations are predicted to decline or even become extinct in the next few decades due to the effects of climatic change (Ostrowski *et al.*, 2016). This reinforces the urgency of safeguarding CWR populations and complementing *in situ* and *ex situ* conservation. Indeed, even if *ex situ* conservation of genetic resources is cost effective and enhances the accessibility to CWR for breeding activities, *in situ* conservation allows CWR populations to evolve in their original range and to retain a higher genetic diversity when compared with seed bank accessions (Thomas *et al.*, 2012).

In situ protected areas meant for the conservation of CWR are generally known as 'genetic reserves'. The first genetic reserves were created in the 1980s, with an increasing attention on CWR *in situ* conservation only in the last 20–25 years (Hunter and Heywood, 2011), and including initiatives such as Farmers' Pride (http://www.farmerspride.eu/), which aims to create a network of protected areas, seed banks and other stakeholders to conserve *in situ* plant genetic resources across Europe. Genetic reserves are often located where protected areas have been established to conserve other aspects of biodiversity, and so the additional resource requirements to conserve CWR may be minimal. Examples of these protected areas are genetic reserves for the conservation of wheat wild relatives and other wild cereals in the Near East and Armenia (Guzzon *et al.*, 2022); protected areas for the conservation of narrow endemic teosinte (wild maize) species in Mexico and Nicaragua (Guzzon *et al.*, 2021); the German network of wild celery genetic reserves (IUCN, n.d.).

Several diversity analyses were made for different countries of the world in order to identify areas with a high CWR diversity that can be considered for the establishment of genetic reserves (e.g. Czech Republic (Taylor *et al.*, 2017), Scotland (Fielder *et al.*, 2016), Sweden (Weibull and Phillips, 2020), Norway (Phillips *et al.*, 2016)). Vincent *et al.* (2019) carried out an extensive global geographical analysis, considering 1261 CWR species from 167 major crop gene pools, to identify key geographical areas for CWR *in situ* conservation worldwide. The authors identified 150 priority sites worldwide where a significant CWR *in situ* diversity, represented by 829 CWR species related to 157 crops, can be systematically conserved *in situ*. Ten top sites for CWR *in situ* conservation, already included in protected areas, were also proposed, with five occurring in the Mediterranean Basin, confirming that the Mediterranean Basin is probably the area with the highest global CWR diversity. Moreover, several gap analyses identified areas of high diversity and gaps in *ex situ* conservation of wild relatives of several crop genepools (e.g. pigeon pea (Khoury *et al.*, 2015a); sweet potato (Khoury *et al.*, 2015b); sunflower (Kantar *et al.*, 2015)). These analyses, beside their great importance for guiding collecting and conservation missions using ecogeographic information, can help identify CWR populations with the potential to contribute agronomic traits of priority to breeders. In future research, it will be important to consider also micro-scale patterns in identifying and conserving CWR populations since microhabitat patterns have a great importance in shaping the adaptive potential of plant populations (Zellweger *et al.*, 2020).

Despite recent initiatives such as the 'Adapting Agriculture to Climate Change' project, it is clear that gaps persist in *ex situ* conservation of many CWR species (Müller *et al.*, 2021), not only in terms of the high percentage of taxa that are underrepresented but also as the lack of knowledge of seed longevity and germination requirements as well as *in vitro* protocols for many species (Flanagan *et al.*, 2018; Gianella *et al.*, 2020). Moreover, as mentioned above, the use of CWR in breeding will also depend on the increasing accessibility of passport data (including novel molecular and physiological data) connected to germplasm accessions. Germplasm banks will therefore need to invest in database systems that can efficiently store and make available these passport data. The *in situ* conservation of CWR appears to be in a very preliminary stage for many taxa and/or centres of CWR diversity. A more proactive conservation approach, specifically targeting endangered CWR, in already existing natural reserves as well as a greater integration of *ex situ* and *in situ* conservation strategies could help to reduce the threats posed to CWR by climatic changes (Vincent *et al.*, 2019; Guzzon *et al.*, 2022).

Conclusion

Agriculture must continue to adapt to maintain and increase food production to cope with the growing global population and increased stresses on agricultural production imposed by climate change. Although often seen as 'marginal', we can show that the use of CWR has played a major role in plant breeding for multiple major crops and at different development stages, improving abiotic and biotic resistance and yield, and their use will necessarily continue to remain important into the future. While the use of CWR in plant breeding is currently often hampered by technical obstacles, the continued development and availability of new techniques such as advances in biotechnology and informatic tools will help overcome these current obstacles. Despite recent advancement through global initiatives, major conservation gaps still exist in the availability of high-quality CWR germplasm material. A significant increase in international collaborative effort on a global scale to identify, collect and make available additional CWR resources is urgently required.

References

Aberkane, H., Amri, A., Belkadi, B., Filali-Maltouf, A., Kehel, Z. *et al.* (2021) Evaluation of durum wheat lines derived from interspecific crosses under drought and heat stress. *Crop Science* 61(1), 119–136. DOI: 10.1002/CSC2.20319.

Allaby, R.G. (2014) Domestication syndrome in plants. In: Smith, C. (ed.) *Encyclopedia of Global Archaeology*. Springer, New York. DOI: 10.1007/978-1-4419-0465-2_2416.

Benlioğlu, B. and Adak, MS. (2019) Importance of crop wild relatives and landraces genetic resources in plant breeding programmes. *Journal of Experimental Agriculture International* 37, 1–8. DOI: 10.9734/bpi/atias/v3.

Bilz, M., Kell, S.P., Maxted, N. and Lansdown, R.V. (2011) *European red list of vascular plants*. Publications Office of the European Union, Luxembourg.

Blair, M.W. and Izquierdo, P. (2012) Use of the advanced backcross-QTL method to transfer seed mineral accumulation nutrition traits from wild to Andean cultivated common beans. *Theoretical and Applied Genetics* 125(5), 1015–1031.

Blair, M.W., Iriarte, G. and Beebe, S. (2006) QTL analysis of yield traits in an advanced backcross population derived from a cultivated Andean x wild common bean (*Phaseolus vulgaris* L.) cross. *Theoretical and Applied Genetics* 112(6), 1149–1163. DOI: 10.1007/s00122-006-0217-2.%20Epub%202006 Jan 24.

Buitrago-Bitar, M.A., Cortés, A.J., López-Hernández, F., Londoño-Caicedo, J.M., Muñoz-Florez, J.E. *et al.* (2021) Allelic diversity at abiotic stress responsive genes in relationship to ecological drought indices for cultivated tepary bean, *Phaseolus acutifolius* A. Gray, and its wild relatives. *Genes* 12, 556. DOI: 10.3390/genes12040556.

Castañeda-Álvarez, N.P., Khoury, C.K., Achicanoy, H.A., Bernau, V., Dempewolf, H. *et al.* (2016) Global conservation priorities for crop wild relatives. *Nature Plants* 2, 1–6. DOI: 10.1038/NPLANTS.2016.22.

Cohen, JE. (2010) Population and climate change. *Proceedings of the American Philosophical Society* 154(2), 158–182.

Cortés, A.J. and Blair, MW. (2021) Genotyping by sequencing and genome-environment associations in wild common bean predict widespread divergent adaptation to drought. *Frontiers in Plant Science* 9, 128. DOI: 10.3389/fpls.2018.00128.

Cortés, A.J., Monserrate, F.A., Ramírez-Villegas, J., Madriñán, S. and Blair, MW. (2021) Drought tolerance in wild plant populations: The case of common beans (*Phaseolus vulgaris* L.). *PLOS ONE* 8(5), e62898. DOI: 10.1371/journal.pone.0062898.

Cortés, A.J. and López-Hernández, F. (2021) Harnessing crop wild diversity for climate change adaptation. *Genes* 12(5), 783. DOI: 10.3390/genes12050783.

Coyne, C.J., Kumar, S. and Von Wettberg, E.J.B. (2020) Potential and limits of exploitation of crop wild relatives for pea, lentil, and chickpea improvement. *Legume Science* 2, e36. DOI: 10.1002/leg3.36.

Crossa, J., Jarquín, D., Franco, J., Pérez-Rodríguez, P., Burgueño, J. *et al.* (2016) Genomic prediction of gene bank wheat landraces. *G3: Genes, Genomes, Genetics* 6(7), 1819–1834.

Dempewolf, H., Eastwood, R.J., Guarino, L., Khoury, C.K., Müller, J.V. *et al.* (2014) Adapting agriculture to climate change: A global initiative to collect, conserve, and use crop wild relatives. *Agroecology and Sustainable Food Systems* 38, 369–377. DOI: 10.1080/21683565.2013.870629.

Dempewolf, H., Baute, G., Anderson, J., Kilian, B., Smith, C. *et al.* (2017) Past and future use of wild relatives in crop breeding. *Crop Science* 57(3), 1070–1082. DOI: 10.2135/CROPSCI2016.10.0885.

De Vitis, M., Hay, F.R., Dickie, J.B., Trivedi, C., Choi, J. *et al.* (2020) Seed storage: Maintaining seed viability and vigor for restoration use. *Restoration Ecology* 28(S3), S249–S255. DOI: 10.1111/rec.13174.

Ebert, AW. (2020) The role of vegetable genetic resources in nutrition security and vegetable breeding. *Plants* 9(6), 736. DOI: 10.3390/PLANTS9060736.

Esposito, S., Carputo, D., Cardi, T. and Tripodi, P. (2019) Applications and trends of machine learning in genomics and phenomics for next-generation breeding. *Plants* 9(1), 34.

Fielder, H., Smith, C., Ford-Lloyd, B. and Maxted, N. (2016) Enhancing the conservation of crop wild relatives in Scotland. *Journal for Nature Conservation* 29, 51–61.

Flanagan, N.S., Chavarriaga, P. and Mosquera-Espinosa, A.T. (2018) Conservation and sustainable use of vanilla crop wild relatives in Colombia. In: *Handbook of Vanilla Science & Technology*. Available at: https://www.researchgate.net/publication/326540798_Conservation_and_Sustainable_Use_of_Vanilla_Crop_Wild_Relatives_in_Colombia

Ford-Lloyd, B.V., Schmidt, M., Armstrong, S.J., Barazani, O., Engels, J. *et al.* (2011) Crop wild relatives—Undervalued, underutilized, and under threat?. *Bioscience* 61(7), 559–565. DOI: 10.1525/bio.2011.61.7.10.

Fuller, D.Q. and Allaby, R. (2009) Seed dispersal and crop domestication: shattering, germination and seasonality in evolution under cultivation. In: Østergaard, L. (ed.) *Annual Plant Reviews*, Vol. 38. Fruit Development and Seed Dispersal, pp. 238–295. DOI: 10.1002/9781444314557.ch7.

Garzón, L.N., Ligarreto, G.A. and Blair, MW. (2008) Molecular marker-assisted backcrossing of anthracnose resistance into Andean climbing beans (*Phaseolus vulgaris* L.). *Crop Science* 48(2), 562–570.

Gasparini, K., Peres, L.E.P. and Zsögön, A. (2021) De novo domestication of wild species to create crops with increased resilience and nutritional value. *Current Opinion in Plant Biology* 60, 102006. DOI: 10.1016/J.PBI.2021.102006.

Gianella, M., Balestrazzi, A., Pagano, A., Müller, J.V., Kyratzis, A.C. *et al.* (2020) Heteromorphic seeds of wheat wild relatives show germination niche differentiation. *Plant Biology* 22, 191–202. DOI: 10.1111/plb.13060.

Guzzon, F., Gianella, M., Giovannini, P. and Payne, T. (2022) Conserving wheat genetic resources. In: Reynold, M. and Braun, H. (eds) *Wheat Improvement. Food Security in a Changing Climate*. Springer Nature, Cham, Switzerland, pp. 299–318.

Guzzon, F., Arandia Rios, L.W., Caviedes Cepeda, G.M., Céspedes Polo, M., Chavez Cabrera, A. *et al.* (2021) Conservation and use of Latin American maize diversity: Pillar of nutrition security and cultural heritage of humanity. *Agronomy* 11, 172.

Hajjar, R. and Hodgkin, T. (2007) The use of wild relatives in crop improvement: A survey of developments over the last 20 years. *Euphytica* 1–13.

Harlan, J.R. and de Wet, J.M.J. (1971) Toward a rational classification of cultivated plants. *Taxon* 20(4), 509–517. DOI: 10.2307/1218252.

Hodgkin, T. and Hajjar, R. (2008) Using crop wild relatives for crop improvement: Trends and perspectives. In: Maxted, N. (ed.) *Crop Wild Relative Conservation and Use*. CABI, Wallingford, United Kingdom, pp. 535–548.

Hunter, D. and Heywood, V. (2011) *Crop wild relatives: A manual of in situ conservation*. Earthscan, London and Washington D.C.

IUCN (n.d.) IUCN SSC crop wild relative specialist group 2020 report.

Kalladan, R., Worch, S., Rolletschek, H., Harshavardhan, V.T., Kuntze, L. *et al.* (2013) Identification of quantitative trait loci contributing to yield and seed quality parameters under terminal drought in barley advanced backcross lines. *Molecular Breeding* 32(1), 71–90. DOI: 10.1007/S11032-013-9853-9.

Kantar, M.B., Sosa, C.C., Khoury, C.K., Castañeda-Álvarez, N.P., Achicanoy, H.A. *et al.* (2015) Ecogeography and utility to plant breeding of the crop wild relatives of sunflower (*Helianthus annuus* L.). *Frontiers in Plant Science* 6, 841.

Kell, S.P., Ford-Lloyd, B.V., Brehm, J.M., Iriondo, J.M. and Maxted, N. (2017) Broadening the base, narrowing the task: Prioritizing crop wild relative taxa for conservation action. *Crop Science* 57, 1042–1058. DOI: 10.2135/cropsci2016.10.0873.

Khoury, C.K., Castañeda-Alvarez, N.P., Achicanoy, H.A., Sosa, C.C., Bernau, V. *et al.* (2015a) Crop wild relatives of pigeonpea [*Cajanus cajan* (L.) millsp.]: Distributions, *ex situ* conservation status, and potential genetic resources for abiotic stress tolerance. *Biological Conservation* 184, 259–270.

Khoury, C.K., Heider, B., Castañeda-Álvarez, N.P., Achicanoy, H.A., Sosa, C.C. *et al.* (2015b) Distributions, *ex situ* conservation priorities, and genetic resource potential of crop wild relatives of sweetpotato [*Ipomoea batatas* (L.) lam., I. series Batatas]. *Frontiers in Plant Science* 6, 841.

Kilian, B., Dempewolf, H., Guarino, L., Werner, P., Coyne, C. *et al.* (2021) Crop science special issue: Adapting agriculture to climate change: A walk on the wild side. *Crop Science* 61(1), 32–36. DOI: 10.1002/csc2.20418.

Kishii, M. (2019) An update of recent use of *Aegilops* species in wheat breeding. *Frontiers in Plant Science* 10, 585. DOI: 10.3389/fpls.2019.00585.

Lei, D., Tan, L., Liu, F., Chen, L. and Sun, C. (2013) Identification of heat-sensitive QTL derived from common wild rice (*Oryza rufipogon* Griff.). *Plant Science* 201–202(1), 121–127. DOI: 10.1016/J.PLANTSCI.2012.12.001.

Levy, A.A. and Feldman, M. (1987) Increase in grain protein percentage in high-yielding common wheat breeding lines by genes from wild tetraploid wheat. *Euphytica* 36, 353–359. DOI: 10.1007/BF00041478.

Li, D. and Pritchard, H. (2009) The science and economics of *ex situ* plant conservation. *Trends in Plant Science* 14, 614–621. DOI: 10.1016/j.

López-Hernández, F. and Cortés, AJ. (2019) Last-generation genome-environment associations reveal the genetic basis of heat tolerance in common bean (*Phaseolus vulgaris* L.). *Frontiers in Genetics* 10, 954. DOI: 10.3389/fgene.2019.00954.

Mammadov, J., Buyyarapu, R., Guttikonda, S.K., Parliament, K., Abdurakhmonov, I.Y. *et al.* (2018) Wild relatives of maize, rice, cotton, and soybean: Treasure troves for tolerance to biotic and abiotic stresses. *Frontiers in Plant Science* 9, 886. DOI: 10.3389/FPLS.2018.00886/BIBTEX.

Maxted, N., Ford-Lloyd, B.V., Jury, S., Kell, S. and Scholten, M. (2006) Towards a definition of a crop wild relative. *Biodiversity and Conservation* 15, 2673–2685. DOI: 10.1007/s10531-005-5409-6.

Maxted, N., Kell, S., Toledo, Á., Dulloo, E., Heywood, V. *et al.* (2010) A global approach to crop wild relative conservation: Securing the gene pool for food and agriculture. *Kew Bulletin* 65(4), 561–576. DOI: 10.1007/S12225-011-9253-4/TABLES/4.

Maxted, N. (2013) *In situ, ex situ* conservation. In: Levin, S.A. (ed.) *Encyclopedia of Biodiversity*, 2nd ed. Academic Press, Amsterdam, pp. 313–323. DOI: 10.1016/B978-0-12-384719-5.00049-6.

Meilleur, B.A. and Hodgkin, T. (2004) *In situ* conservation of crop wild relatives: Status and trends. *Biodiversity and Conservation* 13, 663–684. DOI: 10.1023/B:BIOC.0000011719.03230.17.

Met Office & WFP's Office for Climate Change, Environment and Disaster Risk Reduction (2012) Climate impacts on food security and nutrition. A review of existing knowledge. Exeter, UK: Met Office. Available at: https://lucris.lub.lu.se/ws/portalfiles/portal/129864479/wfp258981.pdf

Mohammed, Y.S.A., Tahir, I.S.A., Kamal, N.M., Eltayeb, A.E., Ali, A.M. *et al.* (2014) Impact of wheat-*Leymus racemosus* added chromosomes on wheat adaptation and tolerance to heat stress. *Breeding Science* 63(5), 450–460. DOI: 10.1270/JSBBS.63.450.

Müller, J.V., Cockel, C.P., Gianella, M. and Guzzon, F. (2021) Treasuring crop wild relative diversity: Analysis of success from the seed collecting phase of the 'Adapting Agriculture to Climate Change' project. *Genetic Resources and Crop Evolution* 68, 2749–2756. DOI: 10.1007/s10722-021-01229-x.

Nass, L.L. and Paterniani, E. (2000) Pre-breeding: A link between genetic resources and maize breeding. *Scientia Agricola* 57(3), 581–587.

Orsenigo, S., Guzzon, F., Abeli, T., Rossi, G., Vagge, I. *et al.* (2017) Comparative germination responses to water potential across different populations of *Aegilops geniculata* and cultivar varieties of *Triticum durum* and *Triticum aestivum*. *Plant Biology* 19(2), 165–171. DOI: 10.1111/PLB.12528.

Ostrowski, M., Prosperi, J. and David, J. (2016) Potential implications of climate change on *Aegilops* species distribution: Sympatry of these crop wild relatives with the major European crop *Triticum aestivum* and conservation issues. *PLoS One* 11, e0153974. DOI: 10.1371/journal.pone.0153974.

Phillips, J., Asdal, Å., Magos Brehm, J., Rasmussen, M. and Maxted, N. (2016) *In situ* and *ex situ* diversity analysis of priority crop wild relatives in Norway. *Diversity and Distribution* 22, 1122–1126. Available at: https://onlinelibrary.wiley.com/doi/10.1111/ddi.12470

Phillips, O.L. and Meelleur, BA. (1998) Usefulness and economic potential of the rare plants of the United States: A statistical survey. *Economic Botany* 52(1), 57–67. DOI: 10.1007/BF02861295.

Pimentel, D., Wilson, C., McCullum, C., Huang, R., Dwen, P. *et al.* (1997) Economic and environmental benefits of biodiversity. *Bioscience* 47(11), 747–757.

Pratap, A., Das, A., Kumar, S. and Gupta, S. (2021) Current perspectives on introgression breeding in food legumes. *Frontiers in Plant Science* 11, 2118. DOI: 10.3389/FPLS.2020.589189/BIBTEX.

Prohens, J., Gramazio, P., Plazas, M., Dempewolf, H., Kilian, B. *et al.* (2017) Introgressiomics: A new approach for using crop wild relatives in breeding for adaptation to climate change. *Euphytica* 213(7), 1–19. DOI: 10.1007/S10681-017-1938-9.

PwC, PricewaterhouseCoopers (2013) *Crop wild relatives: A valuable resource for crop development*. PWC, London, United Kingdom.

Quan, R., Wang, J., Hui, J., Bai, H., Lyu, X. *et al.* (2018) Improvement of salt tolerance using wild rice genes. *Frontiers in Plant Science* 8, 1–11. DOI: 10.3389/fpls.2017.02269.

Raubach, S., Kilian, B., Dreher, K., Amri, A., Bassi, F.M. *et al.* (2021) From bits to bites: Advancement of the germinate platform to support prebreeding informatics for crop wild relatives. *Crop Science* 61(3), 1538–1566.

Rick, CM. (1974) High soluble-solids content in large-fruited tomato lines derived from a wild green-fruited species. *Hilgardia* 42(15), 493–510. DOI: 10.3733/HILG.V42N15P493.

Rossi, G., Montagnani, C., Gargano, D., Peruzzi, L., Abeli, T. *et al.* (2013) Lista Rossa della Flora Italiana. 1. Policy Species e altre specie minacciate. Comitato Italiano IUCN e Ministero Dell'Ambiente e Della Tutela Del Territorio e Del Mare, Rome.

Sharma, S., Upadhyaya, H.D., Varshney, R.K. and Gowda, CLL. (2013) Pre-breeding for diversification of primary gene pool and genetic enhancement of grain legumes. *Frontiers in Plant Science* 4, 309. DOI: 10.3389/fpls.2013.00309.

Simpson, C.E. and Starr, JL. (2001) Registration of 'COAN' peanut. *Crop Science* 41(3), 918.

Singh, D.P., Singh, A.K. and Singh, A. (2021) Plant genetic resources. In: *Plant Breeding and Cultivar Development*. Academic Press, London, United Kingdom, pp. 129–157. DOI: 10.1016/B978-0-12-817563-7.00009-X.

Taylor, N.G., Kell, S.P., Holubec, V., Parra-Quijano, M., Chobot, K. *et al.* (2017) A systematic conservation strategy for crop wild relatives in the Czech Republic. *Diversity and Distribution* 23, 448–462.

Thomas, M., Demeulenaere, E., Dawson, J.C., Khan, A.R., Galic, N. *et al.* (2012) On-farm dynamic management of genetic diversity: The impact of seed diffusions and seed saving practices on a population-variety of bread wheat. *Evolutionary Applications* 5, 779–795.

Tin, H.Q., Loi, N., Labarosa, S.J.E., McNally, K.L., McCouch, S. *et al.* (2021) Phenotypic response of farmer-selected CWR-derived rice lines to salt stress in the Mekong Delta. *Crop Science* 61(1), 201–218.

Ulian, T., Diazgranados, M., Pironon, S., Padulosi, S., Liu, U. *et al.* (2020) Unlocking plant resources to support food security and promote sustainable agriculture. *Plants People Planet* 2(5), 421–445. DOI: 10.1002/ppp3.10145.

UNCC (2021) COP26 the Glasgow Climate Pact. Available at: https://ukcop26.org/wp-content/uploads/2021/11/COP26-Presidency-Outcomes-The-Climate-Pact.pdf

Valkoun, J. (2001) Wheat pre-breeding using wild progenitors. *Euphytica* 119, 17–23. DOI: 10.1023/A:1017562909881.

Veteläinen, M., Negr, V. and Maxted, N. (2009) European landraces: On-farm conservation management and use. In: *Bioversity Technical Bulletin*. Bioversity International, Rome, p. 15.

Vincent, H., Amri, A., Castañeda-Álvarez, N.P., Dempewolf, H., Dulloo, E. *et al.* (2019) Modeling of crop wild relative species identifies areas globally for *in situ* conservation. *Communications Biology* 2, 136.

Voss-Fels, K.P. and Stahl, A. (2019) Modern crop breeding for future food security. *BMC Biology* 17, 18. DOI: 10.1186/s12915-019-0638-4.

Walters, C. and Pence, VC. (2021) The unique role of seed banking and cryobiotechnologies in plant conservation. *Plants People Planet* 3(1), 83–91. DOI: 10.1002/ppp3.10121.

War, A.R., Paulraj, M.G., Ahmad, T., Buhroo, A.A., Hussain, B. *et al.* (2012) Mechanisms of plant defense against insect herbivores. *Plant Signaling & Behavior* 7(10), 1306–1320. DOI: 10.4161/psb.21663.

Warburton, M.L., Rauf, S., Marek, L., Hussain, M., Ogunola, O. *et al.* (2017) The use of crop wild relatives in maize and sunflower breeding. *Crop Science* 57, 1227–1240. DOI: 10.2135/cropsci2016.10.0855.

Weibull, J. and Phillips, J. (2020) Swedish crop wild relatives: Towards a national strategy for *in situ* conservation of CWR. *Genetic Resources* 1(1), 17–23. DOI: 10.46265/genresj.2020.1.17-24.

Zair, W., Maxted, N., Brehm, J.M. and Amri, A. (2021) *Ex situ* and *in situ* conservation gap analysis of crop wild relative diversity in the fertile crescent of the Middle East. *Genetic Resources and Crop Evolution* 68, 693–709. DOI: 10.1007/s10722-020-01017-z.

Zamir, D. (2001) Improving plant breeding with exotic genetic libraries. *Nature Reviews. Genetics* 2, 983–989. DOI: 10.1038/35103590.

Zellweger, F., De Frenne, P., Lenoir, J., Vangansbeke, P., Verheyen, K. *et al.* (2020) Forest microclimate dynamics drive plant responses to warming. *Science* 368(6492), 772–775.

Zia, R., Nawaz, M.S., Siddique, M.J., Hakim, S. and Imran, A. (2021) Plant survival under drought stress: Implications, adaptive responses, and integrated rhizosphere management strategy for stress mitigation. *Microbiological Research* 242, 1–16. DOI: 10.1016/J.MICRES.2020.126626.

Zsögön, A., Cermak, T., Voytas, D. and Peres, LEP. (2016) Genome editing as a tool to achieve the crop ideotype and de novo domestication of wild relatives: Case study in tomato. *Plant Science* 256, 120–130. DOI: 10.1016/J.

21 Food security and livelihood challenges of extensive goat production systems in desertification areas

Pablo Ronda-Borzone[1]*, Guillermo Donoso[2] and José Luis Riveros F.[3]
[1]Facultad de Agronomía e Ingeniería Forestal, Pontificia Universidad Católica de Chile, Santiago, Chile; [2]Departamento de Economía Agraria, Facultad de Agronomía e Ingeniería Forestal, Pontificia Universidad Católica de Chile, Santiago, Chile; [3]Departamento de Ciencias Animales, Facultad de Agronomía e Ingeniería Forestal, Pontificia Universidad Católica de Chile, Santiago, Chile

Abstract
Undernourishment and food insecurity are still challenges threatening poor urban and rural communities. Approximately 12% of the global population faced severe food insecurity in 2020, representing 928 million people. A significant fraction of the world's land surface is covered by drylands, which is expected to increase, affecting the food security and livelihood of 2 billion people due to climate change. Goats have traditionally been a significant asset and source of protein for rural communities inhabiting the drylands. We hypothesize that goats were intentionally selected by communities in dry environments and might be the optimal animal to breed in the future due to their capability to survive in dry environments and to utilize less water than other livestock production systems. However, it must be considered that only a small fraction (1%–2%) of the total water consumed by goats is drunk or utilized directly by the animals. Most of the water consumed is utilized to grow the feed that livestock such as goats consumes, which in most regions of the world is rainfed rangelands. Increasing green water consumption efficiency in goat production is thus critical in the context of increasing water scarcity and desertification. However, there is little research on goat water productivity and the impact of decreasing water security on the sustainability of goat production and, hence, on the livelihoods in small communities in developing countries. Therefore, the impact of increasing water scarcity and desertification on goat production in rainfed lands should be a research priority.

Keywords: goats, livestock, water productivity, water scarcity, food security, climate change

Introduction

Worldwide increasing population, changes in food consumption behavior, and increasing demand for biomass (e.g. bioenergy), combined with the effects of climate change, are likely to lead to increased desertification and water stress scenarios in the future (Hoekstra and Mekonnen, 2011). These already observable patterns are threatening the livelihood and food security of 2 billion people who live in drylands (Feng and Fu, 2013; Prăvălie, 2016). According to the last

*Corresponding author: paronda@uc.cl

IPCC report (2021), there is evidence that human activities are causing climate change and increasing the frequency and intensity of droughts (Masson-Delmotte *et al.*, 2021). Nearly 20% of the global population lives in regions under serious risk from desertification (GEF-IFAD Partnership, 2002). Desertification can be defined as any widespread environmental degradation that reduces the productivity of dryland ecosystems by decreasing plant cover, soil loss, loss of soil organic matter, increasing sand dunes, and increasing run-off (Goudie *et al.*, 2019; Whitford *et al.*, 2020).

As desertification affects drylands, it is relevant to define them to identify the need for public policy. Desertification is usually characterized by the aridity index (AI), a numerical indicator of the degree of climate dryness at a given location. It is usually defined as the ratio between precipitation (P) and potential evapotranspiration (PET), that is, AI = P/PET. Drylands are ecosystems, such as rangelands, grasslands, and woodlands, characterized by high temporal and spatial rainfall variability. Drylands are commonly defined as semiarid regions with an AI <0.65 (UNESCO, UN-Water, 2020).[1] Evidence shows that climate change has been the main contributor to the long-term increasing tendency in AI (UNESCO, UN-Water, 2020; Masson-Delmotte *et al.*, 2021). Thus, climate change adds challenges to the already difficult water management in drylands. Increasing aridity, global warming, and population growth will exacerbate the risk of desertification in the near future (UNESCO, UN-Water, 2020). Approximately 20%–35% of drylands already suffer land degradation, and it is expected to increase significantly, with as much as 80% of this expansion occurring in developing countries, which are typically environmentally and socioeconomically vulnerable and more exposed to food insecurity issues (UNESCO, UN-Water, 2020).

Concomitant with desertification progression, according to Hofste *et al.* (Hofste *et al.*, 2019) 17 countries (that together account for a quarter of the world's population) are currently facing "extremely high" water stress (Hofste *et al.*, 2019). Moreover, the global population is expected to increase to approximately 8.5 billion people in 2030, 10.2 billion in 2050, and 10.8 billion in 2100 (UN, 2019). "By 2025, 1.8 billion people will experience absolute water scarcity, and two-thirds of the world will be living under water-stressed conditions" (UN-Water, 2006). The increasing stress on freshwater to feed a growing population, especially in less-developed countries, is a serious concern (UN-Water, 2006). The rising demand for healthy food and the need to satisfy the nutritional requirements of rural and poor communities facing climate change and desertification challenges ratify the necessity for sustainable production systems. According to the 2020 Sustainable Development Goals Report, hunger and food insecurity have been increasing, and malnutrition still affects millions of children. The situation is likely to worsen, considering the economic slowdowns and disruptions caused by the COVID-19 pandemic-triggered recession (UN, 2021). Faced with this scenario, the need to improve water use and governance has been pinpointed as a critical measure for economic growth, social inclusiveness, and environmental sustainability (OECD, 2018). These consequences of desertification and water scarcity on food security and livelihood are especially relevant in poor rural communities relying on agropastoral production systems. Livestock is critical to the economies of many developing countries. Animals are a source of food, income, and employment. For low-income households, livestock can serve as a store of wealth, provide draught power and organic fertilizer for crop production, and be a means of transport (Pica-Ciamarra Ugo, 2005). Goat production systems are crucial for the subsistence of poor rural communities in these areas that face increasing water scarcity (Akinmoladun *et al.*, 2019), significant climate change impacts, and desertification. First, they play an important role in food security, especially in developing countries under extreme water stress and dry climatic environments (Joy *et al.*, 2020). Several studies have researched the physiological capability of goats to withstand arid and semiarid environments (Akinmoladun *et al.*, 2019). Goats have been recognized for their tolerance to arid environments and their physiological capacity to cope with water deprivation (Silanikove, 2000; Akinmoladun *et al.*, 2019). For these reasons, goats are a significant asset chosen by many producers and communities in desertification areas. Second, goats provide a source of high-quality protein and can improve sustainability as well as employment and income generation opportunities in rural areas, particularly for women (Valdivia, 2001; Peacock, 2005; Escareño *et al.*, 2013). Third, livestock marginal returns in these water-stressed areas have been shown to be higher than rain-fed grain crops (Oweis and Peden, 2008). Pretty *et al.* (Pretty *et al.*, 2006) concluded from a comprehensive evaluation of 144 projects that adopted sustainable agricultural technologies and practices that rain-fed agriculture would be the greatest opportunity for improvement in sustainable water use. Thus, it is important to determine the actual contribution of goats to human food security in arid environments.

Goat production systems generally rely on limited natural resources such as water. Inefficient water use limits community livelihoods in these water-scarce areas. Reducing inefficient water use is thus essential to improve efficiency and sustainability and decrease rural communities' vulnerability to climate change (FAO, 2017). This is particularly important in semiarid and arid areas that are naturally exposed to chronic water shortages and increasing desertification. Additionally, water scarcity is likely to intensify due to several drivers, such as demographic growth, tendency to increase food self-sufficiency by increased domestic agricultural production, urbanization expansion, energy demand, and overall socio-economic development (Roy *et al.*, 2018). The fast-widening divide between the availability and demand of freshwater, the accelerated depletion of groundwater resources, and climate change act as threat multipliers in already fragile regions (Prăvălie, 2016).

Efficiency can be defined as the ratio between the desired output and an input, that is, the number of resources consumed in the process (Willaarts *et al.*, 2014). Three main interpretations of efficiency are usually recognized in the case of water: technical efficiency, water productivity and economic or allocation efficiency. Water productivity (WP) is defined in agricultural systems as the ratio of the output obtained to the amount of water used to produce that output. The numerator of WP, the agricultural output, can be measured in (a) weight (either dry weight or commercial weight); (b) value, or (c) net profit. Improving water productivity means creating more value with less resource consumption. To improve WP, it is necessary to determine the current status of the WP in a given production system and identify its drivers (David *et al.*, 2007).

In a broad sense, WP in livestock production is defined as the ratio of net beneficial outputs from livestock production to the amount of water depleted. The numerator can consider multiple outputs. Goat production systems, for example, may consider products such as meat, milk, manure for fuel and fertilizer use, services such as draught power, social benefits of livestock, including status symbols, and savings (Kijne *et al.*, 2009).

It is critical to promote goat water productivity in these water-scarce environments to improve the production, livelihoods, and sustainability of these production systems under increasing water stress through enhanced management practices. In this context, the assessment of WP in goat production systems provides valuable information to both communities and policy-makers. On the one hand, it creates community awareness about their inefficiency in water use, so they can implement best practices aimed at reducing water consumption. On the other hand, it provides insights to policy-makers to develop guidelines into possible improvements. Nonetheless, livestock production-water interactions in different systems are not totally clear (Peden *et al.*, 2007; Peden *et al.*, 2009; Descheemaeker *et al.*, 2010).

This review analyses the evidence supporting goat production systems may be the animal source of protein to feed subsistence communities in lands suffering from desertification and land degradation. Contrary to popular opinion, livestock economic water productivity compares favorably with marginal returns arising from investments in irrigated horticultural crops and is higher than that observed in rain-fed grain crops (Castillo *et al.*, 2007). Additionally, we discuss the benefits provided by goats to their keepers and the current relationship between goat production, desertification, and food security.

Therefore, the objectives of this review are to (1) strengthen the current understanding of the importance and role of goat production systems for food security and rural livelihood, (2) analyze the resiliency of goat production systems in the context of increasing water scarcity, desertification, and projected impacts of climate change, and (3) define the concept of goat production system water productivity.

Review Methodology

The following databases were searched: CAB Abstracts, Google Scholar, PubMed, and Web of Science (keyword search terms used: 'livestock water productivity', 'goat production systems', 'water scarcity', 'water security', 'water security +goats', 'water productivity +goats', 'desertification +goats', 'goats+food security', 'livestock +food security'). We used references from the articles obtained by this method to check for additional relevant materials.

The Importance of Goat Stock in Food Security and Undernourishment

Food security is a multidimensional phenomenon. These dimensions are (1) availability, the existence of enough food quantity with the appropriate quality supplied through the market or primary production; (2) accessibility, the access of everybody to adequate economic and physical resources to obtain and consume a nutritious diet; (3) utilization, the use of food based on knowledge to meet the nutritional needs of a healthy diet; and (4) stability of the above three dimensions over time (FAO, 2008). Thus, food security can be achieved when all people, at all times, have physical, social, and economic access to sufficient, safe, and nutritious food; that is, their dietary needs and food preferences for an active and healthy life are met (Clay, 2002). Furthermore, food insecurity risk significantly increases under a rising water scarcity scenario (Chen *et al.*, 2021), where the food balance is highly vulnerable to large-scale and regional climatic events (Thalheimer *et al.*, 2019).

The major contribution of foods of animal origin to human health is alleviating malnutrition caused by deficiencies in micronutrients that contribute to poor growth, impaired mental development, and ill health (Ndlovu, 2010). For example, Neumann *et al.* observed that small amounts of meat or milk in the diet of primary school children in Kenya improved their cognitive ability, performance in school, and activity levels (Neumann *et al.*, 2003; Ngambi *et al.*, 2013). Generally, the dietary status in poor farm households improves as animal productivity increases. However, the contribution of animal production to the nutritional status of this group has been difficult to estimate and evaluate (Ndlovu, 2010).

Products from an animal origin are an excellent source of high-quality proteins, essential amino acids, some micronutrients, such as complex B vitamins, and some highly bioavailable minerals, such as zinc and iron (Ndlovu, 2010). This is especially important in pregnant women, lactating mothers, and very young children who have difficulty obtaining sufficient micronutrients from vegetable-based diets (Ndlovu, 2010; Rodríguez *et al.*, 2016). It is important to recall that without the provision of dietary protein (amino acids), the human body cannot synthesize body proteins. Humans consume proteins in foods consisting of animal- and plant-based products, contributing approximately 35% and 65% of the protein in human diets (Wu *et al.*, 2014). In contrast, in the United States, plant and animal foods provide approximately 32% and 68% of the total protein, respectively (Wu *et al.*, 2014). Hence, livestock is the main source of human-consumed animal protein in developed countries such as the United States; however, especially in developing countries, due to the expected increases in population and per capita income, the demand for animal-source foods is expected to substantially increase (Wu *et al.*, 2014). Goat meat consumption per capita varies greatly between countries and is largely supported by local production and tradition. The main goat meat-producing countries are China (39%), India (9%), Pakistan (6%), Nigeria (4%), Bangladesh (4%), and Australia (1%) (Meat and Livestock Australia, 2020). Globally, goat milk production is concentrated by households of low- and medium-income countries such as India, Bangladesh, Sudan, and Pakistan and Mediterranean countries such as France and Spain (Meat and Livestock Australia, 2020).

The animal protein intake is generally lower in villages not keeping goats in the Morogoro Region of Tanzania than in villages that keep them. This suggests that smallholder goat keeping is an effective tool to improve diet protein content (Eik *et al.*, 2008). Goats are a good source of food protein for humans at a relatively low cost, as they graze on lands that are, in general, not suitable for crops and can be fed with farm byproducts. The nutrient value of meat relies mainly on how humans' essential amino acid requirements are met. The amino acid profile of goat meat is comparable to lamb, pork, and beef. Additionally, chevon meat is an excellent source of iron (2.1 mg/g), thiamine (0.10 mg/100 g), riboflavin (0.56 mg/100 g), and niacin (3.6 mg/g) (Webb *et al.*, 2005). Furthermore, the fatty acid composition in goat meat is not inferior to that of meats from other farm animals (Mahgoub *et al.*, 2002). Chevon has lower total fat, saturated fatty acid, and cholesterol content than other meats (Mazhangara *et al.*, 2019). Similarly, goat milk has been recognized as nutritionally acceptable, has an attractive odor and taste, and can be consumed as an alternative to cow milk because it is less allergenic and more digestible (García *et al.*, 2014). The high nutritional value of goat products suggests that goat keeping has the potential to increase the nutritional status of tenants.

Relative to cattle and sheep grazing, rearing goats is more productive in unfavorable arid ecosystems (Tefera, 2007). In the study of Woldu *et al.* (Woldu *et al.*, 2016), a household dietary diversity score was used as an indicator of food security. A significant positive correlation was observed between the gross margin from goats and the score in the arid agropastoral goat production system, indicating the role of goat production in food security (Woldu *et al.*, 2016). Milk and meat goat production systems have been shown to be profitable; however, the production of goat milk has been superior in terms of protein supply to the household and generation of income than goat meat (Safari *et al.*, 2008). For instance, in the Morogoro region of Tanzania, despite the high nutritional value of goat meat, due to the high demand for live animals and the popularity of goat meat at celebrations, many times only a few animals are used for home consumption, and consequently, the actual contribution of meat protein from goats in the family diet is minimal (Eik *et al.*, 2008). Similarly, in rural areas of Ethiopia, products such as meat, eggs, cow's milk, and butter are more important as sources of cash revenue than as means of satisfying the nutritional needs of the family. Nevertheless, goat milk is utilized for home consumption, particularly by family members with more critical protein requirements, such as children, lactating mothers, and sick family members (Tefera, 2007). In rural areas of many countries, such as Ethiopia, women do most of the farm and housework. Providing women with goats enhanced their ability to provide adequate nutrition to their families via the direct use of goat products, such as milk and meat, or using the cash derived from the sale of live animals or their products (Tefera, 2007).

A project in the Harar Highlands (eastern Ethiopia) introduced Anglo-Nubian goats through crossbreeding of the indigenous stock (Somali breed) to increase the income and nutritional status of the poor rural community, especially women and children. The results showed that milk and meat outputs were increased, leading to an average increase in gross per capita income of 19% for the beneficiaries (Animut *et al.*, 2000). Another study focused on women who involved the supply of goats in improving vitamin A supply; researchers concluded that increased milk production and farm income from livestock development projects might improve women and children's nutritional and health status. However, nutrition education must be included to enhance the consumption of goat milk by women and children (Ayalew and Kassa, 1999). Therefore, the development of goat farming in the arid regions of the world can positively impact certain nutrient deficiencies, especially in low-income communities (Peacock, 2005). Accordingly, extensive goat production systems have significantly increased the diet's nutritional value in low-income and rural communities and play an important role in food security, especially in developing countries under extreme water stress and dry climatic environments (Joy *et al.*, 2020). However, there is still a lack of information on the utilization of goat products at the family level and commercialization through informal markets that are not recorded at the national level (Lebbie, 2004).

The Role of Goat Stock in the Local Economy and Rural Livelihood

In 2019, approximately 95% of the goat stock was situated in Africa and Asia (Safari *et al.*, 2008) and was more concentrated in the tropics, dry zones, and developing countries (Animut *et al.*, 2000). Goats are particularly relevant in marginal agricultural lands, especially in arid and semiarid regions, which hold 64% of the goat population (Woldu *et al.*, 2016). Generally, most goats are raised in extensive production systems, with minimal management and low milk or meat production. Goats are often the only asset possessed by poor households. Goats can be sold to buy food or medicine in times of trouble, such as crop failure or family illness. This plays a vital role in ensuring the security of family members (Peacock, 2005). Thus, goats are used to secure food, protein, milk, and cash, especially for women to benefit their children (Silanikove, 2000; Valdivia, 2001; Peacock, 2005).

Additionally, goat rearing provides an opportunity to efficiently utilize family labor (Kumar and Deoghare, 2003). Valdivia studied the role of livestock in poor rural households in Bolivia, Perú, and Kenya, concluding that men are mainly in charge of managing cattle and market transactions, while poultry, sheep, and goats are mainly kept by women (Valdivia, 2001). A study focused on Nigerian women's livelihood showed a positive effect of goat rearing on the standard of living of women in the study area. Approximately 91% of the respondents reported an increase in household income and 83% increased savings. An increase in household food security was reported by 91% of the respondents, and approximately 96% of the women recorded an increase in household assets (Adeleye *et al.*, 2016). These findings agree with those of Wodajo *et al.*, who analyzed why Ethiopian household systems keep different types of animals. In this study, women claimed that small ruminants have a short gestation period, fill gaps in food shortages in the household, and can be sold at short notice during emergencies (Wodajo *et al.*, 2020). Additionally, they represent an important source of milk for children throughout the year, especially during drought seasons when cattle cannot provide milk and goats withstand the drought and survive on low-value feed (Wodajo *et al.*, 2020).

As previously mentioned, goats add value in many ways to production systems. The economic contribution of small ruminants to poor farm households is significant, not only by the consumption of their products but also to household income and survival (Devendra, 2005). For instance, farmers and pastoralists all over Africa who traditionally kept cattle are increasingly turning to goats to survive and increase their incomes. It is hypothesized that this phenomenon is explained by the unique role and potential of goats in sub-Saharan Africa to raise poor people out of poverty and improve their welfare and prosperity (Peacock, 2005). Kaumbata *et al.* (Kaumbata *et al.*, 2020) separate the benefits from goats into tangible and intangible benefits. The tangible benefits included cash, meat, milk, and manure, and the intangible benefits included goats playing roles as a credit buffer and insurance (security) for producers during emergencies. When the intangible benefits were considered, the mean annual net profit and the return on capital increased by 60.3%, indicating the relevance of goats' socio-economic roles in providing current and future economic stability to rural households' economies (Kaumbata *et al.*, 2020). Similarly, in Asia's semiarid and arid areas, goats and sheep provide the main means of survival and security. In these situations, the sale of animals, milk, and manure accounted for between 27.2%–30.7%, 19.7%–84.8%, and 1.0%–4.5% of total farm income, respectively (Devendra, 2005).

In the southwestern semiarid zone of Utter Pradesh, goat rearing was considered the first priority for 39% of respondents, with 57% of the income of goats reared in landless household systems spent on food for the family (Safari *et al.*, 2008). In the same study, the proportion of income from goat rearing of the total income was estimated to be 33.32%, 71.53%, and 81.47% in the small, medium, and large categories, respectively (Safari *et al.*, 2008). Similarly, crop production was ranked as the first economic activity of dairy goat farmers in the Babati and Kongwa regions of Tanzania. It contributes 44.9% and 50% of the household income, respectively. In this study, goat production was ranked second, contributing 30.8% and 25.7% of the total household income in Babati and Kongwa districts, respectively (Adeleye *et al.*, 2016). However, the incomes of dairy goat farmers were higher than those of nondairy goat farmers, indicating that dairy goats contribute to household income by supporting other activities. The difference might be attributed to the use of income from dairy goats in the preparation of crop farms and the purchase of farm inputs. Additionally, manure from dairy goats fertilizes the soil, increasing crop yield and improving household food security and income (Adeleye *et al.*, 2016).

Goat production is also an integral part of farming systems in an arid country such as Jordan and plays an important role in poverty alleviation and improved food security in rural households. In an economic evaluation of goat production systems in Jordan, feed production shared the highest costs (75%) of total variable costs, followed by transportation of water and feed (12%), labor (9%), veterinary services (4%) and water (0.2%) (Wodajo *et al.*, 2020). The productivity of goats and other ruminants in semiarid regions is hindered by many constraints. The major constraints to goat production in southern Africa include high disease and parasite prevalence, low levels of management, limited forage availability, and poor marketing management (Gwaze *et al.*, 2009). Similar constraints were found in the Nigerian study, where the main constraints faced by women rearing goats in order of importance included loss of goats through death and theft; absence of a credit facility; absence of a cooperative society; and absence of hands-on training on goat rearing (Adeleye *et al.*, 2016). Similarly, in six districts of southern Zimbabwe, the high mortality caused by inadequate disease control, extended dry season, feed shortages, and poor nutrition was the major constraint (Homann *et al.*, 2007).

To cope with these concerns and enhance goats' potential in alleviating poverty, it is necessary to develop farmers' awareness and their capacity to effectively use and improve technologies for animal health care, dry season feeding, nutrition, housing, and breeding (Homann *et al.*, 2007). Additionally, collaborative efforts should be made by the government and other institutions to increase the extension of advisory services, credit facilities, high-quality feeds, and veterinary services to stimulate the adoption of recommended practices on goat production for increased productivity and better livelihood (Adeleye *et al.*, 2016).

In conclusion, in arid and semiarid environments, goat rearing is one of the most suitable alternatives for the rural poor's sustenance, livelihood, and poverty alleviation. Furthermore, a reduction in common grazing areas would seriously impair the sustainability of the goat production system of landless rural poor families (Kumar and Deoghare, 2003). A review of studies of goat-based communities and their main research objectives are presented in Table 21.1.

The potential of small ruminants, especially desertic breed goats, to withstand extreme water and therefore contribute to the livelihood of poor household families in areas undergoing desertification is partially determined by their physiological capacity to cope with dehydration. This ability will be discussed in the following section.

Goat Physiological Efficiency to Water Scarcity

As previously pointed out, approximately 90% of the world goat population (~610 million head) is located in Asia and Africa and mostly (80%) in the arid and semiarid regions of the tropical belts exposed to increasing periods of dryness (Silanikove, 2000). Desertification results in water shortages for both livestock and humans (Mdletshe *et al.*, 2018). Camelids are commonly recognized for their adaptation to water scarcity and their ability to survive for long periods of time in deserts without water (Zarrin *et al.*, 2020). This ability relies mainly on their tolerance to dehydration of approximately 30% of their body weight, compared with 10%–12% for most others (Fowler and Bravo, 2010). This capacity has also been described in sheep, goats (Reece, 2015), and cattle (Silanikove, 1994). However, there are differences in the degree of coping with water deprivation among species and breeds. Weight loss under water-deprivation regimens is largely attributed to the fluid content in the gut, mainly the rumen. Desert goats show advantages in utilizing rumen fluid during dehydration and can survive and even sustain

Table 21.1. Review of goat-based community studies and their main research objectives.

Research objectives	Author
Contribution of goat rearing to the household economy, livelihood, and gender issues	Adeleye *et al.* (2016) Animut *et al.* (2000) Devendra (2005) Escareño *et al.* (2013) Gwaze *et al.* (2009) Kaumbata *et al.* (2020) Kumar and Deoghare (2003) Peacock (2005) Safari *et al.* (2008) Valdivia (2001) Peacock (2005) Wodajo *et al.* (2020) Woldu *et al.* (2016)
Contribution of goat food products to people's nutrition and health	Neumann *et al.* (2003) Ngambi *et al.* (2013) Rodríguez *et al.* (2016) Wu *et al.* (2014) Meat and Livestock Australia (2020) Tefera (2007) Safari *et al.* (2008) Animut *et al.* (2000) Webb *et al.* (2005) Mahgoub *et al.* (2002) Mazhangara *et al.* (2019) García *et al.* (2014) Ayalew and Kassa (1999)
Contribution of goats in arid and semiarid environments	Adeleye *et al.* (2016) Escareño *et al.* (2013) Homann *et al.* (2007) Kumar and Deoghare (2003) Mdletshe *et al.* (2018) Safari *et al.* (2008) Wodajo *et al.* (2020) Woldu *et al.* (2016)

production under water-restricted regimens. The main adaptations are related to their large ruminal volume, a better capacity of the kidney to 'desalt' the water absorbed from the gut, and the ability to maintain the salivary flow to the rumen (Silanikove, 2000).

The adaptation capacity of goats to water scarcity can also be seen in the velocity of how body weight is lost in a water deprivation regime. For example, Shankarnarayan *et al.* reported that Marwari goats of the Rajasthan desert lost only 1.5% of their body weight per day compared to 6% loss per day by Marwari sheep, 8% for cattle, and 2% for camel (Shankarnarayan *et al.*, 1985). Additionally, ruminants, especially goats bred for the desert, can rehydrate very quickly. Upon dehydration, the desert black Bedouin goat may drink up to 40% of its body weight in only 3–10 min, cattle 18%, sheep 20%, and camels 25% (Choshniak *et al.*, 1984; Silanikove, 1992). This is mainly due to the ability of the rumen to prevent water from entering the bloodstream too quickly. In non-adapted animals, such amounts of water may cause haemolysis after rehydration (Maltz *et al.*, 1984).

Sheep and goats can concentrate urine under water restrictions, making them superior to cattle in terms of adaptability to water-limited areas and arid environments (Akinmoladun *et al.*, 2019). These findings agree with those of

Choshniak *et al.*, who demonstrated that urine production dropped immediately in Bedouin goats under water deprivation (Choshniak *et al.*, 1984). Furthermore, these animals' kidneys decreased the excretion of water, sodium, and chloride after rehydration in goats previously submitted to a water restriction regimen. In desert sheep and goats, water deprivation also decreases nitrogen losses in urine, increases urea recycling to the gut, and improves nitrogen balance. This is especially relevant during late pregnancy and lactation due to the increased demand for N (Silanikove, 2000).

Decreased metabolism has been described in goats under restricted water regimens. Goats exposed to water restriction have increased body temperature and decreased respiratory rate (Akinmoladun *et al.*, 2019). Evaporative cooling by sweating is one of the major cooling mechanisms during heat stress for goats adapted to hot arid zones; however, it is usually reduced in dehydrated animals (Akinmoladun *et al.*, 2019). Kaliber *et al.* (Kaliber *et al.*, 2016) reported an increase of 0.51°C in water-restricted crossbreed goats (water intake limited to 2 l/day) in comparison with the control group (ad-libitum water intake; average 7.4 l/day). The increase in body temperature allows goats to save water losses by heat dissipation (Kaliber *et al.*, 2016). An increased respiratory rate (panting) is another heat-dissipating mechanism that predominates in small ruminants (Reece, 2015). However, the respiratory rate is reduced during periods of water deprivation in desert goats to reduce water losses (Akinmoladun *et al.*, 2019). During drought, the least harmed animals are browsers such as goats, camels, zebras, and donkeys because they can survive on browse-based diets (Kay, 1997; Silanikove, 2000). Additionally, the ecosystem supports goats longer than other herbivores because of their ability to reduce their metabolism (Silanikove, 2000).

These observable findings, the limited production alternatives, and the multiple benefits of small ruminants described above have influenced farmers' decisions to rear goats in arid and semiarid regions of the world. Some researchers have proposed that goats living in arid areas can maintain their feed intake during water scarcity (Misra and Singh, 2002; Alamer, 2009). This is viewed as an adaptation mechanism to sustain milk production to nourish their offspring (Alamer, 2009). However, this mechanism can counteract various adaptive responses to heat stress usually linked to the desert environment, such as reduced feed intake, rumination time and frequency, and increased water intake to enable small ruminants to deal with high temperatures (Joy *et al.*, 2020). On the other hand, pregnant and lactating goats have 40%–50% higher water turnover rates than nonlactating animals. Additionally, severe dehydration increases antidiuretic hormone (ADH) levels in pregnant goats, resulting in vasoconstriction, which, combined with other factors, such as food deprivation, can result in fetal growth retardation or even abortion (Olsson, 2005). Therefore, goats can sustain production and reproduction under water deprivation. However, subsistence becomes more important than reproduction at some level of water stress.

In conclusion, goats, particularly desert breeds, are physiologically adapted to water deprivation. Additionally, they can survive on browse-based diets. This remarkable resiliency might increase the sustainability of these systems in the context of increasing water scarcity and desertification driven by the climate change scenario.

Efficiency and Water Security of Livestock Systems

Therefore, goats are well adapted to situations of water scarcity and drought. Furthermore, goats are significant assets to their tenants' livelihood and food security in arid and semiarid regions of the world (Peacock, 2005). Hence, it is critical to understand goat water productivity and its drivers in these water-scarce environments to improve the production, livelihoods and sustainability of these production systems under increasing water stress. However, there is a lack of research in terms of the water efficiency of these systems, and, thus, their sustainability is facing increased water scarcity. Thus, due to the limited research on water efficiency in goat production systems, some of the information presented in this review shows data obtained from other rangeland livestock systems. Globally, livestock production is practiced in rangeland areas and in mixed crop-livestock systems that cover approximately 60% of the land area of developing countries (Peden *et al.*, 2007). Agriculture represents 70% of total blue water use and 86% of blue[2] +green water[3] use (Doreau *et al.*, 2012). Animal products consume approximately one-third of the agricultural water (Drastig *et al.*, 2021). Livestock utilizes 15% of irrigated crop evapotranspiration, 33% of that of rain-grown crops, and 68% of that of permanent pastures and rangelands (Doreau *et al.*, 2012). To produce livestock, high-quality water must be ensured. However, livestock production can lead to water pollution (Drastig *et al.*, 2021). To the extent that goat farming is the main source of protein and the main source of income in poor rural communities in drylands, ensuring the sustainability of agropastoral systems in a desertification scenario is essential.

To assess the impact of livestock on water usage and availability, the concept of water productivity has received recent attention. In a broad sense, water productivity is the ratio of the net benefits from the crop, forestry, fishery, livestock, or mixed agricultural systems to the amount of water required to produce those benefits. It widely reflects the goal of obtaining more food, income, livelihoods, and ecological benefits per unit of water used (Molden *et al.*, 2010; Drastig *et al.*, 2021). There are several reasons to increase water productivity, such as meeting the rising demand for food in a growing, wealthier, and increasingly urbanized population in light of water scarcity. Similarly, increasing water productivity may help reduce poverty, especially in rural areas (improving nutrition, employment, and income), and it may relocate water from agriculture to cities or wildlife (Molden *et al.*, 2010).

Although the water for livestock drinking and handling is the most obvious water use in livestock production systems, it is almost negligible compared to the total water consumption. The physical water productivity[4] of animal products is derived mainly from the water required to feed the animals by productive plant transpiration, which is generally approximately 50–100 times the amount needed for drinking (Peden *et al.*, 2007; Descheemaeker *et al.*, 2010). Broadly, 450 m^3 of water is required annually to produce the feed needed to maintain one Tropical Livestock Unit (TLU: measured at 250 kg live weight) (Amede *et al.*, 2009). Nevertheless, water for cleaning, cooling and drinking should not be overlooked (Drastig *et al.*, 2021). The amount of water required to produce 1 kg of animal products varies widely, depending on management practices, the kind of feed, how crop residues are used, the processing system, and how well the animals convert feed and plants into the animal product. Adjusting these factors can improve livestock water productivity (LWP) (Molden *et al.*, 2007).

Measuring water productivity, the denominator is always the unit of water consumed; however, a parallel line of study uses the concept of the water footprint, which is the reciprocal of water productivity (volume of water per unit of product). The practical value of water productivity and the water footprint depends on water use and stress, and both play a major role in local and regional water management and policy (Amarasinghe and Smakhtin, 2014). Mekonnen and Hoekstra estimated the weighted global average water footprint for selected animal products, accounting for 15,415, 10,412, 5988 and 5521 m^3/ton for beef, sheep, pig and goat meat, respectively (Mekonnen and Hoekstra, 2012). Therefore, goats are among the most efficient water users of all domestic species (Peacock, 2005; Mekonnen and Hoekstra, 2012). Additionally, sheep and goats raised in rainfed grassland utilize mainly green water. In the study of Zeng *et al.* (Zeng *et al.*, 2012) in China, sheep and goats had a blue water proportion utilization of 0.3% (Zeng *et al.*, 2012). As previously mentioned, goats are adapted to the arid conditions of drylands, and they are raised mostly by smallholders in extensive production systems, providing many assets to their families. However, little research focuses on this species accounting for the overall impact on water security. Drastig *et al.* (Drastig *et al.*, 2021) analyzed 50 studies accounting for livestock water productivity. These studies were carried out in various regions of the world from 1993 to the present. Of the 50 studies, only four reported goat water productivity. These studies were conducted in China, Ethiopia, Morocco, and the Nile Basin. The differences can be partly explained by the different feed conversion efficiencies of the animals and the feed composition (Mekonnen and Hoekstra, 2012). The average economic livestock water productivity of cattle, sheep, and goats in most areas of the Nile Basin was less than 0.1 USD/m^3, and only a few areas above 0.5 USD/m^3 (van Breugel *et al.*, 2010). Similarly, Haileslassie *et al.* reported water productivity values of USD 0.15–0.69 /m^3 for mixed farming systems (Haileslassie *et al.*, 2009). In Latin America and the Caribbean, the return per unit of water used in livestock products was 0.20 USD/m^3 (Mekonnen *et al.*, 2014). In summary, livestock water productivity is generally low (Mekonnen *et al.*, 2014; van Breugel *et al.*, 2010; Hoekstra and Mekonnen, 2011; Drastig *et al.*, 2021). However, few alternative production systems are suitable for the harsh conditions of arid and semiarid regions. Additionally, water productivity in rainfed agriculture is usually low. Typical values for water productivity for wheat, rice, and maize ranged from 0.03 to 0.30 USD/m^3 (Molden *et al.*, 2007).

Increasing water productivity means more food, more income and lower vulnerability to risks (Cai *et al.*, 2011). Additionally, it liberates water for other users and, more importantly, the environment. Indirectly improving water productivity contributes to alleviating poverty, allowing poor and marginalized people greater access to and more efficient use of water (Peden *et al.*, 2009). Improvements in water productivity indirectly improve food security and reduce poverty (David *et al.*, 2007; Peden *et al.*, 2009; Cai *et al.*, 2011) by increasing sustainable livelihoods in rural and periurban areas, improving human health through better nutrition, lowering agriculture-related pollution, and reducing water-related diseases (Peden *et al.*, 2009). Although LWP is typically low, livestock keeping is one of the most significant means of escaping poverty in rural areas. One of the quickest ways to exacerbate poverty is to deny smallholder livestock keepers of having animals (Peden *et al.*, 2007). Goat keepers represent some of the 'poorest of the poor'; thus, improving the production of their goats is one of the best methods of securing their livelihoods and transforming their lives from one of poverty to relative prosperity (Peacock, 2005). For example, the British NGO FARM-Africa provided goats to the poorest families of the Meru region in Kenya. Some of these families raised their goats so well that they and their descendants improved their welfare and social status. For example, after a while of breeding their goats, one family was able to buy an ox. Then, they rented the ox for plowing, allowing the family to acquire land, build a stone house, and pay the fees to send their sons to secondary school (Peacock, 2005).

Of course, there are also environmental concerns. Extensive goat production in the drylands of developing countries is associated with the progressive degradation of rangelands. This kind of production typically involves the use of communal or state-owned land. The users of this land often do not consider land rehabilitation a priority by the community but rather are perceived as a high cost-low return process (Iñiguez, 2004). On the other hand, there seem to be significant opportunities in goat systems for improving environmental and water management while also improving the livelihoods and food security of poor smallholders (Devendra *et al.*, 2010).

Peden *et al.* (Peden *et al.*, 2009) conclude that there are important opportunities for increasing LWP and, thus, increasing animal production or rehabilitating degraded landscapes. However, there is little research on LWP in extensive animal production systems (Peden *et al.*, 2009). There are a few examples in the literature about how management strategies improve LWP, but these are not focused on goat livestock systems. For instance, increasing milk production from a low to an intermediate level in dairy cows leads to a pronounced increase in WP. Feeding more roughage and less concentrate also increased the water productivity of milk (Krauß *et al.*, 2015). Similarly, the inclusion of crop residues in livestock diets is expected to reduce the water requirement for feed production, resulting in improvements in LWPs. There is a strong and positive relationship between crop biomass production and feed WP. Additionally, LWPs can be impaired when the price of crop residues increases due to the use of crop residues as construction material and fuel and in conservation agriculture for soil fertility enrichment and water conservation management (Bekele *et al.*, 2017).

Discussion and Need for Future Research

Climate change, water scarcity, and overpopulation threaten future livelihood and food security, especially in developing countries. Goat milk and meat consumption have been significant sources of protein and micronutrients to poor households in arid environments. Increased food security and nutritional status have also been achieved due to the higher revenues from goats kept by buying more nutritious products in the market. Thus, goats secure food (high-quality protein), especially for women to support their children. Additionally, goats play a critical socio-economic role in the stability of rural families, adding value to their production systems by providing cash, meat, milk and manure that can be used to fertilize the soil for crops. Furthermore, they provide a credit buffer and insurance (financial security) for producers during emergencies.

Goats are especially valuable in marginal agricultural lands, especially in arid and semiarid regions, which hold almost two-thirds of the goat population. We find that goats are the animal species that are better adapted to thrive in this increased water scarcity scenario and would be the source of animal protein for the upcoming generations affected by desertification. Desertification results in water scarcity for crops, livestock, and humans. Desert goats are extremely resilient to dry environments and have an extraordinary capacity to cope with dehydration, survive with browse-based diets and face water scarcity scenarios. Therefore, goats can sustain the production and livelihood of poor household families living in arid and semiarid regions. These observations likely influenced farmers' decision to rear goats in drylands around the world. Nevertheless, due to the notable differences among goat breeds in terms of water efficiency and dehydration resistance, there is a need for more genetically focused research.

Goat production is one of the most suitable alternatives for rural poor families to sustain their livelihood and increase their nutritional status and income in arid and semiarid environments; however, environmental concerns, such as overgrazing, lead to land degradation. To address these concerns, it is necessary to increase farmers' awareness of these environmental issues and to improve goat feeding management, housing, health care, and breeding. Moreover, policy-makers should stimulate the adoption of recommended practices on goat production for increased productivity and better livelihood.

Therefore, identifying strategies and interventions to increase livestock water efficiency of goat production systems, and thus sustainability in semiarid and arid environments, is critical. Research on water and land utilization, nutritional management, and stocking capacity of drylands are necessary to assess the sustainability and efficiency of goat production systems.

In this review the concept of WP was introduced, which has received less attention in assessing the sustainability of livestock production systems with respect to increased water scarcity. Future research is necessary, especially in the LWP of the complete caprine livestock system in areas undergoing desertification, considering the multiple benefits of goats to rural communities. Finally, increasing the water productivity of goats in areas where resources are becoming relatively scarce is crucial to the food security and livelihood of rural communities that depend on these animals.

Notes

[1] Hyperarid present an AI < 0.05, arid 0.05 < AI < 0.2, semiarid 0.2 < AI < 0.5 and dry subhumid 0.5 < AI < 0.65 (JRC, 2019).

[2] Blue water: refers to the freshwater in rivers, groundwater aquifers, reservoirs, and lakes. It is the main water source for irrigated agriculture.

[3] Green water: refers to the soil moisture generated from rainfall that infiltrates the soil and is available for plants transpiration.

[4] Physical water productivity: when the benefits of water are expressed by the mass of the agricultural output produce (e.g., kg of a crop, milk, etc.) for the amount of water used (e.g., per cubic metre) (David *et al.*, 2007).

References

GEF-IFAD Partnership. (2002) Tackling land degradation and desertification. GEF-IFAD Partnership, Washington, DC, USA. Available at: http://www.ifad.org/events/wssd/ gef/GEF_eng.pdf

Adeleye, O., Alli-Balogun, J., Afiemo, O. and Bako, S. (2016) Effects of goat production on the livelihood of women in Igabi, Chikun and Kajuru local government areas, Kaduna state, Nigeria. *Asian Journal of Agricultural Extension, Economics & Sociology* 11(1), 1–8.

Akinmoladun, O.F., Muchenje, V., Fon, F.N. and Mpendulo, CT. (2019) Small ruminants: Farmers ' hope in a world threatened by water scarcity. *Animals* 9(456), 1–20.

Alamer, M. (2009) Effect of water restriction on lactation performance of aardi goats under heat stress conditions. *Small Ruminant Research* 84, 76–81.

Amarasinghe, U.A. and Smakhtin, V. (2014) Water productivity and water footprint: Misguided concepts or useful tools in water management and policy?. *Water International* 39(7), 1000–1017.

Amede, T., Geheb, K. and Douthwaite, B. (2009) Enabling the uptake of livestockwater productivity interventions in the croplivestock systems of sub-Saharan Africa. *Rangeland Journal* 31(2), 223–230.

Animut, G., Merkel, R. and Sahlu, T. (2000) Increasing food security through improved goat production: A progress report of A UNCF-funded international development partnership between Langston University and Alemaya University. In: Merkel, R.C., Abebe, G. and Goetsch, A.L. (eds) *The opportunities and challenges of enhancing goat production in East Africa. Proceedings of a conference held at Debub University, Awassa, Ethiopia from November 10 to 12, 2000. Langston University*, E (Kika) de la Garza Institute for Goat Research, Langston, OK, pp. 118–126.

Ayalew, W. and Kassa, H. (1999) Reducing vitamin A deficiency in Ethiopia: Linkages with a women-focused dairy goat farming project. Available at: http://citeseerx.ist.psu.edu/viewdoc/download?doi=10.1.1.561.5887&rep=rep1&type=pdf

Bekele, M., Mengistu, A. and Tamir, B. (2017) Livestock and feed water productivity in the mixed crop-livestock system. *Animal* 11(10), 1852–1860. DOI: 10.1017/S1751731117000416.

Cai, X., Molden, D., Mainuddin, M., Sharma, B. and Karimi, P. (2011) Producing more food with less water in a changing world: Assessment of water productivity in 10 major river basins. *Water International* 36(1), 42–62.

Castillo, G.E., Namara, R.E., Ravnborg, H.M., Hanjra, M.A., Smith, L. *et al.* (2007) Reversing the flow: Agricultural water management pathways for poverty reduction.Chapter 4. In: *Water for Food Water for Life: A Comprehensive Assessment of Water Management in Agriculture*. London, UK, Sterling, VA, USA, pp. 278–310.

Chen, L., Chang, J., Wang, Y., Guo, A., Liu, Y. *et al.* (2021) Disclosing the future food security risk of China based on crop production and water scarcity under diverse socioeconomic and climate scenarios. *Science of the Total Environment* 790, 148110. DOI: 10.1016/j.scitotenv.2021.148110.

Choshniak, I., Wittenberg, C., Rosenfeld, J. and Shkolnik, A. (1984) Rapid rehydration and kidney function in the black bedouin goat. *Physiological Zoology* 57(5), 573–579.

Clay, E. (2002) Food security: Concepts and measurement. Paper for FAO expert consultation on trade and food security: Conceptualizing the linkages Rome, 11-12 July 2002. Published as Chapter 2 of Trade Reforms and Food Security: Conceptualizing the linkages. FAO, Rome. Available at: http://www.fao.org/3/y4671e/y4671e06.htm#fnB21

David, M., Jean-Marc, F., Max, F.C., Habiba, G., Joke, M. *et al.* (2007) Chapter 1. Setting the scene. In: *Water for Food Water for Life: A Comprehensive Assessment of Water Management in Agriculture*. Earthscan, London, UK and Sterling, VA, USA, pp. 41–53.

Descheemaeker, K., Amede, T. and Haileslassie, A. (2010) Improving water productivity in mixed crop—livestock farming systems of sub-Saharan Africa. *Agricultural Water Management* 97(5), 579–586. DOI: 10.1016/j.agwat.2009.11.012.

Devendra, C. (2005) Small ruminants in Asia: Contribution to food security, poverty alleviation and opportunities for productivity enhancement. In: *International Workshop on Small Ruminant Production and Development in South East Asia*. Available at: http://hostcambodia.com/mekarn/procsr/Devendra.pdf

Devendra, C., Swanepoel, F., Stroebel, A. and Rooyen, C.J. (2010) Implications and innovative strategies for enhancing the future. In: *The Role of Livestock in Developing Communities: Enhancing Multifunctionality*. Technical Centre for Agricultural and Rural Cooperation (CTA), Westdene, Bloemfontein, South Africa, pp. 77–92.

Doreau, M., Corson, M.S. and Wiedemann, SG. (2012) Water use by livestock: A global perspective for A regional issue?. *Animal Frontiers* 2(2), 9–16.

Drastig, K., Vellenga, L., Qualitz, G., Singh, R., Pfister, S. *et al.* (2021) *Accounting for livestock water productivity: How and why? Land and water discussion papers*. FAO, Rome.

Eik, L., Kifaro, G., Kiango, S., Nordhagen, Ø., Safari, J. *et al.* (2008) Productivity of goats and their contribution to household food security in high potential areas of East Africa: A case of mgeta, Tanzania. *African Journal of Food, Agriculture, Nutrition and Development* 8(3).

Escareño, L., Salinas-Gonzalez, H., Wurzinger, M., Iñiguez, L., Sölkner, J. *et al.* (2013) Dairy goat production systems. *Tropical Animal Health and Production* 45, 17–34.

FAO (2008) An Introduction to the Basic Concepts of Food Security. EC -FAO Food Security Programme. Available at: http://www.foodsec.org/docs /concepts_guide.pdf

FAO (2017) Water for Sustainable Food and Agriculture Water for Sustainable Food and Agriculture. A report produced for the G20 Presidency of Germany. Available at: http://www.fao. org/3/i7959e/i79 59e.pdf

Feng, S. and Fu, Q. (2013) Expansion of global drylands under a warming climate. *Atmospheric Chemistry and Physics* 13(19), 10081–10094.

Fowler, M. and Bravo, P. (2010) *Medicine and surgery of camelids*, 3rd edn. Blackwell Publishing, Inc, Iowa, USA.

García, V., Rovira, S., Boutoial, K. and López, MB. (2014) Improvements in goat milk quality: A review. *Small Ruminant Research* 121(1), 51–57. DOI: 10.1016/j.smallrumres.2013.12.034.

Goudie, A.S., Mirzabaev, A., Wu, J., Evans, J., García-Oliva, F. *et al.* (2019) Desertification. In: *Climate Change and Land: An IPCC Special Report on Climate Change, Desertification, Land Degradation, Sustainable Land Management, Food Security, and Greenhouse Gas Fluxes in Terrestrial Ecosystems*. Cambridge University Press, Cambridge, UK and New York, USA, pp. 46–51.

Gwaze, F.R., Chimonyo, M. and Dzama, K. (2009) Communal goat production in Southern Africa: A review. *Tropical Animal Health and Production* 41(7), 1157–1168.

Haileslassie, A., Peden, D., Gebreselassie, S., Amede, T. and Descheemaeker, K. (2009) Livestock water productivity in mixed crop-livestock farming systems of the Blue Nile Basin: Assessing variability and prospects for improvement. *Agricultural Systems* 102(1–3), 33–40. DOI: 10.1016/j.agsy.2009.06.006.

Hoekstra, A.Y. and Mekonnen, M.M. (2011) Global water scarcity: The monthly blue water footprint compared to blue water availability for the world's major river basins. *Water* (53), 78. Available at: http://www.waterfootprint.org/Reports/Report53-GlobalBlueWater Scarcity.pdf

Hofste, R.W., Reig, P. and Chleifer, L. (2019) 17 Countries, Home to One-Quarter of the World's Population, Face Extremely High Water Stress. [Cited 2021 Jul 20]. Available at: https://www.wri.org/insights/17-countries-home-one-quarter-worlds-population-face-extremely-high-water-stress

Homann, S., van Rooyen, A., Moyo, T. and Nengomasha, Z. (2007) *Goat production and marketing: Baseline information for semi-arid Zimbabwe*. Matopos Research Station, International Crops Research Institute for the Semi-Arid Tropics, Bulawayo, Zimbabwe, p. 84 .

Iñiguez, L. (2004) Goats in resource-poor systems in the dry environments of West Asia, Central Asia and the Inter-Andean valleys. *Small Ruminant Research* 51(2), 137–144.

IPCC, Masson-Delmotte, V., Zhai, P., Pirani, A., Connors, S.L. *et al.* (2021) Climate change 2021: The physical science basis. Contribution of working group I to the sixth assessment report of the intergovernmental panel on climate change. Cambridge University Press, Cambridge, UK and New York, USA.

Mazhangara, I.R., Chivandi, E., Mupangwa, J.F. and Muchenje, V. (2019) The potential of goat meat in the red meat industry. *Sustainability* 11(13), 3671.

Joy, A., Dunshea, F.R., Leury, B.J., Clarke, I.J., Digiacomo, K. *et al.* (2020) Resilience of small ruminants to climate change and increased environmental temperature: A review animals 10(867), 1–18.

JRC (2019) World atlas of desertification. Joint Research Center. Publication Office of the European Union, Luxembourg. Available at: https://wad.jrc.ec.europa.eu/patternsaridity

Kaliber, M., Koluman, N. and Silanikove, N. (2016) Physiological and behavioral basis for the successful adaptation of goats to severe water restriction under hot environmental conditions. *Animal* (2016), 82–88.

Kaumbata, W., Banda, L., Mészáros, G., Gondwe, T., Woodward-Greene, M.J. *et al.* (2020) Tangible and intangible benefits of local goats rearing in smallholder farms in Malawi. *Small Ruminant Research* 187(March). DOI: 10.1016/j.smallrumres.2020.106095.

Kay, RNB. (1997) Responses of African livestock and wild herbivores to drought. *Journal of Arid Environments* 37(4), 683–694.

Kijne, J., Barron, J., Hoff, H., Rockström, J., Karlberg, L. *et al.* (2009) *Opportunities to increase water productivity in agriculture with special reference to Africa and South Asia. A report prepared by Stockholm Environment Institute, for the Swedish Ministry of Environment*. Stockholm Environment Institute, New York, USA, p. 48 .

Krauß, M., Kraatz, S., Drastig, K. and Prochnow, A. (2015) The influence of dairy management strategies on water productivity of milk production. *Agricultural Water Management* 147, 175–186. DOI: 10.1016/j.agwat.2014.07.015.

Kumar, S. and Deoghare, P. (2003) Goat production system and livelihood security of rural landless households. *Indian Journal of Small Ruminants* 9(1), 19–24.

Lebbie, S.H.B. (2004) Goats under household conditions. *Small Ruminant Research* 51(2), 131–136.

Mahgoub, O., Khan, A.J., Al-Maqbaly, R.S., Al-Sabahi, J.N., Annamalai, K. *et al.* (2002) Fatty acid composition of muscle and fat tissues of Omani Jebel Akhdar goats of different sexes and weights. *Meat Science* 61(4), 381–387.

Maltz, E., Olsson, K., Glick, S., Fyhrquist, F., Silanikove, N. *et al.* (1984) Homeostatic responses to water deprivation or hemorrhage in non-lactating Beduin goats. *Comparative Biochemistry and Physiology* 77A(l), 79–84.

Mdletshe, Z.M., Ndlela, S.Z., Nsahlai, I.V. and Chimonyo, M. (2018) Farmer perceptions on factors influencing water scarcity for goats in resource-limited communal farming environments. *Tropical Animal Health and Production* 50(7), 1617–1623.

Meat and Livestock Australia (2020) Glogal Snapshot: Goat Meat. Available at: https://www.mla.com.au/globalassets/mla-corporate/prices--markets/documents/os-markets/export-statistics/jan-2019-snapshots/global-beef-snapshot-jan2019.pdf

Mekonnen, M.M. and Hoekstra, AY. (2012) A global assessment of the water footprint of farm animal products. *Ecosystems* 15(3), 401–415.

Mekonnen, M.M., Pahlow, M., Aldaya, M.M., Zarate, E. and Hoekstra, A.Y. (2014) Water footprint assessment for Latin America and the Caribbean: An analysis of the sustainability, efficiency and equitability of water consumption and pollution. Value of Water Res Report Series No 66.

Misra, A.K. and Singh, K. (2002) Effect of water deprivation on dry matter intake, nutrient utilization and metabolic water production in goats under semi-arid zone of India. *Small Ruminant Research* 46, 159–165.

Molden, D., Oweis, T., Steduto, P., Kijne, J., Hanjra, M.A. *et al.* (2007) Chapter 7: pathways for increasing agricultural water productivity. In: *Water for Food Water for Life: A Comprehensive Assessment of Water Management in Agriculture*. Earthscan, London, UK, Sterling, VA, USA, pp. 278–310.

Molden, D., Oweis, T., Steduto, P., Bindraban, P., Hanjra, M.A. *et al.* (2010) Improving agricultural water productivity: Between optimism and caution. *Agricultural Water Management* 97, 528–535.

Ndlovu, L. (2010) The role of foods of animal origin in human nutrition and health. In: *The Role of Livestock in Developing Communities: Enhancing Multifunctionality*. Technical Centre for Agricultural and Rural Cooperation (CTA), Westdene, Bloemfontein, South Africa, pp. 77–92.

Neumann, C.G., Bwibo, N.O., Murphy, S.P., Sigman, M., Whaley, S. *et al.* (2003) Animal source foods improve dietary quality, micronutrient status, growth and cognitive function in Kenyan school children: Background, study design and baseline findings. *American Society for Nutritional Sciences* 133, 3927S–3931S.

Ngambi, J.W., Alabi, O.J. and Norris, D. (2013) Role of goats in food security, poverty alleviation and prosperity with special reference to sub-saharan africa: A review. *Indian Journal of Animal Research* 47(1), 1–9.

OECD (2018) Implementing the OECD principles on water governance: Indicator framework and evolving practices. OECD studies on water. OECD Publishing, Paris.

Olsson, K. (2005) Fluid balance in ruminants: Adaptation to external and internal challenges. *Annals of the New York Academy of Sciences* 1040, 156–161.

Oweis, T. and Peden, D. (2008) Water and livestock. Proceedings of the international conference on livestock and global climate change. In: *Livestock and Global Climate Change*. Cambridge University Press, Cambridge, UK and New York, USA.

Peacock, C. (2005) Goats—A pathway out of poverty. *Small Ruminant Research* 60(1-2 SPEC. ISS), 179–186.

Peacock, C. (2005) Goats: Unlocking their potential for Africa's farmers. Paper presented at the Seventh Conference of Ministers Responsible for Animal Resources. Available at: https://www.farmafrica.org/wp-content/uploads/2024/07/WP2-Goats-Unlocking-their-Potential.pdf

Peden, D., Taddesse, G. and Haileslassie, A. (2009) Livestock water productivity: Implications for sub-saharan Africa. *Rangeland Journal* 31(2), 187–193.

Peden, D., Tadesse, G., Misra, A.K., Ahmed, F., Astatke, A. *et al.* (2007) Water and livestock for human development. In: *Water for Food Water for Life: A Comprehensive Assessment of Water Management in Agriculture*. Earthscan, London, UK and Sterling, VA, USA, pp. 485–514.

Pica-Ciamarra Ugo (2005) Livestock policies for poverty alleviation: theory and practical evidence (27). Available at: http://www.fao.org/ag/pplpi.html

Prăvălie, R. (2016) Drylands extent and environmental issues. A global approach. *Earth Science Reviews* 161, 259–278.

Pretty, J., Hine, R.E., Morison, J.I.L., Noble, A.D., Bossio, D. *et al.* (2006) Resource-conserving agriculture increases yields in developing countries. *Environmental Science & Technology* 40(4), 1114–1119.

Reece, W. (2015) Body water: Properties and functions. In: Reece, W.O., Erickson, H.H., Goff, J.P. and Uemura, E.E. (eds), 13th edn. Iowa State University, John Wiley and Sons, Inc, Ames, IA, USA, pp. 103–113.

Rodríguez, D.I., Anríquez, G. and Riveros, JL. (2016) Seguridad alimentaria y ganadería: El caso de América Latina y el Caribe. *Ciencia e Investigacion Agraria* 43(1), 5–15.

Roy, J., Tschakert, P., Waisman, H., Abdul Halim, S., Antwi-Agyei, P. *et al.* (2018) Sustainable development, poverty eradication warming, and reducing inequalities. In: *Global Warming of 1.5°C an IPCC Special Report on the Impacts of Global of 1.5°C above Pre-Industrial Levels and Related Global Greenhouse Gas Emission Pathways, in the Context of Strengthening the Global Response to the Threat of Climate Change, Sus*. Cambridge University Press, Cambridge, UK and New York, USA, pp. 445–538.

Safari, J., Mtenga, L., Eik, L., Sundstøl, F. and Johnsen, F. (2008) Analysis of three goat production systems and their contribution to food security in semiarid areas of Morogoro, Tanzania. *Livestock Research for Rural Development* 20, Article #74. Available at: http://www.lrrd.org/lrrd20/5/safa20074.htm (accessed 20 September 2021).

Shankarnarayan, K., Bohra, H. and Ghosh, P. (1985) The goat an appropriate animal for arid and semi-arid regions. *Economic and Political Weekly* 20(45/47), 1965–1967.

Silanikove, N. (1992) Effects of water scarcity and hot environment on appetite and digestion in ruminants: A review. *Livestock Production Science* 30(3131), 175–194.

Silanikove, N. (1994) The struggle to maintain hydration and osmoregulation in animals experiencing severe dehydration and rapid rehydration: the story of ruminants. *Experimental Physiology* 79(3), 281–300.

Silanikove, N. (2000) The physiological basis of adaptation in goats to harsh environments. *Small Ruminant Research* 35, 181–193.

Tefera, T.M. (2007) Case study improving women farmers' welfare through a goat credit project and its implications for promoting food security and rural livelihoods. *Journal of Rural and Community Development* 2(2), 123–129. Available at: http://www.cabdirect.org/ abstracts/20093033088.html%5Cnhttp://www.jrcd.ca/include/getdoc.php?id=245&article=88&mode=pdf

Thalheimer, L., Ficko, A., Roque de Pinho, J., Hashimshony-Yaffe, N., Paschalidou, A. *et al.* (2019) Food security in drylands under a changing climate. In: *COST Action Drylands Facing Change: Interdisciplinary Research on Climate Change, Food Insecurity, Political Instability (CA 16233)*. Arad, Istrael, pp. 1–6.

UN (2021) The Sustainable Development Goals Report 2020. Des Glob Challenges Goals. United Nations Publications, New York, USA.

UN (2019) World population prospects. United Nations. Department of Economic and Social Affairs. Available at: https://population.un.org/wpp/Publications/

UNESCO, UN-Water (2020) United Nations world water development report 2020: Water and climate change. United Nations Educational Scientific and Cultural Organization, Paris, France. Available at: https://en.unesco.org/themes/water-security/wwap/wwdr/2020#download

UN-Water (2006) Coping with water scarcity. Water and Wastewater International, Rome, Italy. Available at: https://www.un.org/waterforlifedecade/pdf/2006 _unwater_coping_with_water_scarcity_eng.pdf

Valdivia, C. (2001) Gender, livestock assets, resource management, and food security: Lessons from the SR-CRSP. *Agriculture and Human Values* 18(1), 27–39.

van Breugel, P., Herrero, M., van de Steeg, J. and Peden, D. (2010) Livestock water use and productivity in the Nile Basin. *Ecosystems* 13(2), 205–221.

Webb, E.C., Casey, N.H. and Simela, L. (2005) Goat meat quality. *Small Ruminant Research* 60(1-2 SPEC. ISS), 153–166.

Whitford, W., Duval, B. and Desertification, D.B. (2020) *Ecology of desert systems*, 2nd edn. Academic Press, Cambridge, Massachusetts, USA, pp. 371–395.

Willaarts, B., Garrido, Al and Llamas, R. (2014) *Water for food security and wellbeing in Latin America and the Caribbean: Social and environmental implications for a globalized economy*. Routledge, London, UK, p. 264 .

Wodajo, H.D., Gemeda, B.A., Kinati, W., Mulem, A.A., van Eerdewijk, A. *et al.* (2020) Contribution of small ruminants to food security for Ethiopian smallholder farmers. *Small Ruminant Research* 184(February), 106064. DOI: 10.1016/j.smallrumres.2020.106064.

Woldu, T., Markemann, A., Reiber, C., Muth, P.C. and Zárate, AV. (2016) Optimising contributions of goat farming to household economic success and food security in three production systems in Ethiopia. *Journal of Agriculture and Rural Development in the Tropics and Subtropics* 117(1), 73–85.

Wu, G., Fanzo, J., Miller, D.D., Pingali, P., Post, M. *et al.* (2014) Production and supply of high-quality food protein for human consumption: Sustainability, challenges, and innovations. *Annals of the New York Academy of Sciences* 1321(1), 1–19.

Zarrin, M., Riveros, J.L., Ahmadpour, A., de Almeida, A.M., Konuspayeva, G. *et al.* (2020) Camelids: New players in the international animal production context. *Tropical Animal Health and Production* 52(3), 903–913.

Zeng, Z., Liu, J., Koeneman, P.H., Zarate, E. and Hoekstra, AY. (2012) Assessing water footprint at river basin level: A case study for the Heihe river Basin in northwest China. *Hydrology and Earth System Sciences* 16(8), 2771–2781.

22 The potential of seaweed for carbon capture

Ellen Ould and Gary S. Caldwell*

School of Natural and Environmental Sciences Reception, 4th Floor, Ridley Building 2, Newcastle University, Newcastle Upon Tyne, NE1 7RU, UK

Abstract

Macroalgae (seaweeds) are drawing interest as potential tools for climate mitigation within a burgeoning blue carbon agenda. With the high rates of net primary productivity, they are one of the main carbon sinks in the global ocean, both as biomass standing stock and as vectors of carbon export through detritus pathways. When coupled with their intrinsic ecological value, their contribution to coastal economies, and their current and future scope for aquaculture production, it is perhaps unsurprising that macroalgae have become a focus for nature-based carbon sequestration. However, the research community must guard against the hyperbole that is beginning to permeate the conversation, such as that which beset the microalgae biofuel movement. Macroalgae undoubtedly have a role to play in combating carbon pollution, and perhaps before significant investment is committed to ramping up macroalgae aquaculture, a more pragmatic view of its contribution to natural capital is justified. This article takes an objective look at the capacity for macroalgae to fill a global carbon sequestration niche, both as managed natural carbon pools and as deployable carbon capture assets.

Review methodology: We searched the following databases: Web of Science, Scopus, Google Scholar using the search terms 'macroalgae OR kelp OR macroalgal AND carbon sequestration OR blue carbon'. In addition, we used the references from the articles obtained by this method to check for additional relevant material. We also spoke to colleagues and checked for any upcoming studies not yet published.

Keywords: Algae, carbon sequestration, carbon sink, climate change, carbon dioxide fixation, detritus, aquaculture

Introduction

Anthropogenic carbon dioxide (CO_2) emissions have severely disrupted global carbon inventories, with atmospheric CO_2 surpassing 416 parts per million (ppm) in February 2021 (Tans and Keeling, 2021). Despite global efforts to reduce emissions at source (e.g. through the enactment of the Kyoto Protocol and the Paris Agreement (Stavins *et al.*, 2014; UNFCCC, 2015)), the upward CO_2 trajectory shows little sign of abating. This has forced policy makers to embrace strategies (both engineered and natural) for CO_2 capture and storage. Biological carbon capture and storage that exploits primary producers (through photosynthetic carbon drawdown) and subsequent conversion to biomass (e.g. afforestation) are the attractive options, although they are not without their drawbacks (Harper *et al.*, 2018). Further, Wang

*Corresponding author: gary.caldwell@ncl.ac.uk

et al. (Wang *et al.*, 2020) showed that the global CO_2 fertilisation effect—the enhancement in vegetation productivity driven by the increased CO_2 concentration in air—has weakened in most terrestrial regions, likely due to changes in key soil nutrients and water resources. Given this scenario, land-based biological mitigation strategies are unlikely to be the panacea for global warming, necessitating a holistic 'one planet' perspective of natural capital.

Macroalgae (seaweeds) are an extensive group of marine plants, comprising over 16,000 species across three phyla: Rhodophyta (red), Chlorophyta (green) and Phaeophyceae (brown) (Algaebase, n.d). Predominantly benthic, they inhabit intertidal and subtidal zones and are often the main primary producers within these habitats. Macroalgae-rich habitats cover between 2 and 6.8 million km^2 of the continental shelf (Trevathan-Tackett *et al.*, 2015) and are one of most productive ecosystems (Krause-Jensen and Duarte, 2016), with estimated global net primary productivity of up to 1.5 Pg C yr^{-1} (Krause-Jensen *et al.*, 2018). Long established as important sources of biomass and biochemicals for diverse applications, viewed through the prism of climate change, macroalgae are attracting much focus as potential carbon pollution mitigation options, both as carbon stores, as renewable feedstock to displace fossil fuels (Raven, 2017), and as regulators of greenhouse gas emissions (e.g. methane from livestock production (Roque *et al.*, 2021)).

Interest in macroalgae carbon sequestration has increased over the past decade (Fig. 22.1), yet considerable research gaps remain. Without addressing these gaps, we risk creating a knowledge vacuum, into which unqualified opinion may prevail (as was evident during the microalgae biofuels bubble (Mascarelli, 2009; Luque, 2010)) that could distort and ultimately derail genuine efforts to develop macroalgae for carbon sequestration. In this article, we present an objective view on the potential for macroalgae to deliver carbon savings. We examine two contrasting yet complementary options: nature-based solutions and those that are more definitively interventionist, for example the aquaculture and bioprocessing of macroalgae (Gao *et al.*, 2018). We discuss these within the remit of carbon sequestration as defined by the IPCC: 'The process of storing carbon in a carbon pool' (IPCC, 2018).

'Blue carbon' refers to carbon sequestered by marine ecosystems such as seagrasses, mangroves and salt marshes, as well as planktonic systems dominated by unicellular phytoplankton (Lovelock and Duarte, 2019) (although there are various definitions with some including or excluding phytoplankton (Nellemann *et al.*, 2009; Macreadie *et al.*, 2019)). Seagrasses, mangroves and salt marshes capture carbon through photosynthesis, burial of detritus from autochthonous and allochthonous sources (Lyimo, 2016). For carbon sequestration, surficial sediments must be considered

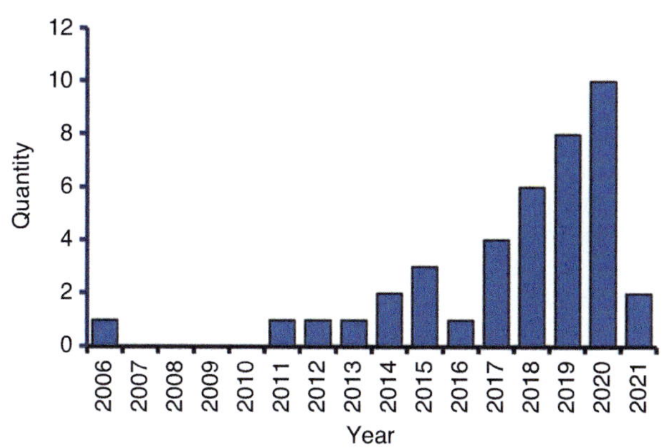

Fig. 22.1. Number of macroalgae carbon sequestration papers published between 2006 and 2021 (Web of Science search using search entry 'macroalgae OR kelp OR macroalgal AND carbon sequestration OR blue carbon', with manual selection of papers based on title and abstract).

as short-term carbon stores as these pools are rapidly remineralised within natural detrital turnover (Githaiga *et al.*, 2019). However, the potential to revert from carbon sink to carbon source is acknowledged, particularly in response to extensive habitat disruption. Sea-level rise, for instance, may lead to blue carbon gains or losses as ecosystems may expand to adjust to new sea-level heights thus promoting new sediment accretion and ultimately increasing carbon sequestration (Lovelock and Reef, 2020) or carbon remineralisation may occur if the new habitats cannot stabilise (Macreadie *et al.*, 2019). Traditionally, macroalgae such as kelp (a group of large brown seaweeds) were considered to have made little contribution to long-term blue carbon stores as most macroalgae attach to hard substrates rather than sediment. While macroalgae store vast quantities of carbon as standing stock, that is the available biomass at a given point in time, this must also be considered a short-term reservoir due to normal carbon turnover through growth, respiration and decomposition pathways (Mcleod *et al.*, 2011).

What is the fate of carbon from macroalgae?

Broadly, macroalgae morphology may be summarised as the holdfast (the section that is affixed to the substratum, loosely equivalent to plant roots), the stipe (the section above the holdfast that provides some rigidity, loosely equivalent to a plant's stalk or stem) and the frond (the main photosynthetically active part of the structure, loosely equivalent to plant leaves, and within which the bulk of the biomass may be found) (Wernberg *et al.*, 2019). Some macroalgae (such as many *Ulva* spp. and some rhodophytes) lack a stipe but have retained the majority of the basic macroalgae structure. The growth of a kelp plant may be described as resembling a 'conveyor belt', with new biomass formed at the meristem (near to the junction between the stipe and frond) and with older frond material found at the distal end. Consequently, there is a more or less continuous export of detritus material to the water column in the form of particulate (POM) or dissolved organic matter (DOM).

Most carbon in the oceans exists in inorganic forms (>2000 µmol/kg seawater), with organic carbon constituting 35–150 µmol/kg seawater. The biological carbon pump fixes carbon into biomass through photosynthesis, which is then transferred by sinking through the water column to depth (Verdugo *et al.*, 2004). Ocean carbon stocks vary seasonally by 5% (McGuire *et al.*, 2009). For macroalgae, ~52% of the exported carbon exists as dissolved organic carbon (DOC) with the remaining ~48% as particulate organic carbon (POC) (Fig. 22.2). Rates of DOC and POC vary across macroalgae species due to chemical composition influencing rate of degradability of detrital material. Carbon from *Palmaria spp.* (red macroalgae) for example is remineralised more readily than *Desmarestia* (brown macroalgae) (Braeckman *et al.*, 2019). A considerable proportion of the POC is recycled within shelf seas (e.g. as beach cast material); however, much of the fraction that is exported beyond the continental shelf sinks to depth.

Organic carbon is considered to be sequestered once it has passed the 'sequestration flux' at a depth of 1000m (Matsumoto *et al.*, 2007). In contrast, much of the labile fraction of the DOC is rapidly remineralised to CO_2 through net respiratory metabolism within the microbial loop and is ultimately returned to the atmospheric carbon inventory; however, a proportion is repackaged as microbial biomass and eventually exported to depth as microbial POC (cf marine snow). The more recalcitrant DOC fraction, while still available to microbial mixotrophs and heterotrophs, is more readily exported to depth. The efficiency of the ocean biological carbon pump is based on a balance between carbon fixation and remineralisation (Passow and Carlson, 2012). When remineralisation occurs at a faster rate, this results in less POC being transported to depth and thus less being buried for long-term storage. As the oceans warm, the remineralisation depth is predicted to increase (thereby the sequestration flux depth will increase), which will reduce overall sequestration capacity; indeed, a deepening in the order of only a few tens of meters can account for ~10–30 ppm changes in atmospheric CO_2 concentrations (Kwon *et al.*, 2009; Chikamoto *et al.*, 2012; Menviel *et al.*, 2012; Segschneider and Bendtsen, 2013; Roth *et al.*, 2014).

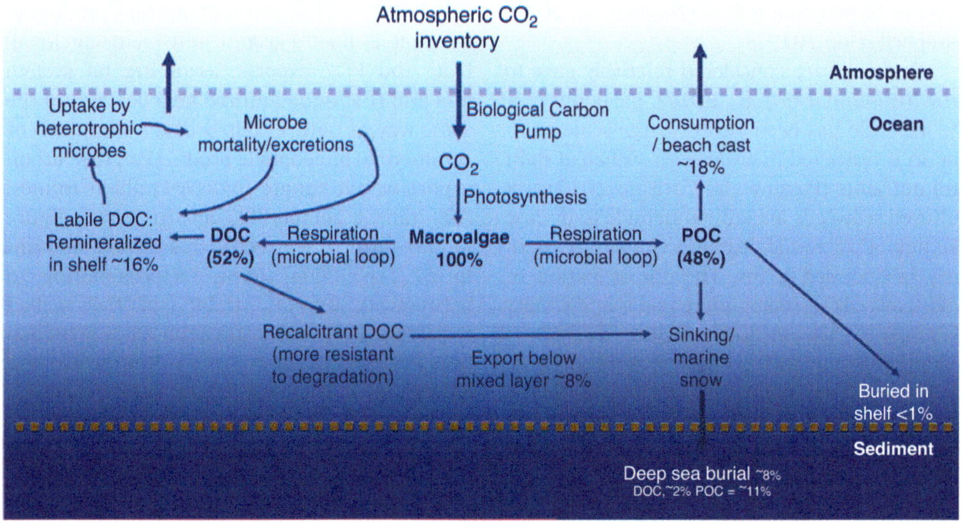

Fig. 22.2. The paths for export and subsequent storage of carbon fixed as macroalgae biomass as influenced by the biological carbon pump. Carbon is exported as dissolved (DOC) and particulate organic carbon (POC). If carbon is exported below a depth of approximately 1000m (the sequestration flux depth), it is considered as entering long-term carbon storage (ultimately as deep sea sediment). The efficiency of the biological carbon pump may be perturbed through ocean warming, which may drive the balance of the process more in favour of carbon remineralisation with consequent loss of some carbon sequestration capacity. Modified from Krause-Jensen and Duarte (2016).

Where these detrital carbon sources are exported to depth (and ultimately assimilated within stable sedimentary stores), this is deemed to constitute long-term carbon storage as remineralisation rates are considerably reduced due to the lower temperatures and oxygen levels. Recent work had estimated that 153 Tg C yr^{-1} (0.153 Gt) of macroalgae carbon is sequestered in deep sea sediments, with 14 Tg C yr^{-1} (0.014 Gt) buried in shelf sediments and a further 6.2 Tg C yr^{-1} (0.0062 Gt) buried from macroalgae growing in soft sediments (Krause-Jensen and Duarte, 2016). These estimates exceed the combined total carbon burial of seagrasses, mangroves and salt marshes although there are significant variations between species and regions; thus, global estimates are difficult to quantify and must be treated conservatively. To meet the 1.5°C limit of global temperature rise, it has been estimated that we need to remove between 640 and 950 $GtCO_2$ from the atmosphere by 2100 (approximating to 8–12 $GtCO_2$ per annum) (Luderer *et al.*, 2018). The capacity for macroalgae to help enable this (0.1732 Gt yr^{-1}, or between 1.44% and 2.17% of the annualised target) should not be dismissed, for example due to growing attached to rock rather than in sediment, or indeed our capacity to scale up biomass production through aquaculture; equally, nor should it be flippantly overblown.

Seaweed aquaculture for CO_2 capture

In 2018, the macroalgae aquaculture sector had a global output of 32.4 million tonnes, of which rhodophytes contributed over 17 million tonnes and phaeophytes almost 15 million tonnes (FAO, 2020). Five genera constitute the bulk of production (~98%)—*Saccharina*, *Undaria*, *Porphyra*, *Eucheuma* and *Gracilaria*. Production is dominated by a small number of nations, notably China, along with several other Asian countries. European production is small in comparison with global production and led by France, Norway and Northern Ireland (Netalgae, n.d.). China accounts for two-thirds of the world's production and is the main producer of brown macroalgae (Zheng *et al.*, 2019; Campbell *et al.*, 2019).

Seaweeds are considered relatively easy to grow (Buschmann *et al.*, 2017), particularly if the life cycle has been closed. Large-scale cultivation is restricted to areas with sufficient light and nutrients (Passow and Carlson, 2012), yet cultured seaweeds are still vulnerable to disease like any other aquacultured organism, particularly in enclosed farms. The domestication of wild seaweeds results in increased susceptibility of crops to disease and parasites through a reduction in genetic diversity (Campbell *et al.*, 2019). Seaweeds can be cultivated onshore, offshore or integrated within existing aquaculture practices. Onshore aquaculture involves growth in tanks and ponds, normally using pumped coastal seawater, whereas offshore cultivation involves floating rafts or ropes with the plants suspended from the surface, or attached to the seafloor (near-shore aquaculture). Offshore operations are more vulnerable to environmental fluctuations and are at greater risk of loss through storm events (García-Poza *et al.*, 2020).

Despite the considerable scale of production, only 0.04% of habitat deemed suitable for farming is currently used. There is undeniably enormous scope to expand seaweed aquaculture, even allowing for potential spatial conflict with other stakeholders (e.g. commercial fishers, offshore renewables, maritime transport), not to mention the geopolitical considerations; yet, attempting to project the scope for growth of seaweed aquaculture is problematic. For instance, Gattuso *et al.* (Gattuso *et al.*, 2006) estimated that 5.71 million km^2 of coastal seas have potential (although this was based solely on light availability in non-polar regions), whereas Froehlich *et al.* (Froehlich *et al.*, 2019), using a broader range of criteria (but arguably still too limited), predicted a maximum area of 48 million km^2. However, utilising 48 million km^2 for seaweed farming is not feasible due to intense competition for marine space with other stakeholders.

Purely as a thought experiment (or perhaps more correctly positioned as a straw man argument), erring on the side of optimism and recognising the sweeping assumptions that we make, if we extrapolate the average kelp detritus export figures estimated within Krumhansl and Scheibling (Krumhansi and Scheibling, 2012) (82% of annual productivity, that is fixed carbon, and factoring in all DOC and POC exports, assuming no system loss and that aquacultured kelp behaves in the same way as natural kelp beds) as a function of commercial phaeophyte production, the carbon export would equate to 13.94 million tonnes. Assuming a comparable detritus export figure for all farmed seaweeds, this would amount to 27.216 million tonnes, approximating to between 0.22% and 0.34% of the carbon sequestration needed to achieve our carbon removal targets. In reality, only a proportion of the detrital carbon would be sequestered. Furthermore, almost all commercially produced kelps are harvested within the year and thus cannot accumulate biomass or export detritus at a comparable level with wild populations. If we further extend this train of thought to estimate the necessary scale of seaweed farming required to fulfil our carbon export targets (to remove between 8 and 12 GtCO_2 per year), this would equate to producing, on an annual basis, between 9.75 and 14.63 Gt of farmed seaweed, which would be a 300- to 450-fold expansion of global seaweed aquaculture. These figures should be caveated that the 82% export figures used are at the upper boundary estimated by Krumhansl and Scheibling, and that not all of the exported carbon will be sequestered. Nonetheless, carrying these figures on, as a very crude estimate based on the scale of the farming operations undertaken in China, this would conservatively equate to the need for a further 120,000 km of shoreline given over to seaweed aquaculture. Clearly, these figures are not intended to be taken seriously as there will be no universal solution to the climate crisis and indeed, and perhaps more importantly, there remain crucial data gaps in our understanding of seaweed carbon export to the deep sea. Nonetheless, they do help frame some of the challenges that society will have to address (economic and political will allowing), for seaweed aquaculture to play a major role in climate mitigation. Evidently, our little thought experiment is an oversimplification and conveniently ignores factors such as the carbon footprint of the logistics of seaweed farming, particularly if conducted offshore (transportation, fuel, drying costs, etc.). However, in lieu of a comprehensive (and future proofed) life cycle assessment, it is sufficient at the stage to acknowledge the intentional naivety of our global scenario.

Up to this point, we have carefully ignored the elephant in the room. The rationale for our article was to explore the raison d'etre for both recognising and accelerating the deployment of macroalgae as a biological route for carbon sequestration. However, consideration must be given to the impacts on the seaweeds of the baked-in ocean temperature rise and associated climatic changes (Chung *et al.*, 2017; Pessarrodona *et al.*, 2018; Straub *et al.*, 2019; Smale *et al.*, 2020). Increased heat stress is likely to result in a restriction of thermal ranges (Harley *et al.*, 2012). Species are beginning to retreat poleward and to greater depths, which is reducing photosynthetic performance and growth, thereby driving the restructuring of food webs (Filbee-Dexter *et al.*, 2020). Intertidal species such as fucoids are particularly susceptible to temperature fluctuations as they generally inhabit areas close to their upper thermal limit; thus, any increase in temperature may significantly impact their abundance and distribution. However, because of their sensitivity, intertidal species can be used as bioindicators for increasing temperatures and may act as an early warning indicator for significant climate change impacts (Smolina *et al.*, 2016). Additionally, macroalgae in cold water habitats are more likely to experience significant stress due to the sensitivity of heat shock protein production in polar and temperate species in comparison with subtropical/tropical species (Jueterbock *et al.*, 2013). Heat shock protein production occurs at a lower temperature in polar/temperate regions than subtropic/tropic regions, and thus, rapid temperature increases would add further stress on the organism. There is some evidence to suggest that cold water species will thrive as temperatures increase closer to their optimum; however, this depends on the rate of warming and other environmental factors such as increased turbidity through glacial melt (Krause-Jensen *et al.*, 2020). Further, the efficiency of the biological carbon pump will potentially be affected by global temperature rise as increased seawater temperature is predicated to increase the rate of organic carbon remineralisation and thus may reduce the export and burial of carbon within the deep sea (Boscolo-Galazzo *et al.*, 2021).

Macroalgae can play a role in our battle against climate change, but how would we achieve this to best effect, and perhaps more importantly, should we even try? Below, we outline some technological steps necessary to dramatically scale aquaculture production. Although given humanities' less than impressive track record with large-scale ecological manipulation, there must be considerable risk that the medicine may be worse than the disease.

Metabolic engineering, as part of a wider systems biology view of macroalgae biology, must be considered alongside conventional thinking around the logistics of scale. Whilst macroalgae are already selected for strains that perform well, by optimising the metabolic characteristics of farmed seaweeds, it should be possible to develop cultivars that are more resistant to abiotic stress, disease and predation, as well as delivering enhanced photosynthetic efficiency, biomass yield and, subsequently, carbon export. Conversations are already taking place concerning genetic modification in relation to fish and invertebrate farming (Stokstad, 2020); however, other than for microalgae and cyanobacteria, the phycological community appears reluctant to broach this subject; perhaps for good reason. Given the scale of farming required (as alluded to in our thought experiment), there is a certain inevitability about the risks posed to wild stock, both genetically—as has been documented in China with *Saccharina japonica* and Japan with *Undaria pinnatifida* (Hu *et al.*, 2021; Uwai *et al.*, 2006) and through increased competition from escaped phenotypes, potentially resulting in local extinction, or at minimum a loss of genomic diversity. Further, seaweed farming is a renowned vector for the transfer of non-native species (Schaffelke and Hewitt, 2007; Eggertsen and Halling, 2021), with initiatives being launched in India and Hawaii to combat escaping seaweed spores through culling juvenile plants in efforts to prevent damage to adjacent coral reefs (Kamalakannan *et al.*, 2014; Neilson *et al.*, 2018; Eggertsen and Halling, 2021).

Offshore systems are difficult in terms of structure. Large-scale farms must be resilient enough to withstand large storm swells. Longlines are considered to be unsuitable for large-scale open water cultivation due to entanglement damage (Mooney-McAuley *et al.*, 2016). New designs in the shape of rings have been found to withstand severe environmental conditions in the North Sea provided the moorings are tended regularly (Buck *et al.*, 2017); however, they present challenges for harvesting. These open opportunities for large-scale cultivation with reduced infrastructure costs and little damage. Integrated multi-trophic aquaculture (IMTA) systems are likely to be the most effective solution due to reduced conflicts and fewer restrictions on farm size; however, these have still proven to be costly and unlikely to be profitable. Van den Burg and co-workers (Heery *et al.*, 2020) used economic modelling to determine profit/loss of cultivating seaweeds using the German ring design and identified that one ring with a life span of 10 year costs US$1380. Using a combination of predicted costs and values, they (van den Burg *et al.*, 2016) modelled that seaweed cultivation integrated within an offshore wind farm (to reflect the diversity of maritime stakeholders) would return a loss of ~US$23,834 per hectare per year. Without sufficient and consistent financial support, increased yield or increased yield prices, these systems are unlikely to come to fruition.

Recent studies have proposed the eco-engineering seaweed farming infrastructure with the concept of 'blue gardens'. These involve incorporating small-scale farms into existing man-made structures such as breakwaters and seawalls. These could support local communities whilst reducing costs for infrastructure (Heery *et al.*, 2020); however, they are unlikely to be suitable on a global scale.

Carbon sequestration: a confusion of terms

As stipulated, we adopt the IPCC definition of carbon sequestration; however, that definition is not without its shortcomings (Bachu *et al.*, 2007; Houghton *et al.*, 2009). Indeed, it may be argued that the interpretation and consequent application of 'carbon sequestration' is at the root of the confusion surrounding the veracity of carbon-saving claims for macroalgae. Curiously, carbon sequestration means different things to different disciplines and, broadly speaking, is delineated by the extent and permanency of carbon removal. For instance, from an engineering perspective, carbon sequestration is conventionally approached from two directions: carbon capture and storage (CCS) and carbon capture and utilisation (CCU).

CCS favours wholesale carbon capture using either oxy-fuel, or pre- and post-combustion technologies, and is typically terminated with geological storage (comprehensive CO_2 removal coupled with permanent storage; (Leung *et al.*, 2014)). CCS processes are expensive and difficult to valorise without incentives such as carbon credits, that is policy-driven change (Yang *et al.*, 2021). CCU on the other hand displaces fossil carbon in industrial and consumer markets with carbon that is recycled from existing industrial practices (market-driven change), in effect, creating short-lived loops in the carbon cycle (restrictive CO_2 removal with temporary storage; (Wang *et al.*, 2011; Centi *et al.*, 2020)). Biomass (or bioenergy) with carbon capture and storage (BECCS; (Kemper, 2015)) is an intermediary position, whereby carbon is fixed as biomass (e.g. lignin in trees) and burnt for heat and power generation, with the ensuing flue gas CO_2 permanently captured through CCS pathways (Fajardy *et al.*, 2020).

In relation to using macroalgae for carbon sequestration, the equivalent to CCS would be akin to farming seaweed with the sole purpose of sinking it to the ocean depths with presumptive storage in deep sea sediments. Some US start-up companies have expanded on this idea; however, they must consider that this proposition may become a financial drain without proper monetary support such as carbon purchase/carbon offset contracts. Further, these proposed 'microfarms' may be more susceptible to damage through adverse environmental conditions and require frequent management (as previously highlighted with offshore seaweed farms) and, unless scaled to a significant extent, will be unlikely to make a major contribution to the carbon sequestration equation (Bever, 2021). These approaches (as is common for new technologies) will invariably require heavy subsidies or strong incentives and should therefore not be considered a realistic option. However, exploiting macroalgae within a BECCS framework opens a range of near to market opportunities that should negate the need for subsidy and could, with appropriate vision and design, develop into sustainable businesses. Such ventures are structured around biorefinery models wherein biomass is sequentially processed to yield numerous biochemicals following a low-volume high-value products philosophy (Flórez-Fernández *et al.*, 2021; Kumar *et al.*, 2021; Ruangrit *et al.*, 2021). After the high-value components have been removed, the remaining biomass residue is typically processed to fuel as both a co-product and a waste management option. Three established water-tolerant biomass-to-fuels conversion technologies with scope for some degree of carbon storage are anaerobic digestion (AD) yielding biomethane and high-solid content digestate that can serve as a fertiliser and soil conditioner (Hinks *et al.*, 2013; Melara *et al.*, 2020; Wang *et al.*, 2021); and hydrothermal liquefaction (Wang *et al.*, 2021) and carbonisation (Shrestha *et al.*, 2021), yielding combustible gases, crude bio-oils (for further refining), nutrient-rich process water and solid residues (biochar) as the carbon store. The biochars generated from both hydrothermal processes are stable and suitable for storage through deep burial. In contrast, AD digestate has poor long-term carbon storage potential but it may stimulate improved carbon drawdown from enhanced soil productivity and may be further stabilised by rerouting it through either of the hydrothermal processes. Equally, the hydrothermal process water can be further valorised by conversion to biogas *via* AD (Brown *et al.*, 2020), which may partly ameliorate issues of parasitic thermal load within the system. A macroalgae biorefinery, assuming that it contains some component of carbon storage, could therefore deliver significant carbon sequestration (proportionate to scale) while stimulating investment in coastal communities.

Concluding Remarks

The carbon storage potential of macroalgae aquaculture is considerable, both as standing inventories (standing stock), long-term storage (export of detritus to depth) and, where suitable, as soil amendments from bioprocessing. In addition to CO_2 absorption, macroalgae provide a vast array of ecosystem services, including coastal protection, and nursery habitats for commercially exploited fish and shellfish, and are biomass sources for the production of biochemicals, biofuels and feed additives. Conversations are shifting towards growth of seaweed farming as a climate mitigation option; however, these conversations must not be had in isolation and must certainly not be at the expense of our natural capital. Nature-based solutions to reduce CO_2 emissions are becoming increasingly popular, with schemes such as Reducing Emissions from Deforestation and Forest Degradation aspiring to preserve natural carbon stocks (UN-REDD, 2021). Such initiatives have fuelled greater appreciation (and valorisation) of the standing carbon potential of global vegetation, including that contained within the global ocean. In stark contrast to seagrass meadows and salt marshes (Environment agency, 2007; Jackson *et al.*, 2012), there is currently no active conservation legislature for macroalgae in the UK and no seaweeds are listed under the protected plants section of the Wildlife and Countryside Act (Schedule 8) (Bailey and Owen, 2014), despite seaweed habitats being at risk from human impacts (Tyler-Walters and Arnold, 2008). There is little effort to conserve seaweeds globally, and only Hawaii (law imposed to prevent overexploitation) (Kim *et al.*, 2019), Australia and North America (both small scale) have attempted active conservation (Layton *et al.*, 2020). If macroalgae are to be used effectively, sensitively and sustainably within a portfolio of climate mitigation tools (biological and engineered), greater appreciation of their contribution to the global carbon cycle must be achieved across the piece, particularly from investors, decision-makers and those outside of the phycological community.

Acknowledgements

We are grateful for constructive and stimulating discussions with Prof Pippa Moore, particularly in relation to the impacts of ocean warming of macroalgae health, range and productivity. We also acknowledge the helpful comments and guidance from the anonymous reviewer.

References

Algaebase. (n.d) Taxonomy browser. Available at: https://www.algaebase.org/search /taxonomy/ (accessed 10 August 2021).

Bachu, S., Bonijoly, D., Bradshaw, J., Burruss, R., Holloway, S. *et al.* (2007) CO_2 storage capacity estimation: methodology and gaps. *International Journal of Greenhouse Gas Control* 1, 430–443.

Bailey, L. and Owen, K. (2014) Seaweed harvesting: Natural England's advice. *Natural England*.

Bever, F. (2021) Maine startup aims to pull carbon out of the atmosphere by growing—and then sinking—kelp farms. WBUR News. Available at: https://www.wbur.org/news /2021/02/16/maine-startup-carbon-kelp (accessed 10 August 2021).

Boscolo-Galazzo, F., Crichton, K.A., Ridgwell, A., Mawbey, E.M., Wade, B.S. *et al.* (2021) Temperature controls carbon cycling and biological evolution in the ocean twilight zone. *Science* 371, 1148–1152.

Braeckman, U., Pasotti, F., Vázquez, S., Zacher, K., Hoffmann, R. *et al.* (2019) Degradation of macroalgal detritus in shallow coastal Antarctic sediments. *Limnology and Oceanography* 64, 1423–1441.

Brown, A.E., Finnerty, G.L., Camargo-Valero, M.A. and Ross, AB. (2020) Valorisation of macroalgae via the integration of hydrothermal carbonisation and anaerobic digestion. *Bioresource Technology* 312, 123539.

Buck, B.H., Nevejan, N., Wille, M., Chambers, M.D. and Chopin, T. (2017) Offshore and multi-use aquaculture with extractive species: seaweeds and bivalves. In: *Aquaculture Perspective Of-Use Sites in the Open Ocean*. Springer, Cham, pp. 23–69.

Buschmann, A.H., Camus, C., Infante, J., Neori, A., Israel, Á. *et al.* (2017) Seaweed production: overview of the global state of exploitation, farming and emerging research activity. *European Journal of Phycology* 52, 391–406.

Campbell, I., Macleod, A., Sahlmann, C., Neves, L., Funderud, J. *et al.* (2019) The environmental risks associated with the development of seaweed farming in Europe—prioritizing key knowledge gaps. *Frontiers in Marine Science* 6, 107.

Centi, G., Perathoner, S., Salladini, A. and Iaquaniello, G. (2020) Economics of CO_2 utilization: A critical analysis. *Frontiers in Energy Research* 8, 567986.

Chikamoto, M.O., Abe-Ouchi, A., Oka, A. and Smith, SL. (2012) Temperature-induced marine export production during glacial period. *Geophysical Research Letters* 39, L21601.

Chung, I.K., Sondak, C.F.A. and Beardall, J. (2017) The future of seaweed aquaculture in a rapidly changing world. *European Journal of Phycology* 52, 495–505.

Eggertsen, M. and Halling, C. (2021) Knowledge gaps and management recommendations for future paths of sustainable seaweed farming in the Western Indian Ocean. *Ambio* 50, 60–73.

Environment agency (2007) Saltmarsh management manual.

Fajardy, M., Morris, J., Gurgel, A., Herzog, H., MacDowell, N. *et al.* (2020) The economics of bioenergy with carbon capture and storage (BECCS) deployment in a 1.5°C or 2°C world.

FAO (2020) The state of worldfisheries and aquaculture: Sustainability in action; Rome.

Filbee-Dexter, K., Wernberg, T., Grace, S.P., Thormar, J., Fredriksen, S. *et al.* (2020) Marine heatwaves and the collapse of marginal north Atlantic kelp forests. *Scientific Reports* 10, 13388.

Flórez-Fernández, N., Illera, M., Sánchez, M., Lodeiro, P., Torres, M.D. *et al.* (2021) Integrated valorization of *Sargassum muticum* in biorefineries. *Chemical Engineering Journal* 404, 125635.

Froehlich, H., Afflerbach, J., Frazier, M. and Halpern, B. (2019) Blue growth potential to mitigate climate change through seaweed offsetting. *Current Biology* 29, 3087–3093.

Gao, G., Clare, A.S., Rose, C. and Caldwell, GS. (2018) *Ulva rigida* in the future ocean: potential for carbon capture, bioremediation and biomethane production. *GCB Bioenergy* 10, 39–51.

García-Poza, S., Leandro, A., Cotas, C., Cotas, J., Marques, J.C. *et al.* (2020) The evolution road of seaweed aquaculture: cultivation technologies and the industry 4.0. *International Journal of Environmental Research and Public Health* 17, 1–42.
Gattuso, J.P., Gentili, B., Duarte, C.M., Kleypas, J.A., Middelburg, J.J. *et al.* (2006) Light availability in the coastal ocean: impact on the distribution of benthic photosynthetic organisms and their contribution to primary production. *Biogeosciences* 3, 489–513.
Githaiga, M.N., Frouws, A.M., Kairo, J.G. and Huxham, M. (2019) Seagrass removal leads to rapid changes in fauna and loss of carbon. *Frontiers in Ecology and Evolution* 7, 62.
Harley, C.D.G., Anderson, K.M., Demes, K.W., Jorve, J.P., Kordas, R.L, *et al.* (2012) Effects of climate change on global seaweed communities. *Journal of Phycology* 48, 1064–1078.
Harper, A.B., Powell, T., Cox, P.M., House, J., Huntingford, C. *et al.* (2018) Land-use emissions play a critical role in land-based mitigation for Paris climate targets. *Nature Communications* 9, 2938.
Heery, E.C., Lian, K.Y., Loke, L.H., Tan, H.T. and Todd, PA. (2020) Evaluating seaweed farming as an eco-engineering strategy for 'blue' shoreline infrastructure. *Ecological Engineering* 152, 105857.
Hinks, J., Edwards, S., Sallis, P.J. and Caldwell, GS. (2013) The steady state anaerobic digestion of *Laminaria hyperborea*—effect of hydraulic residence on biogas production and bacterial community composition. *Bioresource Technology* 143, 221–230.
Houghton, R.A., Hall, F. and Goetz, SJ. (2009) Importance of biomass in the global carbon cycle. *Journal of Geophysical Research: Biogeosciences* 114, G00e03.
Hu, Z.-M., Shan, T.-F., Zhang, J., Zhang, Q.-S., Critchley, A.T. *et al.* (2021) Kelp aquaculture in China: A retrospective and future prospects. *Reviews in Aquaculture* 2021, 12524.
IPCC (2018) Annex I: Glossary.
Jackson, E., Langmead, O., Beaumont, N., Potts, T. and Hattam, C. (2012) Seagrass ecosystem interactions with social and economic systems. Available at: https://randd.defra.gov.uk/ProjectDetails?ProjectId=18148
Jueterbock, A., Tyberghein, L., Verbruggen, H., Coyer, J.A., Olsen, J.L. *et al.* (2013) Climate change impact on seaweed meadow distribution in the north Atlantic rocky intertidal. *Ecology and Evolution* 3, 1356–1373.
Kamalakannan, B., Joyson Joe Jeevamani, J., Arun Nagendran, N., Pandiaraja, D. and Chandrasekaran, S. (2014) Impact of removal of invasive species *Kappaphycus alvarezii* from coral reef ecosystem in Gulf of Mannar, India. *Current Science* (106), 1401–1408.
Kemper, J. (2015) Biomass and carbon dioxide capture and storage: A review. *International Journal of Greenhouse Gas Control* 40, 401–430.
Kim, J., Stekoll, M. and Yarish, C. (2019) Opportunities, challenges and future directions of open-water seaweed aquaculture in the United States. *Phycologia* 58, 446–461.
Krause-Jensen, D. and Duarte, CM. (2016) Substantial role of macroalgae in marine carbon sequestration. *Nature Geoscience* 9, 737–742.
Krause-Jensen, D., Lavery, P., Serrano, O., Marba, N., Masque, P. *et al.* (2018) Sequestration of macroalgal carbon: the elephant in the blue carbon room. *Biology Letters* 14, 20180236.
Krause-Jensen, D., Archambault, P., Assis, J., Bartsch, I., Bischof, K. *et al.* (2020) Imprint of climate change on pan-arctic marine vegetation. *Frontiers in Marine Science* 7, 1129.
Krumhansi, K. and Scheibling, R. (2012) Production and fate of kelp detritus. *Marine Ecology Progress Series* 467, 281–302.
Kumar, M.D., Kavitha, S., Tyagi, V.K., Rajkumar, M., Bhatia, S.K. *et al.* (2021) Macroalgae-derived biohydrogen production: Biorefinery and circular bioeconomy. *Biomass Conversion and Biorefinery* 12, 769–791.
Kwon, E.Y., Primeau, F. and Sarmiento, J.L. (2009) The impact of remineralization depth on the air-sea carbon balance. *Nature Geoscience* 2, 630–635.
Layton, C., Coleman, M.A., Marzinelli, E.M., Steinberg, P.D., Swearer, S.E. *et al.* (2020) Kelp forest restoration in Australia. *Frontiers in Marine Science* 7, 74.
Leung, D.Y.C., Caramanna, G. and Maroto-Valer, MM. (2014) An overview of current status of carbon dioxide capture and storage technologies. *Renewable and Sustainable Energy Reviews* 39, 426–443.
Lovelock, C.E. and Duarte, CM. (2019) Dimensions of blue carbon and emerging perspectives. *Biology Letters* 15, 20180781.
Lovelock, C.E. and Reef, R. (2020) Variable impacts of climate change on blue carbon. *One Earth* 3, 195–211.

Luderer, G., Vrontisi, Z., Bertram, C., Edelenbosch, O.Y., Pietzcker, R.C. *et al.* (2018) Residual fossil CO_2 emissions in 1.5-2 °C pathways. *Nature Climate Change* 8, 626–633.

Luque, R. (2010) Algal biofuels: the eternal promise. *Energy and Environmental Science* 3, 254–257.

Lyimo, LD. (2016) Carbon sequestration processes in tropical seagrass beds. doctoral dissertation, Department of Ecology, Environment and Plant Sciences, Stockholm University.

Macreadie, P.I., Anton, A., Raven, J.A., Beaumont, N., Connolly, R.M. *et al.* (2019) The future of blue carbon science. *Nature Communications* 10, 1–13.

Mascarelli, AL. (2009) Gold rush for algae. *Nature* 461(7263), 460–461.

Matsumoto, K., Hashioka, T. and Yamanaka, Y. (2007) Effect of temperature-dependent organic carbon decay on atmospheric pCO_2. *Journal of Geophysical Research: Biogeosciences* 112, G02007.

McGuire, A., Anderson, L., Christensen, T., Dallimore, S., Guo, L. *et al.* (2009) Sensitivity of the carbon cycle in the arctic to climate change. *Ecological Monographs* 79, 523–555.

Mcleod, E., Chmura, G.L., Bouillon, S., Salm, R., Björk, M. *et al.* (2011) A blueprint for blue carbon: Toward an improved understanding of the role of vegetated coastal habitats in sequestering CO_2. *Frontiers in Ecology and the Environment* 9, 552–560.

Melara, A.J., Singh, U. and Colosi, LM. (2020) Is aquatic bioenergy with carbon capture and storage a sustainable negative emission technology? Insights from a spatially explicit environmental life-cycle assessment. *Energy Conversion and Management* 224, 113300.

Menviel, L., Joos, F. and Ritz, SP. (2012) Simulating atmospheric CO_2, ^{13}C and the marine carbon cycle during the last glacial-interglacial cycle: Possible role for a deepening of the mean remineralization depth and an increase in the oceanic nutrient inventory. *Quatemary Science Reviews* 56, 46–68.

Mooney-McAuley, K.M., Edwards, M.D., Champenois, J. and Gorman, E. (2016) Guidelines for Seaweed Cultivation and Analysis: Public Output report Report [WP1A5.01] of the EnAlgae Project. Available at: https://repository.oceanbestpractices.org/handle/11329/1282

Neilson, B.J., Wall, C.B., Mancini, F.T. and Gewecke, C.A. (2018) Herbivore biocontrol and manual removal successfully reduce invasive macroalgae on coral reefs. *PeerJ* e5332.

Nellemann, C., Corcoran, E., Duarte, C.M., Valdes, L., DeYoung, C. *et al.* (2009) A rapid response assessment. United Nations Environmental Programme: GRID-Arendal.

Netalgae (n.d.) Seaweed industry in Europe. Available at: https://www.seaweed.ie/irish_seaweed_contacts/doc/Filieres_12p_UK.pdf (accessed 10 August 2021).

Passow, U. and Carlson, CA. (2012) The biological pump in a high CO_2 world. *Marine Ecology Progress Series* 470, 249–271.

Pessarrodona, A., Moore, P.J., Sayer, M.D.J. and Smale, DA. (2018) Carbon assimilation and transfer through kelp forests in the NE Atlantic is diminished under a warmer ocean climate. *Global Change Biology* 24, 4386–4398.

Raven, JA. (2017) The possible roles of algae in restricting the increase in atmospheric CO_2 and global temperature. *European Journal of Phycology* 52, 506–522.

Roque, B.M., Venegas, M., Kinley, R.D., Nys, R., Duarte, T.L. *et al.* (2021) Red seaweed (*Asparagopsis taxiformis*) supplementation reduces enteric methane by over 80 percent in beef steers. *PLoS ONE* 16, e0247820.

Roth, R., Ritz, S.P. and Joos, F. (2014) Burial-nutrient feedbacks amplify the sensitivity of atmospheric carbon dioxide to changes in organic matter remineralisation. *Earth System Dynamics* 5, 321–343.

Ruangrit, K., Chaipoot, S., Phongphisutthinant, R., Duangjan, K., Phinyo, K. *et al.* (2021) A successful biorefinery approach of macroalgal biomass as a promising sustainable source to produce bioactive nutraceutical and biodiesel. *Biomass Conversion and Biorefinery* 1–11.

Schaffelke, B. and Hewitt, C.L. (2007) Impacts of introduced seaweeds. *Botanica Marina* 50, 397–417.

Segschneider, J. and Bendtsen, J. (2013) Temperature-dependent remineralization in a warming ocean increases surface pCO_2 through changes in marine ecosystem composition. *Global Biogeochemical Cycles* 27, 1214–1225.

Shrestha, A., Acharya, B. and Farooque, AA. (2021) Study of hydrochar and process water from hydrothermal carbonization of sea lettuce. *Renewable Energy* 163, 589–598.

Smale, D.A., Pessarrodona, A., King, N., Burrows, M.T., Yunnie, A. *et al.* (2020) Environmental factors influencing primary productivity of the forest-forming kelp *Laminaria hyperborea* in the northeast Atlantic. *Scientific Reports* 10, 12161.

Smolina, I., Kollias, S., Jueterbock, A., Coyer, J.A. and Hoarau, G. (2016) Variation in thermal stress response in two populations of the brown seaweed, *Fucus distichus*, from the Arctic and subarctic intertidal. *Royal Society Open Science* 3, 150429.

Stavins, R., Zou, J., Brewer, T., Conte Grand, M., Elzen, M. *et al.* (2014) International cooperation: Agreements and instruments. In: *Climate Change 2014:Mitigation of Climate Change. Contribution of Working Group III to the FifthAssessment Report of the Intergovernmental Panel on Climate Change*. Cambridge, UK, NewYork, USA.

Stokstad, E. (2020) New genetic tools will deliver improved farmed fish, oysters, and shrimp. here's what to expect. In: *ScienceMag.org*. DOI: 10.1126/science.abf7615.

Straub, S.C., Wernberg, T., Thomsen, M.S., Moore, P.J., Burrows, M.T. *et al.* (2019) Resistance, extinction, and everything in between—the diverse responses of seaweeds to marine heatwaves. *Frontiers in Marine Science* 6, 763.

Tans, P. and Keeling, R. (2021). Available at: www.esrl.noaa.gov/gmd/ccgg/trends/

Trevathan-Tackett, S.M., Kelleway, J., Macreadie, P.I., Beardall, J., Ralph, P. *et al.* (2015) Comparison of marine macrophytes for their contributions to blue carbon sequestration. *Ecology* 96, 3043–3057.

Tyler-Walters, H. and Arnold, C. (2008) Sensitivity of intertidal benthic habitats to impacts caused by access to fishing grounds [CCW policy research report no.08/13]. Countryside Council for Wales.

UNFCCC (2015) Report of the Conference of the Parties on its Twenty-First Session, Addendum, at 21.

UN-REDD (2021) UN-REDD programme fact sheet about REDD+.

Uwai, S., Yotsukura, N., Serisawa, Y., Muraoka, D., Hiraoka, M. *et al.* (2006) Intraspecific genetic diversity of *Undaria pinnatifida* in Japan, based on the mitochondrial cox 3 gene and the ITS1 of nrdna. *Hydrobiologia* 553, 345–356.

van den Burg, S., van Duijn, A.P., Bartelings, H., van Krimpen, M. and Poelman, M. (2016) The economic feasibility of seaweed production in the North Sea. *Aquaculture Economics & Management* 20, 235–252.

Verdugo, P., Alldredge, A.L., Azam, F., Kirchman, D.L., Passow, U. *et al.* (2004) The oceanic gel phase: A bridge in the DOM–POM continuum. *Marine Chemistry* 92, 67–85.

Wang, Q., Luo, J., Zhong, Z. and Borgna, A. (2011) CO_2 capture by solid adsorbents and their applications: Current status and new trends. *Energy and Environmental Science* 4, 42–55.

Wang, S., Zhang, Y., Ju, W., Chen, J.M., Ciais, P. *et al.* (2020) Recent global decline of CO_2 fertilization effects on vegetation photosynthesis. *Science* 370(6522), 1295–1300.

Wang, S., Zhao, S., Cheng, X., Qian, L., Barati, B. *et al.* (2021) Study on two-step hydrothermal liquefaction of macroalgae for improving bio-oil. *Bioresource Technology* 319, 124176.

Wernberg, T., Krumhansl, K., Filbee-Dexter, K. and Pedersen, M.F. (2019) Status and trends for the world's kelp forests. In: *World Seas: Anenvironmental Evaluation*. pp. 57–78.

Yang, B., Wei, Y.M., Liu, L.C., Hou, Y.B., Zhang, K. *et al.* (2021) Life cycle cost assessment of biomass co-firing power plants with CO_2 capture and storage considering multiple incentives. *Energy Economics* 96, 105173.

Zheng, Y., Jin, R., Zhang, X., Wang, Q. and Wu, J. (2019) The considerable environmental benefits of seaweed aquaculture in China. *Stochastic Environmental Research and Risk Assessment* 33, 1203–1221.

23 Climate-smart forestry—potential and practicalities

Rasoul Yousefpour[1,2*]

[1]Institute of Forestry and Conservation, John Daniels Faculty of Architecture, Landscape and Design, University of Toronto, 33 Willcocks St., Toronto, ON M5S 3B3, Canada; [2]Forestry Economics and Forest Planning, University of Freiburg, Tennenbacherstr. 4, D-79117 Freiburg, Germany

Abstract

Climate smart forestry (CSF) integrates ecological, economic and social dimensions in forest decision process under climate change. CSF aims to develop adaptive response to global change and optimally contribute to mitigation goals. CSF projects are ecologically effective and economically efficient in reaching the goals and involve society in the process.

Definition

The potential to simultaneously adapt to and mitigate climate change in the managed forest ecosystems is defined as climate-smart forestry (CSF). CSF strategies aim to develop adaptive response to global change and enhance the provision of ecosystem services and optimally contribute to mitigation goals. Recent attention to socio-economic implications of CSF search for comprehensive approaches integrates social dimensions and engages stakeholders in the process.

Review Methodology: After a systematic assessment in the Web of Science database, combinations of three keywords "Climate" AND "Smart" and "Forest*" were used to explore the wealth of knowledge regarding CSF. Finally, the five most cited and most relevant publications were selected for the purpose of this minireview (limited to max. five papers). However, the nature of this minireview is a hybrid of review and opinion paper.

Keywords: climate-smart solutions, carbon forestry, decision analysis

About the Author

The author's main area of research is forestry economics and decision support systems. He applies forest modelling approaches to analyse forest processes and functions involved in the provision of ecosystem services and in response to the management interventions and the economic implications of forest management decisions. He has already published a series of scientific papers adopting methodological approaches from other disciplines for forest decision analysis, for example Bayesian inference, robust optimization, adaptive decision approaches including risk and uncertainty analysis. The studies are quantitative in nature and deal with the novel decision problems and debated solutions in forest sector in local to global scales.

*rasoul.yousefpour@ife.uni-freiburg.de

Approaches

A diversity of methodology and approaches has been introduced in defining the climate-smart forestry CSF strategies and actions. They vary in the extent they integrate ecological process understanding, economic rationale and societal engagement (Bowditch *et al.*, 2020). Essentially, CSF promotes coordinated actions by forest owners, researchers, civil society and policy-makers towards climate change mitigation and resilient pathways. It may differ from 'business-as-usual' approaches by considering the dynamics of climate-ecosystem interactions and fostering innovative policies and new business models (Lipper *et al.*, 2014). It supports efforts from the local to global levels to integrate necessary adaptation measures and capture potential mitigation potentials.

Accordingly, the main three goals of CSF could be defined as follows: (1) sustainable provision of forest ecosystem services for society; (2) adapting and building resilience to climate change from local to national levels; and (3) developing opportunities to reduce GHG emissions and store carbon in forest biomass and woody products. Although it is ideal to reach all three goals optimally, it is not essential to generate triple wins. Most of the studies on CSF, however, focus mainly on forest mitigation potentials (Nabuurs *et al.*, 2017; Yousefpour *et al.*, 2018; Verkerk *et al.*, 2020). Nabuurs *et al.* (2017) explores the forest policies to double the European forests mitigation potentials (i.e. 441 Mt CO_2/year by 2050) by providing concrete forest management examples such as promoting wood use in building sector, afforestation of abandoned farmlands, and establishing large strict forest reserves in remote areas and/or low-productivity areas. The measures are targeted towards mitigation, but implicitly include adaptation issues and other synergetic values. Yousefpour *et al.* (2018) apply a robust optimization approach to find the most ecologically effective and economically low-cost allocation of European forest resources to CSF. The former searches for tree species and regions with a high mitigation potential and the latter allocates the CSF to regions with a lower opportunity cost (defined as the lost income from wood production). They simulate the implementation of CSF for 18 European countries encompassing 68.3 million ha of forest (42.4% of total EU-28 forest area) and found a CSF policy that could sequester 7.3–11.1 billion tons of carbon, projected to be worth 103–141 billion euros in the twenty-first century. Verkerk *et al.* (2020) discusses the relevant policies to expand the total forest area, link mitigation with adaption measures and use wood products for storing carbon and as a substitute for non-renewable products and materials. They highlight the important co-benefits of CSF for the protection of biodiversity and the provision of other ecosystem services essential for sustainable well-being of societies.

Implementation

Policy advices exist on the superiority of CSF to other mitigation strategies in realizing a smarter and higher contribution of forest and wood sector to global change mitigation. Allocating the carbon storage to regions with the fast-growing tree species, lower wood prices and high costs of labour and harvesting has been the backbone of an ecologically effective (high mitigation) and economically efficient (low cost) CSF. Yousefpour *et al.* (2018) studied realization of such an approach in European commercial forests and found fast-growing conifers *Picea abies* and *Pinus sylvestris* are favoured for CSF to broadleaves *Fagus sylvatica* and *Quercus species* elsewhere. Moreover, CSF shall be allocated to northern European countries with higher mitigation potential and eastern countries with a lower opportunity cost. Innovations to establish production and novel business model to substitute fossil-intensive and non-renewable materials by new generation of resource-efficient bio-based products may contribute largely to emission reduction (Verkerk *et al.*, 2020). Similar studies and implementations are currently approaching in a diversity of regions (e.g. in tropics, developing countries) and forestry contexts (afforestation, reforestation, biological conservation) to establish a successful CSF as a responsible approach for global issues particularly climate change mitigation.

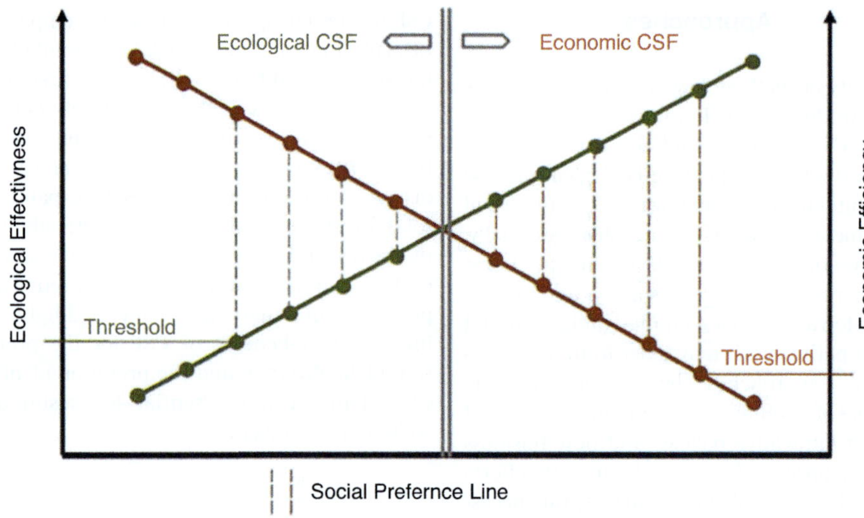

Fig. 23.1. Interaction of ecological effectiveness, economic efficiency and social dimensions in climate-smart forestry (CSF) decision space.

Discussion

CSF is getting recognized as a valid strategy for climate change management in forest and wood sector. However, there are a limited number of studies and implementation cases to conclude about its efficiency (Bowditch *et al.*, 2020; Verkerk *et al.*, 2020). The main strength of CSF remains in the integration of ecological effectiveness, economic efficiency and social inclusiveness in a comprehensive climate change management strategy. As theoretically illustrated in Fig. 23.1, there are trade-offs between ecological and economic dimensions of CSF projects. Decisions on the superiority of dimensions are defined by decision-makers and may be discussed in a socially inclusive (participatory) process. The social CSF decisions may put a threshold for a minimum level of ecological effectiveness (horizontal green line) and for a minimum economic efficiency (horizontal brown line) and eliminate decision options. Social preferences lines may decide among their options (vertical lines in Fig. 23.1) for a higher ecological effectiveness (left side of double grey line), a more economically efficient (right side of double grey line) or a balanced (double reline) CSF strategy. For this, the outcomes may be compared with alternative strategies, for example mitigation potentials in other sectors, which in turn may further limit the decision space.

A socially inclusive CSF decision process facilitates developing bottom-up and nature-based solutions (NBS) in small local scales, for example tree plantations in abandoned farmlands, woodland creation for water-quality improvements, agroforestry and bioenergy projects. These CSF projects may provide a solid platform to realize innovative and socially inclusive projects to foster local bio-economy and thereby contribute to the human well-being. A large-scale CSF, for example European CSF, may target a policy-level engagement and mobilize a series of science-policy interactions and socio-economic impact studies. The latter and similar CSF emphasizes a common forest policy following Paris Agreement commitments and requires fundamental changes in economies and societies to reach the ambitious and simultaneously feasible mitigation impacts and essential adaptation opportunities. Finally, we may conclude that CSF requires substantial changes in land use policy and economy including forestry and holds great potentials for realizing social innovations in developing novel business models (NBS) towards climate change adaptation and mitigation goals.

Finally, CSF is still in its infancy and deserves more and detailed studies and implementation examples to realize its opportune goals. Further research on multiple scales (region—continental) and a diversity of solution-oriented options in line with ongoing debates on; for example, REDD +, in tropics, diversification of forest management in Europe and carbon forestry in plantations are crucial to apply CSF in reaching global GHG mitigation goals.

References

Bowditch, E., Santopuoli, G., Binder, F., Del Río, M., La Porta, N. *et al.* (2020) What is climate-smart forestry? A definition from A multinational collaborative process focused on mountain regions of Europe ecosystem services 43, 101113. DOI: 10.1016/j.ecoser.2020.101113.

Lipper, L., Thornton, P., Campbell, B., Baedeker, T., Braimoh, A. *et al.* (2014) Climate-smart agriculture for food security. *Nature Climate Change* 4(12), 1068–1072. DOI: 10.1038/nclimate2437.

Nabuurs, G.J., Delacote, P., Ellison, D., Hanewinkel, M., Hetemäki, L. *et al.* (2017) By 2050 the mitigation effects of EU forests could nearly double through climate smart forestry. *Forests* 8(12), 484. DOI: 10.3390/f8120484.

Verkerk, P.J., Costanza, R., Hetemäki, L., Kubiszewski, I., Leskinen, P. *et al.* (2020) Climate-smart forestry: The missing link. *Forest Policy and Economics* 115, 102164. DOI: 10.1016/j.forpol.2020.102164.

Yousefpour, R., Augustynczik, A.L.D., Reyer, C.P.O., Lasch-Born, P., Suckow, F. *et al.* (2018) Realizing mitigation efficiency of European commercial forests by climate smart forestry. *Scientific Reports* 8(1), 345. DOI: 10.1038/s41598-017-18778-w.

24 Modelling climate change impacts on maize

Charles Bwalya Chisanga[1]*, Miyanda Moombe[2] and Elijah Phiri[2]
[1]*Department of Plant and Environmental Sciences, Copperbelt University, P.O. Box 21692, Off Jambo Drive, Kitwe 10101, Zambia;* [2]*Department of Soil Science, School of Agriculture Sciences, University of Zambia, P.O. Box 32379, Lusaka 10101, Zambia*

Abstract

Numerous modelling efforts have focused on maize as it is an important cereal crop for both human consumption and livestock production. Crop simulation and multi-linear regressions models can be used to quantify the likely potential impacts of climate change on maize growth and yield. Such models include AquaCrop, Agricultural Production Systems Simulator (APSIM), Decision Support System for Agro-technology Transfer (DSSAT), EPIC, CropSyst, Root Zone Water Quality Model (RZWQM2), SARRA-H, IMPACT+DSSAT, DSSAT, ALAMANC, WOFOST, ADEL, GEPIC, Empirical, MOS, GLAM-Maize, InfoCrop and EcoCrop, among others. Models give predictions utilizing meteorological, soil and crop data in numerical simulations. Various sources of climatic data are available including government meteorological and research departments, world organizations and private institutions. Climate data can also be generated using statistical and dynamical downscaling tools. The review showed that future maize growth and yield would be affected by changes in precipitation, temperature and soil fertility. Rise in temperature is the major factor altering maize yield. Nevertheless, crop simulation models have been observed to give mixed results depending on the region and the crop. Sources of uncertainty in predictions have been attributed to challenges in model parameterization, calibration and validation.

Review Methodology: Review of literature on modelling of climate change impact on maize was carried out by employing keywords to search for relevant peer-reviewed articles on Google Search and Google Scholar. The review was carried out using a semi-systematic approach as outlined (Snyder, 2019). Keywords employed in this search included 'modelling climate change impacts on maize', 'impact of climate change in maize growth and yield', 'impact of precipitation and temperature on maize growth and yield', 'impact of precipitation and temperature on maize yield', 'impact of sowing dates on maize growth and yield', 'impact of carbon on maize growth and yield', 'impact of fertility on maize yield' and 'impact of multimodal ensemble of modelled climate and crop models on maize growth and yield'. The aim of the review was to synthesize the utilization varied models to predict impacts of climate change on maize growth and yield, sources of climatic data utilized in the models and results obtained. Information employed in the review was sourced from a wide range of articles published in peer-reviewed journals.

Keywords: climate change, climate modelling, crop modelling, GCM, maize, yield

*Corresponding author: cbchisanga@gmail.com

Introduction

Agriculture is one of the most sensitive sectors to climate change and variability. Crop productivity is projected to increase or decrease in some areas especially the subtropic and tropics as a result of climate change (Ravindranath and Sathaye, 2003). Modelling of crop production with respect to varying meteorological parameters allows for prediction of associated changes in crop yield (Glotter *et al.*, 2014). Numerous modelling efforts have focused on maize (*Zea mays* L.) among other crops (Rosenzweig *et al.*, 2015; Sultan and Gaetani, 2016). This may be attributed to the fact that maize is an important cereal crop in the world for human consumption and livestock production (NSW, 2009; Beletse *et al.*, 2015; Trnka *et al.*, 2016).

Crop simulation and multi-linear regressions models can be used to quantify the likely potential impacts of climate change on maize growth and yield (Hoogenboom *et al.*, 1999; Falconnier *et al.*, 2020; Chisanga *et al.*, 2020) under both rain-fed and irrigated conditions (Islam *et al.*, 2012; Chisanga *et al.*, 2020). Calibrated and validated maize models have been able to reproduce measured grain yield variations (Islam *et al.*, 2012; Falconnier *et al.*, 2020; Jiang *et al.*, 2021). The projected changes in precipitation and temperature under atmospheric carbon dioxide fertilization may change evapotranspiration (ET) demand and affect soil water available for maize production and productivity (Islam *et al.*, 2012).

The quantitative effects of projected change in precipitation and temperature on maize growth and yield have been simulated using crop models. Crop models are only feasible at predicting the likely impacts of climate change on crop yield at both spatial and temporal scales (Matthews and Stephens, 2002). These models include AquaCrop, Agricultural Production Systems Simulator (APSIM), Decision Support System for Agro-technology Transfer (DSSAT), EPIC, CropSyst, Root Zone Water Quality Model (RZWQM2), SARRA-H, IMPACT +DSSAT, ALAMANC, WOFOST, ADEL, GEPIC, Empirical, MOS, GLAM-Maize, InfoCrop and EcoCrop (Holden and Brereton, 2003; Tingem *et al.*, 2007; Lobell and Burke, 2010; Islam *et al.*, 2012; Robertson *et al.*, 2012; Díaz-Ambrona *et al.*, 2013; Sultan *et al.*, 2013; Monteiro *et al.*, 2017; Gaydon *et al.*, 2017; Luhunga, 2017; Chisanga *et al.*, 2020; Falconnier *et al.*, 2020).

The AquaCrop, APSIM, CropSyst, DSSAT and SARRA-H models have been calibrated and validated for maize yield in many countries (Ouda *et al.*, 2015; Keating *et al.*, 1999; Jones *et al.*, 2003; Iqbal *et al.*, 2014; Beletse *et al.*, 2015; Chisanga *et al.*, 2020). However, comprehensive assessments using multi-models of climate and crop simulation models are insufficient in low-input maize production systems. Furthermore, adequate simulation accuracy uncertainly is currently lacking (Falconnier *et al.*, 2020).

Maize growth and yield can be studied using global climate models, dynamical and statistical downscaling models and crop simulation models (CSMs) (Nkomozepi and Chung, 2013; Falconnier *et al.*, 2020; Chisanga *et al.*, 2020). Studies have shown that an ensemble modelling using a variety of crop models is highly recommended for the quantification of uncertainty (Challinor *et al.*, 2009; Martre *et al.*, 2015; Araya *et al.*, 2015; Wallach *et al.*, 2016; Sultan and Gaetani, 2016). Multi-model ensemble simulations are necessary for providing insights into climate models and CSMs uncertainties as well as developing alternative adaptation and mitigation strategies (Challinor *et al.*, 2009; UNFCCC, 2012; McSweeney *et al.*, 2012; Martre *et al.*, 2015; Chisanga *et al.*, 2017).

Similar to any crop, maize production is vulnerable to climate change (Jiang *et al.*, 2021) and it has been negatively affected in many regions globally (Ravindranath and Sathaye, 2003; IPCC, 2014; Beletse *et al.*, 2015). It is projected that water stress, drought and heat stress would intensify and become more frequent in rain-fed maize growing regions of the world (Hatfield and Prueger, 2015). The extreme climate indices have drastic impact on maize production and productivity (Hatfield and Prueger, 2015).

The projected changes in precipitation and temperature may worsen the current maize production constraints (Falconnier *et al.*, 2020) especially among smallholder farmers in Sub-Saharan Africa (SSA) even with agricultural development (Hillel and Rosenzweig, 2015). Much of this may be due to the fact that smallholder farmers in SSA grow rain-fed maize with limited agriculture inputs (Falconnier *et al.*, 2020). It is predicted that maize yield would decrease by 19% given a 5°C global average warming in SSA (Thornton *et al.*, 2011; IPCC, 2014). A meta-analysis of projected impact of climate change showed a 5% decrease in maize yield by 2050 in West Africa (Knox *et al.*, 2012; Parkes *et al.*, 2018). Reduction in maize yield is also expected as a result of variability in precipitation. Precipitation is the most important meteorological parameter that influences rain-fed maize production (Chisanga, 2019). The mean maize yield reductions of 10% are expected by 2055 at smallholder farms (IPPC, 2007).

Impact of Climate Change on Maize Growth and Yield

Sources of data for modelling maize

Data required by crop simulation models include latitude, longitude, daily precipitation, minimum and maximum air temperature, relative humidity (%), wind speed and solar radiation. Observational meteorological data can be obtained from various institutions and data collectors including Zambia Meteorological Department (ZMD), NASA, NOAA, AgMERRA, WorldClim, Climate Research Unit Timeseries (CRU-TS), TERRACLIMATE, KNMI Climate Explorer, Agricultural Model Inter-comparison and Improvement Project (AgMIP), Climate Information Tool Kit (CLIK), CLIMGEN, MARKSIM, World Climate Research Program's (WCRP's), Climate Change Knowledge Portal, bias corrected and spatially disaggregated (BCSD) climate projections, Climate Change Agriculture and Food Security, PRISM Climate Group and CAIT Climate Data Explorer, among others (Ramirez-Villegas and Jarvis, 2010; Islam *et al.*, 2012; Jarvis and Thornton, 2014; Ruane *et al.*, 2015; Luhunga, 2017).

In the absence of observational data, climate data have been generated using statistical and dynamical downscaling tools. Weather generators such as the Long Ashton Research Weather Generators (LARS-WG), WGEN and SIMMETEO have been used to generate climate data as input into the crop models (Semenov and Brooks, 1999; Rauff and Bello, 2015). Climate data generated by Global Circulation Models (GCMs), Regional Climate Models (RCMs) and Statistical Downscaling Models (SDMs) are also utilized (Lenderink *et al.*, 2007; Jacob, 2009; Flato *et al.*, 2013; Abiodun and Adedoyin, 2016; Lee and Cha, 2020). Nevertheless, crop yield estimates generated employing weather data generated from either GCMs or RCMs require bias correction of climate inputs (Glotter *et al.*, 2014). The use of bias-corrected RCMs to provide input data removes uncertainties associated with GCM data (Parkes *et al.*, 2018).

Requirements for Optimal Maize Growth and Yield

Maize is a C4 crop and its requirements for optimal growth and yield includes water, favourable environmental conditions (air temperature, precipitation, relative humidity and solar radiation), and macro- and micro-nutrients (NSW, 2009). Maize is influenced by changes in maximum and minimum temperature and precipitation, sowing dates and soil fertility (NSW, 2009; Ali *et al.*, 2018). The base temperature for maize growth is between 8 and 10°C (Muchow *et al.*, 1990; Kiniry *et al.*, 1991) and varying numerical models employ either of the two (Yang *et al.*, 2004). The maize crop thrives in temperatures between 28°C and 32°C (Kiniry *et al.*, 1991; Yin *et al.*, 1995) and growth decreases above 34°C (Kiniry *et al.*, 1991), in which maize reduces and consequently ceases (NSW, 2009). For maize to grow optimally, it requires between 450 and 600 mm of water per season mainly extracted from the soil (du Plessis, 2003). The irrigation water requirement of maize is determined by the soil type, cultivar, drained lower limit (DLL), drained upper limit (DUL), plant available soil water (PASW), evaporation, irrigation efficiency and in-crop rainfall (du Plessis, 2003; NSW, 2009). Farmers can manipulate the environment by managing factors such as selection of maize cultivar, sowing dates, tillage, fertilizer application rates and irrigation.

Linear and Crop Simulation Models

Linear models

Crop yield can be estimated using crop simulation models or empirical associative statistical models using explanatory variables such as temperature and rainfall (Choudhury *et al.*, 2015). Linear trends of 50-year time series have been used to investigate the effects of climate change on maize yield in Bologna, Italy (Vitali *et al.*, 2019). A study was undertaken in Georgia to establish the relationship between climate conditions and maize yield (Alexandrov and Hoogenboom, 2001). Statistical crop weather models were developed using a stepwise multilinear regression at 95% probability level. The statistical linear yield model developed (Alexandrov and Hoogenboom, 2001) can be used in assessing the expected anomalies of mean maize yield in Georgia for a particular year. The model can be used to predict maize yield under higher temperatures and drought conditions.

Deep and machine learning has been applied in agriculture to predict crop yield. Outputs from APSIM crop model have been used to develop machine learning meta-model to predict nitrogen loss and maize yield in the United States (U.S.) (Holzworth *et al.*, 2015; Paudel *et al.*, 2021). Random Forests algorithms have been used to predict maize yield in the U.S. (Jeong *et al.*, 2016). The influence of climatic factors on spatiotemporal variation of maize yield in China has been studied (Liu *et al.*, 2021). Stepwise multiple linear regression models were used to ascertain the impact of climate change on maize yield. Results showed that warming trends in China have decreased maize yield in seven provinces from 1978 to 2015. The yield was calculated using crop production, planting area and explanatory variables.

Impact of precipitation and temperature on maize yield

Rise in temperature is expected to be the major factor altering maize yield in the near future (Bassu *et al.*, 2014). At global level, an increase in mean temperature by 1°C is expected to reduce maize yield by 7.4% (Zhao *et al.*, 2017). Maize yield would decrease under conditions of water stress (Kogo *et al.*, 2019). Projected maize yield will decline from 8% to 38% under RCP4.5 and RCP8.5 climate scenarios (Kogo *et al.*, 2019). A review of 50 papers (Knox *et al.*, 2012) revealed that projected maize yield across Africa would reduce by 5% in 2050. Crop yield will decrease by 18% and 22% from 2039 to 2070 in southern Africa and sub-Saharan Africa, respectively (IPCC, 2014; Waongo *et al.*, 2015). In contrast, the modelling of maize yield based on climate change scenarios for 2050 showed a decrease of 5% in East Africa (Waongo *et al.*, 2015). Further, a negative effect of increased temperature on maize growth and yield has been observed (Islam *et al.*, 2012; Zhao *et al.*, 2017). The decline in maize yields under future climate scenarios is due to an increase in temperatures that will reduce the number of days from sowing to physiological maturity (Luhunga, 2017; Chisanga *et al.*, 2020; Kadigi *et al.*, 2020; Tui *et al.*, 2021). The variability in maize yield is also attributed to soil and weather effects (Matthews and Stephens, 2002). An increase in heat stress and dry soils may reduce maize yields to one-third in the subtropical and tropical areas (IPPC, 2007).

The United States Country Studies Program (USCP) in 1994 was used to simulate maize yield using the EPIC model under baseline (1960–1990) and future (2010–2039, 2040–2069 and 2070–2099) climate scenarios (Adejuwon, 2006). The climate scenarios were generated from GCMs. The model predicted increased and decreased maize yield in 2010–2039 and 2040–2069, respectively. Lower yield were predicted due to an increase in air temperature. The EPIC model was used to investigate maize yield response to changes in precipitation, temperature, solar radiation, carbon dioxide and relative humidity (RH) (Adejuwon, 2006).

The DSSAT was used to assess the impact of climate change on maize using four GCMs as input climate data in Philippines. The results showed a decrease in maize yields (Matthews and Stephens, 2002). A study by Makadho (Makadho, 1996) in Zimbabwe using CERES-Maize model and GCMs assessed the potential impact of climate change on maize. The simulated grain yield reduced by 9.5% and the number of days to physiological maturity also reduced. Maize yield in Pakistan would reduce in 2050 due to an increase in temperature (Khaliq, 2008). Using the General Large Area Model (GLAM) in West Africa, (Parkes *et al.* (2018) projected a decrease of 5.95% in maize yield due to an increase in air temperature. Researchers (Lin *et al.*, 2017) using the CERES-Maize model in Northern China revealed that maize yield would decrease in (2.1%) 2020, (12.9%) 2050 and (22.7%) 2080 under RCP4.5. Prediction from InfoCrop model in India showed a decrease in maize biomass and an increase in temperature during 2071 and 2100 (Choudhary *et al.*, 2014).

The nonlinear temperature effects have profound effect on maize yield under climate change in the United States (Schlenkera and Roberts, 2009). Holding the current growing regions fixed, the area-weighted mean maize yields are predicted to decrease under B1 (slowest) and A1FI (rapid warming) by 30%–46% and 63%–82% under the Hadley III model, respectively (Schlenkera and Roberts, 2009). The suitability for maize production has been assessed by Ramirez-Cabral *et al.* (2017). The authors used the CLIMEX distribution model and climate data from GCMs (CSIRO-Mk3.0, MIROC-H) to simulate the loss of suitable areas for maize production. Results indicated that the tropics of Capricorn and Cancer would not be suitable for maize production. However, the regions nearer the pole regions were deemed suitable.

Rain-fed and irrigated maize production (high versus low N fertilizer)

Nitrogen fertilizer application controlled the response to variations in carbon dioxide, temperature and precipitation (Falconnier *et al.*, 2020). Simulating maize yield without N fertilizer input did not benefit from increased atmospheric carbon dioxide fertilization. However, without N fertilizer input, maize yield was less affected by an increase in temperature or reduced rainfall (Chisanga *et al.*, 2020; Falconnier *et al.*, 2020; Chisanga *et al.*, 2020). A study done in Zambia (Chisanga *et al.*, 2020) showed that maize receiving lower nitrogen fertilizer application rates would outperform those receiving higher rates during the 2050s. The effects of reduced rainfall would have an impact on maize growth and yield at higher fertility compared with lower fertility (Chisanga *et al.*, 2020). Similarly, in Zimbabwe maize yield would reduce under both rain-fed and irrigated conditions during 2050s (Makadho, 1996). The reduction in grain yield was attributed to an increase in air temperature, which would shorten the growth period (Makadho, 1996; Khaliq, 2008). Temperatures greater than 35°C can lead to maze growth to fail. The maize yield gaps due to low fertility are projected to widen due to climate change without adaptation options (Falconnier *et al.*, 2020; Siatwiinda *et al.*, 2021).

Maize cultivars that tolerate high temperature with longer growth duration will maintain higher yields under irrigated conditions (Islam *et al.*, 2012). It was observed that projected change in temperature would reduce the growing periods of maize, which would decrease the irrigation water requirement (Islam *et al.*, 2012). The adjustment of crop management practices such as optimizing fertilization and adopting high-yield cultivars enhances maize growth and yield (Jiang *et al.*, 2021). This also improves nutrient use efficiency (NUE) and minimizes environmental risks under the changing climate (Jiang *et al.*, 2021). The simulated maize yield at Bethlehem in South Africa under irrigated conditions reduced the effects of projected changes in rainfall under climate change (Hillel and Rosenzweig, 2015).

The APSIM and DSSAT simulations in India indicated a differential response of irrigated maize productivity to the future climate generated by different GCMs (Ponnusamy *et al.*, 2015). The simulated maize yield using IMPACT-DSSAT indicated a reduction in maize yield under both irrigated and rain-fed conditions at global level (Robertson *et al.*, 2012). The application of sufficient irrigation water can significantly reduce the maize yield losses (Khan and Walker, 2015).

Impact of varying carbon dioxide on maize growth and yield

Predictions (Ravindranath and Sathaye, 2003) from GCMs and the doubling of the current carbon dioxide (CO_2) will increase the average global air temperature between 1.5°C and 4.0°C with accompanying variability in precipitation. The atmospheric carbon dioxide (CO_2) fertilization decreased the evapotranspiration (ET) demand (Islam *et al.*, 2012). According to Rosenzweig and Hillel (2015), high CO_2 concentrations in the atmosphere are unlikely going to outweigh the negative effect of increasing air temperature. Studies on simulated maize yield based on GCMs and crop models projected yield decrease in Uruguay, Brazil, Mexico, Argentina, Chile even with CO_2 fertilization (Matthews and Stephens, 2002; Ravindranath and Sathaye, 2003). Maize yield will be reduced as a result of projected change in temperature. A study by Rosenzweig (1990) using the CERES-Maize model simulated the effects of climate change under 2xCO_2 conditions in the Southern Great Plains of the United States of American. The results indicated that maize yield would increase by 10%. The simulated yield in 2050 relative to the baseline indicated a reduction in grain yield. Lin *et al.* (2017) simulated a reduction in maize yield due to CO_2 fertilization under RCP4.5 (19.5%–22.7%) and RCP8.5 (44.1%–47.5%).

A study (Glotter *et al.*, 2014) compared the NARCCAP dynamically downscaled GCM output and GCM output interpolated to five arc minutes (~10 km) with and without bias correction. The generated climate scenarios were used as inputs into the DSSAT-CERES-Maize model. Climate data corrected for bias eliminated both effects and produced yield distributions consistent with those produced using historical observations (Glotter *et al.*, 2014). The increase in atmospheric CO_2 and effects of climate change are expected to decrease crop production in the Mediterranean, south-west Balkans and south of European Russia (IPCC, 2007; Durdu, 2013).

A study in Greece applied the CERES-Maize to ascertain the impacts of climate change under climate change scenarios and 2xCO_2 from GCMs (Kapetanaki and Rosenzweig, 1997; Khan and Walker, 2015). The study results showed that maize yield would decrease under the current management scenarios due to an increase in air temperature and variation in precipitation. Furthermore, the crop yield would decrease due to reduced number of days from planting to physiological maturity. The study concluded that the impacts of climate change on maize yield would be off-set through earlier sowing dates and the use of drought resistance varieties (Khan and Walker, 2015; Chisanga *et al.*, 2020; Chisanga *et al.*, 2020). Therefore, without CO_2 fertilization, there is need to have effective adaptation and cultivar genetic improvement (Zhao *et al.*, 2017).

Interactive effects of sowing dates, nitrogen fertilizer and cultivars

The crop models have been used to model the response of maize to different sowing dates, nitrogen fertilizer rates using diverse cultivars in temperate, sub-humid and the tropics (Jones and Thornton, 2003; Tovihoudji *et al.*, 2019; Chisanga *et al.*, 2020). Climate change and N input interactions have strong implications for the design of robust adaptation approaches across SSA, because the impact of climate change in low input systems will be modified if farmers intensify maize production with balanced nutrient management (Falconnier *et al.*, 2020).

The impact of climate change will lead to a reduction in future maize yield by 5%–6% relative to baseline in China (Chen *et al.*, 2020). In contrast, simulation studies in North China Plain using high-temperature sensitive maize cultivars showed that maize yield could increase by 1.0%–6.0%, 9.9%–15.2% and 4.1%–5.6% by adopting early sowing, fixing variety growing duration and late planting, respectively (IPCC, 2014). Substantial simulated changes in maize growth and yield during 2050 under RCP4.5 and RCP8.5 scenarios have been reported by researchers (Chisanga *et al.*, 2020). The simulated number of days after planting (DAP) to anthesis and physiological maturity would reduce in 2050. Furthermore, DAP to anthesis and maturity would be affected by an increase in temperature and a reduction in rainfall (Chisanga *et al.*, 2020). The projected changes in temperature and rainfall using a multi-model ensemble of APSIM-Maize and CERES-Maize models simulated reduced maize yield depending on the sowing date (SD) during 2050s (Chisanga *et al.*, 2020).

Applying the future climate scenarios to derive optimized planting date in Burkina Faso for 2020 and 2040, Waongo *et al.* (Waongo *et al.*, 2015) observed that the RCM ensemble mean maize yield from 2011 to 2050 revealed an 8% decrease relative to the baseline. Optimized planting dates achieved an increase of 15% in maize yield regardless of the RCM and scenario (Waongo *et al.*, 2015). The study concluded that the optimized planting dates can be combined with appropriate farming practices to adequately respond to climate change.

Single climate and crop models

A study (Khubaib *et al.*, 2021) in Pakistan assessed the impact of climate change on future distribution of maize production area. The Maxent species distribution model and CMIP5 model (MPI-ESM-LR) were used to predict future production area relative to the baseline. Results showed that the maize production area would reduce by 23%–36% in 2070 depending on the future climate scenarios (RCP4.5 and RCP8.5).

Multi-model ensemble of climate and crop models

The impact of climate change on maize yield has been characterized using the Agricultural Model Inter-comparison and Improvement Project (AgMIP) (Ruane *et al.*, 2013; Díaz-Ambrona *et al.*, 2013; Rosenzweig *et al.*, 2014, 2014; Chisanga *et al.*, 2020). AgMIP Coordinated Global and Regional Assessments of maize yield are characterized by differential impacts across regions and farming systems (Ruane *et al.*, 2018). The AgMIP-coordinated assessments suggested that direct economic and biophysical impacts are substantially larger for regional farming systems than global production changes (Ruane *et al.*, 2018).

Researchers have projected a decrease in maize yield of 10% from HadCM2 climate scenarios for Africa by 2050 (Jones and Thornton, 2003). Further, a projected 20% decrease in maize yield driven by an increase in air temperature in sub-Saharan Africa for the mid-century (2046–2065) has been noted (Schlenker and Lobell, 2010). Maize production will decrease between 10% and 16% as a result of projected climate impacts in South Africa (Cammarano *et al.*, 2020).

It has been shown that rain-fed maize yield in Malawi would reduce by 14% (2050s) and 33% (2090s) (Msowoya *et al.*, 2016). The authors used an ensemble of GCMs and two emission scenarios (SRA1B and SRB1). The decrease in maize yield in future would have effect on the food production and the socio-economic stability. Researchers have projected a decrease in maize yield in Zambia in the mid-century (2040–2069) using an ensemble of crop (APSIM-Maize and CERES-Maize) and climate (CCSM4, HadGEM2-ES, MPIESM-LR) models (Chisanga *et al.*, 2020). The model inter-comparison revealed that simulation of daily soil N supply and N leaching plays a crucial role in simulating climate change impacts for low-input systems (Islam *et al.*, 2012; Falconnier *et al.*, 2020). The use of multi-model ensembles of climate and crop models provides robust estimates of maize growth and yield (Chisanga *et al.*, 2020). Using an ensemble-based projection indicated that maize yield will reduce during 2020s, 2050s and 2080s in North China plain (IPCC, 2014).

The impacts of climate change in West Africa have been studied using multi-model inter-comparison (Waha *et al.*, 2015). The process-based crop models (EPIC, GEPIC, LPJ-GUESS, pDSSAT, PEGASUS) (Deryng, 2015), DSSAT and APSIM (Díaz-Ambrona *et al.*, 2013), SARRA-H and APSIM (Sultan *et al.*, 2014) using the same forcing climate dataset showed similar agreement in the sign of the crop yield response to climate change scenarios. Researchers (Chisanga *et al.*, 2020) have used a multi-model ensemble of APSIM-maize and DSSAT-CERES-maize. An ensemble of crop models has been used to assess the impact of heat stress and rainfall variability on maize yield using historical observations in West African Sudan-Sahel (Salack *et al.*, 2015). Crop water demand, biomass and grain yield were simulated for short-cycle maize cultivars under A1B and RCP8.5. The study projected an increase in temperature (0.6°C–0.8°C) and decrease in precipitation (10%–15%) and reduced grain yield during 2011–2050/1981–2010.

The response of maize growth and yield has been modelled in Sub-Saharan (Tanzania) and South Asia (Rosenzweig *et al.*, 2015). It has been reported that the inter-comparison of seven global gridded crop models (GGCMs) indicated that maize during 2070–2099/1980–2009 grown in lower latitudes has demonstrated greater vulnerability due to rising air temperatures (Rosenzweig *et al.*, 2015; Beletse *et al.*, 2015). Furthermore, global studies with adequate nitrogen fertilizer application showed steady decrease in maize yield in mid-and-high latitude regions with corresponding marginal temperature increase. Climate change will have negative impacts on maize yield for smallholder farmers in Sub-Saharan Africa (Rosenzweig *et al.*, 2013; Rosenzweig *et al.*, 2013). The projected changes in maize yield were analysed for 2010–2040/1980–2010, 2040–2070/1980–2010 and 2070–2100/1980–2010 under RCP4.5 and RCP8.5 using five GCMs (CCSM4, GFDL ESM2, HADGEM2 ES, MIROC5 and MPI ESM) (Beletse *et al.*, 2015). The simulated maize yield in India using DSSAT and APSIM models under RCP8.5 climate conditions (without adaptation) predicted decline in maize yield with varying degree (Ponnusamy *et al.*, 2015). The adaptation strategy proposed (Parkes *et al.*, 2018) is the growing of cultivar resistant to high-temperature stress during silking.

The use of multi-model ensembles of climate models and crop models has been recommended by researchers as they reduce the uncertainties (Kogo *et al.*, 2019). The application of multi-crop model ensemble improves prediction accuracy (Kostková *et al.*, 2021). Multi-model comparisons have been conducted for major staple crops like maize (Bassu *et al.*, 2014; Stöckle and Kemanian, 2020). The standard approach has been to calibrate many models in selected world sites with increasing level of experimental observations made available to modellers (Stöckle and Kemanian, 2020). The comparison of multi-crop model ensemble using mean or median outputs improves predictions (Stöckle and Kemanian, 2020). It has been pointed out that finding the minimum number of required models is not easy (Carter, 2013; Wallach *et al.*, 2018). Further, multi-model ensemble is not a substitute for model improvement (Stöckle and Kemanian, 2020).

Current state and quality of modelling

A meta-analysis of peer reviewed articles by Kogo *et al.* (2019) identified 23 crop simulation models with APSIM and DSSAT-CERES-maize models being the most prominent as shown in Table 24.1. Evaluation of model performance and improvement was considered under AgMIP (Rosenzweig *et al.*, 2013, 2018; Ruane *et al.*, 2018; Falconnier *et al.*, 2020). AgMIP focused on model improvements through protocol-based research on agricultural system components, including crop models, soils, and pests and diseases (Rosenzweig *et al.*, 2018) and evaluated protocol-based multi-model intercomparison at diverse, high-quality sites (Ruane *et al.*, 2017). Results showed mixed areas of positive

Table 24.1. Models used to simulate maize production under climate change. Adapted from Kogo *et al.* (2019) with permission.

Model	No. of articles	Model	No. of articles	Model	No. of articles
CERES-Maize	59	CROPWAT	2	SPACSYS	1
APSIM	30	ISAREG	2	FASSET	1
statistical	12	MAISPROQ	2	SICTOD	1
Cropyst	10	MaxEnt	2	LINTUL5	1
WOFOST	9	4 M crop	1	WINISAREG	1
AquaCrop	8	Agro-IBIS	1	GAEZ	1
DSSAT	8	AgroMetShell	1	MCWLA	1
InfoCrop	8	AirMS	1	SuperEPPS	1
EPIC	4	ARMOSA	1	SIMETAW	1
GLAM	3	BISM	1	CLIMEX	1
CMSM	3	WHCNS	1		

and negative simulated maize yield changes at global scale, with declines in some breadbasket regions, at both 1.5°C and 2.0°C (Rosenzweig *et al.*, 2018). Furthermore, regional simulations for 1.5°C and 2.0°C using site-based crop models had mixed results depending on the region and the crop.

Crop models have inherit limitations (Kogo *et al.*, 2019) and they have a variable skill in simulating crop responses to extreme climatic events (Droutsas *et al.*, 2019). The sources of uncertainty in crop model predictions are model parameterization, calibration, validation and evaluation procedures (Kogo *et al.*, 2019; Manschadi *et al.*, 2021). Furthermore, crop models require complete parameterization for improved accuracy in model predictions. Another challenge in modelling the impact of climate change on maize yield is the unavailability of reliable historical data for model calibrations (Kephe *et al.*, 2021). Scarcity of quality and quantity of input data, such as climate, soil, management practices and cultivar characteristics, hamper the models' effective role for projections. Other limitations of crop models are related to sensitivity to CO_2. This influences the simulated changes and remains an obstacle in assessing the impacts of climate change on crop growth and yield (Ruane *et al.*, 2013).

The abilities of 13 crop models to simulate four different field crops (winter wheat, spring barley, silage maize and winter oilseed rape) have been tested under European conditions (Kostková *et al.*, 2021). Modelling the number of days to anthesis and maturity was well simulated using the ensemble median (EnsMED) compared with the ensemble mean and individual models. The AQUACROP and EPIC models performed best in terms of spread around the line of best fit (RMSE, IA).

Crop growth models simulate crop yield at the field scale by explaining the relationships of crop growth with soil, genetic, management and environmental factors (Maharjan *et al.*, 2019; Leroux *et al.*, 2019; Chen *et al.*, 2021; Manschadi *et al.*, 2021). On the other hand, statistical models are used in estimating crop yield and the input data requirement is low (Pagani *et al.*, 2017; Chen *et al.*, 2021). Therefore, they have relatively high predictive skill with availability of sufficient data (Li *et al.*, 2019). However, empirical models also have shortcomings, such as collinearity between predictors due to sampling extrapolation problems (Schwalbert *et al.*, 2020; Chen *et al.*, 2021).

The major challenges and uncertainties in undertaking modelling studies in certain countries are associated with inadequate information, lack of continuous long-term crop data and facilities that can play a great role in defining research priorities on crop modelling (Di Paola *et al.*, 2016; Kogo *et al.*, 2019). As a result of insufficient modelling, there is inadequate information that can be passed on to stakeholders due to minimal research, lack of access and uncertainties associated with data used in calibration (Whitfield, 2013). Additionally, the accurate and timely maize yield prediction is challenging (Chen *et al.*, 2021).

Major methods, application and possible future directions

The recommended methods for reducing uncertainties include complete model parameterization, utilization of remotely sensed data, machine learning, field and greenhouse experimental data, secondary data and, engaging with farmers to model actual on-farm conditions (Manschadi *et al.*, 2021; Kephe *et al.*, 2021; Chen *et al.*, 2021). Using different methods for model input data calibration and validation can reduce the challenges faced by crop modellers due to non-availability of data (Kephe *et al.*, 2021). Furthermore, the future of modelling depends on the goodness and availability of quality and quantity model input data, the readiness of modellers to cooperate on modularity and standardization, and potential user groups' ability to communicate (Kephe *et al.*, 2021; Chen *et al.*, 2021). The most common adaptation options that have been proposed by crop modellers are changing the sowing dates, crop rotation with legumes, breeding long season cultivars, optimizing N application and crop water management practices (Kogo *et al.*, 2019; Jiang *et al.*, 2021).

Conclusions

Different types of crop simulation models have been used to predict crop growth and yield. These models are statistical, deterministic, dynamic, mechanistic and static in nature. Globally, climate change is affecting the agriculture production and productivity differently. There is need to use improved management practice and germplasm that are tolerant to heat stress and drought as an adaptation strategy. To enhance maize production and productivity, supplemental irrigation, climate smart agriculture, increased fertilizer application, use of improved long-season cultivars and crop diversification are promising adaptation strategies that would improve food security. Crop breeding for high-temperature conditions is a promising option for climate change adaptation. The AgMIP global-gridded crop model showed a decrease in maize yield even with adequate fertilization. Crop simulation models have been observed to give mixed results depending on the region and the crop. Sources of uncertainty in predictions have been attributed to challenges in model parameterization, calibration and validation. Changing sowing dates, crop rotation with legumes, breeding long season cultivars, optimizing N application and crop water management practices have been proposed as adaptation options.

Author contributions

All authors contributed to the conception of the review paper. CBC wrote the paper and incorporated the reviewer comments. MM wrote the introduction, methodology and review of other sections as well. EP edited the paper and made required corrections.

Acknowledgments

The authors would like to thank CABI for being requested to prepare a paper on modelling the impact of climate change on maize yield.

References

Abiodun, B.J. and Adedoyin, A. (2016) A modelling perspective of future climate change. In: *Climate Change*. Elsevier, Amsterdam, Netherlands, pp. 355–371.

Adejuwon, J.O. (2006) Food crop production in Nigeria. II. Potential effects of climate change. *Climate Research* 32, 229–245.

Alexandrov, V.A. and Hoogenboom, G. (2001) Climate variation and crop production in Georgia, USA, during the twentieth century. *Climate Research* 17, 33–43.

Ali, W., Ali, M., Ahmad, Z., Iqbal, J., Anwar, S. *et al.* (2018) Influence of sowing dates on varying maize (*Zea mays* l.) varieties grown under agro-climatic condition of Peshawar. *European Journal of Experimental Biology* 8(6), 36. DOI: 10.21767/2248-9215.100077.

Araya, A., Hoogenboom, G., Luedeling, E., Hadgu, K.M., Kisekka, I. *et al.* (2015) Assessment of maize growth and yield using crop models under present and future climate in southwestern Ethiopia. *Agricultural and Forest Meteorology* 214, 252–265. DOI: 10.1016/j.agrformet.2015.08.259.

Bassu, S., Brisson, N., Durand, J.L., Boote, K., Lizaso, J. *et al.* (2014) How do various maize crop models vary in their responses to climate change factors. *Global Change Biology* 20(7), 2301–2320. DOI: 10.1111/gcb.12520.

Beletse, Y.G., Durand, W. and Nhemachena, C. (2015) Projected impacts of climate change scenarios on the production of maize in Southern Africa: An integrated assessment case study of the Bethlehem District, Central Free State, South Africa. In: *Handbook of Climate Change and Agroecosystems the Agricultural Model Intercomparison and Improvement Project Integrated Crop and Economic Assessments, Part 1*. Joint Publication with American Society of Agronomy, Crop Science Society of America, and Soil Science Society of America, London, UK, pp. 125–155.

Cammarano, D., Valdivia, R.O., Beletse, Y.G., Durand, W., Crespo, O. *et al.* (2020) Integrated assessment of climate change impacts on crop productivity and income of commercial maize farms in northeast South Africa. *Food Security* 12(3), 659–678. DOI: 10.1007/s12571-020-01023-0.

Carter, TR. (2013) Multi-model yield projections. *Nature Climate Change* 3(9), 784–786. DOI: 10.1038/nclimate1995.

Challinor, A.J., Ewert, F., Arnold, S., Simelton, E. and Fraser, E. (2009) Crops and climate change: Progress, trends, and challenges in simulating impacts and informing adaptation. *Journal of Experimental Botany* 60(10), 2775–2789. DOI: 10.1093/jxb/erp062.

Chen, X., Wang, L., Niu, Z., Zhang, M. and Li, J. (2020) The effects of projected climate change and extreme climate on maize and rice in the Yangtze River Basin, China. *Agricultural and Forest Meteorology* 282–283, 107867. DOI: 10.1016/j.agrformet.2019.107867.

Chen, X., Feng, L., Yao, R., Wu, X., Sun, J. *et al.* (2021) Prediction of maize yield at the city level in China using multi-source data. *Remote Sensing* 13(1), 146. DOI: 10.3390/rs13010146.

Chisanga, C.B. (2019) Interactive effects of N fertilization rate, cultivars and planting date under climate change on maize (*Zea mays* L.) yield using crop simulation and statistical downscaling of climate models. Ph.D. thesis, University of Zambia, Lusaka, Zambia.

Chisanga, C.B., Phiri, E., Chinene, V.R. and Chabala, L.M. (2020) Projecting maize yield under local-scale climate change scenarios using crop models: Sensitivity to sowing dates, cultivar, and nitrogen fertilizer rates. *Food and Energy Security* 9(4), 1–17. DOI: 10.1002/fes3.231.

Chisanga, C.B., Phiri, E. and Chinene, V.R.N. (2017) Statistical downscaling of precipitation and temperature using Long Ashton Research Station weather generator in Zambia: A case of Mount Makulu agriculture research station. *American Journal of Climate Change* 06(3), 487–512. DOI: 10.4236/ajcc.2017.63025.

Chisanga, C.B., Phiri, E. and Chinene, V.R.N. (2020) Reliability of rain-fed maize yield simulation using LARS-WG derived CMIP5 climate data at Mount Makulu, Zambia. *Journal of Agricultural Science* 12(11), 275. DOI: 10.5539/jas.v12n11p275.

Choudhary, D., Patel, H.R., Yadav, S.B. and Pandey, V. (2014) Assessment of climate change under A2 scenario and its effect on Rabi maize (*Zea mays* L.) yield using InfoCrop model for Dahod District of Middle Gujarat Agro Climatic Zone. *Annals of Biology* 30, 509–514.

Choudhury, A.H., Jones, J.R. and Spaulding, A.D. (2015) Association of rainfall and detrended crop yield based on piecewise regression. *Journal of Economics & Economic Education Research* 16, 13. Available at: http://www.alliedacademies.org/articles/ association-of-rainfall-and-detrended-crop-yield-based-on-piecewise-regression-for-agricultural-insurance.pdf

Deryng, D. (2015) *Climate Change Impacts on Crop Productivity in Global Semi-arid Areas and Selected Semi-arid Economies Working Paper*. Burkina Faso, Sanmatenga Province.

Di Paola, A., Valentini, R. and Santini, M. (2016) An overview of available crop growth and yield models for studies and assessments in agriculture. *Journal of Science and Food Agriculture* 96, 709–714. DOI: 10.1002/jsfa.7359.

Droutsas, I., Challinor, A.J., Swiderski, M. and Semenov, M.A. (2019) New modelling technique for improving crop model performance—application to the GLAM model. *Environmental Modelling and Software* 118, 187–200. DOI: 10.1016/j.envsoft.2019.05.005.

du Plessis, J. (2003) Maize production. In: *Resource Centre Directorate Agricultural Information Services. Pretoria, South Africa: Directorate Agricultural Information Services*. Directorate Agricultural Information Services, Department of Agriculture , Pretoria, South Africa, pp. 1–38. Available at: http://www.arc.agric.za/arc-gci/FactSheetsLibrary/Maize Production.pdf

Durdu, ÖF. (2013) Evaluation of climate change effects on future corn (*Zea mays* L.) yield in western Turkey. *International Journal of Climatology* 33(2), 444–456. DOI: 10.1002/joc.3435.

Díaz-Ambrona, C.G.H., Gigena, R. and Mendoza, CO. (2013) Climate change impacts on maize and dry bean yields of smallholder farmers in Honduras. *Iberoamerican Journal of Development Studies* 2, 4–22. DOI: 10.26754/ojs_ried/ijds.43.

Falconnier, G.N., Corbeels, M., Boote, K.J., Affholder, F., Adam, M. *et al.* (2020) Modelling climate change impacts on maize yields under low nitrogen input conditions in sub-Saharan Africa. *Global Change Biology* 26(10), 5942–5964. DOI: 10.1111/gcb.15261.

Flato, G., Marotzke, J., Abiodun, B., Braconnot, P., Chou, S.C. *et al.* (2013) Evaluation of climate models. In: Edenhofer, T.F., Qin, D., Plattner, G.-K., Tignor, M., Allen, S.K. *et al.* (eds) *Climate Change 2013: The Physical Science Basis. The Contribution of Working Group I to the Fifth Assessment Report of the Intergovernmental Panel on Climate Change*. Cambridge University Press, Cambridge, UK, pp. 741–866. DOI: 10.1017/CBO9781107415324.

Gaydon, D.S., Wang, E., Poulton, P.L., Ahmad, B., Ahmed, F. *et al.* (2017) Evaluation of the APSIM model in cropping systems of Asia. *Field Crops Research* 204, 52–75. DOI: 10.1016/j.fcr.2016.12.015.

Glotter, M., Elliott, J., McInerney, D., Best, N., Foster, I. *et al.* (2014) Evaluating the utility of dynamical downscaling in agricultural impacts projections. *PNAS* 111(24), 8776–8781. DOI: 10.1073/pnas.1314787111.

Hatfield, J.L. and Prueger, JH. (2015) Temperature extremes: Effect on plant growth and development. *Weather and Climate Extremes* 10, 4–10. DOI: 10.1016/J.WACE.2015.08.001.

Hillel, D. and Rosenzweig, C. (eds) (2015) *Handbook of Climate Change and Agroecosystems. The Agricultural Model Intercomparison and Improvement Project Integrated Crop and Economic Assessments*. Imperial College Press, London.

Holden, N.M. and Brereton, AJ. (2003) Potential impacts of climate change on maize production and the introduction of soybean in Ireland. *Irish Journal of Agricultural and Food Research* 42(1), 1–15. DOI: https://www.jstor.org/stable/25562473.

Holzworth, D.P., Snow, V., Janssen, S., Athanasiadis, I.N., Donatelli, M. *et al.* (2015) Agricultural production systems modelling and software: Current status and future prospects. *Environmental Modelling and Software* 72, 276–286. DOI: 10.1016/j.envsoft.2014.12.013.

Hoogenboom, G., Wilkens, P.W. and Tsuji, G.Y. (eds) (1999) *DSSAT v3*, Vol. 4. University of Hawaii, Honolulu, Hawaii.

IPCC (2014) *Climate Change 2014: Impacts, Adaptation, and Vulnerability. Part A: Global and Sectoral Aspects. Contribution of Working Group II to the Fifth Assessment Report of the Intergovernmental Panel on Climate Change*. Cambridge University Press, Field, C.B. and Barros, V.R. (eds) Cambridge, United Kingdom and New York, NY, USA. DOI: 10.1007/s13398-014-0173-7.2.

IPCC (2007) *Climate Change 2007: The Physical Science Basis. Contribution of Working Group I to the Fourth Assessment Report of the Intergovernmental Panel on Climate Change. Structure*. Solomon, S., Qin, D., Manning, M., Chen, Z., Marquis, M. *et al.* (eds) Cambridge University Press, Cambridge, United Kingdom and New York, NY, USA, p. 976.

IPPC (2007) *Climate Change 2007: Impacts, Adaptation and Vulnerability. Contribution of Working Group II to the Fourth Assessment Report of the Intergovernmental Panel on Climate Change*. Parry, M., Canziani, O.F., Palutikof, J.P., van der Linden, P.J., Hanson, C.E. *et al.* (eds) Cambridge University Press, Cambridge.

Iqbal, M.A., Shen, Y., Stricevic, R., Pei, H., Sun, H. *et al.* (2014) Evaluation of the FAO AquaCrop model for winter wheat on the north China plain under deficit irrigation from field experiment to regional yield simulation. *Agricultural Water Management* 135, 61–72. DOI: 10.1016/j.agwat.2013.12.012.

Islam, A., Ahuja, L.R., Garcia, L.A., Ma, L., Saseendran, A.S. *et al.* (2012) Modeling the impacts of climate change on irrigated corn production in the Central Great Plains. *Agricultural Water Management* 110, 94–108. DOI: 10.1016/j.agwat.2012.04.004.

Jacob, D. (2009) Regional climate models: linking global climate change to local impacts. In: Meyers, R. (ed.) *Encyclopedia of Complexity and Systems Science*. Springer, New York, NY, USA. DOI: 10.1007/978-0-387-30440-3_449.

Jarvis, A. and Thornton, P. (2014) *Beyond the climate science: CCAFS downscaled climate data applied by development agencies around the world*. CIAT, Rome, Italy, p. 22.

Jeong, J.H., Resop, J.P., Mueller, N.D., Fleisher, D.H., Yun, K. *et al.* (2016) Random forests for global and regional crop yield predictions. *PLoS ONE* 11(6), 1–15. DOI: 10.1371/journal.pone.0156571.

Jiang, R., He, W., He, L., Yang, J.Y., Qian, B. *et al.* (2021) Modelling adaptation strategies to reduce adverse impacts of climate change on maize cropping system in Northeast China. *Scientific Reports* 11(810), 1–13. DOI: 10.1038/s41598-020-79988-3.

Jones, P.G. and Thornton, PK. (2003) The potential impacts of climate change on maize production in Africa and Latin America in 2055. *Global Environmental Change* 13(1), 51–59. DOI: 10.1016/S0959-3780(02)00090-0.

Jones, J.W., Hoogenboom, G., Porter, C.H., Boote, K.J., Batchelor, W.D. *et al.* (2003) The DSSAT cropping system model. *European Journal of Agronomy* 18(3–4), 235–265. DOI: 10.1016/S1161-0301(02)00107-7.

Kadigi, I.L., Richardson, J.W., Mutabazi, K.D., Philip, D., Mourice, S.K. *et al.* (2020) The effect of nitrogen-fertilizer and optimal plant population on the profitability of maize plots in the Wami River sub-basin, Tanzania: A bio-economic simulation approach. *Agricultural Systems* 185, 102948. DOI: 10.1016/j.agsy.2020.102948.

Kapetanaki, G. and Rosenzweig, C. (1997) Impact of climate change on maize yield in central and northern Greece: A simulation study with CERES-maize. *Mitigation and Adaptation Strategies for Global Change* 1(3), 251–271. DOI: 10.1007/BF00517806.

Keating, B.A., Robertson, M.J., Muchow, R.C. and Huth, N.I. (1999) Modelling sugarcane production systems I. development and performance of the sugarcane module. *Field Crops Research* 61(3), 253–271.

Kephe, P.N., Ayisi, K.K. and Petja, BM. (2021) Challenges and opportunities in crop simulation modelling under seasonal and projected climate change scenarios for crop production in South Africa. *Agriculture and Food Security* 10(10), 1–24. DOI: 10.1186/s40066-020-00283-5.

Khaliq, T. (2008) Modeling the impact of climate change on maize (*Zea mays* L.). productivity in the Punjab. University of Agriculture, Faisalabad, Pakistan.

Khan, M.I. and Walker, D. (2015) Application of crop growth simulation models in agriculture with special reference to water management planning. *Application of Crop Growth Simulation Models in Agriculture with Special Reference to Water Management Planning* 2(5), 18.

Khubaib, N., Asad, S.A., Khalil, T., Baig, A., Atif, S. *et al.* (2021) Predicting areas suitable for wheat and maize cultivation under future climate change scenarios in Pakistan. *Climate Research* 83, 15–25. DOI: 10.3354/cr01631.

Kiniry, J.R., Rosenthal, W.D., Jackson, B.S. and Hoogenboom, G. (1991) Predicting leaf development of crop plants. In: *Predicting Crop Phenology*. CRC Press, Boca Raton, FL, pp. 29–42. Available at: https://www.ars.usda.gov/ARSUserFiles/30980500/KiniryChapter5.pdf

Knox, J., Hess, T., Daccache, A. and Wheeler, T. (2012) Climate change impacts on crop productivity in Africa and South Asia. *Environmental Research Letters* 7(3), 034032. DOI: 10.1088/1748-9326/7/3/034032.

Kogo, B.K., Kumar, L., Koech, R. and Langat, P. (2019) Modelling impacts of climate change on maize (*Zea mays* L.) growth and productivity: A review of models, outputs and limitations. *Journal of Geoscience and Environment Protection* 07(8), 76–95. DOI: 10.4236/gep.2019.78006.

Kostková, M., Hlavinka, P., Pohanková, E., Kersebaum, K.C., Nendel, C. *et al.* (2021) Performance of 13 crop simulation models and their ensemble for simulating four field crops in central Europe. *Journal of Agricultural Science* 159, 69–89. DOI: 10.1017/S0021859621000216.

Lee, D.-K. and Cha, D.-H. (2020) Regional climate modeling for Asia. *Geoscience Letters* 7(13), 12. DOI: 10.1186/s40562-020-00162-8.

Lenderink, G., Ulden, A., Hurk, B. and Keller, F. (2007) A study on combining global and regional climate model results for generating climate scenarios of temperature and precipitation for the Netherlands. *Climate Dynamics* 29, 157–176. DOI: 10.1007/s00382-007-0227-z.

Leroux, L., Castets, M., Baron, C., Escorihuela, M.J., Bégué, A. *et al.* (2019) Maize yield estimation in West Africa from crop process-induced combinations of multi-domain remote sensing indices. *European Journal of Agronomy* 108, 11–26. DOI: 10.1016/j.eja.2019.04.007.

Li, Y., Guan, K., Yu, A., Peng, B., Zhao, L. *et al.* (2019) Toward building a transparent statistical model for improving crop yield prediction: Modeling rainfed corn in the U.S. *Field Crops Research* 234, 55–65. DOI: 10.1016/j.fcr.2019.02.005.

Lin, Y., Feng, Z., Wu, W., Yang, Y., Zhou, Y. *et al.* (2017) Potential impacts of climate change and adaptation on maize in Northeast China. *Agronomy Journal* 109, 1476–1490. DOI: 10.2134/agronj2016.05.0275.

Liu, Y., Heuvelink, G.B., Bai, Z., He, P., Xu, X. *et al.* (2021) Analysis of spatio-temporal variation of crop yield in China using stepwise multiple linear regression. *Field Crops Research* 264, 108098. DOI: 10.1016/j.fcr.2021.108098.

Lobell, D.B. and Burke, MB. (2010) On the use of statistical models to predict crop yield responses to climate change. *Agricultural and Forest Meteorology* 150, 10. DOI: 10.1016/j.agrformet.2010.07.008.

Luhunga, PM. (2017) Assessment of the impacts of climate change on maize production in the southern and western highlands sub-agro ecological zones of Tanzania. *Frontiers in Environmental Science* 5(51), 1–16. DOI: 10.3389/fenvs.2017.00051.

Maharjan, G.R., Hoffmann, H., Webber, H., Srivastava, A.K., Weihermüller, L. *et al.* (2019) Effects of input data aggregation on simulated crop yields in temperate and mediterranean climates. *European Journal of Agronomy* 103, 32–46. DOI: 10.1016/j.eja.2018.11.001.

Makadho, J.M. (1996) Potential effects of climate change on corn production in Zimbabwe. *Climate Research* Available from, 147–151. Available at: http://libcatalog.cimmyt.org/download/reprints/97135.pdf (accessed 7 July 2016).

Manschadi, A.M., Eitzinger, J., Breisch, M., Fuchs, W., Neubauer, T. *et al.* (2021) Full parameterisation matters for the best performance of crop models: inter-comparison of a simple and a detailed maize model. *International Journal of Plant Production* 15(1), 61–78. DOI: 10.1007/s42106-020-00116-2.

Martre, P., Wallach, D., Asseng, S., Ewert, F., Jones, J.W. *et al.* (2015) Multimodel ensembles of wheat growth: Many models are better than one. *Global Change Biology* 21(2), 911–925. DOI: 10.1111/gcb.12768.

Matthews, R. and Stephens, W. (2002) *Crop-Soil Simulation Models Applications in Developing Countries*. CABI Publishing, UK.

McSweeney, C.F., Jones, R.G. and Booth, BBB. (2012) Selecting ensemble members to provide regional climate change information. *Journal of Climate* 25(20), 7100–7121. DOI: 10.1175/JCLI-D-11-00526.1.

Monteiro, J.E.B.D.A., Assad, E.D., Sentelhas, P.C. and Azevedo, L.D.C. (2017) Modeling of corn yield in Brazil as a function of meteorological conditions and technological level. *Pesquisa Agropecuária Brasileira* 52(3), 137–148. DOI: 10.1590/S0100-204X2017000300001.

Msowoya, K., Madani, K., Davtalab, R., Mirchi, A. and Lund, J.R. (2016) Climate change impacts on maize production in the warm heart of Africa. *Water Resources Management* 30, 5299–5312. DOI: 10.1007/s11269-016-1487-3.

Muchow, R.C., Sinclair, T.R. and Bennett, J.M. (1990) Temperature and solar radiation effects on potential maize yield across locations. *Agronomy Journal* 338–343. Available at: https://www.agronomy.org/publications/aj/pdfs/82/2/AJ0820020338

Nkomozepi, T.D. and Chung, S.-O. (2013) Uncertainty of simulated paddy rice yield using LARS-WG derived climate data in the Geumho River Basin, Korea. *Journal of the Korean Society of Agricultural Engineers* 55(4), 55–63. DOI: 10.5389/KSAE.2013.55.4.055.

NSW (2009) *Maize Growth and Development*. NSW Department of Primary Industries, State of New South Wales. Edwards, J. (ed.). Available at: https://www.dpi.nsw.gov.au/__data/assets/pdf_file/0007/516184/Procrop-maize-growth-and-development.pdf (accessed 8 July 2016).

Ouda, S.A., Noreldin, T., Mounzer, O.H. and Abdelhamid, M.T. (2015) CropSyst model for wheat irrigation water management with fresh and poor quality water. *Journal of Water and Land Development*, 1429–7426. DOI: 10.1515/jwld-2015-0023.

Pagani, V., Guarneri, T., Fumagalli, D., Movedi, E., Testi, L. *et al.* (2017) Improving cereal yield forecasts in Europe—the impact of weather extremes. *European Journal of Agronomy* 89(97–106), 06. DOI: 10.1016/j.eja.2017.

Parkes, B., Defrance, D., Sultan, B., Ciais, P. and Wang, X. (2018) Projected changes in crop yield mean and variability over West Africa in a world 1.5 K warmer than the pre-industrial era. *Earth System Dynamics* 9, 119–134. DOI: 10.5194/esd-9-119-2018.

Parkes, B., Sultan, B. and Ciais, P. (2018) The impact of future climate change and potential adaptation methods on maize yields in West Africa. *Climatic Change* 151(2), 205–217. DOI: 10.1007/s10584-018-2290-3.

Paudel, D., Boogaard, H., Wit, A., Janssen, S., Osinga, S. *et al.* (2021) Machine learning for large-scale crop yield forecasting. *Agricultural Systems* 187, 103016. DOI: 10.1016/j.agsy.2020.103016.

Ponnusamy, P., Vellingiri, G., Danda, R.R., Arunachalam, L., Murthy, D. *et al.* (2015) Integrated assessment of climate change impacts on maize farms and farm household incomes in south India: A case study from Tamil Nadu. In: *Handbook of Climate Change and Agroecosystems*. Imperial College Press, London, UK, pp. 281–313.

Ramirez-Cabral, N.Y.Z., Kumar, L. and Shabani, F. (2017) Global alterations in areas of suitability for maize production from climate change and using a mechanistic species distribution model (CLIMEX). *Scientific Reports* 7, 1–13. DOI: 10.1038/s41598-017-05804-0.

Ramirez-Villegas, J. and Jarvis, A. (2010) Downscaling global circulation model outputs: the delta method. In: *CIAT Decision and Policy Analysis Working Paper. 1*. Cali, Colombia. Available at: http://www.ipcc-data.org/gcm/monthly/SRES_AR4/index.html%5Cnhttp://www.ccafs-climate.org/

Rauff, K.O. and Bello, R. (2015) A review of crop growth simulation models as tools for agricultural meteorology. *Agricultural Sciences* 6, 1098–1105.

Ravindranath, N.H. and Sathaye, J.A. (2003) Climate change and developing countries. In: *Advances in Global Climate Research*, Vol. 11. KluwerAcademic Publishers, NewYork, Boston, Dordrecht, London, Moscow.

Robertson, R., Nelson, G., Thomas, T. and Rosegrant, M. (2012) Incorporating process-based crop simulation models into global economic analyses. *American Journal of Agricultural Economics* 95(2), 228–235. DOI: 10.1093/ajae/aas034.

Rosenzweig, C. (1990) Crop response to climate change in the great plains: A simulation study. *The Professional Geographer* 42(1), 20–37. DOI: 10.1111/j.0033-0124.1990.00020.x.

Rosenzweig, C. and Hillel, D. (2015) Major findings and future activities. In: *Handbook of Climate Change and Agroecosystems*. Imperial College Press, London, UK, pp. 411–426.

Rosenzweig, C., Jones, J.W., Hatfield, J.L., Ruane, A.C., Boote, K.J. *et al.* (2013) The agricultural model intercomparison and improvement project (AgMIP): Protocols and pilot studies. *Agricultural and Forest Meteorology* 170, 166–182. DOI: 10.1016/j.agrformet.2012.09.011.

Rosenzweig, C., Elliott, J., Deryng, D., Ruane, A.C., Müller, C. *et al.* (2013) Assessing agricultural risks of climate change in the 21st century in a global gridded crop model intercomparison. *Proceedings of the National Academy of Sciences of the United States of America* 111, 3268–3273.

Rosenzweig, C., Jones, J.W., Hatfield, J.L., Antle, J.M., Ruane, A.C. *et al.* (2015) The agricultural model intercomparison and improvement project: Phase I activities by a global community of science. In: *Handbook of Climate Change and Agroecosystems*. American Society of Agronomy, Crop Science Society of America, and Soil Science Society of America, London, UK, p. 22.

Rosenzweig, C., Elliott, J., Deryng, D., Ruane, A.C., Müller, C. *et al.* (2014) Assessing agricultural risks of climate change in the 21st century in a global gridded crop model intercomparison. *Proceedings of the National Academy of Sciences* 111(9), 3268–3273. DOI: 10.1073/pnas.1222463110.

Rosenzweig, C., Ruane, A.C., Antle, J., Elliott, J., Ashfaq, M. *et al.* (2018) Coordinating AgMIP data and models across global and regional scales for 1.5°C and 2.0°C assessments. *Philosophical Transactions of the Royal Society A: Mathematical, Physical and Engineering Sciences* 376(2119), 20160455. DOI: 10.1098/rsta.2016.0455.

Rosenzweig, C.E., Jones, J.W., Hatfield, J., Antle, J., Ruane, A. *et al.* (2014) *Guide for Regional Integrated Assessments: Handbook of Methods and Procedures Version 5.1*. Center for Climate Systems Research, Earth Institute, Columbia University, New York.

Ruane, A.C., Goldberg, R. and Chryssanthacopoulos, J. (2015) Climate forcing datasets for agricultural modeling: Merged products for gap-filling and historical climate series estimation. *Agricultural and Forest Meteorology* 200, 233–248. DOI: 10.1016/j.agrformet.2014.09.016.

Ruane, A.C., Cecil, L.D., Horton, R.M., Gordón, R., McCollum, R. *et al.* (2013) Climate change impact uncertainties for maize in Panama: Farm information, climate projections, and yield sensitivities. *Agricultural and Forest Meteorology* 170, 132–145. DOI: 10.1016/j.agrformet.2011.10.015.

Ruane, A.C., Rosenzweig, C., Asseng, S., Boote, K.J., Elliott, J. *et al.* (2017) An AgMIP framework for improved agricultural representation in integrated assessment models. *Environmental Research Letters* 12(12), 125003. DOI: 10.1088/1748-9326/aa8da6.

Ruane, A.C., Antle, J., Elliott, J., Folberth, C., Hoogenboom, G. *et al.* (2018) Biophysical and economic implications for agriculture of +1.5° and +2.0°C global warming using AgMIP coordinated global and regional assessments. *Climate Research* 76(1), 17–39. DOI: 10.3354/cr01520.

Salack, S., Sarr, B., Sangare, S.K., Ly, M., Sanda, I.S. *et al.* (2015) Crop-climate ensemble scenarios to improve risk assessment and resilience in the semi-arid regions of West Africa. *Climate Research* 65, 107–121. DOI: 10.3354/cr01282.

Schwalbert, R.A., Amado, T., Corassa, G., Pott, L.P., Prasad, P.V. *et al.* (2020) Satellite-based soybean yield forecast: Integrating machine learning and weather data for improving crop yield prediction in southern Brazil. *Agricultural and Forest Meteorology* 284, 107886. DOI: 10.1016/j.agrformet.2019.107886.

Schlenkera, W. and Roberts, MJ. (2009) Nonlinear temperature effects indicate severe damages to U.S. crop yields under climate change. *PNAS* 106(37), 15594–15598. DOI: 10.1073/pnas.0906865106.

Schlenker, W. and Lobell, DB. (2010) Robust negative impacts of climate change on African agriculture. *Environmental Research Letters* 5(1), 014010. DOI: 10.1088/1748-9326/5/1/014010.
Semenov, M.A. and Brooks, RJ. (1999) Spatial interpolation of the LARS-WG stochastic weather generator in Great Britain. *Climate Research* 11, 137–148. DOI: 10.3354/cr011137.
Siatwiinda, S.M., Supit, I., van Hove, B., Yerokun, O., Ros, G.H. *et al.* (2021) Climate change impacts on rainfed maize yields in Zambia under conventional and optimized crop management. *Climatic Change* 167(3–4), 39. DOI: 10.1007/s10584-021-03191-0.
Snyder, H. (2019) Literature review as a research methodology: An overview and guidelines. *Journal of Business Research* 104, 333–339.
Stöckle, C.O. and Kemanian, AR. (2020) Can crop models identify critical gaps in genetics, environment, and management interactions. *Frontiers in Plant Science* 11, 1–12. DOI: 10.3389/fpls.2020.00737.
Sultan, B. and Gaetani, M. (2016) Agriculture in West Africa in the twenty-first century: Climate change and impacts scenarios, and potential for adaptation. *Frontiers in Plant Science* 7, 1–20. DOI: 10.3389/fpls.2016.01262.
Sultan, B., Roudier, P., Quirion, P., Alhassane, A., Muller, B. *et al.* (2013) Assessing climate change impacts on sorghum and millet yields in the Sudanian and Sahelian savannas of West Africa. *Environmental Research Letters* 8(1), 014040. DOI: 10.1088/1748-9326/8/1/014040.
Sultan, B., Guan, K., Kouressy, M., Biasutti, M., Piani, C. *et al.* (2014) Robust features of future climate change impacts on sorghum yields in West Africa. *Environmental Research Letters* 10, 104006. DOI: 10.1088/1748-9326/9/10/104006.
Thornton, P.K., Jones, P.G., Ericksen, P.J. and Challinor, A.J. (2011) Agriculture and food systems in sub-Saharan Africa in a 4°C+ world. *Philosophical Transactions of the Royal Society A* 369(1934), 117–136. DOI: 10.1098/rsta.2010.0246.
Tingem, M., Rivington, M., Azam-Ali, S. and Colls, J. (2007) Assessment of the ClimGen stochastic weather generator at Cameroon sites. *African Journal of Environmental Science and Technology* 1(4), 86–92. Available at: http://www.academicjournals.org/AJest (accessed 17 June 2016).
Tovihoudji, P.G., Akponikpè, P.I., Agbossou, E.K. and Bielders, C.L. (2019) Using the DSSAT model to support decision making regarding fertilizer microdosing for maize production in the sub-humid region of Benin. *Frontiers in Environmental Science* 7, 1–15. DOI: 10.3389/fenvs.2019.00013.
Trnka, M., Olesen, J.E., Kersebaum, K.C., Rötter, R.P., Brázdil, R. *et al.* (2016) Changing regional weather – crop yield relationships across Europe between 1901 and 2012. *Climate Research* 70, 195–214. DOI: 10.3354/cr01426.
Tui, S.H.K., Descheemaeker, K., Valdivia, R.O., Masikati, P., Sisito, G. *et al.* (2021) Climate change impacts and adaptation for dryland farming systems in Zimbabwe: A stakeholder-driven integrated multi-model assessment. *Climatic Change* 168(1–2), 10. DOI: 10.1007/s10584-021-03151-8.
UNFCCC (2012) Agriculture. In: *CGE Training Materials for Vulnerability and Adaptation Assessment*. United Nations Framework Convention on Climate Change, Washington, DC, USA, p. 67.
Vitali, G., Rogosin, S. and Baldoni, G. (2019) Climate change and grain production fluctuations. *arXiv Preprint* arXiv:2002.07039.
Waha, K., Huth, N., Carberry, P. and Wang, E. (2015) How model and input uncertainty impact maize yield simulations in West Africa. *Environmental Research Letters* 10, 1–11. DOI: 10.1088/1748-9326/10/2/024017.
Wallach, D., Mearns, L.O., Ruane, A.C., Rötter, R.P. and Asseng, S. (2016) Lessons from climate modeling on the design and use of ensembles for crop modeling. *Climatic Change* 139(3–4), 551–564. DOI: 10.1007/s10584-016-1803-1.
Wallach, D., Martre, P., Liu, B., Asseng, S., Ewert, F. *et al.* (2018) Multimodel ensembles improve predictions of crop–environment–management interactions. *Global Change Biology* 24, 5072–5083. DOI: 10.1111/gcb.14411.
Waongo, M., Laux, P. and Kunstmann, H. (2015) Adaptation to climate change: The impacts of optimized planting dates on attainable maize yields under rainfed conditions in Burkina Faso. *Agricultural and Forest Meteorology* 205, 23–39. DOI: 10.1016/j.agrformet.2015.02.006.
Whitfield, S. (2013) Uncertainty, ignorance and ambiguity in crop modelling for African agricultural adaptation. *Climatic Change* 120(1–2), 325–340. DOI: 10.1007/s10584-013-0795-3.
Yang, H.S., Dobermann, A., Lindquist, J.L., Walters, D.T., Arkebauer, T.J. *et al.* (2004) Hybrid-maize—A maize simulation model that combines two crop modeling approaches. *Field Crops Research* 87, 131–154. DOI: 10.1016/j.fcr.2003.10.003.

Yin, X., Kropff, M.J., McLaren, G. and Visperas, R.M. (1995) A nonlinear model for crop development as a function of temperature. *Agricultural and Forest Meteorology* 77(1–2), 1–16.

Zhao, C., Liu, B., Piao, S., Wang, X., Lobell, D.B. *et al.* (2017) Temperature increase reduces global yields of major crops in four independent estimates. *PNAS* 35, 1–6. DOI: 10.1073/pnas.1701762114.

25 Harnessing soil bacteria and their benefits for sustainable agriculture with changing climate

Rohini Mattoo[1]* and Malali Gowda[2]
[1]Divecha Centre for Climate Change, Indian Institute of Science, Bangalore 560012, India; [2]The University of Trans Disciplinary Health Sciences and Technology, University, Bangalore 560064, India

Abstract

Soil bacteria contribute effectively to key biogeochemical reactions in the soil rhizosphere. They support plants in the rhizosphere to adapt quickly to changing climatic conditions. Differences in root exudates, trace gas chemistry, chemical compounds and nutrient exchange contribute to the recruitment of diverse microorganisms by plant roots. This review highlights the importance of characterizing novel microorganisms to support sustainable agricultural practices. We discuss about tools for characterizing microbes and agricultural practices that influence microbial diversity, and have reviewed how microorganisms may have important but unidentified roles in climate change. Beneficial microbes could improve the turnover of carbon, nitrogen, phosphorus and other minerals thereby avoiding the use of chemical inputs, which are not only causing serious environmental harm but also pose danger to human and animal health

Review Methodology: The review methodology involves identification of manuscripts related to the topic of discussion using existing literature on PubMed, Google Scholar and important reviews along with references taken from important reviews.

Keywords: sustainable climate-resilient agriculture, beneficial soil bacteria, microbial diversity

Introduction

Microorganisms are key players in sustaining life due to their ancient, ubiquitous and important roles across all habitats of the Earth. Microorganisms contribute to an enriched nutrient cycle in the rhizosphere (Ren *et al.*, 2021), aid in disease resistance and improve adaptation to environmental stresses (Tewari and Sharma, 2020; French *et al.*, 2021; Lazcano *et al.*, 2021) through interactions with plant roots and their exudates. Natural and agricultural soils are challenging niches where organisms, the majority of which remain undiscovered, collectively contribute to a variety of soil functions (Wardle *et al.*, 2004; Sahu *et al.*, 2021). *Bacteroidetes*, *Proteobacteria*, *Firmicutes* and *Actinobacteria* are the dominant bacterial groups found in the rhizosphere (Nacke *et al.*, 2011), which play several vital roles in biogeochemical cycles (Nacke *et al.*, 2011), mineral weathering and soil ecosystem regeneration. Further, microbial

*Corresponding author: rohinimattoo@iisc.ac.in

balance and functions are vital to greenhouse gas (GHG) reductions, soil water retention, organic matter cycling and reduction in metal toxicity. For example, in soybean cultivation fields, introduction of *Bradyrhizobium japonicum* species has shown to reduce N_2O, (Sameshima-Saito *et al.*, 2006) a potent GHG. Bacteria are continuously studied for their novel functions and represent the most dominant group of microorganisms studied in soil today (Delgado-Baquerizo *et al.*, 2018). Studies on the beneficial roles of microbes in sustainable agriculture have recently gained momentum and their potential warrants further research.

The aim of this mini review was to highlight the beneficial roles of microbes in assisting crops to adapt to climate fluctuations and highlight how agricultural practices affect microorganisms in soil. We discuss the impact of agricultural practices on microbial diversity and the importance of microbes in supporting plants adapting to changing climatic conditions. Additionally, we discuss the latest techniques to uncover microbes and their novel functions that can be utilized in agroecological research. There is accumulating evidence to show that soil erosion, declining water resource and quality render agricultural ecosystems unstable for sustained crop production (Orgiazzi *et al.*, 2016). We propose that soil microbes warrant investigation to practically utilize their beneficial impact for agricultural purposes.

Microbial diversity in soil

Only one gram of soil, near plant roots, could harbour 90–100 million bacteria and 200,000 fungi (Patten and Glick, 2002; Glick, 2012) with species diversity ranging from 4×10^3 to 5×10^4 species (Raynaud and Nunan, 2014). Soils harbour distinct microorganisms and macrofauna, which differ with increasing depth, composition of soil and microenvironment of the soil. The top soil constitutes a higher concentration of carbon (Jat *et al.*, 2019) an important energy source (Post *et al.*, 1982; Rasse *et al.*, 2005), which accumulates from dead leaves, other decomposing plant parts, exudates and other organic matter. Agricultural soils serve as an important model to study the dynamic carbon flux (due to dead plant and animal/livestock materials, which become part of the soil organic matter [SOM]), which contributes to capturing the carbon dioxide from the atmosphere. Since the carbon content in SOM is higher than that of the atmosphere, SOM is a good indicator of the biodiversity of an ecosystem since it also relates to the availability of carbon and carbon flow between biological communities (Vezzani *et al.*, 2018). The type of soil influences bacteria-to-fungi ratio, which in turn may affect carbon stock in soil (Cozzolino *et al.*, 2021). *Gemmatimonadetes* represent approximately 2% of soil bacterial communities aiding in carbon turnover in soil habitats (Whitman *et al.*, 2016), and methanotrophs (*Methylobacter* and *Methylosinus*) and methanogens are important contributors to the carbon cycles (Aditi and Rohini, 2021). The phototrophic and chemo-/autotrophic microbes capture atmospheric carbon dioxide (CO_2), which leads to generation of organic biomass (Gougoulias *et al.*, 2014), and then, the carbon is liberated back into the atmosphere by auto- and heterotrophic microorganisms (Yahai and Ralf, 2005). The microbes in the carbon cycle are linked to other cycles such as the nitrogen cycle, iron cycle and sulphur cycle (Aditi and Rohini, 2021); hence, it seems worthwhile to speculate that increasing the concentrations of cyanobacteria in soils where their abundance is lower due to biotic or abiotic stresses could help in countering increasing CO_2 levels in the atmosphere.

Apart from carbon, plant roots are able to avail other nutrients such as phosphorus, potassium, iron, nitrogen and even silicon through symbiosis with soil microbes involved in weathering (Raturi *et al.*, 2021). Interestingly, cereal crops have deeper roots (Rich and Watt, 2013), which enable them to attract unique microbes and nutrients since the deeper soil layers constitute important reservoirs of nitrogen, phosphorus, sulphur, iron and soil moisture. Therefore, investigating the nutrient availability, water dynamics in deeper layers of soil and the type of microbial communities in deeper earth would be worthwhile so as to understand which microbial community could be harnessed for benefiting plant growth (Zhang *et al.*, 2014).

Tools to study microbial diversity and abundance

Advances in modern techniques to study microorganisms have revived the curiosity and intensity to decipher microbial functions and their impact on the environment since the 1960s. Primarily, a vast number (~98%) of microorganisms in the soil remain unculturable (Stewart, 2012) and represent an unexplored frontier in modern science. Today, interdisciplinary approaches involving geobiological, biophysical and biochemical tools are being routinely employed to study microbial diversity from different soil habitats because of the hyper-diverse populations of microorganisms present. We have tabulated the advantages of both culture-dependent studies and latest tools for whole genome sequencing in order to understand their independent and interdependent uses towards microbial diversity studies (Table 25.1).

Culture-dependent techniques are routinely employed for isolation, growth and enumeration of microbes on selected media for identification of active heterotrophic microorganisms (Stewart, 2012). Plating methods have helped identify innumerable soil bacteria allowing intense isolation and characterization such as well-studied PGPR, and also *Microbacterium*, *Comamonas* and *Kocuria* from even the composts made from agricultural products (Chandna *et al.*, 2013; Chandna *et al.*, 2013). Previously, microorganisms with low distribution or the slow growers would be missed on culture-dependent approaches (Wade, 2002; Stewart, 2012); however, advances in compositions of nutrient media have allowed isolation and characterization of bacteria with stringent requirements (Lewis *et al.*, 2021). Nutrient media could be formulated for specific pH optima, optimum temperature, special nutrients, particular metals/cofactors, oxygen levels in air (Stewart, 2012) including the addition of selective inhibitors to the media (Lewis *et al.*, 2021). Culture-dependent methods used in synergy with other tools such as optical tweezers, microbioreactors, trans-well plates and diffusion chambers (Pham and Kim, 2012) become highly effective in characterizing specific bacteria. Even the challenging plant endophytes have been isolated using culture-dependent methods especially those from *Azadirachta indica* using specialized media formulations (Agasimundin *et al.*, 2019). Bacteria (~80) and endophytic fungi (~360) have been recovered from at least 10 different parts of neem using novel neem media (Agasimundin *et al.*, 2019), by culture-dependent techniques.

With the advent of successful nucleic acid isolation methods, the vast prokaryote diversity present in soil ecosystems was further unveiled (Borneman *et al.*, 1996; Zhou *et al.*, 1996; Felske and Akkermans, 1998; Felske *et al.*, 1998). Direct soil nucleic acid extractions (Robe *et al.*, 2003) combined with fluorescent *in situ* hybridization (FISH), nucleic acid hybridization, DNA cloning and sequencing, polymer chain reaction (PCR)-based methods (Kirk *et al.*, 2004), etc., are useful in elucidating microbial diversity at the operational taxonomic unit (OTU) level. Biochemical techniques to study the abundance of soil microbial groups include community-level physiological profiles (CLPP), sole carbon source utilization (SCSU) patterns, fatty acid methyl ester (FAME) and phospholipid fatty acid (PLFA) analysis (Willers *et al.*, 2015), which aid in deciphering the abundant taxa, their lipid profiles and ability to thrive in warmer conditions. In addition, isotope methods and advanced microscopy are being incorporated to study the roles of microorganisms in biogeochemistry of soil and paleoclimatology.

Next-generation sequencing (NGS) methods, with the power of multiplexing, such as the whole genome sequencing have been able to generate large quantity of data (Trivedi *et al.*, 2021). The Illumina technique has been particularly advantageous for microbial studies by including multiple reads, which then allows sequencing of hundreds of samples in one run, resulting in the heterogeneous soil sample to be analysed for its vast microbial content. Recently, a new portable DNA sequencer (MinION) from Oxford Nanopore Technologies (ONT) has been developed (Kerkhof *et al.*, 2017), which generates microbial rRNA operon sequence data that obtain vast amount of information on microbes abundance in agricultural soils. Such high-throughput techniques allow elucidation of the genetic potential of microbes that enable them to thrive in various habitats in addition to comparison of cellular functions embedded in

Table 25.1. The tools used in microbial diversity studies and their advantages.

Method	Description with advantages	Disadvantages	References	Applications of the methods
Plating methods	The technique is based on direct viable counts which allow studying of phenotypic and morphological characteristics of single bacterial species	Unculturable and slow-growing bacteria cannot be detected unless the media is specifically designed to isolate certain bacteria	(Koeleman *et al.*, 1998)	1. Separates out bacteria from homogenous mixture 2. Characterizes each bacterium for specific needs, for example pH, carbon source 3. Isolation of beneficial microbes for agricultural purpose, for example, phosphate solubilizers
Fatty acid methyl ester analysis (FAME)	The technique based on grouping of fatty acids where no plating is required	Results could be influenced by many external factors	(Bossio *et al.*, 1998; Bossio and Scow, 1998)	1. Identifies distinct lipids even in extreme habitats, for example, rocks, fossils 2. Used to aid in mineralization of rock in rocks terrain to promote plant growth
Guanine & cytosine (G+C) content	The technique is based on G+C content of the DNA, the technique can be highly quantitative	Large amounts of DNA (up to 50 µg) are needed. Complementary methods for microbial detection are required	(Tiedje *et al.*, 1999)	1. Identifies bacterial responses to environmental adaptations, for example, nutrient poor conditions 2. Could aid in deciphering microbes in poor soils and higher temperatures, and aid in soil remediation processes
Nucleic acid reassociation and hybridization and DGGE	The technique is based on the time needed for half of the DNA to re-associate for use as a diversity index. Allows thousands of genes to be analysed	May not detect bacterial sequences present in low copy numbers	(Muyzer and Smalla, 1998; Cho and Tiedje, 2001)	1. Characterizes predominant bacteria, their role in environment w.r.t gene expression and gene exchange 2. Visualization of bacterial gene expression to characterize upregulation or downregulation of genes in contaminated agricultural soils

Continued

Table 25.1. Continued

Method	Description with advantages	Disadvantages	References	Applications of the methods
DNA microarrays	This is used in conjunction with DNA-DNA hybridization to detect and identify bacterial species. A single array can be useful for species diversity	Detects only highly abundant microbial species, dependent on microbial culturing	(Cho and Tiedje, 2001)	1. Identifies genomic amplifications, deletion, rearrangement, insertion, changes in base-pair in bacteria 2. Allows identification of alterations in dominant bacterial genomes under harsh soil conditions
PCR targeting the 16S rDNA	This technique allows identification of prokaryotes and prediction of phylogenetic relationships	Databases restricted, technical variability can be encountered	(Rodriguez-Brito *et al*., 2010)	1. Specifically identifies bacteria from diverse environment samples containing mixture of microorganisms 2. Distinguishes bacterial communities from macrofauna populations in agricultural soils
18S rDNA and internal transcribed spacer (ITS) region	The technique involves PCR amplified DNA fragments of similar bp. Large number of samples can be analysed simultaneously	PCR biases occurs due to extraction efficiency of samples, less commonly used for proteins	(Muyzer *et al*., 1993; Peters *et al*., 2000)	1. Identifies eukaryotic microorganisms in ecosystems 2. Allows enumeration of eukaryotic microorganisms in soil
Restriction fragment length polymorphism (RFLP)/amplified ribosomal DNA restriction analysis (ARDRA)	The technique is routinely used to detect bacterial communities in environmental samples	Sometimes cannot differentiate between species with a low cut-off	(Vaneechoutte *et al*., 1993)	1. Rapid assessment of microbes with response to changing environment 2. Structural changes in microbial communities can be detected in soil habitats

Continued

Table 25.1. Continued

Method	Description with advantages	Disadvantages	References	Applications of the methods
Terminal restriction fragment length polymorphism (TRFLP)	This technique makes use of the labeled terminal restriction fragment, provides data on diversity since a single band represents one operational taxonomic unit and the band pattern obtained is simpler than RFLP	Dependent on efficiency of sample extraction, and type of restriction enzymes used which may bring bias in data	(Marsh, 1999)	1. Identifies microbes in anaerobic digestion 2. Aids identification of microbes in anoxic horizons of soils
Ribosomal intergenic spacer analysis/ automated ribosomal intergenic spacer analysis (ARISA)	In this technique, intergenic spacer region between the 16S and 23S ribosomal subunits are utilized. This is useful for distinguishing closely related bacteria	Many dominant species often get missed in this analysis	(Ranjard *et al.*, 2001)	1. Allows identification of bacterial at species level 2. Helps in distinguishing members of a functional group in soils, for example, nitrogen fixing bacteria, denitrifiers
Pyrosequencing (first NGS platform)	This was the first commercial NGS instrument developed in 2005. This allows whole genome sequencing, exome, transcriptome, amplicon sequencing, and metagenomics. The FLX+platform efficiency is 99.997% at 15×coverage	For FLX +platform: high-sequencing error rate and time consuming	(Nyren *et al.*, 1993; Ronaghi, 2001)	1. Microbial gene sequencing to identify bacterial species and mutations involved in antimicrobial resistance 2. Allows enumeration of soil microbial diversity and even single-nucleotide polymorphisms over time
Reversible terminator chemistry sequencing (Illumina Solexa, Helicos Biosciences Corporation developed an unblocked terminator named "virtual terminator")	This is an advanced NGS method. At least 99.9% accuracy of Illumina platforms and less time consuming	Substitution errors can get generated on Illumina systems Reduced efficiency due to post-cleavage of blocking group	(Chen *et al.*, 2013)	1. Identifies large samples of bacteria reliably 2. Aids in identification of uncultivable soil communities

Continued

Table 25.1. Continued

Method	Description with advantages	Disadvantages	References	Applications of the methods
Nanopore sequencing	This technique relies on DNA sequence translocation which takes place through pores whose sizes are in nanometers This method is rapid with increased throughput in DNA sequencing with 99.9% accuracy of Illumina platforms	Noise can cause substitution errors in Illumina systems GC bias can occur, which could limit detection	(Kono and Arakawa, 2019)	1. Real-time analysis of single molecules of bacteria, for example, RNA and DNA 2. Aids in identification of uncultivable soil communities
Illumina (MiniSeq, MiSeq, NextSeq, HiSeq (2500), NovaSeq 5000/6000)	This technique is based on clonal amplification and synthesis (SBS) chemistry	Run time could be long	(Besser *et al*., 2018)	1. The method is rapid and reliable for microbial diversity studies 2. Aids in identification of uncultivable soil communities rapidly

the genomes across various habitats. These techniques also detect any alterations in the genomes such as single-nucleotide polymorphism and rearrangements. Coupled with advanced bioinformatics tools, the sequence information obtained through comparative genomics could aid in identification of important gene clusters involved in biogeochemical cycles, climate adaptation, immune response, signalling in plant development and communication with other microbes.

Metagenome studies to decipher the root microbiome of citrus (Xu *et al.*, 2018), banana (Kaushal *et al.*, 2020) and finger millet (Singh and Laxmi, 2015) have revealed important microorganisms associated with the functioning of these crop plants. These studies have identified both the core microbes that associate specifically with the plants and also the dynamic microbial communities which shift under varying environmental influences. Such information and understanding is imperative in engineering the rhizosphere to sustain agricultural productivity and also decipher microbial abundance in forecasting the influence of climate, anthropogenic activities and soil health. Future studies focussing on the specific roles of these identified microorganisms, their gene analysis, cofactors and interactions, which are important in plant growth and development, would be necessary.

Agricultural practices affect soil microbial consortia

Intensive modern agricultural practices involving heavy chemical inputs have led to adverse negative impacts on the environment. Nutrient leaching, soil degradation, greenhouse gases and chemicals found in the atmosphere due to excess chemical inputs in agriculture have severely compromised our environmental quality (Fig. 25.1).

Conservation tillage practices enhance bio-geological properties of soil and microbial activities since they involve reduced soil disturbance. This allows crop residues to settle on the soil and increase the organic matter, which contributes to soil formation and a healthy carbon cycle. Healthy soil composition is additionally necessary for preventing excess water run-off, preservation of soil nutrients and preventing soil erosion (Hobbs *et al.*, 2008; Mathew *et al.*, 2012; Lupwayi *et al.*, 2017; Knapp and van der Heijden, 2018). Conservation tillage restores degraded agricultural soils burdened with chemicals, improving soil physiochemical properties, enhancing carbon sequestration (Bandyopadhyay, 2020) and promoting the carbon biogeochemical cycles (Li *et al.*, 2020).

Practising crop rotations have been demonstrated to be beneficial in reducing plant disease, insects and weeds (Karlen *et al.*, 1994; Karlen *et al.*, 1994; Anderson *et al.*, 1998; Kegode *et al.*, 1999) and supporting active soil microbes, which in turn maintain crop yields. Crops such as pulses support crop rotations by recruiting symbiotic microbes such as Rhizobia to enhance soil nitrogen content (Hirsch *et al.*, 2001; Teng *et al.*, 2015), which naturally increases crop yields and growth. Crop rotation with *Lens culinaris* leads to fixation of 35–100 kg ha^{-1} of nitrogen per year [although this is dependent on the number of nodules per plant] thereby enriching the existing soil nitrogen content (Wiese *et al.*, 2016) and aid in attracting *Rhizobium* species that are the natural symbionts of *Lens culinaris* nodules (Wiese *et al.*, 2016). Rhizobia are also known to confer resistance to plants towards drought and salinity stresses leading to the enhanced production of food grains (Kaushal and Wani, 2016). Pulse crop plants, such as chickpea (*Cicer arietinum*), peas (*Pisum sativum*) or lentils (*Lens culinaris*), harbour microbes, which play roles in biogeochemical cycles (carbon, nitrogen, phosphorus, etc., in the soil). In contrast, long-term monoculture restricts crop growth causing a decline in organic matter in soil, changes in soil pH and a decrease in beneficial microbial population indirectly reducing grain yields and economic problems (Zhao *et al.*, 2018). Since, globally, there is a preferred need to eliminate the dependence on chemical fertilizers and use the naturally available soil nitrogen, incorporating bio-inoculants during crop rotation practices would be desirable.

Controlled slash and burn strategy is another traditional agricultural practice, which allows natural soil regeneration (Altieri, 2002) by increasing nutrient flux in soils, enhancing soil fertility and eradicating crop pest post-harvest. Post-burning, trace soil minerals such as calcium, potassium, phosphorus, magnesium,

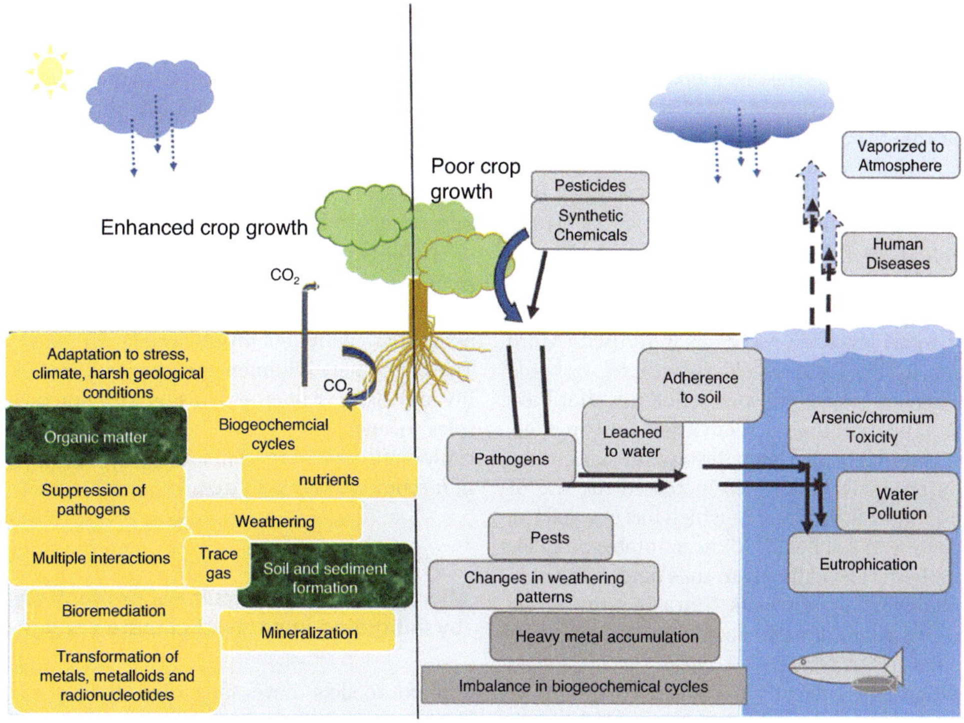

Fig. 25.1. Positive and negative influences of agricultural practices on environment. The left side represents synchronous soil processes such as mineralization, transformation of metals, natural weathering, balanced nutrient and biogeochemical cycles, formation of organic matter, adaptation of plants to climate change due to a harmonious balance in microbial diversity and abundance in soil following sustainable agricultural practices. The right section shows how interference in the natural processes due to excess chemical inputs in agricultural soils initiates a cascade of events leading to a higher abundance of opportunistic pathogens, disruption in rates of weathering processes, heavy metal accumulation due to alterations in microbial diversity. The chemicals get leached into water bodies leading to eutrophication of water and increase in heavy metal toxicity (arsenic, chromium, etc.), thereby posing a risk to not only humans but also aquatic life. Further, the chemicals get vaporized to the atmosphere where they influence cloud formation leading to alterations in greenhouse gases and precipitation patterns.

ammonium-nitrogen and other cations are enriched in ash (Adeduntan and Oyerinde, 2010), which get leached into the top layer of soil resulting in microbial growth and soil enrichment during the rainy season. This effectively suggests that regulated fire management in agricultural lands can impact soil microbial communities positively and contribute to better soil health.

Avoiding synthetic inputs promotes termites and ants as biological engineers in soil that feed on dead and decaying matter (De Bruyn and Conacher, 1990). Additionally, earthworms and other macrofauna modify soil composition through their casts, pellets and other biogenic forms increasing soil carbon (Van Groenigen *et al.*, 2014). Interestingly, the microbes in the gut of macrofauna also contribute effectively towards the biological activity in soil by way of their exudates, mucus and other beneficial decomposing activities (Drake and Horn, 2007). The earthworm gut measures 450 mm^3, and with an average density of 2000 worms/mm^2, there could be a major impact on activities of soil microbes (Drake and Horn, 2007; Pathma and Sakthivel, 2012). Decomposition processes favour organic matter recycling, microbial abundance, and beneficial fungi (Freymann *et al.*, 2008) creating a harmonious balance in biogeochemical cycles by increasing the nutrient content in soil. Core microorganisms identified in a few plant-earthworm interactions further validate their roles in soil function (Jacquiod *et al.*, 2020).

Studies conducted on the surface soils of paddy fields in a district in India revealed that bacterial counts reduced from 6.2×10^5 cfu g^{-1} to 1.8×10^4 cfu g^{-1} due to pesticides (Arora *et al.*, 2019). On the contrary, bacterial counts increased from 5.4×10^4 cfu g^{-1} to 4.1×10^5 cfu g^{-1} when less toxic pesticides were used (Arora *et al.*, 2019). Bacterial diversity, therefore, could be an indicator of various influences that have taken place in the microenvironment (Gu *et al.*, 2019). Due to the Green Revolution, agriculturally productive countries increased the use of pesticides and herbicides, which has now led to a decrease in soil health. Unaccountable pesticide and herbicide additions to soils have resulted in imbalances in nutrients (Purakayastha *et al.*, 2019), which lead to eutrophication of water bodies (Ju *et al.*, 2009). Since TM7 bacteria found in wastewater treatment plants have been associated with the removal of excess phosphate, these could be used to assimilate high concentrations of phosphorus from soil naturally (Bond *et al.*, 1995; Winsley *et al.*, 2014). *Penicillium bilaii* could aid in phosphorus mineralization in soil and its uptake by plant roots (Sharma *et al.*, 2013). Different species of *Pseudomonas, Rhizobium, Bacillus, Enterobacter* (Rodríguez and Fraga, 1999) and species of *Aspergillus* (Narsian and Patel, 2000) have been found to be effective phosphate solubilizers. Phosphate concentrations in soil fluctuate due to chemical inputs and natural mineralization processes. Interestingly, phosphate regulates the interaction of Arabidopsis with *Colletotrichum tofieldiae* wherein an adequate phosphate concentration drives this interaction towards commensalism and phosphate deprivation entails a mutualistic relation (Takahara *et al.*, 2016). Similarly, *Arthrobacter* sp, *Klebsiella* sp., *Bacillus mucilaginosus, Acidithiobacillus ferrooxidans, Bacillus edaphicus, Azotobacter* sp. *Frateuria* sp., *Pseudomonas* sp., *Rhizobium* sp., etc. (Rodríguez and Fraga, 1999; Adesemoye *et al.*, 2008), aid in potassium uptake, the inclusion of which during the plant seedling stage could circumvent the need to add potassium-based chemical fertilizers in the rhizosphere. More naturally occurring microbes warrant identification to support mineralization of metals, which act as important cofactors to promote the plant development.

Recent research in microbial diversity studies has focused on growth-promoting rhizobacteria, molecular techniques to characterize microorganisms, studies on abiotic stresses, improving soil biological quality and improving bio-fertilizer technology. The use of microbes to enhance crop yields is exciting opportunity to sustain agriculture and overcome problems associated with chemical inputs in fields. Therefore, there is immense work ahead to identify and discover microorganisms with practical roles in improving agricultural practices especially with changing climate and a degrading non-renewable soil resource.

Microbes rescue agricultural productivity by mitigating impacts of climate change

Climate models have forecasted that global elevations in temperature and CO_2 levels could occur by 2050 (Naylor *et al.*, 2007; Deutsch *et al.*, 2018; Battisti *et al.*, 2018). There are calculated speculations that increased carbon distribution to the roots from plant leaves influences root exudation (Dusenge *et al.*, 2019), thereby affecting chemo-attractants, signalling molecules and nutrients in soil (Neumann *et al.*, 2014; Kandeler *et al.*, 2006; Rasche *et al.*, 2011; Schreiter *et al.*, 2014). Changes in temperature affect the growth and reproduction of microorganisms, in turn affecting the biological activity of soils (Alexander, 1965). Microorganisms associated with plant roots aid in plant adaptation to extreme heat and insect colonization due to elevated temperatures (Roossinck, 2011; Prasch and Sonnewald, 2013). Using interdisciplinary tools (Table 25.1), the genome of such promising microorganisms could be mapped to identify gene clusters to decipher their molecular functions involved in temperature control, thus paving a way for extending such results to more abundant bacteria (Zhu *et al.*, 2020). Recent research on Actinobacteria adapted to extreme habitats (Hui *et al.*, 2021), or the psychrotrophic soil microbes (Jha *et al.*, 2021), could help tap their potential in supporting plant growth in areas of high temperature or extreme cold. Since microbial activity tends to keep the soil warmer, it additionally aids in buffering of soil when cold conditions prevail. Hence, microbes in agricultural fields could be optimized to support plant growth even during adverse winter conditions. Incorporation of salt-tolerant species of *Bacillus* has been suggested to promote wheat (*Triticum turgidum* subsp. durum) growth in highly saline soils (Ibarra-Villarreal *et al.*, 2021).

Some members of the grass family tend to adapt better to conditions of salinity and high temperatures due to associated endophytes (Rodriguez *et al.*, 2009). Individually, some plants and endophytic fungi cannot grow beyond 38°C but in a symbiotic relationship, they have been demonstrated to grow even when roots were heated to 65°C (Márquez *et al.*, 2007). Endophytes represent an untapped repertoire of phytochemicals to be exploited as alternatives to chemical pesticides in pest control since they act as natural pesticides and fertilizers (Raj, 1950; Rawat, 1995). *Eupenicillium parvum* isolated from Neem (*Azadirachta indica*) produces antibacterials such as azadirachtin A and B under axenic conditions (Kusari *et al.*, 2012). 'Javanicin', a compound isolated from *Chloridium* sp. and found in the roots of *Azadirachta indica A Juss.*, prevents pathogenic *Pseudomonas* (Kharwar *et al.*, 2009). Continued efforts in characterizing plants and microbes of medicinal value under changing climatic conditions would be of major benefit in sustainable agricultural practices and better tools to uncover their potential must be generated in the future. Plants grown in greenhouses have exhibited a decreased intake of water by 20%–30% due to synergy with endophytes (Redman *et al.*, 2011). Therefore, enrichment of natural endophytes may be an important method to extenuate climate change effects such as drought on agricultural crops. Beneficial microorganisms such as *Rhizobium* sp., *A. chroococcum*, *Azospirillum brasilense*, *Pseudomonas corrugata*, *Bacillus subtilis* and *Streptomyces nojiriensis* play roles in promoting sustainable climate adaptable agricultural practices (Bhattacharyya *et al.*, 2016). Incorporation of *Chrococcus*, *Oscillatoria*, *Anabaena*, *Aphanocapra* and *Phormidium* has been shown to fix atmospheric nitrogen thus making this necessary element available for paddy growth (Bagali *et al.*, 2012). Interestingly, *Trichoderma* strains used as bio-control agents increase plant growth by their positive association with different plants (Eslahi *et al.*, 2020). Mechanistically, microbes confer adaptation to climate change such as drought (Singh and Laxmi, 2015; Goel and Singh, 2015; Singh *et al.*, 2015; Vurukonda *et al.*, 2016) by initiating biochemical changes in plants through induced systemic tolerance (IST) pathways as shown in cucumber (Kaushal, 2019) and wheat (Varinderpal-Singh *et al.*, 2020). To tackle the responses of plant microbes to drought studying the microbiomes in different rhizospheres has been suggested as an important avenue (de Vries *et al.*, 2020; Mattoo *et al.*, 2021).

During monsoon, heavy or prolonged rainfall temporarily saturates the soil, reducing oxygen levels and causing bacterial anaerobic respiration to shift towards other electron acceptors to fulfil their metabolic needs (Unger *et al.*, 2009) further altering the functional and total microbial composition of the soil (Elhottová *et al.*, 2002; Mentzer *et al.*, 2006). Increase in humidity in the atmosphere represses the hypersensitive response in plants leading to cellular death and increased pathogen attack during the rainy season (Wang *et al.*, 2005). Additionally, precipitation tends to support decomposition of litter when devoid of plant roots and hence impacting saprophytic microorganisms. In the presence of plant roots, a higher microbial biomass with reduction in respiration and carbon fixation ensues highlighting the influence plants have on surrounding microbes with respect to changing climate (Li *et al.*, 2021). Prolonged anaerobic conditions affect rates of microbial decomposition by microorganisms themselves causing alterations in cycling of nutrients (Schuur *et al.*, 2001; Webster *et al.*, 2004) including fungal populations (Unger *et al.*, 2009). The microbiome of plant species adapted to flood susceptible areas is not well explored and hence needs further investigation.

Increased concentrations of ozone suppress plant growth (Loya *et al.*, 2003), lead to variations in chemical composition of plant tissues (Findlay *et al.*, 1996), reduce carbon distribution to plant roots and decrease plant exudates (Coleman *et al.*, 1995; King *et al.*, 2001; Andersen, 2003). Ozone-led changes in plant exudation in turn impact rhizosphere microbes (McCrady and Andersen, 2000). Molecular biology experiments have revealed alterations in the bacterial genes associated with C, N and S cycling due to increase in ozone levels (Wu *et al.*, 2016; Zytynska *et al.*, 2020). Additionally, pest suppression by introducing bio-inoculants was stronger under eO_3 levels suggesting that despite climate change, microbes when introduced as inoculants could impact plant growth (Zytynska *et al.*, 2020).

Even the symbiotic and plant-microbe commensal interactions are affected by climate. Warmer temperatures supress effector-triggered immunity by dampening the downstream gene WRKY46 (Cheng *et al.*, 2013; Cheng *et al.*, 2013) in Arabidopsis and favour a differential microbial secretion that has been evolutionarily optimized for lower temperatures.

Direct alterations in microbial abundance, alterations in nitrogen fixation and other biogeochemical cycles are linked to warming weather, ozone levels and greenhouse gas emissions due to climate change could influence agricultural productivity and functional activities of rhizosphere microbes (Austin *et al.*, 2009; Castro *et al.*, 2010). Therefore, to address the challenges associated with studying the influence of climatic factors, long-term interdisciplinary experiments coupled with field explorations and computational modelling need to be undertaken.

Perspectives on microbial functions in sustaining agriculture

A diagrammatic representation of how the microbial function lies at core of soil agricultural health with positive and negative influences due to anthropogenic activities and its involvement in biogeochemical cycles has been shown in Fig. 25.1. The soil harbours minerals and nutrients, varying microbial abundance, organic matter and water, which influences trace gases in soil, weathering of rocks and suppression of opportunistic pathogens. Agricultural practices such as incorporation of excessive pesticides and chemical inputs alter the beneficial non-pathogenic soil microbes, biogeochemical cycling of nutrients, heavy metal accumulation causing an increased incidence of opportunistic pathogens, soil degradation, eutrophication of water bodies, which further affects accumulation of heavy metals in water bodies leading to diseases in humans (and aquatic life). The chemicals vaporize to the atmosphere when they interfere with atmosphere gases. Fig. 25.1 also shows the relationship between above-ground and below-ground processes due to the effect of climatic factors on plant-microbe interactions. Various interdisciplinary tools such as whole genome sequencing of soil samples or glaciers, and satellite data for mapping crop pests can help decode the novel functions of microbes for climate-resilient agriculture practices.

Plants and microbes have together evolved compatibilities that support their coexistence (Nihorimbere *et al.*, 2011) by modulating their beneficial features such as uptake of nutrients and overall plant health (Trivedi *et al.*, 2020). Interestingly, it was found that few plants produce exudates rich in citric acid, in which the rhizobacterial strains in the rhizosphere are utilized as their carbon and energy source (Macias-Benitez *et al.*, 2020). Practical efforts can be made to engineer the rhizosphere of agricultural soils to favour the growth of beneficial bacteria, which would mutually benefit their host plants and limit any disease-causing pathogens (Santoyo *et al.*, 2016; Ahkami *et al.*, 2017). Since sterile soils have not been able to confer protection towards disease-causing pathogens (Haas and Défago, 2005), it may be hypothesized that natural organic rich soils may contain numerous factors that may suppress pathogens. Although this generic quality of disease suppression is not currently well deciphered, the presence of microorganisms such as Actinomycetes in that soil may be, additionally, contributing to the suppression of harmful microbes (Buzón-Durán *et al.*, 2020; Chaudhary *et al.*, 2020).

Recent advances in studying soil memory have resulted in emerging concepts of how soil laden with beneficial bacteria could help eradicate pathogens and subsequent restoration of normal soil functions. Even though the exact mechanism of these interactions between pathogenic species and beneficial microbes remains far from being understood, it is clear that composts can help build stronger soil microbial populations and help shape new equilibrium that steers towards populating these beneficial microbes. Growth of many crops including the common bean and soybean crops has positively benefited by association with *Azospirillum brasilense* (Bashan *et al.*, 2004; Pereg *et al.*, 2016). Other beneficial microbes, which have agricultural benefits in soil, include *Azoarcus*, *Azotobacter*, *Arthrobacter*, *Bacillus*, *Clostridium*, *Enterobacter*, *Gluconoacetobacter*, *Pseudomonas*, *Rhizobium* and *Serratia* (Cheng *et al.*, 2009; Mishra *et al.*, 2016). Co-inoculation of laccase-producing *S. melioti* strain with *A. brasilense* grown in low pH soil in Uruguay was shown to enhance rice growth (Castro-Sowinski *et al.*, 2002). Interestingly, different plant species attract unique rhizosphere microbes. Plants with medicinal properties attract microbes, with the ability to produce unique bioactive metabolites (Wang *et al.*, 2012; Köberl *et al.*, 2013). The root-associated microbes of medicinal plant could be tapped as a major phyto-therapeutic resource. *Botryosphaeria rhodina* produces Botryorhodine, only when it is in synergistic association with its host plant *Bidenspilosa* (Abdou *et al.*, 2010).

Even pathogenic microbes tend to relate to several signals from the plants they are associated with, for their own aid and survival. For example, plants whose root exudates contain p-hydroxybenzoic, cinnamic, gallic, salicylic, coumaric and cinnamic acids attract pathogenic fungi and influence their germination (Wu HongSheng *et al.*, 2008; Raza *et al.*, 2015). In contrast, alkaloids produced by the *Veratrum taliense* (Liliaceae) roots were found to prevent the growth of *Phytophthora capsici* and *Rhizoctonia cerealis* (Zhou *et al.*, 2003).

The roles of microbes in promoting a harmonious balance in biogeochemical cycles have been recently summarized (Aditi and Rohini, 2021) where microbes act as central catalysts in the major pathways such as nitrification, denitrification, phosphate solubilisation and uptake. Integrating the knowledge of biogeochemical cycles in the context of climate-adaptable rhizosphere health would be imperative in future.

Concluding points

Given the tremendous diversity of soil microbes, soil flora and fauna, it would continue to be a herculean task to investigate the components of the rhizosphere and their interactions in varying environmental conditions till microbial diversity patterns emerge from different agricultural soils. An understanding of the relationship between microbes involved in biogeochemical cycles and plants in the rhizosphere would help demonstrate that agricultural practices such as unregulated use of chemicals in soils need elimination. Microbial diversity studies are more important than ever for future agricultural research since incorporating beneficial microorganisms aid in increasing plant germination rate, promoting plant growth and allowing plants to acclimatize to harsh environmental conditions. Specific novel microorganisms with specialized biogeochemical functions can be characterized to enhance grain production through mineralization in soil and adaptability of plants to various climatic vagaries. Our understanding is limited to bacteria, but further research is warranted on molecular interactions with other micro-flora such as viruses, fungi, algae. Microbial diversity studies coupled with biogeochemical tools and climate change physics hold enormous promise for sustaining Earth's future agrarian research. We propose that microbes are the major drivers of important ecosystem pathways and could help engineer our agricultural soils to revive the soil health under changing climatic conditions.

Acknowledgements

The authors wish to thank Divecha Centre for Climate Change, Indian Institute of Science, Bangalore, India, for their support.

Review methodology: In particular, we have used PubMed to search for literature for this mini-review. Additional information was gathered from the references of some reviews obtained from the mentioned search. The figure was created using MS Office PowerPoint tool and table in MS office Word.

Conflicts of interest

We have no conflicts of interest to declare. The authors declare that the research was carried out without any commercial or financial associations and hence, there is no potential for any conflict of interest.

References

Abdou, R., Scherlach, K., Dahse, H.-M., Sattler, I. and Hertweck, C. (2010) Botryorhodines A–D, antifungal and cytotoxic depsidones from *Botryosphaeria rhodina*, an endophyte of the medicinal plant *Bidens pilosa*. *Phytochemistry* 71(1), 110–116. DOI: 10.1016/j.phytochem.

Adeduntan, S. and Oyerinde, A. (2010) Evaluation of nutritional and antinutritional characteristics of obeche (*Triplochition scleroxylon scleroxylon*) and several mulberry (*Morus alba*) leaves. *African Journal of Biochemistry Research* 4(7), 175–178.

Adesemoye, A., Obini, M. and Ugoji, E. (2008) Comparison of plant growth-promotion with *Pseudomonas aeruginosa* and *Bacillus subtilis* in three vegetables. *Brazilian Journal of Microbiology* 39, 423–426.

Aditi, G. and Rohini, M. (2021) Microbes in biogeochemical cycles: A perspective on climate resilient agriculture. *CAB Reviews* 16(6), 1–15.

Agasimundin, V.B., Rangiah, K., Sheetal, A. and Gowda, M. (2019) Neem microbiome. In: *The Neem Genome*. Springer, Cham, pp. 111–123.

Ahkami, A.H., White III, R.A., Handakumbura, P.P. and Jansson, C. (2017) Rhizosphere engineering: Enhancing sustainable plant ecosystem productivity. *Rhizosphere* 3, 233–243.

Alexander, M. (1965) Most-probable-number method for microbial populations. In: *Methods of Soil Analysis: Part 2 Chemical and Microbiological Properties*, Vol. 9. pp. 1467–1472.

Altieri, M.A. (2002) Agroecology: The science of natural resource management for poor farmers in marginal environments. *Agriculture, Ecosystems & Environment* 93(1–3), 1–24.

Andersen, C.P. (2003) Source–sink balance and carbon allocation below ground in plants exposed to ozone. *New Phytologist* 157(2), 213–228.

Anderson, R.L., Tanaka, D.L., Black, A.L. and Schweizer, E.E. (1998) Weed community and species response to crop rotation, tillage, and nitrogen fertility. *Weed Technology* 12(3), 531–536.

Arora, S., Sehgal, M., Srivastava, D., Arora, S. and Sarkar, SK. (2019) Rice pest management with reduced risk pesticides in India. *Environmental Monitoring and Assessment* 191(4), 1–20.

Austin, E.E., Castro, H.F., Sides, K.E., Schadt, C.W. and Classen, A.T. (2009) Assessment of 10 years of CO_2 fumigation on soil microbial communities and function in a sweetgum plantation. *Soil Biology and Biochemistry* 41(3), 514–520.

Bagali, A., Patil, H., Guled, M. and Patil, R. (2012) Effect of scheduling of drip irrigation on growth, yield and water use efficiency of onion (*Allium cepa* L.). *Karnataka Journal of Agricultural Sciences* 25(1), 116–119.

Bandyopadhyay, K. (2020) Effect of tillage on soil carbon sequestration. In: *Carbon Management in Tropical and Sub-Tropical Terrestrial Systems*. Springer, Singapore, pp. 213–229.

Bashan, Y., Holguin, G. and De-Bashan, L.E. (2004) Azospirillum-plant relationships: Physiological, molecular, agricultural, and environmental advances (1997-2003). *Canadian Journal of Microbiology* 50(8), 521–577.

Battisti, R., Sentelhas, P. and Boote, K. (2018) Sensitivity and requirement of improvements of four soybean crop simulation models for climate change studies in southern Brazil. *International Journal of Biometeorology* 62(5), 823–832.

Besser, J., Carleton, H.A., Gerner-Smidt, P., Lindsey, R.L. and Trees, E. (2018) Next-generation sequencing technologies and their application to the study and control of bacterial infections. *Clinical Microbiology and Infection* 24, 335–341.

Bhattacharyya, P., Goswami, M. and Bhattacharyya, L. (2016) Perspective of beneficial microbes in agriculture under changing climatic scenario: A review. *Journal of Phytology* 8, 26–41.

Bond, P.L., Hugenholtz, P., Keller, J. and Blackall, L.L. (1995) Bacterial community structures of phosphate-removing and non-phosphate-removing activated sludges from sequencing batch reactors. *Applied and Environmental Microbiology* 61(5), 1910–1916.

Borneman, J., Skroch, P.W., O'Sullivan, K.M., Palus, J.A., Rumjanek, N.G. *et al.* (1996) Molecular microbial diversity of an agricultural soil in Wisconsin. *Applied and Environmental Microbiology* 62(6), 1935–1943.

Bossio, D.A. and Scow, K.M. (1998) Impacts of carbon and flooding on soil microbial communities: Phospholipid fatty acid profiles and substrate utilization patterns. *Microbial Ecology* 35, 265–278.

Bossio, D.A., Scow, K.M., Gunapala, N. and Graham, K.J. (1998) Determinants of soil microbial communities: Effects of agricultural management, season, and soil type on phospholipid fatty acid profiles. *Microbial Ecology* 36, 1–12.

Buzón-Durán, L., Pérez-Lebeña, E., Martín-Gil, J., SánchezBáscones, M. and Martín-Ramos, P. (2020) Applications of streptomyces spp. enhanced compost in sustainable agriculture. In: *Biology of Composts*. Springer, Cham, pp. 257–291.

Castro, H.F., Classen, A.T., Austin, E.E., Norby, R.J. and Schadt, C.W. (2010) Soil microbial community responses to multiple experimental climate change drivers. *Applied and Environmental Microbiology* 76(4), 999–1007.

Castro-Sowinski, S., Martinez-Drets, G. and Okon, Y. (2002) Laccase activity in melanin-producing strains of *Sinorhizobium meliloti*. *FEMS Microbiology Letters* 209(1), 119–125.

Chandna, P., Mallik, S. and Kuhad, R.C. (2013) Assessment of bacterial diversity in agricultural by-product compost by sequencing of cultivated isolates and amplified RDNA restriction analysis. *Applied Microbiology and Biotechnology* 97(15), 6991–7003.

Chandna, P., Nain, L., Singh, S. and Kuhad, R.C. (2013) Assessment of bacterial diversity during composting of agricultural byproducts. *BMC Microbiology* 13(1), 99. DOI: 10.1186/1471-2180-13-99.

Chaudhary, S., Vaish, B., Singh, R.P. and Prasad, V. (2020) Bioactive compost: An approach for managing plant growth in environmentally stressed soils. In: *New Frontiers in Stress Management for Durable Agriculture*. Springer, Singapore, pp. 257–267.

Chen, F., Dong, M., Ge, M., Zhu, L., Ren, L. *et al.* (2013) The history and advances of reversible terminators used in new generations of sequencing technology. *Genomics Proteomics Bioinformatics* 11, 34–40.

Cheng, C., Gao, X., Feng, B., Sheen, J., Shan, L. *et al.* (2013) Plant immune response to pathogens differs with changing temperatures. *Nature Communications* 4(1), 1–9.

Cheng, C., Gao, X., Feng, B., Sheen, J., Shan, L. *et al.* (2013) Differential temperature operation of plant immune responses. *Nature Communications* 4, 2530.

Cheng, F., Cao, G., Wang, X., Zhao, J., Yan, X. *et al.* (2009) Isolation and application of effective nitrogen fixation rhizobial strains on low-phosphorus acid soils in South China. *Chinese Science Bulletin* 54(3), 412–420.

Cho, J.C. and Tiedje, J.M. (2001) Bacterial species determination from DNA-DNA hybridization by using genome fragments and DNA microarrays. *Applied and Environmental Microbiology* 67, 3677–3682.

Coleman, M., Dickson, R., Isebrands, J. and Karnosky, D. (1995) Carbon allocation and partitioning in aspen clones varying in sensitivity to tropospheric ozone. *Tree Physiology* 15(9), 593–604.

Cozzolino, V., Monda, H., Savy, D., Di Meo, V., Vinci, G. *et al.* (2021) Cooperation among phosphate-solubilizing bacteria, humic acids and arbuscular mycorrhizal fungi induces soil microbiome shifts and enhances plant nutrient uptake. *Chemical and Biological Technologies in Agriculture* 8(1), 31. DOI: 10.1186/s40538-021-00230-x.

De Bruyn, L.L. and Conacher, A.J. (1990) The role of termites and ants in soil modification—A review. *Soil Research* 28(1), 55–93.

de Vries, F.T., Griffiths, R.I., Knight, C.G., Nicolitch, O. and Williams, A. (2020) Harnessing rhizosphere microbiomes for drought-resilient crop production. *Science* 368(6488), 270–274.

Delgado-Baquerizo, M., Oliverio, A.M., Brewer, T.E., Benavent-González, A., Eldridge, D.J. *et al.* (2018) A global atlas of the dominant bacteria found in soil. *Science* 359(6373), 320–325.

Deutsch, C.A., Tewksbury, J.J., Tigchelaar, M., Battisti, D.S., Merrill, S.C. *et al.* (2018) Increase in crop losses to insect pests in a warming climate. *Science* 361(6405), 916–919.

Drake, H.L. and Horn, M.A. (2007) As the worm turns: the earthworm gut as a transient habitat for soil microbial biomes. *Annual Review of Microbiology* 61, 169–189.

Dusenge, M.E., Duarte, A.G. and Way, D.A. (2019) Plant carbon metabolism and climate change: Elevated CO_2 and temperature impacts on photosynthesis, photorespiration and respiration. *New Phytologist* 221(1), 32–49.

Elhottová, D., Szili-Kovács, T. and Tříska, J. (2002) Soil microbial community of abandoned sand fields. *Folia Microbiologica* 47(4), 435–440.

Eslahi, N., Kowsari, M., Motallebi, M., Zamani, M.R. and Moghadasi, Z. (2020) Influence of recombinant trichoderma strains on growth of bean (*Phaseolus vulgaris* L) by increased root colonization and induction of root growth related genes. *Scientia Horticulturae* 261, 108932.

Felske, A. and Akkermans, A.D. (1998) Spatial homogeneity of abundant bacterial 16S rRNA molecules in grassland soils. *Microbial Ecology* 36(1), 31–36.

Felske, A., Wolterink, A., Van Lis, R. and Akkermans, A.D. (1998) Phylogeny of the main bacterial 16S rRNA sequences in drentse A grassland soils (The Netherlands). *Applied and Environmental Microbiology* 64(3), 871–879.

Findlay, S., Carreiro, M., Krischik, V. and Jones, C.G. (1996) Effects of damage to living plants on leaf litter quality. *Ecological Applications* 6(1), 269–275.

French, E., Kaplan, I., Iyer-Pascuzzi, A., Nakatsu, C.H. and Enders, L. (2021) Emerging strategies for precision microbiome management in diverse agroecosystems. *Nature Plants* 7(3), 256–267. DOI: 10.1038/s41477-020-00830-9.

Freymann, B.P., Buitenwerf, R., Desouza, O. and Olff, H. (2008) The importance of termites (Isoptera) for the recycling of herbivore dung in tropical ecosystems: A review. *European Journal of Entomology* 105(2), 165–173.

Glick, B.R. (2012) Plant growth-promoting bacteria: Mechanisms and applications. *Scientifica* 2012, 1–15.

Goel, P. and Singh, A.K. (2015) Abiotic stresses downregulate key genes involved in nitrogen uptake and assimilation in *Brassica juncea* L. *PLoS One* 10(11), e0143645.

Gougoulias, C., Clark, J.M. and Shaw, L.J. (2014) The role of soil microbes in the global carbon cycle: Tracking the below-ground microbial processing of plant-derived carbon for manipulating carbon dynamics in agricultural systems. *Journal of the Science of Food and Agriculture* 94(12), 2362–2371. DOI: 10.1002/jsfa.6577.

Gu, L., Deng, H., Ren, Z., Zhao, Y., Yu, S. *et al.* (2019) Dynamic changes in the microbiome and mucosal immune microenvironment of the lower respiratory tract by influenza virus infection. *Frontiers in Microbiology* 10, 2491.

Haas, D. and Défago, G. (2005) Biological control of soil-borne pathogens by fluorescent pseudomonads. *Nature Reviews Microbiology* 3(4), 307–319.

Hirsch, A.M., Lum, M.R. and Downie, J.A. (2001) What makes the rhizobia-legume symbiosis so special. *Plant Physiology* 127(4), 1484–1492.

Hobbs, P.R., Sayre, K. and Gupta, R. (2008) The role of conservation agriculture in sustainable agriculture. *Philosophical Transactions of the Royal Society B: Biological Sciences* 363(1491), 543–555.

Hui, M.L.Y., Tan, L.T.H., Letchumanan, V., He, Y.W., Fang, C.M. *et al.* (2021) The extremophilic actinobacteria: From microbes to medicine. *Antibiotics* 10(6). DOI: 10.3390/antibiotics10060682.

Ibarra-Villarreal, A.L., Gándara-Ledezma, A., Godoy-Flores, A.D., Herrera-Sepúlveda, A., Díaz-Rodríguez, A.M. *et al.* (2021) Salt-tolerant bacillus species as a promising strategy to mitigate the salinity stress in wheat (*Triticum turgidum* subsp. *durum*). *Journal of Arid Environments* 186, 104399. DOI: 10.1016/j.jaridenv.2020.104399.

Jacquiod, S., Puga-Freitas, R., Spor, A., Mounier, A., Monard, C. *et al.* (2020) A core microbiota of the plant-earthworm interaction conserved across soils. *Soil Biology and Biochemistry* 144, 107754.

Jat, H.S., Datta, A., Madhu Choudhary, P.C.S., Yadav, A.K., Vishu Choudhary, M.K.G. *et al.* (2019) Climate smart agriculture practices improve soil organic carbon pools, biological properties and crop productivity in cereal-based systems of North-west India. *Catena* 181, 104059.

Jha, Y., Kulkarni, A. and Subramanian, R.B. (2021) Psychrotrophic soil microbes and their role in alleviation of cold stress in plants. In: Yadav, A.N. (ed.) *Soil Microbiomes for Sustainable Agriculture: Functional Annotation*. Springer International Publishing, Cham, pp. 267–286. DOI: 10.1007/978-3-030-73507-4_10.

Ju, X.T., Xing, G.X., Chen, X.P., Zhang, S.L., Zhang, L.J. *et al.* (2009) Reducing environmental risk by improving N management in intensive Chinese agricultural systems. *Proceedings of the National Academy of Sciences* 106(9), 3041–3046.

Kandeler, E., Deiglmayr, K., Tscherko, D., Bru, D. and Philippot, L. (2006) Abundance of narg, nirs, nirk, and nosz genes of denitrifying bacteria during primary successions of a glacier foreland. *Applied and Environmental Microbiology* 72(9), 5957–5962.

Karlen, D., Varvel, G., Bullock, D.G. and Cruse, R. (1994) Crop rotations for the 21st century. *Advances in Agronomy* 53(1), 45.

Karlen, D.L., Wollenhaupt, N.C., Erbach, D.C., Berry, E.C., Swan, J.B. *et al.* (1994) Long-term tillage effects on soil quality. *Soil and Tillage Research* 32(4), 313–327.

Kaushal, M. (2019) Portraying rhizobacterial mechanisms in drought tolerance: A way forward toward sustainable agriculture. In: *PGPR Amelioration in Sustainable Agriculture*. Elsevier; Woodhead Publishing, pp. 195–216.

Kaushal, M. and Wani, S.P. (2016) Rhizobacterial-plant interactions: Strategies ensuring plant growth promotion under drought and salinity stress. *Agriculture, Ecosystems & Environment* 231, 68–78.

Kaushal, M., Swennen, R. and Mahuku, G. (2020) Unlocking the microbiome communities of banana (*Musa* spp.) under disease stressed (*Fusarium* wilt) and non-stressed conditions. *Microorganisms* 8(3), 443.

Kegode, G.O., Forcella, F. and Clay, S. (1999) Influence of crop rotation, tillage, and management inputs on weed seed production. *Weed Science* 47(2), 175–183.

Kerkhof, L.J., Dillon, K.P., Häggblom, M.M. and McGuinness, L.R. (2017) Profiling bacterial communities by minion sequencing of ribosomal operons. *Microbiome* 5(1), 1–11.

Kharwar, R.N., Verma, V.C., Kumar, A., Gond, S.K., Harper, J.K. *et al.* (2009) Javanicin, an antibacterial naphthaquinone from an endophytic fungus of neem, chloridium sp. *Current Microbiology* 58(3), 233–238.

King, J., Pregitzer, K., Zak, D., Kubiske, M. and Holmes, W. (2001) Correlation of foliage and litter chemistry of sugar maple, *Acer saccharum*, as affected by elevated CO_2 and varying N availability, and effects on decomposition. *Oikos* 94(3), 403–416.

Kirk, J.L., Beaudette, L.A., Hart, M., Moutoglis, P., Klironomos, J.N. *et al.* (2004) Methods of studying soil microbial diversity. *Journal of Microbiological Methods* 58(2), 169–188.

Knapp, S. and van der Heijden, M.G. (2018) A global meta-analysis of yield stability in organic and conservation agriculture. *Nature Communications* 9(1), 1–9.

Köberl, M., Schmidt, R., Ramadan, E.M., Bauer, R. and Berg, G. (2013) The microbiome of medicinal plants: Diversity and importance for plant growth, quality and health. *Frontiers in Microbiology* 4, 400.

Koeleman, J.G., Stoof, J., Biesmans, D.J., Savelkoul, P.H. and Vandenbroucke-Grauls, C.M. (1998) Comparison of amplified ribosomal DNA restriction analysis, random amplified polymorphic DNA analysis, and amplified fragment length polymorphism fingerprinting for identification of acinetobacter genomic species and typing of *Acinetobacter baumannii*. *Journal of Clinical Microbiology* 36, 2522–2529.

Kono, N. and Arakawa, K. (2019) Nanopore sequencing: Review of potential applications in functional genomics. *Development, Growth & Differentiation* 61, 316–326.

Kusari, S., Verma, V.C., Lamshoeft, M. and Spiteller, M. (2012) An endophytic fungus from *Azadirachta indica* A. Juss. that produces Azadirachtin. *World Journal of Microbiology and Biotechnology* 28(3), 1287–1294.

Lazcano, C., Boyd, E., Holmes, G., Hewavitharana, S., Pasulka, A. *et al.* (2021) The rhizosphere microbiome plays a role in the resistance to soil-borne pathogens and nutrient uptake of strawberry cultivars under field conditions. *Scientific Reports* 11(1), 3188. DOI: 10.1038/s41598-021-82768-2.

Lewis, W.H., Tahon, G., Geesink, P., Sousa, D.Z. and Ettema, T.J.G. (2021) Innovations to culturing the uncultured microbial majority. *Nature Reviews Microbiology* 19(4), 225–240. DOI: 10.1038/s41579-020-00458-8.

Li, S.-L., Xu, S., Wang, T.-J., Yue, F.-J., Peng, T. *et al.* (2020) Effects of agricultural activities coupled with karst structures on riverine biogeochemical cycles and environmental quality in the karst region. *Agriculture, Ecosystems & Environment* 303, 107120.

Li, Z., Wang, F., Su, F., Wang, P., Li, S. *et al.* (2021) Climate change drivers alter root controls over litter decomposition in a semi-arid grassland. *Soil Biology and Biochemistry* 158, 108278. DOI: 10.1016/j.soilbio.2021.108278.

Loya, W.M., Pregitzer, K.S., Karberg, N.J., King, J.S. and Giardina, C.P. (2003) Reduction of soil carbon formation by tropospheric ozone under increased carbon dioxide levels. *Nature* 425(6959), 705–707.

Lupwayi, N.Z., Larney, F.J., Blackshaw, R.E., Kanashiro, D.A. and Pearson, D.C. (2017) Phospholipid fatty acid biomarkers show positive soil microbial community responses to conservation soil management of irrigated crop rotations. *Soil and Tillage Research* 168, 1–10.

Macias-Benitez, S., Garcia-Martinez, A.M., Caballero Jimenez, P., Gonzalez, J.M., Tejada Moral, M. *et al.* (2020) Rhizospheric organic acids as biostimulants: monitoring feedbacks on soil microorganisms and biochemical properties. *Frontiers in Plant Science* 11, 633. DOI: 10.3389/fpls.2020.00633.

Márquez, L.M., Redman, R.S., Rodriguez, R.J. and Roossinck, M.J. (2007) A virus in A fungus in A plant: Three-way symbiosis required for thermal tolerance. *Science* 315(5811), 513–515.

Marsh, T.L. (1999) Terminal restriction fragment length polymorphism (t-rflp): An emerging method for characterizing diversity among homologous populations of amplification products. *Current Opinion in Microbiology* 2, 323–327.

Mathew, R.P., Feng, Y., Githinji, L., Ankumah, R. and Balkcom, K.S. (2012) Impact of no-tillage and conventional tillage systems on soil microbial communities. *Applied and Environmental Soil Science* 2012.

Mattoo, R., Umashankar, N. and Raveendra, H. (2021) Contrasting rhizosphere microbial communities between fertilized and bio-inoculated millet. *Rhizosphere* 17, 100273.

McCrady, J. and Andersen, C. (2000) The effect of ozone on below-ground carbon allocation in wheat. *Environmental Pollution* 107(3), 465–472.

Mentzer, J.L., Goodman, R.M. and Balser, T.C. (2006) Microbial response over time to hydrologic and fertilization treatments in a simulated wet prairie. *Plant and Soil* 284(1), 85–100.

Mishra, J., Prakash, J. and Arora, N.K. (2016) Role of beneficial soil microbes in sustainable agriculture and environmental management. *Climate Change and Environmental Sustainability* 4(2), 137–149.

Muyzer, G. and Smalla, K. (1998) Application of denaturing gradient gel electrophoresis (DGGE) and temperature gradient gel electrophoresis (TGGE) in microbial ecology. *Antonie van Leeuwenhoek* 73, 127–141.

Muyzer, G., De Waal, E.C. and Uitterlinden, A.G. (1993) Profiling of complex microbial populations by denaturing gradient gel electrophoresis analysis of polymerase chain reaction-amplified genes coding for 16s rRNA. *Applied and Environmental Microbiology* 59, 695–700.

Nacke, H., Thürmer, A., Wollherr, A., Will, C., Hodac, L. *et al.* (2011) Pyrosequencing-based assessment of bacterial community structure along different management types in German forest and grassland soils. *PloS One* 6(2), e17000.

Narsian, V. and Patel, H. (2000) *Aspergillus aculeatus* as a rock phosphate solubilizer. *Soil Biology and Biochemistry* 32(4), 559–565.

Naylor, R.L., Battisti, D.S., Vimont, D.J., Falcon, W.P. and Burke, M.B. (2007) Assessing risks of climate variability and climate change for Indonesian rice agriculture. *Proceedings of the National Academy of Sciences* 104(19), 7752–7757.

Neumann, G., Bott, S., Ohler, M.A., Mock, H.P., Lippmann, R. *et al.* (2014) Root exudation and root development of lettuce (*Lactuca sativa* L cv. Tizian) as affected by different soils. *Frontiers in Microbiology* 5(2), 2.

Nihorimbere, V., Ongena, M., Smargiassi, M. and Thonart, P. (2011) Beneficial effect of the rhizosphere microbial community for plant growth and health. *Biotechnologie, Agronomie, Société et Environnement* 15(2), 327–337.

Nyren, P., Pettersson, B. and Uhlen, M. (1993) Solid phase DNA minisequencing by an enzymatic luminometric inorganic pyrophosphate detection assay. *Analytical Biochemistry* 208, 171–175.

Orgiazzi, A., Bardgett, R.D. and Barrios, E. (2016) *Global Soil Biodiversity Atlas*. European Commission.

Pathma, J. and Sakthivel, N. (2012) Microbial diversity of vermicompost bacteria that exhibit useful agricultural traits and waste management potential. *SpringerPlus* 1(1), 1–19.

Patten, C.L. and Glick, B.R. (2002) Role of *Pseudomonas putida* indoleacetic acid in development of the host plant root system. *Applied and Environmental Microbiology* 68(8), 3795–3801.

Pereg, L., de-Bashan, L.E. and Bashan, Y. (2016) Assessment of affinity and specificity of azospirillum for plants. *Plant and Soil* 399(1–2), 389–414.

Peters, S., Koschinsky, S., Schwieger, F. and Tebbe, C.C. (2000) Succession of microbial communities during hot composting as detected by PCR-single-strand-conformation polymorphism-based genetic profiles of small-subunit rRNA genes. *Applied and Environmental Microbiology* 66, 930–936.

Pham, V.H. and Kim, J. (2012) Cultivation of unculturable soil bacteria. *Trends in Biotechnology* 30(9), 475–484.

Post, W.M., Emanuel, W.R., Zinke, P.J. and Stangenberger, AG. (1982) Soil carbon pools and world life zones. *Nature* 298(5870), 156–159.

Prasch, C.M. and Sonnewald, U. (2013) Simultaneous application of heat, drought, and virus to arabidopsis plants reveals significant shifts in signaling networks. *Plant Physiology* 162(4), 1849–1866.

Purakayastha, T.J., Pathak, H., Kumari, S., Biswas, S., Chakrabarty, B. *et al.* (2019) Soil health card development for efficient soil management in Haryana, India. *Soil and Tillage Research* 191, 294–305.

Raj, M. (1950) Green leaf manuring and green manuring. *Madras Agricultural Journal* 37, 3–6.

Ranjard, L., Poly, F., Lata, J.C., Mougel, C., Thioulouse, J. *et al.* (2001) Characterization of bacterial and fungal soil communities by automated ribosomal intergenic spacer analysis fingerprints: Biological and methodological variability. *Applied and Environmental Microbiology* 67, 4479–4487.

Rasche, F., Knapp, D., Kaiser, C., Koranda, M., Kitzler, B. *et al.* (2011) Seasonality and resource availability control bacterial and archaeal communities in soils of a temperate beech forest. *The ISME Journal* 5(3), 389–402.

Rasse, D.P., Rumpel, C. and Dignac, M.-F. (2005) Is soil carbon mostly root carbon? Mechanisms for a specific stabilisation. *Plant and Soil* 269(1), 341–356.

Raturi, G., Sharma, Y., Rana, V., Thakral, V., Myaka, B. *et al.* (2021) Exploration of silicate solubilizing bacteria for sustainable agriculture and silicon biogeochemical cycle. *Plant Physiology and Biochemistry* 166, 827–838. DOI: 10.1016/j.plaphy.2021.06.039.

Rawat, J. (1995) Value of a 20-year old irigated teak plantation. *Indian Forester* 121(6), 553–557.

Raynaud, X. and Nunan, N. (2014) Spatial ecology of bacteria at the microscale in soil. *PloS One* 9(1), e87217.

Raza, W., Yuan, J., Ling, N., Huang, Q. and Shen, Q. (2015) Production of volatile organic compounds by an antagonistic strain *Paenibacillus polymyxa* WR-2 in the presence of root exudates and organic fertilizer and their antifungal activity against *Fusarium oxysporum* f. *Niveum*. *Biological Control* 80(89–95), sp. DOI: 10.1016/j.biocontrol.2014.09.004.

Redman, R.S., Kim, Y.O., Woodward, C.J., Greer, C., Espino, L. *et al.* (2011) Increased fitness of rice plants to abiotic stress via habitat adapted symbiosis: A strategy for mitigating impacts of climate change. *PLOS One* 6(7), e14823.

Ren, C., Zhenghu, Z., Yaoxin, G., Gaihe, Y., Fazhu, Z. *et al.* (2021) Contrasting patterns of microbial community and enzyme activity between rhizosphere and bulk soil along an elevation gradient. *CATENA* 196, 104921. DOI: 10.1016/j.catena.2020.104921.

Rich, S.M. and Watt, M. (2013) Soil conditions and cereal root system architecture: Review and considerations for linking Darwin and Weaver. *Journal of Experimental Botany* 64(5), 1193–1208.

Robe, P., Nalin, R., Capellano, C., Vogel, T.M. and Simonet, P. (2003) Extraction of DNA from soil. *European Journal of Soil Biology* 39(4), 183–190.

Rodriguez-Brito, B., Li, L., Wegley, L., Furlan, M., Angly, F. *et al.* (2010) Viral and microbial community dynamics in four aquatic environments. *The Isme Journal* 4, 739.

Rodriguez, R., White, J., Jr, Arnold, A. and Redman, A.R. (2009) Fungal endophytes: Diversity and functional roles. *New Phytologist* 182(2), 314–330.

Rodŕiguez, H. and Fraga, R. (1999) Phosphate solubilizing bacteria and their role in plant growth promotion. *Biotechnology Advances* 17(4–5), 319–339.

Ronaghi, M. (2001) Pyrosequencing sheds light on DNA sequencing. *Genome Research* 11, 3–11.

Roossinck, MJ. (2011) The good viruses: viral mutualistic symbioses. *Nature Reviews Microbiology* 9(2), 99–108.

Sahu, K.P., Kumar, A., Patel, A., Mukesh Kumar, S.G., Prakash, G. *et al.* (2021) Rice blast lesions: An unexplored phyllosphere microhabitat for novel antagonistic bacterial species against *Magnaporthe oryzae*. *Microbial Ecology* 81(3), 731–745. DOI: 10.1007/s00248-020-01617-3.

Sameshima-Saito, R., Chiba, K., Hirayama, J., Itakura, M., Mitsui, H. *et al.* (2006) Symbiotic *Bradyrhizobium japonicum* reduces N_2O surrounding the soybean root system via nitrous oxide reductase. *Applied and Environmental Microbiology* 72(4), 2526–2532. DOI: 10.1128/AEM.72.4.2526-2532.2006.

Santoyo, G., Moreno-Hagelsieb, G., del Carmen OrozcoMosqueda, M. and Glick, B.R. (2016) Plant growth-promoting bacterial endophytes. *Microbiological Research* 183, 92–99.

Schreiter, S., Sandmann, M., Smalla, K. and Grosch, R. (2014) Soil type dependent rhizosphere competence and biocontrol of two bacterial inoculant strains and their effects on the rhizosphere microbial community of field-grown lettuce. *PLoS One* 9(8), e103726.

Schuur, E.A., Chadwick, O.A. and Matson, P.A. (2001) Carbon cycling and soil carbon storage in mesic to wet Hawaiian montane forests. *Ecology* 82(11), 3182–3196.

Sharma, S.B., Sayyed, R.Z., Trivedi, M.H. and Gobi, T.A. (2013) Phosphate solubilizing microbes: Sustainable approach for managing phosphorus deficiency in agricultural soils. *SpringerPlus* 2(1), 1–14.

Singh, D. and Laxmi, A. (2015) Transcriptional regulation of drought response: A tortuous network of transcriptional factors. *Frontiers in Plant Science* 6, 895.

Singh, R., Pandey, N., Naskar, J. and Shirke, P.A. (2015) Physiological performance and differential expression profiling of genes associated with drought tolerance in contrasting varieties of two *Gossypium* species. *Protoplasma* 252(2), 423–438.

Stewart, E.J. (2012) Growing unculturable bacteria. *Journal of Bacteriology* 194(16), 4151–4160.

Takahara, H., Hacquard, S., Kombrink, A., Hughes, H.B., Halder, V. *et al.* (2016) *Colletotrichum higginsianum* extracellular LysM proteins play dual roles in appressorial function and suppression of chitin-triggered plant immunity. *New Phytologist* 211(4), 1323–1337.

Teng, Y., Wang, X., Li, L., Li, Z. and Luo, Y. (2015) Rhizobia and their bio-partners as novel drivers for functional remediation in contaminated soils. *Frontiers in Plant Science* 6, 32.

Tewari, S. and Sharma, S. (2020) Rhizobial exopolysaccharides as supplement for enhancing nodulation and growth attributes of *Cajanus cajan* under multi-stress conditions: A study from lab to field. *Soil and Tillage Research* 198, 104545. DOI: 10.1016/j.still.2019.104545.

Tiedje, J.M., Asuming-Brempong, S., Nüsslein, K., Marsh, T.L. and Flynn, S.J. (1999) Opening the black box of soil microbial diversity. *Applied Soil Ecology* 13, 109–122.

Trivedi, P., Mattupalli, C., Eversole, K. and Leach, J.E. (2021) Enabling sustainable agriculture through understanding and enhancement of microbiomes. *New Phytologist* 230(6), 2129–2147. DOI: 10.1111/nph.17319.

Trivedi, P., Leach, J.E., Tringe, S.G., Sa, T. and Singh, B.K. (2020) Plant–microbiome interactions: From community assembly to plant health. *Nature Reviews Microbiology* 18(11), 607–621. DOI: 10.1038/s41579-020-0412-1.

Unger, I.M., Kennedy, A.C. and Muzika, R.-M. (2009) Flooding effects on soil microbial communities. *Applied Soil Ecology* 42(1), 1–8.

Van Groenigen, J.W., Lubbers, I.M., Vos, H.M., Brown, G.G., De Deyn, G.B. *et al.* (2014) Earthworms increase plant production: A meta-analysis. *Scientific Reports* 4(1), 1–7.

Vaneechoutte, M., De Beenhouwer, H., Claeys, G., Verschraegen, G., De Rouck, A. *et al.* (1993) Identification of mycobacterium species by using amplified ribosomal DNA restriction analysis. *Journal of Clinical Microbiology* 31, 2061–2065.

Varinderpal-Singh, S., Kunal, S., Gosal, S.K., Choudhary, R., Singh, R. *et al.* (2020) Synergistic use of plant growth-promoting rhizobacteria, arbuscular mycorrhizal fungi, and spectral properties for improving nutrient use efficiencies in wheat (triticum aestivum L.). *Communications in Soil Science and Plant Analysis* 51(1), 14–27.

Vezzani, F.M., Anderson, C., Meenken, E., Gillespie, R., Peterson, M. *et al.* (2018) The importance of plants to development and maintenance of soil structure, microbial communities and ecosystem functions. *Soil and Tillage Research* 175, 139–149.

Vurukonda, S., Vardharajula, S., Shrivastava, M. and SkZ, A. (2016) Enhancement of drought stress tolerance in crops by plant growth promoting rhizobacteria. *Microbiological Research* 184, 13–24.

Wade, W. (2002) Unculturable bacteria—the uncharacterized organisms that cause oral infections. *Journal of the Royal Society of Medicine* 95(2), 81–83.

Wang, C., Cai, X. and Zheng, Z. (2005) High humidity represses cf-4/avr4-and cf-9/avr9-dependent hypersensitive cell death and defense gene expression. *Planta* 222(6), 947–956.

Wang, P.L., Zhao, X.Q., Yang, Z.H., Wang, A.X., Wang, C.X. *et al.* (2012) Effect of in-hospital medical complications on case fatality post-acute ischemic stroke: Data from the China national stroke registry. *Chinese Medical Journal* 125(14), 2449–2454.

Wardle, D.A., Bardgett, R.D., Klironomos, J.N., Setälä, H., Van Der Putten, W.H. *et al.* (2004) Ecological linkages between aboveground and belowground biota. *Science* 304(5677), 1629–1633.

Webster, N.S., Smith, L.D., Heyward, A.J., Watts, J.E., Webb, R.I. *et al.* (2004) Metamorphosis of a scleractinian coral in response to microbial biofilms. *Applied and Environmental Microbiology* 70(2), 1213–1221. DOI: 10.1128/AEM.70.2.1213-1221.2004.

Whitman, T., Pepe-Ranney, C., Enders, A., Koechli, C., Campbell, A, *et al.* (2016) Dynamics of microbial community composition and soil organic carbon mineralization in soil following addition of pyrogenic and fresh organic matter. *The ISME Journal* 10(12), 2918–2930.

Wiese, L., Chibanda, C., Chude, V., Vargas, R. and Caon, L. (2016) Promoting sustainable soil management in sub-Saharan Africa through the African Soil Partnership. *Nature & Faune* 30(1), 109–111.

Willers, C., Jansen van Rensburg, P. and Claassens, S. (2015) Phospholipid fatty acid profiling of microbial communities—A review of interpretations and recent applications. *Journal of Applied Microbiology* 119(5), 1207–1218.

Winsley, T.J., Snape, I., McKinlay, J., Stark, J., Dorst, J.M. *et al*. (2014) The ecological controls on the prevalence of candidate division TM7 in polar regions. *Frontiers in Microbiology* 5, 345.

Wu, H., Li, Q., Lu, C., Zhang, L., Zhu, J. *et al*. (2016) Elevated ozone effects on soil nitrogen cycling differ among wheat cultivars. *Applied Soil Ecology* 108, 187–194.

Wu HongSheng, W.H., Liu DongYang, L.D., Ling Ning, L.N., Bao Wei, B.W., Ying RongRong, Y.R. *et al*. (2008) Effects of vanillic acid on the growth and development of *Fusarium oxysporum* f. *niveum*. *Allelopathy Journal* 22, 111–121.

Xu, J., Zhang, Y., Zhang, P., Trivedi, P., Riera, N. *et al*. (2018) The structure and function of the global citrus rhizosphere microbiome. *Nature Communications* 9(1), 1–10.

Yahai, L. and Ralf, C. (2005) *In situ* stable isotope probing of methanogenic archaea in the rice rhizosphere. *Science* 309(5737), 1088–1090. DOI: 10.1126/science.1113435.

Zhang, H.-L., Lal, R., Zhao, X., Xue, J.-F. and Chen, F. (2014) Opportunities and challenges of soil carbon sequestration by conservation agriculture in China. *Advances in Agronomy* 124, 1–36.

Zhao, Q., Xiong, W., Xing, Y., Sun, Y., Lin, X. *et al*. (2018) Long-term coffee monoculture alters soil chemical properties and microbial communities. *Scientific Reports* 8(1), 1–11.

Zhou, C.-X., Liu, J.-Y., Ye, W.-C., Liu, C.-H. and Tan, R.-X. (2003) Neoverataline A and B, two antifungal alkaloids with A novel carbon skeleton from *Veratrum taliense*. *Tetrahedron* 59(30), 5743–5747. DOI: 10.1016/S0040-4020(03)00882-2.

Zhou, J., Bruns, M.A. and Tiedje, J.M. (1996) DNA recovery from soils of diverse composition. *Applied and Environmental Microbiology* 62(2), 316–322.

Zhu, D., Adebisi, W.A., Ahmad, F., Sethupathy, S., Danso, B. *et al*. (2020) Recent development of extremophilic bacteria and their application in biorefinery. *Frontiers in Bioengineering and Biotechnology* 8, 483. DOI: 10.3389/fbioe.2020.00483.

Zytynska, S.E., Eicher, M., Rothballer, M. and Weisser, W.W. (2020) Microbial-mediated plant growth promotion and pest suppression varies under climate change. *Frontiers in Plant Science* 11, 1385.

Index

Printed and bound by CPI Group (UK) Ltd, Croydon, CR0 4YY

11/01/2026

14804883-0003